Algorithms and Computation in Mathematics • Volume 10

Saugata Basu
Richard Pollack
Marie-Françoise Roy

Algorithms in Real Algebraic Geometry

Second Edition

With 37 Figures

Saugata Basu
Georgia Institute of Technology
School of Mathematics
Atlanta, GA 30332-0160
USA
e-mail: saugata@math.gatech.edu

Richard Pollack
Courant Institute of
Mathematical Sciences
251 Mercer Street
New York, NY 10012
USA
e-mail: pollack@cims.nyu.edu

Marie-Françoise Roy
IRMAR Campus de Beaulieu
Université de Rennes I
35042 Rennes cedex
France
e-mail: Marie-Francoise.Roy@univ-rennes1.fr

Mathematics Subject Classification (2000): 14P10, 68W30, 03C10, 68Q25, 52C45

ISSN 1431-1550

ISBN 978-3-642-06964-2 e-ISBN 978-3-540-33099-8

Springer is a part of Springer Science+Business Media

springer.com

Softcover reprint of the hardcover 2nd edition 2006

Cover design: *design & production* GmbH, Heidelberg

Table of Contents

Introduction

Since a real univariate polynomial does not always have real roots, a very natural algorithmic problem, is to design a method to count the number of real roots of a given polynomial (and thus decide whether it has any). The "real root counting problem" plays a key role in nearly all the "algorithms in real algebraic geometry" studied in this book.

Much of mathematics is algorithmic, since the proofs of many theorems provide a finite procedure to answer some question or to calculate something. A classic example of this is the proof that any pair of real univariate polynomials (P, Q) have a greatest common divisor by giving a finite procedure for constructing the greatest common divisor of (P, Q), namely the euclidean remainder sequence. However, different procedures to solve a given problem differ in how much calculation is required by each to solve that problem. To understand what is meant by "how much calculation is required", one needs a fuller understanding of what an algorithm is and what is meant by its "complexity". This will be discussed at the beginning of the second part of the book, in Chapter 8.

The first part of the book (Chapters 1 through 7) consists primarily of the mathematical background needed for the second part. Much of this background is already known and has appeared in various texts. Since these results come from many areas of mathematics such as geometry, algebra, topology and logic we thought it convenient to provide a self-contained, coherent exposition of these topics.

In Chapter 1 and Chapter 2, we study algebraically closed fields (such as the field of complex numbers $\mathbb{C}$) and real closed fields (such as the field of real numbers $\mathbb{R}$). The concept of a real closed field was first introduced by Artin and Schreier in the 1920's and was used for their solution to Hilbert's 17th problem [6, 7]. The consideration of abstract real closed fields rather than the field of real numbers in the study of algorithms in real algebraic geometry is not only intellectually challenging, it also plays an important role in several complexity results given in the second part of the book.

Chapters 1 and 2 describe an interplay between geometry and logic for algebraically closed fields and real closed fields. In Chapter 1, the basic geometric objects are constructible sets. These are the subsets of $\mathbb{C}^n$ which are defined by a finite number of polynomial equations ($P=0$) and inequations ($P \neq 0$). We prove that the projection of a constructible set is constructible. The proof is very elementary and uses nothing but a parametric version of the euclidean remainder sequence. In Chapter 2, the basic geometric objects are the semi-algebraic sets which constitute our main objects of interest in this book. These are the subsets of $\mathbb{R}^n$ that are defined by a finite number of polynomial equations ($P = 0$) and inequalities ($P > 0$). We prove that the projection of a semi-algebraic set is semi-algebraic. The proof, though more complicated than that for the algebraically closed case, is still quite elementary. It is based on a parametric version of real root counting techniques developed in the nineteenth century by Sturm, which uses a clever modification of euclidean remainder sequence. The geometric statement "the projection of a semi-algebraic set is semi-algebraic" yields, after introducing the necessary terminology, the theorem of Tarski that "the theory of real closed fields admits quantifier elimination." A consequence of this last result is the decidability of elementary algebra and geometry, which was Tarski's initial motivation. In particular whether there exist real solutions to a finite set of polynomial equations and inequalities is decidable. This decidability result is quite striking, given the undecidability result proved by Matijacević [113] for a similar question, Hilbert's 10-th problem: there is no algorithm deciding whether or not a general system of Diophantine equations has an integer solution.

In Chapter 3 we develop some elementary properties of semi-algebraic sets. Since we work over various real closed fields, and not only over the reals, it is necessary to reexamine several notions whose classical definitions break down in non-archimedean real closed fields. Examples of these are connectedness and compactness. Our proofs use non-archimedean real closed field extensions, which contain infinitesimal elements and can be described geometrically as germs of semi-algebraic functions, and algebraically as algebraic Puiseux series. The real closed field of algebraic Puiseux series plays a key role in the complexity results of Chapters 13 to 16.

Chapter 4 describes several algebraic results, relating in various ways properties of univariate and multivariate polynomials to linear algebra, determinants and quadratic forms. A general theme is to express some properties of univariate polynomials by the vanishing of specific polynomial expressions in their coefficients. The discriminant of a univariate polynomial P, for example, is a polynomial in the coefficients of P which vanishes when P has a multiple root. The discriminant is intimately related to real root counting, since, for polynomials of a fixed degree, all of whose roots are distinct, the sign of the discriminant determines the number of real roots modulo 4. The discriminant is in fact the determinant of a symmetric matrix whose signature gives an alternative method to Sturm's for real root counting due to Hermite.

Similar polynomial expressions in the coefficients of two polynomials are the classical resultant and its generalization to subresultant coefficients. The vanishing of these subresultant coefficients expresses the fact that the greatest common divisor of two polynomials has at least a given degree. The resultant makes possible a constructive proof of a famous theorem of Hilbert, the Nullstellensatz, which provides a link between algebra and geometry in the algebraically closed case. Namely, the geometric statement 'an algebraic variety (the common zeros of a finite family of polynomials) is empty' is equivalent to the algebraic statement '1 belongs to the ideal generated by these polynomials'. An algebraic characterization of those systems of polynomial equations with a finite number of solutions in an algebraically closed field follows from Hilbert's Nullstellensatz: a system of polynomial equations has a finite number of solutions in an algebraically closed field if and only if the corresponding quotient ring is a finite dimensional vector space. As seen in Chapter 1, the projection of an algebraic set in affine space is constructible. Considering projective space allows an even more satisfactory result: the projection of an algebraic set in projective space is algebraic. This result appears here as a consequence of a quantitative version of Hilbert's Nullstellensatz, following the analysis of its constructive proof. A weak version of Bezout's theorem, bounding the number of simple solutions of polynomials systems is a consequence of this projection theorem.

Semi-algebraic sets are defined by a finite number of polynomial inequalities. On the real line, semi-algebraic sets consist of a finite number of points and intervals. It is thus natural to wonder what kind of geometric finiteness properties are enjoyed by semi-algebraic sets in higher dimensions. In Chapter 5 we study various decompositions of a semi-algebraic set into a finite number of simple pieces. The most basic decomposition is called a cylindrical decomposition: a semi-algebraic set is decomposed into a finite number of pieces, each homeomorphic to an open cube. A finer decomposition provides a stratification, i.e. a decomposition into a finite number of pieces, called strata, which are smooth manifolds, such that the closure of a stratum is a union of strata of lower dimension. We also describe how to triangulate a closed and bounded semi-algebraic set. Various other finiteness results about semi-algebraic sets follow from these decompositions. Among these are:

- a semi-algebraic set has a finite number of connected components each of which is semi-algebraic,
- algebraic sets described by polynomials of fixed degree have a finite number of topological types.

A natural question raised by these results is to find explicit bounds on these quantities now known to be finite.

Chapter 6 is devoted to a self contained development of the basics of elementary algebraic topology. In particular, we define simplicial homology theory and, using the triangulation theorem, show how to associate to semi-algebraic sets certain discrete objects (the simplicial homology vector spaces) which are invariant under semi-algebraic homeomorphisms. The dimensions of these vector spaces, the Betti numbers, are an important measure of the topological complexity of semi-algebraic sets, the first of them being the number of connected components of the set. We also define the Euler-Poincaré characteristic, which is a significant topological invariant of algebraic and semi-algebraic sets.

Chapter 7 presents basic results of Morse theory and proves the classical Oleinik-Petrovsky-Thom-Milnor bounds on the sum of the Betti numbers of an algebraic set of a given degree. The basic technique for these results is the critical point method, which plays a key role in the complexity results of the last chapters of the book. According to basic results of Morse theory, the critical points of a well chosen projection on a line of a smooth hypersurface are precisely the places where a change in topology occurs in the part of the hypersurface inside a half space defined by a hyperplane orthogonal to the line. Counting these critical points using Bezout's theorem yields the Oleinik-Petrovsky-Thom-Milnor bound on the sum of the Betti numbers of an algebraic hypersurface, which is polynomial in the degree and exponential in the number of variables. More recent results bounding the individual Betti numbers of sign conditions defined by a family of polynomials on an algebraic set are described. These results involve a combinatorial part, depending on the number of polynomials considered, which is polynomial in the number of polynomials and exponential in the dimension of the algebraic set, and an algebraic part, given by the Oleinik-Petrovsky-Thom-Milnor bound. The combinatorial part of these bounds agrees with the number of connected components defined by a family of hyperplanes. These quantitative results on the number of connected components and Betti numbers of semi-algebraic sets provide an indication about the complexity results to be hoped for when studying various algorithmic problems related to semi-algebraic sets.

The second part of the book discusses various algorithmic problems in detail. These are mainly real root counting, deciding the existence of solutions for systems of equations and inequalities, computing the projection of a semi-algebraic set, deciding a sentence of the theory of real closed fields, eliminating quantifiers, and computing topological properties of algebraic and semi-algebraic sets.

In Chapter 8 we discuss a few notions of complexity needed to analyze our algorithms and discuss basic algorithms for linear algebra and remainder sequences. We perform a study of a useful tool closely related to remainder sequence, the subresultant sequence. This subresultant sequence plays an important role in modern methods for real root counting in Chapter 9, and also provides a link between the classical methods of Sturm and Hermite seen earlier. Various methods for performing real root counting, and computing the signature of related quadratic forms, as well as an application to counting complex roots in a half plane, useful in control theory, are described.

Chapter 10 is devoted to real roots. In the field of the reals, which is archimedean, root isolation techniques are possible. They are based on Descartes's law of signs, presented in Chapter 2 and properties of Bernstein polynomials, which provide useful constructions in CAD (Computer Aided Design). For a general real closed field, isolation techniques are no longer possible. We prove that a root of a polynomial can be uniquely described by sign conditions on the derivatives of this polynomial, and we describe a different method for performing sign determination and characterizing real roots, without approximating the roots.

In Chapter 11, we describe an algorithm for computing the cylindrical decomposition which had been already studied in Chapter 5. The basic idea of this algorithm is to successively eliminate variables, using subresultants. Cylindrical decomposition has numerous applications among which are: deciding the truth of a sentence, eliminating quantifiers, computing a stratification, and computing topological information of various kinds, an example of which is computing the topology of an algebraic curve. The huge degree bounds (doubly exponential in the number of variables) output by the cylindrical decomposition method give estimates on the number of connected components of semi-algebraic sets which are much worse than those we obtained using the critical point method in Chapter 7.

The main idea developed in Chapters 12 to 16 is that, using the critical point method in an algorithmic way yields much better complexity bounds than those obtained by cylindrical decomposition for deciding the existential theory of the reals, eliminating quantifiers, deciding connectivity and computing connected components.

Chapter 12 is devoted to polynomial system solving. We give a few results about Gröbner bases, and explain the technique of rational univariate representation. Since our techniques in the following chapters involve infinitesimal deformations, we also indicate how to compute the limit of the bounded solutions of a polynomial system when the deformation parameters tend to zero. As a consequence, using the ideas of the critical point method described in Chapter 7, we are able to find a point in every connected components of an algebraic set. Since we deal with arbitrary algebraic sets which are not necessarily smooth, we introduce the notion of a pseudo-critical point in order to adapt the critical point method to this new situation. We compute a point in every semi-algebraically connected component of a bounded algebraic set with complexity polynomial in the degree and exponential in the number of variables. Using a similar technique, we compute the Euler-Poincaré characteristic of an algebraic set, with complexity polynomial in the degree and exponential in the number of variables.

In Chapter 13 we present an algorithm for the existential theory of the reals whose complexity is singly exponential in the number of variables. Using the pseudo-critical points introduced in Chapter 12 and perturbation methods to obtain polynomials in general position, we can compute the set of realizable sign conditions and compute representative points in each of the realizable sign conditions. Applications to the size of a ball meeting every connected component and various real and complex decision problems are provided. Finally we explain how to compute points in realizable sign conditions on an algebraic set taking advantage of the (possibly low) dimension of the algebraic set. We also compute the Euler-Poincaré characteristic of sign conditions defined by a set of polynomials. The complexity results obtained are quite satisfactory in view of the quantitative bounds proved in Chapter 7.

In Chapter 14 the results on the complexity of the general decision problem and quantifier elimination obtained in Chapter 11 using cylindrical decomposition are improved. The main idea is that the complexity of quantifier elimination should not be doubly exponential in the number of variables but rather in the number of blocks of variables appearing in the formula where the blocks of variables are delimited by alternations in the quantifiers $\exists$ and $\forall$. The key notion is the set of realizable sign conditions of a family of polynomials for a given block structure of the set of variables, which is a generalization of the set of realizable sign conditions, corresponding to one single block. Parametrized versions of the methods presented in Chapter 13 give the technique needed for eliminating a whole block of variables.

In Chapters 15 and 16, we compute roadmaps and connected components of algebraic and semi-algebraic sets. Roadmaps can be intuitively described as an one dimensional skeleton of the set, providing a way to count connected components and to decide whether two points belong to the same connected component. A motivation for studying these problems comes from robot motion planning where the free space of a robot (the subspace of the configuration space of the robot consisting of those configurations where the robot is neither in conflict with its environment nor itself) can be modeled as a semi-algebraic set. In this context it is important to know whether a robot can move from one configuration to another. This is equivalent to deciding whether the two corresponding points in the free space are in the same connected component of the free space. The construction of roadmaps is based on the critical point method, using properties of pseudo-critical values. The complexity of the construction is singly exponential in the number of variables, which is a complexity much better than the one provided by cylindrical decomposition. Our construction of parametrized paths gives an algorithm for computing coverings of semi-algebraic sets by contractible sets, which in turn provides a single exponential time algorithm for computing the first Betti number of semi-algebraic sets. Moreover, it gives an efficient algorithm for computing semi-algebraic descriptions of the connected components of a semi-algebraic set in single exponential time.

1 Warning This book is intended to be self contained, assuming only that the reader has a basic knowledge of linear algebra and the rudiments of a basic course in algebra through the definitions and basic properties of groups, rings and fields, and in topology through the elementary properties of closed, open, compact and connected sets.

There are many other aspects of real algebraic geometry that are not considered in this book. The reader who wants to pursue the many aspects of real algebraic geometry beyond the introduction to the small part of it that we provide is encouraged to study other text books [26, 95, 5]. There is also a great deal of material about algorithms in real algebraic geometry that we are not covering in this book. To mention but a few: fewnomials, effective positivstellensatz, semi-definite programming, complexity of quadratic maps and quadratic sets, ...

2 References We have tried to keep our style as informal as possible. Rather than giving bibliographic references and footnotes in the body of the text, we have a section at the end of each chapter giving a brief description of the history of the results with a few of the relevant bibliographic citations. We only try to indicate where, to the best of our knowledge, the main ideas and results appear for the first time, and do not describe the full history and bibliography. We also list below the references containing the material we have used directly.

3 Existing implementations In terms of existing implementation of the algorithms described in the book, the current situation can be roughly summarized as follows: algorithms appearing in Chapters 8 to 12, or more efficient versions based on similar ideas, have been implemented (see a few references below). For most of the algorithms presented in Chapter 13 to 16, there is no implementation at all. The reason for that is that the methods developed are well adapted to complexity results but are not adapted to efficient implementation.

Most algorithms from Chapters 8 to 11 are quite classical and have been implemented several times. We refer to [40] since it is a recent implementation based directly on [20]. It uses in part the work presented in [29]. A very efficient variant of the real root isolation algorithm in the monomial basis in Chapter 10 is described in [138]. Cylindrical algebraic decomposition discussed in Chapter 11 has also been implemented many times, see for example [46, 30, 151]. We refer to [71] for an implementation of an algorithm computing the topology of real algebraic curves close to the one we present in Chapter 11. About algorithms discussed in Chapter 12, most computer algebra systems include Gröbner basis computations. Particularly efficient Gröbner basis computations, based on algorithms not described in the book, can be found in [59]. A very efficient rational univariate representation can be found in [135]. Computing a point in every connected component of an algebraic set based on critical point method techniques is done efficiently in [143], based on the algorithms developed in [8, 144].

4 Comments about the second edition An important change in content between the first edition [20] and the second one is the inversion of the order of Chapter 12 and Chapter 11. Indeed when teaching courses based on the book, we felt that the material on polynomial system solving was not necessary to explain cylindrical decomposition and it was better to make these two chapters independent for teaching purposes. For the same reason, we also made the real root counting technique based on signed subresultant coefficients independent of the signed subresultant polynomials and included it in Chapter 4 rather than in Chapter 9 as before. Some other chapters have been slightly reorganized. Several new topics are included in this second edition: results about normal polynomials and virtual roots in Chapter 2, about discriminants of symmetric matrices in Chapter 4, a new section bounding the Betti numbers of semi-algebraic sets in Chapter 7, an improved complexity analysis of real root isolation, as well as the real root isolation algorithm in the monomial basis, in Chapter 10, the notion of parametrized path in Chapter 15 and the computation of the first Betti number of a semi-algebraic set in single exponential time. We also included a table of notation and completed the bibliography and bibliographical notes at the end of the chapters. Various mistakes and typos have been corrected, and new ones introduced, for sure. As a result of the changes, the numbering of Definitions, Theorems etc. are not identical in the first edition [20] and the second one. Also, Algorithms now have their own numbering.

According to our contract with Springer-Verlag, we have had the right to post updated versions of the first edition of the book on our websites since December 2004. Currently an updated version of the first edition is available online as `bpr-posted1.pdf`. We are going to update on a regular basis this posted version. Here are the various url where these files can be obtained through `http://` at

`www.math.gatech.edu/~saugata/bpr-posted1.html`
`www.math.nyu.edu/faculty/pollack/bpr-posted1.html`
`perso.univ-rennes1.fr/marie-francoise.roy/bpr-posted1.html`

An implementation of algorithms from Chapters 8 to 10 and part of Chapter 11 written in Maxima by Fabrizio Caruso, as well as a version of Jean-Charles Faugère [59] and Fabrice Rouillier [135] software illustrating part of Chapter 12, can also be downloaded at `bpr-posted1-annex`.

Note that the second edition has been prepared inside TeXmacs. The TeXmacs files have been initially produced from classical latex files of the first edition. Even though some manual changes in the latex files have been necessary to obtain correct TeXmacs files, the translation into TeXmacs was made automatically, and it has not been necessary to retype the text and formulas, besides a few exceptions.

After eighteen months of the publication of the current edition of the book, we will post the second edition online and it will be available for downloading from the same url as above.

5 Interactive version of the book Another possibility is to get the book as a $\TeX_{\mathrm{MACS}}$ project by downloading `bpr-posted1-int`. In the $\TeX_{\mathrm{MACS}}$ project version, you are able to travel in the book by clicking on references, to fold/unfold proofs, descriptions of the algorithms and parts of the text. You can use the open-source maxima code corresponding to algorithms of Chapters 8 to 10 and part of Chapter 11 written by Fabrizio Caruso [40]: check examples, read the source code and make your own computations inside the book. You can also use the part of [59] and [135] provided by Jean-Charles Faugère and Fabrice Rouillier to illustrate part of Chapter 12 directly in the book. These functionalities are still experimental. You are welcome to report to the authors' email addresses any problem you might meet in using them.

In the future, $\TeX_{\mathrm{MACS}}$ versions of the book will include other interactive features, such as being able to find all places in the book where a given theorem is quoted.

6 Errors If you find remaining errors in the book, we would appreciate it if you would let us know

email: `saugata.basu@math.gatech.edu`
`pollack@cims.nyu.edu`
`marie-francoise.roy@univ-rennes1.fr`

A list of errors identified in this version will be found at

`www.math.gatech.edu/~saugata/bpr_book/bpr-ed2-errata.html`.

7 Acknowledgment We thank Michel Coste, Greg Friedman, Laureano Gonzalez-Vega, Abdeljaoued Jounaidi, Henri Lombardi, Dimitri Pasechnik, Fabrice Rouillier for their advice and help. We also thank Solen Corvez, Gwenael Guérard, Michael Kettner, Tomas Lajous, Samuel Lelièvre, Mohab Safey, and Brad Weir for studying preliminary versions of the text and helping to improve it. Mistakes or typos in [20] have been identified by Morou Amidou, Emmanuel Briand, Fabrizio Caruso, Fernando Carreras, Keven Commault, Anne Devys, Arno Eigenwillig, Vincent Guenanff, Michael Kettner, Assia Mahboubi, Iona Necula, Adamou Otto, Dimitri Pasechnik, Hervé Perdry, Savvas Perikleous, Moussa Seydou.

Joris Van der Hoeven has provided support for the use of $\TeX_{\mathrm{MACS}}$ and produced several new versions of the software adapted to our purpose. Most figures are the same as in the first edition. However, Henri Lesourd produced some native $\TeX_{\mathrm{MACS}}$ diagrams and figures for us.

At different stages of writing this book the authors received support from CNRS, NSF, Université de Rennes 1, Courant Institute of Mathematical Sciences, University of Michigan, Georgia Institute of Technology, the RIP Program in Oberwolfach, MSRI, DIMACS, RISC, Linz, Centre Emile Borel. Fabrizio Caruso was supported by RAAG during a post doctoral fellowship in Rennes and Santander. The software due to Jean-Charles Faugère [59] and Fabrice Rouillier [135] was developed under the SALSA project at INRIA, CNRS and Université Pierre et Marie Curie, Paris.

8 Sources Our sources for Chapter 2 are: [26] for Section 2.1 and Section 2.4, [140, 98, 49] for Section 2.2, [47] for Section 2.3 and [164, 109] for Section 2.5. Our source for Section 3.1, Section 3.2 and Section 3.3 of Chapter 3 is [26]. Our sources for Chapter 4 are: [63] for Section 4.1, [94] for Theorem 4.47 in Section 4.4, [159, 147] for Section 4.4, [128, 129] for Section 4.6 and [22] for Section 4.7. Our sources for Chapter 5 are [26, 47, 48]. Our source for Chapter 6 is [150]. Our sources for Chapter 7 are [117, 26, 17], and for Section 7.5 [62, 21]. Our sources for Chapter 8 are: [1] for Section 8.2 and [112] for Section 8.3. Our sources for Chapter 9 are [63] and [66, 69, 70, 140, 2] for part of Section 9.1. Our sources for Chapter 10 are: [116] for Section 10.1, [138, 149] for Section 10.2, [141] for Sections 10.3 and [129] for Section 10.4. Our source for Section 11.4 is [52], and for Section 11.6 is [67]. Our sources for Chapter 12 are: for Section 12.1 [51], for Section 12.2 [72], for Section 12.4 [4, 134], for Section 12.5 [13]. The results presented in Section 13.1, Section 13.2 and Section 13.3 of Chapter 13 are based on [13, 15]. Our source for Section 13.4 of Chapter 13 is [18]. Our source for Chapter 14 is [13]. Our sources for Chapter 15 and Chapter 16 are [16, 21].

1

Algebraically Closed Fields

The main purpose of this chapter is the definition of constructible sets and the statement that, in the context of algebraically closed fields, the projection of a constructible set is constructible.

Section 1.1 is devoted to definitions. The main technique used for proving the projection theorem in Section 1.3 is the remainder sequence defined in Section 1.2 and, for the case where the coefficients have parameters, the tree of possible pseudo-remainder sequences. Several important applications of logical nature of the projection theorem are given in Section 1.4.

1.1 Definitions and First Properties

The objects of our interest in this section are sets defined by polynomials with coefficients in an algebraically closed field C.

A field C is **algebraically closed** if any non-constant univariate polynomial $P(X)$ with coefficients in C has a **root** in C, i.e. there exists $x \in \mathrm{C}$ such that $P(x)=0$.

Every field has a minimal extension which is algebraically closed and this extension is called the **algebraic closure** of the field (see Section 2, Chapter 5 of [102]). A typical example of an algebraically closed field is the field $\mathbb{C}$ of complex numbers.

We study the sets of points which are the common zeros of a finite family of polynomials.

If D is a ring, we denote by $\mathrm{D}[X_1, \ldots, X_k]$ the polynomials in k variables $X_1, \ldots, X_k$ with coefficients in D.

Notation 1.1. [Zero set] If $\mathcal{P}$ is a finite subset of $\mathrm{C}[X_1, \ldots, X_k]$ we write the **set of zeros of** $\mathcal{P}$ **in** C^k as

$$\mathrm{Zer}(\mathcal{P}, \mathrm{C}^k) = \{x \in \mathrm{C}^k \mid \bigwedge_{P \in \mathcal{P}} P(x) = 0\}.$$

These are the **algebraic subsets** of C^k.

The set C^k is algebraic since $\mathrm{C}^k = \mathrm{Zer}(\{0\}, \mathrm{C}^k)$. □

Exercise 1.1. Prove that an algebraic subset of C is either a finite set or empty or equal to C.

It is natural to consider the smallest family of sets which contain the algebraic sets and is also closed under the boolean operations (complementation, finite unions, and finite intersections). These are the **constructible sets**. Similarly, the smallest family of sets which contain the algebraic sets, their complements, and is closed under finite intersections is the family of **basic constructible sets**. Such a basic constructible set S can be described as a conjunction of polynomial equations and inequations, namely

$$S = \{x \in \mathrm{C}^k \mid \bigwedge_{P \in \mathcal{P}} P(x) = 0 \wedge \bigwedge_{Q \in \mathcal{Q}} Q(x) \neq 0\}$$

with $\mathcal{P}$, $\mathcal{Q}$ finite subsets of $\mathrm{C}[X_1, \ldots, X_k]$.

Exercise 1.2. Prove that a constructible subset of C is either a finite set or the complement of a finite set.

Exercise 1.3. Prove that a constructible set in C^k is a finite union of basic constructible sets.

The principal goal of this chapter is to prove that the projection from C^{k+1} to C^k that is defined by "forgetting" the last coordinate maps constructible sets to constructible sets. For this, since projection commutes with union, it suffices to prove that the projection

$$\{y \in \mathrm{C}^k \mid \exists\, x \in \mathrm{C} \ \bigwedge_{P \in \mathcal{P}} P(y, x) = 0 \wedge \bigwedge_{Q \in \mathcal{Q}} Q(y, x) \neq 0\}$$

of a basic constructible set,

$$\{(y, x) \in \mathrm{C}^{k+1} \mid \bigwedge_{P \in \mathcal{P}} P(y, x) = 0 \wedge \bigwedge_{Q \in \mathcal{Q}} Q(y, x) \neq 0\}$$

is constructible, i.e. can be described by a boolean combination of polynomial **equations** ($P = 0$) and **inequations** ($P \neq 0$) in $Y = (Y_1, \ldots, Y_k)$.

Some terminology from logic is useful for the study of constructible sets.

We define the language of fields by describing the formulas of this language. The formulas are built starting with atoms, which are polynomial equations and inequations. A formula is written using atoms together with the logical connectives "and", "or", and "negation" ($\wedge$, $\vee$, and $\neg$) and the existential and universal quantifiers ($\exists, \forall$). A formula has free variables, i.e. non-quantified variables, and bound variables, i.e. quantified variables. More precisely, let D be a subring of C. We define the **language of fields with coefficients in D** as follows. An **atom** is $P = 0$ or $P \neq 0$, where P is a polynomial in $\mathrm{D}[X_1, \ldots, X_k]$. We define simultaneously the **formulas** and $\mathrm{Free}(\Phi)$, the set of **free variables of a formula** Φ, as follows

- an atom $P = 0$ or $P \neq 0$, where P is a polynomial in $\mathrm{D}[X_1, \ldots, X_k]$ is a formula with free variables $\{X_1, \ldots, X_k\}$,

- if Φ_1 and Φ_2 are formulas, then $\Phi_1 \wedge \Phi_2$ and $\Phi_1 \vee \Phi_2$ are formulas with
$$\mathrm{Free}(\Phi_1 \wedge \Phi_2) = \mathrm{Free}(\Phi_1 \vee \Phi_2) = \mathrm{Free}(\Phi_1) \cup \mathrm{Free}(\Phi_2),$$
- if Φ is a formula, then $\neg(\Phi)$ is a formula with
$$\mathrm{Free}(\neg(\Phi)) = \mathrm{Free}(\Phi),$$
- if Φ is a formula and $X \in \mathrm{Free}(\Phi)$, then $(\exists X)\ \Phi$ and $(\forall X)\ \Phi$ are formulas with
$$\mathrm{Free}((\exists X)\ \Phi) = \mathrm{Free}((\forall X)\ \Phi) = \mathrm{Free}(\Phi) \setminus \{X\}.$$

If Φ and Ψ are formulas, $\Phi \Rightarrow \Psi$ is the formula $\neg(\Phi) \vee \Psi$.

A **quantifier free formula** is a formula in which no quantifier appears, neither $\exists$ nor $\forall$. A **basic formula** is a conjunction of atoms.

The **C-realization of a formula** Φ with free variables contained in $\{Y_1, ..., Y_k\}$, denoted $\mathrm{Reali}(\Phi, \mathrm{C}^k)$, is the set of $y \in \mathrm{C}^k$ such that $\Phi(y)$ is true. It is defined by induction on the construction of the formula, starting from atoms:

$$\begin{aligned}
\mathrm{Reali}(P=0, \mathrm{C}^k) &= \{y \in \mathrm{C}^k \mid P(y)=0\},\\
\mathrm{Reali}(P \neq 0, \mathrm{C}^k) &= \{y \in \mathrm{C}^k \mid P(y) \neq 0\},\\
\mathrm{Reali}(\Phi_1 \wedge \Phi_2, \mathrm{C}^k) &= \mathrm{Reali}(\Phi_1, \mathrm{C}^k) \cap \mathrm{Reali}(\Phi_2, \mathrm{C}^k),\\
\mathrm{Reali}(\Phi_1 \vee \Phi_2, \mathrm{C}^k) &= \mathrm{Reali}(\Phi_1, \mathrm{C}^k) \cup \mathrm{Reali}(\Phi_2, \mathrm{C}^k),\\
\mathrm{Reali}(\neg\Phi, \mathrm{C}^k) &= \mathrm{C}^k \setminus \mathrm{Reali}(\Phi, \mathrm{C}^k),\\
\mathrm{Reali}((\exists X)\ \Phi, \mathrm{C}^k) &= \{y \in \mathrm{C}^k \mid \exists x \in \mathrm{C} \quad (x, y) \in \mathrm{Reali}(\Phi, \mathrm{C}^{k+1})\},\\
\mathrm{Reali}((\forall X)\ \Phi, \mathrm{C}^k) &= \{y \in \mathrm{C}^k \mid \forall x \in \mathrm{C} \quad (x, y) \in \mathrm{Reali}(\Phi, \mathrm{C}^{k+1})\}
\end{aligned}$$

Two formulas Φ and Ψ such that $\mathrm{Free}(\Phi) = \mathrm{Free}(\Psi) = \{Y_1, ..., Y_k\}$ are **C-equivalent** if $\mathrm{Reali}(\Phi, \mathrm{C}^k) = \mathrm{Reali}(\Psi, \mathrm{C}^k)$.

If there is no ambiguity, we simply write $\mathrm{Reali}(\Phi)$ for $\mathrm{Reali}(\Phi, \mathrm{C}^k)$ and talk about realization and equivalence.

Example 1.2. The formulas $\Phi = ((\exists Y)\ X Y - 1 = 0)$ and $\Psi = (X \neq 0)$ are two formulas of the language of fields with coefficients in $\mathbb{Z}$ and
$$\mathrm{Free}(\Phi) = \mathrm{Free}(\Psi) = \{X\}.$$

Note that the formula Ψ is quantifier free. Moreover, Φ and Ψ are C-equivalent since

$$\begin{aligned}
\mathrm{Reali}(\Phi, \mathrm{C}) &= \{x \in \mathrm{C} \mid \exists y \in \mathrm{C}\ \ x y - 1 = 0\}\\
&= \{x \in \mathrm{C} \mid x \neq 0\}\\
&= \mathrm{Reali}(\Psi, \mathrm{C}).
\end{aligned}$$

□

It is clear that a set is constructible if and only if it can be represented as the realization of a quantifier free formula.

It is easy to see that any formula Φ with $\mathrm{Free}(\Phi) = \{Y_1, ..., Y_k\}$ in the language of fields with coefficients in D is C-equivalent to a a formula

$$(\mathrm{Qu}_1 X_1)...(\mathrm{Qu}_m X_m)\, \mathcal{B}(X_1, ..., X_m, Y_1, ...Y_k)$$

where each $\mathrm{Qu}_i \in \{\forall, \exists\}$ and $\mathcal{B}$ is a quantifier free formula involving polynomials in $\mathrm{D}[X_1, ..., X_m, Y_1, ...Y_k]$. This is called its **prenex normal form** (see Section 10, Chapter 1 of [115]). The variables X_1, ..., X_m are called **bound variables**.

If the formula Φ has no free variables, i.e. $\mathrm{Free}(\Phi) = \emptyset$, then it is called a **sentence**, and it is either C-equivalent to true, when $\mathrm{Reali}(\Phi), \{0\}) = \{0\}$, or C-equivalent to false, when $\mathrm{Reali}(\Phi), \{0\}) = \emptyset$. For example, $0 = 0$ is C-equivalent to true, and $0 = 1$ is C-equivalent to false.

Remark 1.3. Though many statements of algebra can be expressed by a sentence in the language of fields, it is necessary to be careful in the use of this notion. Consider for example the fundamental theorem of algebra: any non constant polynomial with coefficients in $\mathbb{C}$ has a root in $\mathbb{C}$, which is expressed by

$$\forall\, P \in \mathbb{C}[X] \;\; \deg(P) > 0,\; \exists\, X \in \mathbb{C}\; P(X) = 0.$$

This expression is not a sentence of the language of fields with coefficients in $\mathbb{C}$, since quantification over all polynomials is not allowed in the definition of formulas. However, fixing the degree to be equal to d, it is possible to express by a sentence Φ_d the statement: any monic polynomial of degree d with coefficients in $\mathbb{C}$ has a root in $\mathbb{C}$. We write as an example

$$\Phi_2 = ((\forall Y_1)\; (\forall Y_2)\; (\exists X)\; X^2 + Y_1 X + Y_2 = 0).$$

So the definition of an algebraically closed field can be expressed by an infinite list of sentences in the language of fields: the field axioms and the sentences Φ_d, $d \geq 1$. □

Exercise 1.4. Write the formulas for the axioms of fields.

1.2 Euclidean Division and Greatest Common Divisor

We study euclidean division, compute greatest common divisors, and show how to use them to decide whether or not a basic constructible set of C is empty.

In this section, C is an algebraically closed field, D a subring of C and K the quotient field of D. One can take as a typical example of this situation the field $\mathbb{C}$ of complex numbers, the ring $\mathbb{Z}$ of integers, and the field $\mathbb{Q}$ of rational numbers.

Let P be a non-zero polynomial

$$P = a_p X^p + \cdots + a_1 X + a_0 \in \mathrm{D}[X]$$

with $a_p \neq 0$.

We denote the **degree of** P, which is p, by $\deg(P)$. By convention, the degree of the zero polynomial is defined to be $-\infty$. If P is non-zero, we write $\mathrm{cof}_j(P) = a_j$ for the **coefficient of** X^j **in** P (which is equal to 0 if $j > \deg(P)$) and $\mathrm{lcof}(P)$ for its **leading coefficient** $a_p = \mathrm{cof}_{\deg(P)}(P)$. By convention $\mathrm{lcof}(0) = 1$.

Suppose that P and Q are two polynomials in $\mathrm{D}[X]$. The polynomial Q is a **divisor of** P if $P = A Q$ for some $A \in \mathrm{K}[X]$. Thus, while every P divides 0, 0 divides 0 and no other polynomial.

If $Q \neq 0$, the **remainder** in the **euclidean division of** P **by** Q, denoted $\mathrm{Rem}(P, Q)$, is the unique polynomial $R \in \mathrm{K}[X]$ of degree smaller than the degree of Q such that $P = A\,Q + R$ with $A \in \mathrm{K}[X]$. The **quotient** in the euclidean division of P by Q, denoted $\mathrm{Quo}(P, Q)$, is A.

Exercise 1.5. Prove that, if $Q \neq 0$, there exists a unique pair (R, A) of polynomials in $\mathrm{K}[X]$ such that $P = A\,Q + R$, $\deg(R) < \deg(Q)$.

Remark 1.4. Clearly, $\mathrm{Rem}(a\,P, b\,Q) = a\mathrm{Rem}(P, Q)$ for any $a, b \in \mathrm{K}$ with $b \neq 0$. At a root x of Q, $\mathrm{Rem}(P, Q)(x) = P(x)$. □

Exercise 1.6. Prove that x is a root of P in K if and only if $X - x$ is a divisor of P in $\mathrm{K}[X]$.

Exercise 1.7. Prove that if C is algebraically closed, every $P \in \mathrm{C}[X]$ can be written uniquely as

$$P = a\,(X - x_1)^{\mu_1} \cdots (X - x_k)^{\mu_k},$$

with $x_1, \ldots, x_k$ distinct elements of C.

A **greatest common divisor of** P **and** Q, denoted $\gcd(P, Q)$, is a polynomial $G \in \mathrm{K}[X]$ such that G is a divisor of both P and Q, and any divisor of both P and Q is a divisor of G. Observe that this definition implies that P is a greatest common divisor of P and 0. Clearly, any two greatest common divisors (say G_1, G_2) of P and Q must divide each other and have equal degree. Hence $G_1 = a\,G_2$ for some $a \in \mathrm{K}$. Thus, any two greatest common divisors of P and Q are proportional by an element in $\mathrm{K} \setminus \{0\}$. Two polynomials are **coprime** if their greatest common divisor is an element of $\mathrm{K} \setminus \{0\}$.

A **least common multiple of** P **and** Q, $\mathrm{lcm}(P, Q)$ is a polynomial $G \in \mathrm{K}[X]$ such that G is a multiple of both P and Q, and any multiple of both P and Q is a multiple of G. Clearly, any two least common multiples L_1, L_2 of P and Q must divide each other and have equal degree. Hence $L_1 = a\,L_2$ for some $a \in \mathrm{K}$. Thus, any two least common multiple of P and Q are proportional by an element in $\mathrm{K} \setminus \{0\}$.

It follows immediately from the definitions that:

Proposition 1.5. *Let $P \in \mathrm{K}[X]$ and $Q \in \mathrm{K}[X]$, not both zero. Then PQ/G is a least common multiple of P and Q.*

Corollary 1.6.

$$\deg(\mathrm{lcm}(P,Q)) = \deg(P) + \deg(Q) - \deg(\gcd(P,Q)).$$

We now prove that greatest common divisors and least common multiple exist by using euclidean division repeatedly.

Definition 1.7. [Signed remainder sequence] Given $P, Q \in K[X]$, not both 0, we define the **signed remainder sequence of P and Q**,

$$\mathrm{SRemS}(P,Q) = \mathrm{SRemS}_0(P,Q), \mathrm{SRemS}_1(P,Q), \ldots, \mathrm{SRemS}_k(P,Q)$$

by

$$\begin{aligned}
\mathrm{SRemS}_0(P,Q) &= P, \\
\mathrm{SRemS}_1(P,Q) &= Q, \\
\mathrm{SRemS}_2(P,Q) &= -\mathrm{Rem}(\mathrm{SRemS}_0(P,Q), \mathrm{SRemS}_1(P,Q)), \\
&\ \vdots \\
\mathrm{SRemS}_k(P,Q) &= -\mathrm{Rem}(\mathrm{SRemS}_{k-2}(P,Q), \mathrm{SRemS}_{k-1}(P,Q)) \neq 0, \\
\mathrm{SRemS}_{k+1}(P,Q) &= -\mathrm{Rem}(\mathrm{SRemS}_{k-1}(P,Q), \mathrm{SRemS}_k(P,Q)) = 0.
\end{aligned}$$

The signs introduced here are unimportant in the algebraically closed case. They play an important role when we consider analogous problems over real closed fields in Chapter 2. □

In the above, each SRemS_i(P,Q) is the negative of the remainder in the euclidean division of SRemS_{i-2}(P,Q) by $\mathrm{SRemS}_{i-1}(P,Q)$ for $2 \le i \le k+1$, and the sequence ends with $\mathrm{SRemS}_k(P,Q)$when $\mathrm{SRemS}_{k+1}(P,Q) = 0$, for $k \geq 0$.

Proposition 1.8. *The polynomial* $\mathrm{SRemS}_k(P,\ Q)$ *is a greatest common divisor of P and Q.*

Proof: Observe that if a polynomial A divides two polynomials B, C then it also divides $UB + VC$ for arbitrary polynomials U, V. Since

$$\mathrm{SRemS}_{k+1}(P,Q) = -\mathrm{Rem}(\mathrm{SRemS}_{k-1}(P,Q), \mathrm{SRemS}_k(P,Q)) = 0,$$

$\mathrm{SRemS}_k(P,Q)$ divides $\mathrm{SRemS}_{k-1}(P,Q)$ and since,

$$\mathrm{SRemS}_{k-2}(P,Q) = -\mathrm{SRemS}_k(P,Q) + A\,\mathrm{SRemS}_{k-1}(P,Q),$$

$\mathrm{SRemS}_k(P,Q)$ divides $\mathrm{SRemS}_{k-2}(P,Q)$ using the above observation. Continuing this process one obtains that $\mathrm{SRemS}_k(P,Q)$ divides $\mathrm{SRemS}_1(P,Q) = Q$ and $\mathrm{SRemS}_0(P,Q) = P$.

Also, if any polynomial divides $\mathrm{SRemS}_0(P, Q)$, $\mathrm{SRemS}_1(P, Q)$ (that is P, Q) then it divides $\mathrm{SRemS}_2(P, Q)$ and hence $\mathrm{SRemS}_3(P, Q)$ and so on. Hence, it divides $\mathrm{SRemS}_k(P, Q)$. □

Note that the signed remainder sequence of P and 0 is P and when Q is not 0, the signed remainder sequence of 0 and Q is $0, Q$.

Also, note that by unwinding the definitions of the $\mathrm{SRemS}_i(P, Q)$, we can express $\mathrm{SRemS}_k(P, Q) = \gcd(P, Q)$ as $U P + V Q$ for some polynomials U, V in $\mathrm{K}[X]$. We prove bounds on the degrees of U, V by elucidating the preceding remark.

Proposition 1.9. *If G is a greatest common divisor of P and Q, then there exist U and V with*

$$U P + V Q = G.$$

Moreover, if $\deg(G) = g$, *U and V can be chosen so that* $\deg(U) < q - g$, $\deg(V) < p - g$.

The proof uses the extended signed remainder sequence defined as follows.

Definition 1.10. [Extended signed remainder sequence]
Given $P, Q \in K[X]$, not both 0, let

$$\begin{aligned}
\mathrm{SRemU}_0(P, Q) &= 1,\\
\mathrm{SRemV}_0(P, Q) &= 0,\\
\mathrm{SRemU}_1(P, Q) &= 0,\\
\mathrm{SRemV}_1(P, Q) &= 1,\\
A_{i+1} &= \mathrm{Quo}(\mathrm{SRemS}_{i-1}(P, Q), \mathrm{SRemS}_i(P, Q)),\\
\mathrm{SRemS}_{i+1}(P, Q) &= -\mathrm{SRemS}_{i-1}(P, Q) + A_{i+1}\, \mathrm{SRemS}_i(P, Q),\\
\mathrm{SRemU}_{i+1}(P, Q) &= -\mathrm{SRemU}_{i-1}(P, Q) + A_{i+1}\, \mathrm{SRemU}_i(P, Q),\\
\mathrm{SRemV}_{i+1}(P, Q) &= -\mathrm{SRemV}_{i-1}(P, Q) + A_{i+1}\, \mathrm{SRemV}_i(P, Q)
\end{aligned}$$

for $0 \le i \le k$ where k is the least non-negative integer such that $\mathrm{SRemS}_{k+1} = 0$.

The **extended signed remainder sequence** $\mathrm{Ex}(P, Q)$ **of** P **and** Q is $\mathrm{Ex}_0(P, Q), \ldots, \mathrm{Ex}_k(P, Q)$ with

$$\mathrm{Ex}_i(P, Q) = (\mathrm{SRemS}_i(P, Q), \mathrm{SRemU}_i(P, Q), \mathrm{SRemV}_i(P, Q)).$$ □

The proof of Proposition 1.9 uses the following lemma.

Lemma 1.11. *For* $0 \le i \le k+1$,

$$\mathrm{SRemS}_i(P, Q) = \mathrm{SRemU}_i(P, Q) P + \mathrm{SRemV}_i(P, Q) Q.$$

Let $d_i = \deg(\mathrm{SRemS}_i(P, Q))$. *For* $1 \le i \le k$, $\deg(\mathrm{SRemU}_{i+1}(P, Q)) = q - d_i$, *and* $\deg(\mathrm{SRemV}_{i+1}(P, Q)) = p - d_i$.

Proof: It is easy to verify by induction on i that, for $0 \le i \le k+1$,

$$\mathrm{SRemS}_i(P, Q) = \mathrm{SRemU}_i(P, Q)P + \mathrm{SRemV}_i(P, Q)\, Q.$$

Note that $d_i < d_{i-1}$. The proof of the claim on the degrees proceeds by induction. Clearly, since

$$\begin{aligned} \mathrm{SRemU}_2(P, Q) &= -1 \\ \mathrm{SRemU}_3(P, Q) &= -\mathrm{Quo}(\mathrm{SRemS}_1(P, Q), \mathrm{SRemS}_2(P, Q)), \end{aligned}$$

$$\begin{aligned} \deg(\mathrm{SRemU}_2(P, Q)) &= q - d_1, \\ \deg(\mathrm{SRemU}_3(P, Q)) &= q - d_2. \end{aligned}$$

Similarly,

$$\begin{aligned} \deg(\mathrm{SRemV}_2(P, Q)) &= p - d_1, \\ \deg(\mathrm{SRemV}_3(P, Q)) &= p - d_2. \end{aligned}$$

Using the definitions of $\mathrm{SRemU}_{i+1}(P, Q), \mathrm{SRemV}_{i+1}(P, Q)$ and the induction hypothesis, we get

$$\begin{aligned} \deg(\mathrm{SRemU}_{i-1}(P, Q)) &= q - d_{i-2}, \\ \deg(\mathrm{SRemU}_i(P, Q)) &= q - d_{i-1} \\ \deg(A_{i+1}\,\mathrm{SRemU}_i(P, Q)) &= d_{i-1} - d_i + q - d_{i-1} \\ &= q - d_i > q - d_{i-2}. \end{aligned}$$

Hence, $\deg(\mathrm{SRemU}_{i+1}) = q - d_i$. Similarly,

$$\begin{aligned} \deg(\mathrm{SRemV}_{i-1}(P, Q)) &= p - d_{i-2}, \\ \deg(\mathrm{SRemV}_i(P, Q)) &= p - d_{i-1} \\ \deg(A_{i+1}\,\mathrm{SRemV}_i(P, Q)) &= d_{i-1} - d_i + p - d_{i-1} \\ &= p - d_i > p - d_{i-2}. \end{aligned}$$

Hence, $\deg(\mathrm{SRemV}_{i+1}(P, Q)) = p - d_i$. □

Proof of Proposition 1.9: The claim follows by Lemma 1.11 and Proposition 1.8 since $\mathrm{SRemS}_k(P, Q)$ is a gcd of P and Q, taking

$$U = \mathrm{SRemU}_k(P, Q), V = \mathrm{SRemV}_k(P, Q),$$

and noting that $p - d_{k-1} < p - g$, $q - d_{k-1} < q - g$. □

The extended signed remainder sequence also provides a least common multiple of P and Q.

Proposition 1.12. *The equality*

$$\mathrm{SRemU}_{k+1}(P, Q)\, P = -\mathrm{SRemV}_{k+1}(P, Q)\, Q.$$

holds and $\mathrm{SRemU}_{k+1}(P, Q)P = -\mathrm{SRemV}_{k+1}(P, Q)Q$ *is a least common multiple of* P *and* Q.

Proof: Since $d_k = \deg(\gcd(P, Q))$, $\deg(\mathrm{SRemU}_{k+1}(P, Q)) = q - d_k$, $\deg(\mathrm{SRemV}_k(P,Q)) = p - d_k$, and

$$\mathrm{SRemU}_{k+1}(P,Q)\,P + \mathrm{SRemV}_{k+1}(P,Q)Q = 0,$$

it follows that

$$\mathrm{SRemU}_{k+1}(P,Q)P = -\mathrm{SRemV}_{k+1}(P,Q)Q$$

is a common multiple of P and Q of degree $p+q-d_k$, hence a least common multiple of P and Q. □

Definition 1.13. [Greatest common divisor of a family] A **greatest common divisor of a finite family of polynomials** is a divisor of all the polynomials in the family that is also a multiple of any polynomial that divides every polynomial in the family. A greatest common divisor of a family can be obtained inductively on the number of elements of the family by

$$\begin{aligned}\gcd(\emptyset) &= 0,\\ \gcd(\mathcal{P}\cup\{P\}) &= \gcd(P, \gcd(\mathcal{P})).\end{aligned}$$

□

Note that

- $x \in \mathrm{C}$ is a root of every polynomial in $\mathcal{P}$ if and only if it is a root of $\gcd(\mathcal{P})$,
- $x \in \mathrm{C}$ is not a root of any polynomial in $\mathcal{Q}$ if and only if it is not a root of $\prod_{Q\in\mathcal{Q}} Q$ (with the convention that the product of the empty family is 1),
- every root of P in C is a root of Q if and only if $\gcd(P, Q^{\deg(P)}) = P$ (with the convention that $Q^{\deg(0)} = 0$).

With these observations the following lemma is clear:

Lemma 1.14. *If $\mathcal{P}$, $\mathcal{Q}$ are two finite subsets of* $\mathrm{D}[X]$, *then there is an $x \in \mathrm{C}$ such that*

$$\left(\bigwedge_{P\in\mathcal{P}} P(x)=0\right)\wedge\left(\bigwedge_{Q\in\mathcal{Q}} Q(x)\neq 0\right)$$

if and only if

$$\deg(\gcd(\gcd(\mathcal{P}), \prod_{Q\in\mathcal{Q}} Q^d)) \neq \deg(\gcd(\mathcal{P})),$$

where d is any integer greater than $\deg(\gcd(\mathcal{P}))$.

Note that when $\mathcal{Q} = \emptyset$, since $\prod_{Q\in\emptyset} Q = 1$, the lemma says that there is an $x \in \mathrm{C}$ such that $\bigwedge_{P\in\mathcal{P}} P(x) = 0$ if and only if $\deg(\gcd(\mathcal{P})) \neq 0$. Note also that when $\mathcal{P} = \emptyset$, the lemma says that there is an $x \in \mathrm{C}$ such that $\bigwedge_{Q\in\mathcal{Q}} Q(x) \neq 0$ if and only if $\deg(\prod_{Q\in\mathcal{Q}} Q) \geq 0$, i.e. $1 \notin \mathcal{Q}$.

Exercise 1.8. Design an algorithm to decide whether or not a basic constructible set in C is empty.

1.3 Projection Theorem for Constructible Sets

Now that we know how to decide whether or not a basic constructible set in C is empty, we can show that the projection from C^{k+1} to C^k of a basic constructible set is constructible. We shall do this by viewing the multivariate situation as a univariate situation with parameters. Viewing a univariate algorithm parametrically to obtain a multivariate algorithm is among the most important paradigms used throughout this book.

More precisely, the basic constructible set $S \subset C^{k+1}$ can be described as

$$S = \{z \in C^{k+1} \mid \bigwedge_{P \in \mathcal{P}} P(z) = 0 \wedge \bigwedge_{Q \in \mathcal{Q}} Q(z) \neq 0\}$$

with $\mathcal{P}, \mathcal{Q}$ finite subsets of $C[Y_1, \ldots, Y_k, X]$, and its projection $\pi(S)$ (forgetting the last coordinate) is

$$\pi(S) = \{y \in C^k \mid \exists x \in C \ (\bigwedge_{P \in \mathcal{P}} P(y, x) = 0 \wedge \bigwedge_{Q \in \mathcal{Q}} Q(y, x) \neq 0)\}.$$

We can consider the polynomials in $\mathcal{P}$ and $\mathcal{Q}$ as polynomials in the single variable X with the variables $(Y_1, \ldots, Y_k)$ appearing as parameters. For a specialization of Y to $y = (y_1, \ldots, y_k) \in C^k$, we write $P_y(X)$ for $P(y_1, \ldots, y_k, X)$. Hence,

$$\pi(S) = \{y \in C^k \mid \exists x \in C \quad (\bigwedge_{P \in \mathcal{P}} P_y(x) = 0 \wedge \bigwedge_{Q \in \mathcal{Q}} Q_y(x) \neq 0)\},$$

and, for a particular $y \in C^k$ we can decide, using Exercise 1.8, whether or not

$$\exists x \in C \quad (\bigwedge_{P \in \mathcal{P}} P_y(x) = 0 \wedge \bigwedge_{Q \in \mathcal{Q}} Q_y(x) \neq 0)$$

is true.

Defining

$$S_y = \{x \in C \mid \bigwedge_{P \in \mathcal{P}} P_y(x) = 0 \wedge \bigwedge_{Q \in \mathcal{Q}} Q_y(x) \neq 0\},$$

what is crucial now is to partition the parameter space C^k into finitely many parts so that the decision algorithm testing whether S_y is empty or not is the same (is uniform) for all y in any given part. Because of this uniformity, it will turn out that each part of the partition is a constructible set. Since $\pi(S)$ is the union of those parts where $S_y \neq \emptyset$, $\pi(S)$ is constructible being the union of finitely many constructible sets.

We next study the signed remainder sequence of P_y and Q_y for all possible specialization of Y to $y \in C^k$. This cannot be done in a completely uniform way, since denominators appear in the euclidean division process. Nevertheless, fixing the degrees of the polynomials in the signed remainder sequence, it is possible to partition the parameter space, C^k, into a finite number of parts so that the signed remainder sequence is uniform in each part.

Example 1.15. We consider a general polynomial of degree 4. Dividing by its leading coefficient, it is not a loss of generality to take P to be monic. So let $P = X^4 + \alpha X^3 + \beta X^2 + \gamma X + \delta$. Since the translation $X \mapsto X - \alpha/4$ kills the term of degree 3, we can suppose $P = X^4 + a X^2 + b X + c$.

Consider $P = X^4 + a X^2 + b X + c$ and its derivative $P' = 4X^3 + 2a X + b$. Their signed remainder sequence in $\mathbb{Q}(a, b, c)[X]$ is

$$\begin{aligned}
P &= X^4 + a X^2 + b X + c \\
P' &= 4 X^3 + 2 a X + b \\
S_2 &= -\mathrm{Rem}(P, P') \\
&= -\frac{1}{2} a X^2 - \frac{3}{4} b X - c \\
S_3 &= -\mathrm{Rem}(P', S_2) \\
&= \frac{(8 a c - 9 b^2 - 2 a^3) X}{a^2} - \frac{b(12 c + a^2)}{a^2} \\
S_4 &= -\mathrm{Rem}(S_2, S_3) \\
&= \frac{1}{4} \frac{a^2(256 c^3 - 128 a^2 c^2 + 144 a c b^2 - 16 a^4 c - 27 b^4 - 4 b^2 a^3)}{(8 a c - 9 b^2 - 2 a^3)^2}
\end{aligned}$$

Note that when $(a,\ b,\ c)$ are specialized to values in $\mathbb{C}^3$ for which $a = 0$ or $8 a c - 9 b^2 - 2 a^3 = 0$, the signed remainder sequence of P and P' for these special values is not obtained by specializing a, b, c in the signed remainder sequence in $\mathbb{Q}(a, b, c)[X]$. □

In order to take into account all the possible signed remainder sequences that can appear when we specialize the parameters, we introduce the following definitions and notation.

We get rid of denominators appearing in the remainders through the notion of signed pseudo-remainders. Let

$$\begin{aligned}
P &= a_p X^p + \cdots + a_0 \in \mathrm{D}[X], \\
Q &= b_q X^q + \cdots + b_0 \in \mathrm{D}[X],
\end{aligned}$$

where D is a subring of C. Note that the only denominators occurring in the euclidean division of P by Q are b_q^i, $i \le p - q + 1$. The **signed pseudo-remainder** denoted $\mathrm{PRem}(P, Q)$, is the remainder in the euclidean division of $b_q^d\, P$ by Q, where d is the smallest even integer greater than or equal to $p - q + 1$. Note that the euclidean division of $b_q^d P$ by Q can be performed in D and that $\mathrm{PRem}(P, Q) \in \mathrm{D}[X]$. The even exponent is useful in Chapter 2 and later when we deal with signs.

Notation 1.16. [Truncation] Let $Q = b_q X^q + \cdots + b_0 \in \mathrm{D}[X]$. We define for $0 \le i \le q$, the **truncation of Q at i** by

$$\mathrm{Tru}_i(Q) = b_i X^i + \cdots + b_0.$$

The **set of truncations** of a non-zero polynomial $Q \in \mathrm{D}[Y_1, \ldots, Y_k][X]$, where $Y_1, \ldots, Y_k$ are parameters and X is the main variable, is the finite subset of $\mathrm{D}[Y_1, \ldots, Y_k][X]$ defined by

$$\mathrm{Tru}(Q) = \begin{cases} \{Q\} & \text{if } \mathrm{lcof}(Q) \in \mathrm{D} \text{ or } \deg(Q) = 0, \\ \{Q\} \cup \mathrm{Tru}(\mathrm{Tru}_{\deg_X(Q)-1}(Q)) & \text{otherwise.} \end{cases}$$

The **tree of possible signed pseudo-remainder sequences** of two polynomials $P, Q \in \mathrm{D}[Y_1, \ldots, Y_k][X]$, denoted $\mathrm{TRems}(P, Q)$ is a tree whose root R contains P. The children of the root contain the elements of the set of truncations of Q. Each node N contains a polynomial $\mathrm{Pol}(N) \in \mathrm{D}[Y_1, \ldots, Y_k][X]$. A node N is a leaf if $\mathrm{Pol}(N) = 0$. If N is not a leaf, the children of N contain the truncations of $-\mathrm{PRem}(\mathrm{Pol}(p(N)), \mathrm{Pol}(N))$ where $p(N)$ is the parent of N.□

Example 1.17. As in Example 1.15, we consider $P = X^4 + a X^2 + b X + c$ and its derivative $P' = 4X^3 + 2a X + b$. Denoting

$$\begin{aligned}
\overline{S_2} &= -\mathrm{PRem}(P, P') \\
&= -8\,a\,X^2 - 12\,b\,X - 16\,c, \\
\overline{S_3} &= -\mathrm{PRem}(P', \overline{S_2}) \\
&= 64\,((8\,a\,c - 9\,b^2 - 2\,a^3)X - b\,(12\,c + a^2)), \\
\overline{S_4} &= -\mathrm{PRem}(\overline{S_3}, \overline{S_2}) \\
&= 16384\,a^2\left(256\,c^3 - 128\,a^2\,c^2 + 144\,a\,b^2\,c + 16\,a^4\,c - 27\,b^4 - 4\,a^3\,b^2\right), \\
u &= -\mathrm{PRem}(P', (\overline{S_2})) \\
&= 768\,b\left(-27\,b^4 + 72\,a\,c\,b^2 + 256\,c^3\right)
\end{aligned}$$

the tree $\mathrm{TRems}(P, P')$ is the following.

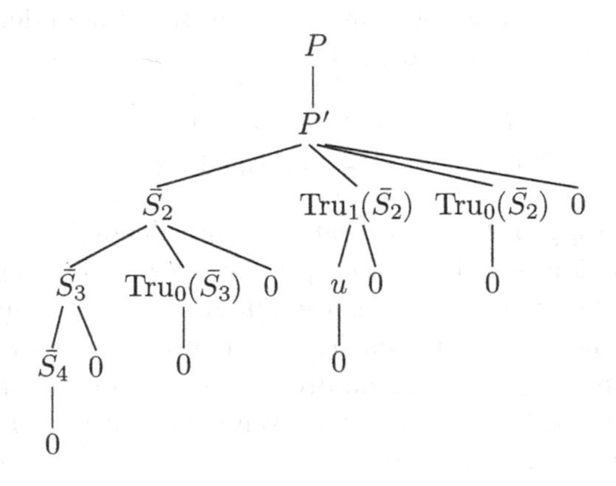

Define

$$\begin{aligned}
s &= 8\,a\,c - 9\,b^2 - 2\,a^3, \\
t &= -b\,(12\,c + a^2) \\
\delta &= 256\,c^3 - 128\,a^2\,c^2 + 144\,a\,b^2\,c + 16\,a^4\,c - 27\,b^4 - 4\,a^3\,b^2.
\end{aligned}$$

The leftmost path in the tree going from the root to a leaf, namely the path $P, P', S_2, S_3, S_4, 0$ can be understood as follows: if $(a, b, c) \in \mathrm{C}^3$ are such that the degree of the polynomials in the remainder sequence of P and P' are $4, 3, 2, 1, 0$, i.e. when $a \neq 0, s \neq 0, \delta \neq 0$ (getting rid of obviously irrelevant factors), then the signed remainder sequence of $P = X^4 + a X^2 + b X + c$ and P' is proportional (up to non-zero squares of elements in C) to $P, P', \overline{S_2}, \overline{S_3}, \overline{S_4}$.□

Notation 1.18. [Degree] For a specialization of $Y = (Y_1, \ldots, Y_k)$ to $y \in \mathrm{C}^k$, and $Q \in \mathrm{D}[Y_1, \ldots, Y_k][X]$, we denote the polynomial in $\mathrm{C}[X]$ obtained by substituting y for Y by Q_y. Given $\mathcal{Q} \subset \mathrm{D}[Y_1, \ldots, Y_k][X]$, we define $\mathcal{Q}_y \subset \mathrm{C}[X]$ as $\{Q_y \mid Q \in \mathcal{Q}\}$.

Let $Q = b_q X^q + \cdots + b_0 \in \mathrm{D}[Y_1, \ldots, Y_k][X]$. We define the basic formula $\deg_X(Q) = i$ as

$$\begin{cases} b_q = 0 \wedge \ldots \wedge b_{i+1} = 0 \wedge b_i \neq 0 & \text{when } 0 \leq i < q, \\ b_q \neq 0 & \text{when } i = q, \\ b_q = 0 \wedge \ldots \wedge b_0 = 0 & \text{when } i = -\infty, \end{cases}$$

so that the sets $\mathrm{Reali}(\deg_X(Q) = i)$ partition C^k and $y \in \mathrm{Reali}(\deg_X(Q) = i)$ if and only if $\deg(Q_y) = i$.

Note that $\mathrm{PRem}(P_y, Q_y) = \mathrm{PRem}(P, \mathrm{Tru}_i(Q))_y$ where $\deg_X(Q_y) = i$.

Given a leaf L of $\mathrm{TRems}(P, Q)$, we denote by $\mathcal{B}_L$ the unique path from the root of $\mathrm{TRems}(P, Q)$ to the leaf L. If N is a node in $\mathcal{B}_L$ which is not a leaf, we denote by $c(N)$ the unique child of N in $\mathcal{B}_L$. We denote by $\mathcal{C}_L$ the basic formula

$$\begin{aligned} \deg_X(Q) &= \deg_X(\mathrm{Pol}(c(R))) \wedge \\ \bigwedge_{N \in \mathcal{B}_L, N \neq R} \deg_X(-\mathrm{PRem}(\mathrm{Pol}(p(N)), \mathrm{Pol}(N))) &= \deg_X(\mathrm{Pol}(c(N))) \end{aligned}$$

□

It is clear from the definitions, since the remainder and pseudo-remainder of two polynomials in $C[X]$ are equal up to a square, that

Lemma 1.19. *The* $\mathrm{Reali}(\mathcal{C}_L)$ *partition* C^k*. Moreover,* $y \in \mathrm{Reali}(\mathcal{C}_L)$ *implies that the signed remainder sequence of* P_y *and* Q_y *is proportional (up to a square) to the sequence of polynomials* $\mathrm{Pol}(N)_y$ *in the nodes along the path* $\mathcal{B}_L$ *leading to* L*. In particular,* $\mathrm{Pol}(p(L))_y$ *is* $\gcd(P_y, Q_y)$*.*

We will now define the set of possible greatest common divisors of a family $\mathcal{P} \subset \mathrm{D}[Y_1, \ldots, Y_k][X]$, called $\mathrm{posgcd}(\mathcal{P})$, which is a finite set containing all the possible greatest common divisors of $\mathcal{P}_y$ which can occur as y ranges over C^k. We define it as a set of pairs $(G, \mathcal{C})$ where $G \in \mathrm{D}[Y_1, \ldots, Y_k][X]$ and $\mathcal{C}$ is a basic formula with coefficients in D so that for each pair $(G, \mathcal{C})$, $y \in \mathrm{Reali}(\mathcal{C})$ implies $\gcd(\mathcal{P}_y) = G_y$. More precisely, we shall make the definition so that the following lemma is true:

Lemma 1.20. *For all* $y \in \mathrm{C}^k$*, there exists one and only one* $(G, \mathcal{C}) \in \mathrm{posgcd}(\mathcal{P})$ *such that* $y \in \mathrm{Reali}(\mathcal{C})$*. Moreover,* $y \in \mathrm{Reali}(\mathcal{C})$ *implies that* G_y *is a greatest common divisors of* $\mathcal{P}_y$*.*

The **set of possible greatest common divisors of a finite family of elements of** $\mathbf{K}[Y_1,\ldots,Y_k][X]$ is defined recursively on the number of elements of the family by

$$\begin{aligned}\mathrm{posgcd}(\emptyset) &= \{(0, 1 \neq 0)\}\\ \mathrm{posgcd}(\mathcal{P} \cup \{P\}) &= \{(\mathrm{Pol}(p(L)), \mathcal{C} \wedge \mathcal{C}_L) \mid (Q, \mathcal{C}) \in \mathrm{posgcd}(\mathcal{P})\\ &\qquad \text{and } L \text{ is a leaf of } \mathrm{TRems}(P, Q)\}.\end{aligned}$$

It is clear from the definitions and Lemma 1.19 that Lemma 1.20 holds.

Example 1.21. Returning to Example 1.17, and using the corresponding notation, the elements of $\mathrm{posgcd}(P, P')$ are (after removing obviously irrelevant factors),

$$\begin{aligned}&(\overline{S}_4, \; a \neq 0 \wedge s \neq 0 \wedge \delta \neq 0),\\ &(\overline{S}_3, \; a \neq 0 \wedge s \neq 0 \wedge \delta = 0),\\ &(\mathrm{Tru}_0(\overline{S}_3), \; a \neq 0 \wedge s = 0 \wedge t \neq 0),\\ &(\overline{S}_2, \; a \neq 0 \wedge s = t = 0),\\ &(u, \; a = 0 \wedge b \neq 0 \wedge u \neq 0),\\ &(\mathrm{Tru}_1(\overline{S}_2), \; a = 0 \wedge b \neq 0 \wedge u = 0),\\ &(\mathrm{Tru}_0(\overline{S}_2), \; a = b = 0 \wedge c \neq 0),\\ &(P', \; a = b = c = 0).\end{aligned}$$

The first pair, which corresponds to the leftmost leaf of $\mathrm{TRems}(P, P')$ can be read as: if $a \neq 0$, $s \neq 0$, and $\delta \neq 0$ (i.e. if the degrees of the polynomials in the remainder sequence are $4, 3, 2, 1, 0$), then $\gcd(P, P') = \overline{S}_4$. The second pair, which corresponds to the next leaf (going left to right) means that if $a \neq 0$, $s \neq 0$, and $\delta = 0$ (i.e. if the degrees of the polynomials in the remainder sequence are $4, 3, 2, 1$), then $\gcd(P, P') = \overline{S}_3$.

If $P = X^4 + a X^2 + b X + c$, the projection of

$$\{(a, b, c, x) \in \mathrm{C}^4 \mid P(x) = P'(x) = 0\}$$

to C^3 is the set of polynomials (where a polynomial is identified with its coefficients (a, b, c)) for which $\deg(\gcd(P, P')) \geq 1$. Therefore, the formula $\exists x \; P(x) = P'(x) = 0$ is equivalent to the formula

$$\begin{aligned}&(a \neq 0 \wedge s \neq 0 \wedge \delta = 0)\\ \vee \; &(a \neq 0 \wedge s = t = 0)\\ \vee \; &(a = 0 \wedge b \neq 0 \wedge u = 0)\\ \vee \; &(a = b = c = 0).\end{aligned}$$

□

The proof of the following projection theorem is based on the preceding constructions of possible gcd.

Theorem 1.22. [Projection theorem for constructible sets] *Given a constructible set in C^{k+1} defined by polynomials with coefficients in* D*, its projection to* C^k *is a constructible set defined by polynomials with coefficients in* D.

Proof: Since every constructible set is a finite union of basic constructible sets it is sufficient to prove that the projection of a basic constructible set is constructible. Suppose that the basic constructible set S in C^{k+1} is

$$\{(y,x)\in \mathrm{C}^k\times \mathrm{C} \mid \bigwedge_{P\in\mathcal{P}} P(y,x)=0 \wedge \bigwedge_{Q\in\mathcal{Q}} Q(y,x)\neq 0\}$$

with $\mathcal{P}$ and $\mathcal{Q}$ finite subsets of $\mathrm{D}[Y_1,\ldots,Y_k,X]$.

Let

$$\mathcal{L}=\mathrm{posgcd}(\{P \mid \exists \mathcal{C}\quad (P,\mathcal{C})\in \mathrm{posgcd}(\mathcal{P})\}\cup\{\prod_{Q\in\mathcal{Q}} Q^d\})$$

where d is the least integer greater than the degree in X of any polynomial in $\mathcal{P}$.

For every $(G,\mathcal{C})\in\mathcal{L}$, there exists a unique $(G_1,\mathcal{C}_1)\in \mathrm{posgcd}(\mathcal{P})$ with $\mathcal{C}_1$ a conjunction of a subset of the atoms appearing in $\mathcal{C}$. Using Lemma 1.14, the projection of S on C^k is the union of the $\mathrm{Reali}(\mathcal{C}\wedge \deg_X(G)\neq \deg_X(G_1))$ for $(G,\mathcal{C})$ in $\mathcal{L}$, and this is clearly a constructible set defined by polynomials with coefficients in D. □

Exercise 1.9.

a) Find the conditions on (a,b,c) for $P=a\,X^2+b\,X+c$ and $P'=2a\,X+b$ to have a common root.

b) Find the conditions on (a,b,c) for $P=a\,X^2+b\,X+c$ to have a root which is not a root of P'.

1.4 Quantifier Elimination and the Transfer Principle

Returning to logical terminology, Theorem 1.22 implies that the theory of algebraically closed fields admits quantifier elimination in the language of fields, which is the following theorem.

Theorem 1.23. [Quantifier Elimination over Algebraically Closed Fields] *Let $\Phi(Y_1,\ldots,Y_\ell)$ be a formula in the language of fields with free variables $\{Y_1,\ldots,Y_\ell\}$, and coefficients in a subring* D *of the algebraically closed field* C. *Then there is a quantifier free formula $\Psi(Y_1,\ldots,Y_\ell)$ with coefficients in* D *which is* C*-equivalent to $\Phi(Y_1,\ldots,Y_\ell)$.*

Notice that an example of quantifier elimination appears in Example 1.2.

The proof of the theorem is by induction on the number of quantifiers, using as base case the elimination of an existential quantifier which is given by Theorem 1.22.

Proof of Theorem 1.23: Given a formula $\Theta(Y) = (\exists X)\ \mathcal{B}(X, Y)$, where $\mathcal{B}$ is a quantifier free formula whose atoms are equations and inequations involving polynomials in $\mathrm{D}[X, Y_1, \ldots, Y_k]$, Theorem 1.22 shows that there is a quantifier free formula $\Xi(Y)$ with coefficients in D that is equivalent to $\Theta(Y)$, since $\mathrm{Reali}(\Theta(Y), \mathrm{C}^k)$, which is the projection of the constructible set $\mathrm{Reali}(\mathcal{B}(X, Y), \mathrm{C}^{k+1})$, is constructible, and constructible sets are realizations of quantifier free formulas. Since $(\forall X)\ \Phi$ is equivalent to $\neg((\exists X)\ \neg(\Phi))$, the theorem immediately follows by induction on the number of quantifiers. □

Corollary 1.24. *Let $\Phi(Y)$ be a formula in the language of fields with coefficients in* C*. The set $\{y \in \mathrm{C}^k | \Phi(y)\}$ is constructible.*

Corollary 1.25. *A subset of* C *defined by a formula in the language of fields with coefficients in* C *is a finite set or the complement of a finite set.*

Proof: By Corollary 1.24, a subset of C defined by a formula in the language of fields with coefficients in C is constructible, and this is a finite set or the complement of a finite set by Exercise 1.2. □

Exercise 1.10. Prove that the sets $\mathbb{N}$ and $\mathbb{Z}$ are not constructible subsets of $\mathbb{C}$. Prove that the sets $\mathbb{N}$ and $\mathbb{Z}$ cannot be defined inside $\mathbb{C}$ by a formula of the language of fields with coefficients in $\mathbb{C}$.

Theorem 1.23 easily implies the following theorem, known as the transfer principle for algebraically closed fields. It is also called the Lefschetz Principle.

Theorem 1.26. [Lefschetz principle] *Suppose that* C' *is an algebraically closed field which contains the algebraically closed field* C*. If* Φ *is a sentence in the language of fields with coefficients in* C*, then it is true in* C *if and only if it is true in* C'*.*

Proof: By Theorem 1.23, there is a quantifier free formula Ψ which is C-equivalent to Φ. It follows from the proof of Theorem 1.22 that Ψ is C'-equivalent to Φ as well. Notice, too, that since Ψ is a sentence, Ψ is a boolean combination of atoms of the form $c = 0$ or $c \neq 0$, where $c \in \mathrm{C}$. Clearly, Ψ is true in C if and only if it is true in C'. □

The **characteristic of a field K** is a prime number p if K contains $\mathbb{Z}/p\mathbb{Z}$ and 0 if K contains $\mathbb{Q}$. The meaning of Lefschetz principle is essentially that a sentence is true in an algebraic closed field if and only if it is true in any other algebraic closed field of the same characteristic.

Let C denote an algebraically closed field and C' an algebraically closed field containing C.

Given a constructible set S in C^k, the **extension of S to C'**, denoted $\mathrm{Ext}(S, \mathrm{C}')$ is the constructible subset of C'^k defined by a quantifier free formula that defines S.

The following proposition is an easy consequence of Theorem 1.26.

Proposition 1.27. *Given a constructible set S in C^k, the set $\mathrm{Ext}(S, \mathrm{C}')$ is well defined (i.e. it only depends on the set S and not on the quantifier free formula chosen to describe it).*

The operation $S \to \mathrm{Ext}(S, \mathrm{C}')$ preserves the boolean operations (finite intersection, finite union and complementation).

If $S \subset T$, then $\mathrm{Ext}(S, \mathrm{C}') \subset \mathrm{Ext}(T, \mathrm{C}')$, where T is a constructible set in C^k.

Exercise 1.11. Prove proposition 1.27.

Exercise 1.12. Show that if S is a finite constructible subset of C^k, then $\mathrm{Ext}(S, \mathrm{C}')$ is equal to S. (Hint: write a formula describing S).

1.5 Bibliographical Notes

Lefschetz's principle (Theorem 1.26) is stated without proof in [105]. Indications for a proof of quantifier elimination over algebraically closed fields (Theorem 1.23) are given in [156] (Remark 16).

2

Real Closed Fields

Real closed fields are fields which share the algebraic properties of the field of real numbers. In Section 2.1, we define ordered, real and real closed fields and state some of their basic properties. Section 2.2 is devoted to real root counting. In Section 2.3 we define semi-algebraic sets and prove that the projection of an algebraic set is semi-algebraic. The main technique used is a parametric version of real root counting algorithm described in the second section. In Section 2.4, we prove that the projection of a semi-algebraic set is semi-algebraic, by a similar method. Section 2.5 is devoted to several applications of the projection theorem, of logical and geometric nature. In Section 2.6, an important example of a non-archimedean real closed field is described: the field of Puiseux series.

2.1 Ordered, Real and Real Closed Fields

Before defining ordered fields, we prove a few useful properties of fields of characteristic zero.

Let K be a field of characteristic zero. The **derivative** of a polynomial

$$P = a_p X^p + \cdots + a_i X^i + \cdots + a_0 \in \mathrm{K}[X]$$

is denoted P' with

$$P' = p\, a_p X^{p-1} + \cdots + i\, a_i X^{i-1} + \cdots + a_1.$$

The i-th derivative of P, $P^{(i)}$, is defined inductively by $P^{(i)} = \left(P^{(i-1)}\right)'$. It is immediate to verify that

$$\begin{aligned}(P+Q)' &= P' + Q',\\ (PQ)' &= P'Q + PQ'.\end{aligned}$$

Taylor's formula holds:

Proposition 2.1. [Taylor's formula] *Let* K *be a field of characteristic zero,*

$$P = a_p X^p + \cdots + a_i X^i + \cdots + a_0 \in \mathrm{K}[X] \text{ and } x \in \mathrm{K}.$$

Then,

$$P = \sum_{i=0}^{\deg(P)} \frac{P^{(i)}(x)}{i!}(X-x)^i.$$

Proof: We prove Taylor's formula holds for monomials X^p by induction on p. The claim is clearly true if $p=0$. Suppose that Taylor's formula holds for $p-1$:

$$X^{p-1} - \sum_{i=0}^{p-1} \frac{(p-1)!}{(p-1-i)!\,i!} x^{p-1-i}(X-x)^i.$$

Then, since $X = x + (X-x)$,

$$\begin{aligned} X^p &= (x+(X-x))\sum_{i=0}^{p-1} \frac{(p-1)!}{(p-1-i)!\,i!} x^{p-1-i}(X-x)^i \\ &= \sum_{i=0}^{p} \frac{p!}{(p-i)!\,i!} x^{p-i}(X-x)^i \end{aligned}$$

since

$$\frac{p!}{(p-i)!\,i!} = \frac{(p-1)}{(p-i)!\,(i-1)!} + \frac{p!}{(p-1-i)!\,(i-1)!}.$$

Hence, Taylor's formula is valid for any polynomial using the linearity of derivation. □

Let $x \in \mathrm{K}$ and $P \in \mathrm{K}[X]$. The **multiplicity** of x as a root of P is the natural number μ such that there exists $Q \in \mathrm{K}[X]$ with $P = (X-x)^\mu Q(X)$ and $Q(x) \neq 0$. Note that if x is not a root of P, the multiplicity of x as a root of P is equal to 0.

Lemma 2.2. *Let* K *be a field of characteristic zero. The element* $x \in \mathrm{K}$ *is a root of* $P \in \mathrm{K}[X]$ *of multiplicity* μ *if and only if*

$$P^{(\mu)}(x) \neq 0, P^{(\mu-1)}(x) = \cdots = P(x) = P'(x) = 0.$$

Proof: Suppose that $P = (X-x)^\mu Q$ and $Q(x) \neq 0$. It is clear that $P(x) = 0$. The proof of the claim is by induction on the degree of P. The claim is obviously true for $\deg(P) = 1$. Suppose that the claim is true for every polynomial of degree $< d$. Since

$$P' = (X-x)^{\mu-1}(\mu Q + (X-x)Q'),$$

and $\mu Q(x) \neq 0$, by induction hypothesis,

$$P'(x) = \cdots = P^{(\mu-1)}(x) = 0, P^{(\mu)}(x) \neq 0.$$

Conversely suppose that

$$P(x) = P'(x) = \cdots = P^{(\mu-1)}(x) = 0, P^{(\mu)}(x) \neq 0.$$

By Proposition 2.1 (Taylor's formula) at x, $P=(X-x)^{\mu}Q$, with

$$Q(x)=P^{(\mu)}(x)/\mu!\neq 0.$$ □

A polynomial $P\in \mathrm{K}[X]$ is **separable** if the greatest common divisor of P and P' is an element of $\mathrm{K}\setminus\{0\}$. A polynomial P is **square-free** if there is no non-constant polynomial $A\in\mathrm{K}[X]$ such that A^2 divides P.

Exercise 2.1. Prove that $P\in\mathrm{K}[X]$ is separable if and only if P has no multiple root in C, where C is an algebraically closed field containing K. If the characteristic of K is 0, prove that $P\in\mathrm{K}[X]$ is separable if and only P is square-free.

A **partially ordered set** $(A, \preceq)$ is a set A, together with a binary relation $\preceq$ that satisfies:

- $\preceq$ is transitive, i.e. $a\preceq b$ and $a\preceq c \Rightarrow a\preceq c$,
- $\preceq$ is reflexive, i.e. $a\preceq a$,
- $\preceq$ is anti-symmetric, i.e. $a\preceq b$ and $b\preceq a \Rightarrow a=b$.

A standard example of a partially ordered set is the power set

$$2^A=\{B \mid B\subseteq A\},$$

the binary relation being the inclusion between subsets of A.

A **totally ordered set** is a partially ordered set $(A, \leq)$ with the additional property that every two elements $a, b\in A$ are comparable, i.e. $a\leq b$ or $b\leq a$. In a totally ordered set, $a<b$ stands for $a\leq b$, $a\neq b$, and $a\geq b$ (resp. $a>b$) for $b\leq a$ (resp. $b<a$).

An **ordered ring** $(\mathrm{A},\leq)$ is a ring, A, together with a total order, $\leq$, that satisfies:

$$x\leq y \;\Rightarrow\; x+z\leq y+z$$
$$0\leq x,\ 0\leq y \;\Rightarrow\; 0\leq x y.$$

An **ordered field** $(\mathrm{F},\leq)$ is a field, F, which is an ordered ring.

An ordered ring $(\mathrm{A},\leq)$ is contained in an ordered field $(\mathrm{F},\leq)$ if $\mathrm{A}\subset\mathrm{F}$ and the inclusion is order preserving. Note that the ordered ring $(\mathrm{A},\leq)$ is necessarily an ordered integral domain.

Exercise 2.2. Prove that in an ordered field $-1<0$.

Prove that an ordered field has characteristic zero.

Prove the law of trichotomy in an ordered field: for every a in the field, exactly one of $a<0$, $a=0$, $a>0$ holds.

Notation 2.3. [Sign] The **sign** of an element a in ordered field $(\mathrm{F},\leq)$ is defined by

$$\begin{cases} \mathrm{sign}(a)=0 & \text{if } a=0,\\ \mathrm{sign}(a)=1 & \text{if } a>0,\\ \mathrm{sign}(a)=-1 & \text{if } a<0.\end{cases}$$

When $a>0$ we say a is positive, and when $a<0$ we say a is negative.

The **absolute value** $|a|$ of a is the maximum of a and $-a$ and is non-negative. □

The fields $\mathbb{Q}$ and $\mathbb{R}$ with their natural order are familiar examples of ordered fields.

Exercise 2.3. Show that it is not possible to order the field of complex numbers $\mathbb{C}$ so that it becomes an ordered field.

In an ordered field, the value at x of a polynomial has the sign of its leading monomial for x sufficiently large. More precisely,

Proposition 2.4. *Let $P = a_p X^p + \cdots + a_0$, $a_p \neq 0$, be a polynomial with coefficients in an ordered field* F. *If $|x|$ is bigger than $2\sum_{0\le i\le p}\frac{|a_i|}{|a_p|}$, then $P(x)$ and $a_p x^p$ have the same sign.*

Proof: Suppose that

$$|x|>2\sum_{0\le i\le p}\left|\frac{a_i}{a_p}\right|,$$

which implies $|x|>2$. Since

$$\begin{aligned}
\frac{P(x)}{a_p x^p} &= 1+\sum_{0\le i\le p-1}\frac{a_i}{a_p}x^{i-p},\\
\frac{P(x)}{a_p x^p} &\ge 1-\left(\sum_{0\le i\le p-1}\frac{|a_i|}{|a_p|}|x|^{i-p}\right)\\
&\ge 1-\left(\sum_{0\le i\le p}\frac{|a_i|}{|a_p|}\right)(|x|^{-1}+|x|^{-2}+\cdots+|x|^{-p})\\
&\ge 1-\frac{1}{2}(1+|x|^{-1}+\cdots+|x|^{-p+1})\\
&= 1-\frac{1}{2}\left(\frac{1-|x|^{-p}}{1-|x|^{-1}}\right)>0.
\end{aligned}$$

□

We now examine a particular way to order the field of rational functions $\mathbb{R}(X)$.

For this purpose, we need a definition: Let $\mathrm{F}\subset\mathrm{F}'$ be two ordered fields. The element $f\in\mathrm{F}'$ is **infinitesimal over F** if it is a positive element smaller than any positive $f\in\mathrm{F}$. The element $f\in\mathrm{F}'$ is **unbounded over F** if it is a positive element greater than any positive $f\in\mathrm{F}$.

Notation 2.5. [Order 0_+] Let F be an ordered field and ε a variable. There is one and only one order on $\mathrm{F}(\varepsilon)$, denoted 0_+, such that ε is infinitesimal over F. If

$$P(\varepsilon)=a_p\,\varepsilon^p+a_{p-1}\,\varepsilon^{p-1}+\cdots+a_{m+1}\,\varepsilon^{m+1}+a_m\,\varepsilon^m$$

with $a_m \neq 0$, then $P(\varepsilon) > 0$ in this order if and only if $a_m > 0$. If $P(\varepsilon)/Q(\varepsilon) \in \mathrm{F}(\varepsilon)$, $P(\varepsilon)/Q(\varepsilon) > 0$ if and only if $P(\varepsilon)\,Q(\varepsilon) > 0$.

Note that the field $\mathrm{F}(\varepsilon)$ with this order contains infinitesimal elements over F, such as ε. The field also contains elements which are unbounded over F such as $1/\varepsilon$. □

Exercise 2.4. Show that 0_+ is an order on $\mathrm{F}(\varepsilon)$ and that it is the only order in which ε is infinitesimal over F.

We define now a cone of a field, which should be thought of as a set of non-negative elements. A **cone** of the field F is a subset $\mathcal{C}$ of F such that:

$$\begin{aligned} x \in \mathcal{C},\ y \in \mathcal{C} &\Rightarrow x + y \in \mathcal{C} \\ x \in \mathcal{C},\ y \in \mathcal{C} &\Rightarrow x\,y \in \mathcal{C} \\ x \in \mathrm{F} &\Rightarrow x^2 \in \mathcal{C}. \end{aligned}$$

The cone $\mathcal{C}$ is **proper** if in addition $-1 \notin \mathcal{C}$.

Let $(\mathrm{F}, \leq)$ be an ordered field. The subset $\mathcal{C} = \{x \in \mathrm{F} \mid x \geq 0\}$ is a cone, the **positive cone** of $(\mathrm{F}, \leq)$.

Proposition 2.6. *Let* $(\mathrm{F}, \leq)$ *be an ordered field. The positive cone* $\mathcal{C}$ *of* $(\mathrm{F}, \leq)$ *is a proper cone that satisfies* $\mathcal{C} \cup -\mathcal{C} = \mathrm{F}$. *Conversely, if* $\mathcal{C}$ *is a proper cone of a field* F *that satisfies* $\mathcal{C} \cup -\mathcal{C} = \mathrm{F}$, *then* F *is ordered by* $x \leq y \Leftrightarrow y - x \in \mathcal{C}$.

Exercise 2.5. Prove Proposition 2.6.

Let K be a field. We denote by $\mathrm{K}^{(2)}$ the set of squares of elements of K and by $\sum \mathrm{K}^{(2)}$ the set of **sums of squares of elements of K**. Clearly, $\sum \mathrm{K}^{(2)}$ is a cone contained in every cone of K.

A field K is a **real field** if $-1 \notin \sum \mathrm{K}^{(2)}$.

Exercise 2.6. Prove that a real field has characteristic 0.

Show that the field $\mathbb{C}$ of complex numbers is not a real field.

Show that an ordered field is a real field.

Real fields can be characterized as follows.

Theorem 2.7. *Let* F *be a field. Then the following properties are equivalent*

a) F *is real.*
b) F *has a proper cone.*
c) F *can be ordered.*
d) For every $x_1, \ldots, x_n$ *in* F, $\sum_{i=1}^n x_i^2 = 0 \Rightarrow x_1 = \cdots = x_n = 0$.

The proof of Theorem 2.7 uses the following proposition.

Proposition 2.8. *Let* $\mathcal{C}$ *be a proper cone of* F, $\mathcal{C}$ *is contained in the positive cone for some order on* F.

The proof of Proposition 2.8 relies on the following lemma.

Lemma 2.9. *Let $\mathcal{C}$ be a proper cone of* F. *If* $-a \notin \mathcal{C}$, *then*

$$\mathcal{C}[a] = \{x + a\,y \mid x, y \in \mathcal{C}\}$$

is a proper cone of F.

Proof: Suppose $-1 = x + a\,y$ with $x, y \in \mathcal{C}$. If $y = 0$ we have $-1 \in \mathcal{C}$ which is impossible. If $y \neq 0$ then $-a = (1/y^2)\,y\,(1+x) \in \mathcal{C}$, which is also impossible. □

Proof of Proposition 2.8: Since the union of a chain of proper cones is a proper cone, Zorn's lemma implies the existence of a maximal proper cone $\overline{\mathcal{C}}$ which contains $\mathcal{C}$. It is then sufficient to show that $\overline{\mathcal{C}} \cup -\overline{\mathcal{C}} = \mathrm{F}$, and to define $x \leq y$ by $y - x \in \overline{\mathcal{C}}$. Suppose that $-a \notin \overline{\mathcal{C}}$. By Lemma 2.9, $\overline{\mathcal{C}}[a]$ is a proper cone and thus, by the maximality of $\overline{\mathcal{C}}$, $\overline{\mathcal{C}} = \overline{\mathcal{C}}[a]$ and thus $a \in \overline{\mathcal{C}}$. □

Proof of Theorem 2.7:

a) ⇒ *b*) since in a real field F, $\sum \mathrm{F}^{(2)}$ is a proper cone.

b) ⇒ *c*) by Proposition 2.8.

c) ⇒ *d*) since in an ordered field, if $x_1 \neq 0$ then $\sum_{i=1}^{n} x_i^2 \geq x_1^2 > 0$.

d) ⇒ *a*), since in a field $0 \neq 1$, so 4 implies that $1 + \sum_{i=1}^{n} x_i^2 = 0$ is impossible. □

A **real closed field** R is an ordered field whose positive cone is the set of squares $\mathrm{R}^{(2)}$ and such that every polynomial in $\mathrm{R}[X]$ of odd degree has a root in R.

Note that the condition that the positive cone of a real closed field R is $\mathrm{R}^{(2)}$ means that R has a unique order as an ordered field, since the positive cone of an order contains necessarily $\mathrm{R}^{(2)}$.

Example 2.10. The field $\mathbb{R}$ of real numbers is of course real closed. The **real algebraic numbers**, i.e. those real numbers that satisfy an equation with integer coefficients, form a real closed field denoted $\mathbb{R}_{\mathrm{alg}}$ (see Exercise 2.11) □

A field R has the **intermediate value property** if R is an ordered field such that, for any $P \in \mathrm{R}[X]$, if there exist $a \in \mathrm{R}$, $b \in \mathrm{R}$, $a < b$ such that $P(a)\,P(b) < 0$, there exists $x \in (a, b)$ such that $P(x) = 0$.

Real closed fields are characterized as follows.

Theorem 2.11. *If* R *is a field then the following properties are equivalent:*

a) R *is real closed.*

b) $\mathrm{R}[i] = \mathrm{R}[T]/(T^2 + 1)$ *is an algebraically closed field.*

c) R *has the intermediate value property.*

d) R *is a real field that has no non-trivial real algebraic extension, that is there is no real field* R_1 *that is algebraic over* R *and different from* R.

The following classical definitions and results about symmetric polynomials are used in the proof of Theorem 2.11.

Let K be a field. A polynomial $Q(X_1, \ldots, X_k) \in \mathrm{K}[X_1, \ldots, X_k]$ is **symmetric** if for every permutation σ of $\{1, \ldots, k\}$,

$$Q(X_{\sigma(1)}, \ldots, X_{\sigma(k)}) = Q(X_1, \ldots, X_k).$$

Exercise 2.7. Denote by $\mathcal{S}_k$ the group of permutations of $\{1, \ldots, k\}$. If $X^\alpha = X_1^{\alpha_1} \ldots X_k^{\alpha_k}$, denote $X_\sigma^\alpha = X_{\sigma(1)}^{\alpha_1} \cdots X_{\sigma(k)}^{\alpha_k}$ and $M_\alpha = \sum_{\sigma \in \mathcal{S}_p} X_\sigma^\alpha$. Prove that every symmetric polynomial can be written as a finite sum $\sum c_\alpha M_\alpha$.

For $i = 1, \ldots, k$, the **i-th elementary symmetric function** is

$$E_i = \sum_{1 \leq j_1 < \cdots < j_i \leq k} X_{j_1} \cdots X_{j_i}.$$

Elementary symmetric functions are related to coefficients of polynomials as follows.

Lemma 2.12. *Let $X_1, \ldots, X_k$ be elements of a field K and*

$$P = (X - X_1) \cdots (X - X_k) = X^k + C_1 X^{k-1} + \cdots + C_k,$$

then $C_i = (-1)^i E_i$.

Proof: Identify the coefficient of X^i on both sides of

$$(X - X_1) \cdots (X - X_k) = X^k + C_1 X^{k-1} + \cdots + C_k. \qquad \square$$

Proposition 2.13. *Let K be a field and let*

$$Q(X_1, \ldots, X_k) \in \mathrm{K}[X_1, \ldots, X_k]$$

be symmetric. There exists a polynomial

$$R(T_1, \ldots, T_k) \in \mathrm{K}[T_1, \ldots, T_k]$$

such that $Q(X_1, \ldots, X_k) = R(E_1, \ldots, E_k)$.

The proof of Proposition 2.13 uses the notion of graded lexicographical ordering. We define first the lexicographical ordering, which is the order of the dictionary and will be used at several places in the book.

We denote by $\mathcal{M}_k$ the set of monomials in k variables. Note that $\mathcal{M}_k$ can be identified with $\mathbb{N}^k$ defining $X^\alpha = X_1^{\alpha_1} \cdots X_k^{\alpha_k}$.

Definition 2.14. [Lexicographical ordering] Let $(B, <)$ be a totally ordered set. The **lexicographical ordering** , $<_{\mathrm{lex}}$, on finite sequences of k elements of B is the total order $<_{\mathrm{lex}}$ defined by induction on k by

$$b <_{\mathrm{lex}} b' \Leftrightarrow b < b'$$
$$(b_1, \ldots, b_k) <_{\mathrm{lex}} (b_1', \ldots, b_k') \Leftrightarrow (b_1 < b_1') \vee (b_1 = b_1' \wedge (b_2, \ldots, b_k) <_{\mathrm{lex}} (b_2', \ldots, b_k')).$$

We denote by $\mathcal{M}_k$ the set of monomials in k variables $X_1, \ldots, X_k$. Note that $\mathcal{M}_k$ can be identified with $\mathbb{N}^k$ defining $X^\alpha = X_1^{\alpha_1} \cdots X_k^{\alpha_k}$. Using this identification defines the lexicographical ordering $<_{\mathrm{lex}}$ on $\mathcal{M}_k$. In thelexicographical ordering, $X_1 >_{\mathrm{grlex}} \ldots >_{\mathrm{grlex}} X_k$. The smallest monomial with respect to the lexicographical ordering is 1, and the lexicographical ordering is compatible with multiplication. Note that the set of monomials less than or equal to a monomial X^α in the lexicographical ordering maybe infinite. □

Exercise 2.8. Prove that a strictly decreasing sequence for the lexicographical ordering is necessarily finite. Hint: by induction on k.

Definition 2.15. [Graded lexicographical ordering] The **graded lexicographical ordering** , $<_{\mathrm{grlex}}$, on the set of monomials in k variables $\mathcal{M}_k$ is the total order $X^\alpha <_{\mathrm{grlex}} X^\beta$ defined by

$$X^\alpha <_{\mathrm{grlex}} X^\beta \Leftrightarrow (\deg(X^\alpha) < \deg(X^\beta)) \vee (\deg(X^\alpha) = \deg(X^\beta) \wedge \alpha <_{\mathrm{lex}} \beta)$$

with $\alpha = (\alpha_1, \ldots, \alpha_k)$, $\beta = (\beta_1, \ldots, \beta_k)$, $X^\alpha = X_1^{\alpha_1} \cdots X_k^{\alpha_k}$, $X^\beta = X_1^{\beta_1} \cdots X_k^{\beta_k}$.

In the graded lexicographical ordering above, $X_1 >_{\mathrm{grlex}} \ldots >_{\mathrm{grlex}} X_k$. The smallest monomial with respect to the graded lexicographical ordering is 1, and the graded lexicographical ordering is compatible with multiplication. Note that the set of monomials less than or equal to a monomial X^α in the graded lexicographical ordering is finite. □

Proof of Proposition 2.13: Since $Q(X_1, \ldots, X_k)$ is symmetric, its leading monomial in the graded lexicographical ordering $c_\alpha X^\alpha = c_\alpha X_1^{\alpha_1} \cdots X_k^{\alpha_k}$ satisfies $\alpha_1 \geq \ldots \geq \alpha_k$. The leading monomial of $c_\alpha E_1^{\alpha_1 - \alpha_2} \cdots E_{k-1}^{\alpha_{k-1} - \alpha_k} E_k^{\alpha_k}$ in the graded lexicographical ordering is also $c_\alpha X^\alpha = c_\alpha X_1^{\alpha_1} \cdots X_k^{\alpha_k}$.

Let $Q_1 = Q(X_1, \ldots, X_k) - c_\alpha E_1^{\alpha_1 - \alpha_2} \cdots E_{k-1}^{\alpha_{k-1} - \alpha_k} E_k^{\alpha_k}$. If $Q_1 = 0$, the proof is over. Otherwise, the leading monomial with respect to the graded lexicographical ordering of Q_1 is strictly smaller than $X_1^{\alpha_1} \cdots X_k^{\alpha_k}$, and it is possible to iterate the construction with Q_1. Since there is no infinite decreasing sequence of monomials for the graded lexicographical ordering, the claim follows. □

Proposition 2.16. *Let $P \in \mathrm{K}[X]$, of degree k, and $x_1, \ldots, x_k$ be the roots of P (counted with multiplicities) in an algebraically closed field C containing K. If a polynomial $Q(X_1, \ldots, X_k) \in \mathrm{K}[X_1, \ldots, X_k]$ is symmetric, then $Q(x_1, \ldots, x_k) \in \mathrm{K}$.*

Proof: Let e_i, for $1 \leq i \leq k$, denote the i-th elementary symmetric function evaluated at x_1, ..., x_k. Since $P \in K[X]$, Lemma 2.12 gives $e_i \in K$. By Proposition 2.13, there exists $R(T_1, \ldots, T_k) \in \mathrm{K}[T_1, \ldots, T_k]$ such that

$$Q(X_1, \ldots, X_k) = R(E_1, \ldots, E_k).$$

Thus, $Q(x_1, \ldots, x_k) = R(e_1, \ldots, e_k) \in \mathrm{K}$. □

With these preliminaries results, it is possible to prove Theorem 2.11.

Proof of Theorem 2.11: $a) \Rightarrow b)$ Let $P \in \mathrm{R}[X]$ a monic separable polynomial of degree $p = 2^m n$ with n odd. We show by induction on m that P has a root in $\mathrm{R}[i]$.

If $m = 0$, then p is odd and P has a root in R, since R is real closed.

Denote by $x_1, \ldots, x_p$ the roots of P in an algebraically closed field C. Let Z be a new indeterminate and $Q(Z, Y)$ the monic polynomial having as roots the $x_i + x_j + Z\, x_i x_j$ where $i < j$.

$$Q(Z,Y) = \prod_{i<j} (Y - (x_i + x_j + Z x_i x_j)).$$

The coefficients of $Q(Z, Y)$ can be explicitly computed as polynomials of the coefficients of P, using Proposition 2.16, thus $Q(Z, Y) \in \mathrm{R}[Z, Y]$. The degree of $Q(Z, Y)$ in Y and Z is $p(p-1)/2$.

Ordering lexicographically the couples (i, j), $i < j$, we define the discriminant of Q as

$$\begin{aligned} D(Z) &= \prod_{\substack{i<j,\, k<\ell \\ (i,j)<(k,\ell)}} ((x_i + x_j + Z\, x_i x_j) - (x_k + x_\ell + Z x_k x_\ell))^2 \\ &= \prod_{\substack{i<j,\, k<\ell \\ (i,j)<(k,\ell)}} (\alpha_{i,j,k,\ell} + Z\, \beta_{i,j,k,\ell})^2 \end{aligned}$$

where $\alpha_{i,j,k,\ell} = (x_i + x_j - x_k + x_\ell)$, $\beta_{i,j,k,\ell} = x_i x_j - x_k x_\ell$. Note that by Proposition 2.16, $D(Z) \in \mathrm{R}[Z]$.

Since all the roots of P are distinct, we get the following implication

$$i < j,\, k < \ell,\, (i,j) < (k,\ell),\, x_i x_j = x_k x_\ell \quad \Rightarrow \quad x_i + x_j \neq x_k + x_\ell.$$

So every factor $\alpha_{i,j,k,\ell} + Z\, \beta_{i,j,k,\ell}$ is nonzero. It follows that $D(Z)$ is not identically zero.

Taking a value $z \in \mathbb{N}$ such that $D(z) \neq 0$, the polynomial $Q(z, Y)$ is a square free polynomial since all its roots are distinct.

We prove now that it is possible to express, for every $1 \leq i < j \leq p$, $x_i + x_j$ and $x_i x_j$ rationally in terms of $\gamma_{i,j} = x_i + x_j + z\, x_i x_j$.

Indeed let

$$\begin{aligned} F(Z,Y) &= \partial Q / \partial Y(Z,Y) \\ &= \sum_{i<j} \prod_{\substack{k<\ell \\ (k,\ell)\neq(i,j)}} (Y - (x_k + x_\ell + Z\, x_k x_\ell)) \\ G(Z,Y) &= \sum_{i<j} (x_i + x_j) \left(\prod_{\substack{k<\ell \\ (k,\ell)\neq(i,j)}} (Y - (x_k + x_\ell + Z x_k x_\ell)) \right), \\ H(Z,Y) &= \sum_{i<j} x_i x_j \left(\prod_{\substack{k<\ell \\ (k,\ell)\neq(i,j)}} (Y - (x_k + x_\ell + Z\, x_k x_\ell)) \right). \end{aligned}$$

Note that by Proposition 2.16, $f(Z, Y), G(Z, Y)$ and $H(Z, Y)$ are elements of $\mathrm{R}[Z, Y]$.

Then, for every $1 \le i < j \le p$,

$$\begin{aligned} F(z,\gamma_{i,j}) &= \prod_{\substack{k<\ell \\ (k,\ell)\neq(i,j)}} (\gamma_{i,j}-\gamma_{k,\ell}), \\ G(z,\gamma_{i,j}) &= (x_i+x_j) \prod_{\substack{k<\ell \\ (k,\ell)\neq(i,j)}} (\gamma_{i,j}-\gamma_{k,\ell}), \\ H(z,\gamma_{i,j}) &= (x_i\, x_j) \prod_{\substack{k<\ell \\ (k,\ell)\neq(i,j)}} (\gamma_{i,j}-\gamma_{k,\ell}). \end{aligned}$$

If follows that

$$\begin{aligned} x_i+x_j &= \frac{G(z,\gamma_{i,j})}{F(z,\gamma_{i,j})}, \\ x_i\, x_j &= \frac{H(z,\gamma_{i,j})}{F(z,\gamma_{i,j})}. \end{aligned}$$

In other words, the roots of the second degree polynomial

$$F(z,\gamma_{i,j})X^2 - G(z,\gamma_{i,j})X + H(z,\gamma_{i,j})$$

are roots of P.

The polynomial $Q(z,Y)$ is of degree $p(p-1)/2$, i.e. of the form $2^{m-1}n'$ with n' odd. By induction hypothesis, it has a root γ in R[i]. Since the classical method for solving polynomials of degree 2 works in $\mathrm{R}[i]$ when R is real closed, the roots of the second degree polynomial

$$F(z,\gamma)X^2 - G(z,\gamma)X + H(z,\gamma)$$

are roots of P that belong to $\mathrm{R}[i]$. We have proved that the polynomial P has a root in $\mathrm{R}[i]$.

For $P = a_p\, X^p + \cdots + a_0 \in \mathrm{R}[i][X]$, we write $\overline{P} = \overline{a}_p\, X^p + \cdots + \overline{a}_0$. Since $P\overline{P} \in \mathrm{R}[X]$, $P\overline{P}$ has a root x in $\mathrm{R}[i]$. Thus $P(x)=0$ or $\overline{P}(x)=0$. In the first case we are done and in the second, $P(\overline{x})=0$.

b) $\Rightarrow$ *c)* Since $\mathrm{C}=\mathrm{R}[i]$ is algebraically closed, P factors into linear factors over C. Since if $c+i\,d$ is a root of P, $c-i\,d$ is also a root of P, the irreducible factors of P are linear or have the form

$$(X-c)^2+d^2=(X-c-i\,d)\,(X-c+i\,d),\ d\neq 0.$$

If $P(a)$ and $P(b)$ have opposite signs, then $Q(a)$ and $Q(b)$ have opposite signs for some linear factor Q of P. Hence the root of Q is in (a,b).

c) $\Rightarrow$ *a)* If y is positive, X^2-y takes a negative value at 0 and a positive value for X big enough, by Proposition 2.4. Thus X^2-y has a root, which is a square root of y. Similarly a polynomial of odd degree with coefficients in R takes different signs for a positive and big enough and b negative and small enough, using Proposition 2.4 again. Thus it has a root in R.

b) $\Rightarrow$ *d)* Since $\mathrm{R}[i]=\mathrm{R}[T]/(T^2+1)$ is a field, T^2+1 is irreducible over R. Hence -1 is not a square in R. Moreover in R, a sum of squares is still a square: let $a,b\in\mathrm{R}$ and $c,d\in\mathrm{R}$ such that $a+i\,b=(c+i\,d)^2$; then $a^2+b^2=(c^2+d^2)^2$. This proves that R is real. Finally, since the only irreducible polynomials of $\mathrm{R}[X]$ of degree >1 are of the form

$$(X-c)^2+d^2=(X-c-i\,d)(X-c+i\,d),\ d\neq 0,$$

and $\mathrm{R}[X]/((X-c)^2+d^2)=\mathrm{R}[i]$, the only non-trivial algebraic extensions of R is $\mathrm{R}[i]$, which is not real.

d) $\Rightarrow$ *a)* Suppose that $a\in\mathrm{R}$. If a is not a square in R, then

$$\mathrm{R}[\sqrt{a}]=\mathrm{R}[X]/(X^2-a)$$

is a non-trivial algebraic extension of R, and thus $\mathrm{R}[\sqrt{a}]$ is not real. Thus,

$$\begin{aligned} -1 &= \sum_{i=1}^{n}(x_i+\sqrt{a}\,y_i)^2 \\ -1 &= \sum_{i=1}^{n}x_i^2+a\sum_{i=1}^{n}y_i^2\in\mathrm{R}. \end{aligned}$$

Since R is real, $-1\neq\sum_{i=1}^{n}x_i^2$ and thus $y=\sum_{i=1}^{n}y_i^2\neq 0$. Hence,

$$\begin{aligned} -a &= \left(\sum_{i=1}^{n}y_i^2\right)^{-1}\left(1+\sum_{i=1}^{n}x_i^2\right) \\ &= \left(\sum_{i=1}^{n}\left(\frac{y_i}{y}\right)^2\right)\left(1+\sum_{i=1}^{n}x_i^2\right)\in\sum\mathrm{R}^{(2)}. \end{aligned}$$

This shows that $\mathrm{R}^{(2)}\cup-\sum\mathrm{R}^{(2)}=\mathrm{R}$ and thus that there is only one possible order on R with $\mathrm{R}^{(2)}=\sum\mathrm{R}^{(2)}$ as positive cone.

It remains to show that if $P\in\mathrm{R}[X]$ has odd degree then P has a root in R. If this is not the case, let P be a polynomial of odd degree $p>1$ such that every polynomial of odd degree $<p$ has a root in R. Since a polynomial of odd degree has at least one odd irreducible factor, we assume without loss of generality that P is irreducible. The quotient $\mathrm{R}[X]/(P)$ is a non-trivial algebraic extension of R and hence $-1=\sum_{i=1}^{n}H_i^2+PQ$ with $\deg(H_i)<p$. Since the term of highest degree in the expansion of $\sum_{i=1}^{n}H_i^2$ has a sum of squares as coefficient and R is real, $\sum_{i=1}^{n}H_i^2$ is a polynomial of even degree $\leq 2p-2$. Hence, the polynomial Q has odd degree $\leq p-2$ and thus has a root x in R. But then $-1=\sum_{i=1}^{n}H_i(x)^2$, which contradicts the fact that R is real. □

Remark 2.17. When $\mathrm{R}=\mathbb{R}$, *a)* $\Rightarrow$ *b)* in Theorem 2.11 is nothing but an algebraic proof of the fundamental theorem of algebra. □

Notation 2.18. [Modulus] If R is real closed, and $\mathrm{R}[i] = \mathrm{R}[T]/(T^2+1)$, we can identify $\mathrm{R}[i]$ with R^2. For $z = a + i\,b \in \mathrm{R}[i]$, $a \in \mathrm{R}$, $b \in \mathrm{R}$, we define the **conjugate** of z by $\overline{z} = a - i\,b$. The **modulus** of $z = a + i\,b \in \mathrm{R}[i]$ is $|z| = \sqrt{a^2+b^2}$. □

Proposition 2.19. *Let* R *be a real closed field,* $P \in \mathrm{R}[X]$. *The irreducible factors of* P *are linear or have the form*

$$(X-c)^2+d^2 = (X-c-i\,d)(X-c+i\,d), d \neq 0$$

with $c, d \in \mathrm{R}$.

Proof: Use the fact that $\mathrm{R}[i]$ is algebraically closed by Theorem 2.11 and that the conjugate of a root of P is a root of P. □

Exercise 2.9. Prove that, in a real closed field, a second degree polynomial

$$P = a\,X^2 + b\,X + c,\ a \neq 0$$

has a constant non-zero sign if and only if its **discriminant** $b^2 - 4\,a\,c$ is negative. Hint: the classical computation over the reals is still valid in a real closed field.

Closed, open and semi-open intervals in R will be denoted in the usual way:

$$\begin{aligned}
(a,b) &= \{x \in \mathrm{R} \mid a < x < b\},\\
[a,b] &= \{x \in \mathrm{R} \mid a \leq x \leq b\},\\
(a,b] &= \{x \in \mathrm{R} \mid a < x \leq b\},\\
(a,+\infty) &= \{x \in \mathrm{R} \mid a < x\},\\
&\ldots
\end{aligned}$$

Proposition 2.20. *Let* R *be a real closed field,* $P \in \mathrm{R}[X]$ *such that* P *does not vanish in* (a,b), *then* P *has constant sign in the interval* (a,b).

Proof: Use the fact that R has the intermediate value property by Theorem 2.11. □

This proposition shows that it makes sense to talk about the sign of a polynomial to the right (resp. to the left) of any $a \in \mathrm{R}$. Namely, the sign of P to the right (resp. to the left) of a is the sign of P in any interval (a,b) (resp. (b, a)) in which P does not vanish. We can also speak of the sign of $P(+\infty)$ (resp. $P(-\infty)$) as the sign of $P(M)$ for M sufficiently large (resp. small) i.e. greater (resp. smaller) than any root of P. This coincides with the sign of $\mathrm{lcof}(P)$ (resp. $(-1)^{\deg(P)}\mathrm{lcof}(P)$) using Proposition 2.4.

Proposition 2.21. *If* r *is a root of* P *of multiplicity* μ *in a real closed field* R *then the sign of* P *to the right of* r *is the sign of* $P^{(\mu)}(r)$ *and the sign of* P *to the left of* r *is the sign of* $(-1)^{\mu}P^{(\mu)}(r)$.

Proof: Write $P = (X - r)^{\mu} Q(x)$ where $Q(r) \neq 0$, and note that

$$\operatorname{sign}(Q(r)) = \operatorname{sign}(P^{(\mu)}(r)).$$

□

We next show that univariate polynomials over a real closed field R share some of the well known basic properties possessed by differentiable functions over $\mathbb{R}$.

Proposition 2.22. [Rolle's theorem] *Let* R *be a real closed field,* $P \in \mathrm{R}[X]$*,* $a, b \in \mathrm{R}$ *with* $a < b$ *and* $P(a) = P(b) = 0$*. Then the derivative polynomial* P' *has a root in* (a, b)*.*

Proof: One may reduce to the case where a and b are two consecutive roots of P, i.e. when P never vanishes on (a, b). Then $P = (X - a)^m (X - b)^n Q$, where Q never vanishes on $[a, b]$. Thus Q has constant sign on $[a, b]$ by Proposition 2.20. Then $P' = (X - a)^{m-1} (X - b)^{n-1} Q_1$, where

$$Q_1 = m\,(X - b)\,Q + n\,(X - a)\,Q + (X - a)\,(X - b)\,Q'.$$

Thus $Q_1(a) = m\ (a - b)\ Q(a)$ and $Q_1(b) = n\ (b - a)\ Q(b)$, and hence $Q_1(a)$ and $Q_1(b)$ have opposite signs. By the intermediate value property, Q_1 has a root in (a, b), and so does P'. □

Corollary 2.23. [Mean Value theorem] *Let* R *be a real closed field,* $P \in \mathrm{R}[X]$*,* $a, b \in \mathrm{R}$ *with* $a < b$*. There exists* $c \in (a, b)$ *such that*

$$P(b) - P(a) = (b - a)\,P'(c).$$

Proof: Apply Rolle's theorem (Proposition 2.22) to

$$Q(X) = (P(b) - P(a))\,(X - a) - (b - a)\,(P(X) - P(a)).$$ □

Corollary 2.24. *Let* R *be a real closed field,* $P \in \mathrm{R}[X]$*,* $a, b \in \mathrm{R}$ *with* $a < b$*. If the derivative polynomial* P' *is positive (resp. negative) over* (a, b)*, then* P *is increasing (resp. decreasing) over* $[a, b]$*.*

The following Proposition 2.28 (Basic Thom's Lemma) which will have important consequences in Chapter 10. We first need a few definitions.

Definition 2.25. Let $\mathcal{Q}$ be a finite subset of $\mathrm{R}[X_1, \ldots, X_k]$. A **sign condition** on $\mathcal{Q}$ is an element of $\{0, 1, -1\}^{\mathcal{Q}}$, i.e. a mapping from $\mathcal{Q}$ to $\{0, 1, -1\}$. A **strict sign condition** on $\mathcal{Q}$ is an element of $\{1, -1\}^{\mathcal{Q}}$, i.e. a mapping from $\mathcal{Q}$ to $\{1, -1\}$. We say that $\mathcal{Q}$ **realizes** the sign condition σ at $x \in \mathrm{R}^k$ if $\bigwedge_{Q \in \mathcal{Q}} \operatorname{sign}(Q(x)) = \sigma(Q)$.

The **realization of the sign condition** σ is

$$\mathrm{Reali}(\sigma) = \{x \in \mathrm{R}^k \mid \bigwedge_{Q \in \mathcal{Q}} \mathrm{sign}(Q(x)) = \sigma(Q)\}.$$

The sign condition σ is **realizable** if $\mathrm{Reali}(\sigma)$ is non-empty. □

Notation 2.26. [Derivatives] Let P be a univariate polynomial of degree p in $\mathrm{R}[X]$. We denote by $\mathrm{Der}(P)$ the list $P, P', \ldots, P^{(p)}$. □

Proposition 2.27. [Basic Thom's Lemma] *Let P be a univariate polynomial of degree p and let σ be a sign condition on* $\mathrm{Der}(P)$ *Then* $\mathrm{Reali}(\sigma)$ *is either empty, a point, or an open interval.*

Proof: The proof is by induction on the degree p of P. There is nothing to prove if $p = 0$. Suppose that the proposition has been proved for $p - 1$. Let $\sigma \in \{0,\ 1,\ -1\}^{\mathrm{Der}(P)}$ be a sign condition on $\mathrm{Der}(P)$, and let σ' be its restriction to $\mathrm{Der}(P')$. If $\mathrm{Reali}(\sigma')$ is either a point or empty, then

$$\mathrm{Reali}(\sigma) = \mathrm{Reali}(\sigma') \cap \{x \in \mathrm{R} \mid \mathrm{sign}(P(x)) = \sigma(P)\}$$

is either a point of empty. If $\mathrm{Reali}(\sigma')$ is an open interval, P' has a constant non-zero sign on it. Thus P is strictly monotone on $\mathrm{Reali}(\sigma')$ so that the claimed properties are satisfied for $\mathrm{Reali}(\sigma)$. □

Proposition 2.27 has interesting consequences. One of them is the fact that a root $x \in \mathrm{R}$ of a polynomial P of degree d with coefficients in R may be distinguished from the other roots of P in R by the signs of the derivatives of P at x.

Proposition 2.28. [Thom encoding] *Let P be a non-zero polynomial of degree d with coefficients in* R. *Let x and x' be two elements of* R, *and denote by σ and σ' the sign conditions on* $\mathrm{Der}(P)$ *realized at x and x'. Then:*

- *If $\sigma = \sigma'$ with $\sigma(P) = \sigma'(P) = 0$ then $x = x'$.*
- *If $\sigma \neq \sigma'$, one can decide whether $x < x'$ or $x > x'$ as follows. Let k be the smallest integer such that $\sigma(P^{(d-k)})$ and $\sigma'(P^{(d-k)})$ are different. Then*
  - $\sigma(P^{(d-k+1)}) = \sigma'(P^{(d-k+1)}) \neq 0.$
  - *If* $\sigma(P^{(d-k+1)}) = \sigma'(P^{(d-k+1)}) = 1,$
  $$x > x' \ \Leftrightarrow\ \sigma(P^{(d-k)}) > \sigma'(P^{(d-k)}).$$
  - *If* $\sigma(P^{(d-k+1)}) = \sigma'(P^{(d-k+1)}) = -1,$
  $$x > x' \ \Leftrightarrow\ \sigma(P^{(d-k)}) < \sigma'(P^{(d-k)}).$$

Proof: The first item is a consequence of Proposition 2.27. The first part of the second item follows from Proposition 2.27 applied to $P^{(d-k+1)}$. The two last parts follow easily since the set

$$\{x \in \mathrm{R} \mid \mathrm{sign}(P^{(i)}(x)) = \sigma(P^{(i)}), i = d-k+1, \cdots, n-1\}$$

is an interval by Proposition 2.28 applied to $P^{(d-k+1)}$, and, on an interval, the sign of the derivative of a polynomial determines whether it is increasing or decreasing. □

Definition 2.29. Let $P \in \mathrm{R}[X]$ and $\sigma \in \{0,1,-1\}^{\mathrm{Der}(P)}$, a sign condition on the set $\mathrm{Der}(P)$ of derivatives of P. The sign condition σ is a **Thom encoding of** $x \in \mathbf{R}$ if $\sigma(P)=0$ and $\mathrm{Reali}(\sigma)=\{x\}$, i.e. σ is the sign condition taken by the set $\mathrm{Der}(P)$ at x. □

Example 2.30. In any real closed field R, $P = X^2 - 2$ has two roots, characterized by the sign of the derivative $2\,X$: one root for which $2\,X > 0$ and one root for which $2\,X < 0$. Note that no numerical information about the roots is needed to characterize them this way. □

Any ordered field can be embedded in a real closed field. More precisely, any ordered field F possesses a unique **real closure** which is the smallest real closed field extending it. The elements of the real closure are algebraic over F (i.e. satisfy an equation with coefficients in F). We refer the reader to [26] for these results.

Exercise 2.10. If F is contained in a real closed field R, the real closure of F consists of the elements of R which are algebraic over F. (Hint: given α and β roots of P and Q in $\mathrm{F}[X]$, find polynomials in $\mathrm{F}[X]$ with roots $\alpha + \beta$ and $\alpha\,\beta$, using Proposition 2.16).

Exercise 2.11. Prove that $\mathbb{R}_{\mathrm{alg}}$ is real closed. Prove that the field $\mathbb{R}_{\mathrm{alg}}$ is the real closure of $\mathbb{Q}$.

The following theorem proves that any algebraically closed field of characteristic zero is the algebraic closure of a real closed field.

Theorem 2.31. *If* C *is an algebraically closed field of characteristic zero, there exists a real closed field* $\mathrm{R} \subset \mathrm{C}$ *such that* $\mathrm{R}[i] = \mathrm{C}$.

Proof: The field C contains a real subfield, the field $\mathbb{Q}$ of rational numbers. Let R be a maximal real subfield of C. The field R is real closed since it has no nontrivial real algebraic extension contained in C (see Theorem 2.11). Note that $\mathrm{C} \setminus \mathrm{R}$ cannot contain a t which is transcendental over R since otherwise $\mathrm{R}(t)$ would be a real field properly containing R. □

An ordered field F is **archimedean** if, whenever a, b are positive elements of F, there exists a natural number $n \in \mathbb{N}$ so that $n\,a > b$.

Real closed fields are not necessarily archimedean and may contain infinitesimal elements. We shall see at the end of this chapter an example of a non-archimedean real closed field when we study the field of Puiseux series.

2.2 Real Root Counting

Although we have a very simple criterion for determining whether a polynomial $P \in \mathrm{C}[X]$ has a root in C (namely, if and only if $\deg(P) \neq 0$), it is much more difficult to decide whether a polynomial $P \in \mathrm{R}[X]$ has a root in R. The first result in this direction was found more than 350 years ago by Descartes. We begin the section with a generalization of this result.

2.2.1 Descartes's Law of Signs and the Budan-Fourier Theorem

Notation 2.32. [Sign variations] The **number of sign variations**, $\mathrm{Var}(a)$, in a sequence, $a = a_0, \cdots, a_p$, of elements in $\mathrm{R} \setminus \{0\}$ is defined by induction on p by:

$$\begin{aligned}\mathrm{Var}(a_0) &= 0\\ \mathrm{Var}(a_0, \cdots, a_p) &= \begin{cases}\mathrm{Var}(a_1, \cdots, a_p) + 1 & \text{if } a_0\, a_1 < 0\\ \mathrm{Var}(a_1, \cdots, a_p) & \text{if } a_0\, a_1 > 0\end{cases}\end{aligned}$$

This definition extends to any finite sequence a of elements in R by considering the finite sequence b obtained by dropping the zeros in a and defining

$$\mathrm{Var}(a) = \mathrm{Var}(b),\ \mathrm{Var}(\emptyset) = 0.$$

For example $\mathrm{Var}(1, -1, 2, 0, 0, 3, 4, -5, -2, 0, 3) = 4.$ □

Let $P = a_p\, X^p + \cdots + a_0$ be a univariate polynomial in $\mathrm{R}[X]$. We write $\mathrm{Var}(P)$ for the number of sign variations in $a_0, \ldots, a_p$ and $\mathrm{pos}(P)$ for the number of positive real roots of P, counted with multiplicity.

Theorem 2.33. [Descartes' law of signs]

- $\mathrm{Var}(P) \geq \mathrm{pos}(P)$
- $\mathrm{Var}(P) - \mathrm{pos}(P)$ *is even.*

We will prove the following generalization of Theorem 2.33 (Descartes's law of signs) due to Budan and Fourier.

Notation 2.34. [Sign variations in a sequence of polynomials at a] Let $\mathcal{P} = P_0, P_1, \ldots, P_d$ be a sequence of polynomials and let a be an element of $\mathrm{R} \cup \{-\infty, +\infty\}$. The **number of sign variations** of $\mathcal{P}$ at a, denoted by $\mathrm{Var}(\mathcal{P}; a)$, is $\mathrm{Var}(P_0(a), \ldots, P_d(a))$ (at $-\infty$ and $+\infty$ the signs to consider are the signs of the leading monomials according to Proposition 2.4).

For example, if $\mathcal{P} = X^5, X^2 - 1, 0, X^2 - 1, X + 2, 1,\ \mathrm{Var}(\mathcal{P}; 1) = 0$.

Given a and b in $\mathrm{R} \cup \{-\infty, +\infty\}$, we denote

$$\mathrm{Var}(\mathcal{P}; a, b) = \mathrm{Var}(\mathcal{P}; a) - \mathrm{Var}(\mathcal{P}; b).$$

□

We denote by $\mathrm{num}(P;(a,b])$ the number of roots of P in $(a,b]$ counted with multiplicities.

Theorem 2.35. [Budan-Fourier theorem] *Let P be a univariate polynomial of degree p in $\mathrm{R}[X]$. Given a and b in $\mathrm{R}\cup\{-\infty,+\infty\}$*

- $\mathrm{Var}(\mathrm{Der}(P);a,b)\geq \mathrm{num}(P;(a,b])$,
- $\mathrm{Var}(\mathrm{Der}(P);a,b)-\mathrm{num}(P;(a,b])$ *is even.*

Theorem 2.33 (Descartes's law of signs) is a particular case of Theorem 2.35 (Budan-Fourier).

Proof of Theorem 2.33 (Descartes' law of signs): The coefficient of degree i of P has the same sign as the $p-i$-th derivative of P evaluated at 0. Moreover, there are no sign variations in the signs of the derivatives at $+\infty$. So that $\mathrm{Var}(P)=\mathrm{Var}(\mathrm{Der}(P);0,+\infty)$. □

The following lemma is the key to the proof of Theorem 2.35 (Budan-Fourier).

Lemma 2.36. *Let c be a root of P of multiplicity $\mu\geq 0$. If no $P^{(k)}$, $0\leq k\leq p$, has a root in $[d,c)\cup(c,d']$, then*

a) $\mathrm{Var}(\mathrm{Der}(P);d,c)-\mu$ *is non-negative and even,*
b) $\mathrm{Var}(\mathrm{Der}(P);c,d')=0$.

Proof: We prove the claim by induction on the degree of P. The claim is true if the degree of P is 1.

Suppose first that $P(c)=0$, and hence $\mu>0$. By induction hypothesis applied to P',

a) $\mathrm{Var}(\mathrm{Der}(P');d,c)-(\mu-1)$ is non-negative and even,
b) $\mathrm{Var}(\mathrm{Der}(P');c,d')=0$.

The sign of P at the left of c is the opposite of the sign of P' at the left of c and the sign of P at the right of c is the sign of P' at the right of c. Thus

$$\begin{aligned}\mathrm{Var}(\mathrm{Der}(P);d) &= \mathrm{Var}(\mathrm{Der}(P');d)+1, \\ \mathrm{Var}(\mathrm{Der}(P);c) &= \mathrm{Var}(\mathrm{Der}(P');c), \\ \mathrm{Var}(\mathrm{Der}(P);d') &= \mathrm{Var}(\mathrm{Der}(P');d'),\end{aligned} \tag{2.1}$$

and the claim follows.

Suppose now that $P(c)\neq 0$, and hence $\mu=0$. Let ν be the multiplicity of c as a root of P'. By induction hypothesis applied to P'

a) $\mathrm{Var}(\mathrm{Der}(P');d,c)-\nu$ is non-negative and even,
b) $\mathrm{Var}(\mathrm{Der}(P');c,d')=0$.

There are four cases to consider.

If ν is odd, and $\text{sign}(P^{(\nu+1)}(c)\,P(c)) > 0$,

$$\begin{aligned}\text{Var}(\text{Der}(P);d) &= \text{Var}(\text{Der}(P');d)+1, \\ \text{Var}(\text{Der}(P);c) &= \text{Var}(\text{Der}(P');c), \\ \text{Var}(\text{Der}(P);d') &= \text{Var}(\text{Der}(P');d').\end{aligned} \tag{2.2}$$

If ν is odd, and $\text{sign}(P^{(\nu+1)}(c)\,P(c)) < 0$,

$$\begin{aligned}\text{Var}(\text{Der}(P);d) &= \text{Var}(\text{Der}(P');d), \\ \text{Var}(\text{Der}(P);c) &= \text{Var}(\text{Der}(P');c)+1, \\ \text{Var}(\text{Der}(P);d') &= \text{Var}(\text{Der}(P');d')+1.\end{aligned} \tag{2.3}$$

If ν is even, and $\text{sign}(P^{(\nu+1)}(c)\,P(c)) > 0$,

$$\begin{aligned}\text{Var}(\text{Der}(P);d) &= \text{Var}(\text{Der}(P');d), \\ \text{Var}(\text{Der}(P);c) &= \text{Var}(\text{Der}(P');c), \\ \text{Var}(\text{Der}(P);d') &= \text{Var}(\text{Der}(P');d').\end{aligned} \tag{2.4}$$

If ν is even, and $\text{sign}(P^{(\nu+1)}(c)\,P(c)) < 0$,

$$\begin{aligned}\text{Var}(\text{Der}(P);d) &= \text{Var}(\text{Der}(P');d)+1, \\ \text{Var}(\text{Der}(P);c) &= \text{Var}(\text{Der}(P');c)+1, \\ \text{Var}(\text{Der}(P);d') &= \text{Var}(\text{Der}(P');d')+1.\end{aligned} \tag{2.5}$$

The claim is true in each of these four cases. □

Proof of Theorem 2.35: It is clear that, for every $c \in (a,b)$,

$$\begin{aligned}\text{num}(P;(a,b]) &= \text{num}(P;(a,c]) + \text{num}(P;(c,b]) \\ \text{Var}(\text{Der}(P);a,b) &= \text{Var}(\text{Der}(P);a,c) + \text{Var}(\text{Der}(P);c,b).\end{aligned}$$

Let $c_1 < \cdots < c_r$ be the roots of all the polynomials $P^{(j)}$, $0 \le j \le p-1$, in the interval (a,b) and let $a = c_0, b = c_{r+1}$, $d_i \in (c_i, c_{i+1})$ so that

$$a = c_0 < d_0 < c_1 < \cdots < c_r < d_r < c_{r+1} = b.$$

Since,

$$\begin{aligned}\text{num}(P;(a,b]) &= \sum_{i=0}^{r} \text{num}(P;(c_i,d_i]) + \text{num}(P;(d_i,c_{i+1}]), \\ \text{Var}(\text{Der}(P);a,b) &= \sum_{i=0}^{r} \text{Var}(\text{Der}(P);c_i,d_i) + \text{Var}(\text{Der}(P);d_i,c_{i+1}),\end{aligned}$$

the claim follows immediately from Lemma 2.36. □

In general it is not possible to conclude much about the number of roots on an interval using only Theorem 2.35 (Descartes's law of signs).

Example 2.37. The polynomial $P = X^2 - X + 1$ has no real root, but $\mathrm{Var}(\mathrm{Der}(P);\ 0,\ 1) = 2$. It is impossible to find $a \in (0,\ 1]$ such that $\mathrm{Var}(\mathrm{Der}(P); 0, a) = 1$ and $\mathrm{Var}(\mathrm{Der}(P); a, 1) = 1$ since otherwise P would have two real roots. This means that however we refine the interval $(0,\ 1]$, we are going to have an interval (the interval $(a, b]$ containing $1/2$) giving 2 sign variations. □

However, there are particular cases where Theorem 2.35 (Budan-Fourier) gives the number of roots on an interval:

Exercise 2.12. Prove that

- If $\mathrm{Var}(\mathrm{Der}(P); a, b) = 0$, then P has no root in $(a, b]$.
- If $\mathrm{Var}(\mathrm{Der}(P);\ a,\ b) = 1$, then P has exactly one root in $(a,\ b]$, which is simple.

Remark 2.38. Another important instance, used in Chapter 8, where Theorem 2.35 (Budan-Fourier) permits a sharp conclusion is the following. When we know in advance that all the roots of a polynomial are real, i.e. when $\mathrm{num}(P; (-\infty, +\infty)) = p$, the number $\mathrm{Var}(\mathrm{Der}(P); a, b)$ is exactly the number of roots counted with multiplicities in $(a, b]$. Indeed the number $\mathrm{Var}(\mathrm{Der}(P); -\infty, +\infty)$, which is always at most p, is here equal to p, hence

$$\begin{aligned} \mathrm{num}(P; (-\infty, a]) &\leq \mathrm{Var}(\mathrm{Der}(P); -\infty, a) \\ \mathrm{num}(P; (a, b]) &\leq \mathrm{Var}(\mathrm{Der}(P); a, b) \\ \mathrm{num}(P; (b, +\infty)) &\leq \mathrm{Var}(\mathrm{Der}(P); b, +\infty) \end{aligned}$$

imply $\mathrm{num}(P, (a, b]) = \mathrm{Var}(\mathrm{Der}(P); a, b)$. □

We are going now to describe situations where the number of sign variations in the coefficients coincides exactly with the number of real roots.

The first case we consider is obvious.

Proposition 2.39. *Let $P \in \mathrm{R}[X]$ be a monic polynomial. If all the roots of P have non-positive real part, then* $\mathrm{Var}(P) = 0$.

Proof: Obvious, using the decomposition of P in products of linear factors and polynomials of degree 2 with complex conjugate roots, since the product of two polynomials whose coefficients are all non-negative have coefficients that are all non-negative. □

The second case we consider is the case of normal polynomials. A polynomial $A = a_p X^p + \cdots + a_0$ with non-negative coefficients is **normal** if

a) $a_p > 0$,
b) $a_k^2 \geqslant a_{k-1} a_{k+1}$ for all index k,
c) $a_h > 0$ and $a_j > 0$ for indices $j < h$ implies $a_{j+1} > 0, \ldots, a_{h-1} > 0$

(with the convention that $a_i = 0$ if $i < 0$ or $i > p$).

Proposition 2.40. *Let $P \in \mathrm{R}[X]$ be a monic polynomial. If all the roots of P belong to the cone $\mathcal{B}$ of the complex plane (see Figure 2.1) defined by $\mathcal{B} = \left\{a + i\,b \,|\, |b| \leqslant -\sqrt{3}\,a\right\}$, then P is normal.*

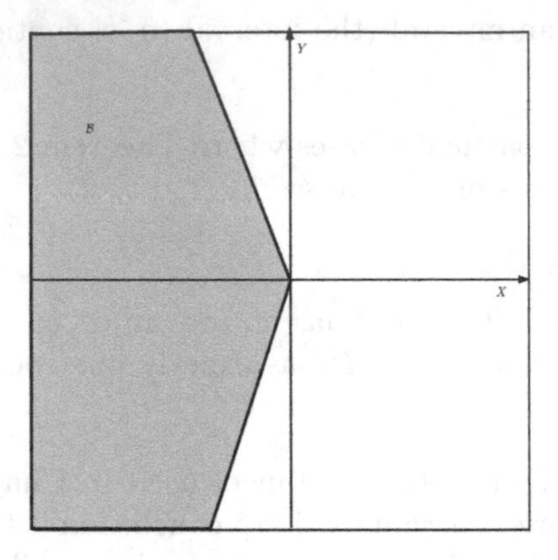

Fig. 2.1. Cone $\mathcal{B}$

The proof of Proposition 2.40 relies on the following lemmas.

Lemma 2.41. *The polynomial $X - x$ is normal if only if $x \leqslant 0$.*

Proof: Follows immediately from the definition of a normal polynomial. □

Lemma 2.42. *A quadratic monic polynomial A with complex conjugate roots is normal if and only if its roots belong to the cone $\mathcal{B}$.*

Proof:

Let $a + i\,b$ and $a - i\,b$ be the roots of A. Then

$$A = X^2 - 2\,a\,X + (a^2 + b^2)$$

is normal if and only if

a) $-2\,a \geqslant 0$,
b) $a^2 + b^2 \geqslant 0$,
c) $(-2\,a)^2 \geqslant a^2 + b^2$.

that is if and only if $a \leqslant 0$ and $4\,a^2 \geqslant a^2 + b^2$, or equivalently $a + i\,b \in \mathcal{B}$. □

Lemma 2.43. *The product of two normal polynomials is normal.*

Proof: Let $A = a_p\,X^p + \cdots + a_0$ and $B = b_q X^q + \cdots + b_0$ be two normal polynomials. We can suppose without loss of generality that 0 is not a root of A and B, i.e. that all the coefficients of A and B are positive.

Let $C = A\,B = c_{p+q} X^{p+q} + \ldots + c_0$. It is clear that all the coefficients of C are positive.

It remains to prove that $c_k^2 \geqslant c_{k-1}\,c_{k+1}$.

Using the partition of $\{(h,j)\in\mathbb{Z}^2 \mid h>j\}$ in $\{(j+1,h-1)\in\mathbb{Z}^2 \mid h\leqslant j\}$ and $\{(h,h-1) \mid h\in\mathbb{Z}\}$.

$$\begin{aligned}
c_k^2 - c_{k-1}c_{k+1} &= \sum_{h\leqslant j} a_h a_j b_{k-h} b_{k-j} + \sum_{h>j} a_h a_j b_{k-h} b_{k-j}\\
&\quad - \sum_{h\leqslant j} a_h a_j b_{k-h+1} b_{k-j-1} - \sum_{h>j} a_h a_j b_{k-h+1} b_{k-j-1}\\
&= \sum_{h\leqslant j} a_h a_j b_{k-h} b_{k-j} + \sum_{h\leqslant j} a_{j+1} a_{h-1} b_{k-j-1} b_{k-h-1}\\
&\quad + \sum_{h} a_h a_{h-1} b_{k-h} b_{k-h+1} - \sum_{h} a_h a_{h-1} b_{k-h+1} b_{k-h}\\
&\quad - \sum_{h\leqslant j} a_h a_j b_{k-h+1} b_{k-j-1} - \sum_{h\leqslant j} a_{j+1} a_{h-1} b_{k-j} b_{k-h}\\
&= \sum_{h\leqslant j} (a_h a_j - a_{h-1} a_{j+1})(b_{k-j} b_{k-h} - b_{k-j-1} b_{k-h+1}).
\end{aligned}$$

Since A is normal and $a_0, \ldots, a_p$ are positive, one has

$$\frac{a_{p-1}}{a_p} \geqslant \frac{a_{p-2}}{a_{p-1}} \geqslant \ldots \geqslant \frac{a_0}{a_1},$$

and $a_h\, a_j - a_{h-1}\, a_{j+1} \geqslant 0$, for all $k \leqslant j$. Similar inequalities hold for the coefficients of B and finally $c_k^2 - c_{k-1}\, c_{k+1}$ is non-negative, being a sum of non-negative quantities. □

Proof of Proposition 2.40: Factor P into linear and quadratic polynomials. By Lemma 2.41 and Lemma 2.42 each of these factors is normal. Now use Lemma 2.43. □

Finally we obtain the following partial reciprocal to Descartes law of signs.

Proposition 2.44. *If A is normal and $x>0$, then* $\mathrm{Var}(A\,(X-x))=1$.

Proof: We can suppose without loss of generality that that 0 is not a root of A, that it that all the coefficients of A are positive.

Then

$$\frac{a_{p-1}}{a_p} \geqslant \frac{a_{p-2}}{a_{p-1}} \geqslant \ldots \geqslant \frac{a_0}{a_1},$$

and

$$\frac{a_{p-1}}{a_p} - x \geqslant \frac{a_{p-2}}{a_{p-1}} - x \geqslant \ldots \geqslant \frac{a_0}{a_1} - x.$$

Since $a_p>0$ and -$a_0\,x<0$, the coefficients of the polynomial

$$(X-x)\,A = a_p X^{p+1} + a_p\left(\frac{a_{p-1}}{a_p} - x\right) X^p + \ldots + a_1\left(\frac{a_0}{a_1} - x\right) X - a_0\, x.$$

have exactly one sign variation. □

A natural question when looking at Budan-Fourier's Theorem (Theorem 2.35), is to interpret the even difference $\mathrm{Var}(\mathrm{Der}(P);a,b) - \mathrm{num}(P;(a,b])$. This can be done through the notion of virtual roots.

The virtual roots of P will enjoy the following properties:

a) the number of virtual roots of P counted with virtual multiplicities is equal to the degree p of P,
b) on an open interval defined by virtual roots, the sign of P is fixed,
c) virtual roots of P and virtual roots of P' are interlaced: if $x_1 \leq \ldots \leq x_p$ are the virtual roots of P and $y_1 \leq \ldots \leq y_{p-1}$ are the virtual roots of P, then

$$x_1 \leq y_1 \leq \ldots \leq x_{p-1} \leq y_{p-1} \leq x_p.$$

Given these properties, in the particular case where P is a polynomial of degree p with all its roots real and simple, virtual roots and real roots clearly coincide.

Definition 2.45. [Virtual roots] The definition of **virtual roots** proceeds by induction on $p = \deg(P)$. We prove simultaneously that properties a), b), c) hold.

If $p = 0$, P has no virtual root and properties a), b), c) hold.

Suppose that properties a), b), c) hold for the virtual roots of P'.

By induction hypothesis the virtual roots of P' are $y_1 \leq \ldots \leq y_{p-1}$. Let

$$I_1 = (-\infty, y_1], \ldots, I_i = [y_{i-1}, y_i], \ldots, I_p = [y_{p-1}, +\infty).$$

By induction hypothesis, the sign of P' is fixed on the interior of each I_i. Let x_i be unique value in I_i such that the absolute value of P on I_i reaches its minimum. The virtual roots of P are $x_1 \leq \ldots \leq x_p$..

According to this inductive definition, properties a), b) and c) are clear for virtual roots of P. Note that the virtual roots of P are always roots of a derivative of P.

The **virtual multiplicity** of x with respect to P, denoted $v(P, x)$ is the number of times x is repeated in the list $x_1 \leq \ldots, \leq x_p$ of virtual roots of P. In particular, if x is not a virtual root of P, its virtual multiplicity is equal to 0. Note that if x is a virtual root of P' with virtual multiplicity ν with respect to P, the virtual multiplicity of x with respect to P' can only be ν, $\nu + 1$ or $\nu - 1$. Moreover, if x is a root of P', the virtual multiplicity of x with respect to P' is necessarily $\nu + 1$. □

Example 2.46. The virtual roots of a polynomial P of degree 2 are

- the two roots of P with virtual multiplicity 1 if P has two distinct real roots,
- the root of P' with virtual multiplicity 2 if P does not have two distinct real roots. □

Given a and b, we denote by $v(P; (a, b])$ the number of virtual roots of P in $(a, b]$ counted with virtual multiplicities.

Theorem 2.47.

$$v(P; (a, b]) = \mathrm{Var}(\mathrm{Der}(P); a, b).$$

The following lemma is the key to the proof of Theorem 2.47.

Lemma 2.48. *Let c be a root of P of virtual multiplicity $v(P,c) \geq 0$. If no $P^{(k)}$, $0 \leq k < p$ has a root in $[d,c)$, then*

$$v(P,c) = \mathrm{Var}(\mathrm{Der}(P); d, c).$$

Proof: The proof of the claim is by induction on $p = \deg(P)$. The claim obviously holds if $p = 0$.

Let $w = v(P,c)$.

- If c is a root of P, the virtual multiplicity of c as a root of P' is $w-1$. By induction hypothesis applied to P', $\mathrm{Var}(\mathrm{Der}(P'); d, c) = w-1$. The claim follows from equation (2.1).
- If c is not a root of P, is a virtual root of P with virtual multiplicity w, and a virtual root of P' with multiplicity ν and virtual multiplicity u, by induction hypothesis applied to P', $\mathrm{Var}(\mathrm{Der}(P'); d, c) = u$.
  - If the sign of P' at the left and at the right of c differ, ν is odd as well as u, using Lemma 2.36 a) and the induction hypothesis for P'.
    - If c is a local minimum of the absolute value of P, $w = u + 1$, $\mathrm{sign}(P^{(\nu+1)}(c)\, P(c)) > 0$, , and the claim follows from (2.2).
    - If c is a local maximum of the absolute value of P, $w = u - 1$, $\mathrm{sign}(P^{(\nu+1)}(c)\, P(c)) < 0$, and the claim follows from (2.3).
  - If the sign of P' at the left and at the right of c coincide, $w = u$, ν is even as well as u using Lemma 2.36 a) and the induction hypothesis for P'. The claim follows from (2.4) and (2.5).

The claim follows in each of these cases. □

It follows clearly from Proposition 2.48 that:

Corollary 2.49. *All the roots of P are virtual roots of P. The virtual multiplicity is at least equal to the multiplicity and the difference is even.*

Proof of Theorem 2.47: It is clear that, for every $c \in (a,b)$,

$$\begin{aligned} v(P;(a,b]) &= v(P;(a,c]) + v(P;(c,b]), \\ \mathrm{Var}(\mathrm{Der}(P); a, b) &= \mathrm{Var}(\mathrm{Der}(P); a, c) + \mathrm{Var}(\mathrm{Der}(P); c, b). \end{aligned}$$

Let $c_1 < \cdots < c_r$ be the roots of all the $P^{(i)}$, $0 \leq i \leq p-1$, in the interval (a,b) and let $c_0 = \infty, c_{r+1} = +\infty$, $d_i \in (c_i, c_{i+1})$ so that $c_0 < d_0 < c_1 < \cdots < c_r < d_r < c_{r+1}$.

Since

$$\begin{aligned} v(P;(a,b) &= \sum_{i=0}^{r} (v(P;(c_i,d_i]) + v(P;(d_i,c_{i+1}])), \\ \mathrm{Var}(\mathrm{Der}(P); a, b) &= \sum_{i=0}^{r} (\mathrm{Var}(\mathrm{Der}(P); c_i, d_i) + \mathrm{Var}(\mathrm{Der}(P); d_i, c_{i+1})), \end{aligned}$$

the claim follows immediately from Lemma 2.36 b) and Lemma 2.48. □

Finally the even number $\text{Var}(\text{Der}(P); a, b) \geq \text{num}(P; (a, b])$ appearing in the statement of Budan-Fourier's Theorem (Theorem 2.35) is the sum of the differences between virtual multiplicities and multiplicities of roots of P in $(a, b]$.

2.2.2 Sturm's Theorem and the Cauchy Index

Let P be a non-zero polynomial with coefficients in a real closed field R. The sequence of signed remainders of P and P', $\text{SRemS}(P, P')$ (see Definition 1.7) is the **Sturm sequence** of P.

We will prove that the number of roots of P in (a, b) can be computed from the Sturm sequence $\text{SRemS}(P, P')$ evaluated at a and b (see Notation 2.34). More precisely the number of roots of P in (a, b) is the difference in the number of sign variations in the Sturm's sequence $\text{SRemS}(P, P')$ evaluated at a and b.

Theorem 2.50. [Sturm's theorem] *Given a and b in $\mathrm{R} \cup \{-\infty, +\infty\}$,*

$$\text{Var}(\text{SRemS}(P, P'); a, b)$$

is the number of roots of P in the interval (a, b).

Remark 2.51. As a consequence, we can decide whether P has a root in R by checking whether $\text{Var}(\text{SRemS}(P, P'); -\infty, +\infty) > 0$. □

Let us first see how to use Theorem 2.50 (Sturm's theorem).

Example 2.52. Consider the polynomial $P = X^4 - 5X^2 + 4$. The Sturm sequence of P is

$$\begin{aligned}
\text{SRemS}_0(P, P') &= P = X^4 - 5X^2 + 4, \\
\text{SRemS}_1(P, P') &= P' = 4X^3 - 10X, \\
\text{SRemS}_2(P, P') &= \frac{5}{2}X^2 - 4, \\
\text{SRemS}_3(P, P') &= \frac{18}{5}X, \\
\text{SRemS}_4(P, P') &= 4.
\end{aligned}$$

The signs of the leading coefficients of the Sturm sequence are $+ + + + +$ and the degrees of the polynomials in the Sturm sequence are $4, 3, 2, 1, 0$. The signs of the polynomials in the Sturm sequence at $-\infty$ are $+ - + - +$, and the signs of the polynomials in the Sturm sequence at $+\infty$ are $+ + + + +$, so $\text{Var}(\text{SRemS}(P, P'); -\infty, +\infty) = 4$. There are indeed 4 real roots: $1, -1, 2$, and -2. □

We are going to prove a statement more general than Theorem 2.50 (Sturm's theorem), since it will be useful not only to determine whether P has a root in R but also to determine whether P has a root at which another polynomial Q is positive.

With this goal in mind, it is profitable to look at the jumps (discontinuities) of the rational function $P'Q/P$. Clearly, these occur only at points c for which $P(c)=0$, $Q(c)\neq 0$. If c occurs as a root of P with multiplicity μ then $P'Q/P=\mu\, Q(c)/(X-c)+R_c$, where R_c is a rational function defined at c. It is now obvious that if $Q(c)>0$, then $P'Q/P$ jumps from $-\infty$ to $+\infty$ at c, and if $Q(c)<0$, then $P'Q/P$ jumps from $+\infty$ to $-\infty$ at c. Thus the number of jumps of $P'Q/P$ from $-\infty$ to $+\infty$ minus the number of jumps of $P'Q/P$ from $+\infty$ to $-\infty$ is equal to the number of roots of P at which Q is positive minus the number of roots of P at which Q is negative. This observation leads us to the following definition. We need first what we mean by a jump from $-\infty$ to $+\infty$.

Definition 2.53. [Cauchy index] Let x be a root of P. The function Q/P **jumps from** $-\infty$ **to** $+\infty$ **at** x if the multiplicity μ of x as a root of P is bigger than the multiplicity ν of x as a root of Q, $\mu-\nu$ is odd and the sign of Q/P at the right of x is positive. Similarly, the function Q/P **jumps from** $+\infty$ **to** $-\infty$ **at** x if if the multiplicity μ of x as a root of P is bigger than the multiplicity ν of x as a root of Q, $\mu-\nu$ is odd and the sign of Q/P at the right of x is negative.

Given $a<b$ in $\mathrm{R}\cup\{-\infty,+\infty\}$ and $P,Q\in\mathrm{R}[X]$, we define the **Cauchy index** of Q/P on (a,b), $\mathrm{Ind}(Q/P;a,b)$, to be the number of jumps of the function Q/P from $-\infty$ to $+\infty$ minus the number of jumps of the function Q/P from $+\infty$ to $-\infty$ on the open interval (a,b). The **Cauchy index** of Q/P on R is simply called the **Cauchy index** of Q/P and it is denoted by $\mathrm{Ind}(Q/P)$, rather than by $\mathrm{Ind}(Q/P;-\infty,+\infty)$. □

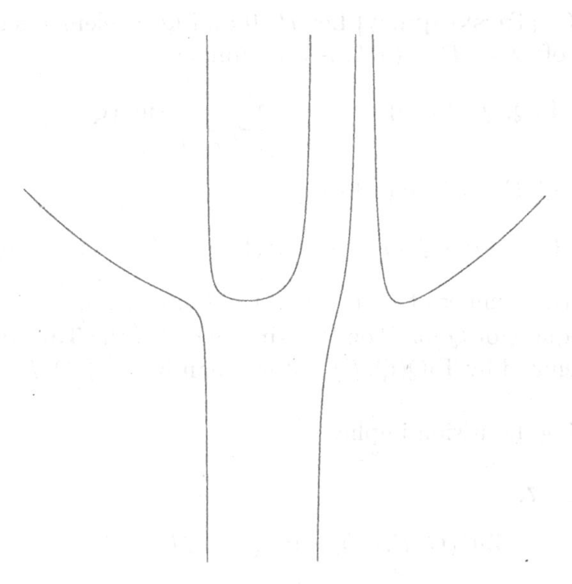

Fig. 2.2. Graph of the rational function Q/P

Example 2.54. Let

$$\begin{aligned} P &= (X-3)^2 (X-1)(X+3), \\ Q &= (X-5)(X-4)(X-2)(X+1)(X+2)(X+4). \end{aligned}$$

The graph of Q/P is depicted in Figure 2.2.

In this example,

$$\begin{aligned} \mathrm{Ind}(Q/P) &= 0 \\ \mathrm{Ind}(Q/P; -\infty, 0) &= 1 \\ \mathrm{Ind}(Q/P; 0, \infty) &= -1 \end{aligned}$$

□

Remark 2.55.

a) Suppose $\deg(P) = p$, $\deg(Q) = q < p$. The Cauchy index $\mathrm{Ind}(Q/P; a, b)$ is equal to p if and only if $q = p-1$, the signs of the leading coefficients of P and Q are equal, all the roots of P and Q are simple and belong to (a, b), and there is exactly one root of Q between two roots of P.
b) If $R = \mathrm{Rem}(Q, P)$, it follows clearly from the definition that

$$\mathrm{Ind}(Q/P; a, b) = \mathrm{Ind}(R/P; a, b).$$

□

Using the notion of Cauchy index we can reformulate our preceding discussion, using the following notation.

Notation 2.56. [Tarski-query] Let $P \neq 0$ and Q be elements of $\mathrm{K}[X]$. The **Tarski-query** of Q for P in (a, b) is the number

$$\mathrm{TaQ}(Q, P; a, b) = \sum_{x \in (a,b), P(x)=0} \mathrm{sign}(Q(x)).$$

Note that $\mathrm{TaQ}(Q, P; a, b)$ is equal to

$$\#(\{x \in (a, b) \mid P(x) = 0 \wedge Q(x) > 0\}) - \#(\{x \in (a, b) \mid P(x) = 0 \wedge Q(x) < 0\})$$

where $\#(S)$ is the number of elements in the finite set S.

The Tarski-query of Q for P on R is simply called the **Tarski-query** of Q for P, and is denoted by $\mathrm{TaQ}(Q, P)$, rather than by $\mathrm{TaQ}(Q, P; -\infty, +\infty)$. □

The preceding discussion implies:

Proposition 2.57.

$$\mathrm{TaQ}(Q, P; a, b) = \mathrm{Ind}(P' Q/P; a, b).$$

In particular the number of roots of P in (a, b) is $\mathrm{Ind}(P'/P; a, b)$.

We now describe how to compute $\mathrm{Ind}(Q/P;\, a,\, b)$. We will see that the Cauchy index is the difference in the number of sign variations in the signed remainder sequence $\mathrm{SRemS}(P,\, Q)$ evaluated at a and b (Definition 1.7 and Notation 2.34).

Theorem 2.58. *Let P, $P \neq 0$, and Q be two polynomials with coefficients in a real closed field* R, *and let a and b (with $a < b$) be elements of $\mathrm{R} \cup \{-\infty, +\infty\}$ that are not roots of P. Then,*

$$\mathrm{Var}(\mathrm{SRemS}(P, Q); a, b) = \mathrm{Ind}(Q/P; a, b).$$

Let $R = \mathrm{Rem}(P, Q)$ and let $\sigma(a)$ be the sign of PQ at a and $\sigma(b)$ be the sign of PQ at b. The proof of Theorem 2.58 proceeds by induction on the length of the signed remainder sequence and is based on the following lemmas.

Lemma 2.59. *If a and b are not roots of a polynomial in the signed remainder sequence,*

$$\begin{aligned} &\mathrm{Var}(\mathrm{SRemS}(P, Q); a, b) \\ = &\begin{cases} \mathrm{Var}(\mathrm{SRemS}(Q, -R); a, b) + \sigma(b) & \text{if } \sigma(a)\,\sigma(b) = -1, \\ \mathrm{Var}(\mathrm{SRemS}(Q, -R); a, b) & \text{if } \sigma(a)\,\sigma(b) = 1. \end{cases} \end{aligned}$$

Proof: The claim follows from the fact that at any x which is not a root of P and Q (and in particular at a and b)

$$\mathrm{Var}(\mathrm{SRemS}(P, Q); x) = \begin{cases} \mathrm{Var}(\mathrm{SRemS}(Q, -R); x) + 1 & \text{if } P(x)\,Q(x) < 0, \\ \mathrm{Var}(\mathrm{SRemS}(Q, -R); x) & \text{if } P(x)\,Q(x) > 0, \end{cases}$$

looking at all possible cases. □

Lemma 2.60. *If a and b are not roots of a polynomial in the signed remainder sequence,*

$$\mathrm{Ind}(Q/P; a, b) = \begin{cases} \mathrm{Ind}(-R/Q; a, b) + \sigma(b) & \text{if } \sigma(a)\,\sigma(b) = -1, \\ \mathrm{Ind}(-R/Q; a, b) & \text{if } \sigma(a)\,\sigma(b) = 1. \end{cases}$$

Proof: We can suppose without loss of generality that Q and P are coprime. Indeed if D is a greatest common divisor of P and Q and

$$P_1 = P/D, Q_1 = Q/D, R_1 = \mathrm{Rem}(P_1, Q_1) = R/D,$$

then P_1 and Q_1 are coprime,

$$\mathrm{Ind}(Q/P; a, b) = \mathrm{Ind}(Q_1/P_1; a, b), \mathrm{Ind}(-R/Q; a, b) = \mathrm{Ind}(-R_1/Q_1; a, b),$$

and the signs of $P(x)Q(x)$ and $P_1(x)Q_1(x)$ coincide at any point which is not a root of PQ.

Let n_{-+} (resp. n_{+-}) denote the number of sign variations from -1 to 1 (resp. from 1 to -1) of PQ when x varies from a to b. It is clear that

$$n_{-+} - n_{+-} = \begin{cases} \sigma(b) & \text{if } \sigma(a)\,\sigma(b) = -1 \\ 0 & \text{if } \sigma(a)\,\sigma(b) = 1. \end{cases}$$

It follows from the definition of Cauchy index that

$$\mathrm{Ind}(Q/P; a, b) + \mathrm{Ind}(P/Q; a, b) = n_{-+} - n_{+-}.$$

Noting that

$$\mathrm{Ind}(R/Q; a, b) = \mathrm{Ind}(P/Q; a, b),$$

the claim of the lemma is now clear. □

Proof of Theorem 2.58: We can assume without loss of generality that a and b are not roots of a polynomial in the signed remainder sequence. Indeed if $a < a' < b' < b$ with $(a, a']$ and $[b', b)$ containing no root of the polynomials in the signed remainder sequence, it is clear that

$$\mathrm{Ind}(Q/P; a, b) = \mathrm{Ind}(Q/P; a', b').$$

We prove now that

$$\mathrm{Var}(\mathrm{SRemS}(P, Q); a, b) = \mathrm{Var}(\mathrm{SRemS}(P, Q); a', b').$$

We omit $(P,\ Q)$ in the notation in the following lines. First notice that since a is not a root of P, a is not a root of the greatest common divisor of P and Q, and hence a is not simultaneously a root of SRemS_j and SRemS_{j+1} (resp. SRemS_{j-1} and SRemS_j). So, if a is a root of SRemS_j, $j \neq 0$, $\mathrm{SRemS}_{j-1}(a)\,\mathrm{SRemS}_{j+1}(a) < 0$, since

$$\mathrm{SRemS}_{j+1} = -\mathrm{SRemS}_{j-1} + \mathrm{Quo}(\mathrm{SRemS}_j, \mathrm{SRemS}_{j-1})\,\mathrm{SRemS}_j$$

(see Remark 1.4) so that

$$\begin{aligned} & \mathrm{Var}(\mathrm{SRemS}_{j-1}, \mathrm{SRemS}_j, \mathrm{SRemS}_{j+1}; a) \\ = \ & \mathrm{Var}(\mathrm{SRemS}_{j-1}, \mathrm{SRemS}_j, \mathrm{SRemS}_{j+1}; a') \\ = \ & 1. \end{aligned}$$

This implies $\mathrm{Var}(\mathrm{SRemS}(P,\ Q);\ a) = \mathrm{Var}(\mathrm{SRemS}(P,\ Q);\ a')$, and similarly $\mathrm{Var}(\mathrm{SRemS}(P, Q); b) = \mathrm{Var}(\mathrm{SRemS}(P, Q); b')$.

The proof of the theorem now proceeds by induction on the number $n \geq 2$ of elements in the signed remainder sequence. The base case $n = 2$ corresponds to $R = 0$ and follows from Lemma 2.59 and Lemma 2.60. Let us suppose that the Theorem holds for $n - 1$ and consider P and Q such that their signed remainder sequence has n elements. The signed remainder sequence of Q and $-R$ has $n - 1$ elements and, by the induction hypothesis,

$$\mathrm{Var}(\mathrm{SRemS}(Q, -R); a, b) = \mathrm{Ind}(-R/Q; a, b).$$

So, by Lemma 2.59 and Lemma 2.60,

$$\mathrm{Var}(\mathrm{SRemS}(P, Q); a, b) = \mathrm{Ind}(Q\,P; a, b).$$ □

As a consequence of the above we derive the following theorem.

Theorem 2.61. [Tarski's theorem] *If $a < b$ are elements of $\mathrm{R} \cup \{-\infty, +\infty\}$ that are not roots of P, with $P, Q \in \mathrm{R}[X]$, then*

$$\mathrm{Var}(\mathrm{SRemS}(P, P'\,Q); a, b) = \mathrm{TaQ}(Q, P; a, b).$$

Proof: This is immediate from Theorem 2.58 and Proposition 2.57. □

Theorem 2.50 (Sturm's theorem) is a particular case of Theorem 2.61, taking $Q = 1$.

Proof of Theorem 2.50: The proof is immediate by taking $Q = 1$ in Theorem 2.61. □

2.3 Projection Theorem for Algebraic Sets

Let R be a real closed field. If $\mathcal{P}$ is a finite subset of $\mathrm{R}[X_1, \ldots, X_k]$, we write the **set of zeros** of $\mathcal{P}$ in R^k as

$$\mathrm{Zer}(\mathcal{P}, \mathrm{R}^k) = \{x \in \mathrm{R}^k \mid \bigwedge_{P \in \mathcal{P}} P(x) = 0\}.$$

These are the **algebraic sets** of $\mathrm{R}^k = \mathrm{Zer}(\{0\}, \mathrm{R}^k)$.

An important way in which this differs from the algebraically closed case is that the common zeros of $\mathcal{P}$ are also the zeros of a single polynomial $Q = \sum_{P \in \mathcal{P}} P^2$.

The smallest family of sets of R^k that contains the algebraic sets and is closed under the boolean operations (complementation, finite unions, and finite intersections) is the **constructible sets**.

We define the **semi-algebraic sets** of R^k as the smallest family of sets in R^k that contains the algebraic sets as well as sets defined by polynomial **inequalities** i.e. sets of the form $\{x \in \mathrm{R}^k | P(x) > 0\}$ for some polynomial $P \in \mathrm{R}[X_1, \ldots, X_k]$, and which is also closed under the boolean operations (complementation, finite unions, and finite intersections). If the coefficients of the polynomials defining S lie in a subring $\mathrm{D} \subset \mathrm{R}$, we say that the semi-algebraic set S is **defined over** D.

It is obvious that any semi-algebraic set in R^k is the finite union of sets of the form $\{x \in \mathrm{R}^k | P(x) = 0 \wedge \bigwedge_{Q \in \mathcal{Q}} Q(x) > 0\}$. These are the **basic semi-algebraic sets**.

Notice that the constructible sets are semi-algebraic as the basic constructible set

$$S = \{x \in \mathrm{R}^k | P(x) = 0 \wedge \bigwedge_{Q \in \mathcal{Q}} Q(x) \neq 0\}$$

is the basic semi-algebraic set

$$\{x \in \mathrm{R}^k | P(x) = 0 \wedge \bigwedge_{Q \in \mathcal{Q}} Q^2(x) > 0\}.$$

The goal of the next pages is to show that the projection of an algebraic set in R^{k+1} is a semi-algebraic set of R^k if R is a real closed field.

This is a new example of the paradigm described in Chapter 1 for extending an algorithm from the univariate case to the multivariate case by viewing the univariate case parametrically. The algebraic set $Z \subset \mathrm{R}^{k+1}$ can be described as

$$Z = \{(y, x) \in \mathrm{R}^{k+1} \mid P(y, x) = 0\}$$

with $P \in \mathrm{R}[X_1, \ldots, X_k, X_{k+1}]$, and its projection $\pi(Z)$ (forgetting the last coordinate) is

$$\pi(Z) = \{y \in \mathrm{R}^k \mid \exists x \in \mathrm{R}\ P(y, x) = 0)\}.$$

For a particular $y \in \mathrm{R}^k$ we can decide, using Theorem 2.50 (Sturm's theorem) and Remark 2.51, whether or not $\exists x \in \mathrm{R}\ P_y(x) = 0$ is true.

Defining $Z_y = \{x \in \mathrm{R} \mid P_y(x) = 0\}$, (see Notation 1.18) what is crucial here is to partition the parameter space R^k into finitely many parts so that each part is either contained in $\{y \in \mathrm{R}^k \mid Z_y = \emptyset\}$ or in $\{y \in \mathrm{R}^k \mid Z_y \neq \emptyset\}$. Moreover, the algorithm used for constructing the partition ensures that the decision algorithm testing whether Z_y is empty or not is the same (is uniform) for all y in any given part. Because of this uniformity, it turns out that each part of the partition is a semi-algebraic set. Since $\pi(Z)$ is the union of those parts where $Z_y \neq \emptyset$, $\pi(Z)$ is semi-algebraic being the union of finitely many semi-algebraic sets.

We first introduce some terminology from logic which is useful for the study of semi-algebraic sets.

We define the language of ordered fields by describing the formulas of this language. The definitions are similar to the corresponding notions in Chapter 1, the only difference is the use of inequalities in the atoms. The formulas are built starting with atoms, which are polynomial equations and inequalities. A formula is written using atoms together with the logical connectives "and", "or", and "negation" ($\wedge$, $\vee$, and $\neg$) and the existential and universal quantifiers ($\exists, \forall$). A formula has free variables, i.e. non-quantified variables, and bound variables, i.e. quantified variables. More precisely, let D be a subring of R. We define the **language of ordered fields with coefficients in D** as follows. An **atom** is $P = 0$ or $P > 0$, where P is a polynomial in $\mathrm{D}[X_1, \ldots, X_k]$. We define simultaneously the **formulas** and the set $\mathrm{Free}(\Phi)$ of **free variables of a formula** Φ as follows

- an atom $P = 0$ or $P > 0$, where P is a polynomial in $\mathrm{D}[X_1, \ldots, X_k]$ is a formula with free variables $\{X_1, \ldots, X_k\}$,

- if Φ_1 and Φ_2 are formulas, then $\Phi_1 \wedge \Phi_2$ and $\Phi_1 \vee \Phi_2$ are formulas with $\mathrm{Free}(\Phi_1 \wedge \Phi_2) = \mathrm{Free}(\Phi_1 \vee \Phi_2) = \mathrm{Free}(\Phi_1) \cup \mathrm{Free}(\Phi_2)$,
- if Φ is a formula, then $\neg(\Phi)$ is a formula with $\mathrm{Free}(\neg(\Phi)) = \mathrm{Free}(\Phi)$,
- if Φ is a formula and $X \in \mathrm{Free}(\Phi)$, then $(\exists X)\ \Phi$ and $(\forall X)\ \Phi$ are formulas with $\mathrm{Free}((\exists X)\ \Phi) = \mathrm{Free}((\forall X)\ \Phi) = \mathrm{Free}(\Phi) \setminus \{X\}$.

If Φ and Ψ are formulas, $\Phi \Rightarrow \Psi$ is the formula $\neg(\Phi) \vee \Psi$.

A **quantifier free formula** is a formula in which no quantifier appears, neither $\exists$ nor $\forall$. A **basic formula** is a conjunction of atoms.

The **R-realization of a formula** Φ with free variables contained in $\{Y_1, \ldots, Y_k\}$, denoted $\mathrm{Reali}(\Phi, \mathrm{R}^k)$, is the set of $y \in \mathrm{R}^k$ such that $\Phi(y)$ is true. It is defined by induction on the construction of the formula, starting from atoms:

$$\begin{aligned}
\mathrm{Reali}(P=0,\mathrm{R}^k) &= \{y \in \mathrm{R}^k \mid P(y)=0\},\\
\mathrm{Reali}(P>0,\mathrm{R}^k) &= \{y \in \mathrm{R}^k \mid P(y)>0\},\\
\mathrm{Reali}(P<0,\mathrm{R}^k) &= \{y \in \mathrm{R}^k \mid P(y)<0\},\\
\mathrm{Reali}(\Phi_1 \wedge \Phi_2,\mathrm{R}^k) &= \mathrm{Reali}(\Phi_1,\mathrm{R}^k) \cap \mathrm{Reali}(\Phi_2,\mathrm{R}^k),\\
\mathrm{Reali}(\Phi_1 \vee \Phi_2,\mathrm{R}^k) &= \mathrm{Reali}(\Phi_1,\mathrm{R}^k) \cup \mathrm{Reali}(\Phi_2,\mathrm{R}^k),\\
\mathrm{Reali}(\neg\Phi,\mathrm{R}^k) &= \mathrm{R}^k \setminus \mathrm{Reali}(\Phi,\mathrm{R}^k),\\
\mathrm{Reali}((\exists X)\ \Phi,\mathrm{R}^k) &= \{y \in \mathrm{R}^k \mid \exists x \in \mathrm{R} \quad (x,y) \in \mathrm{Reali}(\Phi,\mathrm{R}^{k+1})\},\\
\mathrm{Reali}((\forall X)\ \Phi,\mathrm{R}^k) &= \{y \in \mathrm{R}^k \mid \forall x \in \mathrm{R} \quad (x,y) \in \mathrm{Reali}(\Phi,\mathrm{R}^{k+1})\}
\end{aligned}$$

Two formulas Φ and Ψ such that $\mathrm{Free}(\Phi) = \mathrm{Free}(\Psi) = \{Y_1, \ldots, Y_k\}$ are **R-equivalent** if $\mathrm{Reali}(\Phi, \mathrm{R}^k) = \mathrm{Reali}(\Psi, \mathrm{R}^k)$. If there is no ambiguity, we simply write $\mathrm{Reali}(\Phi)$ for $\mathrm{Reali}(\Phi, \mathrm{R}^k)$ and talk about realization and equivalence.

It is clear that a set is semi-algebraic if and only if it can be represented as the realization of a quantifier free formula. It is also easy to see that any formula in the language of fields with coefficients in D is R-equivalent to a formula

$$\Phi(Y) = (\mathrm{Qu}_1 X_1) \ldots (\mathrm{Qu}_m X_m)\ \mathcal{B}(X_1, \ldots, X_m, Y_1, \ldots Y_k)$$

where each $\mathrm{Qu}_i \in \{\forall, \exists\}$ and $\mathcal{B}$ is a quantifier free formula involving polynomials in $\mathrm{D}[X_1, \ldots, X_m, Y_1, \ldots Y_k]$. This is called its **prenex normal form** (see Section 10, Chapter 1 of [115]). The variables $X_1, \ldots, X_m$ are called **bound variables** . If a formula has no free variables, then it is called a **sentence**, and is either R-equivalent to true, when $\mathrm{Reali}(\Phi, \{0\}) = \{0\}$, or R-equivalent to false, when $\mathrm{Reali}(\Phi, \{0\}) = \emptyset$. For example, $1 > 0$ is R-equivalent to true, and $1 < 0$ is R-equivalent to false.

We now prove that the projection of an algebraic set is semi-algebraic.

Theorem 2.62. *Given an algebraic set of* R^{k+1} *defined over* D, *its projection to* R^k *is a semi-algebraic set defined over* D.

Before proving Theorem 2.62, let us explain the mechanism of its proof on an example.

Example 2.63. We describe the projection of the algebraic set

$$\{(a,b,c,X)\in \mathrm{R}^4 \mid X^4+a\,X^2+b\,X+c=0\}$$

to R^3, i.e. the set

$$\{(a;b;c)\in \mathrm{R}^3 \mid \exists X\in \mathrm{R}\quad X^4+a\,X^2+b\,X+c=0\},$$

as a semi-algebraic set.

We look at all leaves of $\mathrm{TRems}(P,P')$ and at all possible signs for leading coefficients of all possible signed pseudo-remainders (using Example 1.15). We denote by n the difference between the number of sign variations at $-\infty$ and $+\infty$ in the Sturm sequence of $P=X^4+a\,X^2+b\,X+c$ for each case. We indicate for each leaf L of $\mathrm{TRems}(P,P')$ the basic formula $\mathcal{C}_L$ and the degrees occurring in the signed pseudo-remainder sequence of P and P' along the path $\mathcal{B}_L$.

$$(a\neq 0\wedge s\neq 0\wedge \delta\neq 0,(4,3,2,1,0))$$

a	$-$	$-$	$-$	$-$	$+$	$+$	$+$	$+$
s	$+$	$+$	$-$	$-$	$+$	$+$	$-$	$-$
δ	$+$	$-$	$+$	$-$	$+$	$-$	$+$	$-$
n	4	2	0	2	0	-2	0	2

The first column can be read as follows: for every polynomial

$$P=X^4+a\,X^2+b\,X+c$$

satisfying $a<0$, $s>0$, $\delta>0$, the number of real roots is 4. Indeed the leading coefficients of the signed pseudo-remainder sequence of P and P' are $1,4,-a,64\,s,16384\,a^2\,\delta$ (see Example 1.17) and the degrees of the polynomials in the signed pseudo-remainder sequence of P and P' are $4,3,2,1,0$, the signs of the signed pseudo-remainder sequence of P and P' at $-\infty$ are $+-+-+$ and at $+\infty$ are $+++++$. We can apply Theorem 2.50 (Sturm's Theorem).

The other columns can be read similarly. Notice that n can be negative (for $a > 0$, $s > 0$, $\delta < 0$). Though this looks paradoxical, Sturm's theorem is not violated. It only means that there is no polynomial $P \in \mathrm{R}[X]$ with $P = X^4 + a X^2 + b X + c$ and $a > 0$, $s > 0$, $\delta < 0$. Notice that even when n is non-negative, there might be no polynomial $P \in \mathrm{R}[X]$ with $P = X^4 + a X^2 + b X + c$ and (a, s, δ) satisfying the corresponding sign condition.

Similarly, for the other leaves of $\mathrm{TRems}(P, P')$

$$(a \neq 0 \wedge s \neq 0 \wedge \delta = 0, (4, 3, 2, 1))$$

a	$-$	$-$	$+$	$+$
s	$+$	$-$	$+$	$-$
n	3	1	-1	1

$$(a \neq 0 \wedge s = 0 \wedge t \neq 0, (4, 3, 2, 0))$$

a	$-$	$-$	$+$	$+$
t	$+$	$-$	$+$	$-$
n	2	2	0	0

$$(a \neq 0 \wedge s = t = 0, (4, 3, 2))$$

a	$-$	$+$
n	2	0

$$(a = 0 \wedge b \neq 0 \wedge u \neq 0, (4, 3, 1, 0))$$

b	$+$	$+$	$-$	$-$
u	$+$	$-$	$+$	$-$
n	2	0	0	2

$$(a = 0 \wedge b \neq 0 \wedge u = 0, (4, 3, 1))$$

b	$+$	$-$
n	1	1

$$(a = b = 0 \wedge c \neq 0, (4, 3, 0))$$

c	$+$	$-$
n	0	2

$$(a = b = c = 0, (4, 3))$$

$$n = 1$$

Finally, the formula $\exists X \quad X^4 + a\,X^2 + b\,X + c = 0$ is R-equivalent to the quantifier-free formula $\Phi(a, b, c)$:

$$
\begin{aligned}
& (a<0 \wedge s>0) \\
\vee\ & (a<0 \wedge s<0 \wedge \delta<0) \\
\vee\ & (a>0 \wedge s<0 \wedge \delta<0) \\
\vee\ & (a<0 \wedge s\neq 0 \wedge \delta=0) \\
\vee\ & (a>0 \wedge s<0 \wedge \delta=0) \\
\vee\ & (a<0 \wedge s=0 \wedge t\neq 0) \\
\vee\ & (a<0 \wedge s=0 \wedge t=0) \\
\vee\ & (a=0 \wedge b<0 \wedge u<0) \\
\vee\ & (a=0 \wedge b>0 \wedge u>0) \\
\vee\ & (a=0 \wedge b\neq 0 \wedge u=0) \\
\vee\ & (a=0 \wedge b=0 \wedge c<0) \\
\vee\ & (a=0 \wedge b=0 \wedge c=0),
\end{aligned}
$$

by collecting all the sign conditions with $n \geq 1$. Thus, we have proven that the projection of the algebraic set

$$\{(x, a, b, c) \in \mathrm{R}^4 \mid x^4 + a\,x^2 + b\,x + c\}$$

into R^3 is the semi-algebraic subset $\mathrm{Reali}(\Phi, \mathrm{R}^3)$. □

The proof of Theorem 2.62 follows closely the method illustrated in the example.

Proof of Theorem 2.62: Let $Z = \{z \in \mathrm{R}^{k+1} \mid P(z) = 0\}$. Let Z' be the intersection of Z with the subset of $(y, x) \in \mathrm{R}^{k+1}$ such that P_y is not identically zero.

Let L be a leaf of $\mathrm{TRems}(P, P')$, and let $\mathcal{A}(L)$ be the set of non-zero polynomials in $\mathrm{D}[Y_1, \ldots, Y_k]$ appearing in the basic formula $\mathcal{C}_L$, (see Notation 1.18).

Let $\mathcal{L}$ be the set of all leaves of $\mathrm{TRems}(P, P')$, and

$$\mathcal{A} = \bigcup_{L \in \mathcal{L}} \mathcal{A}(L) \subset \mathrm{D}[Y_1, \ldots, Y_k].$$

If $\tau \in \{0, 1, -1\}^{\mathcal{A}}$, we define the realization of τ by

$$\mathrm{Reali}(\tau) = \{y \in \mathrm{R}^k \mid \bigwedge_{A \in \mathcal{A}} \mathrm{sign}(A(y)) = \tau(A)\}.$$

Let $Z_y = \{x \in \mathrm{R} \mid P(y, x) = 0\}$. Note that $\mathrm{Reali}(\tau) \subset \{y \in \mathrm{R}^k \mid Z_y \neq \emptyset\}$ or $\mathrm{Reali}(\tau) \subset \{y \in \mathrm{R}^k \mid Z_y = \emptyset\}$, by Theorem 2.50 (Sturm's theorem) and Remark 2.51. Let

$$\Sigma = \{\tau \in \{0, 1, -1\}^{\mathcal{A}} \mid \forall y \in \mathrm{Reali}(\tau) \quad Z_y \neq \emptyset\}.$$

It is clear that the semi-algebraic set $\bigcup_{\tau \in \Sigma} \mathrm{Reali}(\tau)$ coincides with the projection of S'.

The fact that the projection of the intersection of Z with the subset of $(y, x) \in \mathrm{R}^{k+1}$ such that P_y is identically zero is semi-algebraic is obvious.

Thus the whole projection of $Z = Z' \cup (Z \setminus Z')$ is semi-algebraic since it is a union of semi-algebraic sets. □

2.4 Projection Theorem for Semi-Algebraic Sets

We are going to prove by a similar method that the projection of a semi-algebraic set is semi-algebraic. We start with a decision algorithm deciding if a given sign condition has a non-empty realization at the zeroes of a univariate polynomial.

When P and Q have no common roots, we can find the number of roots of P at each possible sign of Q in terms of the Tarski-querics of 1 and Q for P.

We denote

$$\begin{aligned} Z &= \mathrm{Zer}(P, \mathrm{R}) \\ &= \{x \in \mathrm{R} \mid P(x) = 0\}, \\ \mathrm{Reali}(Q = 0, Z) &= \{x \in Z \mid \mathrm{sign}(Q(x)) = 0\} = \{x \in Z \mid Q(x) = 0\}, \\ \mathrm{Reali}(Q > 0, Z) &= \{x \in Z \mid \mathrm{sign}(Q(x)) = 1\} = \{x \in Z \mid Q(x) > 0\}, \\ \mathrm{Reali}(Q < 0, Z) &= \{x \in Z \mid \mathrm{sign}(Q(x)) = -1\} = \{x \in Z \mid Q(x) < 0\}, \end{aligned}$$

and $c(Q = 0, Z)$, $c(Q > 0, Z)$, $c(Q < 0, Z)$ are the cardinalities of the corresponding sets.

Proposition 2.64. *If P and Q have no common roots in* R, *then*

$$\begin{aligned} c(Q > 0, Z) &= (\mathrm{TaQ}(1, P) + \mathrm{TaQ}(Q, P))/2, \\ c(Q < 0, Z) &= (\mathrm{TaQ}(1, P) - \mathrm{TaQ}(Q, P))/2. \end{aligned}$$

Proof: We have

$$\begin{aligned} \mathrm{TaQ}(1, P) &= c(Q > 0, Z) + c(Q < 0, Z), \\ \mathrm{TaQ}(Q, P) &= c(Q > 0, Z) - c(Q < 0, Z). \end{aligned}$$

Now solve. □

With a little more effort, we can find the number of roots of P at each possible sign of Q in terms of the Tarski-queries of 1, Q, and Q^2 for P.

Proposition 2.65. *The following holds*

$$\begin{aligned} c(Q=0,Z) &= \mathrm{TaQ}(1,P) - \mathrm{TaQ}(Q^2,P), \\ c(Q>0,Z) &= (\mathrm{TaQ}(Q^2,P) + \mathrm{TaQ}(Q,P))/2, \\ c(Q<0,Z) &= (\mathrm{TaQ}(Q^2,P) - \mathrm{TaQ}(Q,P))/2. \end{aligned}$$

Proof: Indeed, we have

$$\begin{aligned} \mathrm{TaQ}(1,P) &= c(Q=0,Z) + c(Q>0,Z) + c(Q<0,Z), \\ \mathrm{TaQ}(Q,P) &= c(Q>0,Z) - c(Q<0,Z), \\ \mathrm{TaQ}(Q^2,P) &= c(Q>0,Z) + c(Q<0,Z). \end{aligned}$$

Now solve. □

We want to extend these results to the case of many polynomials.

We consider a $P \in \mathrm{R}[X]$ with P not identically zero, $\mathcal{Q}$ a finite subset of $\mathrm{R}[X]$, and the finite set $Z = \mathrm{Zer}(P, \mathrm{R}) = \{x \in \mathrm{R} \mid P(x) = 0\}$.

We will give an expression for the number of elements of Z at which $\mathcal{Q}$ satisfies a given sign condition σ.

Let σ be a sign condition on $\mathcal{Q}$ i.e. an element of $\{0,1,-1\}^{\mathcal{Q}}$. The **realization of the sign condition** σ **over** Z is

$$\mathrm{Reali}(\sigma, Z) = \{x \in \mathrm{R} \mid P(x) = 0 \wedge \bigwedge_{Q \in \mathcal{Q}} \mathrm{sign}(Q(x)) = \sigma(Q)\}.$$

Its cardinality is denoted $c(\sigma, Z)$.

Given $\alpha \in \{0,1,2\}^{\mathcal{Q}}$, and $\sigma \in \{0,1,-1\}^{\mathcal{Q}}$ we write σ^α for $\prod_{Q \in \mathcal{Q}} \sigma(Q)^{\alpha(Q)}$, and $\mathcal{Q}^\alpha$ for $\prod_{Q \in \mathcal{Q}} Q^{\alpha(Q)}$. When $\mathrm{Reali}(\sigma, Z) \neq \emptyset$, the sign of $\mathcal{Q}^\alpha$ is fixed on $\mathrm{Reali}(\sigma, Z)$ and is equal to σ^α, with the convention that $0^0 = 1$.

We number the elements of $\mathcal{Q}$ so that $\mathcal{Q} = \{Q_1, \ldots, Q_s\}$ and use the lexicographical orderings on $\{0, 1, 2\}^{\mathcal{Q}}$ (with $0<1<2$) and $\{0, 1, -1\}^{\mathcal{Q}}$ (with $0 \prec 1 \prec -1$) (see Definition 2.14).

Given a list of elements $A = \alpha_1, \ldots, \alpha_m$ of $\{0,1,2\}^{\mathcal{Q}}$ with $\alpha_1 <_{\mathrm{lex}} \ldots <_{\mathrm{lex}} \alpha_m$, we define

$$\begin{aligned} \mathcal{Q}^A &= \mathcal{Q}^{\alpha_1}, \ldots, \mathcal{Q}^{\alpha_m} \\ \mathrm{TaQ}(\mathcal{Q}^A, P) &= \mathrm{TaQ}(\mathcal{Q}^{\alpha_1}, P), \ldots, \mathrm{TaQ}(\mathcal{Q}^{\alpha_m}, P). \end{aligned}$$

Given a list of elements $\Sigma = \sigma_1, \ldots, \sigma_n$ of $\{0,1,-1\}^{\mathcal{Q}}$, with $\sigma_1 <_{\mathrm{lex}} \ldots <_{\mathrm{lex}} \sigma_n$, we define

$$\begin{aligned} \mathrm{Reali}(\Sigma, Z) &= \mathrm{Reali}(\sigma_1, Z), \ldots, \mathrm{Reali}(\sigma_n, Z) \\ c(\Sigma, Z) &= c(\sigma_1, Z), \ldots, c(\sigma_n, Z). \end{aligned}$$

Definition 2.66. The **matrix of signs of** $\mathcal{Q}^A$ **on** Σ is the $m \times n$ matrix $\mathrm{Mat}(A, \Sigma)$ whose i, j-th entry is $\sigma_j^{\alpha_i}$. □

Example 2.67. If $\mathcal{Q} = \{Q_1, Q_2\}$ and $A = \{0, 1, 2\}^{\{Q_1, Q_2\}}$, $\{Q_1, Q_2\}^A$ is the list $1, Q_2, Q_2^2, Q_1, Q_1Q_2, Q_1Q_2^2, Q_1^2, Q_1^2Q_2, Q_1^2Q_2^2$. Taking $\Sigma = \{0, 1, -1\}^{\{Q_1, Q_2\}}$, i.e. the list

$$\begin{array}{l} Q_1 = 0 \wedge Q_2 = 0, Q_1 = 0 \wedge Q_2 > 0, Q_1 = 0 \wedge Q_2 < 0, \\ Q_1 > 0 \wedge Q_2 = 0, Q_1 > 0 \wedge Q_2 > 0, Q_1 > 0 \wedge Q_2 < 0, \\ Q_1 < 0 \wedge Q_2 = 0, Q_1 < 0 \wedge Q_2 > 0, Q_1 < 0 \wedge Q_2 < 0, \end{array}$$

the matrix of signs of these nine polynomials on these nine sign conditions is

$$\mathrm{Mat}(A, \Sigma) = \begin{bmatrix} 1 & 1 & 1 & 1 & 1 & 1 & 1 & 1 & 1 \\ 0 & 1 & -1 & 0 & 1 & -1 & 0 & 1 & -1 \\ 0 & 1 & 1 & 0 & 1 & 1 & 0 & 1 & 1 \\ 0 & 0 & 0 & 1 & 1 & 1 & -1 & -1 & -1 \\ 0 & 0 & 0 & 0 & 1 & -1 & 0 & -1 & 1 \\ 0 & 0 & 0 & 0 & 1 & 1 & 0 & -1 & -1 \\ 0 & 0 & 0 & 1 & 1 & 1 & 1 & 1 & 1 \\ 0 & 0 & 0 & 0 & 1 & -1 & 0 & 1 & -1 \\ 0 & 0 & 0 & 0 & 1 & 1 & 0 & 1 & 1 \end{bmatrix}.$$

For example, the 5-th row of the matrix reads as follows: the signs of the 5-th polynomial of $\mathcal{Q}^A$ which is Q_1Q_2 on the 9 sign conditions of Σ are

$$\begin{bmatrix} 0 & 0 & 0 & 0 & 1 & -1 & 0 & -1 & 1 \end{bmatrix}.$$

□

Proposition 2.68. *If* $\bigcup_{\sigma \in \Sigma} \mathrm{Reali}(\sigma, Z) = Z$ *then*

$$\mathrm{Mat}(A, \Sigma) \cdot c(\Sigma, Z) = \mathrm{TaQ}(\mathcal{Q}^A, P).$$

Proof: It is obvious since the (i, j) – th entry of $\mathrm{Mat}(A, \Sigma)$ is $\sigma_j^{\alpha_i}$. □

Note that when $\mathcal{Q} = \{Q\}$, $A = \{0, 1, 2\}^{\{Q\}}$ and $\Sigma = \{0, 1, -1\}^{\{Q\}}$ the conclusion of Proposition 2.68 is

$$\begin{bmatrix} 1 & 1 & 1 \\ 0 & 1 & -1 \\ 0 & 1 & 1 \end{bmatrix} \cdot \begin{bmatrix} c(Q = 0, Z) \\ c(Q > 0, Z) \\ c(Q < 0, Z) \end{bmatrix} = \begin{bmatrix} \mathrm{TaQ}(1, P) \\ \mathrm{TaQ}(Q, P) \\ \mathrm{TaQ}(Q^2, P) \end{bmatrix} \tag{2.6}$$

which was hidden in the proof of Proposition 2.65.

It follows from Proposition 2.68 that when the matrix $M(\mathcal{Q}^A, \Sigma)$ is invertible, we can express $c(\Sigma, Z)$ in terms of $\mathrm{TaQ}(\mathcal{Q}^A, P)$. This is the case when $A = \{0, 1, 2\}^{\mathcal{Q}}$ and $\Sigma = \{0, 1, -1\}^{\mathcal{Q}}$, as we will see now.

Notation 2.69. [Tensor product] Let M and $M' = [\, m'_{ij} \,]$ be two matrices with respective dimensions $n \times m$ and $n' \times m'$. The matrix $M \otimes M'$ is the $n\,n' \times m\,m'$ matrix

$$[\, m_{ij} M' \,].$$

The matrix $M \otimes M'$ is the **tensor product** of M and M'. □

Example 2.70. If

$$M = M' = \begin{bmatrix} 1 & 1 & 1 \\ 0 & 1 & -1 \\ 0 & 1 & 1 \end{bmatrix},$$

$$M \otimes M' = \begin{bmatrix} 1 & 1 & 1 & 1 & 1 & 1 & 1 & 1 & 1 \\ 0 & 1 & -1 & 0 & 1 & -1 & 0 & 1 & -1 \\ 0 & 1 & 1 & 0 & 1 & 1 & 0 & 1 & 1 \\ 0 & 0 & 0 & 1 & 1 & 1 & -1 & -1 & -1 \\ 0 & 0 & 0 & 0 & 1 & -1 & 0 & -1 & 1 \\ 0 & 0 & 0 & 0 & 1 & 1 & 0 & -1 & -1 \\ 0 & 0 & 0 & 1 & 1 & 1 & 1 & 1 & 1 \\ 0 & 0 & 0 & 0 & 1 & -1 & 0 & 1 & -1 \\ 0 & 0 & 0 & 0 & 1 & 1 & 0 & 1 & 1 \end{bmatrix}.$$

Notice that $M \otimes M'$ coincides with the matrix of signs of $A = \{0,1,2\}^{\{Q_1,Q_2\}}$ on $\Sigma = \{0,1,-1\}^{\{Q_1,Q_2\}}$. □

Notation 2.71. Let M_s be the $3^s \times 3^s$ matrix defined inductively by

$$M_1 = \begin{bmatrix} 1 & 1 & 1 \\ 0 & 1 & -1 \\ 0 & 1 & 1 \end{bmatrix}$$

$$M_{t+1} = M_t \otimes M_1.$$

□

Exercise 2.13. Prove that M_s is invertible using induction on s.

Proposition 2.72. *Let $\mathcal{Q}$ be a finite set of polynomials with s elements, $A = \{0,1,2\}^{\mathcal{Q}}$ and $\Sigma = \{0,1,-1\}^{\mathcal{Q}}$, ordered lexicographically. Then*

$$\mathrm{Mat}(A,\Sigma) = M_s.$$

Proof: The proof is by induction on s. If s=1, the claim is Equation (2.6). If the claim holds for s, it holds also for $s+1$ given the definitions of M_{s+1}, of $\mathrm{Mat}(A,\Sigma)$, and the orderings on $A = \{0,1,2\}^{\mathcal{Q}}$ and $\Sigma = \{0,1,-1\}^{\mathcal{Q}}$. □

So, Proposition 2.68 and Proposition 2.72 imply

Corollary 2.73. $M_s \cdot c(\Sigma, Z) = \mathrm{TaQ}(\mathcal{Q}^A, P)$.

We now have all the ingredients needed to decide whether a subset of R defined by a sign condition is empty or not, with the following two lemmas.

Lemma 2.74. *Let $Z = \mathrm{Zer}(P, \mathrm{R})$ be a finite set and let σ be a sign condition on $\mathcal{Q}$. Whether or not $\mathrm{Reali}(\sigma, Z) = \emptyset$ is determined by the degrees of the polynomials in the signed pseudo-remainder sequences of P, $P'\mathcal{Q}^\alpha$ and the signs of their leading coefficients for all $\alpha \in A = \{0,1,2\}^{\mathcal{Q}}$.*

Proof: For each $\alpha \in \{0, 1, 2\}^{\mathcal{Q}}$, the degrees and the signs of the leading coefficients of all of the polynomials in the signed pseudo-remainder sequences $\mathrm{SRemS}(P,\ P'\mathcal{Q}^{\alpha})$ clearly determine the number of sign variations of $\mathrm{SRemS}(P,\ P'\mathcal{Q}^{\alpha})$ at $-\infty$ and $+\infty$, i.e. $\mathrm{Var}(\mathrm{SRemS}(P, P'\mathcal{Q}^{\alpha}); -\infty)$ and $\mathrm{Var}(\mathrm{SRemS}(P,\ P'\mathcal{Q}^{\alpha});\ +\infty)$, and their difference is $\mathrm{TaQ}(\mathcal{Q}^{\alpha},\ P)$ by Theorem 2.61. Using Propositions 2.72, Proposition 2.68, and Corollary 2.73

$$M_s^{-1} \cdot \mathrm{TaQ}(\mathcal{Q}^A, P) = c(\Sigma, Z).$$

Denoting the row of M_s^{-1} that corresponds to the row of σ in $c(\Sigma, Z)$ by r_σ, we see that $r_\sigma \cdot \mathrm{TaQ}(\mathcal{Q}^A, P) = c(\sigma, Z)$. Finally,

$$\mathrm{Reali}(\sigma, Z) = \{x \in \mathrm{R} | P(x) = 0 \wedge \bigwedge_{Q \in \mathcal{Q}} \mathrm{sign}(Q(x)) = \sigma(Q)\}$$

is non-empty if and only if $c(\sigma, Z) > 0$. □

Lemma 2.75. *Let σ be a strict sign condition on $\mathcal{Q}$. Whether or not $\mathrm{Reali}(\sigma) = \emptyset$ is determined by the degrees and the signs of the leading coefficients of the polynomials in $\mathrm{Var}(\mathrm{SRemS}(C, C'))$ (with $C = \prod_{Q \in \mathcal{Q}} Q$) and the signs of the leading coefficients of the polynomials in $\mathrm{Var}(\mathrm{SRemS}(C', C''\mathcal{Q}^{\alpha}))$ for all $\alpha \in A = \{0, 1, 2\}^{\mathcal{Q}}$.*

Proof: Recall (Theorem 2.50) that the number of roots of C is determined by the signs of the leading coefficients of $\mathrm{Var}(\mathrm{SRemS}(C, C'))$.

- If C has no roots, then each $Q \in \mathcal{Q}$ has constant sign which is the same as the sign of its leading coefficient.
- If C has one root, then the possible sign conditions on $\mathcal{Q}$ are determined by the sign conditions on $\mathcal{Q}$ at $+\infty$ and at $-\infty$.
- If C has at least two roots, then all intervals between two roots of C contain a root of C' and thus all sign conditions on $\mathcal{Q}$ are determined by the sign conditions on $\mathcal{Q}$ at $+\infty$ and at $-\infty$ and by the sign conditions on $\mathcal{Q}$ at the roots of C'. This is covered by Lemma 2.74. □

The goal of the remainder of the section is to show that the semi-algebraic sets in R^{k+1} are closed under projection if R is a real closed field. The result is a generalization of Theorem 2.62 and the proof is based on a similar method.

Let us now describe our algorithm for proving that the projection of a semi-algebraic set is semi-algebraic. Using how to decide whether or not a basic semi-algebraic set in R is empty (see Lemmas 2.74 and 2.75), we can show that the projection from R^{k+1} to R^k of a basic semi-algebraic set is semi-algebraic. This is a new example of our paradigm for extending an algorithm from the univariate case to the multivariate case by viewing the univariate case parametrically. The basic semi-algebraic set $S \subset \mathrm{R}^{k+1}$ can be described as

$$S = \{x \in \mathrm{R}^{k+1} | \bigwedge_{P \in \mathcal{P}} P(x) = 0 \wedge \bigwedge_{Q \in \mathcal{Q}} Q(x) > 0\}$$

with $\mathcal{P}$, $\mathcal{Q}$ finite subsets of $\mathrm{R}[X_1, ..., X_k, X_{k+1}]$, and its projection $\pi(S)$ (forgetting the last coordinate) is

$$\pi(S) = \{y \in \mathrm{R}^k \mid \exists x \in \mathrm{R} \; (\bigwedge_{P \in \mathcal{P}} P_y(x) = 0 \bigwedge_{Q \in \mathcal{Q}} Q_y(x) > 0).$$

For a particular $y \in \mathrm{R}^k$ we can decide, using Lemmas 2.74 and 2.75, whether or not

$$\exists x \in \mathrm{R} \; (\bigwedge_{P \in \mathcal{P}} P_y(x) = 0 \bigwedge_{Q \in \mathcal{Q}} Q_y(x) > 0)$$

is true.

What is crucial here is to partition the parameter space R^k into finitely many parts so that each part is either contained in $\{y \in \mathrm{R}^k \mid S_y = \emptyset\}$ or in $\{y \in \mathrm{R}^k \mid S_y \neq \emptyset\}$, where

$$S_y = \{x \in \mathrm{R} \mid \bigwedge_{P \in \mathcal{P}} P_y(x) = 0 \wedge \bigwedge_{Q \in \mathcal{Q}} Q_y(x) > 0\}.$$

Moreover, the algorithm used for constructing the partition ensures that the decision algorithm testing whether S_y is empty or not is the same (is uniform) for all y in any given part. Because of this uniformity, it turn out that each part of the partition is a semi-algebraic set. Since $\pi(S)$ is the union of those parts where $S_y \neq \emptyset$, $\pi(S)$ is semi-algebraic being the union of finitely many semi-algebraic sets.

Theorem 2.76. [Projection theorem for semi-algebraic sets] *Given a semi-algebraic set of* R^{k+1} *defined over* D, *its projection to* R^k *is a semi-algebraic set defined over* D.

Proof: Since every semi-algebraic set is a finite union of basic semi-algebraic sets it is sufficient to prove that the projection of a basic semi-algebraic set is semi-algebraic. Suppose that the basic semi-algebraic set S in R^{k+1} is

$$\mathrm{Reali}(\sigma, Z) = \{(y, x) \in \mathrm{R}^k \times \mathrm{R} \mid P(y, x) = \wedge \bigwedge_{Q \in \mathcal{Q}} \mathrm{sign}(\mathcal{Q}(y, x)) = \sigma(Q)\},$$

with $Z = \{z \in \mathrm{R}^{k+1} \mid P(z) = 0\}$. Let S' be the intersection of S with the subset of $(y, x) \in \mathrm{R}^{k+1}$ such that P_y is not identically zero.

Let L be a function on $\{0, 1, 2\}^{\mathcal{Q}}$ associating to each $\alpha \in \{0, 1, 2\}^{\mathcal{Q}}$ a leaf L_α of $\mathrm{TRems}(P, P'\mathcal{Q}^\alpha)$, and let $\mathcal{A}(L_\alpha)$ be the set of non-zero polynomials in $\mathrm{D}[Y_1, ..., Y_k]$ appearing in the quantifier free formula $\mathcal{C}_{L_\alpha}$, (see Notation 1.18).

Let $\mathcal{L}$ be the set of all functions L on $\{0, 1, 2\}^{\mathcal{Q}}$ associating to each α a leaf L_α of $\mathrm{TRems}(P, P'\mathcal{Q}^\alpha)$, and

$$\mathcal{A} = \bigcup_{L \in \mathcal{L}} \bigcup_{\alpha \in \{0,1,2\}^{\mathcal{Q}}} \mathcal{A}(L_\alpha) \subset \mathrm{D}[Y_1, ..., Y_k].$$

Note that since $\mathcal{A}$ contains the coefficients of P',the signs of the coefficients of P are fixed as soon as the signs of the polynomials in $\mathcal{A}$ are fixed.

If $\tau \in \{0,1,-1\}^{\mathcal{A}}$, we define the realization of τ by

$$\mathrm{Reali}(\tau)=\{y\in \mathrm{R}^k \mid \bigwedge_{A\in\mathcal{A}} \mathrm{sign}(A(y))=\tau(A)\}.$$

Let $Z_y=\{x\in \mathrm{R} \mid P(y,x)=0\}$, $\sigma_y(Q_y)=\sigma(Q)$, and note that either

$$\mathrm{Reali}(\tau)\subset\{y\in \mathrm{R}^k \mid \mathrm{Reali}(\sigma_y,Z_y)\neq\emptyset\}$$

or

$$\mathrm{Reali}(\tau)\subset\{y\in \mathrm{R}^k \mid \mathrm{Reali}(\sigma_y,Z_y)=\emptyset\},$$

by Lemma 2.74. Let

$$\Sigma=\{\tau\in\{0,1,-1\}^{\mathcal{A}} \mid \forall y\in \mathrm{Reali}(\tau)\quad \mathrm{Reali}(\sigma_y,Z_y)\neq\emptyset\}.$$

It is clear that the semi-algebraic set $\bigcup_{\tau\in\Sigma}\mathrm{Reali}(\tau)$ coincides with the projection of S'.

The fact that the projection of the intersection of S with the subset of $(y,x)\in \mathrm{R}^{k+1}$ such that P_y is identically zero is semi-algebraic follows in a similar way, using Lemma 2.75.

Thus the whole projection $S=S'\cup(S\setminus S')$ is semi-algebraic as a union of semi-algebraic sets. □

Exercise 2.14. Find the conditions on a, b such that $X^3+a\,X+b$ has a strictly positive real root.

2.5 Applications

2.5.1 Quantifier Elimination and the Transfer Principle

As in Chapter 1, the projection theorem (Theorem 2.76) implies that the theory of real closed fields admits quantifier elimination in the language of ordered fields, which is the following theorem.

Theorem 2.77. [Quantifier Elimination over Real Closed Fields] *Let $\Phi(Y)$ be a formula in the language of ordered fields with coefficients in an ordered ring* D *contained in the real closed field* R*. Then there is a quantifier free formula $\Psi(Y)$ with coefficients in* D *such that for every $y\in \mathrm{R}^k$, the formula $\Phi(y)$ is true if and only if the formula $\Psi(y)$ is true.*

The proof of the theorem is by induction on the number of quantifiers, using as base case the elimination of an existential quantifier which is given by Theorem 2.76.

Proof: Given a formula $\Theta(Y) = (\exists X)\, \mathcal{B}(X, Y)$, where $\mathcal{B}$ is a quantifier free formula whose atoms are equations and inequalities involving polynomials in $\mathrm{D}[X, Y_1, \ldots, Y_k]$, Theorem 2.76 shows that there is a quantifier free formula $\Xi(Y)$ whose atoms are equations and inequalities involving polynomials in $\mathrm{D}[X, Y_1, \ldots, Y_k]$ and that is equivalent to $\Theta(Y)$. This is because $\mathrm{Reali}(\Theta(Y), \mathrm{R}^k)$ which is the projection of the semi-algebraic set $\mathrm{Reali}(\mathcal{B}(X,Y), \mathrm{R}^{k+1})$ defined over D is semi-algebraic and defined over D, and semi-algebraic sets defined over D are realizations of quantifier free formulas with coefficients in D. Since $(\forall X)\, \Phi$ is equivalent to $\neg((\exists X)\, \neg(\Phi))$, the theorem immediately follows by induction on the number of quantifiers. □

Corollary 2.78. *Let $\Phi(Y)$ be a formula in the language of ordered fields with coefficients in* D. *The set $\{y \in \mathrm{R}^k | \Phi(y)\}$ is semi-algebraic.*

Corollary 2.79. *A subset of* R *defined by a formula in the language of ordered fields with coefficients in* R *is a finite union of points and intervals.*

Proof: By Theorem 2.77 a subset of R defined by a formula in the language of ordered fields with coefficients in R is semi-algebraic and this is clearly a finite union of points and intervals. □

Exercise 2.15. Show that the set $\{(x, y) \in \mathbb{R}^2 | \exists n \in \mathbb{N} \quad y = n x\}$ is not a semi-algebraic set.

Theorem 2.77 immediately implies the following theorem known as the Tarski-Seidenberg Principle or the Transfer Principle for real closed fields.

Theorem 2.80. [Tarski-Seidenberg principle] *Suppose that* R' *is a real closed field that contains the real closed field* R. *If* Φ *is a sentence in the language of ordered fields with coefficients in* R, *then it is true in* R *if and only if it is true in* R'.

Proof: By Theorem 2.77, there is a quantifier free formula Ψ R-equivalent to Φ. It follows from the proof of Theorem 2.76 that Ψ is R'-equivalent to Φ as well. Notice, too, that Ψ is a boolean combination of atoms of the form $c = 0, c > 0$, or $c < 0$, where $c \in \mathrm{R}$. Clearly, Ψ is true in R if and only if it is true in R'. □

Since any real closed field contains the real closure of $\mathbb{Q}$, a consequence of Theorem 2.80 is

Theorem 2.81. *Let* R *be a real closed field. A sentence in the language of fields with coefficients in* $\mathbb{Q}$ *is true in* R *if and only if it is true in any real closed field.*

The following application of quantifier elimination will be useful later in the book.

Proposition 2.82. *Let* F *be an ordered field and* R *its real closure. A semi-algebraic set* $S \subset \mathrm{R}^k$ *can be defined by a quantifier free formula with coefficients in* F.

Proof: Any element $a \in \mathrm{R}$ is algebraic over F, and is thus a root of a polynomial $P_a(X) \in \mathrm{F}[X]$. Suppose that $a = a_j$ where $a_1 < \cdots < a_\ell$ are the roots of P_a in R.

Let $\Delta_a(Y)$ be the formula

$$(\exists Y_1)...(\exists Y_\ell)\ [Y_1 < Y_2 < \cdots < Y_\ell \wedge (P_a(Y_1) = \cdots = P_a(Y_\ell) = 0)$$
$$\wedge ((\forall X)\ P_a(X) = 0 \Rightarrow (X = Y_1 \vee \cdots \vee X = Y_\ell)) \wedge Y = Y_j].$$

Then, for $y \in \mathrm{R}$, $\Delta_a(y)$ is true if and only if $y = a$.

Let A be the finite set of elements of $\mathrm{R} \setminus \mathrm{F}$ appearing in a quantifier free formula Φ with coefficients in R such that $S = \{x \in \mathrm{R}^k \mid \Phi(x)\}$. For each $a \in A$, replacing each occurrence of a in Φ by new variables Y_a gives a formula $\Psi(X, Y)$, with $Y = (Y_a, a \in A)$. Denoting $n = \#(A)$, it is clear that $S = \{x \in \mathrm{R}^k \mid \forall y \in \mathrm{R}^n\ (\bigwedge_{a \in A} \Delta_a(y_a) \Rightarrow \Psi(x, y))\}$.

The conclusions follows from Theorem 2.77 since the formula

$$\forall Y \left(\bigwedge_{a \in A} \Delta_a(Y_a) \Rightarrow \Psi(X, Y) \right)$$

is equivalent to a quantifier free formula with coefficients in F. □

2.5.2 Semi-Algebraic Functions

Since the main objects of our interest are the semi-algebraic sets we want to introduce mappings which preserve semi-algebraicity. These are the semi-algebraic functions. Let $S \subset \mathrm{R}^k$ and $T \subset \mathrm{R}^\ell$ be semi-algebraic sets. A function $f: S \rightarrow T$ is **semi-algebraic** if its graph $\mathrm{Graph}(f)$ is a semi-algebraic subset of $\mathrm{R}^{k+\ell}$.

Proposition 2.83. *Let* $f: S \rightarrow T$ *be a semi-algebraic function. If* $S' \subset S$ *is semi-algebraic, then its image* $f(S')$ *is semi-algebraic. If* $T' \subset T$ *is semi-algebraic, then its inverse image* $f^{-1}(T')$ *is semi-algebraic.*

Proof: The set $f(S')$ is the image of $(S' \times T) \cap \mathrm{Graph}(f)$ under the projection from $S \times T$ to T and is semi-algebraic by Theorem 2.76.

The set $f^{-1}(T'))$ is the image of $(S \times T') \cap \mathrm{Graph}(f))$ under the projection, $S \times T \rightarrow S$ and is semi-algebraic, again by Theorem 2.76 □

Proposition 2.84. *If* A, B, C *are semi-algebraic sets in* R^k, R^ℓ, *and* R^m, *resp., and* $f: A \rightarrow B$, $g: B \rightarrow C$ *are semi-algebraic functions, then the composite function* $g \circ f: A \rightarrow C$ *is semi-algebraic.*

Proof: Let $F \subset \mathrm{R}^{k+\ell}$ be the graph of f and $G \subset \mathrm{R}^{\ell+m}$ the graph of g. The graph of $g \circ f$ is the projection of $(F \times \mathrm{R}^m) \cap (\mathrm{R}^k \times G)$ to R^{k+m} and hence is semi-algebraic by Theorem 2.76. □

Proposition 2.85. *Let A be a semi-algebraic set of R^k. The semi-algebraic functions from A to R form a ring.*

Proof: Follows from Proposition 2.84 by noting that $f + g$ is the composition of $(f, g)\colon A \to \mathrm{R}^2$ with $+ : \mathrm{R}^2 \to \mathrm{R}$, and $f \times g$ is the composition of $(f, g)\colon A \to \mathrm{R}^2$ with $\times : \mathrm{R}^2 \to \mathrm{R}$. □

Proposition 2.86. *Let $S \subset \mathrm{R}$ be a semi-algebraic set, and $\varphi\colon S \to \mathrm{R}$ a semi-algebraic function. There exists a non-zero polynomial $P \in \mathrm{R}[X, Y]$ such that for every x in S, $P(x, \varphi(x)) = 0$.*

Proof: The graph Γ of φ is the finite union of non-empty semi-algebraic sets of the form

$$\Gamma_i = \{(x, y) \in \mathrm{R} \times \mathrm{R} \mid P_i(x, y) = 0 \wedge Q_{i,1}(x, y) > 0 \wedge \ldots \wedge Q_{i,m_i}(x, y) > 0\}$$

with P_i not identically zero, for otherwise, given $(x, y) \in \Gamma_i$, the graph of φ intersected with the line $X = x$ would contain a non-empty interval of this line. We can then take P as the product of the P_i. □

2.5.3 Extension of Semi-Algebraic Sets and Functions

In the following paragraphs, R denotes a real closed field and R' a real closed field containing R. Given a semi-algebraic set S in R^k, the **extension** of S to R', denoted $\mathrm{Ext}(S, \mathrm{R}')$, is the semi-algebraic subset of R'^k defined by the same quantifier free formula that defines S.

The following proposition is an easy consequence of Theorem 2.80.

Proposition 2.87. *Let $S \subset \mathrm{R}^k$ be a semi-algebraic set. The set* $\mathrm{Ext}(S, \mathrm{R}')$ *is well defined (i.e. it only depends on the set S and not on the quantifier free formula chosen to describe it).*

The mapping $S \to \mathrm{Ext}(S, \mathrm{R}')$ preserves the boolean operations (finite intersection, finite union, and complementation).

If $S \subset T$, with $T \subset \mathrm{R}^k$ semi-algebraic, then $\mathrm{Ext}(S, \mathrm{R}') \subset \mathrm{Ext}(T, \mathrm{R}')$.

Of course $\mathrm{Ext}(S, \mathrm{R}') \cap \mathrm{R}^k = S$. But $\mathrm{Ext}(S, \mathrm{R}')$ may not be the only semi-algebraic set of R'^k with this property: if $S = [0, 4] \subset \mathbb{R}_{\mathrm{alg}}$ (the real algebraic numbers), $\mathrm{Ext}(S, \mathbb{R}) = [0, 4] \subset \mathbb{R}$; but also $([0, \pi) \cup (\pi, 4]) \cap \mathbb{R}_{\mathrm{alg}} = S$, where $\pi = 3.14...$ is the area enclosed by the unit circle.

Exercise 2.16. Show that if S is a finite semi-algebraic subset of R^k, then $\mathrm{Ext}(S, \mathrm{R}')$ is equal to S.

For any real closed field R, we denote by π the projection mapping

$$\pi\colon \mathrm{R}^{k+1} \to \mathrm{R}^k$$

that "forgets" the last coordinate.

Proposition 2.88. *If* R *is a real closed field and* $S \subset \mathrm{R}^{k+1}$ *is a semi-algebraic set then* $\pi(S)$ *is semi-algebraic. Moreover, if* R′ *is an arbitrary real closed extension of* R, *then* $\pi(\mathrm{Ext}(S, \mathrm{R}')) = \mathrm{Ext}(\pi(S), \mathrm{R}')$.

Proof: We use Theorem 2.80. Since the projection of the semi-algebraic set S is the semi-algebraic set B, $B = \pi(S)$ is true in R. This is expressed by a formula which is thus also true in R′. □

Let $S \subset \mathrm{R}^k$ and $T \subset \mathrm{R}^\ell$ be semi-algebraic sets, and let $f: S \to T$ be a semi-algebraic function whose graph is $G \subset S \times T$.

Proposition 2.89. *If* R′ *is a real closed extension of* R, *then* $\mathrm{Ext}(G, \mathrm{R}')$ *is the graph of a semi-algebraic function* $\mathrm{Ext}(f, \mathrm{R}'): \mathrm{Ext}(S, \mathrm{R}') \to \mathrm{Ext}(T, \mathrm{R}')$.

Proof: Let Φ, Ψ and Γ be quantifier free formulas such that

$$\begin{aligned} S &= \{x \in \mathrm{R}^k \mid \Phi(x)\} \\ T &= \{y \in \mathrm{R}^\ell \mid \Psi(y)\} \\ G &= \{(x, y) \in \mathrm{R}^{k+\ell} \mid \Gamma(x, y)\}. \end{aligned}$$

The fact that G is the graph of a function from S to T can be expressed by the sentence $\forall X\ A$, with

$$\begin{aligned} A = \ & ((\Phi(X) \Leftrightarrow (\exists Y\, \Gamma(X, Y)) \wedge (\forall Y\, \Gamma(X, Y) \Rightarrow \Psi(Y)) \\ & \wedge (\forall Y \forall Y' (\Gamma(X, Y) \wedge \Gamma(X, Y')) \Rightarrow Y = Y')), \end{aligned}$$

with $X = (X_1, \ldots, X_k)$, $Y = (Y_1, \ldots, Y_k)$ and $Y' = (Y_1', \ldots, Y_\ell')$.

Applying Theorem 2.80, $\forall X\ A$ is therefore true in R′, which expresses the fact that $\mathrm{Ext}(G, \mathrm{R}')$ is the graph of a function from $\mathrm{Ext}(S, \mathrm{R}')$ to $\mathrm{Ext}(T, \mathrm{R}')$, since

$$\begin{aligned} \mathrm{Ext}(S, \mathrm{R}') &= \{x \in \mathrm{R}'^k \mid \Phi(x)\} \\ \mathrm{Ext}(T, \mathrm{R}') &= \{y \in \mathrm{R}'^\ell \mid \Psi(y)\} \\ \mathrm{Ext}(G, \mathrm{R}') &= \{(x, y) \in \mathrm{R}'^{k+\ell} \mid \Gamma(x, y)\}. \end{aligned}$$

□

The semi-algebraic function $\mathrm{Ext}(f, \mathrm{R}')$ of the previous proposition is called the **extension of** f **to** R′.

Proposition 2.90. *Let* S' *be a semi-algebraic subset of* S. *Then*

$$\mathrm{Ext}(f(S'), \mathrm{R}') = \mathrm{Ext}(f, \mathrm{R}')(\mathrm{Ext}(S', \mathrm{R}')).$$

Proof: The semi-algebraic set $f(S')$ is the projection of $G \cap (S' \times \mathrm{R}^\ell)$ onto R^ℓ, so the conclusion follows from Proposition 2.88. □

Exercise 2.17.

a) Show that the semi-algebraic function f is injective (resp. surjective, resp. bijective) if and only if $\mathrm{Ext}(f,\mathrm{R}')$ is injective (resp. surjective, resp. bijective).
b) Let T' be a semi-algebraic subset of T. Show that
$$\mathrm{Ext}(f^{-1}(T'),\mathrm{R}')=\mathrm{Ext}(f,\mathrm{R}')^{-1}(\mathrm{Ext}(T',\mathrm{R}'))\,.$$

2.6 Puiseux Series

The field of Puiseux series provide an important example of a non-archimedean real closed field.

The collection of Puiseux series in ε with coefficients in R will be a real closed field containing the field $\mathrm{R}(\varepsilon)$ of rational functions in the variable ε ordered by 0_+ (see Notation 2.5). In order to include in our field roots of equations such as $X^2-\varepsilon=0$, we introduce rational exponents such as $\varepsilon^{1/2}$. This partially motivates the following definition of Puiseux series.

Let K be a field and ε a variable. The **ring of formal power series in ε with coefficients in K**, denoted $\mathrm{K}[[\varepsilon]]$, consists of series of the form $a=\sum_{i\geq 0} a_i\varepsilon^i$ with $i\in\mathbb{N}$, $a_i\in\mathrm{K}$.

Its field of quotients, denoted $\mathrm{K}((\varepsilon))$, is the **field of Laurent series in ε with coefficients in K** and consists of series of the form $\overline{a}=\sum_{i\geq k} a_i\varepsilon^i$ with $k\in\mathbb{Z}$, $i\in\mathbb{Z}$, $a_i\in\mathrm{K}$.

Exercise 2.18. Prove that $\mathrm{K}((\varepsilon))$ is a field, and is the quotient field of $\mathrm{K}[[\varepsilon]]$.

A **Puiseux series in ε with coefficients in K** is a series of the form $a=\sum_{i\geq k} a_i\varepsilon^{i/q}$ with $k\in\mathbb{Z}$, $i\in\mathbb{Z}$, $a_i\in\mathrm{K}$, q a positive integer. Puiseux series are formal Laurent series in the indeterminate $\varepsilon^{1/q}$ for some positive integer q. The **field of Puiseux series in ε with coefficients in K** is denoted $\mathrm{K}\langle\langle\varepsilon\rangle\rangle$.

These series are formal in the sense that there is no assertion of convergence; ε is simply an indeterminate. We assume that the different symbols ε^r, $r\in\mathbb{Q}$, satisfy

$$\begin{aligned}\varepsilon^{r_1}\varepsilon^{r_2} &= \varepsilon^{r_1+r_2},\\ (\varepsilon^{r_1})^{r_2} &= \varepsilon^{r_1 r_2},\\ \varepsilon^0 &= 1.\end{aligned}$$

Hence any two Puiseux series, $\overline{a}=\sum_{i\geq k_1} a_i\,\varepsilon^{i/q_1}$, $\overline{b}=\sum_{j\geq k_2} b_j\,\varepsilon^{j/q_2}$ can be written as formal Laurent series in $\varepsilon^{1/q}$, where q is the least common multiple of q_1 and q_2. Thus, it is clear how to add and multiply two Puiseux series. Also, any finite number of Puiseux series can be written as formal Laurent series in $\varepsilon^{1/q}$ with a common q.

If $\overline{a}=a_1\varepsilon^{r_1}+a_2\varepsilon^{r_2}+\cdots\in \mathrm{K}\langle\langle\varepsilon\rangle\rangle$, (with $a_1\neq 0$ and $r_1<r_2<\ldots$), then the **order** of $\overline{a}$, denoted $o(\overline{a})$, is r_1 and the **initial coefficient** of $\overline{a}$, denoted $\mathrm{In}(\overline{a})$ is a_1. By convention, the order of 0 is ∞. The order is a function from $\mathrm{K}\langle\langle\varepsilon\rangle\rangle$ to $\mathbb{Q}\cup\{\infty\}$ satisfying

- $o(\overline{a}\overline{b})=o(\overline{a})+o(\overline{b})$,
- $o(\overline{a}+\overline{b})\geq \min(o(\overline{a}),o(\overline{b}))$, with equality if $o(\overline{a})\neq o(\overline{b})$.

Exercise 2.19. Prove that $\mathrm{K}\langle\langle\varepsilon\rangle\rangle$ is a field.

When K is an ordered field, we make $\mathrm{K}\langle\langle\varepsilon\rangle\rangle$ an ordered field by defining a Puiseux series $\overline{a}$ to be positive if $\mathrm{In}(\overline{a})$ is positive. It is clear that the field of rational functions $\mathrm{K}(\varepsilon)$ equipped with the order 0_+ is a subfield of the ordered field of Puiseux series $\mathrm{K}\langle\langle\varepsilon\rangle\rangle$, using Laurent's expansions about 0.

In the ordered field $\mathrm{K}\langle\langle\varepsilon\rangle\rangle$, ε is infinitesimal over K (Definition page 32), since it is positive and smaller than any positive $r\in\mathrm{K}$, since $r-\varepsilon>0$. Hence, the field $\mathrm{K}\langle\langle\varepsilon\rangle\rangle$ is non-archimedean. This is the reason why we have chosen to name the indeterminate ε rather than some more neutral X.

The remainder of this section is primarily devoted to a proof of the following theorem.

Theorem 2.91. *Let* R *be a real closed field. Then, the field* $\mathrm{R}\langle\langle\varepsilon\rangle\rangle$ *is real closed.*

As a corollary

Theorem 2.92. *Let* C *be an algebraically closed field of characteristic* 0*. The field* $\mathrm{C}\langle\langle\varepsilon\rangle\rangle$ *is algebraically closed.*

Proof: Apply Theorem 2.31, Theorem 2.11 and Theorem 2.91, noticing that $\mathrm{R}[i]\langle\langle\varepsilon\rangle\rangle=\mathrm{R}\langle\langle\varepsilon\rangle\rangle[i]$. □

The first step in the proof of Theorem 2.91 is to show is that positive elements of $\mathrm{R}\langle\langle\varepsilon\rangle\rangle$ are squares in $\mathrm{R}\langle\langle\varepsilon\rangle\rangle$.

Lemma 2.93. *A positive element of* $\mathrm{R}\langle\langle\varepsilon\rangle\rangle$ *is the square of an element in* $\mathrm{R}\langle\langle\varepsilon\rangle\rangle$.

Proof: Suppose that $\overline{a}=\sum_{i\geq k}a_i\,\varepsilon^{i/q}\in\mathrm{R}\langle\langle\varepsilon\rangle\rangle$ with $a_k>0$. Defining $\overline{b}=\sum_{i\geq k+1}(a_i/a_k)\,\varepsilon^{(i-k)/q}$, we have $\overline{a}=a_k\,\varepsilon^{k/q}(1+\overline{b})$ and $o(\overline{b})>0$.

The square root of $1+\overline{b}$ is obtained by taking the Taylor series expansion of $(1+\overline{b})^{1/2}$ which is

$$\overline{c}=1+\frac{1}{2}\overline{b}+\cdots+\frac{1}{n!}\frac{1}{2}\left(\frac{1}{2}-1\right)\cdots\left(\frac{1}{2}-(n-1)\right)\overline{b}^n+\cdots.$$

In order to check that $\overline{c}^2=1+\overline{b}$, just substitute. Since $a_k>0$ and R is real closed, $\sqrt{a_k}\in\mathrm{R}$. Hence, $\sqrt{a_k}\,\varepsilon^{k/2q}\,\overline{c}$ is the square root of $\overline{a}$. □

In order to complete the proof of Theorem 2.91, it remains to prove that an odd degree polynomial in $\mathrm{R}\langle\langle\varepsilon\rangle\rangle[X]$ has a root in $\mathrm{R}\langle\langle\varepsilon\rangle\rangle$. Given

$$P(X) = \overline{a}_0 + \overline{a}_1 X + \cdots + \overline{a}_p X^p \in \mathrm{R}\langle\langle\varepsilon\rangle\rangle[X]$$

with p odd, we will construct an $\overline{x} \in \mathrm{R}\langle\langle\varepsilon\rangle\rangle$ such that $P(\overline{x}) = 0$. We may assume that $\overline{a}_0 \neq 0$, since otherwise 0 is a root of P. Furthermore, we may assume without loss of generality that

$$o(\overline{a}_i) = \frac{m_i}{m}$$

with the same m for every $0 \leq i \leq p$. Our strategy is to consider an unknown

$$\overline{x} = x_1 \varepsilon^{\xi_1} + x_2 \varepsilon^{\xi_1 + \xi_2} + \cdots + x_i \varepsilon^{\xi_1 + \cdots + \xi_i} + \cdots \tag{2.7}$$

with $\xi_2 > 0$, ..., $\xi_j > 0$ and determine, one after the other, the unknown coefficients x_i and the unknown exponents ξ_i so that $\overline{x} \in \mathrm{R}\langle\langle\varepsilon\rangle\rangle$ and satisfies $P(\overline{x}) = 0$.

Natural candidates for the choice of ξ_1 and x_1 will follow from the geometry of the exponents of P, that we study now. The polynomial $P(X)$ can be thought of as a formal sum of expressions $X^i \varepsilon^r$ ($i \in \mathbb{Z}$, $r \in \mathbb{Q}$) with coefficients in R. The points (i, r) for which $X^i \varepsilon^r$ occurs in $P(X)$ with non-zero coefficient constitute the **Newton diagram** of P. Notice that the points of the Newton diagram are arranged in columns and that the points $M_i = (i, o(a_i))$, $i = 0, \ldots, p$, for which $\overline{a}_i \neq 0$ are the lowest points in each column.

The **Newton polygon** of P is the sequence of points

$$M_0 = M_{i_0}, \ldots, M_{i_\ell} = M_p$$

satisfying:

- All points of the Newton diagram of P lie on or above each of the lines joining $M_{i_{j-1}}$ to M_{i_j} for $j = 1, \ldots, \ell$.
- The ordered triple of points $M_{i_{j-1}}$, M_{i_j}, $M_{i_{j+1}}$ is oriented counter-clockwise, for $j = 1, \ldots, \ell - 1$. This is saying that the edges joining adjacent points in the sequence $M_0 = M_{i_0}, \ldots, M_{i_\ell} = M_p$ constitute a **convex chain**.

In such a case the **slope** of $[M_{i_{j-1}}, M_{i_j}]$ is $\frac{o(a_{i_j}) - o(a_{i_{j-1}})}{i_j - i_{j-1}}$, and its **horizontal projection** is the interval $[i_{j-1}, i_j]$.

Notice that the Newton polygon of P is the lower convex hull of the Newton diagram of P.

To the segment $E = [M_{i_{j-1}}, M_{i_j}]$ with horizontal projection $[i_{j-1}, i_j]$, we associate its **characteristic polynomial**

$$Q(P, E, X) = \sum a_h X^h \in \mathrm{R}[X],$$

where the sum is over all h for which

$$M_h = (h, o(\overline{a}_h)) = \left(h, \frac{m_h}{m}\right) \in E \text{ and } a_h = \mathrm{In}(\overline{a}_h).$$

Note that if $-\xi$ is the slope of E, then $o(\overline{a}_h)+h\,\xi$ has a constant value β for all M_h on E.

Example 2.94. Let

$$P(X)=\varepsilon-2\,\varepsilon^2\,X^2-X^3+\varepsilon\,X^4+\varepsilon\,X^5.$$

The Newton diagram of P is

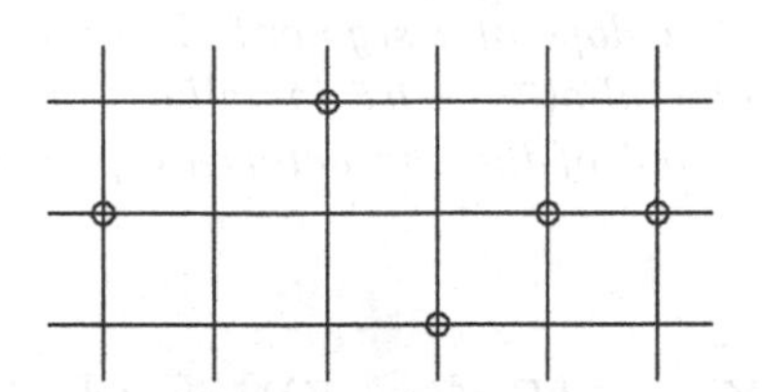

Fig. 2.3. Newton diagram

The Newton polygon of P consists of two segments $E=[M_0,\ M_3]$ and $F=[M_3,M_5]$. The segment E has an horizontal projection of length 3 and the segment F has an horizontal projection of length 2

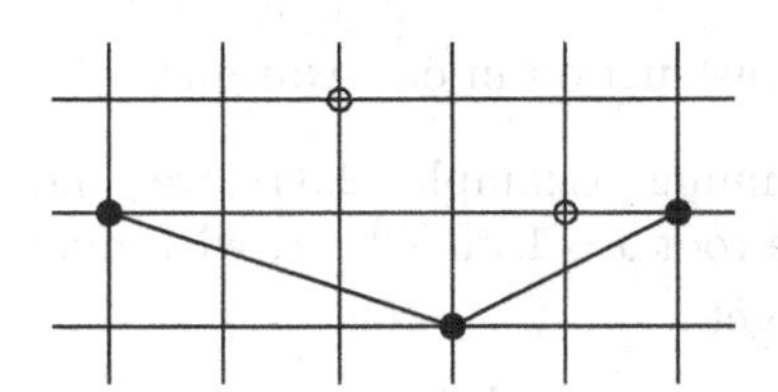

Fig. 2.4. Newton polygon

We have

$$\begin{aligned} Q(P,E,X) &= 1-X^3 \\ Q(P,F,X) &= X^3\,(X^2-1). \end{aligned}$$

The two slopes are $-1/3$ and $1/2$ and the corresponding values of ξ are $1/3$ and $-1/2$. The common value β of $o(\overline{a}_h)+h\xi$ on the two segments are 1 and $-3/2$. □

If x is a non-zero root of multiplicity r of the characteristic polynomial of a segment E of the Newton polygon with slope $-\xi$, we construct a root of P which is a Puiseux series starting with $x\varepsilon^{\xi}$. In other words we find

$$\overline{x}=x\,\varepsilon^{\xi}+x_2\,\varepsilon^{\xi+\xi_2}+\cdots+x_i\,\varepsilon^{\xi+\xi_2+\cdots+\xi_i}+\cdots \tag{2.8}$$

with $\xi_2>0,\ldots,\xi_j>0$ such that $P(\overline{x})=0$.

The next lemma is a key step in this direction. The result is the following: if we replace in P X by $\varepsilon^{\xi}(x+X)$ and divide the result by $\varepsilon^{-\beta}$, where β is the common value of $o(\overline{a}_h)+h\xi$ on E, we obtain a new Newton polygon with a part having only negative slopes, whose horizontal projection is $[0, r]$. A segment of this part of the Newton polygon will be used to find the second term of the series.

Lemma 2.95. *Let*

- ξ *be the opposite of the slope of a segment E of the Newton polygon of P,*
- β *be the common value of $o(\overline{a}_h)+h\xi$ for all q_h on E,*
- $x\in \mathrm{R}$ *be a non-zero root of the characteristic polynomial $Q(P, E, X)$ of multiplicity r.*

a) The polynomial

$$R(P,E,x,Y)=\varepsilon^{-\beta}P(\varepsilon^{\xi}(x+Y))=\bar{b}_0+\bar{b}_1 Y+\cdots\bar{b}_p Y^p$$

satisfies

$$\begin{array}{ll} o(\bar{b}_i)\geq 0, & i=0,\dots,p,\\ o(\bar{b}_i)>0, & i=0,\dots,r-1,\\ o(\bar{b}_r)=0. & \end{array}$$

b) For every $\overline{x}\in\mathrm{R}\langle\langle\varepsilon\rangle\rangle$ such that $\overline{x}=\varepsilon^{\xi}(x+\overline{y})$ with $o(\overline{y})>0$, $o(P(\overline{x}))>\beta$.

We illustrate the construction in our example.

Example 2.96. Continuing Example 2.94, we choose the segment E, with $\xi=1/3$, chose the root $x=1$ of X^3-1, with multiplicity 1, and replace X by $\varepsilon^{1/3}(1+X)$ and get

$$\begin{aligned} P_1(X) &= \varepsilon^{-1}P(\varepsilon^{1/3}(1+X))\\ &= \varepsilon^{5/3}X^5+\left(\varepsilon^{4/3}+5\,\varepsilon^{5/3}\right)X^4\\ &+ \left(-1+4\,\varepsilon^{4/3}+10\,\varepsilon^{5/3}\right)X^3\\ &+ \left(-3+8\,\varepsilon^{5/3}+6\,\varepsilon^{4/3}\right)X^2\\ &+ \left(\varepsilon^{5/3}-3+4\,\varepsilon^{4/3}\right)X-\varepsilon^{5/3}+\varepsilon^{4/3}. \end{aligned}$$

The Newton polygon of p_1 is

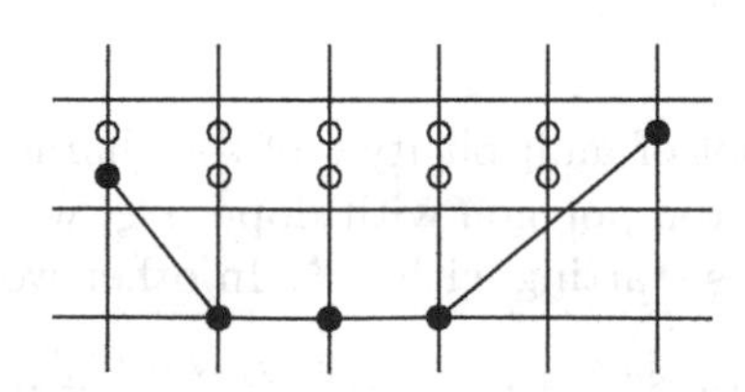

Fig. 2.5. Newton polygon of p_1

We chose the negative slope with corresponding characteristic polynomial $-3X+1$ and make the change of variable $X=\varepsilon^{4/3}(1/3+Y)$.

We have obtained this way the two first terms $\varepsilon^{1/3}+(1/3)\,\varepsilon^{1/3+4/3}+\cdots$ of a Puiseux series $\overline{x}$ satisfying $P(\overline{x})=0$. □

The proof of Lemma 2.95 uses the next lemma which describes a property of the characteristic polynomials associated to the segments of the Newton polygon.

Lemma 2.97. *The slope $-\xi$ of E has the form $-c/(m\ q)$ with $q>0$ and* $\gcd(c,q)=1$. *Moreover, $Q(P,E,X)=X^j\,\phi(X^q)$, where $\phi\in \mathrm{R}[X]$, $\phi(0)\neq 0$, and $\deg\phi=(k-j)/q$.*

Proof: The slope of $E=[M_j,M_k]$ is

$$\frac{o(\bar{a}_k)-o(\bar{a}_j)}{k-j}=\frac{m_k-m_j}{m\,(k-j)}=-\frac{c}{m\,q},$$

where $q>0$ and $\gcd(c,q)=1$. If $(h,o(\overline{a}_h))=(h,\frac{m_h}{m})$ is on E then

$$\frac{c}{m\,q}=\frac{o(\bar{a}_j)-o(\bar{a}_h)}{h-j}=\frac{m_j-m_h}{m\,(h-j)}.$$

Hence, q divides $h-j$, and there exists a non-negative s such that $h=j+s\,q$. The claimed form of $Q(P,E,X)$ follows. □

Proof of Lemma 2.95: For a) since x is a root of $\phi(X^q)$ of multiplicity r, we have

$$\phi(X^q)=(X-x)^r\,\psi(X),\quad \psi(x)\neq 0.$$

Thus,

$$\begin{aligned}R(P,E,x,Y) &= \varepsilon^{-\beta}P(\varepsilon^{\xi}(x+Y))\\ &= \varepsilon^{-\beta}(\bar{a}_0+\overline{a}_1\varepsilon^{\xi}(x+Y)+\cdots+\bar{a}_p\varepsilon^{p\xi}(x+Y)^p)\\ &= A(Y)+B(Y),\end{aligned}$$

where

$$\begin{aligned}A(Y) &= \varepsilon^{-\beta}\sum_{h,q_h\in E} a_h\,\varepsilon^{o(a_h)+h\xi}\,(x+Y)^h\\ B(Y) &= \varepsilon^{-\beta}\left(\sum_{h,q_h\in E}(\overline{a}_h-a_h\,\varepsilon^{o(a_h)})\,\varepsilon^{h\xi}\,(x+Y)^h+\sum_{\ell,q_\ell\notin E}\overline{a}_\ell\,\varepsilon^{\ell\xi}\,(x+Y)^\ell\right).\end{aligned}$$

Since $o(\overline{a}_h)+h\,\xi=\beta$,

$$\begin{aligned}A(Y) &= Q(P,E,x+Y)\\ &= (x+Y)^j\,\phi((x+Y)^q)\\ &= Y^r(x+Y)^j\,\psi(x+Y)\\ &= c_r\,Y^r+c_{r+1}\,Y^{r+1}+\cdots+c_p\,Y^p,\end{aligned}$$

with $c_r=x^j\,\psi(x)\neq 0$ and $c_i\in\mathrm{R}$.

Since $o((\overline{a}_h - a_h\,\varepsilon^{o(a_h)})\,\varepsilon^{h\xi}) > \beta$ and $o(\overline{a}_\ell\,\varepsilon^{\ell\xi}) > \beta$,

$$R(P,E,x,Y) = B(Y) + c_r\,Y^r + c_{r+1}\,Y^{r+1} + \cdots + c_p\,Y^p,$$

where every coefficient of $B(Y) \in \mathrm{R}\langle\langle\varepsilon\rangle\rangle[Y]$ has positive order. The conclusion follows.

For b), since $o(\overline{y}) > 0$, $o(R(P,E,x,\overline{y})) > 0$ is an easy consequence of a). The conclusion follows noting that $P(x) = \varepsilon^{\beta}\,R(P,E,x,y)$. □

It is now possible to proceed with the proof of Theorem 2.91.

Proof of Theorem 2.91: Consider P with odd degree. Hence, we can choose a segment E_1 of the Newton polygon of P which has a horizontal projection of odd length. Let the slope of E_1 be $-\xi_1$. It follows from Lemma 2.97 that the corresponding characteristic polynomial $Q(P,E_1,X)$ has a non-zero root x_1 in R of odd multiplicity r_1, since R is real closed. Define $P_1(X) = R(P,E_1,x_1,X)$ using this segment and the root x_1.

Note that $(r_1,\ 0)$ is a vertex of the Newton polygon of $R(P,\ E_1,\ x_1,\ X)$, and that all the slopes of segments $[M_j,\ M_k]$ of the Newton polygon of $R(P,E_1,x_1,X)$ for $k \le r_1$ are negative: this is an immediate consequence of Lemma 2.95.

Choose recursively a segment E_{i+1} of the Newton polygon of P_i with negative slope $-\xi_{i+1}$, and horizontal projection of odd length, so that the corresponding characteristic polynomial $Q(P_i,E_{i+1},X)$ has a non-zero root x_{i+1} in R of odd multiplicity r_{i+1}, and take $P_{i+1}(X) = R(P_i,E_{i+1},x_{i+1},X)$. The only barrier to continuing this process is if we cannot choose a segment with negative slope over the interval $[0,r_i]$ and this is the case only if 0 is a root of $P_i(X)$. But in this exceptional case $x_1\,\varepsilon^{\xi_1} + \cdots + x_i\,\varepsilon^{\xi_1+\cdots+\xi_i}$ is clearly a root of P.

Suppose we have constructed x_i, ξ_i for $i \in \mathbb{N}$ and let

$$\overline{x} = x_1\,\varepsilon^{\xi_1} + x_2\,\varepsilon^{\xi_1+\xi_2} + \cdots.$$

Then from the definition of the $P_i(X)$, it follows by induction that $o(P(x)) > \beta_1 + \cdots + \beta_j$ for all j. To complete the proof, we need to know that $\overline{x} \in \mathrm{R}\langle\langle\varepsilon\rangle\rangle$ and that the sums $\beta_1 + \cdots + \beta_j$ are unbounded. Both these will follow if we know that the q in Lemma 2.97 is eventually 1. Note that the multiplicities of the chosen roots x_i are non-increasing and hence are eventually constant, at which point they have the value r. This means that from this point on, the Newton polygon has a single segment with negative slope, and horizontal projection of length r. Therefore all subsequent roots chosen also have multiplicity r. It follows (since $Q_j(X)$ must also have degree r) that $Q_j(X) = c\,(X - x_j)^r$ with $x_j \neq 0$, from which it follows that the corresponding q is equal to 1, since the coefficient of degree 1 of ϕ_j is $-r\,c\,x_j^{r-1}$, which is not zero. □

If K is a field, we denote by $\mathrm{K}\langle\varepsilon\rangle$ the subfield of $\mathrm{K}\langle\langle\varepsilon\rangle\rangle$ of **algebraic Puiseux series**, which consists of those elements that are algebraic over $\mathrm{K}(\varepsilon)$, i.e. that satisfy a polynomial equation with coefficients in $\mathrm{K}(\varepsilon)$.

Corollary 2.98. *When* R *is real closed,* $\mathrm{R}\langle\varepsilon\rangle$ *is real closed. The field* $\mathrm{R}\langle\varepsilon\rangle$ *is the real closure of* $\mathrm{R}(\varepsilon)$ *equipped with the order* 0_+.

Proof: Follows immediately from Theorem 2.91 and Exercise 2.10. □

Similarly, if $\mathrm{C}=\mathrm{R}[i]$, then $\mathrm{C}\langle\varepsilon\rangle=\mathrm{R}\langle\varepsilon\rangle[i]$ is an algebraic closure of $\mathrm{C}(\varepsilon)$.

We shall see in Chapter 3 that algebraic Puiseux series with coefficients in R can be interpreted as of germs semi-algebraic and continuous functions at the right of the origin.

A **valuation ring** of a field F is a subring of F such that either x or its inverse is in the ring for every non-zero x.

Proposition 2.99. *The elements of* $\mathrm{K}\langle\varepsilon\rangle$ *with non-negative order constitute a valuation ring denoted* $\mathrm{K}\langle\varepsilon\rangle_b$. *The elements of* $\mathrm{R}\langle\varepsilon\rangle_b$ *are exactly the elements of* $\mathrm{R}\langle\varepsilon\rangle$ *bounded over* R *(i.e. their absolute value is less than a positive element of* R*). The elements of* $\mathrm{C}\langle\varepsilon\rangle_b$ *are exactly the elements of* $\mathrm{C}\langle\varepsilon\rangle$ *bounded over* R *(i.e. their modulus is less than a positive element of* R*).*

Notation 2.100. [Limit] We denote by $\lim_\varepsilon$ the ring homomorphism from $\mathrm{K}\langle\varepsilon\rangle_b$ to K which maps $\sum_{i\in\mathrm{N}} a_i\,\varepsilon^{i/q}$ to a_0. The mapping $\lim_\varepsilon$ simply replaces ε by 0 in a bounded Puiseux series. □

2.7 Bibliographical Notes

The theory of real closed fields was developed by Artin and Schreier [7] and used by Artin [6] in his solution to Hilbert's 17-th problem. The algebraic proof of the fundamental theorem of algebra is due to Gauss [65].

Real root counting began with Descartes's law of sign [53], generalized by Budan [34] and Fourier [60], and continued with Sturm [152]. The connection between virtual roots [68] and Budan-Fourier's theorem comes from [49]. The notion of Cauchy index appears in [41]. Theorem 2.58 is already proved in two particular cases (when $Q=P'$ and when P is square-free) in [152]. The partial converse to Descartes's law of sign presented here appears in [126].

Quantifier elimination for real closed fields is a fundamental result. It was known to Tarski before 1940 (it is announced without a proof in [154]) and published much later [156]. The version of 1940, ready for publication in Actualités Scientifiques et Industrielles (Hermann), was finally not published at that time, "as a result of war activities", and has appeared in print much later [155]. The proof presented here follows the original procedure of Tarski. Theorem 2.61 is explicitely stated in [155, 156], and the sign determination algorithm is sketched.

There are many different proofs of quantifier elimination for real closed fields, in particular by Seidenberg [148], Cohen [43] and Hormander [92].

Puiseux series have been considered for the first time by Newton [123].

3

Semi-Algebraic Sets

In Section 3.1 and Section 3.2, we define the topology of semi-algebraic sets and study connectedness in a general real closed field. In order to study the properties of closed and bounded semi-algebraic sets in Section 3.4, we introduce semi-algebraic germs in Section 3.3. The semi-algebraic germs over a real closed field constitute a real closed field containing infinitesimal elements, closely related to the field of Puiseux series seen in Chapter 2, and play an important role throughout the whole book. We end the chapter with Section 3.5 on semi-algebraic differentiable functions.

3.1 Topology

Let R be a real closed field. Since R is an ordered field, we can define the topology on R^k in terms of open balls in essentially the same way that we define the topology on $\mathbb{R}^k$. The **euclidean norm, open balls, closed balls,** and **spheres** are defined as follows:

With $x=(x_1,\ldots,x_k)\in \mathrm{R}^k$, $r\in \mathrm{R}$, $r>0$, we denote

$$\begin{array}{rcll}
\|x\| &=& \sqrt{x_1^2+\cdots+x_k^2} & \text{(euclidean norm of } x\text{)},\\
B_k(x,r) &=& \{y\in \mathrm{R}^k \mid \|y-x\|^2<r^2\} & \text{(open ball)},\\
\overline{B}_k(x,r) &=& \{y\in \mathrm{R}^k \mid \|y-x\|^2\le r^2\} & \text{(closed ball)},\\
S^{k-1}(x,r) &=& \{y\in \mathrm{R}^k \mid \|y-x\|^2=r^2\} & \text{(}(k-1)\text{-sphere)}.
\end{array}$$

Note that $B_k(x,r), \overline{B}_k(x,r)$, and $S^{k-1}(x,r)$ are semi-algebraic sets.

We omit both x and r from the notation when x is the origin of R^k and $r=1$, i.e. for the unit ball and sphere centered at the origin. We also omit the subscript k when it leads to no ambiguity.

We recall the definitions of the basic notions of open, closed, closure, interior, continuity, etc.

A set $U \subset \mathrm{R}^k$ is **open** if it is the union of open balls, i.e. if every point of U is contained in an open ball contained in U. A set $F \subset \mathrm{R}^k$ is **closed** if its complement is open. Clearly, the arbitrary union of open sets is open and the arbitrary intersection of closed sets is closed. The **closure** of a set S, denoted $\overline{S}$, is the intersection of all closed sets containing S. The **interior** of S, denoted S°, is the union of all open subsets of S and thus is also the union of all open balls in S. We also have a notion of subsets of S being open or closed relative to S. A subset of S is called **open in** S if it is the intersection of an open set with S. It is **closed in** S if it is the intersection of a closed set with S. A function from S to T is **continuous** if the inverse image of any set open in T is open in S. It is easy to prove that polynomial maps from R^k to R^ℓ are continuous in the Euclidean topology: one proves first that $+$ and $\times$ are continuous, then that the composite of continuous functions is continuous.

These definitions are clearly equivalent to the following formulations:

- U is open if and only if $\forall x \in U\ \exists r \in \mathrm{R}, r > 0\ B(x,r) \subset U$.
- $\overline{S} = \{x \in \mathrm{R}^k \mid \forall r > 0\ \exists y \in S\ \|y - x\|^2 < r^2\}$.
- $S^\circ = \{x \in S \mid \exists r > 0, \forall y\ \|y - x\|^2 < r^2 \Rightarrow y \in S\}$.
- If $S \subset \mathrm{R}^k$ and $T \subset \mathrm{R}^\ell$, a function $f: S \to T$ is continuous if and only if it is continuous at every point of S, i.e.
$$\forall x \in S\ \forall r > 0\ \exists \delta > 0, \forall y \in S\ \|y - x\| < \delta \Rightarrow \|f(y) - f(x)\| < r.$$

Note that if U, S, T, f are semi-algebraic, these definitions are expressed by formulas in the language of ordered fields. Indeed, it is possible to replace in these definitions semi-algebraic sets and semi-algebraic functions by quantifier-free formulas describing them. For example let $\Psi(X_1, \ldots, X_k)$ be a quantifier free formula such that

$$S = \{(x_1, \ldots, x_k) \in \mathrm{R}^k \mid \Psi(x_1, \ldots, x_k)\}.$$

Then, if $\Phi(X_1, \ldots, X_k, Y_1, \ldots, Y_\ell)$ is a formula, $\forall x \in S\ \ \Phi(x, y)$ can be replaced by

$$(\forall x_1) \ldots (\forall x_k)\ (\Psi(x_1, \ldots, x_k) \Rightarrow \Phi(x_1, \ldots, x_k, y_1, \ldots, y_\ell)),$$

and $\exists x \in S\ \Phi(x, y_1, \ldots, y_\ell)$ can be replaced by

$$(\exists x_1) \ldots (\exists x_k)\ (\Psi(x_1, \ldots, x_k) \wedge \Phi(x_1, \ldots, s_k, , y_1, \ldots, y_\ell)).$$

An immediate consequence of these observations and of Theorem 2.77 (Quantifier elimination) (more precisely Corollary 2.78) is

Proposition 3.1. *The closure and the interior of a semi-algebraic set are semi-algebraic sets.*

Remark 3.2. It is tempting to think that the closure of a semi-algebraic set is obtained by relaxing the strict inequalities describing the set, but this idea is mistaken. Take $S = \{x \in \mathrm{R} \mid x^3 - x^2 > 0\}$. The closure of S is not $T = \{x \in \mathrm{R} \mid x^3 - x^2 \geq 0\}$ but is $\overline{S} = \{x \in \mathrm{R} \mid x^3 - x^2 \geq 0 \ \wedge\ x \geq 1\}$, as 0 is clearly in T and not in $\overline{S}$. □

We next consider semi-algebraic and continuous functions. The following proposition is clear, noting that Proposition 2.85 and Proposition 2.84 take care of the semi-algebraicity:

Proposition 3.3. *If A, B, C are semi-algebraic sets and $f\colon A \rightarrow B$ and $g\colon B \rightarrow C$ are semi-algebraic continuous functions, then the composite function $g \circ f\colon A \rightarrow C$ is semi-algebraic and continuous.*

Let A be a semi-algebraic set of R^k. The semi-algebraic continuous functions from A to R form a ring.

Exercise 3.1. Let R' be a real closed field containing R.

a) Show that the semi-algebraic set $S \subset \mathrm{R}^k$ is open (resp. closed) if and only if $\mathrm{Ext}(S, \mathrm{R}')$ is open (resp. closed). Show that

$$\mathrm{Ext}(\overline{S}, \mathrm{R}') = \overline{\mathrm{Ext}(S, \mathrm{R}')}.$$

b) Show that a semi-algebraic function f is continuous if and only if $\mathrm{Ext}(f, \mathrm{R}')$ is continuous.

The intermediate value property is valid for semi-algebraic continuous functions.

Proposition 3.4. *Let f be a semi-algebraic and continuous function defined on $[a, b]$. If $f(a)\, f(b) < 0$, then there exists x in (a, b) such that $f(x) = 0$.*

Proof: Suppose,without loss of generality, that $f(a) > 0$, $f(b) < 0$. Let $A = \{x \in [a, b] \mid f(x) > 0\}$. The set A is semi-algebraic, non-empty, and open. So, by Corollary 2.79, A is the union of a finite non-zero number of open subintervals of $[a, b]$. Let $A = [a, b_1) \cup \ldots \cup (a_\ell, b_\ell)$. Then $f(b_1) = 0$ since f is continuous, thus $f(b_1) \leq 0$. □

Proposition 3.5. *Let f be a semi-algebraic function defined on the semi-algebraic set S. Then f is continuous if and only if for every $x \in S$ and every $y \in \mathrm{Ext}(S, \mathrm{R}\langle\varepsilon\rangle)$ such that $\lim_\varepsilon (y) = x$, $\lim_\varepsilon (\mathrm{Ext}(f, \mathrm{R}\langle\varepsilon\rangle)(y)) = f(x)$.*

Proof: Suppose that f is continuous. Then

$$\forall x \in S\, \forall a > 0\, \exists b(a)\, \forall y \in S\, |x - y| < b(a) \Rightarrow |f(x) - f(y)| < a.$$

holds in R. Taking $y \in \mathrm{Ext}(S, \mathrm{R}\langle\varepsilon\rangle)$ such that $\lim_\varepsilon (y) = x$, for every positive $a \in \mathrm{R}$, $|x - y| < b(a)$, thus $|f(x) - \mathrm{Ext}(f, \mathrm{R}\langle\varepsilon\rangle)(y)| < a$, using Tarski-Seidenberg principle (Theorem 2.80).

In the other direction, suppose that f is not continuous. It means that

$$\exists x \in S\, \exists a > 0 \forall b\, \exists y \in S\, |x - y| < b \wedge |f(x) - f(y)| > a$$

holds in R. Taking $b=\varepsilon$, there exists $y \in \mathrm{Ext}(S, \mathrm{R}\langle\varepsilon\rangle)$ such that $\lim_\varepsilon (y) = x$, while $| f(x) - \mathrm{Ext}(f, \mathrm{R}\langle\varepsilon\rangle)(y) | > a$, using again Tarski-Seidenberg principle (Theorem 2.80), which implies that $f(x)$ and $\lim_\varepsilon (\mathrm{Ext}(f, \mathrm{R}\langle\varepsilon\rangle)(y))$are not infinitesimally close.

□

A **semi-algebraic homeomorphism** f from a semi-algebraic set S to a semi-algebraic set T is a semi-algebraic bijection which is continuous and such that f^{-1} is continuous.

Exercise 3.2. Let R' be a real closed field containing R. Prove that if f is a semi-algebraic homeomorphism from a semi-algebraic set S to a semi-algebraic set T, then $\mathrm{Ext}(f, \mathrm{R}')$ is a semi-algebraic homeomorphism from $\mathrm{Ext}(S, \mathrm{R}')$ to $\mathrm{Ext}(T, \mathrm{R}')$.

3.2 Semi-algebraically Connected Sets

Recall that a set $S \subset \mathbb{R}^k$ is connected if S is not the disjoint union of two non-empty sets which are both closed in S. Equivalently, S does not contain a non-empty strict subset which is both open and closed in S.

Unfortunately, this definition is too general to be suitable for R^k with R an arbitrary real closed field, as it allows R to be disconnected.

For example, consider $\mathbb{R}_{\mathrm{alg}}$, the field of real algebraic numbers. The set $(-\infty, \pi) \cap \mathbb{R}_{\mathrm{alg}}$ is both open and closed (with $\pi = 3.14...$), and hence $\mathbb{R}_{\mathrm{alg}}$ is not connected. However, the set $(-\infty, \pi) \cap \mathbb{R}_{\mathrm{alg}}$ is not a semi-algebraic set in $\mathbb{R}_{\mathrm{alg}}$, since π is not an algebraic number.

Since semi-algebraic sets are the only sets in which we are interested, we restrict our attention to these sets.

A semi-algebraic set $S \subset \mathrm{R}^k$ is **semi-algebraically connected** if S is not the disjoint union of two non-empty semi-algebraic sets that are both closed in S. Or, equivalently, S does not contain a non-empty semi-algebraic strict subset which is both open and closed in S.

A semi-algebraic set S in R^k is **semi-algebraically path connected** when for every x, y in S, there exists a **semi-algebraic path** from x to y, i.e. a continuous semi-algebraic function φ: $[0, 1] \to S$ such that $\varphi(0) = x$ and $\varphi(1) = y$.

We shall see later, in Chapter 5 (Theorem 5.23), that the two notions of being semi-algebraically connected and semi-algebraically path connected agree for semi-algebraic sets. We shall see also (Theorem 5.22) that the two notions of being connected and semi-algebraically connected agree for semi-algebraic subsets of $\mathbb{R}$.

Exercise 3.3. Prove that if A is semi-algebraically connected, and the semi-algebraic set B is semi-algebraically homeomorphic to A then B is semi-algebraically connected.

Since the semi-algebraic subsets of the real closed field R are the finite unions of open intervals and points, the following proposition is clear:

Proposition 3.6. *A real closed field* R *(as well as all its intervals) is semi-algebraically connected.*

A subset C of R^k is **convex** if $x, y \in C$ implies that the segment

$$[x, y] = \{(1-\lambda)\, x + \lambda\, y \mid \lambda \in [0, 1] \subset \mathrm{R}\}$$

is contained in C.

Proposition 3.7. *If C is semi-algebraic and convex then C is semi-algebraically connected.*

Proof: Suppose that C is the disjoint union of two non-empty sets F_1 and F_2 which are closed in C. Let $x_1 \in F_1$ and $x_2 \in F_2$. The segment $[x_1, x_2]$ is the disjoint union of $F_1 \cap [x_1, x_2]$ and $F_2 \cap [x_1, x_2]$, which are closed, semi-algebraic, and non-empty. This contradicts the fact that $[x_1, x_2]$ is semi-algebraically connected (Proposition 3.6). □

Since the open cube $(0, 1)^k$ is convex, the following proposition is clear:

Proposition 3.8. *The open cube $(0, 1)^k$ is semi-algebraically connected.*

The following useful property holds for semi-algebraically connected sets.

Proposition 3.9. *If S is a semi-algebraically connected semi-algebraic set and $f: S \to \mathrm{R}$ is a locally constant semi-algebraic function (i.e. given $x \in S$, there is an open $U \subset S$ such that for all $y \in U$, $f(y) = f(x)$), then f is a constant.*

Proof: Let $d \in f(S)$. Since f is locally constant $f^{-1}(d)$ is open. If f is not constant, $f(S) \setminus \{d\}$ is non-empty and $f^{-1}(f(S) \setminus \{d\})$ is open. Clearly, $S = f^{-1}(d) \cup f^{-1}(f(S) \setminus \{d\})$. This contradicts the fact that S is semi-algebraically connected, since $f^{-1}(d)$ and $f^{-1}(f(S) \setminus \{d\}$ are non-empty open and disjoint semi-algebraic sets. □

3.3 Semi-algebraic Germs

We introduce the field of germs of semi-algebraic continuous functions at the right of the origin and prove that it provides another description of the real closure $\mathrm{R}\langle\varepsilon\rangle$ of $\mathrm{R}(\varepsilon)$ equipped with the order 0_+. We saw in Chapter 2 that $\mathrm{R}\langle\varepsilon\rangle$ is the field of algebraic Puiseux series (Corollary 2.98). The field $\mathrm{R}\langle\varepsilon\rangle$ is used in Section 3.4 to prove results in semi-algebraic geometry, and it will also play an important role in the second part of the book, which is devoted to algorithms.

In order to define the field of germs of semi-algebraic continuous functions at the right of the origin, some preliminary work on semi-algebraic and continuous functions is necessary.

Proposition 3.10. *Let S be a semi-algebraic set and let P be a univariate polynomial with coefficients semi-algebraic continuous functions defined on S. Then if y is a simple root of $P(x, Y)$ for a given $x \in S$, there is a semi-algebraic and continuous function f defined on a neighborhood of x in S such that $f(x) = y$ and for every $x' \in U$, $f(x')$ is a simple root of $P(x', Y)$.*

Proof: Let $m > 0$ such that for every $m' \in (0, m)$,

$$P(x, y - m')\, P(x, y + m') < 0.$$

Such an m exists because, y being a simple root of $P(x, Y)$, $P(x, Y)$ is either increasing or decreasing on an interval $(y - m,\ y + m)$. Note that y is the only root of $P(x, Y)$ in $(y - m,\ y + m)$. Suppose without loss of generality, that $\partial P / \partial Y(x,\ y) > 0$ and let V be a neighborhood of $(x,\ y)$ in $S \times \mathrm{R}$ where $\partial P / \partial Y$ is positive. For every m', $0 < m' < m$, the set

$$\{u \in S \mid P(u, y - m') P(u, y + m') < 0 \wedge [(u, y - m'), (u, y + m')] \subset V\}$$

is an open semi-algebraic subset of S containing x. This proves that $P(u, Y)$ has a simple root $y(u)$ on $(y - m', y + m')$ and that the function associating to $u \in U$ the value $y(u)$ is continuous. □

The set of **germs of semi-algebraic continuous functions at the right of the origin** is the set of semi-algebraic continuous functions with values in R which are defined on an interval of the form $(0, t)$, $t \in \mathrm{R}_+$, modulo the equivalence relation

$$f_1 \simeq f_2 \Leftrightarrow \exists t > 0 \quad \forall t' \ 0 < t' < t \ \ f_1(t') = f_2(t').$$

Proposition 3.11. *The germs of semi-algebraic continuous functions at the right of the origin form a real closed field.*

Proof: Let φ and φ' be two germs of semi-algebraic continuous functions at the right of the origin, and consider semi-algebraic continuous functions f and f' representing φ and φ', defined without loss of generality on a common interval $(0, t)$. The sum (resp. product) of φ and φ' is defined as the germ at the right of the origin of the sum (resp. product) of the semi-algebraic and continuous function $f + f'$ (resp. $f f'$) defined on $(0, t)$. It is easy to check that equipped with this addition and multiplication, the germs of semi-algebraic continuous functions at the right of the origin form a ring. The 0 (resp. 1) element of this ring is the germ of semi-algebraic continuous function at the right of the origin with representative the constant function with value 0 (resp. 1).

Consider a germ φ of semi-algebraic continuous function at the right of the origin and a representative f of φ defined on $(0,\ t)$. The set $A=\{x\in(0,t)\mid f(x)=0\}$ is a semi-algebraic set, and thus a finite union of points and intervals (Corollary 2.79). If A contains an interval $(0,\ t')$, then $\varphi=0$. Otherwise, denoting by t' the smallest element of A (defined as t is A is empty), the restriction of f to $(0,\ t')$ is everywhere non-zero, and hence $1/f$ is a semi-algebraic and continuous function defined on $(0,t')$ with associated germ $1/\varphi$. Thus the germs of semi-algebraic continuous functions at the right of the origin form a field.

Consider a germ φ of semi-algebraic continuous function at the right of the origin and a representative f of φ defined on $(0,t)$. The sets

$$\begin{aligned} A &= \{x\in(0,t)\mid f(x)=0\},\\ B &= \{x\in(0,t)\mid f(x)>0\},\\ C &= \{x\in(0,t)\mid f(x)<0\}. \end{aligned}$$

are semi-algebraic and partition $(0,\ t)$ into a finite number of points and intervals. One and only one of these three sets contains an interval of the form $(0,t')$. Thus, the sign of a germ φ of a semi-algebraic continuous function at the right of the origin is well defined. It is easy to check that equipped with this sign function, the germs of semi-algebraic continuous functions at the right of the origin form an ordered field.

It remains to prove that the germs of semi-algebraic continuous functions at the right of the origin have the intermediate value property, by Theorem 2.11.

It is sufficient to prove the intermediate value property for $\overline{P}$ separable, by Lemma 3.12.

Lemma 3.12. *The property* $(I(P,a,b))$

$$P(a)\,P(b)<0\Rightarrow\exists x\quad a<x<b\quad P(x)=0$$

holds for any $P\in\mathrm{R}[X]$ *if and only if it holds for any* $P\in\mathrm{R}[X]$*, with* P *separable.*

Proof of Lemma 3.12: It is clear that if $(I(P,a,b))$ holds for any $P\in\mathrm{R}[X]$, it holds for any $P\in\mathrm{R}[X]$, with P separable. In the other direction, if P is separable, there is nothing to prove. So, suppose that $P(a)\ P(b)<0$. If $P_1=\gcd(P(X),P'(X))\neq1,\ P(X)=P_1(X)\,P_2(X)$ with

$$\deg(P_1(X))<\deg(P(X)),\deg(P_2(X))<\deg(P(X)),$$

and either $P_1(a)\ P_1(b)<0$ or $P_2(a)\ P_2(b)<0$. This process can be continued up to the moment where a divisor Q of P, with $\gcd(Q(X),\ Q'(X))=1$, $Q(a)\,Q(b)<0$ is found. Applying property $(I(Q,a,b))$ gives a root of P. □

So, let $\overline{P}(Y) = \alpha_p Y^p + \cdots + \alpha_0, \alpha_p \neq 0$, be a separable polynomial, where the α_i are germs of semi-algebraic continuous functions at the right of the origin, and let φ_1 and φ_2 be such that $\overline{P}(\varphi_1)\,\overline{P}(\varphi_2) < 0$. Let $a_p, \ldots, a_0, f_1, f_2$ be representatives of $\alpha_p, \ldots, \alpha_0, \varphi_1, \varphi_2$ defined on $(0, t_0)$. For every $t \in (0, t_0)$, let $P(t, Y) = a_p(t)\, Y^p + \cdots + a_0(t)$. Shrinking $(0, t_0)$, if necessary, so that all the coefficients appearing in the signed remainder sequence of $\overline{P}$, $\overline{P}'$ have representatives defined on $(0, t_0)$, we can suppose that for every $t \in (0, t_0)$, $\deg(P(t, Y)) = p$, $P(t, f_1(t)) P(t, f_2(t)) < 0$, and $\gcd(P(t, Y), \overline{P}'(t, Y)) = 1$. It is clear that, for every $t \in (0, t_0)$, $P(t, Y)$ has a root in $(f_1(t), f_2(t))$. Consider, for every $0 \leq r \leq p$, the set $A_r \subset (0, t_0)$ of those t such that $P(t, Y)$ has exactly r distinct roots in R. Since A_r can be described by a formula, it is a semi-algebraic subset of $(0, t_0)$. The A_r partition $(0, t_0)$ into a finite union of points and intervals, and exactly one of the A_r contains an interval of the form $(0, t_1)$. We are going to prove that for $0 \leq i \leq r$, the function g_i associating to $t \in (0, t_1)$ the i-th root of $P(t, Y)$ is semi-algebraic and continuous and that one of them lies between f_1 and f_2.

Let $t \in (0, t_1)$ and consider the $g_i(t)$. By Proposition 3.10, there exists an open interval $(t - m, t + m)$ and semi-algebraic continuous functions h_i defined on $(t - m, t + m)$ such that $h_i(u)$ is a simple root of $P(u, Y)$ for every $u \in (t - m, t + m)$. This root is necessarily $g_i(u)$ because the number of roots of $P(t, Y)$ on S is fixed. Thus, g_i is continuous.

Since for every $t \in (0, t_1)$, $P(t, f_1(t))\, P(t, f_2(t)) < 0$, the graph of g_i does not intersect the graphs of f_1 and f_2. So there is at least one g_i lying between f_1 and f_2. □

Proposition 3.13. *The germs of semi-algebraic continuous functions at the right of the origin is the real closure of* R(ε) *equipped with the unique order making* ε *infinitesimal. The element* ε *is sent to the germ of the identity map at the right of the origin.*

Proof: By Proposition 3.11, the germs of semi-algebraic continuous functions at the right of the origin form a real closed field. By Proposition 2.86, a germ of semi-algebraic function at the right of the origin is algebraic over R(ε). □

Using Corollary 2.98 and Proposition 3.13,

Theorem 3.14. *The real closed field of germs of semi-algebraic continuous functions at the right of the origin is isomorphic to the field of algebraic Puiseux series* R$\langle\varepsilon\rangle$.

Using germs of semi-algebraic continuous functions at the right of the origin, the extension of a semi-algebraic set from R to R$\langle\varepsilon\rangle$ has a particularly simple meaning. Before explaining this, we need a notation.

Notation 3.15. [Composition with germs] Consider a germ φ of semi-algebraic continuous functions at the right of the origin and f defined on $(0, t)$ representing φ. If g is a continuous semi-algebraic function defined on the image of f, we denote by $g \circ \varphi$ the germ of semi-algebraic continuous functions at the right of the origin associated to the semi-algebraic continuous function $g \circ f$ defined on $(0, t)$. Note that $g \circ \varphi$ is independent of the choice of the representative f of φ. Note also that if f represents φ, $f \circ \varepsilon = \varphi$, since ε is the germ of the identity map at the right of the origin. □

Proposition 3.16. *Let $S \subset \mathrm{R}^k$ be a semi-algebraic set and $\varphi = (\varphi_1, \ldots, \varphi_k) \in \mathrm{R}\langle\varepsilon\rangle^k$. Let $f_1, \ldots, f_k$ be continuous semi-algebraic functions defined on $(0, t)$ and representing $\varphi_1, \ldots, \varphi_k$ and let $f = (f_1, \ldots, f_k)$. Then*

$$\varphi \in \mathrm{Ext}(S, \mathrm{R}\langle\varepsilon\rangle) \Leftrightarrow \exists\, t > 0 \quad \forall t' \quad 0 < t' < t \quad f(t') \in S.$$

Suppose that $\varphi \in \mathrm{Ext}(S, \mathrm{R}\langle\varepsilon\rangle)$ and let g be a semi-algebraic function defined on S. Then $\mathrm{Ext}(g, \mathrm{R}\langle\varepsilon\rangle)(\varphi) = g \circ \varphi$.

In particular, $\mathrm{Ext}(f, \mathrm{R}\langle\varepsilon\rangle)(\varepsilon) = \varphi$.

Proof: The first part of the proposition is clear since, as we have seen above in the proof of Proposition 3.11, if $P \in \mathrm{R}[X_1, \ldots, X_k]$ and $\varphi_1, \ldots, \varphi_k$ are germs of semi-algebraic continuous functions at the right of the origin with representatives $f_1, \ldots, f_k$ defined on a common $(0, t)$,

- $P(\varphi_1, \ldots, \varphi_k) = 0$ in $\mathrm{R}\langle\varepsilon\rangle$ if and only if there is an interval $(0, t) \subset \mathrm{R}$ such that $\forall\, t' \in (0, t) \quad P(f_1(t'), \ldots, f_k(t')) = 0$
- $P(\varphi_1, \ldots, \varphi_k) > 0$ in $\mathrm{R}\langle\varepsilon\rangle$ if and only if there is an interval $(0, t) \subset \mathrm{R}$ such that $\forall\, t' \in (0, t) \quad P(f_1(t'), \ldots, f_k(t')) > 0$.

The second part is clear as well by definition of the extension. The last part is a consequence of the second one, taking $S = \mathrm{R}\langle\varepsilon\rangle$, $\varphi = \varepsilon$, $f = \mathrm{Id}$, $g = f$ and using the remark at the end of Notation 3.15. □

An important property of $\mathrm{R}\langle\varepsilon\rangle$ is that sentences with coefficients in $\mathrm{R}[\varepsilon]$ which are true in $\mathrm{R}\langle\varepsilon\rangle$ are also true on a sufficiently small interval $(0, r) \subset \mathrm{R}$. Namely:

Proposition 3.17. *If Φ is a sentence in the language of ordered fields with coefficients in $\mathrm{R}[\varepsilon]$ and $\Phi'(t)$ is the sentence obtained by substituting $t \in \mathrm{R}$ for ε in Φ, then Φ is true in $\mathrm{R}\langle\varepsilon\rangle$ if and only if there exists t_0 in R such that $\Phi'(t)$ is true for every $t \in (0, t_0) \cap \mathrm{R}$.*

Proof: The semi-algebraic set $A = \{t \in \mathrm{R} \mid \Phi'(t)\}$ is a finite union of points and intervals. If A contains an interval $(0, t_0)$ with t_0 a positive element of R, then the extension of A to $\mathrm{R}\langle\varepsilon\rangle$ contains $(0, t_0) \subset \mathrm{R}\langle\varepsilon\rangle$, so that $\varepsilon \in \mathrm{Ext}(A, \mathrm{R}\langle\varepsilon\rangle)$ and $\Phi = \Phi'(\varepsilon)$ is true in $\mathrm{R}\langle\varepsilon\rangle$.

On the other hand, if A contains no interval $(0,t)$ with t a positive element of R, there exists t_0 such that $(0,t_0)\cap A=\emptyset$ and thus $\mathrm{Ext}((0,t_0)\cap A,\mathrm{R}\langle\varepsilon\rangle)=\emptyset$ and $\varepsilon\notin\mathrm{Ext}(A,\mathrm{R}\langle\varepsilon\rangle)$, which means that Φ is not true in $\mathrm{R}\langle\varepsilon\rangle$. □

The subring of germs of semi-algebraic continuous functions at the right of the origin which are bounded by an element of R coincides with the valuation ring $\mathrm{R}\langle\varepsilon\rangle_b$ defined in Chapter 2 (Notation 2.100). Indeed, is clear by Proposition 3.17 that a germ φ of semi-algebraic continuous functions at the right of the origin is **bounded** by an element of R if and only if φ has a representative f defined on $(0,t)$ which is bounded. Note that this property is independent of the representative f chosen for φ.

The ring homomorphism $\lim_\varepsilon$ defined on $\mathrm{R}\langle\varepsilon\rangle_b$ in Notation 2.100 has a useful consequence for semi-algebraic functions.

Proposition 3.18. *Let $f\colon(0,a)\to\mathrm{R}$ be a continuous bounded semi-algebraic function. Then f can be continuously extended to a function $\overline{f}$ on $[0,a)$.*

Proof: Let M bound the absolute value of f on $(0,a)$. Thus M bounds the germ of semi-algebraic continuous function $\varphi\in\mathrm{R}\langle\varepsilon\rangle$ associated to f using Proposition 3.16 and $\lim_\varepsilon(\varphi)$ is well-defined. Let $b=\lim_\varepsilon(\varphi)$. Defining

$$\overline{f}(t)=\begin{cases} b & \text{if } t=0,\\ f(t) & \text{if } t\in(0,a)\end{cases}$$

we easily see that $\overline{f}$ is continuous at 0. Indeed for every $r>0$ in R, the extension of the set $\{t\in\mathrm{R}\mid |f(t)-b|\le r\}$ to $\mathrm{R}\langle\varepsilon\rangle$ contains ε, since $\mathrm{Ext}(f,\mathrm{R}\langle\varepsilon\rangle)(\varepsilon)-b=\varphi-b$ is infinitesimal, and therefore there is a positive δ in R such that it contains the interval $(0,\delta)$ by Proposition 3.17. □

We can now prove a more geometric result. Note that its statement does not involve Puiseux series, while the proof we present does.

Theorem 3.19. [Curve selection lemma] *Let $S\subset\mathrm{R}^k$ be a semi-algebraic set. Let $x\in\overline{S}$. Then there exists a continuous semi-algebraic mapping $\gamma\colon[0,1)\to\mathrm{R}^k$ such that $\gamma(0)=x$ and $\gamma((0,1))\subset S$.*

Proof: Let $x\in\overline{S}$. For every $r>0$ in R, $B(x,r)\cap S$ is non-empty, hence $B(x,\varepsilon)\cap\mathrm{Ext}(S,\mathrm{R}\langle\varepsilon\rangle)$ is non-empty by the Transfer principle (Theorem 2.80). Let $\varphi\in B(x,\varepsilon)\cap\mathrm{Ext}(S,\mathrm{R}\langle\varepsilon\rangle)$. By Proposition 3.16 there exists a representative of φ which is a semi-algebraic continuous function f defined on $(0,t)$ such that for every t', $0<t'<t$, $f(t')\in B(x,r)\cap S$. Using Proposition 3.18 and scaling, we get $\gamma\colon[0,1)\to\mathrm{R}^k$ such that $\gamma(0)=x$ and $\gamma((0,1))\subset S$. It is easy to check that γ is continuous at 0. □

3.4 Closed and Bounded Semi-algebraic Sets

In $\mathbb{R}^k$, a closed bounded set S is compact, i.e. has the property that whenever S is covered by a family of sets open in S, it is also covered by a finite subfamily of these sets. This is no longer true for a general real closed field R, as can be seen by the following examples.

a) The interval $[0,1] \subset \mathbb{R}_{\mathrm{alg}}$ is not compact since the family

$$\{[0,r) \cup (s,1] \mid 0 < r < \pi/4 < s < 1, r \in \mathbb{R}_{\mathrm{alg}}\}$$

(where $\pi = 3.14...$), is an open cover of $[0,1]$ which has no finite subcover.

b) The interval $[0,1] \subset \mathbb{R}_{\mathrm{alg}}$ is not compact since the family

$$\{[0,r) \cup (s,1] \mid 0 < r < \pi/4 < s < 1, r \in \mathbb{R}_{\mathrm{alg}}\}$$

(where $\pi = 3.14...$), is an open cover of $[0,1]$ which has no finite subcover.

c) The interval $[0,1] \subset \mathbb{R}\langle\varepsilon\rangle$ is not compact since the family

$$\{[0,f) \cup (r,1] \mid f > 0 \text{and infinitesimal over } \mathbb{R}, r \in \mathbb{R}, 0 < r < 1\}$$

is an open cover with no finite subcover.

However, closed and bounded semi-algebraic sets do enjoy properties of compact subsets, as we see now. We are going to prove the following result.

Theorem 3.20. *Let S be a closed, bounded semi-algebraic set and g a semi-algebraic continuous function defined on S. Then $g(S)$ is closed and bounded.*

Though the statement of this theorem is geometric, the proof we present uses the properties of the real closed extension $\mathrm{R}\langle\varepsilon\rangle$ of R.

The proof of the theorem uses the following lemma:

Lemma 3.21. *Let g be a semi-algebraic continuous function defined on a closed, bounded semi-algebraic set $S \subset \mathrm{R}^k$. If $\varphi \in \mathrm{Ext}(S, \mathrm{R}\langle\varepsilon\rangle)$, then $g \circ \varphi$ is bounded over* R *and*

$$g(\lim_\varepsilon(\varphi)) = \lim_\varepsilon(g \circ \varphi).$$

Proof: Let $f = (f_1, ..., f_k)$ be a semi-algebraic function defined on $(0,t)$ and representing $\varphi = (\varphi_1, ..., \varphi_k) \in \mathrm{R}\langle\varepsilon\rangle^k$ and let $\overline{f}$ its extension to $[0,t)$, using Proposition 3.18. By definition of $\lim_\varepsilon$,

$$\overline{f}(0) = b = \lim_\varepsilon(\varphi)$$

since $\varphi - b$ is infinitesimal. Since S is closed $b \in S$. Thus g is continuous at b. Hence, for every $r > 0 \in \mathrm{R}$, there is an η such that if $z \in S$ and $\|z - b\| < \eta$ then $\|g(z) - g(b)\| < r$. Using the Transfer Principle (Theorem 2.80) together with the fact that $\varphi \in \mathrm{Ext}(S, \mathrm{R}\langle\varepsilon\rangle)$ and $\varphi - b$ is infinitesimal over R we see that $\|g \circ \varphi - g(b)\|$ is smaller than any $r > 0$. Thus $g \circ \varphi$ is bounded over R and infinitesimally close to $g(b)$, and hence $g(\lim_\varepsilon(\varphi)) = \lim_\varepsilon(g \circ \varphi)$. □

Proof of Theorem 3.20: We first prove that $g(S)$ is closed. Suppose that x is in the closure of $g(S)$. Then $B(x, r) \cap g(S)$ is not empty, for any $r \in \mathrm{R}$. Hence, by the Transfer principle (Theorem 2.80), $B(x, \varepsilon) \cap \mathrm{Ext}(g(S), \mathrm{R}\langle\varepsilon\rangle)$ is not empty. Thus, there is a $\varphi \in \mathrm{Ext}(g(S), \mathrm{R}\langle\varepsilon\rangle)$ for which $\lim_\varepsilon(\varphi) = x$. By Proposition 2.90, there is a $\varphi' \in \mathrm{Ext}(S, \mathrm{R}\langle\varepsilon\rangle)$ such that $g \circ \varphi' = \varphi$. Since S is closed and bounded and φ' has a representative f' defined on $(0, t)$ which can be extended continuously to $\overline{f'}$ at 0, $\lim_\varepsilon(\varphi') = \overline{f'}(0) \in S$, and we conclude that $g(\lim_\varepsilon(\varphi')) = \lim_\varepsilon(\varphi) = x$. Hence $x \in g(S)$.

We now prove that $g(S)$ is bounded. The set

$$A = \{t \in \mathrm{R} \mid \exists x \in S \quad \|g(x)\| = t\}$$

is semi-algebraic and so it is a finite union of points and intervals. For every $\varphi \in \mathrm{Ext}(S, \mathrm{R}\langle\varepsilon\rangle)$, $g \circ \varphi$ is bounded over R by Lemma 3.21. Thus $\mathrm{Ext}(A, \mathrm{R}\langle\varepsilon\rangle)$ does not contain $1/\varepsilon$. This implies that A contains no interval of the form $(M, +\infty)$, and thus A is bounded. □

3.5 Implicit Function Theorem

The usual notions of differentiability over $\mathbb{R}$ can be developed over an arbitrary real closed field R. We do this now.

Let f be a semi-algebraic function from a semi-algebraic open subset U of R^k to R^p, and let $x_0 \in U$. We write $\lim_{x \to x_0} f(x) = y_0$ for

$$\forall r > 0\ \exists \delta\ \forall x\ \|x - x_0\| < \delta \Rightarrow \|f(x) - y_0\| < r$$

and $f(x) = o(\|x - x_0\|)$ for

$$\lim_{x \to x_0} \frac{f(x)}{\|x - x_0\|} = 0.$$

If M is a semi-algebraic subset of U, we write $\lim_{x \in M, x \to x_0} f(x) = y_0$ for

$$\forall r > 0\ \exists\, \delta\ \forall x \in M\ \|x - x_0\| < \delta \Rightarrow \|f(x) - y_0\| < r.$$

The function f: $(a, b) \to \mathrm{R}$ is **differentiable at** $x_0 \in (a, b)$ with derivative $f'(x_0)$ if

$$\lim_{x \to x_0} \frac{f(x) - f(x_0)}{x - x_0} = f'(x_0).$$

We consider only semi-algebraic functions. Theorem 3.20 implies that a semi-algebraic function continuous on a closed and bounded interval is bounded and attains its bounds.

Exercise 3.4. Prove that Rolle's Theorem and the Mean Value Theorem hold for semi-algebraic differentiable functions.

Proposition 3.22. *Let f: $(a, b) \to \mathrm{R}$ be a semi-algebraic function differentiable on the interval (a, b). Then its derivative f' is a semi-algebraic function.*

Proof: Describe the graph of f' by a formula in the language of ordered fields with parameters in R, and use Corollary 2.78. □

Exercise 3.5. Provide the details of the proof of Proposition 3.22.

Partial derivatives of multivariate semi-algebraic functions are defined in the usual way and have the usual properties. In particular let $U \subset \mathrm{R}^k$ be a semi-algebraic open set and $f\colon U \to \mathrm{R}^p$, and suppose that the partial derivatives of the coordinate functions of f with respect to $X_1, \ldots, X_k$ exist on U and are continuous. These partial derivatives are clearly semi-algebraic functions.

For every $x_0 \in U$, let $\mathrm{d}f(x_0)$ denote the **derivative of** f **at** x_0, i.e. the linear mapping from R^k to R^p sending $(h_1, \ldots, h_k)$ to

$$\left(\sum_{j=1,\ldots,k} \frac{\partial f_1}{\partial X_j}(x_0)\, h_j, \ldots, \sum_{j=1,\ldots,k} \frac{\partial f_p}{\partial X_j}(x_0)\, h_j\right).$$

The matrix of $\mathrm{d}f(x_0)$ is the **Jacobian matrix of** f **at** x_0 and its determinant is the **Jacobian of** f **at** x_0. Following the usual arguments from a calculus course, It is clear that

$$f(x) - f(x_0) - \mathrm{d}f(x_0)(x - x_0) = o(\|x - x_0\|).$$

As in the univariate case, one can iterate the above definition to define higher derivatives.

Let $U \subset \mathrm{R}^k$ be a semi-algebraic open set and $B \subset \mathrm{R}^p$ a semi-algebraic set. The set of semi-algebraic functions from U to B for which all partial derivatives up to order ℓ exist and are continuous is denoted $\mathcal{S}^\ell(U, B)$, and the class $\mathcal{S}^\infty(U, B)$ is the intersection of $\mathcal{S}^\ell(U, B)$ for all finite ℓ. The ring $\mathcal{S}^\ell(U, R)$ is abbreviated $\mathcal{S}^\ell(U)$, and the ring $\mathcal{S}^\infty(U, \mathrm{R})$ is also called the ring of **Nash functions.**

We present a semi-algebraic version of the implicit function theorem whose proof is essentially the same as the classical proofs.

Given a linear mapping $F\colon \mathrm{R}^k \to \mathrm{R}^p$, we define the **norm** of F by $\|F\| = \sup(\{\|F(x)\| \mid \|x\| = 1\})$. This is a well-defined element of R by Theorem 3.20, since $x \mapsto \|F(x)\|$ is a continuous semi-algebraic function and $\{x \mid \|x\| = 1\}$ is a closed and bounded semi-algebraic set.

Proposition 3.23. *Let x and y be two points of R^k, U an open semi-algebraic set containing the segment $[x, y]$, and $f \in \mathcal{S}^1(U, \mathrm{R}^\ell)$. Then*

$$\|f(x) - f(y)\| \le M\, \|x - y\|,$$

where $M = \sup(\{\|\mathrm{d}f(z)\| \mid z \in [x, y]\})$ (M is well defined, by Theorem 3.20).

Proof: Define $g(t) = f((1 - t)x + t y)$ for $t \in [0, 1]$. Then $\|g'(t)\| \le M\, \|x - y\|$ for $t \in [0, 1]$. For any positive $c \in \mathrm{R}$, we define

$$A_c = \{t \in [0, 1] \mid \|g(t) - g(0)\| \le M\, \|x - y\| t + c t\}$$

which is a closed semi-algebraic subset of $[0,\,1]$ containing 0. Let t_0 be the largest element in A_c. Suppose $t_0 \neq 1$. We have

$$\|g(t_0) - g(0)\| \leq M\,\|x - y\|t_0 + c\,t_0\,.$$

Since $\|g'(t_0)\| \leq M\,\|x - y\|$, we can find $r > 0$ in R such that if $t_0 < t < t_0 + r$,

$$\|g(t) - g(t_0)\| \leq M\,\|x - y\|(t - t_0) + c(t - t_0)\,.$$

So, for $t_0 < t < t_0 + r$, by summing the two displayed inequalities, we have

$$\|g(t) - g(0)\| \leq M\,\|x - y\|t + c t,$$

which contradicts the maximality of t_0. Thus $1 \in A_c$ for every c, which gives the result. □

Proposition 3.24. [Inverse Function Theorem] *Let U' be a semi-algebraic open neighborhood of the origin 0 of R^k, $f \in \mathcal{S}^\ell(U', \mathrm{R}^k)$, $\ell \geq 1$, such that $f(0) = 0$ and that $\mathrm{d}f(0)\colon \mathrm{R}^k \to \mathrm{R}^k$ is invertible. Then there exist semi-algebraic open neighborhoods U, V of 0 in R^k, $U \subset U'$, such that $f|_U$ is a homeomorphism onto V and $(f|_U)^{-1} \in \mathcal{S}^\ell(V, U)$.*

Proof: We can suppose that $\mathrm{d}f(0)$ is the identity Id of R^k (by composing with $\mathrm{d}f(0)^{-1}$). Take $g = f - \mathrm{Id}$. Then $\mathrm{d}g(0) = 0$, and there is $r_1 \in \mathrm{R}$ such that $\|\mathrm{d}g(x)\| \leq \frac{1}{2}$ if $x \in B_k(0, r_1)$. By Proposition 3.23, if $x, y \in B_k(0, r_1)$, then:

$$\|f(x) - f(y) - (x - y)\| \leq \frac{1}{2}\|x - y\|$$

and thus

$$\frac{1}{2}\|x - y\| \leq \|f(x) - f(y)\| \leq \frac{3}{2}\|x - y\|\,,$$

using the triangle inequalities. This implies that f is injective on $B_k(0, r_1)$. We can find $r_2 < r_1$ with $\mathrm{d}f(x)$ invertible for $x \in B_k(0,\ r_2)$.Now we prove that $f(B_k(0,\ r_2)) \supset B_k(0,\ r_2/4)$. For y^0 with $\|y^0\| < r_2/4$, define $h(x) = \|f(x) - y^0\|^2$. Then h reaches its minimum on $\overline{B}_k(0,\ r_2)$ and does not reach it on the boundary $S^{k-1}(0,\ r_2)$ since if $\|x\| = r_2$, one has $\|f(x)\| \geq r_2/2$ and thus $h(x) > (r_2/4)^2 > h(0)$. Therefore, this minimum is reached at a point $x^0 \in B_k(0, r_2)$. One then has, for $i = 1, \ldots, n$,

$$\frac{\partial h}{\partial x_i}(x^0) = 0\,, \quad \text{i.e.} \sum_{j=1}^{k} (f_j(x^0) - y_j^0)\frac{\partial f_j}{\partial x_i}(x^0) = 0\,.$$

Since $\mathrm{d}f(x^0)$ is invertible, we have $f(x^0) = y^0$. We then define $V = B_k(0, r_2/4)$, $U = f^{-1}(V) \cap B_k(0, r_2)$. The function f^{-1} is continuous because

$$\|f^{-1}(x) - f^{-1}(y)\| \leq 2\|x - y\|$$

for $x, y \in V$, and we easily get $\mathrm{d}(f^{-1})(x) = (\mathrm{d}f(f^{-1}(x)))^{-1}$. □

Theorem 3.25. [Implicit Function Theorem] *Let $(x^0, y^0) \in \mathrm{R}^{k+\ell}$, and let $f_1, \ldots, f_\ell$ be semi-algebraic functions of class $\mathcal{S}^m$ on an open neighborhood of (x^0, y^0) such that $f_j(x^0, y^0) = 0$ for $j = 1, \ldots, \ell$ and the Jacobian matrix of $f = (f_1, \ldots, f_\ell)$ at (x^0, y^0) with respect to the variables $y_1, \ldots, y_\ell$ is invertible. Then there exists a semi-algebraic open neighborhood U (resp. V) of x^0 (resp. y^0) in R^k (resp. R^ℓ) and a function $\varphi \in \mathcal{S}^m(U, V)$ such that $\varphi(x^0) = y^0$, and, for every $(x, y) \in U \times V$, we have*

$$f_1(x, y) = \cdots = f_\ell(x, y) = 0 \Leftrightarrow y = \varphi(x) .$$

Proof: Apply Proposition 3.24 to the function $(x, y) \mapsto (x, f(x, y))$. □

We now have all the tools needed to develop "semi-algebraic differential geometry".

The notion of an $\mathcal{S}^\infty$-diffeomorphism between semi-algebraic open sets of R^k is clear. The semi-algebraic version of $\mathcal{C}^\infty$ submanifolds of R^k is as follows.

An **$\mathcal{S}^\infty$-diffeomorphism** φ from a semi-algebraic open U of R^k to a semi-algebraic open Ω of R^k is a bijection from U to Ω that is $\mathcal{S}^\infty$ and such that $\varphi^{(-1)}$ is $\mathcal{S}^\infty$.

A semi-algebraic subset M of R^k is an **$\mathcal{S}^\infty$ submanifold of R^k of dimension ℓ** if for every point x of M, there exists a semi-algebraic open U of R^k and an $\mathcal{S}^\infty$-diffeomorphism φ from U to a semi-algebraic open neighborhood Ω of x in R^k such that $\varphi(0) = x$ and

$$\varphi(U \cap (\mathrm{R}^\ell \times \{0\})) = M \cap \Omega$$

(where $\mathrm{R}^\ell \times \{0\} = \{(a_1, \ldots, a_\ell, 0, \ldots, 0) | (a_1, \ldots, a_\ell) \in \mathrm{R}^\ell\}$).

A semi-algebraic map from M to N, where M (resp. N) is an $\mathcal{S}^\infty$ submanifold of R^m (resp. R^n), is an **$\mathcal{S}^\infty$ map** if it is locally the restriction of an $\mathcal{S}^\infty$ map from R^m to R^n.

A point x of a semi-algebraic set $S \subset \mathrm{R}^k$ is a **smooth point of dimension ℓ** if there is a semi-algebraic open subset U of S containing x which is an $\mathcal{S}^\infty$ submanifold of R^k of dimension ℓ.

Let x be a smooth point of dimension ℓ of an $\mathcal{S}^\infty$ submanifold M of R^k and let Ω be a semi-algebraic open neighborhood of x in R^k and $\varphi: U \to \Omega$ as in the definition of a submanifold. Let $X_1, \ldots, X_k$ be the coordinates of the domain of $\varphi = (\varphi_1, \ldots, \varphi_k)$. We call the set $T_x(M) = x + \mathrm{d}\varphi(0)(\mathrm{R}^\ell \times \{0\})$ the **tangent space to M at x.** Clearly, the tangent space contains x and is a translate of an ℓ dimensional linear subspace of R^k, i.e. an **ℓ-flat**. More concretely, note that the tangent space $T_x(M)$ is the translate by x of the linear space spanned by the first ℓ columns of the Jacobian matrix.

We next prove the usual geometric properties of tangent spaces.

Proposition 3.26. *Let x be a point of an $\mathcal{S}^\infty$ submanifold M of R^k having dimension ℓ and let π denote orthogonal projection onto the ℓ-flat $T_x(M)$. Then, $\lim_{y \in M, y \to x} \frac{\|y - \pi(y)\|}{\|y - x\|} = 0$.*

Proof: Let Ω be a semi-algebraic open neighborhood of x in R^k and $\varphi: U \to \Omega$ as in the definition of a submanifold. Let $X_1, \ldots, X_k$ be the coordinates of the domain of $\varphi = (\varphi_1, \ldots, \varphi_k)$. Then,

$$T_x(M) = x + \mathrm{d}\varphi(0)(\mathrm{R}^\ell \times \{0\}).$$

From elementary properties of derivatives (see Equation (3.5)), it is clear that for $u \in \mathrm{R}^\ell \times \{0\}$, $\varphi(u) - \mathrm{d}\varphi(0)(u) = o(\|u\|)$.

Now, for $y \in M \cap \Omega$, let $u = \varphi^{-1}(y)$. Then, since π is an orthogonal projection,

$$\|y - \pi(y)\| \leq \|\varphi(u) - \mathrm{d}\varphi(0)(u)\| = o(\|u\|).$$

Since, φ^{-1} is an S^∞ map, for any bounded neighborhood of x there is a constant C such that $\|\varphi^{-1}(y)\| \leq C\|y - x\|$ for all y in the neighborhood. Since $\|u\| = \|\varphi^{-1}(y)\| \leq C\|y - x\|$,

$$\|\varphi(u) - \mathrm{d}\varphi(0)(u)\| = o(\|y - x\|),$$

and the conclusion follows. □

We next prove that the tangent vector at a point of a curve lying on an $\mathcal{S}^\infty$ submanifold M of R^k is contained in the tangent space to M at that point.

Proposition 3.27. *Let x be a point of the $\mathcal{S}^\infty$ submanifold M in R^k having dimension ℓ, and let γ: $[-1, 1] \to \mathrm{R}^k$ be an $\mathcal{S}^\infty$ curve contained in M with $\gamma(0) = x$. Then the tangent vector $x + \gamma'(0)$ is contained in the tangent space $T_x(M)$.*

Proof: Let $\gamma(t) = (\gamma_1(t), \ldots, \gamma_k(t))$. Let Ω, φ be as in the definition of submanifold, and consider the composite map $\varphi^{-1} \circ \gamma: [-1, 1] \to \mathrm{R}^k$. Applying the chain rule, $\mathrm{d}(\varphi^{-1} \circ \gamma)(0) = \mathrm{d}\varphi^{-1}(x)(\gamma'(0))$. Since $\gamma([-1, 1]) \subset M$, it follows that $\varphi^{-1}(\gamma([-1, 1])) \subset \mathrm{R}^\ell \times \{0\}$, and $\mathrm{d}(\varphi^{-1} \circ \gamma)(t) \in \mathrm{R}^\ell \times \{0\}$ for all $t \in [-1, 1]$. Thus, $\mathrm{d}\varphi^{-1}(x)(\gamma'(0)) \in \mathrm{R}^\ell \times \{0\}$. Since $\mathrm{d}\varphi^{-1}(x) = (\mathrm{d}\varphi(0))^{-1}$, applying $\mathrm{d}\varphi(0)$ to both sides we have $\gamma'(0) \in \mathrm{d}\varphi(0)(\mathrm{R}^\ell \times \{0\})$, and finally $x + \gamma'(0) \in T_x(M)$. □

The notion of derivatives defined earlier for multivariate functions can now be extended to $\mathcal{S}^\infty$ submanifolds.

Let f: $M \to N$ be an $\mathcal{S}^\infty$ map, where M (resp. N) is a m' (resp. n') dimensional $\mathcal{S}^\infty$ submanifold of R^m (resp. R^n).

Let $x \in M$ and let Ω (resp. Ω') be a neighborhood of x (resp. $f(x)$) in R^m (resp. R^n) and φ (resp. ψ) a semi-algebraic diffeomorphism from U to Ω (resp. U' to Ω') such that $\varphi(0) = x$ (resp. $\psi(0) = f(x)$) and

$$\varphi(\mathrm{R}^{m'} \times \{0\}) = M \cap \Omega \text{ (resp. } \psi(\mathrm{R}^{n'} \times \{0\}) = N \cap \Omega' \text{)}.$$

Clearly, $\psi^{-1} \circ f \circ \varphi: \mathrm{R}^m \to \mathrm{R}^n$ is an $\mathcal{S}^\infty$ map, and its restriction to $\mathrm{R}^{m'} \times \{0\}$ is an $\mathcal{S}^\infty$ map to $\mathrm{R}^{n'} \times \{0\}$.

The derivative $\mathrm{d}(\psi^{-1} \circ f \circ \varphi)(0)$ restricted to $\mathrm{R}^{m'} \times \{0\}$ maps $\mathrm{R}^{m'} \times \{0\}$ into $\mathrm{R}^{n'} \times \{0\}$.

The linear map $\mathrm{d}f(x): T_x(M) \to T_{f(x)}(N)$ defined by

$$\mathrm{d}f(x)(v) = f(x) + \mathrm{d}\psi(0)(\mathrm{d}(\psi^{-1} \circ f \circ \varphi)(0)(\mathrm{d}\varphi^{-1}(x)(v - x))),$$

is called the **derivative** of f at x.

Proposition 3.28.

a) *A semi-algebraic open subset of an $\mathcal{S}^\infty$ submanifold V of dimension i is an $\mathcal{S}^\infty$ submanifold of dimension i.*
b) *If V' is an $\mathcal{S}^\infty$ submanifold of dimension j contained in an $\mathcal{S}^\infty$ submanifold V of dimension i, then $j \le i$.*

Proof: a) is clear. b) follows from the fact that the tangent space to V' at $x \in V'$ is a subspace of the tangent space to V at x. □

3.6 Bibliographical Notes

Semi-algebraic sets appear first in a logical context in Tarski's work [154]. They were studied from a geometrical and topological point of view by Brakhage [28], in his unpublished thesis. The modern study of semi-algebraic sets starts with Lojasiewicz, as a particular case of semi-analytic sets [110, 111].

4

Algebra

We start in Section 4.1 with the discriminant, and the related notion of subdiscriminant. In Section 4.2, we define the resultant and signed subresultant coefficients of two univariate polynomials an indicate how to use them for real root counting. We describe in Section 4.3 an algebraic real root counting technique based on the signature of a quadratic form. We then give a constructive proof of Hilbert's Nullstellensatz using resultants in Section 4.4. In Section 4.5, we algebraically characterize systems of polynomials with a finite number of solutions and prove that the corresponding quotient rings are finite dimensional vector spaces. In Section 4.6, we give a multivariate generalization of the real root counting technique based on the signature of a quadratic form described in Section 4.3. In Section 4.7, we define projective space and prove a weak version of Bézout's theorem.

Throughout Chapter 4, K is a field of characteristic zero and C is an algebraically closed field containing it. We will also denote by R a real closed field containing K when K is an ordered field.

4.1 Discriminant and Subdiscriminant

Notation 4.1. [Discriminant] Let $P \in \mathrm{R}[X]$ be a monic polynomial of degree p,

$$P = X^p + a_{p-1}X^{p-1} + \cdots + a_0,$$

and let $x_1, \ldots, x_p$ be the roots of P in C (repeated according to their multiplicities). The **discriminant** of P, $\mathrm{Disc}(P)$, is defined by

$$\mathrm{Disc}(P) = \prod_{p \geq i > j \geq 1} (x_i - x_j)^2. \qquad \square$$

Remark 4.2. The discriminant played a key role in the algebraic proof of the fundamental theorem of algebra (proof of $a) \Rightarrow b)$) in Theorem 2.11, see Remark 2.17). $\square$

Proposition 4.3. $\mathrm{Disc}(P) = 0$ *if and only if* $\deg(\gcd(P, P')) > 0$.

Proof: It is clear from the definition that $\mathrm{Disc}(P)=0$ if and only if P has a multiple root in C. □

Remark 4.4. When all the roots of P are in R and distinct, $\mathrm{Disc}(P)>0$. □

The sign of the discriminant counts the number of real roots modulo 4.

Proposition 4.5. *Let $P \in \mathrm{R}[X]$ be monic with* R *real closed, of degree p, and with p distinct roots in* C*; Denoting by t the number of roots of P in* R*,*

$$\begin{aligned} \mathrm{Disc}(P)>0 &\Leftrightarrow t \equiv p \bmod 4, \\ \mathrm{Disc}(P)<0 &\Leftrightarrow t \equiv p-2 \bmod 4. \end{aligned}$$

Proof: Let $y_1, \ldots, y_t$ be the roots of P in R and $z_1, \overline{z_1}, \ldots, z_s, \overline{z_s}$ the roots of P in $\mathrm{C} \setminus \mathrm{R}$, with $\mathrm{C} = \mathrm{R}[i]$.

The conclusion is clear since

$$\begin{aligned} \mathrm{sign}\Big(\prod_{i=1}^{s} (z_i - \overline{z_i})^2\Big) &= (-1)^s, \\ (y_i - y_j)^2 &> 0, 1 \le i < j \le t, \\ \big((z_i - z_j)(z_i - \overline{z_j})(\overline{z_i} - z_j)(\overline{z_i} - \overline{z_j})\big)^2 &> 0, 1 \le i < j \le s, \\ \big((y_i - z_j)(y_i - \overline{z_j})\big)^2 &> 0, 1 \le i \le t, 1 \le j \le s. \end{aligned}$$

Thus, $\mathrm{Disc}(P)>0$ if and only if s is even, and $\mathrm{Disc}(P)<0$ if and only if s is odd. □

The $p-k$**-subdiscriminant of** P , $1 \le k \le p$, is by definition

$$\mathrm{sDisc}_{p-k}(P) = \sum_{\substack{I \subset \{1,\ldots,p\} \\ \#(I)=k}} \prod_{\substack{(j,\ell) \in I \\ \ell > j}} (x_j - x_\ell)^2.$$

Note that $\mathrm{sDisc}_{p-1}(P)=p$. The discriminant is the 0-th subdiscriminant:

$$\mathrm{sDisc}_0(P) = \mathrm{Disc}(P) = \prod_{p \ge j > \ell \ge 1} (x_j - x_\ell)^2.$$

Remark 4.6. It is clear that when all the roots of P are in R

$$\mathrm{sDisc}_0(P) = \ldots = \mathrm{sDisc}_{j-1}(P) = 0, \mathrm{sDisc}_j(P) \neq 0$$

if and only if P has $p-j$ distinct roots. We shall see later in Proposition 4.29 that this property is true in general.

□

The subdiscriminants are intimately related to the Newton sums of P.

Definition 4.7. The i**-th Newton sum** of the polynomial P, denoted N_i, is $\sum_{x \in \mathrm{Zer}(P,\mathrm{C})} \mu(x) x^i$, where $\mu(x)$ is the multiplicity of x. □

The Newton sums can be obtained from the coefficients of P by the famous Newton identities.

Proposition 4.8. *Let* $P = a_p X^p + a_{p-1} X^{p-1} + \cdots + a_1 X + a_0$. *For any* i

$$(p-i)\, a_{p-i} = a_p N_i + \cdots + a_0 N_{i-p}, \tag{4.1}$$

with the convention $a_i = N_i = 0$ *for* $i < 0$.

Proof: We have

$$\begin{aligned} P &= a_p \prod_{x \in \mathrm{Zer}(P,\mathrm{C})} (X-x)^{\mu(x)}, \\ \frac{P'}{P} &= \sum_{x \in \mathrm{Zer}(P,\mathrm{C})} \frac{\mu(x)}{(X-x)}. \end{aligned}$$

Using

$$\frac{1}{X-x} = \sum_{i=0}^{\infty} \frac{x^i}{X^{i+1}},$$

we get

$$\begin{aligned} \frac{P'}{P} &= \sum_{i=0}^{\infty} \frac{N_i}{X^{i+1}}, \\ P' &= \left(\sum_{i=0}^{\infty} \frac{N_i}{X^{o+1}} \right) P. \end{aligned}$$

Equation (4.1) follows by equating the coefficients of X^{p-i-1} on both sides of the last equality. □

Consider the square matrix

$$\mathrm{Newt}_{p-k}(P) = \begin{bmatrix} N_0 & N_1 & \iddots & & \iddots & N_{k-1} \\ N_1 & \iddots & & \iddots & N_{k-1} & N_k \\ \iddots & & \iddots & N_{k-1} & N_k & \iddots \\ & \iddots & N_{k-1} & N_k & \iddots & \\ \iddots & N_{k-1} & N_k & \iddots & & \iddots \\ N_{k-1} & N_k & \iddots & & \iddots & N_{2k-2} \end{bmatrix}$$

with entries the Newton sums of the monic polynomial P of degree p.

We denote as usual by $\det(M)$ the determinant of a square matrix M.

Proposition 4.9. *For every* k, $1 \le k \le p$,

$$\mathrm{sDisc}_{p-k}(P) = \det(\mathrm{Newt}_{p-k}(P)).$$

The proof of Proposition 4.9 uses the Cauchy-Binet formula.

Proposition 4.10. **[Cauchy-Binet]** *Let A be a $n \times m$ matrix and B be a $m \times n$ matrix, $m \geq n$. For every $I \subset \{1, \ldots, m\}$ of cardinality n, denote by A_I the $n \times n$ matrix obtained by extracting from A the columns with indices in I. Similarly let B^I be the $n \times n$ matrix obtained by extracting from B the rows with indices in I.*

$$\det(A\,B) = \sum_{\substack{I \subset \{1, \ldots, m\} \\ \#(I) = n}} \det(A_I) \det(B^I).$$

Proof:

We introduce an m-dimensional diagonal matrix D_λ with diagonal entries the variables $\lambda_1, \ldots, \lambda_m$ and study $\det(A\,D_\lambda\,B)$. Since the entries of the matrix $A\,D_\lambda\,B$ are homogeneous linear forms in the λ_i, $\det(A\,D_\lambda\,B)$ is a homogeneous polynomial of degree n in the λ_i.

We are going to prove that the only monomials with non-zero coefficients of $\det(A\,D_\lambda\,B)$ are of the form $\lambda_I = \prod_{i \in I} \lambda_i$ for a subset $I \subset \{1, \ldots, m\}$, $\#(I) = n$.

Indeed if we consider $I \subset \{1, \ldots, m\}$, $\#(I) < n$, the specialization of $\det(A\,D_\lambda\,B)$ obtained by sending λ_j to 0 for $j \notin I$ is identically null. This implies that the coefficients of all the monomials where a variable is repeated are 0.

If we choose $I \subset \{1, \ldots, m\}$, $\#(I) = n$, and specialize the variables λ_i, $i \in I$ to 1 and the variables λ_i, $i \notin I$ to 0, we get the coefficient of $\lambda_I = \prod_{i \in I} \lambda_i$ in $\det(A\,D_\lambda\,B)$, which is $\det(A_I)\det(B^I)$.

Specializing finally all the λ_i to 1, we get the required identity. □

The proof of Proposition 4.9 makes also use of the classical Vandermonde determinant. Let $x_1, \ldots, x_r$ be elements of a field K. The **Vandermonde determinant** of $x_1, \ldots, x_r$ is $\det(V(x_1, \ldots, x_r))$ with

$$V(x_1, \ldots, x_{r-1}, x_r) = \begin{bmatrix} 1 & \cdots & 1 & 1 \\ x_1 & \cdots & x_{r-1} & x_r \\ \vdots & & \vdots & \vdots \\ x_1^{r-1} & \cdots & x_{r-1}^{r-1} & x_r^{r-1} \end{bmatrix}$$

the **Vandermonde matrix**.

Lemma 4.11.

$$\det(V(x_1, \ldots, x_r)) = \prod_{r \geq i > j \geq 1} (x_i - x_j).$$

Proof: The claim is true when $x_1, \ldots, x_r$ are not all distinct since both sides are 0. The proof when $x_1, \ldots, x_r$ are all distinct is by induction on r. The claim is obviously true for $r = 2$. Suppose that the claim is true for $r - 1$ and consider

$$V(x_1, \ldots, x_{r-1}, X) = \begin{bmatrix} 1 & \cdots & 1 & 1 \\ x_1 & \cdots & x_{r-1} & X \\ \vdots & & \vdots & \vdots \\ x_1^{r-1} & \cdots & x_{r-1}^{r-1} & X^{r-1} \end{bmatrix}.$$

The polynomial $\det(V(x_1, ..., x_{r-1}, X))$ has degree at most $r-1$, with $r-1$ distinct roots $x_1, ..., x_{r-1}$ because, replacing X by x_i in $V(x_1, ..., x_{r-1}, X)$, we get a matrix with two equal columns. So

$$\det(V(x_1, ..., x_{r-1}, X)) = c \prod_{r-1 \geq j \geq 1} (X - x_j).$$

The coefficient of $\det(V(x_1, ..., x_{r-1}, X))$ is the Vandermonde determinant of $x_1, ..., x_{r-1}$, $\det(V(x_1, ..., x_{r-1}))$ is equal to

$$\prod_{r-1 \geq i > j \geq 1} (x_i - x_j),$$

by the induction hypothesis. So

$$\det(V(x_1, ..., x_{r-1}, X)) = \prod_{r-1 \geq i > j \geq 1} (x_i - x_j) \prod_{r-1 \geq j \geq 1} (X - x_j).$$

Now substitute x_r for X to get the claim. □

Proof of Proposition 4.9: Define

$$V_k = \begin{bmatrix} 1 & \cdots & \cdots & \cdots & 1 \\ x_1 & \cdots & \cdots & \cdots & x_p \\ \vdots & & & & \vdots \\ x_1^{k-1} & \cdots & \cdots & \cdots & x_p^{k-1} \end{bmatrix}.$$

It is clear that $V_k V_k^t = \mathrm{Newt}_{p-k}(P)$. Now apply Binet-Cauchy formula, noting that, if $I \subset \{1, ..., p\}$, $\#(I) = k$, and V_{kI} is the $k \times k$ matrix obtained by extracting from V_k the columns with indices in I

$$\det(V_{kI}) = \prod_{\substack{(j,\ell) \in I \\ \ell > j}} (x_j - x_\ell),$$

by Lemma 4.11. □

4.2 Resultant and Subresultant Coefficients

4.2.1 Resultant

Let P and Q be two non-zero polynomials of degree p and q in $\mathrm{D}[X]$, where D is a ring. When D is a domain, its fraction field is denoted by K. Let

$$\begin{aligned} P &= a_p X^p + a_{p-1} X^{p-1} + \cdots + a_0, \\ Q &= b_q X^q + b_{q-1} X^{q-1} + \cdots + b_0. \end{aligned}$$

We define the Sylvester matrix associated to P and Q and the resultant of P and Q.

Notation 4.12. [Sylvester matrix] The **Sylvester matrix** of P and Q, denoted by $\mathrm{Syl}(P,Q)$, is the matrix

$$\begin{bmatrix} a_p & \cdots & \cdots & \cdots & \cdots & a_0 & 0 & \cdots & 0 \\ 0 & \ddots & & & & & \ddots & \ddots & \vdots \\ \vdots & \ddots & \ddots & & & & & \ddots & 0 \\ 0 & \cdots & 0 & a_p & \cdots & \cdots & \cdots & \cdots & a_0 \\ b_q & \cdots & \cdots & \cdots & b_0 & 0 & \cdots & \cdots & 0 \\ 0 & \ddots & & & & \ddots & \ddots & & \vdots \\ \vdots & \ddots & \ddots & & & & \ddots & \ddots & \vdots \\ \vdots & & \ddots & \ddots & & & & \ddots & 0 \\ 0 & \cdots & \cdots & 0 & b_q & \cdots & \cdots & \cdots & b_0 \end{bmatrix}.$$

It has $p+q$ columns and $p+q$ rows. Note that its rows are

$$X^{q-1}P, \ldots, P, X^{p-1}Q, \ldots, Q$$

considered as vectors in the basis $X^{p+q-1}, \ldots, X, 1$.

The **resultant** of P and Q, denoted $\mathrm{Res}(P,\ Q)$, is the determinant of $\mathrm{Syl}(P,Q)$. □

Remark 4.13. This matrix comes about quite naturally since it is the transpose of the matrix of the linear mapping $U,\ V \mapsto U\,P + V\,Q$, where $(U,\ V)$ is identified with

$$(u_{q-1}, \ldots, u_0, v_{p-1}, \ldots, v_0),$$

and $U = u_{q-1}X^{q-1} + \cdots + u_0$, $V = v_{p-1}X^{p-1} + \cdots + v_0$. □

The following lemma is clear from this remark.

Lemma 4.14. *Let* D *be a domain. Then* $\mathrm{Res}(P,\ Q) = 0$ *if and only if there exist non-zero polynomials* $U \in \mathrm{K}[X]$ *and* $V \in \mathrm{K}[X]$, *with* $\deg(U) < q$ *and* $\deg(V) < p$, *such that* $UP + VQ = 0$.

We can now prove the well-known proposition.

Proposition 4.15. *Let* D *be a domain. Then* $\mathrm{Res}(P,Q) = 0$ *if and only if* P *and* Q *have a common factor in* $\mathrm{K}[X]$.

Proof: The proposition is an immediate consequence of the preceding lemma and of Proposition 1.5, since the least common multiple of P and Q has degree $< p+q$ if and only if there exist non-zero polynomials U and V with $\deg(U) < q$ and $\deg(V) < p$ such that $UP + VQ = 0$. □

If D is a domain, with fraction field K, $a_p \neq 0$ and $b_q \neq 0$, the resultant can be expressed as a function of the roots of P and Q in an algebraically closed field C containing K.

Theorem 4.16. *Let*

$$\begin{aligned} P &= a_p \prod_{i=1}^{p} (X - x_i) \\ Q &= b_q \prod_{j=1}^{q} (X - y_j), \end{aligned}$$

in other words x_1, ..., x_p are the roots of P (counted with multiplicities) and $y_1, \ldots, y_q$ are the roots of Q (counted with multiplicities).

$$\mathrm{Res}(P, Q) = a_p^q b_q^p \prod_{i=1}^{p} \prod_{j=1}^{q} (x_i - y_j).$$

Proof: Let

$$\Theta(P, Q) = a_p^q b_q^p \prod_{i=1}^{p} \prod_{j=1}^{q} (x_i - y_j).$$

If P and Q have a root in common, $\mathrm{Res}(P, Q) = \Theta(P, Q) = 0$, and the theorem holds. So we suppose now that P and Q are coprime. The theorem is proved by induction on the length n of the remainder sequence of P and Q.

When $n = 2$, Q is a constant b, and $\mathrm{Res}(P, Q) = \Theta(P, Q) = b^p$.

The induction step is based on the following lemma.

Lemma 4.17. *Let R be the remainder of the Euclidean division of P by Q and let r be the degree of R. Then,*

$$\begin{aligned} \mathrm{Res}(P, Q) &= (-1)^{pq} b_q^{p-r} \mathrm{Res}(Q, R), \\ \Theta(P, Q) &= (-1)^{pq} b_q^{p-r} \Theta(Q, R). \end{aligned}$$

Proof of Lemma 4.17: Let $R = c_r X^r + \cdots + c_0$. Replacing the rows of coefficients of the polynomials $X^{q-1} P, \ldots, P$ by the rows of coefficients of the polynomials $X^{q-1} R, \ldots, R$ in the Sylvester matrix of P and Q gives the matrix

$$M = \begin{bmatrix} 0 & 0 & c_r & \cdots & \cdots & c_0 & 0 & \cdots & 0 \\ \vdots & & \ddots & \ddots & & & \ddots & \ddots & \vdots \\ \vdots & & & \ddots & \ddots & & & \ddots & 0 \\ 0 & \cdots & \cdots & \cdots & 0 & c_r & \cdots & \cdots & c_0 \\ b_q & \cdots & \cdots & \cdots & b_0 & 0 & \cdots & \cdots & 0 \\ 0 & \ddots & & & & \ddots & \ddots & & \vdots \\ \vdots & \ddots & \ddots & & & & \ddots & \ddots & \vdots \\ \vdots & & \ddots & \ddots & & & & \ddots & 0 \\ 0 & \cdots & \cdots & 0 & b_q & \cdots & \cdots & \cdots & b_0 \end{bmatrix}.$$

such that

$$\det(M) = \mathrm{Res}(P, Q).$$

Indeed,

$$R = P - \sum_{i=0}^{p-q} d_i (X^i Q),$$

where $C = \sum_{i=0}^{p-q} d_i X^i$ is the quotient of P in the euclidean division of P by Q, and adding to a row a multiple of other rows does not change the determinant.

Denoting by N the matrix whose rows are $X^{p-1} Q, \ldots, X^{r-1} Q, \ldots, Q, X^{q-1} R, \ldots, R$, we note that

$$N = \begin{bmatrix} b_q & \cdots & \cdots & \cdots & b_0 & 0 & \cdots & \cdots & 0 \\ 0 & b_q & & & & b_0 & \ddots & & \vdots \\ \vdots & 0 & b_q & \cdots & \cdots & \cdots & b_0 & \ddots & \vdots \\ \vdots & \vdots & \ddots & \ddots & & & & \ddots & 0 \\ \vdots & \vdots & \cdots & 0 & b_q & \cdots & \cdots & \cdots & b_0 \\ \vdots & \vdots & c_r & \cdots & \cdots & c_0 & 0 & \cdots & 0 \\ \vdots & \vdots & 0 & \ddots & & & \ddots & \ddots & \vdots \\ \vdots & \vdots & \vdots & \ddots & \ddots & & & \ddots & 0 \\ 0 & 0 & 0 & \cdots & 0 & c_r & \cdots & \cdots & c_0 \end{bmatrix}.$$

is obtained from M by exchanging the order of rows, so that

$$\det(N) = (-1)^{pq} \det(M).$$

It is clear, developing the determinant of N by its $p - r$ first columns, that

$$\det(N) = b_q^{p-r} \mathrm{Res}(Q, R).$$

On the other hand, since $P = CQ + R$, $P(y_j) = R(y_j)$ and

$$\Theta(P, Q) = a_p^q \prod_{i=1}^{p} Q(x_i) = (-1)^{pq} b_q^p \prod_{j=1}^{q} P(y_j),$$

we have

$$\begin{aligned} \Theta(P, Q) &= (-1)^{pq} b_q^p \prod_{j=1}^{q} P(y_j) \\ &= (-1)^{pq} b_q^p \prod_{j=1}^{q} R(y_j) \\ &= (-1)^{pq} b_q^{p-r} \Theta(Q, R). \end{aligned}$$

□ □

For any ring D, the following holds:

Proposition 4.18. *If $P, Q \in \mathrm{D}[X]$, then there exist $U, V \in \mathrm{D}[X]$ such that $\deg(U) < q$, $\deg(V) < p$, and $\mathrm{Res}(P, Q) = UP + VQ$.*

Proof: Let Syl(P, Q)* be the matrix whose first $p+q-1$ columns are the first $p+q-1$ first columns of Syl(P, Q) and such that the elements of the last column are the polynomials $X^{q-1}P,\ldots,P,X^{p-1}Q,\ldots,Q$. Using the linearity of det(Syl(P, Q)*) as a function of its last column it is clear that

$$\det(\mathrm{Syl}(P,Q)^*) = \mathrm{Res}(P,Q) + \sum_{j=1}^{p+q-1} d_j X^j,$$

where d_j is the determinant of the matrix Syl(P, Q)$_j$ whose first $p+q-1$ columns are the first $p+q-1$ columns of Syl(P, Q) and such that the last column is the $p+q-j$-th column of Syl(P, Q). Since Syl(P, Q)$_j$ has two identical columns, $d_j=0$ for $j=1,\ldots,p+q-1$ and

$$\det(\mathrm{Syl}(P,Q)^*) = \mathrm{Res}(P,Q).$$

Expanding the determinant of Syl(P, Q)* by its last column, we obtain the claimed identity. □

The Sylvester matrix and the resultant also have the following useful interpretation. Let C be an algebraically closed field. Identify a monic polynomial

$$X^q + b_{q-1}X^{q-1} + \cdots + b_0 \in \mathrm{C}[X]$$

of degree q with the point $(b_{q-1},\ldots,b_0) \in \mathrm{C}^q$. Let

$$\begin{array}{rcl} m\colon \mathrm{C}^q \times \mathrm{C}^p & \longrightarrow & \mathrm{C}^{q+p} \\ (Q,P) & \longmapsto & QP \end{array}$$

be the mapping defined by the multiplication of monic polynomials. The map m sends

$$(b_{q-1},\ldots,b_0,a_{p-1},\ldots,a_0)$$

to the vector whose entries are $(m_{p+q-1},\ldots,m_0)$, where

$$m_j = \sum_{q-i+p-k=j} b_{q-i}a_{p-k} \text{for } j=p+q-1,\ldots,0$$

(with $b_q=a_p=1$). The following proposition is thus clear:

Proposition 4.19. *The Jacobian matrix of m is the Sylvester matrix of P and Q and the Jacobian of m is the resultant.*

Finally, the definition of resultants as determinants implies that:

Proposition 4.20. *If P is monic,* $\deg(Q) \le \deg(P)$, *and* $f\colon \mathrm{D} \to \mathrm{D}'$ *is a ring homomorphism, then* $f(\mathrm{Res}(P,Q)) = \mathrm{Res}(f(P),f(Q))$ *(denoting by f the induced homomorphism from* D[X] *to* D′[X]*).*

4.2.2 Subresultant Coefficients

We now define the Sylvester-Habicht matrices and the signed subresultant coefficients of P and Q when $p = \deg(P) > q = \deg(Q)$.

Notation 4.21. [Sylvester-Habicht matrix] Let $0 \le j \le q$ The **j-th Sylvester-Habicht matrix** of P and Q, denoted $\mathrm{SyHa}_j(P,Q)$, is the matrix whose rows are $X^{q-j-1}P, \dots, P, Q, \dots, X^{p-j-1}Q$ considered as vectors in the basis $X^{p+q-j-1}, \dots, X, 1$:

$$\begin{bmatrix} a_p & \cdots & \cdots & \cdots & \cdots & a_0 & 0 & 0 \\ 0 & \ddots & & & & & \ddots & 0 \\ \vdots & \ddots & a_p & \cdots & \cdots & \cdots & \cdots & a_0 \\ \vdots & & 0 & b_q & \cdots & \cdots & \cdots & b_0 \\ \vdots & ⋰ & ⋰ & & & & ⋰ & 0 \\ 0 & ⋰ & & & & ⋰ & ⋰ & \vdots \\ b_q & \cdots & \cdots & \cdots & b_0 & 0 & \cdots & 0 \end{bmatrix}.$$

It has $p+q-j$ columns and $p+q-2j$ rows.

The **j-th signed subresultant coefficient** denoted $\mathrm{sRes}_j(P,Q)$ or sRes_j is the determinant of the square matrix $\mathrm{SyHa}_{j,j}(P,Q)$obtained by taking the first $p+q-2j$ columns of $\mathrm{SyHa}_j(P,Q)$.

By convention, we extend these definitions for $q < j \le p$ by

$$\begin{aligned} \mathrm{sRes}_p(P,Q) &= \mathrm{sign}(a_p), \\ \mathrm{sRes}_j(P,Q) &= 0, \; q < j < p. \end{aligned}$$

□

Remark 4.22. The matrix $\mathrm{SyHa}_j(P,Q)$ comes about quite naturally since it is the transpose of the matrix of the mapping $U, V \mapsto UP + VQ$, where (U,V) is identified with

$$(u_{q-j-1}, \dots, u_0, v_0, \dots, v_{p-j-1}),$$

with $U = u_{q-j-1}X^{q-j-1} + \dots + u_0$, $V = v_{p-j-1}X^{p-j-1} + \dots + v_0$.

The peculiar order of rows is adapted to the real root counting results presented later, in Chapter 8. □

The following lemma is clear from this remark:

Lemma 4.23. *Let* D *be a domain and* $0 \le j \le \min(p,\, q)$ *if* $p \ne q$ *(resp.* $0 \le j \le p-1$ *if* $p = q$*). Then* $\mathrm{sRes}_j(P,\, Q) = 0$ *if and only if there exist non-zero polynomials* $U \in \mathrm{K}[X]$ *and* $V \in \mathrm{K}[X]$*, with* $\deg(U) < q - j$ *and* $\deg(V) < p - j$*, such that* $\deg(UP + VQ) < j$.

The following proposition will be useful for the cylindrical decomposition in Chapter 5.

Proposition 4.24. *Let* D *be a domain and* $0 \le j \le \min(p, q)$ *if* $p \ne q$ *(resp.* $0 \le j \le p-1$ *if* $p = q$*). Then* $\deg(\gcd(P, Q)) \ge j$ *if and only if*

$$\mathrm{sRes}_0(P, Q) = \cdots = \mathrm{sRes}_{j-1}(P, Q) = 0.$$

Proof: Suppose that $\deg(\gcd(P, Q)) \ge j$. Then, the least common multiple of P and Q,

$$\mathrm{lcm}(P, Q) = \frac{PQ}{\gcd(P, Q)}$$

(see Proposition 1.5) has degree $\le p + q - j$. This is clearly equivalent to the existence of polynomials U and V, with $\deg(U) \le q - j$ and $\deg(V) \le p - j$, such that $U\,P = -V\,Q = \mathrm{lcm}(P,\ Q)$. Or, equivalently, that there exist polynomials U and V with $\deg(U) \le q - j$ and $\deg(V) \le p - j$ such that $UP + VQ = 0$. This implies that

$$\mathrm{sRes}_0 = \cdots = \mathrm{sRes}_{j-1} = 0$$

using Lemma 4.23.

The reverse implication is proved by induction on j. If $j = 1$, $\mathrm{sRes}_0 = 0$ implies, using Lemma 4.23, that there exist U and V with $\deg(U) < q$ and $\deg(V) < p$ satisfying $UP + VQ = 0$. Hence $\deg(\gcd(P, Q)) \ge 1$. If

$$\mathrm{sRes}_0(P, Q) = \cdots = \mathrm{sRes}_{j-2}(P, Q) = 0,$$

the induction hypothesis implies that $\deg(\gcd(P, Q)) \ge j - 1$. If in addition $\mathrm{sRes}_{j-1} = 0$ then, by Lemma 4.23, there exist U and V with $\deg(U) \le q - j$ and $\deg(V) \le p - j$ such that $\deg(UP + VQ) < j - 1$. Since the greatest common divisor of P and Q divides $UP + VQ$ and has degree $\ge j - 1$, we have $UP + VQ = 0$, which implies that $\deg(\mathrm{lcm}(P, Q)) \le p + q - j$ and hence $\deg(\gcd(P, Q)) \ge j$. □

The following consequence is clear, using Lemma 4.23 and Proposition 4.24.

Proposition 4.25. *Let* D *be a domain and* $0 \le j \le \min(p, q)$ *if* $p \ne q$ *(resp.* $0 \le j \le p-1$ *if* $p = q$*). Then* $\deg(\gcd(P, Q)) = j$ *if and only if*

$$\mathrm{sRes}_0(P, Q) = \cdots = \mathrm{sRes}_{j-1}(P, Q) = 0, \mathrm{sRes}_j(P, Q) \ne 0.$$

Notation 4.26. [Reversing rows] We denote by ε_i the signature of the permutation reversing the order of i consecutive rows in a matrix, i.e. $\varepsilon_i = (-1)^{i(i-1)/2}$. For every natural number $i \ge 1$,

$$\varepsilon_{4i} = 1, \varepsilon_{4i-1} = -1, \varepsilon_{4i-2} = -1, \varepsilon_{4i-3} = 1. \qquad (4.2)$$

In particular, $\varepsilon_{i-2j} = (-1)^j \varepsilon_i$. □

Thus, it is clear from the definitions that

$$\mathrm{sRes}_0(P, Q) = \varepsilon_p \mathrm{Res}(P, Q). \qquad (4.3)$$

Note that, as a consequence Proposition 4.15 is a special case of Proposition 4.24.

Let us make the connection between subresultant coefficients and subdiscriminants.

We first define subdiscriminants of non-monic polynomials. Let

$$\begin{aligned} P &= a_p X^p + \cdots + a_0, \\ \mathrm{sDisc}_{p-k}(P) &= a_p^{2k-2} \mathrm{sDisc}_{p-k}(P/a_p) \\ &= a_p^{2k-2} \sum_{\substack{I \subset \{1,\ldots,p\} \\ \#(I)=k}} \prod_{(j,\ell)\in I, \ell>j} (x_j - x_\ell)^2 \end{aligned}$$

Proposition 4.27.

$$a_p \,\mathrm{sDisc}_{p-k}(P) = \mathrm{sRes}_{p-k}(P, P'). \tag{4.4}$$

Proof: Indeed if

$$D_k = \begin{bmatrix} 1 & 0 & \cdots & 0 & 0 & \cdots & \cdots & \cdots & 0 \\ 0 & 1 & \ddots & \vdots & \vdots & & & & \vdots \\ \vdots & \ddots & \ddots & 0 & \vdots & & & & \vdots \\ \vdots & & \ddots & 1 & 0 & \cdots & \cdots & \cdots & 0 \\ 0 & \cdots & \cdots & 0 & N_0 & N_1 & \cdots & \cdots & N_{k-1} \\ \vdots & & ⋰ & ⋰ & ⋰ & & & ⋰ & \vdots \\ \vdots & ⋰ & ⋰ & ⋰ & & & ⋰ & & \vdots \\ 0 & N_0 & N_1 & & & N_{k-1} & & & N_{2k-3} \\ N_0 & N_1 & \cdots & \cdots & N_{k-1} & \cdots & \cdots & \cdots & N_{2k-2} \end{bmatrix},$$

and

$$D_k' = \begin{bmatrix} a_p & \cdots & \cdots & \cdots & \cdots & \cdots & \cdots & \cdots & a_{p-2k+2} \\ 0 & a_p & \ddots & & & & & & \vdots \\ \vdots & \ddots & \ddots & \ddots & & & & & \vdots \\ \vdots & & 0 & a_p & \ddots & & & & \vdots \\ \vdots & & & 0 & a_p & \ddots & & & \vdots \\ \vdots & & & & \ddots & \ddots & \ddots & & \vdots \\ \vdots & & & & & \ddots & \ddots & \ddots & \vdots \\ \vdots & & & & & & \ddots & a_p & \vdots \\ 0 & \cdots & \cdots & \cdots & \cdots & \cdots & \cdots & 0 & a_p \end{bmatrix},$$

it is a easy to see that $\mathrm{SyHa}_{p-k,p-k}(P, P') = D_k \cdot D_k'$, using the relations (4.1). Since $\det(D_k') = a_p^{2k-1}$,

$$\begin{aligned} \det(\mathrm{SyHa}_{p-k,p-k}(P, P')) &= a_p^{2k-1} \mathrm{sDisc}_{p-k}(P/a_p) \\ &= a_p \,\mathrm{sDisc}_{p-k}(P). \\ \det(\mathrm{SyHa}_{p-k,p-k}(P, P')) &= a_p^{2k-1} \mathrm{sDisc}_{p-k}(P/a_p) \\ &= a_p \,\mathrm{sDisc}_{p-k}(P). \end{aligned}$$

On the other hand $\det(D_k) = \mathrm{sRes}_{p-k}(P, P')$. The claim follows by Proposition 4.18. □

Remark 4.28. Note that if $P \in \mathrm{D}[X]$, then $\mathrm{sDisc}_i(P) \in \mathrm{D}$ for every $i \leq p$. □

Proposition 4.29. *Let* D *be a domain. Then* $\deg(\gcd(P, P')) = j$, $0 \leq j < p$ *if and only if*

$$\mathrm{sDisc}_0(P) = \ldots = \mathrm{sDisc}_{j-1}(P) = 0, \mathrm{sDisc}_j(P) \neq 0.$$

Proof: Follows immediately from Proposition 4.27 and Proposition 4.25 □

4.2.3 Subresultant Coefficients and Cauchy Index

We indicate how to compute the Cauchy index by using only the signed subresultant coefficients. We need a definition:

Notation 4.30. [Generalized Permanences minus Variations] Let $s = s_p, \ldots, s_0$ be a finite list of elements in an ordered field K such that $s_p \neq 0$. Let $q < p$ such that $s_{p-1} = \cdots = s_{q+1} = 0$, and $s_q \neq 0$, and $s' = s_q, \ldots, s_0$. (if there exist no such q, s' is the empty list). We define inductively

$$\mathrm{PmV}(s) = \begin{cases} 0 & \text{if } s' = \emptyset, \\ \mathrm{PmV}(s') + \varepsilon_{p-q}\,\mathrm{sign}(s_p\, s_q) & \text{if } p-q \text{ is odd,} \\ \mathrm{PmV}(s') & \text{if } p-q \text{ is even.} \end{cases}$$

where $\varepsilon_{p-q} = (-1)^{(p-q)(p-q-1)/2}$, using Notation 4.26.

Note that when all elements of s are non-zero, $\mathrm{PmV}(s)$ is the difference between the number of sign permanence and the number of sign variations in $s_p, \ldots, s_0$. Note also that when s is the sequence of coefficients of polynomials $\mathcal{P} = P_p, \ldots, P_0$ with $\deg(P_i) = i$, then

$$\mathrm{PmV}(s) = \mathrm{Var}(\mathcal{P}; -\infty, +\infty)$$

(see Notation 2.32). □

Let P and Q be two polynomials with:

$$\begin{aligned} P &= a_p X^p + a_{p-1} X^{p-1} + \cdots + a_0 \\ Q &= b_{p-1} X^{p-1} + \cdots + b_0, \end{aligned}$$

$\deg(P) = p, \deg(Q) = q \leq p-1$.

We denote by $\mathrm{sRes}(P, Q)$ the sequence of $\mathrm{sRes}_j(P, Q)$, $j = p, \ldots, 0$.

Theorem 4.31. $\mathrm{PmV}(\mathrm{sRes}(P, Q)) = \mathrm{Ind}(Q/P)$.

Before proving Theorem 4.31 let us list some of its consequences.

Theorem 4.32. *Let P and Q be polynomials in* $\mathrm{D}[X]$ *and R the remainder of $P'Q$ and P. Then* $\mathrm{PmV}(\mathrm{sRes}(P, R)) = \mathrm{TaQ}(Q, P)$.

Proof: Apply Theorem 4.31 and Proposition 2.57, since

$$\mathrm{Ind}(P'Q/P) = \mathrm{Ind}(R/P)$$

by Remark 2.55. □

Theorem 4.33. *Let P be a polynomial in $\mathrm{D}[X]$. Then*

$$\mathrm{PmV}(\mathrm{sDisc}_{p-1}(P), \ldots, \mathrm{sDisc}_0(P))$$

is the number of roots of P in R.

Proof: Apply Theorem 4.31 and Proposition 4.27. □

The proof of Theorem 4.31 uses the following two lemmas.

Lemma 4.34.

$$\mathrm{Ind}(Q/P) = \begin{cases} \mathrm{Ind}(-R/Q) + \mathrm{sign}(a_p b_q) & \text{if } p\text{-}q \text{ is odd,} \\ \mathrm{Ind}(-R/Q) & \text{if } p-q \text{ is even.} \end{cases}$$

Proof: The claim is an immediate consequence of Lemma 2.60. □

Lemma 4.35.

$$\mathrm{PmV}(\mathrm{sRes}(P,Q)) = \begin{cases} \mathrm{PmV}(\mathrm{sRes}(Q,-R)) + \mathrm{sign}(a_p b_q) & \text{if } p-q \text{ is odd,} \\ \mathrm{PmV}(\mathrm{sRes}(Q,-R)) & \text{if } p-q \text{ is even.} \end{cases}$$

The proof of Lemma 4.35 is based on the following proposition.

Proposition 4.36. *Let r be the degree of $R = \mathrm{Rem}(P, Q)$.*

$$\mathrm{sRes}_j(P,Q) \;=\; \varepsilon_{p-q} b_q^{p-r}\, \mathrm{sRes}_j(Q,-R) \;\; \text{if } j \le r,$$

where $\varepsilon_i = (-1)^{i(i-1)/2}$.

Moreover, $\mathrm{sRes}_j(P,Q) = \mathrm{sRes}_j(Q,-R) = 0$ *if f* $r < j < q$.

Proof: Replacing the polynomials $X^{q-j-1}P, \ldots, P$ by the polynomials $X^{q-j-1}R, \ldots, R$ in $\mathrm{SyHa}_j(P, Q)$ does not modify the determinant based on the $p+q-2j$ first columns. Indeed,

$$R = P - \sum_{i=0}^{p-q} c_i X^i Q,$$

where $C = \sum_{i=0}^{p-q} c_i X^i$ is the quotient of P in the euclidean division of P by Q, and adding to a polynomial of a sequence a multiple of another polynomial of the sequence does not change the determinant based on the $p+q-2j$ first columns.

Reversing the order of the polynomials multiplies the determinant based on the $p+q-2j$ first columns. by ε_{p+q-2j}. Replacing R by $-R$ multiplies the determinant based on the $p+q-2j$ first columns by $(-1)^{q-j}$, and

$$(-1)^{q-j}\varepsilon_{p+q-2j}=\varepsilon_{p-q}$$

(see Notation 4.26). Denoting by D_j the determinant obtained by taking the $p+q-2j$ first columns of the matrix the rows corresponding to the coefficients of $X^{p-j-1}Q,\ldots,Q,-R,\ldots,-X^{q-j-1}R$,

$$\mathrm{sRes}_j(P,Q)=\varepsilon_{p-q}D_j.$$

If $j\leq r$, it is clear that

$$D_j=b_q^{p-r}\,\mathrm{sRes}_j(Q,-R).$$

If $r<j<q$, it is clear that

$$D_j=\mathrm{sRes}_j(P,Q)=\mathrm{sRes}_j(Q,-R)=0.$$

using the convention in Notation 4.20 and noting that the $q-j$-th row of the determinant D_j is null. □

Proof of Lemma 4.35: Using Proposition 4.36,

$$\mathrm{PmV}(\mathrm{sRes}_r(P,Q),\ldots,\mathrm{sRes}_0(P,Q))=\mathrm{PmV}(\mathrm{sRes}_r(Q,-R),\ldots,\mathrm{sRes}_0(Q,-R)).$$

If $q-r$ is even

$$\begin{aligned}
&\mathrm{PmV}(\mathrm{sRes}_q(P,Q),\ldots,\mathrm{sRes}_0(P,Q))\\
=\;&\mathrm{PmV}(\mathrm{sRes}_r(P,Q),\ldots,\mathrm{sRes}_0(P,Q))\\
=\;&\mathrm{PmV}(\mathrm{sRes}_r(Q,-R),\ldots,\mathrm{sRes}_0(Q,-R))\\
=\;&\mathrm{PmV}(\mathrm{sRes}_q(Q,-R),\ldots,\mathrm{sRes}_0(Q,-R)).
\end{aligned}$$

If $q-r$ is odd, since

$$\begin{aligned}
\mathrm{sRes}_q(P,Q) &= \varepsilon_{p-q}b_q^{p-q},\\
\mathrm{sRes}_q(Q,-R) &= \mathrm{sign}(b_q),\\
\mathrm{sRes}_r(P,Q) &= \varepsilon_{p-q}b_q^{p-r}\mathrm{sRes}_r(Q,-R),
\end{aligned}$$

denoting $d_r=\mathrm{sRes}_r(Q,-R)$,

$$\begin{aligned}
&\mathrm{PmV}(\mathrm{sRes}_q(P,Q),\ldots,\mathrm{sRes}_0(P,Q))\\
=\;&\mathrm{PmV}(\mathrm{sRes}_r(P,Q),\ldots,\mathrm{sRes}_0(P,Q))+\varepsilon_{q-r}\mathrm{sign}(b_q d_r)\\
=\;&\mathrm{PmV}(\mathrm{sRes}_q(Q,-R),\ldots,\mathrm{sRes}_0(Q,-R)).
\end{aligned}$$

Thus in all cases

$$\begin{aligned}
&\mathrm{PmV}(\mathrm{sRes}_q(P,Q),\ldots,\mathrm{sRes}_0(P,Q))\\
=\;&\mathrm{PmV}(\mathrm{sRes}_q(Q,-R),\ldots,\mathrm{sRes}_0(Q,-R)).
\end{aligned}$$

If $p-q$ is even

$$\begin{aligned} & \mathrm{PmV}(\mathrm{sRes}_p(P,Q),\ldots,\mathrm{sRes}_0(P,Q)) \\ = & \mathrm{PmV}(\mathrm{sRes}_q(P,Q),\ldots,\mathrm{sRes}_0(P,Q)) \\ = & \mathrm{PmV}(\mathrm{sRes}_q(Q,-R),\ldots,\mathrm{sRes}_0(Q,-R)). \end{aligned}$$

If $p-q$ is odd, since

$$\begin{aligned} \mathrm{sRes}_p(P,Q) &= \mathrm{sign}(a_p), \\ \mathrm{sRes}_q(P,Q) &= \varepsilon_{p-q} b_q^{p-q}, \end{aligned}$$

$$\begin{aligned} & \mathrm{PmV}(\mathrm{sRes}_p(P,Q),\ldots,\mathrm{sRes}_0(P,Q)) \\ = & \mathrm{PmV}(\mathrm{sRes}_q(P,Q),\ldots,\mathrm{sRes}_0(P,Q)) + \mathrm{sign}(a_p b_q) \\ = & \mathrm{PmV}(\mathrm{sRes}_q(Q,-R),\ldots,\mathrm{sRes}_0(Q,-R)) + \mathrm{sign}(a_p b_q). \end{aligned}$$

□

Proof of Theorem 4.31: The proof proceeds by induction on the number n of elements with distinct degrees in the signed subresultant sequence.

If $n=2$, Q divides P. We have

$$\mathrm{Ind}(Q/P) = \begin{cases} \mathrm{sign}(a_p b_q) & \text{if } p\text{-q is odd,} \\ 0 & \text{if } p-q \text{ is even.} \end{cases}$$

by Lemma 4.34 and

$$\mathrm{PmV}(\mathrm{sRes}(P,Q)) = \begin{cases} \mathrm{sign}(a_p b_q) & \text{if } p\text{-q is odd,} \\ 0 & \text{if } p-q \text{ is even.} \end{cases}$$

by Lemma 4.35.

Let us suppose that the theorem holds for $n-1$ and consider P and Q such that their signed subresultant sequence has n elements with distinct degrees. The signed subresultant sequence of Q and $-R$ has $n-1$ elements with distinct degrees. By the induction hypothesis,

$$\mathrm{PmV}(\mathrm{sRes}(Q,-R)) = \mathrm{Ind}(-R/Q).$$

So, by Lemma 4.34 and Lemma 4.35,

$$\mathrm{PmV}(\mathrm{sRes}(P,Q)) = \mathrm{Ind}(Q/P).$$

□

Example 4.37. Consider again $P = X^4 + a X^2 + b X + c$,

$$\begin{aligned} \mathrm{sDisc}_3(P) &= 4, \\ \mathrm{sDisc}_2(P) &= -8a, \\ \mathrm{sDisc}_1(P) &= 4(8ac - 9b^2 - 2a^3) \\ \mathrm{sDisc}_0(P) &= 256c^3 - 128a^2c^2 + 144ab^2c + 16a^4c - 27b^4 - 4a^3b^2. \end{aligned}$$

As in Example 1.15, let

$$
\begin{aligned}
s &= 8\,a\,c-9\,b^2-2\,a^3,\\
\delta &= 256\,c^3-128\,a^2\,c^2+144\,a\,b^2\,c+16\,a^4\,c-27\,b^4-4\,a^3\,b^2.
\end{aligned}
$$

We indicate in the following tables the number of real roots of P (computed using Theorem 4.31) in the various cases corresponding to all the possible signs for a, s, δ:

1	+	+	+	+	+	+	+	+	+
4	+	+	+	+	+	+	+	+	+
$-a$	+	+	+	+	+	+	+	+	+
s	+	+	+	−	−	−	0	0	0
δ	+	−	0	+	−	0	+	−	0
n	4	2	3	0	2	1	2	2	2

1	+	+	+	+	+	+	+	+	+
4	+	+	+	+	+	+	+	+	+
$-a$	−	−	−	−	−	−	−	−	−
s	+	+	+	−	−	−	0	0	0
δ	+	−	0	+	−	0	+	−	0
n	0	−2	−1	0	2	1	0	0	0

1	+	+	+	+	+	+	+	+	+
4	+	+	+	+	+	+	+	+	+
$-a$	0	0	0	0	0	0	0	0	0
s	+	+	+	−	−	−	0	0	0
δ	+	−	0	+	−	0	+	−	0
n	2	0	1	0	2	1	0	2	1

Note that when $a=s=0$, according to the definition of PmV when there are two consecutive zeroes,

$$
\begin{cases}
\mathrm{PmV}(\mathrm{sRes}(P,P'))=0 & \text{if } \delta>0\\
\mathrm{PmV}(\mathrm{sRes}(P,P'))=2 & \text{if } \delta<0\\
\mathrm{PmV}(\mathrm{sRes}(P,P'))=1 & \text{if } \delta=0.
\end{cases}
$$

Notice that the only sign conditions on a, s, δ for which all the roots of P are real is $a<0$, $s>0$, $\delta>0$, according to Corollary 9.8. Remark that, according to Corollary 4.3, when $\delta<0$ there are always two distinct real roots. This looks incompatible with the tables we just gave. In fact, the sign conditions with $\delta<0$ giving a number of real roots different from 2, and the sign conditions with $\delta>0$ giving a number of real roots equal to 2 have empty realizations.

We represent in Figure 4.1 the set of polynomials of degree 4 in the plane $a=-1$ and the zero sets of s, δ.

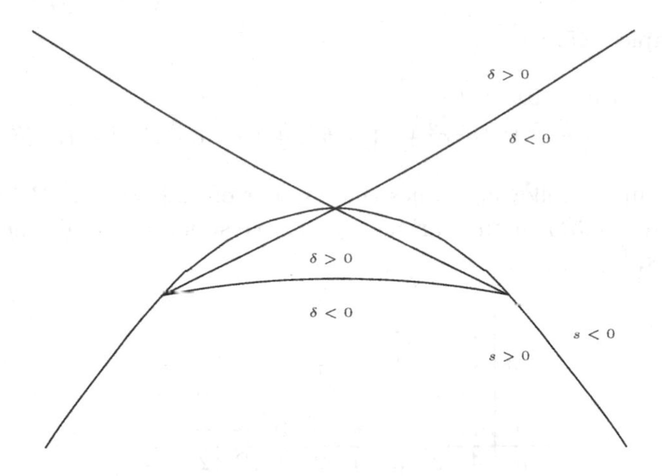

Fig. 4.1. $a = -1, s = \delta = 0$

Finally, in Figure 4.2 we represent the set of polynomials of degree 4 in a, b, c space and the zero sets of s, δ.

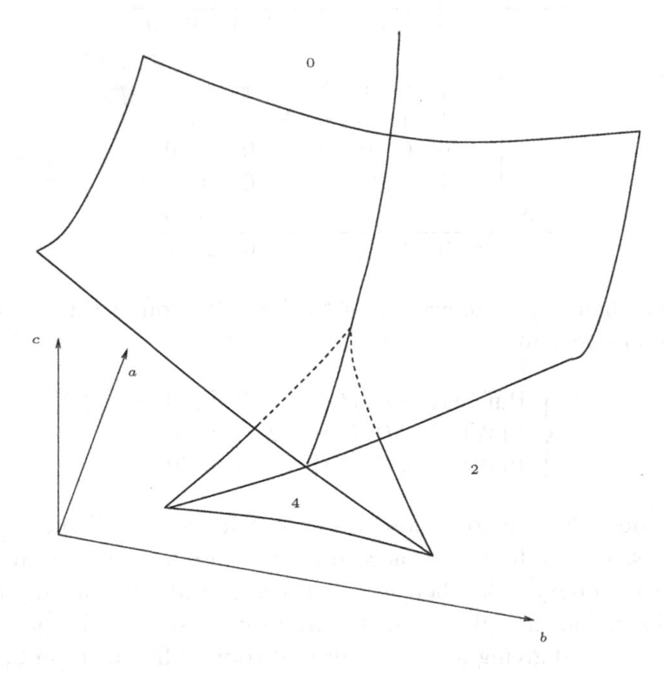

Fig. 4.2. The set defined by $\delta = 0$ and the different regions labelled by the number of real roots

Exercise 4.1. Find all sign conditions on a, s, δ with non-empty realizations.

As a consequence, the formula $\exists X \quad X^4 + a X^2 + b X + c = 0$ is equivalent to the quantifier-free formula

$$\begin{array}{c}(a<0 \wedge s \geq 0 \wedge \delta > 0) \vee \\ (a<0 \wedge \delta \leq 0) \vee \\ (a>0 \wedge s<0 \wedge \delta \leq 0) \vee \\ (a=0 \wedge s>0 \wedge \delta \geq 0) \vee \\ (a=0 \wedge s \leq 0 \wedge \delta \leq 0).\end{array}$$

collecting all sign conditions giving $n \geq 1$. It can be checked easily that the realization of the sign conditions $(a=0 \wedge s>0 \wedge \delta \geq 0)$ and $(a<0 \wedge s=0 \wedge \delta > 0)$ are empty. So that $(\exists X)\ X^4 + a X^2 + b X + c = 0$ is finally equivalent to

$$\begin{array}{c}(a<0 \wedge s>0 \wedge \delta > 0) \vee \\ (a<0 \wedge \delta \leq 0) \vee \\ (a>0 \wedge s<0 \wedge \delta \leq 0) \vee \\ (a=0 \wedge s \leq 0 \wedge \delta \leq 0).\end{array}$$

It is interesting to compare this result with Example 2.63: the present description is more compact and involves only sign conditions on the principal subresultants a, s, δ. □

4.3 Quadratic Forms and Root Counting

4.3.1 Quadratic Forms

The **transpose** of an $n \times m$ matrix $A = [a_{i,j}]$ is the $m \times n$ matrix $A^t = [b_{j,i}]$ defined by $b_{j,i} = a_{i,j}$. A square matrix A is **symmetric** if $A^t = A$.

A **quadratic form** with coefficients in a field K of characteristic 0 is a homogeneous polynomial of degree 2 in a finite number of variables of the form

$$\Phi(f_1, \ldots, f_n) = \sum_{i,j=1}^{n} m_{i,j} f_i f_j$$

with $M = [m_{i,j}]$ a symmetric matrix of size n. If $f = (f_1, \ldots, f_n)$, then $\Phi = f \cdot M \cdot f^t$, where f^t is the transpose of f. The **rank** of Φ, denoted by $\mathrm{Rank}(\Phi(f_1, \ldots, f_n))$, is the rank of the matrix M.

A **diagonal expression** of the quadratic form $\Phi(f_1, \ldots, f_n)$ is an identity

$$\Phi(f_1, \ldots, f_n) = \sum_{i=1}^{r} c_i L_i(f_1, \ldots, f_n)^2$$

with $c_i \in K, c_i \neq 0$ and the $L_i(f_1, \ldots, f_n)$ are linearly independent linear forms with coefficients in K. The elements c_i, $i = 1, \ldots, r$ are the **coefficients** of the diagonal expression. Note that $r = \mathrm{Rank}(\Phi(f_1, \ldots, f_n))$.

Theorem 4.38. [Sylvester's law of inertia]

- *A quadratic form $\Phi(f_1, \ldots, f_n)$ of dimension n has always a diagonal expression.*
- *If K is ordered, the difference between the number of positive coefficients and the number of negative coefficients in a diagonal expression of $\Phi(f_1, \ldots, f_n)$ is a well defined quantity.*

Proof: Let $\Phi(f_1, \ldots, f_n) = \sum_{i,j=1}^{n} m_{i,j} f_i f_j$.

The first claim is proved by induction on n. The result is obviously true if $n=1$. It is also true when $M=0$.

If some diagonal entry $m_{i,i}$ of M is not zero, we can suppose without loss of generality (reordering the variables) that $m_{n,n}$ is not 0. Take

$$L(f_1, \ldots, f_n) = \sum_{k=1}^{n} m_{k,n} f_k.$$

The quadratic form

$$\Phi(f_1, \ldots, f_n) - \frac{1}{m_{n,n}} L(f_1, \ldots, f_n)^2$$

does not depend on the variable f_n, and we can apply the induction hypothesis to

$$\Phi_1(f_1, \ldots, f_{n-1}) = \Phi(f_1, \ldots, f_n) - \frac{1}{m_{n,n}} L(f_1, \ldots, f_n)^2.$$

Since $L(f_1, \ldots, f_n)$ is a linear form containing f_n, it is certainly linearly independent from the linear forms in the decomposition of $\Phi_1(f_1, \ldots, f_{n-1})$.

If all diagonal entries are equal to 0, but $M \neq 0$, we can suppose without loss of generality (reordering the variables) that $m_{n-1,n} \neq 0$. Performing the linear change of variable

$$\begin{aligned} g_i &= f_i, 1 \leq i \leq n-2, \\ g_{n-1} &= \frac{f_n + f_{n-1}}{2}, \\ g_n &= \frac{f_n - f_{n-1}}{2}, \end{aligned}$$

we get

$$\Phi(g_1, \ldots, g_n) = \sum_{i,j=1}^{n} r_{i,j} g_i g_j$$

with $r_{n,n} = 2 m_{n,n-1} \neq 0$, so we are in the situation where some diagonal entry is not zero, and we can apply the preceding transformation.

So we have decomposed

$$\Phi(f_1, \ldots, f_n) = \sum_{i=1}^{r} c_i L_i(f_1, \ldots, f_n)^2,$$

where r is the rank of M, and the $L_i(f_1, \ldots, f_n)$'s are linearly independent linear forms, since the rank of M and the rank of the diagonal matrix with entries c_i are equal.

For the second claim, suppose that we have a second diagonal expression

$$\Phi(f_1, \ldots, f_n) = \sum_{i=1}^{r} c_i' L_i'(f_1, \ldots, f_n)^2,$$

with $c_i' \neq 0$, and the $L_i'(f_1, \ldots, f_n)$ are linearly independent forms, and, without loss of generality, assume that

$$\begin{aligned} &c_1 > 0, \ldots, c_s > 0, c_{s+1} < 0, \ldots, c_r < 0, \\ &c_1' > 0, \ldots, c_{s'}' > 0, c_{s'+1}' < 0, \ldots, c_r' < 0, \end{aligned}$$

with $0 \leq s \leq s' \leq r$. If $s' > s$, choose values of $f = (f_1, \ldots, f_n)$ such that the values at f of the $r - (s' - s)$ forms

$$L_1(f), \ldots, L_s(f), L_{s'+1}'(f), \ldots, L_r'(f)$$

are zero and the value at f of one of the forms

$$L_{s+1}(f), \ldots, L_r(f)$$

is not zero.

To see that this is always possible observe that the vector subspace V_1 defined by

$$L_1(f) = \cdots = L_s(f) = L_{s+1}'(f) = \cdots = L_r'(f) = 0$$

has dimension $\geq n - r + s' - s > n - r$, while the vector subspace V_2 defined by

$$L_1(f) = \cdots = L_s(f) = L_{s+1}(f) = \cdots = L_r(f) = 0$$

has dimension $n - r$, since the linear forms $L_i(f)$ are linearly independent, and thus there is a vector $f = (f_1, \ldots, f_n) \in V_1 \setminus V_2$ which satisfies

$$L_1(f) = \cdots = L_s(f) = 0,$$

and $L_i(f) \neq 0$ for some i, $s < i \leq r$.

For this value of $f = (f_1, \ldots, f_n)$, $\sum_{i=1}^{r} c_i L_i(f)^2$ is strictly negative while $\sum_{i=1}^{r} c_i' L_i'(f)^2$ is non-negative. So the hypothesis $s' > s$ leads to a contradiction. □

If K is ordered, the **signature** of Φ, $\mathrm{Sign}(\Phi)$, is the difference between the numbers of positive c_i and negative c_i in its diagonal form.

The preceding theorem immediately implies

Corollary 4.39. *There exists a basis B such that, denoting also by B the matrix of B in the canonical basis,*

$$B D B^t = M$$

where D is a diagonal matrix with r_+ positive entries, r_- negative entries, with $\mathrm{Rank}(\Phi) = r_+ + r_-$, $\mathrm{Sign}(\Phi) = r_+ - r_-$.

Let R be a real closed field. We are going to prove that a symmetric matrix with coefficients in R can be diagonalized in a basis of orthogonal vectors.

We denote by $u \cdot u'$ the **inner product** of vectors of R^n

$$u \cdot u' = \sum_{k=1}^{n} u_k u'_k,$$

where $u = (u_1, \ldots, u_n)$, $u' = (u'_1, \ldots, u'_n)$. The **norm** of u is $\|u\| = \sqrt{u.u}$.Two vectors u and u' are **orthogonal** if $u \cdot u' = 0$.

A basis $v_1, \ldots, v_n$ of vectors of R^n is orthogonal if

$$v_i \cdot v_j = \sum_{k=1}^{n} v_{i,k} v_{k,j} = 0$$

for all $i = 1, \ldots, n$, $j = 1, \ldots, n$, $j \neq i$.

A basis $v_1, \ldots, v_n$ of vectors of R^n is **orthonormal** if is is orthogonal and morevoer $\|u\| = 1$, for all $i = 1, \ldots, n$.

Two linear forms

$$L = \sum_{i=1}^{n} u_i f_i, \; L' = \sum_{i=1}^{n} u'_i f_i$$

are orthogonal if $u \cdot u' = 0$.

We first describe the Gram-Schmidt orthogonalization process.

Proposition 4.40. [Gram-Schmidt orthogonalization] *Let v_1, ..., v_n be linearly independent vectors with coefficients in* R. *There is a family of linearly independent orthogonal vectors $w_1, \ldots, w_n$ with coefficients in* R *such that for every $i = 1, \ldots, n$, $w_i - v_i$ belong to the vector space spanned by $v_1, \ldots$, v_{i-1}.*

Proof: The construction proceeds by induction, starting with $w_1 = v_1$ and continuing with

$$w_i = v_i - \sum_{j=1}^{i-1} \mu_{i,j} w_j,$$

where

$$\mu_{i,j} = \frac{v_i \cdot w_j}{\|w_j\|^2}.$$

□

Let M be a symmetric matrix of dimension n with entries in R.

If $f = (f_1, \ldots, f_n)$, $g = (g_1, \ldots, g_n)$, let

$$\begin{aligned} \Phi_M(f) &= f \cdot M \cdot f^t, \\ B_M(f, g) &= g \cdot M \cdot f^t, \\ u_M(f) &= M \cdot f. \end{aligned}$$

The quadratic form Φ is **non-negative** if for every $f \in \mathrm{R}^n$, $\Phi_M(f) \geq 0$.

Proposition 4.41. [Cauchy-Schwarz inequality] *If Φ is non-negative,*

$$B_M(f,g)^2 \leq \Phi_M(f)\,\Phi_M(g).$$

Proof: Fix f and g and consider the second degree polynomial

$$P(T) = \Phi_M(f + Tg) = \Phi_M(f) + 2TB_M(f,g) + T^2\,\Phi_M(g).$$

For every $t \in \mathrm{R}$, $P(t)$ is non-negative since Φ_M is non-negative. So P can be

- of degree 0 if $\Phi_M(g) = B_M(f,g) = 0$, in this case the inequality claimed holds
- of degree 2 with negative discriminant if $\Phi_M(g) \neq 0$. Since the discriminant of P is

$$4B_M(f,g)^2 - 4\Phi_M(f)\,\Phi_M(g),$$

the inequality claimed holds in this case too. □

Our main objective in the end of the section is to prove the following result.

Theorem 4.42. *Let M be a symmetric matrix with entries in* R. *The eigenvalues of M are in* R, *and there is an orthonormal basis of eigenvectors for M with coordinates in* R.

As a consequence, since positive elements in R are squares, there exists an orthogonal basis B such that,denoting also by B the matrix of B in the canonical basis,

$$B\,D\,B^t = M$$

where D is the diagonal matrix with r_+ entries 1, r_- entries -1, and $n-r$ entries 0, $r = r_+ + r_-$:

Corollary 4.43. *A quadratic form Φ with coefficients in* R *can always be written as*

$$\Phi = \sum_{i=1}^{r_+} L_i^2 - \sum_{i=r_++1}^{r_++r_-} L_i^2$$

where the L_i are independent orthogonal linear forms with coefficients in R, *and $r = r_+ + r_-$ is the rank of Φ.*

Corollary 4.44. *Let r_+, r_-, and r_0 be the number of >0, <0, and $=0$ eigenvalues of the symmetric matrix associate to the quadratic form Φ, counted with multiplicities. Then*

$$\begin{aligned}\mathrm{Rank}(\Phi) &= r_+ + r_-,\\ \mathrm{Sign}(\Phi) &= r_+ - r_-.\end{aligned}$$

Proof of Theorem 4.42: The proof is by induction on n. The Theorem is obviously true for $n=1$.

Let $M=[m_{i,j}]_{i,j=1\ldots n}$, $N=[m_{i,j}]_{i,j=1\ldots n-1}$. By induction hypothesis, there exists an orthonormal matrix $\mathcal{B}$ with entries in R such that

$$\mathcal{B}^t N \mathcal{B} = D(y_1, \ldots, y_{n-1})$$

where $D(y_1, \ldots, y_{n-1})$ is a diagonal matrix with entries

$$y_1 \leqslant \ldots \leqslant y_{n-1}.$$

Note that the column vectors of $\mathcal{B}$, $w_1, \ldots, w_{n-1}$, form a basis of eigenvectors of the quadratic form associated to N. We can suppose without loss of generality that $N w_i = y_i w_i$. Let v_i be the vector of R^n whose first coordinates coincide with w_i and whose last coordinate is 0 and let $\mathcal{C}$ be an orthonormal basis completing $v_1 \ldots, v_{n-1}$ by Proposition 4.40. We have

$$\mathcal{C}^t M \mathcal{C} = \begin{bmatrix} y_1 & 0 & 0 & 0 & b_1 \\ 0 & \ddots & 0 & 0 & \vdots \\ 0 & 0 & \ddots & 0 & \vdots \\ 0 & 0 & 0 & y_{n-1} & b_{n-1} \\ b_1 & \ldots & \ldots & b_{n-1} & a \end{bmatrix}$$

Let ε be a variable. Define $b_i' = b_i$ if $b_i \neq 0$, and $b_i' = \varepsilon$ otherwise, and if

$$y_{i-1} < y_i = \ldots = y_j < y_{j+1},$$

$y_k' = y_i + (k-i)\varepsilon$, for $0 \leqslant k \leqslant j-i$. We define the symmetric matrix M' with entries in $\mathrm{R}\langle\varepsilon\rangle$ by

$$\mathcal{C}^t M' \mathcal{C} = \begin{bmatrix} y_1' & 0 & 0 & 0 & b_1' \\ 0 & \ddots & 0 & 0 & \vdots \\ 0 & 0 & \ddots & 0 & \vdots \\ 0 & 0 & 0 & y_{n-1}' & b_{n-1}' \\ b_1' & \ldots & \ldots & b_{n-1}' & a \end{bmatrix}.$$

Note that $\lim_\varepsilon (y_i') = y_i$, $\lim_\varepsilon (b_i') = b_i$, hence $\lim_\varepsilon (M') = M$. Developing the characteristic polynomial P of $\mathcal{C}^t M' \mathcal{C}$, which is equal to the characteristic polynomial of M, on the last column and the last row we get

$$P = \prod_{i=1}^{n-1} (X - y_i')(X - a') - \sum_{i=1}^{n-1} b_i'^2 \prod_{j \neq i} (X - y_j').$$

Evaluating at y_i', we get

$$\mathrm{sign}(P(y_i')) = \mathrm{sign}\left(b_i'^2 \prod_{j \neq i} (y_i' - y_j') \right) = \mathrm{sign}(-1)^{n-i}.$$

Since the sign of P at $-\infty$ is $(-1)^n$, and the sign of P at $+\infty$ is 1, the polynomial P has n real roots satisfiying

$$x_1' < y_1' < x_2' < \ldots < x_{n-1}' < y_{n-1}' < x_n'.$$

Taking eigenvectors of norm 1 defines an orthonormal matrix $\mathcal{D}'$ such that

$$\mathcal{D}'^t M' \mathcal{D}' = D(x'_1, \ldots, x'_n).$$

Applying $\lim_\varepsilon$ on both sides we obtain an orthonormal matrix such that

$$\mathcal{D}^t M \mathcal{D} = D(x'_1, \ldots, x'_n),$$

noting that x_1 and x_n are bounded by an element of R by Proposition 2.4. Note that $x_1 \leqslant \ldots \leqslant x_n$ are the eigenvalues of M. □

We now prove that the subdiscriminants of characteristic polynomials of symmetric matrices are sums of squares. Let M is a symmetric $p \times p$ matrix with coefficients in a field K and $\mathrm{Tr}(M)$ its trace. The k-th subdiscriminant of the characteristic polynomial of M $\mathrm{sDisc}_k(M)$ is the determinant of the matrix $\mathrm{Newt}_k(M)$ whose (i,j)-th entry is $\mathrm{Tr}(M^{i+j-2})$, $i, j = 1, \ldots, pk$. Indeed, the Newton sum N_i of $\mathrm{CharPol}(M)$ is $\mathrm{Tr}(M^i)$, the trace of the matrix M^i. If M is a symmetric $p \times p$ matrix with coefficients in a ring D, we also define $\mathrm{sDisc}_k(M)$ as the determinant of the matrix $\mathrm{Newt}_k(M)$ whose (i,j)-th entry is $\mathrm{Tr}(M^{i+j-2})$, $i, j = 1, \ldots p - k$.

We define a linear basis $E_{j,\ell}$ of the space $\mathrm{Sym}(p)$ of symmetric matrices of size p as follows. First define $F_{j,\ell}$ as the matrix having all zero entries except 1 at (j, ℓ). Then take $E_{j,j} = F_{j,j}$, $E_{j,\ell} = 1/\sqrt{2}(F_{j,\ell} + F_{\ell,j})$, $\ell > j$. Define E as the ordered set $E_{j,\ell}$ $p \geq \ell \geq j \geq 0$, indices being taken in the order

$$(1,1), \ldots, (p,p), (1,2), \ldots, (1,p), \ldots, (p-1,p).$$

For simplicity, we index elements of E pairs (j,ℓ), $\ell \geq j$.

Proposition 4.45. *The map associating to* $(A, B) \in \mathrm{Sym}(p) \times \mathrm{Sym}(p)$ *the value* $\mathrm{Tr}(A\,B)$ *is a scalar product on* $\mathrm{Sym}(p)$ *with orthogonal basis* E.

Proof: Simply check. □

Let A_k be the $(p - k) \times p(p + 1)/2$ matrix with $(i, (j, \ell))$-th entry the (j,ℓ)-th component of M^{i-1} in the basis E.

Proposition 4.46. $\mathrm{Newt}_k(M) = A_k\, A_k^t$.

Proof: Immediate since $\mathrm{Tr}(M^{i+j})$ is the scalar product of M^i by M^j in the basis E. □

We consider a generic symmetric matrix $M = [m_{i,j}]$ whose entries are $p(p+1)/2$ independent variables $m_{j,\ell}, \ell \geq j$. We are going to give an explicit expression of $\mathrm{sDisc}_k(M)$ as a sum of products of powers of 2 by squares of elements of the ring $\mathbb{Z}[m_{j,\ell}]$.

Let A_k be the $(p - k) \times p(p + 1)/2$ matrix with $(i, (j, \ell))$-th entry the (j,ℓ)-th component of M^{i-1} in the basis E.

Theorem 4.47. $\mathrm{sDisc}_k(M)$ *is the sum of squares of the* $(p-k)\times(p-k)$ *minors of* A_k.

Proof: Use Proposition 4.46 and Proposition 4.10 (Cauchy-Binet formula). □

Noting that the square of a $(p-k)\times(p-k)$ minor of A_k is a power of 2 multiplied by a square of an element of $\mathbb{Z}[m_{j,\ell}]$, we obtain an explicit expression of $\mathrm{sDisc}_k(M)$ as a sum of products of powers of 2 by squares of elements of the ring $\mathbb{Z}[m_{j,\ell}]$.

As a consequence the k-th subdiscriminant of the characteristic polynomial of a symmetric matrix with coefficients in a ring D is a sum of products of powers of 2 by squares of elements in D.

Let us take a simple example and consider

$$M=\begin{bmatrix} m_{11} & m_{12} \\ m_{12} & m_{22} \end{bmatrix}.$$

The characteristic polynomial of M is $X^2-(m_{11}+m_{22})X+m_{11}m_{22}-m_{12}^2$, and its discriminant is $(m_{11}+m_{22})^2-4(m_{11}m_{22}-m_{12}^2)$. On the other hand the sum of the squares of the 2 by 2 minors of

$$A_0=\begin{bmatrix} 1 & 1 & 0 \\ m_{11} & m_{22} & \sqrt{2}m_{12} \end{bmatrix}$$

is

$$(m_{22}-m_{11})^2+(\sqrt{2}m_{12})^2+(\sqrt{2}m_{12})^2.$$

It is easy to check the statement of Proposition 4.46 in this particular case.

Proposition 4.48. *Given a symmetric matrix* M, *there exists* $k, n-1\geq k\geq 0$ *such that the signs of the subdiscriminants of the characteristic polynomial of* M *are given by*

$$\bigwedge_{p-1\geq i\geq k} \mathrm{sDisc}_i(M)>0 \wedge \bigwedge_{0\leq i<k} \mathrm{sDisc}_i(M)=0.$$

Proof: First note that, by Proposition 4.46, $\mathrm{sDisc}_i(M)\geq 0$. Moreover, it follows from Proposition 4.46 that $\mathrm{sDisc}_i(M)=0$ if only if the rank of A_i is less than $n-i$. So, $\mathrm{sDisc}_{k-1}(M)=0$ implies $\mathrm{sDisc}_i(M)=0$ for every $0\leq i<k$ and $\mathrm{sDisc}_k(M)>0$ implies $\mathrm{sDisc}_i(M)>0$ for every $n-1\geq i\geq k$. In other words, for every symmetric matrix M, there exists $k, n-1\geq k\geq 0$ such that the signs of the subdiscriminants of M are given by

$$\bigwedge_{p-1\geq i\geq k} \mathrm{sDisc}_i(M)>0 \wedge \bigwedge_{0\leq i<k} \mathrm{sDisc}_i(M)=0.$$

□

As a corollary, we obtain an algebraic proof of a part of Theorem 4.42.

Proposition 4.49. *Let M be a symmetric matrix with entries in* R*. The eigenvalues of M are in* R.

Proof: The number of roots in R of the characteristic polynomial $\mathrm{CharPol}(M)$ is $p - k$, using Proposition 4.48, and Theorem 4.33, while the number of distinct roots of $\mathrm{CharPol}(M)$ in C is $p-k$ using Proposition 4.25. □

Proposition 4.50. *Let P be a polynomial in* R$[X]$*, $P = a_p X^p + \cdots + a_0$. All the roots of P are in* R *if and only if there exists $p > k \geq 0$ such that* $\mathrm{sDisc}_i(P) > 0$ *for all i from p to k and* $\mathrm{sDisc}_i(P) = 0$ *for all i from $k-1$ to* 0

Proof: Since it is clear that every polynomial having all its roots in R is the characteristic polynomial of a diagonal symmetric matrix with entries in R, Proposition 4.49 implies that the set of polynomials having all their roots in R is contained in the set described

$$\bigvee_{k=p-1,\ldots,0}\left(\bigwedge_{p-1\geq i\geq k} \mathrm{sDisc}_i(P) > 0 \wedge \bigwedge_{0\leq i<k} \mathrm{sDisc}_i(P) = 0\right).$$

The other inclusion follows immediately from Theorem 4.31. □

Remark 4.51. Note that the sign condition

$$\mathrm{sDisc}_{p-2}(P) \geq 0 \wedge \ldots \wedge \mathrm{sDisc}_0(P) \geq 0$$

does not imply that P has all its roots in R: the polynomials $X^4 + 1$ has no real root (its four roots are $\pm\sqrt{2}/2 \pm i\sqrt{2}/2$, and it is immediate to check that is satisfies $\mathrm{sDisc}_2(P) = \mathrm{sDisc}_1(P) = 0, \mathrm{sDisc}_0(A) > 0$.

In fact, the set of polynomials having all their roots in R is the closure of the set defined by

$$\mathrm{sDisc}_{p-2}(P) > 0 \wedge \ldots \wedge \mathrm{sDisc}_0(P) > 0,$$

but does not coincide with the set defined by

$$\mathrm{sDisc}_{p-2}(A) \geq 0 \wedge \ldots \wedge \mathrm{sDisc}_0(A) \geq 0.$$ □

This is a new occurrence of the fact that the closure of a semi-algebraic set is not necessarily obtained by relaxing sign conditions defining it (see Remark 3.2).

4.3.2 Hermite's Quadratic Form

We define Hermite's quadratic form and indicate how its signature is related to real root counting.

Let R be a real closed field, D an ordered integral domain contained in R, K the field of fractions of D, and C = R$[i]$ (with $i^2 = -1$).

We consider P and Q, two polynomials in $\mathrm{D}[X]$, with P monic of degree p and Q of degree $q < p$:

$$\begin{aligned} P &= X^p + a_{p-1}X^{p-1} + \cdots + a_1 X + a_0 \\ Q &= b_q X^q + b_{q-1}X^{q-1} + \cdots + b_1 X + b_0. \end{aligned}$$

We define the **Hermite quadratic form** $\mathrm{Her}(P,\ Q)$ depending of the p variables $f_1, \ldots, f_p$ in the following way:

$$\mathrm{Her}(P,Q)(f_1,\ldots,f_p) - \sum_{x \in \mathrm{Zer}(P,\mathrm{C})} \mu(x) Q(x) (f_1 + f_2 x + \cdots + f_p x^{p-1})^2,$$

where $\mu(x)$ is the multiplicity of x. Note that

$$\mathrm{Her}(P,Q) = \sum_{k=1}^{p} \sum_{j=1}^{p} \sum_{x \in \mathrm{Zer}(P,\mathrm{C})} \mu(x)\, Q(x)\, x^{k+j-2}\, f_k\, f_j.$$

When $Q = 1$, we get:

$$\begin{aligned} \mathrm{Her}(P,1) &= \sum_{k=1}^{p} \sum_{j=1}^{p} \sum_{x \in \mathrm{Zer}(P,\mathrm{C})} \mu(x)\, x^{k+j-2} f_k\, f_j \\ &= \sum_{k=1}^{p} \sum_{j=1}^{p} N_{k+j-2} f_k\, f_j \end{aligned}$$

where N_n is the n-th Newton sum of P (see Definition 4.7). So the matrix associated to $\mathrm{Her}(P,Q)$ is $\mathrm{Newt}_0(P)$.

Since the expression of $\mathrm{Her}(P,Q)$ is symmetric in the x's, the quadratic form $\mathrm{Her}(P,Q)$ has coefficients in K by Proposition 2.13. In fact, the coefficients of $\mathrm{Her}(P,Q)$ can be expressed in terms of the trace map.

We define $\mathrm{A} = \mathrm{K}[X]/(P)$. The ring A is a K-vector space of dimension p with basis $1, X, \ldots, X^{p-1}$. Indeed any $f \in \mathrm{K}[X]$ has a representative $f_1 + f_2 X + \cdots + f_p X^{p-1}$ obtained by taking its remainder in the euclidean division by P, and if f and g are equal modulo P, their remainder in the euclidean division by P are equal.

We denote by Tr the usual **trace** of a linear map from a finite dimensional vector space A to A, which is the sum of the entries on the diagonal of its associated matrix in any basis of A.

Notation 4.52. [Multiplication map] For $f \in \mathrm{A}$, we denote by $L_f\colon \mathrm{A} \to \mathrm{A}$ the linear map of multiplication by f, sending any $g \in \mathrm{A}$ to the remainder of fg in the euclidean division by P. □

Proposition 4.53. *The quadratic form* $\mathrm{Her}(P,\ Q)$ *is the quadratic form associating to*

$$f = f_1 + f_2 X \cdots + f_p X^{p-1} \in A = \mathrm{K}[X]/(P)$$

the expression $\mathrm{Tr}(L_{Qf^2})$.

The proof of Proposition 4.53 relies on the following results.

Proposition 4.54.

$$\mathrm{Tr}(L_f) = \sum_{x \in \mathrm{Zer}(P,C)} \mu(x) f(x).$$

Proof: The proof proceeds by induction on the number of distinct roots of P.

When $P = (X - x)^{\mu(x)}$, since x is root of $f - f(x)$,

$$(f - f(x))^{\mu(x)} = 0 \text{ modulo } P$$

and $L_{f - f(x)}$ is nilpotent, with characteristic polynomial $X^{\mu(x)}$. Thus $L_{f - f(x)}$ has a unique eigenvalue 0 with multiplicity $\mu(x)$. So $\mathrm{Tr}(L_{f - f(x)}) = 0$ and $\mathrm{Tr}(L_f) = \mu(x) f(x)$.

If $P = P_1 P_2$ with P_1 and P_2 coprime, by Proposition 1.9 there exists U_1 and U_2 with $U_1 P_1 + U_2 P_2 = 1$. Let

$$e_1 = U_2 P_2 = 1 - U_1 P_1,\ e_2 = U_1 P_1 = 1 - U_2 P_2.$$

It is easy to verify that

$$e_1^2 = e_1, e_2^2 = e_2, e_1 e_2 = 0, e_1 + e_2 = 1$$

in A. It is also easy to check that the mapping from $\mathrm{K}[X]/(P_1) \times \mathrm{K}[X]/(P_2)$ to $\mathrm{K}[X]/(P)$ associating to (Q_1, Q_2) the polynomial $Q = Q_1 e_1 + Q_2 e_2$ is an isomorphism. Moreover, if $f_1 = f \bmod P_1$ and $f_2 = f \bmod P_2$, $\mathrm{K}[X]/(P_1)$ and $\mathrm{K}[X]/(P_2)$ are stable by L_f and L_{f_1} and L_{f_2} are the restrictions of L_f to $\mathrm{K}[X]/(P_1)$ and $\mathrm{K}[X]/(P_2)$. Then $\mathrm{Tr}(L_f) = \mathrm{Tr}(L_{f_1}) + \mathrm{Tr}(L_{f_2})$. This proves the proposition by induction, since the number of roots of P_1 and P_2 are smaller than the number of roots of P. □

Proposition 4.55. *Let* $C = \mathrm{Quo}(P'Q, P)$, *then*

$$\frac{P'Q}{P} = C + \sum_{n=0}^{\infty} \frac{\mathrm{Tr}(L_{QX^n})}{X^{n+1}}.$$

Proof: As already seen in the proof of Proposition 4.8

$$\frac{P'}{P} = \sum_{x \in \mathrm{Zer}(P,\mathrm{C})} \frac{\mu(x)}{(X - x)}.$$

Dividing Q by $X - x$ and letting C_x be the quotient,

$$Q = Q(x) + (X - x) C_x,$$

and thus

$$\frac{P'Q}{P} = \sum_{x \in \mathrm{Zer}(P,\mathrm{C})} \mu(x) \left(C_x + \frac{Q(x)}{(X - x)} \right).$$

Since

$$\frac{1}{X-x}=\sum_{n=0}^{\infty}\frac{x^n}{X^{n+1}},$$

the coefficient of $1/X^{n+1}$ in the development of $P'Q/P$ in powers of $1/X$ is thus,

$$\sum_{x\in \mathrm{Zer}(P,\mathrm{C})}\mu(x)Q(x)x^n.$$

Now apply Proposition 4.54 □

Proof of Proposition 4.53: By Proposition 4.55,

$$\mathrm{Tr}(L_{QX^{k+j}})=\sum_{x\in \mathrm{Zer}(P,\mathrm{C})}\mu(x)Q(x)x^{k+j}.$$

In other words, $\mathrm{Tr}(L_{QX^{k+j}})$ is the $j+1$, $k+1$-th entry of the symmetric matrix associated to Hermite's quadratic form $\mathrm{Her}(P,Q)$ in the basis $1,X,\ldots,X^{p-1}$. □

Note that Proposition 4.55 implies that the coefficients of $\mathrm{Her}(P, Q)$ belong to D, since L_f expressed in the canonical basis has entries in D.

Remark 4.56. As a consequence of Proposition 4.53, the quadratic form $\mathrm{Her}(P,1)$ is the quadratic form associating to

$$f=f_1+f_2X\cdots+f_pX^{p-1}\in \mathrm{A}=\mathrm{K}[X]/(P)$$

the expression $\mathrm{Tr}(L_{f^2})$. So the $j+1,k+1$-th entry of the symmetric matrix associated to Hermite's quadratic form $\mathrm{Her}(P,1)$ in the basis $1,X,\ldots,X^{p-1}$ is $\mathrm{Tr}(L_{X^{j+k}})=N_{k+j}$. Note that Proposition 4.55 is a generalization of Proposition 4.8. □

The main result about Hermite's quadratic form is the following theorem. We use again the notation

$$\mathrm{TaQ}(Q,P)=\sum_{x\in \mathrm{R},\,P(x)=0}\mathrm{sign}(Q(x)).$$

Theorem 4.57. [Hermite]

$$\begin{aligned}\mathrm{Rank}(\mathrm{Her}(P,Q)) &= \#\{x\in \mathrm{C}\mid P(x)=0\wedge Q(x)\neq 0\},\\ \mathrm{Sign}(\mathrm{Her}(P,Q)) &= \mathrm{TaQ}(Q,P).\end{aligned}$$

As an immediate consequence

Theorem 4.58. *The rank of* $\mathrm{Her}(P,1)$ *is equal to the number of roots of* P *in* C. *The signature of* $\mathrm{Her}(P,1)$ *is equal to the number of roots of* P *in* R.

Proof of Theorem 4.57: For $x \in \mathrm{C}$, let $L(x, -)$ be the linear form on C^p defined by:

$$L(x, f) = f_1 + f_2 x + \dots + f_p x^{p-1}.$$

Let $\{x \in \mathrm{C} \mid P(x) = 0 \wedge Q(x) \neq 0\} = \{x_1, \dots, x_r\}$. Thus,

$$\mathrm{Her}(P, Q) = \sum_{i=1}^{r} \mu(x_i)\, Q(x_i) L(x_i, f)^2.$$

The linear forms $L(x_i, f)$ are linearly independent since the roots are distinct and the Vandermonde determinant

$$\det(V(x_1, \dots, x_r)) = \prod_{r \geq i > j \geq 1} (x_i - x_j).$$

is non-zero. Thus the rank of $\mathrm{Her}(P, Q)$ is r.

Let

$$\{x \in \mathrm{R} \mid P(x) = 0 \wedge Q(x) \neq 0\} = \{y_1, \dots, y_s\}.$$
$$\{x \in \mathrm{C} \setminus \mathrm{R} \mid P(x) = 0 \wedge Q(x) \neq 0\} = \{z_1, \overline{z_1}, \dots, z_t, \overline{z_t}\}.$$

The quadratic form $\mathrm{Her}(P, Q)$ is equal to

$$\sum_{i=1}^{s} \mu(y_i) Q(y_i) L(y_i, f)^2 + \sum_{j=1}^{t} \mu(z_j)(Q(z_j) L(z_j, f)^2 + Q(\overline{z_j}) L(\overline{z_j}, f)^2),$$

with the $L(y_i, f)$, $L(z_j, f)$, $L(\overline{z_j}, f)$ ($i = 1, \dots, s$, $j = 1, \dots, t$) linearly independent.

Writing $\mu(z_j) Q(z_j) = (a(z_j) + i\, b(z_j))^2$ with $a(z_j), b(z_j) \in \mathrm{R}$ and denoting by $s_i(z_j)$ and $t_i(z_j)$ the real and imaginary part of z_j^i,

$$L_1(z_j) = \sum_{i=1}^{p} (a(z_j) s_i(z_j) - b(z_j) t_i(z_j)) f_i$$
$$L_2(z_j) = \sum_{i=1}^{p} (a(z_j) t_i(z_j) + b(z_j) s_i(z_j)) f_i$$

are linear forms with coefficients in R such that

$$\mu(z_j)(Q(z_j) L(z_j, f)^2 + Q(\overline{z_j}) L(\overline{z_j}, f)^2) = 2 L_1(z_j)^2 - 2 L_2(z_j)^2.$$

Moreover the $L(y_i, f)$, $L_1(z_j)$, $L_2(z_j)$ ($i = 1, \dots, s$, $j = 1, \dots, t$) are linearly independent linear forms. So, using Theorem 4.38 (Sylvester's inertia law), the signature of $\mathrm{Her}(P, Q)$ is the signature of $\sum_{i=1}^{s} \mu(y_i) Q(y_i) L(y_i, f)^2$. Since the linear forms $L(y_i, f)$ are linearly independent, the signature of $\mathrm{Her}(P, Q)$ is $\mathrm{TaQ}(Q, P)$. □

Remark 4.59. Note that it follows from Theorem 4.58 and Theorem 4.33 that the signature of $\text{Her}(P,1)$,which is the number of roots of P in R can be computed from the signs of the principal minors $\text{sDisc}_{p-k}(P)$,$k=1,\ldots,p$ of the symmetric matrix $\text{Newt}_0(P)$ defining $\text{Her}(P,1)$. This is a general fact about Hankel matrices that we shall define and study in Chapter 9.

□

4.4 Polynomial Ideals

4.4.1 Hilbert's Basis Theorem

An **ideal** I of a ring A is a subset $I\subset \mathrm{A}$ containing 0 that is closed under addition and under multiplication by any element of A. To an ideal I of A is associated an equivalence relation on A called **congruence modulo** I. We write $a=b \bmod I$ if and only if $a-b\in I$. It is clear that if $a_1-b_1\in I, a_2-b_2\in I$ then $(a_1+a_2)-(b_1+b_2)\in I, a_1 a_2-b_1 b_2=a_1(a_2-b_2)+b_2(a_1-b_1)\in I$.

The **quotient ring** A/I is the set of equivalence classes equipped with the natural ring structure obtained by defining the sum or product of two classes as the class of the sum or product of any members of the classes. Observation 4.9 shows that this is well defined.

The set of those elements a such that a power of a belongs to the ideal I is an ideal called the **radical of** I:

$$\sqrt{I}=\{a\in\mathrm{A}\mid\exists m\in\mathbb{N}\quad a^m\in I\}.$$

A **prime ideal** is an ideal such that $x\,y\in I$ implies $x\in I$ or $y\in I$.

To a finite set of polynomials $\mathcal{P}\subset\mathrm{K}[X_1,\ldots,X_k]$ is associated $\text{Ideal}(\mathcal{P},\mathrm{K})$, the **ideal generated by** $\mathcal{P}$ **in**],$\mathrm{K}[X_1,\ldots,X_k]$ i.e.,

$$\text{Ideal}(\mathcal{P},\mathrm{K})=\left\{\sum_{P\in\mathcal{P}} A_P P\mid A_P\in\mathrm{K}[X_1,\ldots,X_k]\right\}.$$

A polynomial in $\text{Ideal}(\mathcal{P},\mathrm{K})$ vanishes at every point of $\text{Zer}(\mathcal{P},\mathrm{C}^k)$.

Note that when $k=1$, the ideal generated by $\mathcal{P}$ in $\mathrm{K}[X_1]$ is **principal** (i.e. generated by a single polynomial) and generated by the greatest common divisor of the polynomials in $\mathcal{P}$ (Definition, page 13).

This is no longer true for a general k, but the following finiteness theorem holds.

Theorem 4.60. [Hilbert's basis theorem] *Any ideal* $I\subset\mathrm{K}[X_1,\ldots,X_k]$ *is finitely generated, i.e. there exists a finite set* $\mathcal{P}$ *such that* $I=\text{Ideal}(\mathcal{P},\mathrm{K})$.

The proof uses the partial order of divisibility on the set $\mathcal{M}_k$ of monomials in k variables $X_1,\ldots,X_k$, which can be identified with $\mathbb{N}^k$, partially ordered by

$$\alpha=(\alpha_1,\ldots,\alpha_k)\prec\beta=(\beta_1,\ldots,\beta_k)\Leftrightarrow\alpha_1\le\beta_1,\ldots,\alpha_k\le\beta_k.$$

If $\alpha=(\alpha_1,\ldots,\alpha_{k-1})\in\mathbb{N}^{k-1}$ and $n\in\mathbb{N}$, we denote by $(\alpha,n)=(\alpha_1,\ldots,\alpha_{k-1},n)$.

Lemma 4.61. [Dickson's lemma] *Every subset of $\mathcal{M}_k$ closed under multiplication has a finite number of minimal elements with respect to the partial order of divisibility.*

Proof: The proof is by induction on the number k of variables. If $k=1$, the result is clear. Suppose that the property holds for $k-1$. Let $B\subset\mathcal{M}_k$ and

$$A=\left\{X^\alpha\in\mathcal{M}_{k-1}\mid\exists n\in\mathbb{N}\, X^{(\alpha,n)}\in B\right\}.$$

By induction hypothesis, A has a finite set of minimal elements for the partial order of divisibility

$$\left\{X^{\alpha(1)},\ldots,X^{\alpha(N)}\right\}.$$

Let n be such that for every $i=1,\ldots,N$, $X^{(\alpha(i),n)}\in B$. For every $m<n$,

$$C_m=\left\{X^\alpha\in\mathcal{M}_{k-1}\mid\, X^{(\alpha,m)}\in B\right\}$$

has a finite set of minimal elements with respect to the partial order of divisibility

$$\left\{X^{\gamma(m,1)},\ldots,X^{\gamma(m,\ell(m))}\right\},$$

using again the induction hypothesis. Consider the finite set

$$D=\left\{X^{(\alpha(i),n)}\mid i=1,\ldots,N\right\}\bigcup_{m=0}^{n}\left\{X^{(\gamma(m,i),m)}\mid i=1,\ldots,\ell(m)\right\}.$$

Let $X^\beta\in B$, with $\beta=(\alpha,\,r)$. If $r\geq n$, X^β is multiple of $X^{(\alpha(i),n)}$ for some $i=1,\ldots,N$. On the other hand, if $r<n$, X^β is multiple of $X^{(\gamma(r,i),r)}$ for some $i=1,\ldots,\ell(r)$. So every element of B is multiple of an element in D. It is clear that a finite number of minimal elements for the partial order of divisibility can be extracted from D. □

In order to prove Theorem 4.60, the notion of monomial ordering is useful.

Definition 4.62. [Monomial ordering] A total ordering on the set $\mathcal{M}_k$ of monomials in k variables is a **monomial ordering** if the following properties hold

a) $X^\alpha>1$ for every $\alpha\in\mathbb{N}^k$, $\alpha\neq(0,\ldots,0)$
b) $X_1>\ldots>X_k$,
c) $X^\alpha>X^\beta\Longrightarrow X^{\alpha+\gamma}>X^{\beta+\gamma}$, for every α,β,γ elements of $\mathbb{N}^k$,
d) every decreasing sequence of monomials for the monomial order $<$ is finite. □

The lexicographical ordering defined in Notation 2.14 and the graded lexicographical ordering defined in Notation 2.15 are examples of monomial orderings. Another important example of monomial ordering is the reverse lexicographical ordering defined above.

Definition 4.63. [Reserve lexicographical ordering] The **reverse lexicographical ordering** , $<_{\mathrm{revlex}}$, on the set $\mathcal{M}_k$ of monomials in k variables is the total order $X^\alpha <_{\mathrm{grlex}} X^\beta$ defined by

$$X^\alpha <_{\mathrm{grlex}} X^\beta \Leftrightarrow (\deg(X^\alpha) < \deg(X^\beta)) \vee \left(\deg(X^\alpha) = \deg(X^\beta) \wedge \overline{\beta} <_{\mathrm{lex}} \overline{\alpha}\right)$$

with $\alpha = (\alpha_1, \ldots, \alpha_k)$, $\beta = (\beta_1, \ldots, \beta_k)$, $\overline{\alpha} = (\alpha_k, \ldots, \alpha_1)$, $\overline{\beta} = (\beta_k, \ldots, \beta_1)$, $X^\alpha = X_1^{\alpha_1} \cdots X_k^{\alpha_k}$, $X^\beta = X_1^{\beta_1} \cdots X_k^{\beta_k}$, and $<_{\mathrm{lex}}$ is the lexicographical ordering defined in Notation 2.14.

In the reverse lexicographical ordering above, $X_1 >_{\mathrm{revlex}} \ldots >_{\mathrm{revlex}} X_k$. The smallest monomial with respect to the reverse lexicographical ordering is 1, and the reverse lexicographical ordering order is compatible with multiplication. Note that the set of monomials less than or equal to a monomial X^α in the reverse lexicographical ordering is finite. □

Definition 4.64. Given a polynomial $P \in K[X_1, \ldots, X_k]$ we write $\mathrm{cof}(X^\alpha, P)$ for the coefficient of the monomial X^α in the polynomial P. The monomial X^α is a **monomial of** P if $\mathrm{cof}(X^\alpha, P) \neq 0$, and $\mathrm{cof}(X^\alpha, P) X^\alpha$ is a **term of** P.

Given a monomial ordering $<$ on $\mathcal{M}_k$, we write $\mathrm{lmon}(P)$ for the **leading monomial** of P with respect to $<$ i.e. the largest monomial of P with respect to $<$. The **leading coefficient** of P is $\mathrm{lcof}(P) = \mathrm{cof}(\mathrm{lmon}(P), P)$, and the **leading term** of P is $\mathrm{lt}(P) = \mathrm{lcof}(P)\mathrm{lmon}(P)$. Let X^α be a monomial of P, and let G be another polynomial. The **reduction** of (P, X^α) by G is defined by

$$\begin{aligned}&\mathrm{Red}(P, X^\alpha, G)\\ = &\begin{cases} P - (\mathrm{cof}(X^\alpha, P)/\mathrm{lcof}(G))\, X^\beta G & \text{if } \exists\, \beta \in \mathbb{N}^k\ X^\alpha = X^\beta\, \mathrm{lmon}(G),\\ P & \text{otherwise.}\end{cases}\end{aligned}$$

Given a finite set of polynomials, $\mathcal{G} \subset \mathrm{K}[X_1, \ldots, X_k]$, Q is a **reduction** of P modulo $\mathcal{G}$ if there is a $G \in \mathcal{G}$ and a monomial X^α of P such that $Q = \mathrm{Red}(P, X^\alpha, G)$. We say that P is **reducible** to Q modulo $\mathcal{G}$ if there is a finite sequence of reductions modulo $\mathcal{G}$ starting with P and ending at Q.

□

Remark 4.65. Note that if P is reducible to Q modulo $\mathcal{G}$, it follows that $(P - Q) \in \mathrm{Ideal}(\mathcal{G}, \mathrm{K})$. Note also that if P is reducible to 0 modulo $\mathcal{G}$, then

$$\exists\, G_1 \in \mathcal{G} \ldots \exists\, G_s \in \mathcal{G}\ \ P = A_1 G_1 + \cdots + A_s G_s,$$

with $\mathrm{lmon}(A_i G_i) \leq \mathrm{lmon}(P)$ for all $i = 1, \ldots, s$. □

Definition 4.66. A **Gröbner basis** of an ideal $I \subset K[X_1, ..., X_k]$ for the monomial ordering $<$ on $\mathcal{M}_k$ is a finite set, $\mathcal{G} \subset I$, such that

- the leading monomial of any element in I is a multiple of the leading monomial of some element in $\mathcal{G}$,
- the leading monomial of any element of $\mathcal{G}$ is not a multiple of the leading monomial of another element in $\mathcal{G}$.

A **Gröbner basis** for the monomial ordering $<$ on $\mathcal{M}_k$ is a finite set $\mathcal{G} \subset K[X_1, ..., X_k]$ which is a Gröbner basis of the ideal $\mathrm{Ideal}(\mathcal{G}, \mathrm{K})$. □

Deciding whether an element belongs to an ideal I is easy given a Gröbner basis of I.

Proposition 4.67. *If $\mathcal{G}$ is a Gröbner basis of I for the monomial ordering $<$ on $\mathcal{M}_k$, $P \in I$ if and only if P is reducible to 0 modulo $\mathcal{G}$.*

Proof: It is clear that if P is reducible to 0 modulo $\mathcal{G}$, $P \in I$. Conversely, let $P \neq 0 \in I$. Then, the leading monomial of P is a multiple of the leading monomial of some $G \in \mathcal{G}$, so that defining $Q = \mathrm{Red}(P, \mathrm{lmon}(P), G)$ either $Q = 0$ or $\mathrm{lmon}(Q) < \mathrm{lmon}(P)$, $Q \in I$. Since there is non infinite decresing sequence for $<$, this process must terminate at zero after a finite number of steps. □

As a consequence

Proposition 4.68. *A Gröbner basis of I for the monomial ordering $<$ on $\mathcal{M}_k$ is a set of generators of I.*

Proof: Let $P \in I$. By Proposition 4.67, P is reducible to 0 by $\mathcal{G}$ and $P \in I(\mathcal{G}, \mathrm{K})$. □

Proposition 4.69. *Every ideal of $\mathrm{K}[X_1, ..., X_k]$ has a Gröbner basis for any monomial ordering $<$ on $\mathcal{M}_k$.*

Proof: Let $I \subset \mathrm{K}[X_1, ..., X_k]$ be an ideal and let $\mathrm{lmon}(I)$ be the set of leading monomials of elements of I. By Lemma 4.61, there is a finite set of minimal elements in $\mathrm{lmon}(I)$ for the partial order of divisibility, denoted by $\{X^{\alpha(1)}, ..., X^{\alpha(N)}\}$. Let $\mathcal{G} = \{G_1, ..., G_N\}$ be elements of I with leading monomials $\{X^{\alpha(1)}, ..., X^{\alpha(N)}\}$. By definition of $\mathcal{G}$, the leading monomial of any polynomial in I is a multiple of the leading monomial of some polynomial in $\mathcal{G}$, and no leading monomial of $\mathcal{G}$ is divisible by another leading monomial of $\mathcal{G}$. □

Proof of Theorem 4.60: The claim is an immediate corollary of Proposition 4.69 since a Gröbner basis of an ideal is a finite number of generators, by Proposition 4.68. □

Corollary 4.70. *Let $I_1 \subset I_2 \subset \cdots \subset I_n \subset \cdots$ be an ascending chain of ideals of $\mathrm{K}[X_1, ..., X_k]$. Then $\exists\, n \in \mathbb{N}\ \forall m \in \mathbb{N}\ (m > n \Rightarrow I_m = I_n)$.*

Proof: It is clear that $I = \bigcup_{i \geq 0} I_i$ is an ideal and has a finite set of generators according to Theorem 4.60. This finite set of generators belongs to some I_N and so $I_N = I$. □

If $I \subset \mathrm{K}[X_1, \ldots, X_k]$ is an ideal and L is a field containing K, we denote by $\mathrm{Zer}(I, \mathrm{L}^k)$ the set of common zeros of I in L^k,

$$\mathrm{Zer}(I, \mathrm{L}^k) = \{x \in \mathrm{L}^k \mid \forall P \in I \; P(x) = 0\}.$$

When $\mathrm{L} = \mathrm{K}$, this defines the **algebraic sets** contained in K^k. Note that Theorem 4.60 implies that every algebraic set contained in K^k is of the form

$$\mathrm{Zer}(\mathcal{P}, \mathrm{K}^k) = \{x \in \mathrm{K}^k \mid \bigwedge_{P \in \mathcal{P}} P(x) = 0\},$$

where $\mathcal{P}$ is a finite set of polynomials, so that the definition of algebraic sets given here coincides with the definition of algebraic sets given in Chapter 1 (Definition page 11) when $\mathrm{K} = \mathrm{C}$ and in Chapter 2 (Definition page 57) when $\mathrm{K} = \mathrm{R}$.

4.4.2 Hilbert's Nullstellensatz

Hilbert's Nullstellensatz (weak form) is the following result.

Theorem 4.71. [Weak Hilbert's Nullstellensatz] *Let $\mathcal{P} = \{P_1, \ldots, P_s\}$ be a finite subset of $\mathrm{K}[X_1, \ldots, X_k]$ then $\mathrm{Zer}(\mathcal{P}, \mathrm{C}^k) = \emptyset$ if and only if there exist $A_1, \ldots, A_s \in K[X_1, \ldots, X_k]$ such that*

$$A_1 P_1 + \cdots + A_s P_s = 1.$$

We develop several tools and technical results before proving it.

The **degree** of a monomial $X^\alpha = X_k^{\alpha_k} \cdots X_1^{\alpha_1}$ in k variables is the sum of the degrees with respect to each variable and the **degree** of a polynomial P in k variables, denoted $\deg(Q)$, is the maximum degree of its monomials. A polynomial is **homogeneous** if all its monomials have the same degree.

Definition 4.72. A non-zero polynomial $P \in \mathrm{K}[X_1, \ldots, X_{k-1}][X_k]$ is **quasi-monic** with respect to X_k if its leading coefficient with respect to X_k is an element of K. A set of polynomials $\mathcal{P}$ is quasi-monic with respect to X_k if each polynomial in $\mathcal{P}$ is quasi-monic with respect to X_k. □

If v is a linear automorphism $\mathrm{K}^k \to \mathrm{K}^k$ and $[v_{i,j}]$ is its matrix in the canonical basis, we write

$$v(X) = \left(\sum_{j=1}^{k} v_{1,j} X_j, \ldots, \sum_{j=1}^{k} v_{k,j} X_j \right).$$

Lemma 4.73. *Let $\mathcal{P} \subset \mathrm{K}[X_1, \ldots, X_k]$ be a finite subset. Then, there exists a linear automorphism v: $\mathrm{K}^k \to \mathrm{K}^k$ such that for all $P \in \mathcal{P}$, the polynomial $P(v(X))$ is quasi-monic in X_k.*

Proof: Choose a linear automorphism of the form

$$v(X_1, \ldots, X_k) = (X_1 + a_1 X_k,\ X_2 + a_2 X_k, \ldots,\ X_{k-1} + a_{k-1} X_k,\ X_k)$$

with $a_i \in \mathrm{K}$. Writing $P(X) = \Pi(X) + \cdots$, where Π is the homogeneous part of highest degree (say d) of P, we have

$$P(v(X)) = \Pi(a_1, \ldots, a_{k-1}, 1)\, X_k^d + Q$$

(where Q has smaller degree in X_k); it is enough to choose $a_1, \ldots, a_{k-1}$ such that none of the $\Pi(a_1, \ldots, a_{k-1}, 1)$ is zero. This can be done by taking the product of the Π and using the following Lemma 4.74. □

Lemma 4.74. *If a polynomial $B(Z_1, \ldots, Z_k)$ in $\mathrm{K}[Z_1, \ldots, Z_k]$ is not identically zero and has degree d, there are elements $(z_1, \ldots, z_k)$ in $\{0, \ldots, d\}^k$ such that $B(z_1, \ldots, z_k)$ is a non-zero element of* K.

Proof: The proof is by induction on k. It is true for a polynomial in one variable since a non-zero polynomial of degree d has at most d roots in a field, so it does not vanish on at least one point of $\{0, \ldots, d\}$. Suppose now that it is true for $k-1$ variables, and consider a polynomial $B(Z_1, \ldots, Z_k)$ in k variables of degree d that is not identically zero. Thus, if we consider B as a polynomial in Z_k with coefficients in $\mathrm{K}[Z_1, \ldots, Z_{k-1}]$, one of its coefficients is not identically zero in $\mathrm{K}[Z_1, \ldots, Z_{k-1}]$. Hence, by the induction hypothesis, there exist $(z_1, \ldots, z_{k-1})$ in $\{0, \ldots, d\}^{k-1}$ with $B(z_1, \ldots, z_{k-1}, Z_k)$ not identically zero. The degree of $B(z_1, \ldots, z_{k-1}, Z_k)$ is at most d, so we have reduced to the case of one variable, which we have already considered. □

Let $\mathcal{P} \subset \mathrm{K}[X_1, \ldots, X_k]$ and $\mathcal{Q} \subset \mathrm{K}[X_1, \ldots, X_{k-1}]$ be two finite sets of polynomials. The projection π from C^k to C^{k-1} forgetting the last coordinate is a **finite mapping from** $\mathrm{Zer}(\mathcal{P}, \mathrm{C}^k)$ **onto** $\mathrm{Zer}(\mathcal{Q}, \mathrm{C}^{k-1})$ if its restriction to $\mathrm{Zer}(\mathcal{P}, \mathrm{C}^k)$ is surjective on $\mathrm{Zer}(\mathcal{Q}, \mathrm{C}^{k-1})$ and if $\mathcal{P}$ contains a polynomial quasi-monic in X_k, denoted by P. Since P is quasi-monic in X_k, for every $y \in \mathrm{Zer}(\mathcal{Q}, \mathrm{C}^{k-1})$, $\mathrm{Zer}(P(y, X_k), \mathrm{C})$ is finite. Thus

$$\pi^{-1}(y) \cap \mathrm{Zer}(\mathcal{P}, \mathrm{C}^k) \subset \mathrm{Zer}(P(y, X_k), \mathrm{C})$$

is finite.

Proposition 4.75. *Let $\mathcal{P} = \{P_1, \ldots, P_s\} \subset \mathrm{K}[X_1, \ldots, X_k]$ with P_1 quasi-monic in X_k. There exists a finite set*

$$\mathrm{Proj}_{X_k}(\mathcal{P}) \subset \mathrm{K}[X_1, \ldots, X_{k-1}] \cap \mathrm{Ideal}(\mathcal{P}, \mathrm{K})$$

such that

$$\pi(\mathrm{Zer}(\mathcal{P}, \mathrm{C}^k)) = \mathrm{Zer}(\mathrm{Proj}_{X_k}(\mathcal{P}), \mathrm{C}^{k-1})$$

(where π is the projection from C^k to C^{k-1} forgetting the last coordinate) and π is a finite mapping from $\mathrm{Zer}(\mathcal{P}, \mathrm{C}^k)$ to $\mathrm{Zer}(\mathrm{Proj}_{X_k}(\mathcal{P}), \mathrm{C}^{k-1})$.

Proof: If $s=1$, take $\mathrm{Proj}_{X_k}(\mathcal{P})=\{0\}$. Since P_1 is quasi-monic, the conclusion is clear.

If $s>1$, then let U be a new indeterminate, and let

$$R(U, X_1, \ldots, X_k) = P_2(X) + U P_3(X) + \cdots + U^{s-2} P_s(X).$$

The resultant of P_1 and R with respect to X_k (apply definition in page 106), belongs to $\mathrm{K}[U, X_1, \ldots, X_{k-1}]$ and is written

$$\mathrm{Res}_{X_k}(P_1, R) = Q_t U^{t-1} + \cdots + Q_1,$$

with $Q_i \in \mathrm{K}[X_1, \ldots, X_{k-1}]$. Let $\mathrm{Proj}_{X_k}(\mathcal{P}) = \{Q_1, \ldots, Q_t\}$.

By Proposition 4.18, there are polynomials M and N in $\mathrm{K}[U, X_1, \ldots, X_k]$ such that

$$\mathrm{Res}_{X_k}(P_1, R) = M P_1 + N R.$$

Identifying the coefficients of the powers of U in this equality, one sees that there are for $i = 1, \ldots, t$ identities $Q_i = M_i P_1 + N_{i,2} P_2 + \cdots + N_{i,s} P_s$ with M_i and $N_{i,2} \ldots N_{i,s}$ in $\mathrm{K}[X_1, \ldots, X_k]$ so that $Q_1, \ldots, Q_t$ belong to $\mathrm{Ideal}(\mathcal{P}, \mathrm{K}) \cap \mathrm{K}[X_1, \ldots, X_{k-1}]$.

Since

$$\mathrm{Proj}_{X_k}(\mathcal{P}) \subset \mathrm{Ideal}(\mathcal{P}, \mathrm{K}) \cap \mathrm{K}[X_1, \ldots, X_{k-1}],$$

it follows that

$$\pi(\mathrm{Zer}(\mathcal{P}, \mathrm{C}^k)) \subset \mathrm{Zer}(\mathrm{Proj}_{X_k}(\mathcal{P}), \mathrm{C}^{k-1}).$$

In the other direction, suppose $x' \in \mathrm{Zer}(\mathrm{Proj}_{X_k}(\mathcal{P}), \mathrm{C}^{k-1})$. Then for every $u \in \mathrm{C}$, we have $\mathrm{Res}_{X_k}(P_1, R)(u, x') = 0$. Since P_1 is quasi-monic with respect to X_k,

$$\mathrm{Res}(P_1(x', X_k), R(u, x', X_k)) = \mathrm{Res}_{X_k}(P_1, R)(u, x') = 0,$$

using Proposition 4.20. For every $u \in \mathrm{C}$, by Proposition 4.15 the polynomials $P(x', X_k)$ and $R(u, x', X_k)$ have a common factor in $\mathrm{K}[X_k]$, hence a common root in C. Since $P(x', X_k)$ has a finite number of roots in C, one of them, say x_k, is a root of $R(u, x', X_k)$ for infinitely many $u \in \mathrm{C}$. Choosing $s-1$ such distinct elements $u_1, \ldots, u_{s-1}$, we get that the polynomial $R(U, x', x_k)$ of degree $\leq s-2$ in U has $s-1$ distinct roots, which is possible only if $R(U, x', x_k)$ is identically zero. So one has $P_2(x', x_k) = \cdots = P_s(x', x_k) = 0$. We have proved that for any $x' \in \mathrm{Zer}(\mathrm{Proj}_{X_k}(\mathcal{P}), \mathrm{C}^{k-1})$, there exist a finite non-zero number of x_k such that $(x', x_k) \in \mathrm{Zer}(\mathcal{P}, \mathrm{C}^k)$, so that

$$\mathrm{Zer}(\mathrm{Proj}_{X_k}(\mathcal{P}), \mathrm{C}^{k-1}) \subset \pi(\mathrm{Zer}(\mathcal{P}, \mathrm{C}^k)).$$

Since P_1 is monic, $\mathrm{Zer}(\mathcal{P}, \mathrm{C}^k)$ is finite on $\mathrm{Zer}(\mathrm{Proj}_{X_k}(\mathcal{P}), \mathrm{C}^{k-1})$. □

Let $\mathcal{P} \subset \mathrm{K}[X_1, \ldots, X_k]$, $\mathcal{T} \subset \mathrm{K}[X_1, \ldots, X_{k'}]$ be two finite sets of polynomials and $k > k'$. The projection Π from C^k to $\mathrm{C}^{k'}$ forgetting the last $(k - k')$ coordinates is a **finite mapping from** $\mathrm{Zer}(\mathcal{P}, \mathbf{C}^k)$ **to** $\mathrm{Zer}(\mathcal{T}, \mathbf{C}^{k'})$ if for each i, $0 \le i \le k - k'$ there exists a finite set of polynomials $\mathcal{Q}_{k-i} \subset \mathrm{K}[X_1, \ldots, X_{k-i}]$ with $\mathcal{P} = \mathcal{Q}_k, \mathcal{T} = \mathcal{Q}_{k'}$ such that for every i, $0 \le i \le k - k' - 1$, the projection from C^{k-i} to C^{k-i-1} forgetting the last coordinate is a finite mapping from $\mathrm{Zer}(\mathcal{Q}_{k-i}, \mathrm{C}^{k-i})$ to $\mathrm{Zer}(\mathcal{Q}_{k-i-1}, \mathrm{C}^{k-i-1})$.

Proposition 4.76. *Let* $\mathcal{P} = \{P_1, \ldots, P_s\} \subset \mathrm{K}[X_1, \ldots, X_k]$. *Then*

- *either* $1 \in \mathrm{Ideal}(\mathcal{P}, \mathrm{K})$,
- *or there exists a linear automorphism* $v: \mathrm{K}^k \rightarrow \mathrm{K}^k$ *and a natural number* k', $0 \le k' \le k$, *such that the canonical projection* Π *from* C^k *to* $\mathrm{C}^{k'}$ *forgetting the last* $k - k'$ *coordinates is a finite mapping from* $v(\mathrm{Zer}(\mathcal{P}, \mathrm{C}^k))$ *to* $\mathrm{C}^{k'}$ *(the linear automorphism* v *being extended to* C^k*).*

Proof: The proof is by induction on k.

When $k = 1$, consider the greatest common divisor Q of $\mathcal{P}$, which generates the ideal generated by $\mathcal{P}$. If $Q = 0$ take $k' = 1$, and if $Q \neq 0$, take $k' = 0$.

If $\deg(Q) = 0$, then Q is a non-zero constant and $1 \in \mathrm{Ideal}(\mathcal{P}, \mathrm{K})$. If $\deg(Q) > 0$, then $\mathrm{Zer}(\mathcal{P}, \mathrm{C}) = \mathrm{Zer}(Q, \mathrm{C}^k)$ is non-empty and finite, so the projection to $\{0\}$ is finite and the result holds insthis ca

Suppose now that $k > 1$ and that the theorem holds for $k - 1$.
If $\mathrm{Ideal}(\mathcal{P}, \mathrm{K}) = \{0\}$, the theorem obviously holds by taking $k' = 0$.
If $\mathrm{Ideal}(\mathcal{P}, \mathrm{K}) \neq \{0\}$, it follows from Lemma 4.73 that we can perform a linear change of variables w and assume that $P_1(w(X))$ is quasi-monic with respect to X_k.

Let $\mathcal{P}_w = \{P_1(w(X)), \ldots, P_s(w(X))\}$.

Applying the induction hypothesis to $\mathrm{Proj}_{X_k}(\mathcal{P}_w)$,

- either $1 \in \mathrm{Ideal}(\mathrm{Proj}_{X_k}(\mathcal{P}_w, \mathrm{K}))$,
- or there exists a linear automorphism v': $\mathrm{K}^{k-1} \rightarrow \mathrm{K}^{k-1}$ and a natural number $k', 0 \le k' \le k - 1$, such that the canonical projection Π' from C^{k-1} to $\mathrm{C}^{k'}$ is a finite mapping from $v'(\mathrm{Zer}(\mathrm{Proj}_{X_k}(\mathcal{P}_w, \mathrm{C}^{k-1}))$ to $\mathrm{C}^{k'}$.

Since $\mathrm{Proj}_{X_k}(\mathcal{P}_w) \subset \mathrm{Ideal}(\mathcal{P}, \mathrm{K})$, it is clear that if $1 \in \mathrm{Ideal}(\mathrm{Proj}_{X_k}(\mathcal{P}_w), \mathrm{K}))$, then $1 \in \mathrm{Ideal}(\mathcal{P}_w, \mathrm{K})$, which implies $1 \in \mathrm{Ideal}(\mathcal{P}, \mathrm{K})$.

We now prove that if there exists a linear automorphism $v': \mathrm{K}^{k-1} \rightarrow \mathrm{K}^{k-1}$ and a natural number $k', 0 \le k' \le k - 1$, such that the canonical projection Π' from C^{k-1} to $\mathrm{C}^{k'}$ is a finite mapping from $v'(\mathrm{Zer}(\mathrm{Proj}_{X_k}(\mathcal{P}_w), \mathrm{C}^{k-1}))$ to $\mathrm{C}^{k'}$, there exists a linear automorphism $v: \mathrm{K}^k \rightarrow \mathrm{K}^k$ such that the canonical projection Π from C^k to $\mathrm{C}^{k'}$ is a finite mapping from $v(\mathrm{Zer}(\mathcal{P}, \mathrm{C}^k))$ to $\mathrm{C}^{k'}$. By Proposition 4.75, $w^{-1}(\mathrm{Zer}(\mathcal{P}, \mathrm{C}^k)) = \mathrm{Zer}(\mathcal{P}_w, \mathrm{C}^k)$ is finite on $\mathrm{Zer}(\mathrm{Proj}_{X_k}(\mathcal{P}_w), \mathrm{C}^{k-1})$, so $v = (v', \mathrm{Id}) \circ w^{-1}$ is a linear automorphism from K^k to K^k such that the canonical projection Π from C^k to $\mathrm{C}^{k'}$ is a finite mapping from $v(\mathrm{Zer}(\mathcal{P}, \mathrm{C}^k))$ to $\mathrm{C}^{k'}$. □

We are now ready for the proof of Theorem 4.71 (Weak Hilbert's Nullstellensatz).

Proof of Theorem 4.71: The existence of $A_1, \ldots, A_s \in K[X_1, \ldots, X_k]$ such that $A_1 P_1 + \cdots + A_s P_s = 1$clearly implies $\mathrm{Zer}(\mathcal{P}, \mathrm{C}^k) = \emptyset$.

On the other hand, by Proposition 4.76, if $\mathrm{Zer}(\mathcal{P}, \mathrm{C}^k) = \emptyset$, there cannot exist a linear automorphism $v: \mathrm{K}^k \to \mathrm{K}^k$ and a natural number k', $0 \leq k' \leq k$ such that the canonical projection Π from C^k to $\mathrm{C}^{k'}$ is a finite mapping from $v(\mathrm{Zer}(\mathcal{P}, \mathrm{C}^k))$ to $\mathrm{C}^{k'}$, since such a map must be surjective by definition.

So, using Proposition 4.76, $1 \in \mathrm{Ideal}(\mathcal{P}, \mathrm{K})$ which means that there exist $A_1, \ldots, A_s \in K[X_1, \ldots, X_k]$ such that $A_1 P_1 + \cdots + A_s P_s = 1$. □

Hilbert's Nullstellensatz is derived from the weak form of Hilbert's Nullstellensatz (Theorem 4.71) using what is commonly known as Rabinovitch's trick.

Theorem 4.77. [Hilbert's Nullstellensatz] *Let $\mathcal{P}$ be a finite subset of* $\mathrm{K}[X_1, \ldots, X_k]$. *If a polynomial P with coefficients in* K *vanishes on* $\mathrm{Zer}(\mathcal{P}, \mathrm{C}^k)$, *then $P^n \in \mathrm{Ideal}(\mathcal{P}, \mathrm{K})$ for some n.*

Proof: The set of polynomials $\mathcal{P} \cup \{T P - 1\}$ in the variables $X_1, \ldots, X_k, T$ has no common zeros in C^{k+1} so according to Theorem 4.71 if $\mathcal{P} = \{P_1, \ldots, P_s\}$, there exist polynomials

$$A_1(X_1, \ldots, X_k, T), \ldots, A_s(X_1, \ldots, X_k, T), A(X_1, \ldots, X_k, T)$$

in $\mathrm{K}[X_1, \ldots, X_k, T]$ such that $1 = A_1 P_1 + \cdots + A_s P_s + A\,(T P - 1)$. Replacing everywhere T by $1P$ and clearing denominators by multiplying by an appropriate power of P, we see that a power of P is in the ideal $\mathrm{Ideal}(\mathcal{P}, \mathrm{K})$. □

Another way of stating Hilbert's Nullstellensatz which follows immediately from the above is:

Theorem 4.78. *Let $\mathcal{P}$ be a finite subset of* $\mathrm{K}[X_1, \ldots, X_k]$. *The radical of* $\mathrm{Ideal}(\mathcal{P}, \mathrm{K})$ *coincides with the set of polynomials in* $\mathrm{K}[X_1, \ldots, X_k]$ *vanishing on* $\mathrm{Zer}(\mathcal{P}, \mathrm{C}^k)$ *i.e.*

$$\sqrt{\mathrm{Ideal}(\mathcal{P}, \mathrm{K})} = \{P \in \mathrm{K}[X_1, \ldots, X_k] \mid \forall x \in \mathrm{Zer}(\mathcal{P}, \mathrm{C}^k), P(x) = 0\}.$$

Corollary 4.79. [Homogeneous Hilbert's Nullstellensatz]
Let $\mathcal{P} = \{P_1, \ldots, P_s\} \subset \mathrm{K}[X_1, \ldots, X_k]$ be a finite set of homogeneous polynomials with $\deg(P_i) = d_i$. If a homogeneous polynomial $P \in \mathrm{K}[X_1, \ldots, X_k]$ of degree p vanishes on the common zeros of $\mathcal{P}$ in C^k, then there exists $n \in \mathbb{N}$ and homogeneous polynomials $H_1, \ldots, H_s$ in $\mathrm{K}[X_1, \ldots, X_k]$ of degrees $c_1, \ldots, c_s$ such that

$$\begin{aligned} P^n &= H_1 P_1 + \cdots + H_s P_s, \\ n\,p &= c_1 + d_1 = \cdots = c_s + d_s. \end{aligned}$$

Proof: According to Hilbert's Nullstellensatz, there exist $n \in \mathbb{N}$ and polynomials $B_1, \ldots, B_s$ in $\mathrm{K}[X_1, \ldots, X_k]$ such that $P^n = B_1 P_1 + \cdots + B_s P_s$.

Decompose B_i as $H_i + C_i$ with H_i homogeneous of degree $n p - d_i$, and notice that no monomial of $C_i P_i$ has degree $n p$. So $P^n = H_1 P_1 + \cdots + H_s P_s$. □

Remark 4.80. Let us explain the statement claimed in the introduction of the chapter that our proof of Hilbert's Nullstellensatz is constructive.

Indeed, the method used for proving Theorem 4.71 (Weak Hilbert's Nullstellensatz) provides an algorithm for deciding, given a finite set $\mathcal{P} \subset \mathrm{K}[X_1, \ldots, X_k]$, whether $\mathrm{Zer}(\mathcal{P}, \mathrm{C}^k)$ is empty and if it is empty, computes an algebraic identity certifying that 1 belongs to the ideal generated by $\mathcal{P}$.

The algorithms proceeds by eliminating variables one after the other. Given a finite family $\mathcal{P} \neq \{0\}$ of polynomials in k variables, we check whether it contains a non-zero constant. If this is the case we conclude that $\mathrm{Zer}(\mathcal{P}, \mathrm{C}^k)$ is empty and that 1 belongs to the ideal generated by $\mathcal{P}$. Otherwise, we perform a linear change of coordinates so that one of the polynomials of the family gets monic and replace the initial $\mathcal{P}$ by this new family. Then we compute $\mathrm{Proj}_{X_k}(\mathcal{P})$, which is a family of polynomials in $k-1$ variables together with algebraic identities expressing that the elements of $\mathrm{Proj}_{X_k}(\mathcal{P})$ belong to the ideal generated by $\mathcal{P}$. If $\mathrm{Proj}_{X_k}(\mathcal{P}) = \{0\}$ we conclude that $\mathrm{Zer}(\mathcal{P}, \mathrm{C}^k)$ is not empty. If $\mathrm{Proj}_{X_k}(\mathcal{P}) \neq \{0\}$ we continue the process replacing k by $k-1$ and $\mathcal{P}$ by $\mathrm{Proj}_{X_k}(\mathcal{P})$. After eliminating all the variables, we certainly have that either the family of polynomials is $\{0\}$ or it contains a non-zero constant, and we can conclude in both cases. □

Let us illustrate the algorithm outlined in the preceding remark by two examples.

Example 4.81. a) Let $\mathcal{P} = \{X_2,\ X_2 + X_1,\ X_2 + 1\}$. The first polynomial is monic with respect to X_2. We consider the resultant with respect to X_2 of X_2 and $(U+1)\, X_2 + U\, X_1 + 1$, which is equal to $U\, X_1 + 1$. Thus $\mathrm{Proj}_{X_2}(\mathcal{P}) = \{X_1, 1\} \neq \{0\}$, and contains a non-zero constant. Moreover $1 = (X_2 + 1) - X_2$. So we already proved that 1 belongs to the ideal generated by $\mathcal{P}$ and $\mathrm{Zer}(\mathcal{P}, \mathrm{C}^2) = \emptyset$.

b) Let $\mathcal{P} = \{X_2,\ X_2 + X_1,\ X_2 + 2\, X_1\}$. The first polynomial is monic with respect to X_2. The resultant with respect to X_2 of X_2 and $(U+1)\, X_2 + (U+2)\, X_1$ is equal to $(U+2)\, X_1$. Thus $\mathrm{Proj}_{X_2}(\mathcal{P}) = \{X_1\} \neq \{0\}$, contains a single element, and $\mathrm{Proj}_{X_1}(\mathrm{Proj}_{X_2}(\mathcal{P})) = 0$. Thus 1 does not belong to the ideal generated by $\mathcal{P}$ and $\mathrm{Zer}(\mathcal{P}, \mathrm{C}^2) \neq \emptyset$. In fact, $\mathrm{Zer}(\mathcal{P}, \mathrm{C}^2) = \{(0,0)\}$. □

Since the proof of Theorem 4.71 (Weak Hilbert's Nullstellensatz) is constructive, it is not surprising that it produces a bound on the degrees of the algebraic identity output. More precisely we have the following quantitative version of Hilbert's Nullstellensatz.

Theorem 4.82. [Quantitative Hilbert's Nullstellensatz]
Let $\mathcal{P}=\{P_1,\ldots,P_s\}\subset \mathrm{K}[X_1,\ldots,X_k]$ be a finite set of less than d polynomials, of degrees bounded by d. If a polynomial $P\in \mathrm{K}[X_1,\ldots,X_k]$ of degree at most d in k variables vanishes on the common zeros of $\mathcal{P}$ in C^k, then there exists $n\le d\,(2\,d)^{2^{k+1}}$ and s polynomials $B_1,\ldots,B_s$ in $\mathrm{K}[X_1,\ldots,X_k]$ each of degree $\le d\,(2\,d)^{2^{k+1}}$ such that $P^n=B_1\,P_1+\cdots+B_s\,P_s$.

Proof: The proof of the theorem follows from a close examination of the proofs of Proposition 4.76 and Theorem 4.77, using the notation of these proofs.

Suppose that $\mathcal{P}=\{P_1,\ldots,P_s\}$ has no common zeros in C^k.

We consider first the case of 1 variable X. Since $\mathrm{Zer}(\mathcal{P},\mathrm{C})=\emptyset$,

$$\mathrm{Res}_X(P_1,P_2+U\,P_3+\cdots+U^{s-2}\,P_s)\in \mathrm{K}[U]$$

is not the zero polynomial, and we can find $u\in\mathrm{K}$ such that

$$\mathrm{Res}_X(P_1,P_2+u\,P_3+\cdots+u^{s-2}\,P_s)$$

is a non-zero element of K. By Proposition 1.9, there exist U and V of degree at most d such that

$$U\,P_1+V\,(P_2+u\,P_3+\cdots+u^{s-2}\,P_s)=1,$$

which gives the identity

$$1=U\,P_1+V\,P_2+u\,V\,P_3+\cdots+u^{s-2}\,V\,P_s$$

with $\deg(U),\deg(V)\le d$.

Consider now the case of k variables. Since $\mathrm{Res}_{X_k}(P_1,\,R)$ is the determinant of the Sylvester matrix, which is of size at most 2 d, the degree of $\mathrm{Res}_{X_k}(P_1,R)$ with respect to $X_1,\ldots,X_{k-1},U$ is at most $2d^2$ (the entries of the Sylvester matrix are polynomials of degrees at most d in $X_1,\ldots,X_{k-1},U$). So there are at most 2 d^2 polynomials of degree 2 d^2 in $k-1$ variables to which the induction hypothesis is applied. Thus, the function g defined by

$$\begin{aligned} g(d,1) &= d \\ g(d,k) &= g(2\,d^2,k-1) \end{aligned}$$

bounds the degree of the polynomials A_i. It is clear that $(2\;d)^{2^k}$ is always bigger than $g(d,k)$.

For $P\neq 1$, using Rabinovitch's trick and apply the preceding bound to $P_1,\ldots,P_s,PT-1$, we get an identity

$$A_1\,P_1+\cdots+A_s\,P_s+A\,(P\,T-1)=1,$$

with $A_1,\ \ldots,\ A_s,\ A$ of degree at most $(2\ d)^{2^{k+1}}$. Replacing T by $1/P$ and removing denominators gives an identity

$$P^n=B_1\,P_1+\cdots+B_s\,P_s$$

with $n \le (2\,d)^{2^{k+1}}$ and $\deg(B_i) \le d\,(2\,d)^{2^{k+1}}$. □

The following corollary follows from Theorem 4.82 using the proof of Corollary 4.79.

Corollary 4.83. *Let $\mathcal{P} = \{P_1 ..., P_s\} \subset \mathrm{K}[X_1, ..., X_k]$ be a finite set of less than d homogeneous polynomials of degrees d_i bounded by d. If a homogeneous polynomial $P \in \mathrm{K}[X_1, ..., X_k]$ of degree p vanishes on the common zeros of $\mathcal{P}$ in C^k, there exist $n \in \mathbb{N}$, $n \le (2\ d)^{2^{k+1}}$, and homogeneous polynomials $H_1, ..., H_s$ in $\mathrm{K}[X_1, ..., X_k]$ of respective degrees $c_1, ..., c_s$ such that*

$$\begin{aligned} P^n &= H_1 P_1 + \cdots + H_s P_s \\ n\,p &= c_1 + d_1 = \cdots = c_s + d_s. \end{aligned}$$

Remark 4.84. Note that the double exponential degree bound in the number of variables obtained in Theorem 4.82 comes from the fact that the elimination of one variable between polynomials of degree d using resultant produces polynomials of degree d^2. A similar process occurs in Chapter 11 when we study cylindrical decomposition. □

4.5 Zero-dimensional Systems

Let $\mathcal{P}$ be a finite subset of $\mathrm{K}[X_1, ..., X_k]$. The set of zeros of $\mathcal{P}$ in C^k

$$\mathrm{Zer}(\mathcal{P}, \mathrm{C}^k) = \{x \in \mathrm{C}^k \mid \bigwedge_{P \in \mathcal{P}} P(x) = 0\}$$

is also called the set of **solutions** in C^k of the polynomial system of equations $\mathcal{P} = 0$. Abusing terminology, we speak of the solutions of a polynomial system $\mathcal{P}$. A system of polynomial equations $\mathcal{P}$ is **zero-dimensional** if it has a finite number of solutions in C^k, i.e. if $\mathrm{Zer}(\mathcal{P}, \mathrm{C}^k)$ is a non-empty finite set. We denote by $\mathrm{Ideal}(\mathcal{P}, \mathrm{K})$ the ideal generated by $\mathcal{P}$ in $\mathrm{K}[X_1, ..., X_k]$.

We are going to prove that a system of polynomial equations

$$\mathcal{P} \subset \mathrm{K}[X_1, ..., X_k]$$

is zero-dimensional if and only if the K-vector space

$$\mathrm{A} = \mathrm{K}[X_1, ..., X_k]/\mathrm{Ideal}(\mathcal{P}, \mathrm{K})$$

is finite dimensional. The proof of this result relies on Hilbert's Nullstellensatz.

Theorem 4.85. *Let K be a field, C an algebraically closed field containing K, and $\mathcal{P}$ a finite subset of $\mathrm{K}[X_1, ..., X_k]$.*

The vector space $A = \mathrm{K}[X_1, ..., X_k]/\mathrm{Ideal}(\mathcal{P}, \mathrm{K})$ is of finite dimension $m > 0$ if and only if $\mathcal{P}$ is zero-dimensional.

The number of elements of $\mathrm{Zer}(\mathcal{P}, \mathrm{C}^k)$ *is less than or equal to the dimension of* A *as a* K*-vector space.*

Note that the fact that C is algebraically closed is essential in the statement, since otherwise there exist univariate polynomials of positive degree with no root.

The proof of the theorem will use the following definitions and results.

We consider the ideal $\mathrm{Ideal}(\mathcal{P}, \mathrm{C})$ generated by $\mathcal{P}$ in $\mathrm{C}[X_1, \ldots, X_k]$ and define $\overline{A} = \mathrm{C}[X_1, \ldots, X_k]/\mathrm{Ideal}(\mathcal{P}, \mathrm{C})$. Given $x = (x_1, \ldots, x_k) \in \mathrm{Zer}(\mathcal{P}, \mathrm{C}^k)$ and $Q \in \overline{A}$, $Q(x) \in \mathrm{C}$ is well-defined, since two polynomials in $\mathrm{C}[X_1, \ldots, X_k]$ having the same image in $\overline{A}$ have the same value at x.

The following result makes precise the relationship between A and $\overline{A}$.

Lemma 4.86. $A \subset \overline{A}$.

Proof: If a and b are elements of $\mathrm{K}[X_1, \ldots, X_k]$ equal modulo $\mathrm{Ideal}(\mathcal{P}, \mathrm{C})$, then there exists for each $P \in \mathcal{P}$ a polynomial A_P of some degree d_P in $\mathrm{C}[X_1, \ldots, X_k]$ such that $a - b = \sum A_P P$. Since the various polynomials $A_P P$ are linear combinations of a finite number of monomials, this identity can be seen as the fact that a system of linear equations with coefficients in K has a solution in C (the unknowns being the coefficients of the A_P). We know from elementary linear algebra that this system of linear equations must then also have solutions in K, which means that there are $C_P \in \mathrm{K}[X_1, \ldots, X_k]$ such that $a - b = \sum_{P \in \mathcal{P}} C_P P$. Thus $a = b$ in A. This implies that the inclusion morphism $\mathrm{A} \subset \overline{A}$ is well-defined. □

We also have

Lemma 4.87. *Let* $\mathcal{P}$ *be a finite set of polynomials in* $\mathrm{K}[X_1, \ldots, X_k]$*. Then* $A = \mathrm{K}[X_1, \ldots, X_k]/\mathrm{Ideal}(\mathcal{P}, \mathrm{K})$ *is a finite dimensional vector space of dimension* m *over* K *if and only if* $\overline{A} = \mathrm{C}[X_1, \ldots, X_k]/\mathrm{Ideal}(\mathcal{P}, \mathrm{C})$ *is a finite dimensional vector space of dimension* m *over* C.

Proof: Suppose that A has finite dimension m and consider any finite set of $m' > m$ monomials in $\mathrm{K}[X_1, \ldots, X_k]$. Since the images in A of these monomials are linearly dependent in A over K, the images in $\overline{A}$ of these monomials are linearly dependent in $\overline{A}$ over C. Therefore the dimension of $\overline{A}$ is finite and no greater than the dimension of A, since both A and $\overline{A}$ are spanned by monomials.

For the other direction, if $\overline{A}$ has finite dimension m then we consider any family $B_1, \ldots, B_m$ of $m' > m$ elements in $\mathrm{K}[X_1, \ldots, X_k]$ and denote by $b_1, \ldots, b_{m'}$ their images in A and by $b_1', \ldots, b_{m'}'$ their images in $\overline{A}$. Since $b_1', \ldots, b_{m'}'$ are linearly dependent, there exist $(\lambda_1, \ldots, \lambda_{m'})$ in $\mathrm{C}^{m'}$ which are not all zero and for each $P \in \mathcal{P}$ a polynomial A_P of some degree d_P in $\mathrm{C}[X_1, \ldots, X_k]$ such that

$$\lambda_1 B_1 + \cdots + \lambda_{m'} B_{m'} = \sum A_P P. \tag{4.5}$$

Since the various polynomials $A_P P$ are linear combinations of a finite number of monomials, the identity 4.11 can be seen as the fact that a system of linear equations with coefficients in K has a solution in C (the unknowns being the λ_i and the coefficients of the A_P). We know from elementary linear algebra that this system of linear equations must then also have solutions in K which means that there are $\mu_i \in \mathrm{K}$ not all zero and $C_P \in \mathrm{K}[X_1, \ldots, X_k]$ such that

$$\mu_1 B_1 + \cdots + \mu_{m'} B_{m'} = \sum C_P P.$$

Thus, $b_1, \ldots, b_{m'}$ are linearly dependent over K and hence the dimension of A is no greater than the dimension of $\overline{A}$. □

Definition 4.88. An element a of A is **separating for** $\mathcal{P}$ if a has distinct values at distinct elements of $\mathrm{Zer}(\mathcal{P}, \mathrm{C}^k)$. □

Separating elements always exist when $\mathcal{P}$ is zero-dimensional.

Lemma 4.89. *If* $\#\mathrm{Zer}(\mathcal{P}, \mathrm{C}^k) = n$, *then there exists* i, $0 \le i \le (k-1)\binom{n}{2}$, *such that*

$$a_i = X_1 + i\, X_2 + \cdots + i^{k-1} X_k$$

is separating.

Proof: Let $x = (x_1, \ldots, x_k), y = (y_1, \ldots, y_k)$ be two distinct points of $\mathrm{Zer}(\mathcal{P}, \mathrm{C}^k)$ and let $\ell(x, y)$ be the number of i, $0 \le i \le (k-1)\binom{n}{2}$, such that $a_i(x) = a_i(y)$. Since the polynomial

$$(x_1 - y_1) + (x_2 - y_2)\, T + \cdots + (x_k - y_k)\, T^{k-1}$$

is not identically zero, it has no more than $k-1$ distinct roots. If follows that $\ell(x, y) \le k-1$. As the number of 2-element subsets of $\mathrm{Zer}(\mathcal{P}, \mathrm{C}^k)$ is $\binom{n}{2}$, the lemma is proved. □

An important property of separating elements is the following lemma:

Lemma 4.90. *If* a *is separating and* $\mathrm{Zer}(\mathcal{P}, \mathrm{C}^k)$ *has* n *elements, then* $1, a, \ldots, a^{n-1}$ *are linearly independent in* A.

Proof: Suppose that there exist $c_i \in \mathrm{K}$ such that $\sum_{i=0}^{n-1} c_i a^i = 0$ in A, whence the polynomial $c_0 + c_1\, a + \cdots + c_{n-1}\, a^{n-1}$ is in $\mathrm{Ideal}(\mathcal{P}, \mathrm{K})$. Thus for all $x \in \mathrm{Zer}(\mathcal{P}, \mathrm{C}^k)$, $\sum_{i=0}^{n-1} c_i\, a^i(x) = 0$. The univariate polynomial $\sum_{i=0}^{n-1} c_i\, T^i = 0$ has n distinct roots and is therefore identically zero. □

Proof of Theorem 4.85: If A is a finite dimensional vector space of dimension N over K, then $1, X_1, \ldots, X_1^N$ are linearly dependent in A. Consequently, there is a polynomial $p_1(X_1)$ of degree at most N in the ideal $\mathrm{Ideal}(\mathcal{P}, \mathrm{C})$. It follows that the first coordinate of any $x \in \mathrm{Zer}(\mathcal{P}, \mathrm{C}^k)$ is a root of p_1. Doing the same for all the variables, we see that $\mathrm{Zer}(\mathcal{P}, \mathrm{C}^k)$ is a finite set.

Conversely, if $\operatorname{Zer}(\mathcal{P}, \mathrm{C}^k)$ is finite, take a polynomial $p_1(X_1) \in \mathrm{C}[X_1]$ whose roots are the first coordinates of the elements of $\operatorname{Zer}(\mathcal{P}, \mathrm{C}^k)$. According to Hilbert's Nullstellensatz (Theorem 4.77) a power of p_1 belongs to the ideal $\operatorname{Ideal}(\mathcal{P}, \mathrm{C})$. Doing the same for all variables, we see that for every i, there exists a polynomial of degree d_i in $\mathrm{C}[X_i]$ in the ideal $\operatorname{Ideal}(\mathcal{P}, \mathrm{C})$. It follows that $\overline{A}$ has a basis consisting of monomials whose degree in X_i is less than d_i. Thus, $\overline{A}$ is finite dimensional over C. We conclude that A is finite dimensional over K using Lemma 4.87.

Part b) of the theorem follows from Lemma 4.89 and Lemma 4.90. □

We now explain how the quotient ring $\overline{A}$ splits into a finite number of local factors, one for each $x \in \operatorname{Zer}(\mathcal{P}, \mathrm{C}^k)$). These local factors are used to define the multiplicities of the solutions of the system of polynomial equations. In the case where all the multiplicities are one these local factors will be nothing but the field C itself, and the projection onto the factor corresponding to an $x \in \operatorname{Zer}(\mathcal{P}, \mathrm{C}^k)$ consists in sending an element of $\overline{A}$ to its value at x.

We need a definition. A **local ring** B is a ring, such that for every $a \in B$, either a is invertible or $1 + a$ is invertible. A field is always a local ring.

Exercise 4.2. A ring B is local if and only if has a unique maximal (proper) ideal which is the set of non-invertible elements.

Given a multiplicative subset S of a ring A (i.e. a subset of A closed under multiplication), we define an equivalence relation on ordered pairs (a, s) with $a \in A$ and $s \in S$ by $(a, s) \sim (a', s')$ if and only if there exist $t \in S$ such that $t(a\, s' - a' s) = 0$. The class of (a, s) is denoted a/s. The **ring of fractions** $S^{-1}A$ is the set of classes a/s equipped with the following addition and multiplication

$$a/s + a'/s' = (a\, s' + a' s)/(s\, s'),$$
$$(a/s)(a'/s') = (a\, a')/(s\, s').$$

The **localization of $\overline{A}$ at** $x \in \operatorname{Zer}(\mathcal{P}, \mathbf{C}^k)$ is denoted $\overline{A}_x$. It is the ring of fractions associated to the multiplicative subset S_x consisting of elements of $\overline{A}$ not vanishing at x. The ring $\overline{A}_x$ is local: an element P/Q of $\overline{A}_x$ is invertible if and only if $P(x) \neq 0$, and it is clear that either P/Q is invertible or $1 + P/Q = (Q + P)/Q$ is invertible.

We will prove that the ring $\overline{A}$ is isomorphic to the product of the various $\overline{A}_x$ for $x \in \operatorname{Zer}(\mathcal{P}, \mathrm{C}^k)$. The proof relies on the following result.

Proposition 4.91. *If* $\operatorname{Zer}(\mathcal{P}, \mathrm{C}^k)$ *is finite, then, for every* $x \in \operatorname{Zer}(\mathcal{P}, \mathrm{C}^k)$, *there exists an element* e_x *of* $\overline{A}$ *such that*

- $\sum_{x \in \operatorname{Zer}(\mathcal{P}, C^k)} e_x = 1$,
- $e_x\, e_y = 0$ *for* $y \neq x$ *with* $y, x \in \operatorname{Zer}(\mathcal{P}, \mathrm{C}^k)$,
- $e_x^2 = e_x$,
- $e_x(x) = 1$ *for* $x \in \operatorname{Zer}(\mathcal{P}, \mathrm{C}^k)$,
- $e_x(y) = 0$ *for* $x, y \in \operatorname{Zer}(\mathcal{P}, \mathrm{C}^k)$ *and* $x \neq y$.

Proof: We first prove that, for every $x \in \mathrm{Zer}(\mathcal{P}, \mathrm{C}^k)$, there exists an element s_x of $\overline{A}$ with $s_x(x) = 1$, $s_x(y) = 0$ for every $y \in \mathrm{Zer}(\mathcal{P}, \mathrm{C}^k)$, $y \neq x$. Making if necessary a linear change of variables, we suppose that the variable X_1 is separating. The classical Lagrange interpolation formula gives polynomials in X_1 with the required properties. Namely, writing each $x \in \mathrm{Zer}(\mathcal{P}, \mathrm{C}^k)$ as $(x_1, ..., x_k)$, we set

$$s_x = \prod_{\substack{y \in \mathrm{Zer}(\mathcal{P}, \mathrm{C}^k) \\ y \neq x}} \frac{X_1 - y_1}{x_1 - y_1}.$$

Since $s_x s_y$ vanishes at every common zero of $\mathcal{P}$, Hilbert's Nullstellensatz (Theorem 4.77) implies that there exists a power of each s_x, denoted t_x, such that $t_x\, t_y = 0$ in $\overline{A}$ for $y \neq x$, and $t_x(x) = 1$. The family of polynomials $\mathcal{P} \cup \{t_x \mid x \in \mathrm{Zer}(\mathcal{P}, \mathrm{C}^k)\}$ has no common zero so, according to Hilbert's Nullstellensatz, there exist polynomials r_x such that $\sum t_x\, r_x = 1$ in $\overline{A}$. Take $e_x = t_x\, r_x$. It is now easy to verify the claimed properties. □

The element e_x is called the **idempotent associated to** x. Since $e_x^2 = e_x$, the set $e_x \overline{A}$ equipped with the restriction of the addition and multiplication of $\overline{A}$ is a ring with identity (namely e_x).

Proposition 4.92. *The ring $e_x \overline{A}$ is isomorphic to the localization $\overline{A}_x$ of $\overline{A}$ at x.*

Proof: Note that if $Q(x) \neq 0$, $e_x\, Q$ is invertible in $e_x\, \overline{A}$. Indeed, we can decompose $Q = Q(x)\ (1 + v)$ with $v(x) = 0$. Since $\forall y \in \mathrm{Zer}(\mathcal{P}, \mathrm{C}^k)$, we have $v\, e_x(y) = 0$, $(v\, e_x)^N = 0$ for some $N \in \mathbb{N}$ by Hilbert's Nullstellensatz and thus $e_x\,(1 + v)$ is invertible in $e_x\, \overline{A}$. Its inverse is

$$(1 - e_x\, v + \cdots + (-v)^{N-1})\, e_x,$$

and it follows that $e_x\, Q$ is invertible as well.

So, denoting by $(e_x\, Q)^{-1}$ the inverse of $e_x\, Q$ in $e_x \overline{A}$, consider the mapping from $\overline{A}_x$ to $\overline{A}$ which sends P/Q to $P\,(e_x\, Q)^{-1} = e_x\, P\,(e_x\, Q)^{-1}$. It is easy to check that this is a ring homomorphism. Conversely, to $P e_x$ is associated $P/1$, which is a ring homomorphism from $e_x\, \overline{A}$ to $\overline{A}_x$.

To see that these two ring homomorphisms are inverses to each other, we need only prove that $(P(e_x Q)^{-1})/1 = P/Q$ in $\overline{A}_x$. This is indeed the case since

$$P e_x((e_x\, Q)\,(e_x\, Q)^{-1} - 1) = 0$$

and $e_x(x) = 1$. □

We now prove that $\overline{A}$ is the product of the $\overline{A}_x$.

Theorem 4.93. *For each $x \in \mathrm{Zer}(\mathcal{P}, \mathrm{C}^k)$ there exists an idempotent e_x such that $e_x \overline{A} = \overline{A}_x$ and*

$$\overline{A} \cong \prod_{x \in \mathrm{Zer}(\mathcal{P}, C^k)} \overline{A}_x.$$

Proof: Since $\sum_{x \in \mathrm{Zer}(\mathcal{P}, \mathrm{C}^k)} e_x = 1$, $\overline{A} \cong \prod_{x \in \mathrm{Zer}(\mathcal{P}, \mathrm{C}^k)} \overline{A}_x$. The canonical surjection of $\overline{A}$ onto $\overline{A}_x$ coincides with multiplication by e_x. □

We denote by $\mu(x)$ the dimension of $\overline{A}_x$ as a C-vector space. We call $\mu(x)$ the **multiplicity** of the zero $x \in \mathrm{Zer}(\mathcal{P}, \mathrm{C}^k)$.

If the multiplicity of x is 1 we say that x is **simple** . Then $\overline{A}_x = \mathrm{C}$ and the canonical surjection $\overline{A}$ onto $\overline{A}_x$ coincides with the homomorphism from $\overline{A}$ to C sending P to its value at x. Indeed, suppose that $P(x) = 0$. Then $P e_x(y) = 0$ for every $y \in \mathrm{Zer}(\mathcal{P}, \mathrm{C}^k)$ and hence by Hilbert's Nullstellensatz there exists $N \in \mathbb{N}$ such that $(P e_x)^N = 0$. Since e_x is idempotent this implies that $P^N e_x = 0$, and thus $P^N = 0$ in $\overline{A}_x$ which is a field. Hence $P = 0$ in $\overline{A}_x$.

When the system of polynomial equations $\mathcal{P} = \{P_1, \ldots, P_k\}$ is zero-dimensional, simple zeros coincide with non-singular zeros as we see now.

Let $P_1, \ldots, P_k$ be polynomials in $\mathrm{C}[X_1, \ldots, X_k]$. A **non-singular zero** of

$$P_1(X_1, \ldots, X_k), \ldots, P_k(X_1, \ldots, X_k)$$

is a k-tuple $x = (x_1, \ldots, x_k)$ of elements of C^k such that

$$P_1(x_1, \ldots, x_k) = \cdots = P_k(x_1, \ldots, x_k) = 0$$

and $\det\left(\left[\frac{\partial P_i}{\partial X_j}(x)\right]\right) \neq 0$.

Proposition 4.94. *Let $\mathcal{P} = \{P_1, \ldots, P_k\} \subset K[X_1, \ldots, X_k]$ be a zero dimensional system and x a zero of $\mathcal{P}$ in C^k. Then the following are equivalent*

a) x is a non-singular zero of $\mathcal{P}$,
b) x is simple, i.e. the multiplicity of x is 1 and $\overline{A}_x = \mathrm{C}$,
c) $M_x \subset \mathrm{Ideal}(\mathcal{P}, \mathrm{C}) + (M_x)^2$, denoting by M_x the ideal of elements of $\mathrm{C}[X_1, \ldots, X_k]$ vanishing at x.

Proof: *a)* $\Rightarrow$ *c)* Using Taylor's formula at x,

$$P_j = \sum_i \frac{\partial P_j}{\partial X_i}(x)(X_i - x_i) + B_j$$

with $B_j \in (M_x)^2$. So

$$\sum_i \frac{\partial P_j}{\partial X_i}(x)(X_i - x_i) \in \mathrm{Ideal}(\mathcal{P}, \mathrm{K}) + (M_x)^2.$$

Since the matrix $\left[\frac{\partial P_j}{\partial X_i}(x)\right]$ is invertible, for every i

$$(X_i - x_i) \in \mathrm{Ideal}(\mathcal{P}, \mathrm{K}) + (M_x)^2.$$

c) $\Rightarrow$ *b)* Since $(X_i - x_i) e_x$ vanishes on $\mathrm{Zer}(\mathcal{P})$ for every i, and $e_x^2 = e_x$, according to Hilbert's Nullstellensatz, there exists N_i such that

$$(X_i - x_i)^{N_i} e_x \in \mathrm{Ideal}(\mathcal{P}, \mathrm{K}).$$

So there exist N such that $(M_x)^N \cdot e_x \subset \mathrm{Ideal}(\mathcal{P}, \mathrm{K})$. Using repeatedly $M_x \subset \mathrm{Ideal}(\mathcal{P}, \mathrm{K}) + (M_x)^2$, we get

$$(M_x)^{N-1} \cdot e_x \subset \mathrm{Ideal}(\mathcal{P}, \mathrm{K}), \ldots, M_x \cdot e_x \subset \mathrm{Ideal}(\mathcal{P}, \mathrm{K}).$$

This implies $\overline{A}_x = \mathrm{C}$.

b) $\Rightarrow$ *a)* If $\overline{A}_x = \mathrm{C}$, then for every i, $(X_i - x_i)\, e_x \in \mathrm{Ideal}(\mathcal{P}, \mathrm{K})$. Indeed $(X_i - x_i)\, e_x = 0$ in $\overline{A}_x = C$ and $(X_i - x_i)\, e_x\, e_y = 0$ in $\overline{A}_y$ for $y \neq x$ and $y \in \mathrm{Zer}(\mathcal{P}, \mathrm{C})$. So, for every i there exist $A_{i,j}$ such that

$$(X_i - x_i)\, e_x = \sum_j A_{i,j} P_j.$$

Differentiating with respect to X_i and $X_\ell, \ell \neq i$ and evaluating at x we get

$$\begin{aligned} 1 &= \sum_j A_{i,j}(x) \frac{\partial P_j}{\partial X_i}(x), \\ 0 &= \sum_j A_{i,j}(x) \frac{\partial P_j}{\partial X_\ell}(x), \ell \neq i, \end{aligned}$$

so the matrix $\left[\frac{\partial P_j}{\partial X_i}(x)\right]$ is invertible. □

4.6 Multivariate Hermite's Quadratic Form

We consider a zero dimensional system $\mathcal{P}$ and its set of solutions in C^k

$$\mathrm{Zer}(\mathcal{P}, \mathrm{C}^k) = \{x \in \mathrm{C}^k \mid \bigwedge_{P \in \mathcal{P}} P(x) = 0\}.$$

We indicate first the relations between $\mathrm{Zer}(\mathcal{P}, \mathrm{C}^k)$ and the eigenvalues of certain linear maps on the finite dimensional vector spaces

$$\begin{aligned} A &= \mathrm{K}[X_1, \ldots, X_k]/\mathrm{Ideal}(\mathcal{P}, \mathrm{K}) \quad \text{and} \\ \overline{A} &= \mathrm{C}[X_1, \ldots, X_k]/\mathrm{Ideal}(\mathcal{P}, \mathrm{C}). \end{aligned}$$

Notation 4.95. [Multiplication map] If $f \in \mathrm{A}$, we denote by $L_f \colon \mathrm{A} \to \mathrm{A}$ the linear map of multiplication by f defined by $L_f(g) = f g$ for $g \in \mathrm{A}$. Similarly, if $f \in \overline{A}$, we also denote by' $L_f \colon \overline{A} \mapsto \overline{A}$ the linear map of multiplication by f defined by $L_f(g) = f g$ for $g \in \overline{A}$. By Lemma 4.86, $\mathrm{A} \subset \overline{A}$, so we denote also by $L_f \colon \overline{A} \mapsto \overline{A}$ the linear map of multiplication by $f \in \mathrm{A}$ defined by $L_f(g) = f g$ for $g \in \overline{A}$ and for $f \in \mathrm{A}$. □

We denote as above by $\overline{A}_x$ the localization at a zero x of $\mathcal{P}$ and by $\mu(x)$ its multiplicity. We denote by $L_{f,x}$ the linear map of multiplication by f from $\overline{A}_x$ to $\overline{A}_x$ defined by $L_{f,x}(P/Q) = f P/Q$. Note that $L_{f,x}$is well-defined since if $P_1/Q_1 = P/Q$,then $f P_1/Q_1 = f P/Q$. Considering $\overline{A}_x$ as a sub-vector space of $\overline{A}$, $L_{f,x}$ is the restriction of L_f to $\overline{A}_x$.

Theorem 4.96. *The eigenvalues of L_f are the $f(x)$, with multiplicity $\mu(x)$, for $x \in \mathrm{Zer}(\mathcal{P}, \mathrm{C}^k)$.*

Proof: As $e_x(f - f(x))$ vanishes on $\mathrm{Zer}(\mathcal{P}, \mathrm{C}^k)$, Hilbert's Nullstellensatz (Theorem 4.77) implies that there exists $m \in \mathbb{N}$ such that $(e_x(f - f(x)))^m = 0$ in $\overline{A}$, which means that $L_{e_x(f - f(x))}$ is nilpotent and has a unique eigenvalue 0 with multiplicity $\mu(x)$. Thus $L_{f,x}$ has only one eigenvalue $f(x)$ with multiplicity $\mu(x)$. Using Theorem 4.93 completes the proof. □

It follows immediately:

Theorem 4.97. [Stickelberger] *For $f \in \overline{A}$, the linear map L_f has the following properties:*

The trace of L_f is

$$\mathrm{Tr}(L_f) = \sum_{x \in \mathrm{Zer}(\mathcal{P}, C^k)} \mu(x) f(x). \tag{4.6}$$

The determinant of L_f is

$$\det(L_f) = \prod_{x \in \mathrm{Zer}(\mathcal{P}, C^k)} f(x)^{\mu(x)}. \tag{4.7}$$

The characteristic polynomial $\chi(\mathcal{P}, f, T)$ of L_f is

$$\chi(\mathcal{P}, f, T) = \prod_{x \in \mathrm{Zer}(\mathcal{P}, C^k)} (T - f(x))^{\mu(x)}. \tag{4.8}$$

Note that the statement on the trace is a generalization of Proposition 4.54.

Remark 4.98. Note that if $f \in \mathrm{A}$, $\mathrm{Tr}(L_f)$ and $\det(L_f)$ are in K and $\chi(\mathcal{P}, f, T) \in \mathrm{K}[T]$. Moreover, if the multiplication table of A in the basis $\mathcal{B}$ has entries in a ring D contained in K and f has coefficients in D in the basis $\mathcal{B}$, $\mathrm{Tr}(L_f)$ and $\det(L_f)$ are in D and $\chi(\mathcal{P}, f, T) \in \mathrm{D}[T]$. □

A consequence of Theorem 4.97 (Stickelberger) is a multivariate generalization of the univariate Hermite's theorem seen earlier in this chapter (Theorem 4.57).

We first define the multivariate generalization of Hermite quadratic form. For every $Q \in \mathrm{A}$, we define the **Hermite's bilinear map** as the bilinear map:

$$\begin{array}{rccc} \mathrm{her}(\mathcal{P}, Q): & \mathrm{A} \times \mathrm{A} & \longrightarrow & K \\ & (f, g) & \longmapsto & \mathrm{Tr}(L_{fgQ}) \end{array}.$$

The corresponding quadratic form associated to $\mathrm{her}(\mathcal{P}, Q)$ is called the **Hermite's quadratic form**,

$$\begin{array}{rccc} \mathrm{Her}(\mathcal{P}, Q): & \mathrm{A} & \longrightarrow & K \\ & f & \longmapsto & \mathrm{Tr}(L_{f^2 Q}) \end{array}.$$

When $Q=1$ we simply write $\mathrm{her}(\mathcal{P})=\mathrm{her}(\mathcal{P},1)$ and $\mathrm{Her}(\mathcal{P})=\mathrm{Her}(\mathcal{P},1)$.

We shall write $\mathrm{A}_{\mathrm{rad}}$ to denote the ring $\mathrm{K}[X_1,\ldots,X_k]/\sqrt{\mathrm{Ideal}(\mathcal{P},\mathrm{K})}$.

The next theorem gives the connection between the radical of $\mathrm{Ideal}(\mathcal{P},\mathrm{K})$ and the radical of the quadratic form $\mathrm{Her}(\mathcal{P})$:

$$\mathrm{Rad}(\mathrm{Her}(\mathcal{P}))=\{f\in \mathrm{A}\mid \forall g\in \mathrm{A}\ \mathrm{her}(\mathcal{P})(f,g)=0\}.$$

Theorem 4.99.

$$\sqrt{\mathrm{Ideal}(\mathcal{P},\mathrm{K})}=\mathrm{Rad}(\mathrm{Her}(\mathcal{P})).$$

Proof: Let f be an element of $\sqrt{\mathrm{Ideal}(\mathcal{P},\mathrm{K})}$. Then f vanishes on every element of $\mathrm{Zer}(\mathcal{P},\mathrm{C}^k)$. So, applying Corollary 4.97, we obtain the following equality for every $g\in K[X_1,\ldots,X_k]$:

$$\mathrm{her}(\mathcal{P})(f,g)=\mathrm{Tr}(L_{fg})=\sum_{x\in \mathrm{Zer}(\mathcal{P},\mathrm{C}^k)}\mu(x)f(x)g(x)=0.$$

Conversely, if f is an element such that $\mathrm{her}(\mathcal{P})(f,g)=0$ for any g in A, then Corollary 4.97 gives:

$$\forall g\in \mathrm{A}\ \mathrm{her}(\mathcal{P})(f,g)=\mathrm{Tr}(L_{fg})=\sum_{x\in \mathrm{Zer}(\mathcal{P},\mathrm{C}^k)}\mu(x)f(x)g(x)=0. \tag{4.9}$$

Let a be a separating element (Definition 4.88). If $\mathrm{Zer}(\mathcal{P},\mathrm{C}^k)=\{x_1,\ldots,x_n\}$, Equality (4.15) used with each of $g=1,\ldots,a^{n-1}$ gives,

$$\begin{bmatrix} 1 & \ldots & 1\\ \vdots & & \vdots\\ a(x_1)^{n-1} & \ldots & a(x_n)^{n-1}\end{bmatrix}\cdot\begin{bmatrix}\mu(x_1)\,f(x_1)\\ \vdots\\ \mu(x_n)\,f(x_n)\end{bmatrix}=\begin{bmatrix}0\\ \vdots\\ 0\end{bmatrix}$$

so that $f(x_1)=\cdots=f(x_n)=0$, since a is separating and the matrix at the left hand side is a Vandermonde matrix, hence invertible. Using Hilbert's Nullstellensatz 4.71, we obtain $f\in\sqrt{\mathrm{Ideal}(\mathcal{P},\mathrm{K})}$ as desired. □

The following result generalizes Hermite's Theorem (Theorem 4.57) and has a very similar proof.

We define the Tarski-query of Q for $\mathcal{P}$ as

$$\mathrm{TaQ}(Q,\mathcal{P})=\sum_{x\in \mathrm{Zer}(\mathcal{P},\mathrm{R}^k)}\mathrm{sign}(Q(x))$$

Theorem 4.100. [Multivariate Hermite]

$$\begin{aligned}\mathrm{Rank}(\mathrm{Her}(\mathcal{P},Q)) &= \#\{x\in \mathrm{Zer}(\mathcal{P},\mathrm{C}^k)\mid Q(x)\neq 0\},\\ \mathrm{Sign}(\mathrm{Her}(\mathcal{P},Q)) &= \mathrm{TaQ}(Q,\mathcal{P}).\end{aligned}$$

Proof: Consider a separating element a. The elements 1, a, ..., a^{n-1} are linearly independent in A by Lemma 4.90 and can be completed to a basis $\omega_1=1,\omega_2=a,\ldots,\omega_n=a^{n-1},\omega_{n+1},\ldots,\omega_N$ of the K-vector space A.

Corollary 4.97 provides the following expression for the quadratic form $\mathrm{Her}(\mathcal{P}, Q)$: if $f = \sum_{j=1}^{N} f_j \omega_j \in \mathrm{A}$

$$\mathrm{Her}(\mathcal{P}, Q)(f) = \sum_{x \in \mathrm{Zer}(\mathcal{P}, \mathrm{C}^k)} \mu(x) Q(x) \left(\sum_{j=1}^{N} f_j \omega_j(x) \right)^2.$$

Consequently, denoting $\mathrm{Zer}(\mathcal{P}, \mathrm{C}^k) = \{x_1, \ldots, x_n\}$, $\mathrm{Her}(\mathcal{P}, Q)$ is the map

$$f \mapsto (f_1, \ldots, f_N) \cdot \Gamma^t \cdot \Delta(\mu(x_1)\, Q(x_1), \ldots, \mu(x_n)\, Q(x_n)) \cdot \Gamma \cdot (f_1, \ldots, f_N)^t,$$

where

$$\Gamma = \begin{bmatrix} 1 & a(x_1) & \ldots & a(x_1)^{n-1} & \omega_{n+1}(x_1) & \ldots & \omega_N(x_1) \\ \vdots & \vdots & & \vdots & \vdots & & \vdots \\ 1 & a(x_n) & \ldots & a(x_n)^{n-1} & \omega_{n+1}(x_n) & \ldots & \omega_N(x_n) \end{bmatrix}$$

and Δ denotes a diagonal matrix with the indicated diagonal entries. Therefore it suffices to prove that the rank of Γ is equal to n. But a is separating and the principal minor of the matrix Γ is a Vandermonde determinant.

Given $(f_1, \ldots, f_N)$, let $f = \sum_{i=1}^{N} f_i\, \omega_i$. According to Corollary 4.97, the quadratic form $\mathrm{Her}(\mathcal{P}, Q)$ is given in this basis by

$$\sum_{y \in \mathrm{Zer}(\mathcal{P}, \mathrm{R}^k)} \mu(y) Q(y) \left(\sum_{i=1}^{N} f_i\, \omega_i(y) \right)^2 + \sum_{z \in \mathrm{Zer}(\mathcal{P}, \mathrm{C}^k) \setminus \mathrm{Zer}(\mathcal{P}, \mathrm{R}^k)} \mu(z) Q(z) \left(\sum_{i=1}^{N} f_i\, \omega_i(z) \right)^2$$

as a quadratic form in the variables f_i. We have already seen in the first part of the proof that the n rows of Γ are linearly independent over C. Moreover, if z and $\overline{z}$ are complex conjugate solutions of $\mathcal{P}$, with $Q(z) \neq 0$,

$$\mu(z) Q(z) \left(\sum_{i=1}^{N} f_i\, \omega_i(z) \right)^2 + \mu(\overline{z}) Q(\overline{z}) \left(\sum_{i=1}^{N} f_i\, \omega_i(\overline{z}) \right)^2$$

is easily seen to be a difference of two squares of real linear forms. Indeed, writing $\mu(z)\, Q(z) = (a(z) + i\, b(z))^2$,

$$(a(z) + i\, b(z)) \left(\sum_{i=1}^{N} f_i\, \omega_i(x) \right) = L_{1,z} + i\, L_{2,z},$$

with $s_i(z)$ and $i\, t_i(z)$ the real and imaginary part of $\omega_i(z)$,

$$L_{1,z} = \sum_{i=1}^{N} (a(z)\, s_i(z) - b(z)\, t_i(z))\, f_i$$

$$L_{2,x} = \sum_{i=1}^{N} (a(z)\, t_i(z) + b(z)\, s_i(z))\, f_i$$

are real linear forms in $f_1, \ldots, f_N$ with coefficients in R so that

$$\mu(z)Q(z)\left(\sum_{i=1}^{N} f_i\omega_i(z)\right)^2 + \mu(\overline{z})Q(\overline{z})\left(\sum_{i=1}^{N} f_i\omega_i(\overline{z})\right)^2 = 2\,L_{1,z}^2 - 2\,L_{2,z}^2.$$

Moreover, $L(y,f), L_1(z), L_2(z)$ $(y \in \mathrm{Zer}(\mathcal{P}, R^k),\ z, \overline{z} \in \mathrm{Zer}(\mathcal{P}, \mathrm{C}^k) \setminus \mathrm{Zer}(\mathcal{P}, \mathrm{R}^k))$ are linearly independent linear forms.

So the signature of $\mathrm{Her}(\mathcal{P}, Q)$ is the signature of

$$\sum_{y \in \mathrm{Zer}(\mathcal{P}, \mathrm{R}^k)} \mu(y)Q(y)L(y,f)^2.$$

Since the linear forms $L(y, f)$, are linearly independent the signature of $\mathrm{Her}(\mathcal{P}, Q)$ is $\sum_{y \in \mathrm{Zer}(\mathcal{P},\mathrm{R}^k)} \mathrm{sign}(Q(y)) = \mathrm{TaQ}(Q, \mathcal{P})$. □

4.7 Projective Space and a Weak Bézout's Theorem

Let R be a real closed field and $\mathrm{C} = \mathrm{R}[i]$. The **complex projective space of dimension** k, $\mathbb{P}_k(\mathrm{C})$, is the set of lines of C^{k+1} through the origin.

A $(k+1)$-tuple $x = (x_0, x_1, \ldots, x_k) \neq (0, 0, \ldots, 0)$ of elements of C defines a line $\overline{x}$ through the origin. This is denoted by $\overline{x} = (x_0 \colon x_1 \colon \ldots \colon x_k)$ and $(x_0, x_1, \ldots, x_k)$ are **homogeneous coordinates** of $\overline{x}$. Clearly,

$$(x_0 \colon x_1 \colon \ldots \colon x_k) = (y_0 \colon y_1 \colon \ldots \colon y_k)$$

if and only if there exists $\lambda \neq 0$ in C with $x_i = \lambda y_i$.

A polynomial P in $\mathrm{C}[X_{1,0}, \ldots, X_{1,k_1}, \ldots, X_{m,0}, \ldots, X_{m,k_m}]$ is **multi-homogeneous** of multidegree $d_1, \ldots, d_m$ if it is homogeneous of degree d_i in the block of variables $X_{i,0}, \ldots, X_{i,k_i}$ for every $i \leq m$.

For example $T(X^2 + Y^2)$ is homogeneous of degree 1 in the variable T and homogeneous of degree 2 in the variables$\{X, Y\}$.

If P is multi-homogeneous of multidegree $d_1, \ldots, d_m$, a **zero** of P in $\mathbb{P}_{k_1}(\mathrm{C}) \times \ldots \times \mathbb{P}_{k_m}(\mathrm{C})$ is a point

$$x = (\overline{x_1}, \ldots, \overline{x_m}) \in \mathbb{P}_{k_1}(\mathrm{C}) \times \ldots \times \mathbb{P}_{k_m}(\mathrm{C})$$

such that $P(x_1, \ldots, x_m) = 0$, and this property denoted by $P(x) = 0$ depends only on x and not on the choice of the homogeneous coordinates. An **algebraic set of** $\mathbb{P}_{k_1}(\mathbf{C}) \times \ldots \times \mathbb{P}_{k_m}(\mathbf{C})$ is a set of the form

$$\mathrm{Zer}(\mathcal{P}, \prod_{i=1}^{m} \mathbb{P}_{k_i}(\mathrm{C})) = \{x \in \prod_{i=1}^{m} \mathbb{P}_{k_i}(\mathrm{C}) \mid \bigwedge_{P \in \mathcal{P}} P(x) = 0\},$$

where $\mathcal{P}$ is a finite set of multi-homogeneous polynomials in

$$\mathrm{C}[X_1, \ldots, X_m] = \mathrm{C}[X_{1,0}, \ldots, X_{1,k_1}, \ldots, X_{m,0}, \ldots, X_{m,k_m}].$$

Lemma 4.101. *An algebraic subset of* $\mathbb{P}_1(\mathrm{C})$ *is either* $\mathbb{P}_1(\mathrm{C})$ *or a finite set of points.*

Proof: Let $\mathcal{P} = \{P_1, \ldots, P_s\}$ with P_i homogeneous of degree d_i. If all the P_i are zero, $\mathrm{Zer}(\mathcal{P}, \mathbb{P}_1(\mathrm{C})) = \mathbb{P}_1(\mathrm{C})$. Otherwise, $\mathrm{Zer}(\mathcal{P}, \mathbb{P}_1(\mathrm{C}))$ contains $(0{:}\,1)$ if and only if

$$P_1(0, X_1) = \cdots = P_s(0, X_1) = 0.$$

The other elements of $\mathrm{Zer}(\mathcal{P}, \mathbb{P}_1(\mathrm{C}))$ are the points of the form $(1{:}\,x_1)$, where x_1 is a solution of

$$P_1(1, X_1) = \cdots = P_s(1, X_1) = 0,$$

which is a finite number of points since the $P_i(1, X_1)$ are not all zero. □

Theorem 4.102. *If* $V \subset \mathbb{P}_{k_1}(\mathrm{C}) \times \mathbb{P}_{k_2}(\mathrm{C})$ *is algebraic, its projection on* $\mathbb{P}_{k_2}(\mathrm{C})$ *is algebraic.*

Proof: We first introduce some notation. With $X = (X_0, \ldots, X_k)$, we denote the set of homogeneous polynomials of degree d in X by $\mathrm{C}[X]_d$. Let $\mathcal{P} = \{P_1, \ldots, P_s\}$ be a finite set of homogeneous polynomials with P_i of degree d_i in X. For $d \geq d_i$, let $M_d(\mathcal{P})$ be the mapping

$$\mathrm{C}[X]_{d-d_1} \times \ldots \times \mathrm{C}[X]_{d-d_s} \to C[X]_d$$

sending $(H_1, \ldots, H_s)$ to $H_1\, P_1 + \cdots + H_s\, P_s$. Identifying a homogeneous polynomial with its vector of coefficients in the basis of monomials, $M_d(\mathcal{P})$ defines a matrix $\mathcal{M}_d(\mathcal{P})$.

The projection of $\mathrm{Zer}(\mathcal{P}, \mathbb{P}_{k_1}(\mathrm{C}) \times \mathbb{P}_{k_2}(\mathrm{C}))$ on $\mathbb{P}_{k_2}(\mathrm{C})$ is

$$\pi(\mathrm{Zer}(\mathcal{P}, \mathbb{P}_{k_1}(\mathrm{C}) \times \mathbb{P}_{k_2}(\mathrm{C})) = \{\overline{y} \in \mathbb{P}^{k_2}(\mathrm{C}) \mid \exists \overline{x} \in \mathbb{P}_{k_1}(\mathrm{C}) \bigwedge_{P \in \mathcal{P}} P(x, y) = 0\}$$

Consider $\overline{y} \notin \pi(V)$, i.e. $\overline{y} \in \mathbb{P}_{k_2}(\mathrm{C})$ and such that

$$\{\overline{x} \in \mathbb{P}^{k_1}(\mathrm{C}) \mid \bigwedge_{P \in \mathcal{P}} P(x, y) = 0\} = \emptyset.$$

Then

$$\{x \in C^{k_1+1} \mid \bigwedge_{P \in \mathcal{P}} P(x, y) = 0\} = \{0\}.$$

According to Corollary 4.83, there exists for every $i = 0, \ldots, k_i$ an integer $n_i \leq (2d)^{2^{k_1+2}}$ and polynomials $A_{i,j} \in \mathrm{C}[X]_{n_i - d_i}$ such that

$$X_i^{n_i} = A_{i,1}(X)\, P_1(X, y) + \cdots + A_{i,s}(X)\, P_s(X, y).$$

Since any monomial of degree $N = \sum_{i=0}^{k_1} (2d)^{2^{k_1+2}}$ is a multiple of $X_i^{n_i}$ for some $1 \leq i \leq k_1$, for every polynomial $P \in \mathrm{C}[X]_N$ there exist polynomials $H_1, \ldots, H_s$ with $H_i \in \mathrm{C}[X]_{N-d_i}$ such that

$$P = H_1(X)\, P_1(X, y) + \cdots + H_s(X)\, P_s(X, y).$$

We have proved that $\overline{y} \notin \pi(V)$ if and only if $M_N(\{P_1(X,y),...,P_s(X,y)\})$ is surjective. This can be expressed by a finite disjunction of conditions $M_i(y) \neq 0$, where the $M_i(Y)$ are the maximal minors extracted from the matrix $\mathcal{M}_N(\{P_1(X, Y, ..., P_s(X, Y)\})$ in which $Y = (Y_0, ..., Y_{k_2})$ appear as variables. Hence, $\pi(V) = \{\overline{y} \mid \bigwedge M_i(y) = 0\}$, which is an algebraic set of $\mathbb{P}_{k_2}(\mathrm{C})$. □

The remainder of the chapter is devoted to proving a weak version of Bézout's theorem, estimating the number of non-singular projective zeros of a polynomial system of equations. The proof of this theorem is quite simple. The basic idea is that we look at a polynomial system which has exactly the maximum number of such zeros and move continuously from this polynomial system to the one under consideration in such a way that the number of non-singular projective zeros cannot increase. In order to carry out this simple idea, we define projective zeros and elaborate a little on the geometry of $\mathbb{P}_k(\mathrm{C})$.

If P_1, ..., P_k are homogeneous polynomials in $\mathrm{C}[X_0, \ldots, X_k]$, we say that $x = (x_0 : x_1 : \ldots : x_k) \in \mathbb{P}_k(\mathrm{C})$ is a **non-singular projective zero** of $P_1, \ldots, P_k$ if $P_i(x) = 0$ for $i = 1, \ldots, k$ and

$$\mathrm{rank}\left(\left[\frac{\partial P_i}{\partial X_j}(x)\right]\right) = k,$$

for $i = 1, \ldots, k$, $j = 0, \ldots, k$. Note that $(x_1, \ldots, x_k)$ is a non-singular zero of

$$P_1(1, X_1, \ldots, X_k), \ldots, P_k(1, X_1, \ldots, X_k)$$

if and only if $(1 : x_1 : \ldots : x_k)$ is a non-singular projective zero of

$$P_1, \ldots, P_k.$$

For $i = 0, \ldots, k$, let ϕ_i be the map from C^k to $\mathbb{P}_k(\mathrm{C})$ which maps $(x_1, \ldots, x_k)$ to $(x_1 : \ldots : x_{i-1} : 1 : x_i : \ldots : x_k)$, and let $\mathcal{U}_i = \phi_i(\mathrm{C}^k)$. Note that

$$\begin{aligned} \mathcal{U}_i &= \{\overline{x} \in \mathbb{P}_k(\mathrm{C}) \mid x_i \neq 0\}, \\ \phi_i^{-1}(x_0 : x_{i-1} : x_i : x_{i+1} : \ldots : x_k) &= \left(\frac{x_0}{x_i}, \ldots, \frac{x_{i-1}}{x_i}, \frac{x_{i+1}}{x_i}, \ldots, \frac{x_k}{x_i}\right). \end{aligned}$$

It is clear that $\cup_{i=0,\ldots,k} \mathcal{U}_i = \mathbb{P}_k(\mathrm{C})$. It is also clear that $\phi_i^{-1}(\mathcal{U}_i \cap \mathcal{U}_j)$ is a semi-algebraic open subset of $\mathrm{C}^k = R^{2k}$ and that $\phi_j^{-1} \circ \phi_i$ is a semi-algebraic bijection from $\phi_i^{-1}(\mathcal{U}_i \cap \mathcal{U}_j)$ to $\phi_j^{-1}(\mathcal{U}_i \cap \mathcal{U}_j)$.

We define the euclidean topology and semi-algebraic sets of $\mathbb{P}_k(\mathrm{C})$ as follows:

- a subset U of $\mathbb{P}_k(\mathrm{C})$ is **open** in the euclidean topology if only if for every $i = 0, \ldots, k$, $\phi_i^{-1}(U \cap \mathcal{U}_i)$ is an open set in the euclidean topology of $\mathrm{C}^k = \mathrm{R}^{2k}$,
- a subset S of $\mathbb{P}_k(\mathrm{C})$ is **semi-algebraic** if only if for every $i = 0, \ldots, k$, $\phi_i^{-1}(S \cap \mathcal{U}_i)$ is semi-algebraic in $\mathrm{C}^k = \mathrm{R}^{2k}$.

Note that the $\mathcal{U}_i$ are semi-algebraic open subsets of $\mathbb{P}_k(\mathrm{C})$.

Similarly, it is easy to define the semi-algebraic sets of $\mathbb{P}_k(\mathrm{C}) \times \mathbb{P}_\ell(\mathrm{C})$. A semi-algebraic mapping from $\mathbb{P}_k(\mathrm{C})$ to $\mathbb{P}_\ell(\mathrm{C})$ is a mapping whose graph is semi-algebraic.

Since every point of $\mathbb{P}_k(\mathrm{C})$ has a neighborhood that is contained in some $\mathcal{U}_i$, the local properties of $\mathbb{P}_k(\mathrm{C})$ are the same as the local properties of $\mathrm{C}^k = \mathrm{R}^{2k}$. In particular the notion of differentiability and the classes $\mathcal{S}^m$ and $\mathcal{S}^\infty$ can be defined in a similar way and the corresponding implicit function theorem remains valid.

Theorem 4.103. [Projective Implicit Function Theorem]
Let $(x^0, y^0) \in \mathbb{P}_k(\mathrm{C}) \times \mathbb{P}_\ell(\mathrm{C})$, and let $f_1, \ldots, f_\ell$ be semi-algebraic functions of class $\mathcal{S}^m$ on an open neighborhood of $(\overline{x^0}, \overline{y^0})$ such that $f_j(\overline{x^0}, \overline{y^0}) = 0$ for $j = 1, \ldots, \ell$ and the Jacobian matrix

$$\left[\frac{\partial f_j}{\partial y_i}(\overline{x^0}, \overline{y^0})\right]$$

is invertible. Then there exists a semi-algebraic open neighborhood U (resp. V) of $\overline{x^0}$ (resp. $\overline{y^0}$) in $\mathbb{P}_k(\mathrm{C})$ (resp. $\mathbb{P}_\ell(\mathrm{C})$) and a function $\varphi \in \mathcal{S}^m(U, V)$ such that $\varphi(\overline{x^0}) = \overline{y^0}$ and such that for every $(\overline{x}, \overline{y}) \in U \times V$, we have

$$f_1(\overline{x}, \overline{y}) = \cdots = f_\ell(\overline{x}, \overline{y}) = 0 \Leftrightarrow \overline{y} = \varphi(\overline{x}) .$$

Our final observation is the following lemma showing that the complement of a finite subset of $\mathbb{P}_1(\mathrm{C})$ is semi-algebraically path connected.

If S is a semi-algebraic subset of $\mathbb{P}_k(\mathrm{C})$, we say that S is **semi-algebraically path connected** if for every x and y in S, there exists a continuous path from x to y, i.e. a continuous mapping γ from $[0, 1]$ to S such that $\gamma(0) = x, \gamma(1) = y$ and the graph of γ is semi-algebraic.

Lemma 4.104. *If Δ is a finite subset of $\mathbb{P}_1(\mathrm{C})$, then $\mathbb{P}_1(\mathrm{C}) \setminus \Delta$ is semi-algebraically path connected.*

Proof: If x and y both belong to $\mathcal{U}_0$ (resp. $\mathcal{U}_1$), it is clear that there is a semi-algebraic continuous path from $\phi_0^{-1}(x)$ to $\phi_0^{-1}(y)$ (resp. $\phi_1^{-1}(x)$ to $\phi_1^{-1}(y)$) avoiding $\phi_0^{-1}(\Delta \cap \mathcal{U}_0)$ (resp. $\phi_1^{-1}(\Delta \cap \mathcal{U}_0)$). If $x \in \mathcal{U}_0$, $y \in \mathcal{U}_1$, take $z \in (\mathbb{P}_1(\mathrm{C}) \setminus \Delta) \cap \mathcal{U}_0 \cap \mathcal{U}_1$ and connect x to z and then z to y outside Δ by semi-algebraic and continuous paths. □

Proposition 4.105. *Let $P_1, \ldots, P_k \in \mathrm{C}[X_0, \ldots, X_k]$ be homogeneous polynomials of degrees $d_1, \ldots, d_k$. Then the number of non-singular projective zeros of $P_1, \ldots, P_k$ is at most $d_1 \cdots d_k$.*

Proof: For $i = 1, \ldots, k$, let

$H_{i,\lambda,\mu}(X_0, \ldots, X_k) = \lambda\, P_i + \mu\, (X_i - X_0)\, (X_i - 2X_0) \cdots (X_i - d_i X_0)$, for $(\lambda, \mu) \in \mathrm{C}^2 \setminus \{0\}$.

We denote by $\mathcal{S}_{(\lambda:\mu)}$ the polynomial system

$$H_{1,\lambda,\mu}, \ldots, H_{k,\lambda,\mu}.$$

Note that the polynomial system $\mathcal{S}_{(0:1)}$ has $d_1 \cdots d_k$ non-singular projective zeros and $\mathcal{S}_{(1:0)}$ is $P_1, \ldots, P_k$. The subset of $(x, (\lambda: \mu)) \in \mathbb{P}_k(\mathrm{C}) \times \mathbb{P}_1(\mathrm{C})$ such that x is a singular projective zero of the polynomial system $\mathcal{S}_{(\lambda:\mu)}$ is clearly algebraic. Therefore, according to Theorem 4.102, its projection Δ on $\mathbb{P}_1(\mathrm{C})$ is an algebraic subset of $\mathbb{P}_1(\mathrm{C})$. Since $(0:1) \notin \Delta$, the set Δ consists of finitely many points, using Lemma 4.101. Since $\mathbb{P}_1(\mathrm{C}) \setminus \Delta$ is semi-algebraically connected, there is a semi-algebraic continuous path γ: $[0, 1] \subset \mathrm{R} \to \mathbb{P}_1(\mathrm{C})$ such that $\gamma(0) = (1{:}\ 0)$, $\gamma(1) = (0{:}\ 1)$, and $\gamma((0, 1]) \subset \mathbb{P}_1(\mathrm{C}) \setminus \Delta$. Note that $(\lambda: \mu) \in \mathbb{P}_1(\mathrm{C}) \setminus \Delta$ if and only if all projective zeros of $\mathcal{S}_{(\lambda:\mu)}$ are non-singular. By the implicit function theorem, for every non-singular projective zero x of $\mathcal{S}_{(1:0)}$, there exists a continuous path σ_x: $[0, 1] \to \mathbb{P}_k(\mathrm{C})$ such that $\sigma_x(0) = x$ and, for every $t \in (0, 1]$, $\sigma_x(t)$ is a non-singular projective zero of $\mathcal{S}_{\gamma(t)}$. Moreover, if y is another non-singular projective zero of $\mathcal{S}_{(1:0)}$, then $\sigma_x(t) \neq \sigma_y(t)$ for every $t \in [0, 1]$. From this we conclude that the number of non-singular projective zeros of $\mathcal{S}_{(1:0)}$: $P_1 = \cdots = P_k = 0$ is less than or equal to the number of projective zeros of $\mathcal{S}_{(0:1)}$, which is $d_1 \cdots d_k$. □

Theorem 4.106. [Weak Bézout] *Let $P_1, \ldots, P_k \in \mathrm{C}[X_1, \ldots, X_k]$ be polynomials of degrees $d_1, \ldots, d_k$. Then the number of non-singular zeros of $P_1, \ldots, P_k$ is at most $d_1 \cdots d_k$.*

Proof: Define

$$P_i^h = X_0^{d_i} P_i\left(\frac{X_1}{X_0}, \ldots, \frac{X_k}{X_0}\right), i = 1, \ldots, k,$$

and apply Proposition 4.105. The claim follows, noticing that any non-singular zero of $P_1, \ldots, P_k$ is a non-singular projective zero of $P_1^h, \ldots, P_k^h$. □

4.8 Bibliographical Notes

Resultants were introduced by Euler [56] and Bézout [24] and have been studied by many authors, particularly Sylvester [153]. Subresultant coefficients are discussed already in Gordan's textbook [74].

The use of quadratic forms for real root counting, in the univariate and multivariate case, is due to Hermite [89].

Hilbert's Nullstellensatz appears in [91] and a constructive proof giving doubly exponential degrees can be found in [88]. Much better degree bounds have been proved more recently, and are not included in our book [31, 35, 97].

5

Decomposition of Semi-Algebraic Sets

In this chapter, we decompose semi-algebraic sets in various ways and study several consequences of these decompositions. In Section 5.1 we introduce the cylindrical decomposition which is a key technique for studying the geometry of semi-algebraic sets. In Section 5.2 we use the cylindrical decomposition to define and study the semi-algebraically connected components of a semi-algebraic set. In Section 5.3 we define the dimension of a semi-algebraic set and obtain some basic properties of dimension. In Section 5.4 we get a semi-algebraic description of the partition induced by the cylindrical decomposition using Thom's lemma. In Section 5.5 we decompose semi-algebraic sets into smooth manifolds, called strata, generalizing Thom's lemma in the multivariate case. In Section 5.6 we define simplicial complexes, and establish the existence of a triangulation for a closed and bounded semi-algebraic set in Section 5.7. This triangulation result is used in Section 5.8 to prove Hardt's triviality theorem which has several important consequences, notably among them the finiteness of topological types of algebraic sets defined by polynomials of fixed degrees. We conclude the chapter with a semi-algebraic version of Sard's theorem in Section 5.9.

5.1 Cylindrical Decomposition

We first define what is a cylindrical decomposition: a decomposition of R^k into a finite number of semi-algebraically connected semi-algebraic sets having a specific structure with respect to projections.

Definition 5.1. A cylindrical decomposition of R^k is a sequence $\mathcal{S}_1, \ldots, \mathcal{S}_k$ where, for each $1 \leq i \leq k$, $\mathcal{S}_i$ is a finite partition of R^i into semi-algebraic subsets, called the **cells of level** i, which satisfy the following properties:

- Each cell $S \in \mathcal{S}_1$ is either a point or an open interval.
- For every $1 \leq i < k$ and every $S \in \mathcal{S}_i$, there are finitely many continuous semi-algebraic functions

$$\xi_{S,1} < \cdots < \xi_{S,\ell_S} \colon S \longrightarrow \mathrm{R}$$

such that the cylinder $S \times \mathrm{R} \subset \mathrm{R}^{i+1}$ is the disjoint union of cells of $\mathcal{S}_{i+1}$ which are:

- either the graph of one of the functions $\xi_{S,j}$, for $j=1,\ldots,\ell_S$:
$$\{(x', x_{j+1}) \in S \times \mathrm{R} \mid x_{j+1} = \xi_{S,j}(x')\},$$
- or a band of the cylinder bounded from below and from above by the graphs of the functions $\xi_{S,j}$ and $\xi_{S,j+1}$, for $j = 0, \ldots, \ell_S$, where we take $\xi_{S,0} = -\infty$ and $\xi_{i,\ell_S+1} = +\infty$:
$$\{(x', x_{j+1}) \in S \times \mathrm{R} \mid \xi_{S,j}(x') < x_{j+1} < \xi_{S,j+1}(x')\}. \qquad \square$$

Remark 5.2. Denoting by π_ℓ the canonical projection of R^k to R^ℓ, it follows immediately from the definition that for every cell T of $\mathcal{S}_i$, $i \geq \ell$, $S = \pi_\ell(T)$ is a cell of $\mathcal{S}_\ell$. We say that the cell T lies above the cell S. It is also clear that if S is a cell of $\mathcal{S}_i$, denoting by $T_1, \ldots, T_m$ the cells of $\mathcal{S}_{i+1}$ lying above S, we have $S \times \mathrm{R} = \bigcup_{j=1}^m T_j$. $\square$

Proposition 5.3. *Every cell of a cylindrical decomposition is semi-algebraically homeomorphic to an open i-cube* $(0,1)^i$ *(by convention,* $(0, 1)^0$ *is a point) and is semi-algebraically connected.*

Proof: We prove the proposition for the cells of $\mathcal{S}_i$ by induction on i.

If $i = 0$, the cells are clearly either points or open intervals and the claim holds.

Observe that if S is a cell of $\mathcal{S}_i$, the graph of $\xi_{S,j}$ is semi-algebraically homeomorphic to S and every band

$$\{(x', x_{j+1}) \in S \times \mathrm{R} \mid \xi_{S,j}(x') < x_{j+1} < \xi_{S,j+1}(x')\}$$

is semi-algebraically homeomorphic to $S \times (0, 1)$. In the case of the graph of $\xi_{S,j}$, the homeomorphism simply maps $x' \in S$ to $(x', \xi_{S,j}(x'))$.

For $\{(x', x_{j+1}) \in S \times \mathrm{R} \mid \xi_{S,j}(x') < x_{j+1} < \xi_{S,j+1}(x')\}$, $0 < j < \ell_S$ we map $(x', t) \in S \times (0,1)$ to $(x', (1-t)\xi_{S,j}(x') + t\xi_{S,j+1}(x'))$.

In the special case $j = 0$, $j = \ell_S$, we take

$$\begin{aligned} &\left(x', \frac{t-1}{t} + \xi_{S,j}(x')\right) && \text{if } j=0,\ \ell_S \neq 0, \\ &\left(x', \frac{t}{1-t} + \xi_{S,\ell_S}(x')\right) && \text{if } j=\ell_S \neq 0, \\ &\left(x', -\frac{1}{t} + \frac{1}{1-t}\right) && \text{if } j=\ell_S=0. \end{aligned}$$

These mappings are clearly bijective, bicontinuous and semi-algebraic, noting that the mappings sending $t \in (0,1)$ to

$$\frac{t-1}{t} + a, \frac{t}{1-t} + a, -\frac{1}{t} + \frac{1}{1-t},$$

are semi-algebraic bijections from $(0,1)$ to $(-\infty, a)$, $(a, +\infty)$, $(-\infty, +\infty)$. $\square$

A **cylindrical decomposition adapted to a finite family of semi-algebraic sets** $T_1, \ldots, T_\ell$ is a cylindrical decomposition of R^k such that every T_i is a union of cells.

Example 5.4. We illustrate this definition by presenting a cylindrical decomposition of R^3 adapted to the unit sphere.

Note that the projection of the sphere on the X_1, X_2 plane is the unit disk. The intersection of the sphere and the cylinder above the open unit disk consists of two hemispheres. The intersection of the sphere and the cylinder above the unit circle consists of a circle. The intersection of the sphere and the cylinder above the complement of the unit disk is empty. Note also that the projection of the unit circle on the line is the interval $[-1, 1]$.

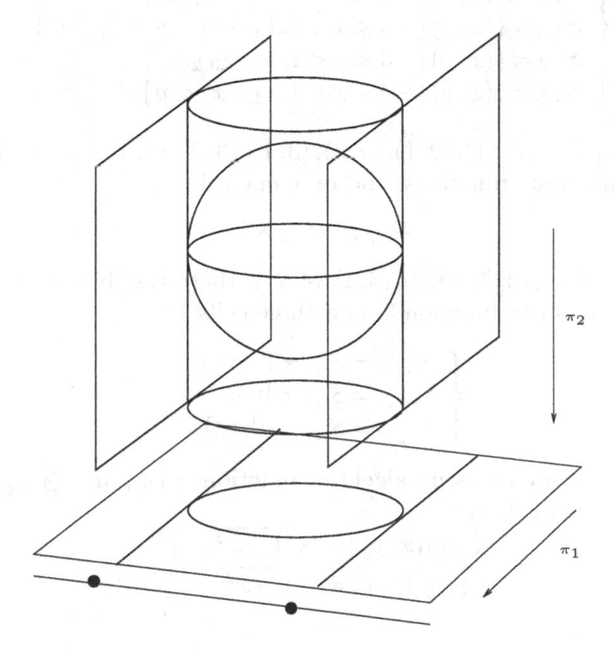

Fig. 5.1. A cylindrical decomposition adapted to the sphere in R^3

The decomposition of R consists of five cells of level 1 corresponding to the points -1 and 1 and the three intervals they define.

$$\begin{cases} S_1 = (-\infty, -1) \\ S_2 = \{-1\} \\ S_3 = (-1, 1) \\ S_4 = \{1\} \\ S_5 = (1, \infty). \end{cases}$$

Above S_i $i = 1, 5$, there are no semi-algebraic functions, and only one cell $S_{i,1} = S_i \times \mathrm{R}$.

Above S_i, $i=2,4$ there is only one semi-algebraic function associating to -1 and 1 the constant value 0, and there are three cells.

$$\begin{cases} S_{i,1}=S_i\times(-\infty,0)\\ S_{i,2}=S_i\times\{0\}\\ S_{i,3}=S_i\times(0,\infty)\end{cases}$$

Above S_3, there are two semi-algebraic functions $\xi_{3,1}$ and $\xi_{3,2}$ associating to $x\in S_3$ the values $\xi_{3,1}(x)=-\sqrt{1-x^2}$ and $\xi_{3,2}(x)=\sqrt{1-x^2}$. There are 5 cells above S_3, the graphs of $\xi_{3,1}$ and $\xi_{3,2}$ and the bands they define

$$\begin{cases} S_{3,1}=\{(x,y)\mid -1<x<1, y<\xi_{3,1}(x)\}\\ S_{3,1}=\{(x,y)\mid -1<x<1, y=\xi_{3,1}(x)\}\\ S_{3,3}=\{(x,y)\mid -1<x<1, \xi_{3,1}(x)<y<\xi_{3,2}(x)\}\\ S_{3,4}=\{(x,y)\mid -1<x<1, y=\xi_{3,2}(x)\}\\ S_{3,5}=\{(x,y)\mid -1<x<1, \xi_{3,2}(x)<y\}.\end{cases}$$

Above $S_{i,j}$, $(i,j)\in\{(1,1),(2,1),(2,3),(3,1),(3,5),(4,1),(4,3),(5,1)\}$, there are no semi-algebraic functions, and only one cell:

$$S_{i,j,1}=S_{i,j}\times\mathrm{R}$$

Above $S_{i,j}$, $(i,j)\in\{(2,2),(3,2),(3,4),(4,2)\}$, there is only one semi-algebraic function, the constant function 0, and three cells:

$$\begin{cases} S_{i,j,1}=S_{i,j}\times(-\infty,0)\\ S_{i,j,2}=S_{i,j}\times\{0\}\\ S_{i,j,3}=S_{i,j}\times(0,\infty)\end{cases}$$

Above $S_{3,3}$, there are two semi-algebraic functions $\xi_{3,3,1}$ and $\xi_{3,3,2}$ associating to $(x,y)\in S_{3,3}$ the values

$$\begin{aligned}\xi_{3,3,1}(x,y)&=-\sqrt{1-x^2-y^2}\\ \xi_{3,3,2}(x,y)&=\sqrt{1-x^2-y^2},\end{aligned}$$

and five cells

$$\begin{cases} S_{3,3,1}=\{(x,y,z)\mid (x,y)\in S_{3,3}, z<\xi_{3,3,1}(x,y)\}\\ S_{3,3,2}=\{(x,y,z)\mid (x,y)\in S_{3,3}, z=\xi_{3,3,1}(x,y)\}\\ S_{3,3,3}=\{(x,y,z)\mid (x,y)\in S_{3,3}, \xi_{3,3,1}(x,y)<z<\xi_{3,3,2}(x,y)\}\\ S_{3,3,4}=\{(x,y,z)\mid (x,y)\in S_{3,3}, z=\xi_{3,3,2}(x,y)\}\\ S_{3,3,5}=\{(x,y,z)\mid (x,y)\in S_{3,3}, \xi_{3,3,2}(x,y)<z\}.\end{cases}$$

□

Note that a cylindrical decomposition has a recursive structure. Let S be a cell of level i of a cylindrical decomposition $\mathcal{S}$ and $x\in S$. Denoting by π_i the canonical projection of R^k to R^i and identifying $\pi_i^{-1}(x)$ with R^{k-i}, the finite partition of R^{k-i} obtained by intersecting the cells of $\mathcal{S}_{i+j}$ above S with $\pi_i^{-1}(x)$ is a cylindrical decomposition of R^{k-i} called the **cylindrical decomposition induced by $\mathcal{S}$ above x.**

Definition 5.5. Given a finite set $\mathcal{P}$ of polynomials in $\mathrm{R}[X_1, \ldots, X_k]$, a subset S of R^k is **$\mathcal{P}$-semi-algebraic** if S is the realization of a quantifier free formula with atoms $P=0$, $P>0$ or $P<0$ with $P\in\mathcal{P}$. It is clear that for every semi-algebraic subset S of R^k, there exists a finite set $\mathcal{P}$ of polynomials in $\mathrm{R}[X_1, \ldots, X_k]$ such that S is $\mathcal{P}$-semi-algebraic.

A subset S of R^k is **$\mathcal{P}$-invariant** if every polynomial $P\in\mathcal{P}$ has a constant sign (>0, <0, or $=0$) on S.

A **cylindrical decomposition of R^k adapted to** $\mathcal{P}$ is a cylindrical decomposition for which each cell $C\in\mathcal{S}_k$ is $\mathcal{P}$-invariant. It is clear that if S is $\mathcal{P}$-semi-algebraic, a cylindrical decomposition adapted to $\mathcal{P}$ is a cylindrical decomposition adapted to S. □

The main purpose of the next few pages is to prove the following result.

Theorem 5.6. [Cylindrical decomposition] *For every finite set $\mathcal{P}$ of polynomials in $\mathrm{R}[X_1, \ldots, X_k]$, there is a cylindrical decomposition of R^k adapted to $\mathcal{P}$.*

The theorem immediately implies:

Corollary 5.7. *For every finite family of semi-algebraic sets $S_1, \ldots, S_\ell$ of R^k, there is a cylindrical decomposition of R^k adapted to $S_1, \ldots, S_\ell$.*

Since we intend to construct a cylindrical decomposition adapted to $\mathcal{P}$ it is convenient if for $S\in\mathcal{S}_{k-1}$ we choose each $\xi_{S,j}$ to be a root of a polynomial $P\in\mathcal{P}$, as a function of $(x_1, \ldots, x_{k-1})\in S$. To this end, we shall prove that the real and complex roots (those in $\mathrm{R}[i]=\mathrm{C}$) of a univariate polynomial depend continuously on its coefficients.

Notation 5.8. [Disk] We write $D(z,r)=\{w\in\mathrm{C}\mid |z-w|<r\}$ for the **open disk** centered at z with radius r. □

First we need the following bound on the modulus of the roots of a polynomial.

Proposition 5.9. *Let $P=a_pX^p+\cdots+a_1X+a_0\in\mathrm{C}[X]$, with $a_p\neq 0$. If $x\in\mathrm{C}$ is a root of P, then*

$$|x|\leq \max_{i=1,\ldots,p}\left(p\left|\frac{a_{p-i}}{a_p}\right|\right)^{1/i}=M\,.$$

Proof: If $z\in\mathrm{C}$ is such that $|z|>M$, then $|a_{p-i}|<|a_p|\,|z|^i/p$, $i=1,\ldots,p$. Hence,

$$|a_{p-1}\,z^{p-1}+\cdots+a_0|\leq|a_{p-1}|\,|z|^{p-1}+\cdots+|a_0|<|a_p\,z^p|$$

and $P(z)\neq 0$. □

We identify the monic polynomial $X^q+b_{q-1}\,X^{q-1}+\cdots+b_0\in\mathrm{C}[X]$ of degree q with the point $(b_{q-1},\ldots,b_0)\in\mathrm{C}^q$.

Lemma 5.10. *Given $r > 0$, there is an open neighborhood U of $(X - z)^\mu$ in C^μ such that every monic polynomial in U has its roots in $D(z, r)$.*

Proof: Without loss of generality, we can suppose that $z = 0$ and apply Proposition 5.9. □

Consider the map

$$\begin{aligned} m: \mathrm{C}^q \times \mathrm{C}^r &\longrightarrow \mathrm{C}^{q+r} \\ (Q, R) &\longmapsto Q R \end{aligned}$$

defined by the multiplication of monic polynomials of degrees q and r respectively.

Lemma 5.11. *Let $P_0 \in \mathrm{C}^{q+r}$ be a monic polynomial such that $P_0 = Q_0 R_0$, where Q_0 and R_0 are coprime monic polynomials of degrees q and r, respectively. There exist open neighborhoods U of P_0 in C^{q+r}, U_1 of Q_0 in C^q and U_2 of R_0 in C^r such that any $P \in U$ is uniquely the product of coprime monic polynomials $Q R$ with $Q \in U_1$ and $R \in U_2$,.*

Proof: The Jacobian matrix of m is the Sylvester matrix associated to

$$X^{q-1} R_0, \ldots, R_0, X^{r-1} Q_0, \ldots, Q_0$$

(Proposition 4.19). Hence the Jacobian of m is equal, up to sign, to the resultant of R_0 and Q_0 and is therefore different from zero by Proposition 4.15, since R_0 and Q_0 have no common factor. The conclusion follows using the implicit function theorem (Theorem 3.25). □

We can now prove

Theorem 5.12. [Continuity of Roots] *Let $P \in \mathrm{R}[X_1, \ldots, X_k]$ and let S be a semi-algebraic subset of R^{k-1}. Assume that $\deg(P(x', X_k))$ is constant on S and that for some $a' \in S$, $z_1, \ldots, z_j$ are the distinct roots of $P(a', X_k)$ in $\mathrm{C} = \mathrm{R}[i]$, with multiplicities $\mu_1, \ldots, \mu_j$, respectively.*

If the open disks $D(z_i, r) \subset \mathrm{C}$ are disjoint then there is an open neighborhood V of a' such that for every $x' \in V \cap S$, the polynomial $P(x', X_k)$ has exactly μ_i roots, counted with multiplicity, in the disk $D(z_i, r)$, for $i = 1, \ldots, j$.

Proof: Let $P_0 = (X - z_1)^{\mu_1} \cdots (X - z_j)^{\mu_j}$. By Lemma 5.11 there exist open neighborhoods U of P_0 in $\mathrm{C}^{\mu_1 + \cdots + \mu_j}$, U_1 of $(X - z_1)^{\mu_1}$ in C^{μ_1}, ..., U_j of $(X - z_j)^{\mu_j}$ in C^{μ_j} such that every $P \in U$ can be factored uniquely as $P = Q_1 \cdots Q_j$, where the Q_i are monic polynomials in U_i.

Using Lemma 5.10, it is clear that there is a neighborhood V of a' in S so that for every $x' \in V$ the polynomial $P(x', X_k)$ has exactly μ_i roots counted with multiplicity in $D(z_i, r)$, for $i = 1, \ldots, j$. □

We next consider the conditions which ensure that the zeros of two polynomials over a connected semi-algebraic set define a cylindrical structure.

Proposition 5.13. *Let $P, Q \in \mathrm{R}[X_1, \ldots X_k]$ and S a semi-algebraically connected subset of R^{k-1}. Suppose that P and Q are not identically 0 over S and that $\deg(P(x', X_k))$, $\deg(Q(x', X_k))$, $\deg(\gcd(P(x', X_k), Q(x', X_k)))$, the number of distinct roots of $P(x', X_k)$ in C and the number of distinct roots of $Q(x', X_k)$ in C are constant as x' varies over S. Then there exists ℓ continuous semi-algebraic functions $\xi_1 < \cdots < \xi_\ell$: $S \to \mathrm{R}$ such that, for every $x' \in S$, the set of real roots of $(PQ)(x', X_k)$ is exactly $\{\xi_1(x'), \ldots, \xi_\ell(x')\}$.*

Moreover for $i = 1, \ldots, \ell$, the multiplicity of the root $\xi_i(x')$ of $P(x', X_k)$ (resp. $Q(x', X_k)$) is constant for $x' \in S$. (If a is not a root, its multiplicity is zero, see Definition page 30).

Proof: Let $a' \in S$ and let $z_1, \ldots, z_j$ be the distinct roots in C of the product $(P\,Q)(a', X_k)$. Let μ_i (resp. ν_i) be the multiplicity of z_i as a root of $P(a', X_k)$ (resp. $Q(a', X_k)$). The degree of $\gcd(P(a', X_k), Q(a', X_k))$ is $\sum_{i=1}^{j} \min(\mu_i, \nu_i)$, and each z_i has multiplicity $\min(\mu_i, \nu_i)$ as a root of this greatest common divisor. Choose $r > 0$ such that all disks $D(z_i, r)$ are disjoint.

Using Theorem 5.12 and the fact that the number of distinct complex roots stays constant over S, we deduce that there exists a neighborhood V of a' in S such that for every $x' \in V$, each disk $D(z_i, r)$ contains one root of multiplicity μ_i of $P(x', X_k)$ and one root of multiplicity ν_i of $Q(x', X_k)$. Since the degree of $\gcd(P(x', X_k), Q(x', X_k))$ is equal to $\sum_{i=1}^{j} \min(\mu_i, \nu_i)$, this greatest common divisor must have exactly one root ζ_i, of multiplicity $\min(\mu_i, \nu_i)$, in each disk $D(z_i, r)$ such that $\min(\mu_i, \nu_i) > 0$. So, for every $x' \in V$, and every $i = 1, \ldots, j$, there is exactly one root ζ_i of $(PQ)(x', X_k)$ in $D(z_i, r)$ which is a root of $P(x', X_k)$ of multiplicity μ_i and a root of $Q(x', X_k)$ of multiplicity ν_i. If z_i is real, ζ_i is real (otherwise, its conjugate $\overline{\zeta_i}$ would be another root of $(P\,Q)(x', X_k)$ in $D(z_i, r)$). If z_i is not real, ζ_i is not real, since $D(z_i, r)$ is disjoint from its image by conjugation. Hence, if $x' \in V$, the polynomial $(P\,Q)(x', X_k)$ has the same number of distinct real roots as $(P\,Q)(a', X_k)$. Since S is semi-algebraically connected, the number of distinct real roots of $(PQ)(x', X_k)$ is constant for $x' \in S$ according to Proposition 3.9. Let ℓ be this number. For $1 \le i \le \ell$, denote by ξ_i: $S \to \mathrm{R}$ the function which sends $x' \in S$ to the i-th real root (in increasing order) of $(PQ)(x', X_k)$. The argument above, with arbitrarily small r also shows that the functions ξ_i are continuous. It follows from the fact that S is semi-algebraically connected that each $\xi_i(x')$ has constant multiplicity as a root of $P(x', X_k)$ and as a root of $Q(x', X_k)$ (cf Proposition 3.9). If S is described by the formula $\Theta(X')$, the graph of ξ_i is described by the formula

$$\begin{aligned}
&\Theta(X') \\
\wedge\ &((\exists Y_1)\ldots(\exists Y_\ell)\ (Y_1 < \cdots < Y_\ell \wedge (PQ)(X', Y_1) = 0 \wedge \ldots \wedge (PQ)(X', Y_\ell) = 0) \\
\wedge\ &((\forall Y)\ (PQ)(X', Y) = 0 \Rightarrow (Y = Y_1 \vee \ldots \vee Y = Y_\ell)) \wedge X_k = Y_i),
\end{aligned}$$

which shows that ξ_i is semi-algebraic, by quantifier elimination (Corollary 2.78). □

We have thus proved:

Proposition 5.14. *Let $\mathcal{P}$ be a finite subset of* $\mathrm{R}[X_1, \ldots X_k]$ *and S a semi-algebraically connected semi-algebraic subset of* R^{k-1}. *Suppose that, for every $P \in \mathcal{P}$,* $\deg(P(x', X_k))$ *and the number of distinct real roots of P are constant over S and that, for every pair $P, Q \in \mathcal{P}$,* $\deg(\gcd(P(x', X_k), Q(x', X_k))$ *is also constant over S. Then there are ℓ continuous semi-algebraic functions $\xi_1 < \cdots < \xi_\ell$: $S \to \mathrm{R}$ such that, for every $x' \in S$, the set of real roots of $\prod_{P \in \mathcal{P}'} P(x', X_k)$, where $\mathcal{P}'$ is the subset of $\mathcal{P}$ consisting of polynomials not identically 0 over S, is exactly $\{\xi_1(x'), \ldots, \xi_\ell(x')\}$. Moreover, for $i = 1, \ldots, \ell$ and for every $P \in \mathcal{P}'$, the multiplicity of the root $\xi_i(x')$ of $P(x', X_k)$ is constant for $x' \in S$.*

It follows from Chapter 4 (Proposition 4.24) that the number of distinct roots of P, of Q and the degree of the greatest common divisor of P and Q are determined by whether the signed subresultant coefficients $\mathrm{sRes}_i(P, P')$ and $\mathrm{sRes}_i(P, Q)$ are zero or not, as long as the degrees (with respect to X_k) of P and Q are fixed.

Notation 5.15. [Elimination] Using Notation 1.16, with parameters $X_1, \ldots, X_{k-1}$ and main variable X_k, let

$$\mathrm{Tru}(\mathcal{P}) = \{\mathrm{Tru}(P) \mid P \in \mathcal{P}\}.$$

We define $\mathrm{Elim}_{X_k}(\mathcal{P})$ to be the set of polynomials in $\mathrm{R}[X_1, \ldots, X_{k-1}]$ defined as follows:

- If $R \in \mathrm{Tru}(\mathcal{P})$, $\deg_{X_k}(R) \geq 2$, $\mathrm{Elim}_{X_k}(\mathcal{P})$ contains all $\mathrm{sRes}_j(R, \partial R / \partial X_k)$ which are not in R, $j = 0, \ldots, \deg_{X_k}(R) - 2$.
- If $R \in \mathrm{Tru}(\mathcal{P})$, $S \in \mathrm{Tru}(\mathcal{P})$,
  - if $\deg_{X_k}(R) > \deg_{X_k}(S)$, $\mathrm{Elim}_{X_k}(\mathcal{P})$ contains all $\mathrm{sRes}_j(R, S)$ which are not in R, $j = 0, \ldots, \deg_{X_k}(S) - 1$,
  - if $\deg_{X_k}(R) < \deg_{X_k}(S)$, $\mathrm{Elim}_{X_k}(\mathcal{P})$ contains all $\mathrm{sRes}_j(S, R)$ which are not in R, $j = 0, \ldots, \deg_{X_k}(R) - 1$,
  - if $\deg_{X_k}(R) = \deg_{X_k}(S)$, $\mathrm{Elim}_{X_k}(\mathcal{P})$ contains all $\mathrm{sRes}_j(S, \overline{R})$, with $\overline{R} = \mathrm{lcof}(S) R - \mathrm{lcof}(R) S$ which are not in R, $j = 0, \ldots, \deg_{X_k}(\overline{R}) - 1$.
- If $R \in \mathrm{Tru}(\mathcal{P})$, and $\mathrm{lcof}(R)$ is not in R, $\mathrm{Elim}_{X_k}(\mathcal{P})$ contains $\mathrm{lcof}(R)$. □

Theorem 5.16. *Let $\mathcal{P}$ be a set of polynomials in* $\mathrm{R}[X_1, \ldots, X_k]$, *and let S be a semi-algebraically connected semi-algebraic subset of* R^{k-1} *which is $\mathrm{Elim}_{X_k}(\mathcal{P})$-invariant. Then there are continuous semi-algebraic functions $\xi_1 < \cdots < \xi_\ell$: $S \to \mathrm{R}$ such that, for every $x' \in S$, the set $\{\xi_1(x'), \ldots, \xi_\ell(x')\}$ is the set of all real roots of all non-zero polynomials $P(x', X_k)$, $P \in \mathcal{P}$. The graph of each ξ_i (resp. each band of the cylinder $S \times \mathrm{R}$ bounded by these graphs) is a semi-algebraically connected semi-algebraic set semi-algebraically homeomorphic to S (resp. $S \times (0, 1)$) and is $\mathcal{P}$-invariant.*

Proof: For P in $\mathcal{P}$, $R \in \mathrm{Tru}(P)$, consider the constructible set $A \subset \mathrm{R}^{k-1}$ defined by $\mathrm{lcof}(R) \neq 0, \deg(P) = \deg(R)$. By Proposition 4.24, for every $a' \in A$, the vanishing or non-vanishing of the $\mathrm{sRes}_j(R, \partial R/\partial X_k)(a')$ determines the number of distinct roots of $P(a', X_k)$ in C, which is

$$\deg(R(a', X_k)) - \deg(\gcd(R(a', X_k), \partial R/\partial X_k(a', X_k))$$

Similarly, for $R \in \mathrm{Tru}(P)$, $S \in \mathrm{Tru}(Q)$, consider the constructible set B defined by

$$\mathrm{lcof}(R) \neq 0, \deg(P) = \deg(R), \mathrm{lcof}(S) \neq 0, \deg(Q) = \deg(S).$$

For every $a' \in B$, which of the $\mathrm{sRes}_j(R, S)(a')$ (resp. $\mathrm{sRes}_j(S, R)(a')$, $\mathrm{sRes}_j(S, \overline{R})(a')$) vanish, determine $\deg(\gcd(P(a', X_k), Q(a', X_k)))$, by Proposition 4.24. Thus, the assumption that a connected semi-algebraic subset of R^{k-1} is $\mathrm{Elim}_{X_k}(\mathcal{P})$-invariant implies that the hypotheses of Proposition 5.14 are satisfied. □

We are finally ready for the proof of Theorem 5.6.

Proof of Theorem 5.6 The proof is by induction on the dimension of the ambient space.

Let $\mathcal{Q} \subset \mathrm{R}[X_1]$ be finite. It is clear that there is a cylindrical decomposition of R adapted to $\mathcal{Q}$ since the real roots of the polynomials in $\mathcal{Q}$ decompose the line into finitely many points and open intervals which constitute the cells of a cylindrical decomposition of R adapted to $\mathcal{Q}$.

Let $\mathcal{Q} \subset \mathrm{R}[X_1, \ldots, X_i]$ be finite. Starting from a cylindrical decomposition of R^{i-1} adapted to $\mathrm{Elim}_{X_i}(\mathcal{Q})$, and applying to the cells of this cylindrical decomposition Proposition 5.16, yields a cylindrical decomposition of R^i adapted to $\mathcal{Q}$.

This proves the theorem. □

Example 5.17. We illustrate this result by presenting a cylindrical decomposition of R^3 adapted to the polynomial $P = X_1^2 + X_2^2 + X_3^2 - 1$. The 0-th Sylvester-Habicht matrix of P and $\partial P/\partial X_3$ is

$$\begin{bmatrix} 1 & 0 & X_1^2 + X_2^2 - 1 \\ 0 & 2 & 0 \\ 2 & 0 & 0 \end{bmatrix}.$$

Hence, $\mathrm{sRes}_0(P, \partial P/\partial X_3) = -4(X_1^2 + X_2^2 - 1)$ and $\mathrm{sRes}_1(P, \partial P/\partial X_3) = 2$. Getting rid of irrelevant constant factors, we obtain

$$\mathrm{Elim}_{X_3}(P) = \{X_1^2 + X_2^2 - 1\}.$$

Similarly,

$$\mathrm{Elim}_{X_2}(\mathrm{Elim}_{X_3}(P)) = \{X_1^2 - 1\}.$$

The associated cylindrical decomposition is precisely the one described in Example 5.4. □

Remark 5.18. The proof of Theorem 5.6 provides a method for constructing a cylindrical decomposition adapted to $\mathcal{P}$. In a projection phase, we eliminate the variables one after the other, by computing $\mathrm{Elim}_{X_k}(\mathcal{P})$, then $\mathrm{Elim}_{X_{k-1}}(\mathrm{Elim}_{X_k}(\mathcal{P}))$ etc. until we obtain a finite family of univariate polynomials.

In a lifting phase, we decompose the line in a finite number of cells which are the points and intervals defined by the family of univariate polynomials. Then we decompose the cylinder contained in R^2 above each of these points and intervals in a finite number of cells consisting of graphs and bands between these graphs. Then we decompose the cylinder contained in R^2 above each of plane cells in a finite number of cells consisting of graphs and bands between these graphs etc.

Note that the projection phase of the construction provides in fact an algorithm computing explicitly a family of polynomials in one variable. The complexity of this algorithm will be studied in Chapter 12. □

Theorem 5.19. *Every semi-algebraic subset S of R^k is the disjoint union of a finite number of semi-algebraic sets, each of them semi-algebraically homeomorphic to an open i-cube $(0,1)^i \subset \mathrm{R}^i$ for some $i \leq k$ (by convention $(0,1)^0$ is a point).*

Proof: According to Corollary 5.7, there exists a cylindrical decomposition adapted to S. Since these cells are homeomorphic to an open i-cube $(0,1)^i \subset \mathrm{R}^i$ for some $i \leq k$, the conclusion follows immediately. □

An easy consequence is the following which asserts the piecewise continuity of semi-algebraic functions.

Proposition 5.20. *Let S be a semi-algebraic set and let f: $S \to \mathrm{R}^k$ be a semi-algebraic function. There is a partition of S in a finite number of semi-algebraic sets $S_1, \ldots, S_n$ such that the restriction f_i of f to S_i is semi-algebraic and continuous.*

Proof: By Theorem 5.19, the graph G of f is the union of open i-cubes of various dimensions, which are clearly the graphs of semi-algebraic continuous functions. □

5.2 Semi-algebraically Connected Components

Theorem 5.21. *Every semi-algebraic set S of R^k is the disjoint union of a finite number of semi-algebraically connected semi-algebraic sets $C_1, \ldots, C_\ell$ that are both closed and open in S.*

The C_1, ..., C_ℓ are called the **semi-algebraically connected components** of S.

Proof of Theorem 5.21: By Theorem 5.19, S is the disjoint union of a finite number of semi-algebraic sets S_i semi-algebraically homeomorphic to open $d(i)$-cubes $(0,1)^{d(i)}$ and hence semi-algebraically connected by Proposition 3.8 Consider the smallest equivalence relation $\mathcal{R}$ on the set of the S_i containing the relation "$S_i \cap \overline{S_j} \neq \emptyset$". Let $C_1, \ldots, C_\ell$ be the unions of the equivalence classes for $\mathcal{R}$. The C_j are semi-algebraic, disjoint, closed in S, and their union is S. We show now that each C_j is semi-algebraically connected. Suppose that we have $C_j = F_1 \cup F_2$ with F_1 and F_2 disjoint, semi-algebraic and closed in C_j. Since each S_i is semi-algebraically connected, $S_i \subset C_j$ implies that $S_i \subset F_1$ or $S_i \subset F_2$. Since F_1 (resp. F_2) is closed in C_j, if $S_i \subset F_1$ (resp. F_2) and $S_i \cap \overline{S_{i'}} \neq \emptyset$ then $S_{i'} \subset F_1$ (resp. F_2). By the definition of the C_j, we have $C_j = F_1$ or $C_j = F_2$. So C_j is semi-algebraically connected. □

Theorem 5.22. *A semi-algebraic subset S of $\mathbb{R}^k$ is semi-algebraically connected if and only if it is connected. Every semi-algebraic set (and in particular every algebraic subset) of $\mathbb{R}^k$ has a finite number of connected components, each of which is semi-algebraic.*

Proof: It is clear that if S is connected, it is semi-algebraically connected.

If S is not connected then there exist open sets O_1 and O_2 (not necessarily semi-algebraic) with

$$S \subset O_1 \cup O_2,\ O_1 \cap S \neq \emptyset, O_2 \cap S \neq \emptyset$$

and $(S \cap O_1) \cap (S \cap O_2) = \emptyset$. By Theorem 5.19, we know that S is a union of a finite number $C_1, \ldots, C_\ell$ of semi-algebraic sets homeomorphic to open cubes of various dimensions. If $O_1 \cap S$ and $O_2 \cap S$ are unions of a finite number of semi-algebraic sets among C_1, ..., C_ℓ, $O_1 \cap S$ and $O_2 \cap S$ are semi-algebraic and S is not semi-algebraically connected. Otherwise, some C_i is disconnected by O_1 and O_2, which is impossible since C_i is homeomorphic to an open cube.

Hence a semi-algebraic subset S of $\mathbb{R}^k$ is semi-algebraically connected if and only if it is connected. The remainder of the theorem follows from Theorem 5.21. □

Theorem 5.23. *A semi-algebraic set is semi-algebraically connected if and only if it is semi-algebraically path connected.*

Proof: Since $[0,1]$ is semi-algebraically connected, it is clear that semi-algebraic path connectedness implies semi-algebraic connectedness. We prove the converse by using Theorem 5.19 and the proof of Theorem 5.21. It is obvious that an open d-cube is semi-algebraically path connected. It is then enough to show that if S_i and S_j are semi-algebraically homeomorphic to open d-cubes, with $S_i \cap \overline{S_j} \neq \emptyset$, then $S_i \cup S_j$ is semi-algebraically path connected. But this is a straightforward consequence of the Curve Selection Lemma (Theorem 3.19). □

Let R' be a real closed extension of the real closed field R.

Proposition 5.24. *The semi-algebraic set S is semi-algebraically connected if and only if* $\mathrm{Ext}(S, \mathrm{R}')$ *is semi-algebraically connected.*

More generally, if C_1, ..., C_ℓ are the semi-algebraically connected components of S, then $\mathrm{Ext}(C_1, \mathrm{R}')$, ..., $\mathrm{Ext}(C_\ell, \mathrm{R}')$ *are the semi-algebraically connected components of* $\mathrm{Ext}(S, \mathrm{R}')$.

Proof: Given a decomposition $S = \bigcup_{i=1}^m S_i$, with, for each i, a semi-algebraic homeomorphism φ_i: $(0,1)^{d(i)} \to S_i$, the extension gives a decomposition

$$\mathrm{Ext}(S, \mathrm{R}') = \bigcup_{i=1}^{m} \mathrm{Ext}(S_i, \mathrm{R}'),$$

and semi-algebraic homeomorphisms

$$\mathrm{Ext}(\varphi_i, \mathrm{R}') \colon (\mathrm{Ext}((0,1), \mathrm{R}')^{d_i} \to (\mathrm{Ext}(S_i, \mathrm{R}').$$

The characterization of the semi-algebraically connected components from a decomposition (cf. Theorem 5.21) then gives the result. □

5.3 Dimension

Let S be a semi-algebraic subset of R^k. Take a cylindrical decomposition of R^k adapted to S. A naive definition of the dimension of S is the maximum of the dimension of the cells contained in S, the dimension of a cell semi-algebraically homeomorphic to $(0,1)^d$ being d. But this definition is not intrinsic. We would have to prove that the dimension so defined does not depend on the choice of a cylindrical decomposition adapted to S. Instead, we introduce an intrinsic definition of dimension and show that it coincides with the naive one.

The **dimension** $\dim(S)$ of a semi-algebraic set S is the largest d such that there exists an injective semi-algebraic map from $(0,1)^d$ to S. By convention, the dimension of the empty set is -1. Note that the dimension of a set is clearly invariant under semi-algebraic bijections. Observe that it is not obvious for the moment that the dimension is always $< +\infty$. It is also not clear that this definition of dimension agrees with the intuitive notion of dimension for cells.

We are going to prove the following result.

Theorem 5.25. *Let $S \subset R^k$ be semi-algebraic and consider a cylindrical decomposition of R^k adapted to S. Then the dimension of S is finite and is the maximum dimension of the cells contained in S.*

The key ingredient for proving this result is the following lemma.

Lemma 5.26. *Let S be a semi-algebraic subset of R^k with non-empty interior. Let f: $S \to R^k$ be an injective semi-algebraic map. Then $f(S)$ has non-empty interior.*

Proof: We prove the lemma by induction on k. If $k=1$, S is semi-algebraic and has infinite cardinality, hence $f(S) \subset \mathrm{R}$ is semi-algebraic and infinite and must therefore contain an interval.

Assume that $k > 1$ and that the lemma is proved for all $\ell < k$. Using the piecewise continuity of semi-algebraic functions (Proposition 5.20), we can assume moreover that f is continuous. Take a cylindrical decomposition of R^k adapted to $f(S)$. If $f(S)$ has empty interior, it contains no cell open in R^k. Hence $f(S)$ is the union of cells C_1, ..., C_n that are not open in R^k and, for $i=1,\ldots,n$, there is a semi-algebraic homeomorphism $C_i \to (0,1)^{\ell_i}$ with $\ell_i < k$. Take a cylindrical decomposition of R^k adapted to the $f^{-1}(C_i)$. Since $S = \bigcup_{i=1}^n f^{-1}(C_i)$ has non-empty interior, one of the $f^{-1}(C_i)$, say $f^{-1}(C_1)$, must contain a cell C open in R^k. The restriction of f to C gives an injective continuous semi-algebraic map $C \to C_1$.

Since C is semi-algebraically homeomorphic to $(0,1)^k$ and C_1 semi-algebraically homeomorphic to $(0,1)^\ell$ with $\ell < k$, we obtain an injective continuous semi-algebraic map g from $(0,1)^k$ to $(0,1)^\ell$. Set $a = (\frac{1}{2}, \ldots, \frac{1}{2}) \in R^{k-\ell}$ and consider the mapping g_a from $(0,1)^\ell$ to $(0,1)^\ell$ defined by $g_a(x) = g(a,x)$. We can apply the inductive assumption to g_a. It implies that $g_a((0,1)^\ell)$ has non-empty interior in $(0,1)^\ell$. Choose a point $c = g_a(b)$ in the interior of $g_a((0,1)^\ell)$. Since g is continuous, all points close enough to (a,b) are mapped by g to the interior of $g_a((0,1)^\ell)$. Let (x,b) be such a point with $x \neq a$. Since g_a is onto the interior of $g_a((0,1)^\ell)$ there is $y \in (0,1)^\ell$ such that $g(x,b) = g_a(y) = g(a,y)$, which contradicts the fact that g is injective. Hence, $f(S)$ has non-empty interior. □

Proposition 5.27. *The dimension of* $(0,1)^d$ *is* d. *The dimension of a cell semi-algebraically homeomorphic to* $(0,1)^d$ *is* d.

Proof: There is no injective semi-algebraic map from $(0,1)^e$ to $(0,1)^d$ if $e > d$. Otherwise, the composition of such a map with the embedding of $(0,1)^d$ in $\mathrm{R}^e = \mathrm{R}^d \times \mathrm{R}^{e-d}$ as $(0,1)^d \times \{0\}$ would contradict Lemma 5.26. This shows the first part of the corollary. The second part follows, using the fact that the dimension, according to its definition, is invariant under semi-algebraic bijection. □

Proposition 5.28. *If* $S \subset T$ *are semi-algebraic sets,* $\dim(S) \leq \dim(T)$.

If S *and* T *are semi-algebraic subsets of* R^k, $\dim(S \cup T) = \max(\dim(S), \dim(T))$.

If S *and* T *are semi-algebraic sets,* $\dim(S \times T) = \dim(S) + \dim(T)$.

Proof: That $\dim(S) \leq \dim(T)$ is clear from the definition. The inequality $\dim(S \cup T) \geq \max(\dim S, \dim T)$ follows from 1. Now let $f\colon (0,1)^d \to S \cup T$ be a semi-algebraic injective map. Taking a cylindrical decomposition of R^d adapted to $f^{-1}(S)$ and $f^{-1}(T)$, we see that $f^{-1}(S)$ or $f^{-1}(T)$ contains a cell of dimension d. Since f is injective, we have $\dim(S) \geq d$ or $\dim(T) \geq d$. This proves the reverse inequality $\dim(S \cup T) \leq \max(\dim(S), \dim(T))$.

Since $\dim(S \cup T) = \max(\dim(S), \dim(T))$, it is sufficient to consider the case where S and T are cells.

Since $S \times T$ is semi-algebraically homeomorphic to $(0,1)^{\dim S} \times (0,1)^{\dim T}$, the assertion in this case follows from Proposition 5.27. □

Proof of Theorem 5.25: The result follows immediately from Proposition 5.27 and Proposition 5.28. □

Proposition 5.29. *Let S be a semi-algebraic subset of R^k, and let $f: S \to \mathrm{R}^\ell$ a semi-algebraic mapping. Then $\dim(f(S)) \le \dim(S)$. If f is injective, then $\dim(f(S)) = \dim(S)$.*

The proof uses the following lemma.

Lemma 5.30. *Let $S \subset \mathrm{R}^{k+\ell}$ be a semi-algebraic set, π the projection of $\mathrm{R}^{k+\ell}$ onto R^ℓ. Then $\dim(\pi(S)) \le \dim(S)$. If, moreover, the restriction of π to S is injective, then $\dim(\pi(S)) = \dim S$.*

Proof: When $\ell = 1$ and S is a graph or a band in a cylindrical decomposition of R^{k+1}, the result is clear. If S is any semi-algebraic subset of R^{k+1}, it is a union of such cells for a decomposition, and the result is still true. The case of any ℓ follows by induction. □

Proof of Proposition 5.30: Let $G \subset \mathrm{R}^{k+\ell}$ be the graph of f. Lemma 5.30 tells us that $\dim(S) = \dim(G)$ and $\dim(f(S)) \le \dim(S)$, with equality if f is injective. □

Finally the following is clear:

Proposition 5.31. *Let V be an $\mathcal{S}^\infty$ submanifold of dimension d of R^k (as a submanifold of R^k, see Definition 3.25). Then the dimension of V as a semi-algebraic set is d.*

5.4 Semi-algebraic Description of Cells

In the preceding sections, we decomposed semi-algebraic sets into simple pieces, the cells, which are semi-algebraically homeomorphic to open i-cubes. We have also explained how to produce such a decomposition adapted to a finite set of polynomials $\mathcal{P}$. But the result obtained is not quite satisfactory, as we do not have a semi-algebraic description of the cells by a boolean combination of polynomial equations and inequalities. Since the cells are semi-algebraic, this description certainly exists. It would be nice to have the polynomials defining the cells of a cylindrical decomposition adapted to $\mathcal{P}$. This will be possible with the help of the derivatives of the polynomials.

We need to introduce a few definitions.

Definition 5.32. A **weak sign condition** is an element of

$$\{\{0\},\{0,1\},\{0,-1\}\}.$$

Note that

$$\begin{cases} \operatorname{sign}(x)\in\{0\} & \text{if and only if } x=0,\\ \operatorname{sign}(x)\in\{0,1\} & \text{if and only if } x\geq 0,\\ \operatorname{sign}(x)\in\{0,-1\} & \text{if and only if } x\leq 0.\end{cases}$$

A **weak sign condition** on $\mathcal{Q}$ is an element of $\{\{0\},\{0,1\},\{0,-1\}\}^{\mathcal{Q}}$. If $\sigma\in\{0,1,-1\}^{\mathcal{Q}}$, its **relaxation** $\bar{\sigma}$ is the weak sign condition on $\mathcal{Q}$ defined by $\bar{\sigma}(Q)=\overline{\sigma(Q)}$. The **realization of the weak sign condition** τ is

$$\mathrm{Reali}(\tau)=\{x\in\mathrm{R}^k \mid \bigwedge_{Q\in\mathcal{Q}} \operatorname{sign}(Q(x))\in\tau(Q)\}.$$

The weak sign condition τ is **realizable** if $\mathrm{Reali}(\tau)$ is non-empty.

A set of polynomials $\mathcal{Q}\subset\mathrm{R}[X]$ is **closed under differentiation** if $0\notin\mathcal{Q}$ and if for each $Q\in\mathcal{Q}$ then $Q'\in\mathcal{Q}$ or $Q'=0$. □

The following result is an extension of the Basic Thom's lemma (Lemma 2.28) seen in Chapter 2. It implies that if a family of polynomials is stable under differentiation, the cells it defines on a line are described by sign conditions on this family.

Lemma 5.33. [Thom's lemma] *Let $\mathcal{Q}\subset\mathrm{R}[X]$ be a finite set of polynomials closed under differentiation and let σ be a sign condition on the set $\mathcal{Q}$. Then*

- $\mathrm{Reali}(\sigma)$ *is either empty, a point, or an open interval.*
- *If* $\mathrm{Reali}(\sigma)$ *is empty, then* $\mathrm{Reali}(\bar{\sigma})$ *is either empty or a point.*
- *If* $\mathrm{Reali}(\sigma)$ *is a point, then* $\mathrm{Reali}(\bar{\sigma})$ *is the same point.*
- *If* $\mathrm{Reali}(\sigma)$ *is an open interval then* $\mathrm{Reali}(\bar{\sigma})$ *is the corresponding closed interval.*

Proof: The proof is by induction on s, the number of polynomials in $\mathcal{Q}$. There is nothing to prove if $s=0$. Suppose that the proposition has been proved for s and that Q has maximal degree in $\mathcal{Q}$, which is closed under differentiation and has $s+1$ elements. The set $\mathcal{Q}\setminus\{Q\}$ is also closed under differentiation. Let $\sigma\in\{0,1,-1\}^{\mathcal{Q}}$ be a sign condition on $\mathcal{Q}$, and let σ' be its restriction to $\mathcal{Q}\setminus\{Q\}$. If $\mathrm{Reali}(\sigma')$ is either a point or empty, then

$$\mathrm{Reali}(\sigma)=\mathrm{Reali}(\sigma')\cap\{x\in\mathrm{R}\mid \operatorname{sign}(Q(x))=\sigma(Q)\}$$

is either a point or empty. If $\mathrm{Reali}(\sigma')$ is an open interval, the derivative of Q (which is among $\mathcal{Q}\setminus\{Q\}$), has a constant non-zero sign on it (except if Q is a constant, which is a trivial case). Thus Q is strictly monotone on $\mathrm{Reali}(\bar{\sigma}')$ so that the claimed properties are satisfied for $\mathrm{Reali}(\sigma)$. □

By alternately applying the operation Elim and closing under differentiation we obtain a set of polynomials whose realizable sign conditions define the cells of a cylindrical decomposition adapted to $\mathcal{P}$.

Theorem 5.34. *Let $\mathcal{P}^\star = \bigcup_{i=1,\ldots,k} \mathcal{P}_i$ be a finite set of non-zero polynomials such that:*

- *$\mathcal{P}_k$ contains $\mathcal{P}$,*
- *for each i, $\mathcal{P}_i$ is a subset of $\mathrm{R}[X_1,\ldots,X_i]$ that is closed under differentiation with respect to X_i,*
- *for $i \le k$, $\mathrm{Elim}_{X_i}(\mathcal{P}_i) \subset \mathcal{P}_{i-1}$.*

Writing $\mathcal{P}_{\le i} = \bigcup_{j \le i} \mathcal{P}_j$, the families $\mathcal{S}_i$, for $i = 1, \ldots, k$, consisting of all $\mathrm{Reali}(\sigma)$ with σ a realizable sign condition on $\mathcal{P}_{\le i}$ constitute a cylindrical decomposition of R^k adapted to $\mathcal{P}$.

Proof: The case $k=1$ is covered by Lemma 5.33. The proof of the general case is clear by induction on k, again using Lemma 5.33. □

Remark 5.35. Since $\mathcal{P}_{\le i+1}$ is closed under differentiation, for every cell $S \subset \mathrm{R}^i$ and every semi-algebraic function $\xi_{S,j}$ of the cylindrical decomposition described in the theorem, there exists $P \in \mathcal{P}_{\le i+1}$ such that, for every $x \in S$, $\xi_{S,j}(x)$ is a simple root of $P(x, X_{i+1})$. □

5.5 Stratification

We do not have so far much information concerning which cells of a cylindrical decomposition are adjacent to others, for cells which are not above the same cell.

In the case of the cylindrical decomposition adapted to the sphere, it is not difficult to determine the topology of the sphere from the cell decomposition. Indeed, the two functions on the disk defined by $X_1^2 + X_2^2 + X_3^2 < 1$, whose graphs are the two open hemispheres, have an obvious extension by continuity to the closed disk.

Example 5.36. We give an example of a cylindrical decomposition where it is not the case that the functions defined on the cells have an extension by continuity to boundary of the cell. Take $P = (X_1 X_2 X_3) - (X_1^2 + X_2^2)$. In order to visualize the corresponding zero set, it is convenient to fix the value of x_3.

The zero set Z of $(X_1 X_2 x_3) - (X_1^2 + X_2^2)$ can be described as follows.

- If $-2 < x_3 < 2$, Z consists of the isolated point $(0, 0, x_3)$.
- If $x_3 = -2$ or $x_3 = 2$, Z consists of one double line through the origin in the plane $X_3 = x_3$.

– If $x_3 > -2$ or $x_3 < 2$, Z consists of two distinct lines through the origin in the plane $X_3 = x_3$.

Note that the set of zeroes of P in the ball of center 0 and radius of 1 is the segment $(-1, 1)$ of the X_3 axis, so that $\mathrm{Zer}(P, \mathrm{R}^3)$ has an open subset which is a semi-algebraic set of dimension 1.

– When $X_1 X_2 \neq 0$, $P = 0$ is equivalent to

$$X_3 = \frac{X_1^2 + X_2^2}{X_1 X_2}.$$

– When $X_1 = 0$, $X_2 \neq 0$, the polynomial P is $-X_2^2$.
– When $X_2 = 0$, $X_1 \neq 0$, the polynomial P is $-X_1^2$.
– When $X_1 = 0$, $X_2 = 0$, P is identically zero.

The function $(X_1^2 + X_2^2)/(X_1 X_2)$ does not have a well defined limit when X_1 and X_2 tend to 0. The function describing the zeros of P on each open quadrant cannot be extended continuously to the closed quadrant.

The main difference with the example of the sphere is the fact that the polynomial P is not monic as polynomial in X_3: the leading coefficient $X_1 X_2$ vanishes, and P is even identically zero for $X_1 = X_2 = 0$. □

We explain now that the information provided by the cylindrical decomposition is not sufficient to determine the topology.

Example 5.37. We describe two surfaces having the same cylindrical decomposition and a different topology, namely the two surfaces defined as the zero sets of

$$\begin{aligned} P_1 &= (X_1 X_2 X_3)^2 - (X_1^2 + X_2^2)^2 \\ P_2 &= P_2 = (X_1 X_2 X_3 - (X_1 - X_2)^2)\,(X_1 X_2 X_3 - (X_1 + X_2)^2). \end{aligned}$$

Consider first $P_1 = (X_1 X_2 X_3)^2 - (X_1^2 + X_2^2)^2$. In order to visualize the zero sct of P_1, it is convenient to fix the value of x_3.

The zero set of $P_1 = (X_1 X_2 x_3)^2 - (X_1^2 + X_2^2)^2$ is the union of the zero set Z_1 of $(X_1 X_2 x_3) + (X_1^2 + X_2^2)$ and the zero set Z_2 of $(X_1 X_2 x_3) - (X_1^2 + X_2^2)$ in the plane $X_3 = x_3$.

– If $-2 < x_3 < 2$, $Z_1 = Z_2$ consists of the isolated point $(0, 0, x_3)$.
– If $x_3 = -2$ or $x_3 = 2$, $Z_1 \cup Z_2$ consists of two distinct lines through the origin in the plane $X_3 = x_3$.
– If $x_3 > -2$ or $x_3 < 2$, $Z_1 \cup Z_2$ consists of four distinct lines through the origin in the plane $X_3 = x_3$.

Note that the set of zeroes of P_1 in the ball of center 0 and radius of 1 is the segment $(-1, 1)$ of the X_3 axis, so that $\mathrm{Zer}(P_1, \mathrm{R}^3)$ has an open subset which is a semi-algebraic set of dimension 1.

It is easy to see that the 9 cells of R^2 defined by the signs of X_1 and X_2 together with the 3 cells of R defined by the sign of X_1 determine a cylindrical decomposition adapted to $\{P_1\}$.

- When $X_1 X_2 \neq 0$, $P_1 = 0$ is equivalent to

$$X_3 = \frac{X_1^2 + X_2^2}{X_1 X_2} \text{ or } X_3 = -\frac{X_1^2 + X_2^2}{X_1 X_2}.$$

  So the zeroes of P_1 are described by two graphs of functions over each open quadrant, and the cylindrical decomposition of P_1 has five cells over each open quadrant. The sign of P_1 in these five cells is $1, 0, -1, 0, 1$.
- When $X_1 = 0$, $X_2 \neq 0$, the polynomial P_1 is $-X_2^4$. The cylinders over each open half-axis have one cell on which P_1 is negative.
- When $X_1 \neq 0$, $X_2 = 0$, the polynomial P_1 is $-X_1^4$. The cylinders over each open half-axis have one cell on which P_1 is negative.
- When $X_1 = 0$, $X_2 = 0$, P_1 is identically zero. The cylinder over the origin has one cell on which P_1 is zero.

The function $(X_1^2 + X_2^2)/(X_1 \ X_2)$ does not have a well defined limit when X_1 and X_2 tend to 0. Moreover, the closure of the graph of the function $(X_1^2 + X_2^2)/(X_1 X_2)$ on $X_1 > 0$, $X_2 > 0$ intersected with the line above the origin is $[2, +\infty)$, which is not a cell of the cylindrical decomposition.

Consider now $P_2 = (X_1 X_2 X_3 - (X_1 - X_2)^2)\,(X_1 X_2 X_3 - (X_1 + X_2)^2)$,

In order to visualize the corresponding zero set, it is convenient to fix the value of x_3.

The zero set of $(X_1 \ X_2 \ x_3 - (X_1 - X_2)^2)\ (X_1 \ X_2 \ x_3 - (X_1 + X_2)^2)$ is the union of the zero set Z_1 of $X_1 \ X_2 \ x_3 - (X_1 - X_2)^2$ and the zero set Z_2 of $X_1 X_2 x_3 - (X_1 + X_2)^2$ in the plane $X_3 = x_3$.

It can be easily checked that:

- If $-4 < x_3$, or $x_3 > 4$, the zeroes of P_2 in the plane $X_3 = x_3$ consist of four lines through the origin.
- If $x_3 = -4$ or $x_3 = 4$, the zeroes of P_2 in the plane $X_3 = x_3$ consists of three lines through the origin.
- If $x_3 = 0$, the zeroes of P_2 in the plane $X_3 = x_3$ consists of two lines through the origin.
- If $-4 < x_3 < 0$ or $0 < x_3 < 4$, the zeroes of P_2 in the plane $X_3 = x_3$ consists of two lines through the origin.

It is also easy to see that the 9 cells of R^2 defined by the signs of X_1 and X_2 and the 3 cells of R defined by the sign of X_1 determine a cylindrical decomposition adapted to $\{P_2\}$.

- When $X_1 X_2 \neq 0$, $P_2 = 0$ is equivalent to

$$X_3 = \frac{(X_1 - X_2)^2}{X_1 X_2} \text{ or } X_3 = \frac{(X_1 + X_2)^2}{X_1 X_2}.$$

So the zeroes of P_2 are described by two graphs of functions over each open quadrant, and the cylindrical decomposition of P_2 has five cells over each open quadrant. The sign of P_2 in these five cells is $1, 0, -1, 0, 1$.

- When $X_1 = 0$, $X_2 \neq 0$, the polynomial P_2 is $-X_2^4$. The cylinders over each open half-axis have one cell on which P_2 is negative.
- When $X_1 \neq 0$, $X_2 = 0$, the polynomial P_2 is $-X_1^4$. The cylinders over each open half-axis have one cell on which P_2 is negative.
- When $X_1 = 0$, $X_2 = 0$, P_2 is identically zero. The cylinder over the origin has one cell on which P_2 is zero.

Finally, while the descriptions of the cylindrical decompositions of P_2 and P_2 are identical, $\mathrm{Zer}(P_1, \mathrm{R}^3)$ and $\mathrm{Zer}(P_2, \mathrm{R}^3)$ are not homeomorphic: $\mathrm{Zer}(P_1, \mathrm{R}^3)$ has an open subset which is a semi-algebraic set of dimension 1, and it is not the case for $\mathrm{Zer}(P_2, \mathrm{R}^3)$. □

A **semi-algebraic stratification** of a finite family $S_1, \ldots, S_\ell$ of semi-algebraic sets is a finite partition of each S_i into semi-algebraic sets $S_{i,j}$ such that

- every $S_{i,j}$ is a $\mathcal{S}^\infty$ submanifold,
- the closure of $S_{i,j}$ in S_i is the union of $S_{i,j}$ with some $S_{i,j'}$'s where the dimensions of the $S_{i,j'}$'s are less than the dimension of $S_{i,j}$.

The $S_{i,j}$ are called **strata** of this stratification. A **cell stratification of R^k adapted to $\mathcal{P}$** is a stratification of R^k for which every stratum S_i is $\mathcal{S}^\infty$ diffeomorphic to an open cube $(0,1)^{d_i}$ and is also $\mathcal{P}$-invariant. A cell stratification of R^k adapted to $\mathcal{P}$ induces a stratification of $S_1, \ldots, S_\ell$ for every finite family $S_1, \ldots, S_\ell$ of $\mathcal{P}$-semi-algebraic sets.

Theorem 5.38. *For every finite set $\mathcal{P} \subset \mathrm{R}[X_1, \ldots, X_k]$, there exists a cell stratification of R^k adapted to $\mathcal{P}$.*

In Thom's lemma, the closures of the different "pieces" (points and open intervals) are obtained by relaxing strict inequalities. The key technique to prove Theorem 5.38 is to extend these properties to the case of several variables. In the cylindrical decomposition, when the polynomials are quasi-monic, we can control what happens when we pass from a cylinder $S \times \mathrm{R}$ to another $T \times \mathrm{R}$ such that $T \subset \overline{S}$. The quasi-monicity is needed to avoid the kind of bad behavior described in Example 5.37.

The following result can be thought of as a multivariate version of Thom's lemma.

Suppose that

- $\mathcal{P} \subset \mathrm{R}[X_1, \ldots, X_k]$ is closed under differentiation with respect to X_k and each $P \in \mathcal{P}$ is quasi-monic with respect to X_k (see Definition 4.72),
- S and S' are semi-algebraically connected semi-algebraic subsets of R^{k-1}, both $\mathrm{Elim}_{X_k}(\mathcal{P})$-invariant, and S' is contained in the closure of S.

It follows from Proposition 5.16 that there are continuous semi-algebraic functions $\xi_1 < \cdots < \xi_\ell : S \to \mathrm{R}$ and $\xi'_1 < \cdots < \xi'_{\ell'} : S' \to \mathrm{R}$ which describe, for all $P \in \mathcal{P}$, the real roots of the polynomials $P(x, X_k)$ as functions of $x = (x_1, \ldots, x_{k-1})$ in S or in S'. Denote by Γ_j and Γ'_j the graphs of ξ_j and ξ'_j, respectively. Since $\mathcal{P}$ is closed under differentiation, there is a polynomial $P \in \mathcal{P}$ such that, for every $x \in S$ (resp. $x \in S'$), $\xi_j(x)$ (resp. $\xi'_j(x)$) is a simple root of $P(x, X_k)$ for $P \in \mathcal{P}$ (see Remark 5.35). Denote by B_j and B'_j the bands of the cylinders $S \times \mathrm{R}$ and $S' \times \mathrm{R}$, respectively, which are bounded by these graphs.

Proposition 5.39. [Generalized Thom's Lemma]

- *Every function ξ_j can be continuously extended to S', and this extension coincides with one of the functions $\xi'_{j'}$.*
- *For every function $\xi'_{j'}$, there is a function ξ_j whose extension by continuity to S' is $\xi'_{j'}$.*
- *For every $\sigma \in \{0, 1, -1\}^{\mathcal{P}}$, the set*

$$\mathrm{Reali}(\sigma, S \times \mathrm{R}) = \{(x, x_k) \in S \times \mathrm{R} \mid \mathrm{sign}(\mathcal{P}(x, x_k)) = \sigma\}$$

is either empty or one of the Γ_j or one of the B_j. Let $\mathrm{Reali}(\overline{\sigma}, S \times \mathrm{R})$ be the subset of $S \times \mathrm{R}$ obtained by relaxing the strict inequalities:

$$\mathrm{Reali}(\overline{\sigma}, S \times \mathrm{R}) = \{(x, x_k) \in S \times \mathrm{R} \mid \mathrm{sign}(\mathcal{P}(x, x_k)) \in \overline{\sigma}\},$$

and let

$$\mathrm{Reali}(\overline{\sigma}; S' \times \mathrm{R}) = \{(x, x_k) \in S' \times \mathrm{R} \mid \mathrm{sign}(\mathcal{P}(x, x_k)) \in \overline{\sigma}\}.$$

If $\mathrm{Reali}(\sigma, S \times \mathrm{R}) \neq \emptyset$, we have $\overline{\mathrm{Reali}(\sigma, S \times \mathrm{R})} \cap (S \times \mathrm{R}) = \mathrm{Reali}(\overline{\sigma}, S \times \mathrm{R}))$ and $\overline{\mathrm{Reali}(\sigma, S \times \mathrm{R})} \cap (S' \times \mathrm{R}) = \mathrm{Reali}(\overline{\sigma}, S' \times \mathrm{R})$. Moreover, $\mathrm{Reali}(\sigma, S' \times \mathrm{R})$ is either a graph $\Gamma'_{j'}$ or the closure of one of the bands $B'_{j'}$ in $S' \times \mathrm{R}$.

Proof: Let $x' \in S'$. Consider one of the functions ξ_j. Since $\mathcal{P}$ is closed under differentiation, there is a polynomial $P \in \mathcal{P}$ such that, for every $x \in S$, $\xi_j(x)$ is a simple root of

$$P(x, X_k) = a_p X_k^p + a_{p-1}(x) X_k^{p-1} + \cdots + a_0(x),$$

(see Remark 5.35). Moreover, a_p is a non-zero constant. Let

$$M(x') = \max_{i=1,\ldots,p} \left(p \left| \frac{a_{p-i}(x')}{a_p} \right| \right)^{1/i}.$$

By Proposition 5.9, and the continuity of M, there is a neighborhood U of x' in R^{k-1} such that, for every $x \in S \cap U$, we have

$$\xi_j(x) \in [-M(x') - 1, M(x') + 1]$$

Choose a continuous semi-algebraic path γ such that

$$\gamma((0,1]) \subset S \cap U, \ \gamma(0) = x'.$$

The semi-algebraic function $f = \xi_j \circ \gamma$ is bounded and therefore has, by Proposition 3.18, a continuous extension $\overline{f}$ with

$$\overline{f}(0) \in [-M(x') - 1, M(x') + 1].$$

Let $\tau_1 = \mathrm{sign}(P'(x, \xi_j(x))), \ldots, \tau_p = \mathrm{sign}(P^{(p)}(x, \xi_j(x)))$, for $x \in S$ (observe that these signs are constant for $x \in S$). Since every point in the graph of ξ_j satisfies

$$P(x', x_k') = 0, \ \mathrm{sign}(P'(x', x_k')) = \tau_1, \ldots, \ \mathrm{sign}(P^{(p)}(x', x_k')) = \tau_d \,,$$

every point (x', x_k') in the closure of the graph of ξ_j must satisfy

$$P(x', x_k') = 0, \ \mathrm{sign}(P'(x', x_k')) \in \overline{\tau}_1, \ldots, \ \mathrm{sign}(P^{(p)}(x', x_k')) \in \overline{\tau}_d \,.$$

By Lemma 5.33 (Thom's lemma), there is at most one x_k' satisfying these inequalities. Since $(x', \overline{f}(0))$ is in the closure of the graph of ξ_j, it follows that ξ_j extends continuously to x' with the value $\overline{f}(0)$. Hence, it extends continuously to S', and this extension coincides with one of the functions $\xi_{j'}'$. This proves the first item.

We now prove the second item. Choose a function $\xi_{j'}'$. Since $\xi_{j'}'$ is a simple root of some polynomial P in the set, by Proposition 3.10 there is a function ξ_j, also a root of P, whose continuous extension to S' is $\xi_{j'}'$.

We now turn to the third item. The properties of $\mathrm{Reali}(\sigma, S \times \mathrm{R})$ and $\mathrm{Reali}(\overline{\sigma}, S \times \mathrm{R})$ are straightforward consequences of Thom's lemma, since $P \in \mathcal{P}$ has constant sign on each graph Γ_j and each band B_j. The closure of B_j in $S \times \mathrm{R}$ is $\Gamma_j \cup B_j \cup \Gamma_{j+1}$, where $\Gamma_0 = \Gamma_{\ell+1} = \emptyset$ and therefore it is obvious that $\overline{\mathrm{Reali}(\sigma, S \times \mathrm{R})} \cap (S' \times \mathrm{R}) \subset \mathrm{Reali}(\overline{\sigma}, S' \times \mathrm{R})$. It follows from 1 and 2 that $\overline{\mathrm{Reali}(\sigma, S \times \mathrm{R})} \cap (S' \times \mathrm{R})$ is either a graph Γ_j' or the closure of one of the bands $B_{j'}'$ in $S' \times \mathrm{R}$.

By Thom's lemma, this is also the case for $\mathrm{Reali}(\overline{\sigma}, S' \times \mathrm{R})$. It remains to check that it cannot happen that $\mathrm{Reali}(\overline{\sigma}, S' \times \mathrm{R})$ is the closure of a band $B_{j'}'$ and $\overline{\mathrm{Reali}(\sigma, S \times \mathrm{R})} \cap (S' \times \mathrm{R})$ is one of the graphs $\Gamma_{j'}'$ or $\Gamma_{j'+1}'$. In this case, all $\sigma(P)$ should be different from zero and the sign of P should be $\sigma(P)$ on every sufficiently small neighborhood V of a point x' of $B_{j'}'$. This implies that $V \cap (S \times \mathrm{R}) \subset R(\sigma, S \times \mathrm{R})$ and, hence, $x' \in \overline{R(\sigma, S \times \mathrm{R})}$, which is impossible. □

Proposition 5.40. *Let $\mathcal{P}^\star = \bigcup_{i=1}^{k} \mathcal{P}_i$ be finite sets of non-zero polynomials such that:*

- *$\mathcal{P}_k$ contains $\mathcal{P}$,*
- *or each i, $\mathcal{P}_i$ is a subset of $\mathrm{R}[X_1, \ldots, X_i]$ that is closed under differentiation and quasi-monic with respect to X_i,*

– *for* $i \leq k$, $\mathrm{Elim}_{X_i}(\mathcal{P}_i) \subset \mathcal{P}_{i-1}$.

Writing $\mathcal{P}_{\leq i} = \bigcup_{j \leq i} \mathcal{P}_j$, *the families* $\mathcal{S}_i$, *for* $i = 1, \ldots, k$, *consisting of all* $\mathrm{Reali}(\sigma)$ *with* σ *a realizable sign conditions on* $\mathcal{P}_{\leq i}$ *constitute a cylindrical decomposition of* R^k *that is a cell stratification of* R^k *adapted to* $\mathcal{P}$.

Proof: The proof of the proposition is a simple induction on k. The preceding Proposition 5.39 (Generalized Thom's Lemma) provides the induction step and Thom's lemma the base case for $k = 1$. To show that the dimension condition is satisfied, observe that if $\sigma \in \{0, 1, -1\}^{\mathcal{P}}$ and $\mathrm{Reali}(\sigma) \neq \emptyset$, then $\mathrm{Reali}(\overline{\sigma})$ is the union of $\mathrm{Reali}(\sigma)$ and some $\mathrm{Reali}(\sigma')$, $\sigma' \neq \sigma$.

Since $\mathrm{Reali}(\sigma)$ (resp. $\mathrm{Reali}(\sigma')$) is a cell of a cylindrical decomposition, $\mathrm{Reali}(\sigma)$ (resp. $\mathrm{Reali}(\sigma')$) is semi-algebraically homeomorphic to $(0,1)^{d(\sigma)}$ (resp. $(0,1)^{d(\sigma')}$). That $d(\sigma') < d(\sigma)$ is easily seen by induction. □

A family $\mathcal{P}$ is a **stratifying family** if it satisfies the hypothesis of Proposition 5.40.

The theorem above holds for a stratifying family of polynomials. But we shall now see that it is always possible to convert a finite set of polynomials to a quasi-monic set by making a suitable linear change of coordinates. By successively converting to quasi-monic polynomials, closing under differentiation and applying Elim, we arrive at a stratifying family.

Proposition 5.41. *Let* $\mathcal{P} \subset \mathrm{R}[X_1, \ldots, X_k]$. *There is a linear automorphism* $u: \mathrm{R}^k \to \mathrm{R}^k$ *and a stratifying family of polynomials* $\mathcal{Q}^\star = \cup_{i=1,\ldots k} \mathcal{Q}_i$ *such that* $P(u(X)) \in \mathcal{Q}_k$ *for all* $P \in \mathcal{P}$ *(where* $X = (X_1, \ldots, X_k)$*).*

Proof: By Lemma 4.73, there is a linear change of variables v such that, for all $P \in \mathcal{P}$, the polynomial $P(v(X))$ is quasi-monic with respect to X_k.

Let $\mathcal{Q}_k$ consist of all polynomials $P(v(X))$ for $P \in \mathcal{P}$ together with all their non-zero derivatives of every order with respect to X_k. Using induction, applied to $\mathrm{Elim}_{X_k}(\mathcal{Q}_k)$, there is a linear automorphism $u': \mathrm{R}^{k-1} \to \mathrm{R}^{k-1}$ and a stratifying family of polynomials $\cup_{1 \leq i \leq k-1} \mathcal{R}_i$ such that $Q(u'(X')) \in \mathcal{R}_{k-1}$ for every $Q \in \mathrm{Elim}_{X_k}(\mathcal{Q}_k)$, where $X' = (X_1, \ldots, X_{k-1})$. Finally, set $u = (u' \times \mathrm{Id}) \circ v$ (where $u' \times \mathrm{Id}(X', X_k) = (u'(X'), X_k)$), $\mathcal{Q}_j = \{R(X) \mid R \in \mathcal{R}_j\}$ for $j \leq k-1$. □

We are now ready for the proof of Theorem 5.38.

Proof of Theorem 5.38: Use Proposition 5.41 to get a linear automorphism $u: \mathrm{R}^k \to \mathrm{R}^k$ and a stratifying family $\mathcal{Q}^\star$ that contains

$$\mathcal{Q} = \{P(u(X)) \mid P \in \mathcal{P}\}$$

in order to obtain, by Proposition 5.40, a cell stratification adapted to $\mathcal{Q}$. Clearly, u^{-1} converts this cell stratification to one adapted to $\mathcal{P}$. □

Theorem 5.38 has consequences for the dimension of semi-algebraic sets.

Theorem 5.42. *Let $S \subset \mathrm{R}^k$ be a semi-algebraic set. Then,*

$$\begin{aligned}\dim(\overline{S}) &= \dim(S),\\ \dim(\overline{S}\setminus S) &< \dim(S).\end{aligned}$$

Proof: This is clear from Proposition 5.28 and Theorem 5.38, since the closure of a stratum is the union of this stratum and of a finite number of strata of smaller dimensions. □

5.6 Simplicial Complexes

We first recall some basic definitions and notations about simplicial complexes.

Let $a_0, \ldots, a_d$ be points of R^k that are **affinely independent** (which means that they are not contained in any affine subspace of dimension $d-1$). The d-**simplex** with vertices $a_0, \ldots, a_d$ is

$$[a_0, \ldots, a_d] = \{\lambda_0 a_0 + \cdots + \lambda_d a_d \mid \sum_{i=0}^{d} \lambda_i = 1 \text{and } \lambda_0, \ldots, \lambda_d \geq 0\}.$$

Note that the dimension of $[a_0, \ldots, a_d]$ is d.

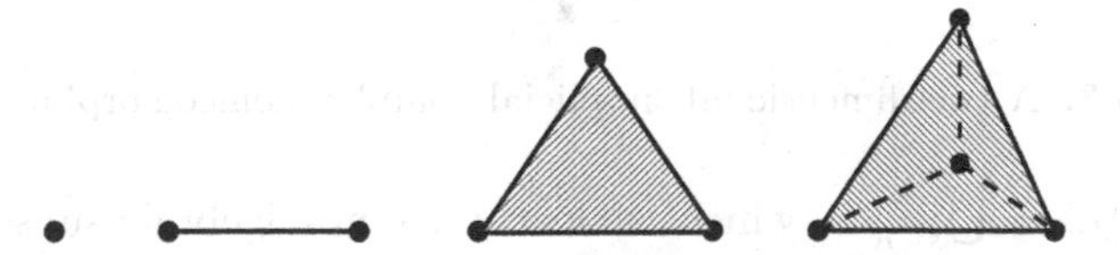

Fig. 5.2. Zero, one, two, and three dimensional simplices

An $e-$ **face** of the $d-$ simplex $s = [a_0, \ldots, a_d]$ is any simplex $s' = [b_0, \ldots, b_e]$ such that

$$\{b_0, \ldots, b_e\} \subset \{a_0, \ldots, a_d\}.$$

The face s' is a **proper face** of s if $\{b_0, \ldots, b_e\} \neq \{a_0, \ldots, a_d\}$. The $0-$ faces of a simplex are its **vertices**, the $1-$ faces are its **edges**, and the $(d-1)-$ faces of a $d-$ simplex are its **facets**. We also include the empty set as a simplex of dimension -1, which is a face of every simplex. If s' is a face of s we write $s' \prec s$.

The open simplex corresponding to a simplex s is denoted s° and consists of all points of s which do not belong to any proper face of s:

$$s^\circ = (a_0, \ldots, a_d) = \{\lambda_0 a_0 + \cdots + \lambda_d a_d \mid \sum_{i=0}^{d} \lambda_i = 1 \text{and } \lambda_0 > 0, \ldots, \lambda_d > 0\}.$$

which is the interior of $[a_0, ..., a_d]$. By convention, if s is a $0-$simplex then $s^\circ = s$.

The **barycenter** of a $d-$simplex $s = [a_0, \ldots, a_d]$ in R^k is the point $\mathrm{ba}(s) \in \mathrm{R}^k$ defined by $\mathrm{ba}(s) = 1/(d+1)\sum_{0 \le i \le d} a_i$.

A **simplicial complex** K in R^k is a finite set of simplices in R^k such that $s, s' \in K$ implies

- every face of s is in K,
- $s \cap s'$ is a common face of both s and s'.

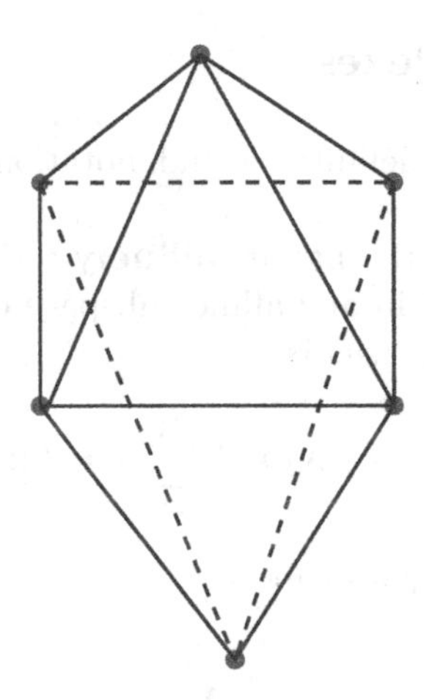

Fig. 5.3. A two dimensional simplicial complex homeomorphic to S^2

The set $|K| = \bigcup_{s \in K} s$, which is clearly a semi-algebraic subset of R^k, is called the the **realization of** K. Note that the realization of K is the disjoint union of its open simplices. A **polyhedron** in R^k is a subset P of R^k such that there exists a simplicial complex K in R^k with $P = |K|$. Such a K is called a **simplicial decomposition** of P.

Let K and L be two simplicial complexes. Then L is called a **subdivision** of K if

- $|L| = |K|$,
- for every simplex $s \in L$ there is a simplex $s' \in K$ such that $s \subset s'$.

Given a simplicial complex K, an **ascending sequence of simplices** is a collection of simplices $\{s_0, s_1, \ldots, s_j\}$ such that $s_0 \prec s_1 \prec \cdots \prec s_j$.

Let K be a simplicial complex. Let $\mathrm{ba}(K)$ denote the set of simplices that are spanned by the barycenters of some ascending sequence of simplices of K. Thus for every ascending sequence of simplices in K, $s_0 \prec s_1 \prec \cdots \prec s_j$, we include in K' the simplex $[\mathrm{ba}(s_0), \ldots, \mathrm{ba}(s_j)]$, and we call $\mathrm{ba}(s_j)$ the **leading vertex** of $[\mathrm{ba}(s_0), ..., \mathrm{ba}(s_j)]$. It is easy to check that $\mathrm{ba}(K)$ is a simplicial complex, called the **barycentric subdivision** of K.

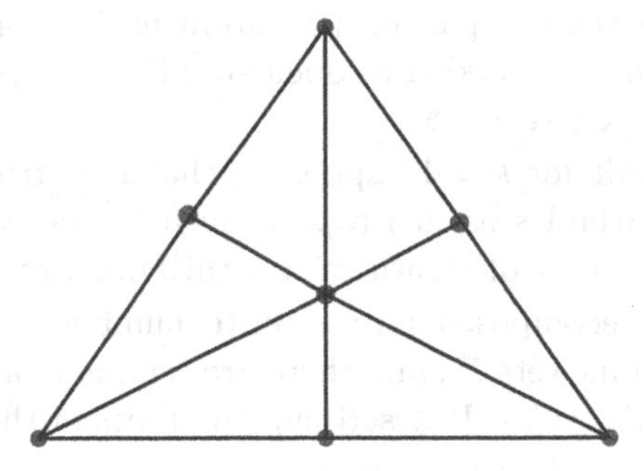

Fig. 5.4. The barycentric subdivision of a two dimensional simplex

5.7 Triangulation

A **triangulation** of a semi-algebraic set S is a simplicial complex K together with a semi-algebraic homeomorphism h from $|K|$ to S. We next prove that any closed and bounded semi-algebraic set can be triangulated. In fact, we prove a little more, which will be useful for technical reasons. The triangulation will also be a stratification of S which respects any given finite collection of semi-algebraic subsets of S, i.e. the images of the open simplices will be the strata and each of the specified subsets of S will be stratified as well.

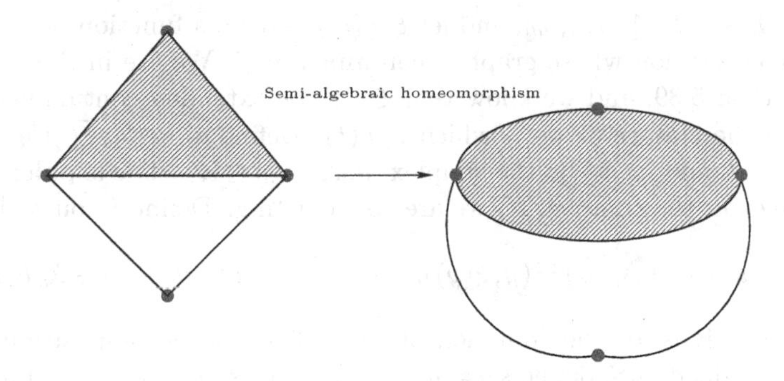

Fig. 5.5. Semi-algebraic triangulation

A triangulation of S **respecting a finite family of semi-algebraic sets** $S_1, \ldots, S_q$ contained in S is a triangulation K, h such that each S_j is the union of images by h of open simplices of K.

Theorem 5.43. [Triangulation] *Let $S \subset \mathrm{R}^k$ be a closed and bounded semi-algebraic set, and let S_1, ..., S_q be semi-algebraic subsets of S. There exists a triangulation of S respecting $S_1, \ldots, S_q$. Moreover, the vertices of K can be chosen with rational coordinates.*

Proof: We first prove the first part of the statement. The proof is by induction on k. For $k=1$, let $|K|=S$, taking as open simplices the points and bounded open intervals which constitute S.

We prove the result for $k>1$ supposing that it is true for $k-1$. After a linear change of variables as in Proposition 5.41, we can suppose that S and the S_j are the union of strata of a stratifying set of polynomials $\mathcal{P}$. Thus R^{k-1} can be decomposed into a finite number of semi-algebraically connected semi-algebraic sets C_i, and there are semi-algebraic and continuous functions $\xi_{i,1}<\cdots<\xi_{i,\ell_i}: C_i \to \mathrm{R}$ describing the roots of the non-zero polynomials among $P(x, X_k)$, $P \in \mathcal{P}_k$, as functions of $x \in C_i$. We know that S, and the S_j, are unions of some graphs of $\xi_{i,j}$ and of some bands of cylinders $C_i \times \mathrm{R}$ between these graphs. Denote by $\pi: \mathrm{R}^k \to \mathrm{R}^{k-1}$ the projection which forgets the last coordinate. The set $\pi(S)$ is closed and bounded, semi-algebraic, and the union of some C_i; also, each $\pi(S_j)$ is the union of some C_i. By the induction hypothesis, there is a triangulation $g: |L| \to \pi(S)$ (where g is a semi-algebraic homeomorphism, L a simplicial complex in R^{k-1}) such that each $C_i \subset \pi(S)$ is a union of images by g of open simplices of L. Thus, at the top level, R^k is decomposed into cylinders over sets of the form $g(t^\circ)$ for t a simplex of L.

We next extend the triangulation of $\pi(S)$ to one for S. For every t in L we construct a simplicial complex K_t and a semi-algebraic homeomorphism

$$h_t: |K_t| \longrightarrow \overline{S \cap (g(t^\circ) \times \mathrm{R})}.$$

Fix a t in L, say $t=[b_0, \ldots, b_d]$, and let $\xi: g(t^\circ) \to \mathrm{R}$ be a function of the cylindrical decomposition whose graph is contained in S. We are in the situation of Proposition 5.39, and we know that ξ can be extended continuously to $\overline{\xi}$ defined on the closure of $g(t^\circ)$ which is $g(t)$. Define $a_i=(b_i, \overline{\xi}(g(b_i))) \in \mathrm{R}^k$ for $i=0, \ldots, d$, and let s_ξ be the simplex $[a_0, \ldots, a_d] \subset \mathrm{R}^k$. The simplex s_ξ will be a simplex of the complex K_t we are constructing. Define h_t on s_ξ by

$$h_t(\lambda_0 a_0+\cdots+\lambda_d a_d)=(y, \overline{\xi}(y)), \quad \text{where } y=g(\lambda_0 b_0+\cdots+\lambda_d b_d).$$

If $\xi': g(t^\circ) \to \mathrm{R}$ is another function of the cylindrical decomposition whose graph is contained in S, define $s_{\xi'}=[a_0', \ldots, a_d']$ in the same way. It is important that $s_{\xi'}$ not coincide with s_ξ. At least one of the a_i' must differ from the corresponding a_i. Similarly when the restrictions of $\overline{\xi}$ and $\overline{\xi'}$ to a face r of t are not the same, we require that on at least one vertex b_i of r, the values of $\overline{\xi}$ and $\overline{\xi'}$ are different (so that the corresponding a_i and a_i' are distinct). Thus we require that or every simplex t of L, if ξ and ξ' are two distinct functions $g(t^\circ) \to \mathrm{R}$ of the cylindrical decomposition then there exists a vertex b of t such that $\overline{\xi}(g(b)) \neq \overline{\xi'}(g(b))$. It is clear that this requirement will be satisfied if we replace L by its barycentric division $\mathrm{ba}(L)$. Hence, after possibly making this replacement, we can assume that our requirement is satisfied by L.

Now consider a band between two graphs of successive functions

$$\xi < \xi' : g(t^\circ) \longrightarrow \mathrm{R}$$

contained in S (note that an unbounded band cannot be contained in S, since S is closed and bounded). Let P be the polyhedron above t whose bottom face is s_ξ and whose top face is $s_{\xi'}$. This polyhedron P has a simplicial decomposition

$$P = \bigcup_{i=0}^{d} [a_0', \ldots, a_i', a_i, \ldots, a_d].$$

Note that it may happen that $a_i' = a_i$ in which case we understand

$$[a_0', \ldots, a_i', a_i, \ldots, a_d]$$

to be the $d-1$-simplex

$$[a_0', \ldots, a_i', a_{i+1}, \ldots, a_d].$$

The complex K_t we are constructing contains the simplices (and their faces) of this simplicial decomposition of P. We define h_t on P by the condition that the segment $[\lambda_0 a_0 + \cdots + \lambda_d a_d, \lambda_0 a_0' + \cdots + \lambda_d a_d']$ is sent by an affine function to $[(y, \overline{\xi}(y)), (y, \overline{\xi'}(y))]$, where

$$y = g(\lambda_0 b_0 + \cdots + \lambda_d b_d).$$

Having constructed K_t and h_t for each simplex t of L, it remains to prove that these K_t and h_t can be glued together to give K and h as a triangulation of S. We next show that it is possible if we first choose a total order for all vertices of L and then label the simplices of L compatibly with this total order.

It is enough to check this for a simplex t and one of its faces r. The first thing to notice is that if we have a simplex s_η in K_r that is sent by h_r onto the closure of the graph $\eta: g(r^\circ) \to \mathrm{R}$, a function of the algebraic decomposition, and if s_η meets $|K_t|$, then it is a simplex of K_t: indeed in this case η coincides with one of the $\overline{\xi}$ on $g(r^\circ)$ by point 2 of Proposition 5.39 (for $\xi: g(t^\circ) \to \mathrm{R}$, s_ξ simplex of K_t and s_η a facet of s_ξ). For this reason, it is also the case that h_t and h_r coincide on $|K_t| \cap |K_r|$. What remains to verify is that the simplicial complex of the polyhedron P in $t \times \mathrm{R}$ (see above) induces the simplicial decomposition of the polyhedron $P \cap (r \times \mathrm{R})$. This is the case if the simplicial decomposition $P = \bigcup_{i=0}^{d} [a_0', \ldots, a_i', a_i, \ldots, a_d]$ is compatible with a fixed total order on the vertices of L.

It remains to prove that there exists a simplicial complex L with rational coordinates such that $|K|$ and $|L|$ are semi-algebraically homeomorphic. The proof is by induction on k. When $k=1$, the semi-algebraic subsets $S, S_1, \ldots, S_q$ are a finite number of points and intervals and the claim is clear. The inductive steps uses the cylindrical structure of the constructed triangulation. □

The following corollary of Theorem 5.43 will be used in the proof of Theorem 5.46.

Proposition 5.44. *Let $S \subset \mathrm{R}^k$ be a closed and bounded semi-algebraic set, and let S_1, ..., S_q be semi-algebraic subsets of S, such that S and the S_j are given by boolean combinations of sign conditions over polynomials that are all either monic in the variable X_k or independent of X_k. Let π be the projection of R^k to R^{k-1} that forgets the last factor. There are semi-algebraic triangulations*

$$\Phi: |K| = \bigcup_{p=1}^{s'} |s_p| \to S, \quad |K| \subset \mathrm{R}^k$$

and

$$\Psi: |L| = \bigcup_{\ell=1}^{s} |t_\ell| \to \pi(S), \quad |L| \subset \mathrm{R}^{k-1}$$

such that $\pi \circ \Phi(x) = \Psi \circ \pi(x)$ for $x \in |K|$, and each S_j is the union of some $\Phi(s_i^\circ)$.

5.8 Hardt's Triviality Theorem and Consequences

Hardt's triviality theorem is the following.

Theorem 5.45. [Hardt's triviality theorem] *Let $S \subset \mathrm{R}^m$ and $T \subset \mathrm{R}^k$ be semi-algebraic sets. Given a continuous semi-algebraic function $f: S \to T$, there exists a finite partition of T into semi-algebraic sets $T = \bigcup_{i=1}^r T_i$, so that for each i and any $x_i \in T_i$, $T_i \times f^{-1}(x_i)$ is semi-algebraically homeomorphic to $f^{-1}(T_i)$.*

For technical reasons, we prove the slightly more general:

Theorem 5.46. [Semi-algebraic triviality] *Let $S \subset \mathrm{R}^m$ and $T \subset \mathrm{R}^k$ be semi-algebraic sets. Given a continuous semi-algebraic function $f: S \to T$ and S_1, ..., S_q semi-algebraic subsets of S, there exists a finite partition of T into semi-algebraic sets $T = \bigcup_{i=1}^r T_i$, so that for each i and any $x_i \in T_i$, $T_i \times f^{-1}(x_i)$ is semi-algebraically homeomorphic to $f^{-1}(T_i)$. More precisely, writing $F_i = f^{-1}(x_i)$, there exist semi-algebraic subsets $F_{i,1}, ...F_{i,q}$ of F_i and a semi-algebraic homeomorphism $\theta_i: T_i \times F_i \to f^{-1}(T_i)$ such that $f \circ \theta_i$ is the projection mapping $T_i \times F_i \to T_i$ and such that*

$$\theta_i(T_i \times F_{i,j}) = S_j \cap f^{-1}(T_i).$$

Proof: We may assume without loss of generality that

- S and T are both bounded (using if needed homeomorphisms of the form $x \mapsto x/(1 + \|x\|)$, which are obviously semi-algebraic),
- S is a semi-algebraic subset of R^{m+k} and f is the restriction to S of the projection mapping Π: $R^{m+k} \to R^k$ that forgets the first m coordinates, (replacing S by the graph of f which is semi-algebraically homeomorphic to S).

The proof proceeds by induction on the lexicographic ordering of the pairs (m, k).

The sets S and S_j are given by boolean combinations of sign conditions over a finite number of polynomials $\mathcal{P} \subset \mathrm{R}[X, Y]$, where $X = (X_1, \ldots, X_m)$ and $Y = (Y_1, \ldots, Y_k)$. Making, if needed, a linear substitution of the variables of the Y's only as in Proposition 5.41, one may suppose that each $P \in \mathcal{P}$ can be written

$$g_{P,0}(X)\, Y_k^{d(P)} + g_{P,1}(X, Y')\, Y_k^{d(P)-1} + \cdots + g_{P,d(P)}(X, Y'),$$

where $Y' = (Y_1, \ldots, Y_{k-1})$, with $g_{P,0}(X)$ not identically zero. Let

$$A(X) = \prod_{P \in \mathcal{P}} g_{P,0}(X).$$

The dimension of the semi-algebraic set $T'' = \{x \in T \mid A(x) = 0\}$ is strictly smaller than m. By Theorem 5.19, this set can be written as the finite union of sets of the form $\varphi((0, 1)^d)$ where φ is a semi-algebraic homeomorphism and $d < m$. Taking the inverse image under φ, we have to deal with a subset of R^d, and our induction hypothesis takes care of this case.

It remains to handle what happens above $T' = T \setminus T''$. We multiply each polynomial in $\mathcal{P}$ by a convenient product of powers of $g_{Q,0}(X)$, $Q \in \mathcal{P}$, so that the leading coefficient of P becomes $(A(X) Y_k)^{d(P)}$. Replacing $A(X) Y_k$ by Z_k defines a semi-algebraic homeomorphism from $S \cap (T' \times \mathrm{R}^k)$ onto a bounded semi-algebraic set $S' \subset \mathrm{R}^{m+k}$. Denote by S_j' the image of $S_j \cap (T' \times \mathrm{R}^k)$ under this homeomorphism. Now, the sets S' and S_j' are both given by boolean combinations of sign conditions over polynomials that are all either quasi-monic in the variable Z_k or independent of Z_k. Up to a linear substitution of the variables involving only the variables X and $(Y_1, \ldots, Y_{k-1})$, one may suppose that S' and the S_j' are given by boolean combinations of sign conditions over polynomials of a stratifying family. By Proposition 5.40, $\overline{S'}$ is also given by a boolean combination of sign conditions over the same polynomials.

One can now apply Corollary 5.45 to $\overline{S'}$ and $S_0' = S', S_1', \ldots, S_q'$: there are semi-algebraic triangulations

$$\Phi: |K| = \bigcup_{p=1}^{s'} |s_p| \to \overline{S'}, \quad |K| \subset \mathrm{R}^{m+k}$$

and

$$\Psi: |L| = \bigcup_{\ell=1}^{s} |t_\ell| \to \pi(\overline{S'}), \quad |L| \subset \mathrm{R}^{m+k-1}$$

such that $\pi \circ \Phi(x) = \Psi \circ \pi(x)$ if $x \in |K|$, and each S_j' $(j = 0, \ldots, q)$ is the union of some $\Phi(s_i^\circ)$.

We now apply the induction hypothesis to $\pi(\overline{S'})$, with the subsets $\Psi(t_\ell^\circ)$ and the projection mapping Π': $\mathrm{R}^{m-1+k} \to \mathrm{R}^k$. We obtain a finite partition of R^k into semi-algebraic sets $(T_i')_{i=1,\ldots,r}$. We also obtain semi-algebraic sets $G_i, G_{i,0}, G_{i,1}, \ldots, G_{i,s}$ with $G_{i,\ell} \subset G_i \subset \mathrm{R}^{m-1}$ and semi-algebraic homeomorphisms ρ_i: $T_i' \times G_i \to \Pi'^{-1}(T_i') \cap \pi(\overline{S'})$ such that $\Pi' \circ \rho_i$ is the projection mapping $T_i' \times G_i \to T_i'$. Moreover, for every ℓ, $\rho_\ell(T_i' \times G_{i,\ell}) = \Pi'^{-1}(T_i') \cap \Psi(t_\ell^\circ)$. Let us fix i, and let x_1 be a point of T_i'. One may suppose that

$$G_i = \Pi'^{-1}(x_1) \cap \pi(\overline{S'})$$

and that if $(x_1, y') \in G_i$, then $\rho_i(x_1, (x_1, y')) = (x_1, y')$. Let us then set $F_i' = \Pi^{-1}(x_1) \cap S'$ and $F_{i,j}' = \Pi^{-1}(x_1) \cap S_j'$. It remains to build

$$\theta_i: T_i' \times F_i' \to \Pi^{-1}(T_i') \cap \overline{S'}.$$

Let $x \in T_i'$ and $(x_1, y') \in G_i$; (x_1, y') belongs to one of the $\Psi(t_\ell^\circ)$, say $\Psi(t_1^\circ)$, and $\rho_i(x, (x_1, y')) \in \Psi(t_1^\circ)$. By the properties of the triangulations Φ and Ψ, the intersections with the $\Phi(s_p)$ decompose

$$\pi^{-1}(x_1, y') \cap S' \text{ and } \pi^{-1}(\rho_\ell(x, (x_1, y'))) \cap \overline{S'}$$

in the same way: θ_i maps affinely the segment

$$\{x\} \times (\pi^{-1}(x_1, y') \cap \Phi(s_p)) \subset T_i' \times F_i'$$

(which is possibly either a point or empty) onto the segment

$$\pi^{-1}(\rho_i(x, (x_1, y'))) \cap \Phi(s_p).$$

We leave it to the reader to verify that the θ_i built in this way is a semi-algebraic homeomorphism and that $\theta_i(T_i' \times F_{i,j}') = \Pi^{-1}(T_i') \cap S_j'$. □

Theorem 5.46 (Semi-algebraic triviality) makes it possible to give an easy proof that the number of topological types of algebraic subsets of R^k is finite if one fixes the maximum degree of the polynomials.

Theorem 5.47. [Finite topological types] *Let k and d be two positive integers. Let $\mathcal{M}(k, d)$ be the set of algebraic subsets $V \subset \mathrm{R}^k$ such that there exists a finite set $\mathcal{P} \subset \mathrm{R}[X_1, \ldots, X_k]$ with $V = \mathrm{Zer}(\mathcal{P}, \mathrm{R}^k)$ and $\deg(P) \le d$ for every $P \in \mathcal{P}$. There exist a finite number of algebraic subsets $V_1, \ldots, V_s$ of R^k in $\mathcal{M}(k, d)$ such that for every V in $\mathcal{M}(k, d)$ there exist i, $1 \le i \le s$, and a semi-algebraic homeomorphism φ: $\mathrm{R}^k \to \mathrm{R}^k$ with $\varphi(V_i) = V$.*

Proof: The set $\mathcal{M}(k, d)$ is contained in the set $\mathcal{F}$ of algebraic subsets of R^k given by a single equation of degree $\le 2d$ because

$$\mathrm{Zer}(\mathcal{P}, \mathrm{R}^k) = \mathrm{Zer}(\sum_{P \in \mathcal{P}} P^2, \mathrm{R}^k).$$

One parametrizes the set $\mathcal{F}$ by the space R^N of coefficients of the equation: abusing notation, P denotes the point of R^N whose coordinates are the coefficients of P. Let $S = \{(P, x) \in \mathrm{R}^N \times \mathrm{R}^k \mid P(x) = 0\}$. The set S is algebraic. Let $\Pi: \mathrm{R}^N \times \mathrm{R}^k \to \mathrm{R}^N$ be the canonical projection mapping. One has $\Pi^{-1}(P) \cap S = \{P\} \times \mathrm{Zer}(P, \mathrm{R}^k)$. Theorem 5.46 applied to the projection mapping $\Pi: \mathrm{R}^N \times \mathrm{R}^k \to \mathrm{R}^N$ and to the subset S of $\mathrm{R}^N \times \mathrm{R}^k$ gives the result. □

Another consequence of Theorem 5.46 (Semi-algebraic triviality) is the theorem of local conic structure of the semi-algebraic sets.

Theorem 5.48. [Local conic structure] *Let E be a semi-algebraic subset of R^k and x a non-isolated point of E. Then there exist $r \in \mathrm{R}$, $r > 0$, and for every r', $0 < r' \le r$, a semi-algebraic homeomorphism $\varphi: \overline{B}_k(x, r') \to \overline{B}_k(x, r')$ such that:*

- $\|\varphi(y) - x\| = \|y - x\|$ *for every* $y \in \overline{B}_k(x, r')$,
- $\varphi \mid S^{k-1}(x, r')$ *is the identity mapping,*
- $\varphi(E \cap \overline{B}_k(x, r'))$ *is a cone with vertex x and base* $E \cap S^{k-1}(x, r')$.

Proof: Apply Theorem 5.46 (Semi-algebraic triviality) with $S = \mathrm{R}^k$, $S_1 = E$, and $f: S \to \mathrm{R}$ defined by $f(y) = \|y - x\|$ to deduce that there exists $r > 0$ and for every r', $0 < r' \le r$, a semi-algebraic homeomorphism

$$\theta: (0, r'] \times S^{k-1}(x, r') \to \overline{B}_k(x, r') \setminus \{x\}$$

such that, for every y in $S^{k-1}(x, r')$, $\|\theta(t, y) - x\| = t$ for $t \in (0, r']$, $\theta(r', y) = y$, and $\theta((0, r'] \times (E \cap S^{k-1}(x, r'))) = E \cap \overline{B}_k(x, r) \setminus \{x\}$. It is then easy to build φ. □

Let S be a closed semi-algebraic set and T a closed semi-algebraic subset of S. A **semi-algebraic deformation retraction** from S to T, is a continuous semi-algebraic function h: $S \times [0, 1] \to S$ such that $h(-, 0)$ is the identity mapping of S, such that $h(-, 1)$ has its values in T and such that for every $t \in [0, 1]$ and every x in T, $h(x, t) = x$.

Proposition 5.49. [Conic structure at infinity] *Let S be a closed semi-algebraic subset of R^k. There exists $r \in \mathrm{R}$, $r > 0$, such that for every r', $r' \ge r$, there is a semi-algebraic deformation retraction from S to $S_{r'} = S \cap \overline{B_k}(0, r')$ and a semi-algebraic deformation retraction from $S_{r'}$ to S_r.*

Proof: Let us suppose that S is not bounded. Through an inversion mapping $\varphi: \mathrm{R}^k \setminus \{0\} \to \mathrm{R}^k \setminus \{0\}$, $\varphi(x) = x / \|x\|^2$, which is obviously semi-algebraic, one can reduce to the property of local conic structure for $\varphi(S) \cup \{0\}$ at 0. □

Proposition 5.50. *Let f: $S \to T$ be a semi-algebraic function that is a local homeomorphism. There exists a finite cover $S = \bigcup_{i=1}^n U_i$ of S by semi-algebraic sets U_i that are open in S and such that $f \mid U_i$ is a homeomorphism for every i.*

Proof: We assume, as in the proof of Theorem 5.46, that T is bounded and that the partition $T = \bigcup_{\ell=1}^{r} T_\ell$ is induced by a semi-algebraic triangulation Φ: $|K| = \bigcup_{\ell=1}^{s} |s_\ell| \rightarrow \overline{T}$ such that $T_\ell = \Phi(s_\ell^0)$. We then replace T by $Z = \bigcup_{\ell=1}^{r} |s_\ell^0|$ and set $g = \Phi^{-1} \circ f$. There are semi-algebraic homeomorphisms θ_ℓ: $s_\ell^0 \times F_\ell \rightarrow g^{-1}(s_\ell^0)$ such that $g \circ \theta_\ell$ is the projection mapping $s_\ell^0 \times F_\ell \rightarrow s_\ell^0$ by Theorem 5.46 (Semi-algebraic triviality). Each F_ℓ consists of a finite number of points since g is a local homeomorphism and F_ℓ is semi-algebraic. Let $x_{\ell,1}, \ldots, x_{\ell,p_\ell}$ denote these points. Note that if $s_\ell^0 \subset Z$, then $s_{\ell'}^0 \subset Z$, s_ℓ is a face of $s_{\ell'}$ and $x_{\ell,\lambda} \in F_\ell$, then there exists a unique point $x_{\ell',\lambda'} = \beta_{\ell,\ell'}(x_{\ell,\lambda}) \in F_{\ell'}$ such that $\theta_\ell(s_\ell^0 \times \{x_{\ell,\lambda}\})$ is equal to the closure of $(\theta_{\ell'}(s_{\ell'}^0 \times \{x_{\ell',\lambda'}\})) \cap g^{-1}(s_\ell^0)$. Fix ℓ and λ and set

$$V_{\ell,\lambda} = \bigcup \{\theta_{\ell'}(s_{\ell'}^0 \times \{\beta_{\ell,\ell'}(x_{\ell,\lambda})\}) \mid s_{\ell'}^0 \subset Z \text{and } s_\ell \text{is a face of } s_{\ell'}\}.$$

By the previous remark, $g \mid V_{\ell,\lambda}$ is a homeomorphism over the union of the $s_{\ell'}^0 \subset Z$ such that s_ℓ is a face of $s_{\ell'}$. The proposition is then proved, since the $V_{\ell,\lambda}$ form a finite open cover of S. □

Corollary 5.51. *Let M be an $\mathcal{S}^\infty$ submanifold of R^k of dimension d. There exists a finite cover of M by semi-algebraic open sets M_i such that, for each M_i, one can find $j_1, \ldots, j_d \in \{1, \ldots, k\}$ in such a way that the restriction to M_i of the projection mapping $(x_1, \ldots, x_k) \mapsto (x_{j_1}, \ldots, x_{j_d})$ from R^k onto R^d is an $\mathcal{S}^\infty$ diffeomorphism onto its image (stated differently, over each M_i one can express $k-d$ coordinates as $\mathcal{S}^\infty$ functions of the other d coordinates).*

Proof: Let Π: $\mathrm{R}^k \rightarrow \mathrm{R}^d$ be the projection mapping that forgets the last $k-d$ coordinates, and let $M' \subset M$ be the set of points x such that Π induces an isomorphism from the tangent space $T_x(M)$ onto R^d. The function $\Pi \mid M'$ is a local homeomorphism, hence, by Proposition 5.50, one can cover M' by the images of a finite number of semi-algebraic continuous sections (i.e. local inverses) of $\Pi \mid M'$, defined over semi-algebraic open sets of R^d; these sections are $\mathcal{S}^\infty$ functions by Theorem 3.25 (Implicit Function Theorem). We do the same with projections onto all other $k-d$-coordinates, thereby exhausting the manifold. □

We now introduce a notion of local dimension.

Proposition 5.52. *Let $S \subset \mathrm{R}^k$ be a semi-algebraic set, and let a be a point of S. There exists a semi-algebraic open neighborhood U of x in S such that, for any other semi-algebraic open neighborhood U' of x in S contained in U, one has* $\dim(U) = \dim(U')$.

Proof: Clear by the properties of the dimension and Theorem 5.48 (local conic structure). □

Let $S \subset \mathrm{R}^k$ be a semi-algebraic set and x a point of S. With U as in Proposition 5.52, one calls $\dim(U)$ the **local dimension of S at x**, denoted $\dim(S_x)$.

A point $x \in S$ is a **smooth point of dimension** d of S if there exists a semi-algebraic open neighborhood U of R^k such that $S \cap U$ is an $\mathcal{S}^\infty$ manifold of dimension d. Note that a smooth point of S of dimension d has local dimension d.

Proposition 5.53. *Let S be a semi-algebraic set of dimension d. There exists a non-empty semi-algebraic subset $T \subset S$ such that every point of T is a smooth point of dimension d and $S^{(d)} = \{x \in S \mid \dim(S_x) = d\}$ is a non-empty closed semi-algebraic subset S, which is the closure of T. Moreover* $\dim(S \setminus S^{(d)}) < d$.

Proof: By Theorem 5.38, the set S is a finite union of semi-algebraic sets S_i, each $\mathcal{S}^\infty$ diffeomorphic to $(0,1)^{d(i)}$. Let T be the union of the S_i such that $d(i) = d$ (there are such S_i since $d = \sup(d(i))$). It is clear that every point of T is a smooth point of dimension d. Let S' be the closure in S of T. Of course $S' \subset S^{(d)}$. If $x \notin S'$, there is a sufficiently small open neighborhood U of x such that $S_i \cap U \neq \emptyset$ implies $d(i) < d$ hence $x \notin S^{(d)}$. Therefore $\mathrm{R}^k \setminus S^{(d)}$ is open. Note that $S \setminus S^{(d)}$ contains no stratum of dimension d. This proves the claim. □

Proposition 5.54. *Let S be a semi-algebraic set. There exist non-empty semi-algebraic subsets of S, $S_1, \ldots, S_\ell$ such that every point of S_i is a smooth point of dimension $d(i)$ and S is the union of the closure of the S_i.*

Proof: The proof is by induction on the dimension of S. The claim is obviously true for $\dim(S) = 1$, since S is a finite number of points and intervals, and a closed interval is the closure of the corresponding open interval.

Suppose by induction hypothesis that the claim holds for all semi-algebraic sets of dimension $< d$ and consider a semi-algebraic set S of dimension d. By Proposition 5.53, the set $S^{(d)} = \{x \in S \mid \dim(S_x) = d\}$ is the closure of a semi-algebraic subset T_1 such that every point of T_1 is a smooth point of dimension d. Define $S_1 = S \setminus S^{(d)}$. It follows from Proposition 5.53 that the dimension of S_1 is $< d$, and the claim follows by applying the induction hypothesis to S_1. □

5.9 Semi-algebraic Sard's Theorem

Definition 5.55. [Critical point] If $f: N \to M$ is an $\mathcal{S}^\infty$ function between two $\mathcal{S}^\infty$ submanifolds N and M, then a **critical point** of f is a point x of N where the rank of the differential $D f(x): T_x(N) \to T_{f(x)}(M)$ is strictly smaller than the dimension of M; a **critical value** of f is the image of a critical point under f. A **regular point** of f on N is a point which is not critical and a **regular value** of f on M is a value of f which is not critical. □

We now give the semi-algebraic version of Sard's Theorem.

Theorem 5.56. [Sard's theorem] *Let $f: N \to M$ be an $\mathcal{S}^\infty$ function between two $\mathcal{S}^\infty$ submanifolds. The set of critical values of f is a semi-algebraic subset of M whose dimension is strictly smaller than the dimension of M.*

The proof of Theorem 5.56 uses the constant rank theorem which can be proved from the inverse function theorem for $\mathcal{S}^\infty$ functions.

Theorem 5.57. [Constant Rank] *Let f be a $\mathcal{S}^\infty$ function from a semi-algebraic open set A of R^k into R^m such that the rank of the derivative $\mathrm{d}f(x)$ is constant and equal to p over A, and a be a point in A.*

There exists a semi-algebraic open neighborhood U of a which is contained in A, an $\mathcal{S}^\infty$ diffeomorphism u: $U \to (-1, 1)^k$, a semi-algebraic open set $V \supset f(U)$, and an $\mathcal{S}^\infty$ diffeomorphism $v: (-1,1)^m \to V$ such that $f|_U = v \circ g \circ u$, where $g: (-1,1)^k \to (-1,1)^m$ is the mapping $(x_1, \ldots, x_k) \mapsto (x_1, \ldots, x_p, 0, \ldots, 0)$.

Proof: Without loss of generality let $a = O$ be the origin and $f(a) = O$. Then, $\mathrm{d}f(O): \mathrm{R}^k \to \mathrm{R}^m$ is a linear map of rank p. Let $M \subset \mathrm{R}^k$ be the kernel of $\mathrm{d}f(O)$ and $N \subset \mathrm{R}^m$ be its image. It is clear that $\dim(M) = k - p$ and $\dim(N) = p$.

Without loss of generality we can assume that M is spanned by the last $k-p$ coordinates of R^k, and we denote by M' the subspace spanned by the first p coordinates.

We also assume without loss of generality that N is spanned by the first p coordinates of R^m and we denote by N' the subspace spanned by the last $m-p$ coordinates. We will denote by π_M (resp. π_N, $\pi_{N'}$) the projection maps from R^k to M (resp. R^m to N, N').

Let U' be a neighborhood of O in A and consider the map $\tilde{u}_1: U' \to N \times M$ defined by

$$\tilde{u}_1(x) = (\pi_N(f(x)), \pi_M(x)).$$

Clearly, $\mathrm{d}\tilde{u}_1(O)$ is invertible, so by Proposition 3.24 (Inverse Function Theorem), there exists a neighborhood U'' of the origin such that $\tilde{u}_1|_{U''}$ is an $\mathcal{S}^\infty$ diffeomorphism and $\tilde{u}_1(U'')$ contains the set $I_k(r) = (-r, r)^{k-p} \times (-r, r)^p$ for some sufficiently small $r > 0$. Let $U = \tilde{u}_1^{-1}(I_k(r))$ and $u_1 = \tilde{u}_1|_U$.

Let V' be a neighborhood of the origin in R^m containing $f(U)$ and define $\tilde{v}_1: V' \to N \times N'$ by

$$\tilde{v}_1(y) = (\pi_N(y), \pi_{N'}(y - f(u_1^{-1}(\pi_N(y), O)))).$$

Shrinking U and r if necessary, we can choose a neighborhood $V'' \subset V'$ containing $f(U)$, such that $\tilde{v}_1|_{V''}$ is an $\mathcal{S}^\infty$ diffeomorphism. To see this observe that $\mathrm{d}\tilde{v}_1(O)$ is invertible, and apply Proposition 3.24 (Inverse Function Theorem). Shrink r if necessary so that $\tilde{v}_1(V'')$ contains

$$I_m(r) = (-r, r)^p \times (-r, r)^{m-p}.$$

Let $V = \tilde{v}_1^{-1}(I_m(r))$ and $v_1 = \tilde{v}_1|_V$. Finally, let u: $U \to I_k(1)$ be defined by $u(x) = u(x)/r$ and let v: $I_m(1) \to V$ be the $\mathcal{S}^\infty$ diffeomorphism defined by $v(y) = v_1^{-1}(r\,y)$.

We now prove that $f\,|_U = v \circ g \circ u$, where g: $(-1,1)^k \to (-1,1)^m$ is the projection mapping $(x_1, \ldots, x_k) \mapsto (x_1, \ldots, x_p, 0, \ldots, 0)$.

Since the rank of the derivative $\mathrm{d}f(x)$ is constant and equal to p for all $x \in U$, we have that for each $x \in U$ the image N_x of $\mathrm{d}f(x)$ is a p-dimensional linear subspace of R^m. Also, choosing r small enough we can ensure that π_N restricted to N_x is a bijection. We let L_x: $N \to N_x$ denote the inverse of this bijection.

Now, consider the $\mathcal{S}^\infty$ map f_1: $(-r,r)^k \to \mathrm{R}^m$ defined by,

$$f_1(z_1, z_2) = f(u_1^{-1}(\pi_N(z_1), z_2)).$$

We first show that $f_1(z_1, z_2)$ is in fact independent of z_2.

Clearly,

$$f(x) = f_1(u_1^{-1}(\pi_N(f(x)), \pi_M(x))).$$

Differentiating using the chain rule, denoting $g = \mathrm{d}u^{-1}(\pi_N(f(x)), \pi_M(x))$, for all $t \in \mathrm{R}^k$,

$$\begin{aligned} \mathrm{d}f(x)(t) &= \mathrm{d}_1 f_1(u_1^{-1}(\pi_N(f(x)), \pi_M(x))) \circ g \circ \pi_N \circ \mathrm{d}f(x)(t) \\ &\quad + \mathrm{d}_2 f_1(u_1^{-1}(\pi_N(f(x)), \pi_M(x))) \circ g \circ \pi_M(t), \end{aligned}$$

where d_i is the derivative with respect to z_i. Note also that,

$$\mathrm{d}f(x)(t) = L_x \circ \pi_N \circ \mathrm{d}f(x)(t).$$

Hence, with $L = (L_x - \mathrm{d}_1 f_1(u_1^{-1}(\pi_N(f(x)), \pi_M(x)))$

$$\begin{aligned} &\mathrm{d}_2 f_1(u_1^{-1}(\pi_N(f(x)), \pi_M(x))) \circ \mathrm{d}u^{-1}(\pi_N(f(x)), \pi_M(x)) \circ \pi_M(t) \\ =\ & L \circ \mathrm{d}u^{-1}(\pi_N(f(x)), \pi_M(x))) \circ \pi_N \circ \mathrm{d}f(x)(t). \end{aligned}$$

Let S_x denote the linear map

$$L_x - \mathrm{d}_1 f_1(u_1^{-1}(\pi_N(f(x)), \pi_M(x))) \circ \mathrm{d}u^{-1}(\pi_N(f(x)), \pi_M(x)) \colon N \to N_x.$$

For $t \in M'$, $\pi_M(t) = 0$ and hence, $S_x \circ \pi_N \circ \mathrm{d}f(x)(t) = 0$. Since, $\pi_N \circ \mathrm{d}f(x)$ is a bijection onto N, this implies that $S_x = 0$. Therefore, we get that

$$\mathrm{d}_2 f_1(u_1^{-1}(\pi_N(f(x)), \pi_M(x))) \circ \mathrm{d}u^{-1}(\pi_N(f(x)), \pi_M(x)) \circ \pi_M(t) = 0$$

for all $t \in \mathrm{R}^k$ implying that

$$\mathrm{d}_2 f_1(u_1^{-1}(\pi_N(f(x)), \pi_M(x))) = 0$$

for all $x \in U$. This shows that $f_1(z_1, z_2)$ is in fact independent of z_2.

Suppose now that $v_1(y) \in N$ for some $y \in V$. This means that,

$$\pi_{N'}(y - f(u_1^{-1}(\pi_N(y), O))) = 0.$$

Let $u_1^{-1}(\pi_N(y), O) = x$. It follows from the definition of u_1 and from our assumption that $\pi_N(f(x)) = \pi_N(y)$ and $\pi_{N'}(y) = \pi_{N'}(f(x))$. Hence, $y = f(x)$.

Conversely, suppose that $y = f(x)$. Then using the fact that $f_1(z_1, z_2)$ does not depend on z_2 and the fact that u_1 is injective we get that

$$f(u_1^{-1}(\pi_N(y), O)) = f(u_1^{-1}(\pi_N(f(x)), \pi_M(x))) = f(x) = y$$

and hence $\pi_{N'}(y - f(u_1^{-1}(\pi_N(y), O))) = 0$. Thus, $v_1(y) \in N$. □

Proof of Theorem 5.56: By Corollary 5.51, one may suppose that M is a semi-algebraic open set of R^m. Let $S \subset N$ be the set of critical points of f. The set S is semi-algebraic since the partial derivatives of f are $\mathcal{S}^\infty$ functions. By Proposition 5.40, S is a finite union of semi-algebraic sets S_i that are the images of $\mathcal{S}^\infty$ embeddings φ_i: $(0, 1)^{d(i)} \to N$. The rank of the composite function $f \circ \varphi_i$ is $< m$. It remains to prove that the dimension of the image of $f \circ \varphi_i$ is $< m$. This is done in the following lemma.

Lemma 5.58. *Let g: $(0, 1)^d \to \mathrm{R}^m$ be an $\mathcal{S}^\infty$ function such that the rank of the differential $\mathrm{d}g(x)$ is everywhere $< m$. Then, the dimension of the image of g is $< m$.*

Proof of Lemma 5.58: Let us suppose that $\dim(g((0,1)^d)) = m$. By applying Corollary 5.51 to g, one can find a semi-algebraic open set U of R^m that is contained in $g((0,1)^d)$ and a semi-algebraic homeomorphism θ: $U \times F \to g^{-1}(U)$ such that $g \circ \theta$ is the projection of $U \times F$ onto U. If $x \in g^{-1}(U)$, then the image under g of every semi-algebraic open neighborhood of x is a semi-algebraic open neighborhood of $g(x)$ and is thus of dimension m. If for x one chooses a point where the rank of $\mathrm{d}g(x)$ is maximal (among the values taken over $g^{-1}(U)$), then one obtains a contradiction with Theorem 5.57 (Constant Rank). □ □

5.10 Bibliographical Notes

The geometric technique underlying the cylindrical decomposition method can be found already in [160], for algebraic sets. The specific cylindrical decomposition method using subresultant coefficients comes from Collins [45].

Triangulation of semi-algebraic sets seems to appear for the first time in [28].

Hardt's triviality theorem appears originally in [83].

6

Elements of Topology

In this chapter, we introduce basic concepts of algebraic topology adapted to semi-algebraic sets. We show how to associate to semi-algebraic sets discrete objects (the homology and cohomology groups) that are invariant under semi-algebraic homeomorphisms. In Section 6.1, we develop a combinatorial theory for homology and cohomology that applies only to simplicial complexes. In Section 6.2 we show how to extend this theory to closed semi-algebraic sets using the triangulation theorem proved in Chapter 5. In Section 6.3 we define homology groups, Borel-Moore homology groups and Euler-Poincaré characteristic for special cases of locally closed semi-algebraic sets.

6.1 Simplicial Homology Theory

6.1.1 The Homology Groups of a Simplicial Complex

We define the simplicial homology groups of a simplicial complex K in a combinatorial manner. We use the notions and notation introduced in Section 5.6.

Given a simplicial complex K, let K_i be the set of i-dimensional simplices of K. In particular, K_0 is the set of vertices of K.

6.1.1.1 Chain Groups

Let $p \in \mathbb{N}$. A non-degenerate **oriented p-simplex** is a p-simplex $[a_0, \ldots, a_p]$ together with an equivalence class of total orderings on the set of vertices $\{a_0, \ldots, a_p\}$, two orderings are equivalent if they differ by an even permutation of the vertices. Thus, a simplex has exactly two orientations. If $a_0, \ldots, a_p$ are not affinely independent, we set $[a_0, \ldots, a_p] = 0$, which is a degenerate oriented p-simplex.

Abusing notation, if $s=[a_0,\ldots,a_p]$ is a p-simplex, we denote by $[a_0,\ldots,a_p]$ the oriented simplex corresponding to the order $a_0 < a_1 < \cdots < a_p$ on the vertices. So, $s=[a_0,\ldots,a_p]$ is an oriented simplex and $-s=[a_1,a_0,a_2,\ldots,a_p]$ is the oppositely oriented simplex.

Given a simplicial complex K, the $\mathbb{Q}$-vector space generated by the p-dimensional oriented simplices of K is called the **p-chain group** of K and is denoted $\mathrm{C}_p(K)$. The elements of $\mathrm{C}_p(K)$ are called the **p-chains** of K. Notice that if K contains no p-simplices then $\mathrm{C}_p(K)$ is a $\mathbb{Q}$-vector space generated by the empty set, which is $\{0\}$. Since K_p is finite, $\mathrm{C}_p(K)$ is finite dimensional. An element of $\mathrm{C}_p(K)$ can be written $c=\sum_i n_i s_i, n_i \in \Lambda, s_i \in K_p$. For $p<0$, we define $\mathrm{C}_p(K)=0$. When s is the oriented p-simplex $[a_0,\ldots,a_p]$, we define $[b,s]$ to be the oriented $p+1$-simplex $[b,a_0,\ldots,a_p]$. If $c=\sum_i n_i s_i$, (with $n_i \in \Lambda$) is a p-chain, then we define $[b,c]$ to be $\sum_i n_i[b,s_i]$.

Given an oriented p-simplex $s=[a_0,\ldots,a_p]$, $p>0$, the **boundary** of s is the $(p-1)$-chain

$$\partial_p(s)=\sum_{0\le i\le p}(-1)^i[a_0,\ldots,a_{i-1},\hat{a_i},a_{i+1},\ldots,a_p],$$

where the hat ˆ means that the corresponding vertex is omitted.

The map ∂_p extends linearly to a homomorphism ∂_p: $\mathrm{C}_p(K)\to \mathrm{C}_{p-1}(K)$ by the rule

$$\partial_p\Bigl(\sum_i n_i s_i\Bigr)=\sum_i n_i\partial_p(s_i).$$

Note that, if c is a p-chain, $\partial_{p+1}([b,c])=c-[b,\partial_p(c)]$.

For $p\le 0$, we define $\partial_p=0$. Thus, we have the following sequence of vector space homomorphisms,

$$\cdots\longrightarrow \mathrm{C}_p(K)\xrightarrow{\partial_p}\mathrm{C}_{p-1}(K)\xrightarrow{\partial_{p-1}}\mathrm{C}_{p-2}(K)\xrightarrow{\partial_{p-2}}\cdots\xrightarrow{\partial_1}\mathrm{C}_0(K)\xrightarrow{\partial_0}0.$$

Using the definition of ∂_p and expanding, it is not too difficult to show that, for all p

$$\partial_{p-1}\circ\partial_p=0.$$

The sequence of pairs $\{(\mathrm{C}_p(K),\partial_p)\}_{p\in\mathbb{N}}$ is denoted $\mathrm{C}_\bullet(K)$.

Given two simplicial complexes K, L, a map ϕ: $|K|\to|L|$ is a **simplicial map** if it is the piecewise linear extension to each simplex of a map ϕ_0: $K_0\to L_0$ that maps the vertices of every simplex in K to the vertices of a simplex in L (not necessarily of the same dimension). A simplicial map ϕ defines a sequence of homomorphisms $\mathrm{C}_p(\phi)$ from $\mathrm{C}_p(K)$ to $\mathrm{C}_p(L)$ by

$$\mathrm{C}_p(\phi)[a_0,\ldots,a_p]=[\phi_0(a_0),\ldots,\phi_0(a_p)].$$

Notice that the right hand side is automatically zero if ϕ_0 is not injective on the set $\{a_0, ..., a_p\}$, in which case $[\phi_0(a_0), ..., \phi_0(a_p)]$ is a degenerate simplex. Also note that a simplicial map is automatically semi-algebraic.

6.1.1.2 Chain Complexes and Chain Maps

The chain groups obtained from a simplicial complex are a special case of more general abstract algebraic objects called chain complexes. The homomorphisms between the chain groups obtained from simplicial maps are then special cases of the more general chain homomorphisms, which we introduce below.

A sequence $\{\mathrm{C}_p\}$, $p \in \mathbb{Z}$, of vector spaces together with a sequence $\{\partial_p\}$ of homomorphisms $\partial_p: \mathrm{C}_p \to \mathrm{C}_{p-1}$ for which $\partial_{p-1} \circ \partial_p = 0$ for all p is called a **chain complex**. Given two chain complexes, $\mathrm{C}_\bullet = (\mathrm{C}_p, \partial_p)$ and $\mathrm{C}'_\bullet = (\mathrm{C}'_p, \partial'_p)$, a **chain homomorphism** $\phi_\bullet: \mathrm{C}_\bullet \to \mathrm{C}'_\bullet$ is a sequence of homomorphisms ϕ_p: $\mathrm{C}_p \to \mathrm{C}'_p$ for which $\partial'_p \circ \phi_p = \phi_{p-1} \circ \partial_p$ for all p.

In other words, the following diagram is commutative.

$$\begin{array}{ccccccc} \cdots & \longrightarrow & \mathrm{C}_p & \xrightarrow{\partial_p} & \mathrm{C}_{p-1} & \longrightarrow & \cdots \\ & & \downarrow \phi_p & & \downarrow \phi_{p-1} & & \\ \cdots & \longrightarrow & \mathrm{C}'_p & \xrightarrow{\partial'_p} & \mathrm{C}'_{p-1} & \longrightarrow & \cdots \end{array}$$

Notice that if $\phi: K \to K'$ is a simplicial map, then $\mathrm{C}_\bullet(\phi): \mathrm{C}_\bullet(K) \to \mathrm{C}_\bullet(K')$ is a chain homomorphism between the chain complexes $\mathrm{C}_\bullet(K)$ and $\mathrm{C}_\bullet(K')$.

6.1.1.3 Homology of Chain Complexes

Given a chain complex $\mathrm{C}_\bullet$, the elements of $B_p(\mathrm{C}_\bullet) = \mathrm{Im}(\partial_{p+1})$ are called p-**boundaries** and those of $Z_p(\mathrm{C}_\bullet) = \mathrm{Ker}(\partial_p)$ are called p-**cycles**. Note that, since $\partial_{p-1} \circ \partial_p = 0$, $B_p(\mathrm{C}_\bullet) \subset Z_p(\mathrm{C}_\bullet)$. The **homology groups** $\mathrm{H}_p(\mathrm{C}_\bullet)$ are defined by $\mathrm{H}_p(\mathrm{C}_\bullet) = Z_p(\mathrm{C}_\bullet)/B_p(\mathrm{C}_\bullet)$.

Note that, by our definition, the homology groups $\mathrm{H}_p(C_\bullet)$ are all $\mathbb{Q}$-vector spaces (finite dimensional if the vector spaces C_p's are themselves finite dimensional). We still refer to them as groups as this is standard terminology in algebraic topology where more general rings of coefficients, for instance the integers, are often used in the definition of the chain complexes. In such situations, the homology groups are not necessarily vector spaces over a field, but rather modules over the corresponding ring.

This sequence of groups together with the sequence of homomorphisms which sends each $\mathrm{H}_p(C_\bullet)$ to $0 \in \mathrm{H}_{p-1}(C_\bullet)$ constitutes a chain complex $(\mathrm{H}_p(C_\bullet), 0)$ which is denoted by $\mathrm{H}_\star(C_\bullet)$.

Lemma 6.1. *Given two chain complexes* $\mathrm{C}_\bullet$ *and* $\mathrm{C}'_\bullet$*, a chain homomorphism* $\phi_\bullet\colon \mathrm{C}_\bullet \to \mathrm{C}'_\bullet$ *induces a homomorphism* $\mathrm{H}_\star(\phi_\bullet)\colon \mathrm{H}_\star(\mathrm{C}_\bullet) \to \mathrm{H}_\star(\mathrm{C}'_\bullet)$ *which respects composition. In other words, given another chain homomorphism* $\psi_\bullet$: $\mathrm{C}'_\bullet \to \mathrm{C}''_\bullet$,

$$\mathrm{H}_\star(\psi_\bullet \circ \phi_\bullet) = \mathrm{H}_\star(\psi_\bullet) \circ \mathrm{H}_\star(\phi_\bullet)$$

and $\mathrm{H}_\star(\mathrm{Id}_{C_\bullet}) = \mathrm{Id}_{\mathrm{H}_\star(C_\bullet)}$.

Proof: Using the fact that the diagram of a chain homomorphism commutes, we see that a chain homomorphism carries cycles to cycles and boundaries to boundaries. Thus, the chain homomorphism ϕ induces homomorphisms

$$Z_p(\phi_\bullet)\colon Z_p(C_\bullet) \to Z_p(C'_\bullet),$$

$$B_p(\phi_\bullet)\colon B_p(C_\bullet) \to B_p(C'_\bullet).$$

Thus, it also induces a homomorphism

$$\mathrm{H}_p(\phi_\bullet)\colon \mathrm{H}_p(C_\bullet) \to \mathrm{H}_p(C'_\bullet).$$

The remaining claims follow easily. □

6.1.1.4 Homology of a Simplicial Complex

Definition 6.2. Given a simplicial complex K, $\mathrm{H}_p(K) = \mathrm{H}_p(\mathrm{C}_\bullet(K))$ is the **p-th simplicial homology group** of K. As a special case of Lemma 6.1, it follows that a simplicial map from K to L induces homomorphisms between the homology groups $\mathrm{H}_p(K)$ and $\mathrm{H}_p(L)$.

We denote by $\mathrm{H}_\star(K)$ the chain complex $(\mathrm{H}_p(K), 0)$ and call it the **homology of** K.

It is clear from the definition that $\mathrm{H}_p(K)$ is a finite dimensional $\mathbb{Q}$-vector space. The dimension of $\mathrm{H}_p(K)$ as a $\mathbb{Q}$-vector space is called the **p-th Betti number** of K and denoted $\mathrm{b}_p(K)$.

The **Euler-Poincaré characteristic** of K is

$$\chi(K) = \sum_i (-1)^i \mathrm{b}_i(K).$$

□

Proposition 6.3. *Let* $n_i(K)$ *be the number of simplexes of dimension* i *of* K. *Then*

$$\chi(K) = \sum_i (-1)^i n_i(K).$$

Proof: Recall from the definition of $\mathrm{H}_i(K)$ that,

$$\mathrm{b}_i(K) = \dim\ \mathrm{H}_i(K) = \dim \operatorname{Ker}(\partial_i) - \dim \operatorname{Im}(\partial_{i+1}).$$

Moreover,

$$n_i(K) = \dim \mathrm{C}_i(K) = \dim \mathrm{Ker}(\partial_i) + \dim \mathrm{Im}(\partial_i).$$

An easy calculation now shows that,

$$\begin{aligned} \chi(K) &= \sum_i (-1)^i \, \mathrm{b}_i(K) \\ &= \sum_i (-1)^i \, (\dim \mathrm{Ker}(\partial_i) - \dim \mathrm{Im}(\partial_{i+1})) \\ &= \sum_i (-1)^i \, (\dim \mathrm{Ker}(\partial_i) + \dim \mathrm{Im}(\partial_i)) \\ &= \sum_i (-1)^i \, n_i. \end{aligned}$$

□

6.1.2 Simplicial Cohomology Theory

We have defined the homology groups of a simplicial complex K in the previous section. We now define a dual notion – namely that of cohomology groups. One reason for defining cohomology groups is that in many situations, it is more convenient and intuitive to reason with the cohomology groups than with the homology groups.

Given a simplicial complex K, we will denote by $\mathrm{C}^p(K)$ the vector space dual to $\mathrm{C}_p(K)$, and the sequence of homomorphisms,

$$0 \to \mathrm{C}^0(K) \xrightarrow{\delta^0} \mathrm{C}^1(K) \xrightarrow{\delta^1} \mathrm{C}^2(K) \cdots \mathrm{C}^p(K) \xrightarrow{\delta^p} \mathrm{C}^{p+1}(K) \xrightarrow{\delta^{p+1}} \cdots$$

is called the **cochain complex** of K. Here, δ^p is the homomorphism dual to ∂_{p+1} in the chain complex $\mathrm{C}^\bullet(K)$. The sequence of pairs $\{(\mathrm{C}^p(K), \delta^p)\}_{p \in \mathbb{N}}$ is denoted by by $\mathrm{C}^\bullet(K)$. Notice that each $\phi \in \mathrm{C}^p(K)$ is a linear functional on the vector space $\mathrm{C}^p(K)$, and thus ϕ is determined by the values it takes on each i-simplex of K.

6.1.2.1 Cochain Complexes

The dual notion for chain complexes is that of cochain complexes. A sequence $\{\mathrm{C}^p\}$, $p \in \mathbb{Z}$, of vector spaces together with a sequence $\{\delta^p\}$ of homomorphisms $\delta^p \colon \mathrm{C}^p \to \mathrm{C}^{p+1}$ for which $\delta^{p+1} \circ \delta^p = 0$ for all p is called a **cochain complex**. Given two cochain complexes, $\mathrm{C}^\bullet = (\mathrm{C}^p, \delta^p)$ and $\mathrm{D}^\bullet = (\mathrm{D}^p, \delta'^p)$, a **cochain homomorphism** $\phi^\bullet \colon \mathrm{C}^\bullet \to \mathrm{D}^\bullet$ is a sequence of homomorphisms $\phi^p \colon \mathrm{C}^p \to \mathrm{D}^p$ for which $\partial'^p \circ \phi^p = \phi^{p+1} \circ \partial^p$ for all p.

In other words, the following diagram is commutative.

$$\begin{array}{ccccccc}
\cdots & \longrightarrow & \mathrm{C}^p & \xrightarrow{\delta^p} & \mathrm{C}^{p+1} & \longrightarrow & \cdots \\
 & & \downarrow \phi^p & & \downarrow \phi^{p+1} & & \\
\cdots & \longrightarrow & \mathrm{D}^p & \xrightarrow{\delta'^p} & \mathrm{D}^{p+1} & \longrightarrow & \cdots
\end{array}$$

It is clear that given a chain complex $\mathrm{C}_\bullet = \{(\mathrm{C}_p, \partial_p)\}_{p\in\mathbb{N}}$, we can obtain a corresponding cochain complex $\mathrm{C}^\bullet = \{(\mathrm{C}^p, \delta^p)\}_{p\subset\mathbb{N}}$ by taking duals of each term and homomorphisms. Doing so reverses the direction of every arrow in the corresponding diagram.

6.1.2.2 Cohomology of Cochain Complexes

The elements of $B^p(\mathrm{C}^\bullet) = \mathrm{Im}(\delta^{p-1})$ are called ***p*-coboundaries** and those of $Z^p(\mathrm{C}^\bullet) = \mathrm{Ker}(\delta^p)$ are called ***p*-cocycles**. It is easy to verify that $B^p(\mathrm{C}^\bullet) \subset Z^p(\mathrm{C}^\bullet)$. The **cohomology groups**, $\mathrm{H}^p(C^\bullet)$, are defined by

$$\mathrm{H}^p(\mathrm{C}^\bullet) = \frac{Z^p(\mathrm{C}^\bullet)}{B^p(\mathrm{C}^\bullet)}.$$

This sequence of groups together with the sequence of homomorphisms which sends each $\mathrm{H}^p(\mathrm{C}^\bullet)$ to $0 \in \mathrm{H}^{p+1}(\mathrm{C}^\bullet)$ constitutes a chain complex $(\mathrm{H}^p(\mathrm{C}^\bullet), 0)$ which is denoted by $\mathrm{H}^\star(\mathrm{C}^\bullet)$.

It is an easy exercise in linear algebra to check that:

Proposition 6.4. *Let $\mathrm{C}_\bullet$ be a chain complex and $\mathrm{C}^\bullet$ the corresponding cochain complex. Then, for every $p \geq 0$, $\mathrm{H}_p(\mathrm{C}_\bullet) \cong \mathrm{H}^p(\mathrm{C}^\bullet)$.*

Given a simplicial complex K, the p-th cohomology group $\mathrm{H}^p(\mathrm{C}^\bullet(K))$ will be denoted by $\mathrm{H}^p(K)$. The cohomology group $\mathrm{H}^0(K)$ has a particularly natural interpretation. It is the vector space of locally constant functions on $|K|$.

Proposition 6.5. *Let K be a simplicial complex such that $K_0 \neq \emptyset$. The cohomology group $\mathrm{H}^0(K)$ is the vector space of locally constant functions on $|K|$. As a consequence, the number of connected components of $|K|$ is $\mathrm{b}_0(K)$.*

Proof: Clearly, $\mathrm{H}^0(K)$ depends only on the 1-skeleton of K, that is the subcomplex of K consisting of the zero and one-dimensional simplices.

Let $z \in \mathrm{C}^0(K)$ be a cocycle, that is such that $\mathrm{d}^0(z) = 0$. This implies that for any $e = [u, v] \in K_1$, $z(u) - z(v) = 0$. Hence z takes a constant value on vertices in a connected component of $|K|$. Since $B^0(\mathrm{C}^\bullet(K))$ is 0, this shows that $\mathrm{H}^0(K)$ is the vector space of locally constant functions on $|K|$.

Using Proposition 6.4, the last part of the claim follows since the dimension of the vector space of locally constant functions on $|K|$ is the number of connected components of $|K|$. □

It follows immediately from Lemma 6.1 that

Lemma 6.6. *Given two cochain complexes* $\mathrm{C}^\bullet$ *and* $\mathrm{C}'^\bullet$, *a cochain homomorphism* $\phi^\bullet: \mathrm{C}^\bullet \to \mathrm{C}'^\bullet$ *induces a homomorphism* $\mathrm{H}^\star(\phi^\bullet): \mathrm{H}^\star(\mathrm{C}^\bullet) \to \mathrm{H}^\star(\mathrm{C}'^\bullet)$ *which respects composition. In other words, given another chain homomorphism* $\psi^\bullet$: $\mathrm{C}'^\bullet \to \mathrm{C}''^\bullet$,

$$\mathrm{H}^\star(\psi^\bullet \circ \phi^\bullet) = \mathrm{H}^\star(\psi^\bullet) \circ \mathrm{H}^\star(\phi^\bullet) \text{ and } \mathrm{H}^\star(\mathrm{Id}_{\mathrm{C}^\bullet}) = \mathrm{Id}_{\mathrm{H}^\star(\mathrm{C}^\bullet)}.$$

6.1.3 A Characterization of H^1 in a Special Case.

Let A be a simplicial complex and $A^1, \ldots, A^s$ sub-complexes of A such that, each A^i is connected and

$$\begin{aligned} A &= A^1 \cup \cdots \cup A^s, \\ \mathrm{H}^1(A^i) &= 0, 1 \le i \le s. \end{aligned}$$

For $1 \le i < j < \ell \le s$, we denote by A^{ij} the sub-complex $A^i \cap A^j$, and by $A^{ij\ell}$ the sub-complex $A^i \cap A^j \cap A^\ell$. We will denote by C_α^{ij} the sub-complexes corresponding to connected components of A^{ij}, and by $C_\beta^{ij\ell}$ the sub-complexes corresponding to connected components of $A^{ij\ell}$.

We will show that the simplicial cohomology group, $\mathrm{H}^1(A)$, is isomorphic to the first cohomology group of a certain complex defined in terms of $\mathrm{H}^0(A^i)$, $\mathrm{H}^0(A^{ij})$ and $\mathrm{H}^0(A^{ij\ell})$. This result will be the basis of an efficient algorithm for computing the first Betti number of semi-algebraic sets, which will be developed in Chapter 16.

Let

$$N^\bullet = N^0 \longrightarrow N^1 \longrightarrow N^2 \to 0$$

denote the complex

$$\mathrm{C}^0(A) \xrightarrow{d^0} \mathrm{C}^1(A) \xrightarrow{d^1} \mathrm{C}^2(A) \to 0.$$

Note that $N^\bullet$ is just a truncated version of the cochain complex of A. The coboundary homomorphisms d^0, d^1 are identical to the ones in $\mathrm{C}^\bullet(A)$.

For each $h \ge 0$, we define

$$\delta_0: \bigoplus_{1 \le i \le s} \mathrm{C}^h(A^i) \longrightarrow \bigoplus_{1 \le i < j \le s} \mathrm{C}^h(A^{ij})$$

as follows.

For $\phi \in \oplus_{1 \le i \le s} \mathrm{C}^0(A^i)$, $1 \leqslant i < j \leqslant s$, and each oriented h-simplex $\sigma \in A_h^{ij}$,

$$\delta_0^h \phi_{i,j}(\sigma) = \phi_i(\sigma) - \phi_j(\sigma).$$

Similarly, we define

$$\delta_1^h\colon \bigoplus_{1\le i<j\le}^{h} \mathrm{C}^h(A^{ij}) \longrightarrow \bigoplus_{1\le i<j<\ell\le s} \mathrm{C}^h(A^{ij\ell})$$

by defining for $\psi \in \oplus_{1\le i<j\le s} \mathrm{C}^h(A^{ij})$, $1 \leqslant i < j < \ell \leqslant s$ and each oriented h-simplex $\sigma \in A_h^{ij}$,

$$(\delta_1^h \psi)_{ij\ell}(\sigma) = \psi_{j\ell}(\sigma) - \psi_{i\ell}(\sigma) + \psi_{ij}(\sigma).$$

Let

$$M^\bullet = M^0 \longrightarrow M^1 \longrightarrow M^2 \to 0$$

denote the complex

$$\bigoplus_{1\le i\le s} \mathrm{C}^0(A^i) \xrightarrow{D_0} \bigoplus_{1\le i\le s} \mathrm{C}^1(A^i) \ \bigoplus_{1\le i<j\le s} \mathrm{C}^0(A^{ij}) \xrightarrow{D_1} \bigoplus_{\ell+n=2} \bigoplus_{J_n} \mathrm{C}^\ell(A^{i_1\ldots i_n}) \to 0.$$

where $J_n = \{i_1 \ldots i_n \mid 1 \leqslant i_1 < \ldots < i_n \leqslant s.\}$

The homomorphism D_0 is defined by

$$D_0(\phi) = d^0(\phi) \oplus \delta_0^0(\phi),\ \phi \in \bigoplus_{1\le i\le s} \mathrm{C}^0(A^i)$$

and D_1 is defined by

$$D_1(\phi \oplus \psi) = d^1(\phi) \oplus (-\delta_0^1(\phi) + d^0(\psi)) \oplus -\delta_1^0(\psi),$$

$$\phi \in \bigoplus_{1\le i\le s} \mathrm{C}^1(A^i),\ \psi \in \bigoplus_{1\le i<j\le s} \mathrm{C}^0(A^{ij}).$$

Finally let

$$L^\bullet = L^0 \longrightarrow L^1 \longrightarrow L^2 \to 0$$

denote the complex

$$\bigoplus_{1\le i\le s} \mathrm{H}^0(A^i) \xrightarrow{\delta_0} \bigoplus_{1\le i<j\le s} \mathrm{H}^0(A^{ij}) \xrightarrow{\delta_1} \bigoplus_{1\le i<j<\ell\le s} \mathrm{H}^0(A^{ij\ell}) \to 0.$$

Recall that $\mathrm{H}^0(X)$ can be identified as the vector space of locally constant functions on the simplicial complex X, and is thus a vector space whose dimension equals the number of connected components of X. The homomorphisms δ_i in the complex $L^\bullet$ are generalized restriction homomorphisms. Thus, for $\phi \in \bigoplus_{1\le i\le s} \mathrm{H}^0(A^i)$,

$$(\delta_0 \phi)_{ij} = \phi_i|_{A^{ij}} - \phi_j|_{A^{ij}}$$

and for $\psi \in \bigoplus_{1\le i<j\le s} \mathrm{H}^0(A^{ij})$,

$$(\delta_1 \psi)_{ij\ell} = \psi_{j\ell}|_{A^{ij\ell}} - \psi_{i\ell}|_{A^{ij\ell}} + \psi_{ij}|_{A^{ij\ell}}.$$

We now define a homomorphism of complexes,

$$F^\bullet: L^\bullet \to M^\bullet,$$

as follows:

For $\phi \in L^0$ and $u \in A_0^i$,

$$F^1(\phi)_i(u) = \phi_i(A^i).$$

For $\psi \in L^2$ and $e \in A_1^{ij}$,

$$F^2(\psi)_i = 0,$$

and

$$F^2(\psi)_{ij}(e) = \psi_{ij}(C_\alpha^{ij}),$$

where C_α^{ij} is the connected component of A^{ij} containing e.

For $\theta \in L^3$ and $\sigma \in A_2^{ij\ell}$

$$F^3(\theta)_i = F^3(\theta)_{ij} = 0,$$

and

$$F^3(\theta)_{ij\ell} = \psi_{ij\ell}(C_{ij\ell}^\beta),$$

where $C_\beta^{ij\ell}$ is the connected component of $A^{ij\ell}$ containing σ. It is easy to verify that $F^\bullet$ is a homomorphism of complexes, and thus induces an homomorphism

$$\mathrm{H}^\star(F^\bullet): \mathrm{H}^\star(L^\bullet) \to \mathrm{H}^\star(M^\bullet).$$

We now prove that,

Proposition 6.7. *The induced homomorphism,*

$$\mathrm{H}^1(F^\bullet): \mathrm{H}^1(L^\bullet) \to \mathrm{H}^1(M^\bullet)$$

is an isomorphism.

Proof: We first prove that, $\mathrm{H}^1(F^\bullet)$: $\mathrm{H}^1(L^\bullet) \to \mathrm{H}^1(M^\bullet)$ is surjective. Let $z = \phi \oplus \psi \in M^1$ be a cocycle, where

$$\phi \in \bigoplus_{1 \le i \le s} \mathrm{C}^1(A^i)$$

and

$$\psi \in \bigoplus_{1 \le i < j \le s} \mathrm{C}^0(A^{ij})$$

Since z is a cocycle, that is $D_0(z) = 0$, we have from the definition of D_0 that,

$$\begin{aligned} d^1\phi &= 0, \\ \delta_1^0\psi &= 0, \\ \delta_0^1\phi + d^0\psi &= 0. \end{aligned}$$

From the first property, and the fact $\mathrm{H}^1(A^i)=0$ for each $i, 1\le i\le \ell$, we deduce that there exists

$$\theta \in M^0 = \bigoplus_{1\le i\le \ell} \mathrm{C}^0(A^i)$$

such that, for $e=[u,v]\in A_1^i$,

$$\phi_i(e)=\theta_i(u)-\theta_i(v). \tag{6.1}$$

As a consequence of the second property, we have that for $1\le i<j<\ell\le s$, and $u\in A_0^{ij\ell}$,

$$\psi_{j\ell}(u)-\psi_{i\ell}(u)+\psi_{ij}(u)=0. \tag{6.2}$$

Finally, from the third property we get that, for $1\le i<j\le s$ and $e=[u, v]\in A_1^{ij}$, and θ defined above,

$$\theta_i(u)-\theta_i(v)-\theta_j(u)+\theta_j(v)+\psi_{ij}(u)-\psi_{ij}(v)=0. \tag{6.3}$$

We now define $z'=0\oplus\gamma\in M^1$ by defining, for $1\le i<j\le s$ and $u\in A_0^{ij}$,

$$\gamma_{ij}(u)=\theta_i(u)-\theta_j(u)+\psi_{ij}(u).$$

From (6.3) it follows that $\gamma_{ij}(u)$ is constant for all vertices u in any connected component of A^{ij}. Thus, $z'\in F^1(L^1)$. Next, for $1\le i<j\le s$, and $e=[u,v]\in A_1^{ij}$

$$\begin{aligned} d\gamma_{ij}(e) &= \theta_i(u)-\theta_j(u)+\psi_{ij}(u)-(\theta_i(v)-\theta_j(v)+\psi_{ij}(v)) \\ &= 0. \end{aligned}$$

where we again use (6.3). This shows that z' is a cycle.

Finally, it is easy to check, using the facts that $\psi-\gamma=\delta\theta$, and $\phi=d^0\theta$, that, $z-z'=(d^0+\delta_0^0)\theta$ is a coboundary in $M^\bullet$. This proves the surjectivity of $\mathrm{H}^1(F^\bullet)$.

We now prove that $\mathrm{H}^1(F^\bullet)$ is injective by proving that for any $z\in L^1$, if $F^1(z)$ is a coboundary in M^1 then z must be a coboundary in L^1. Let

$$F^1(z)=(d^0+\delta_0^0)\theta$$

for $\theta\in\oplus_{1\le i\le s} C^0(A^i)$. We define $\gamma\in\bigoplus_{1\le i\le s}\mathrm{H}^0(A^i)$ by defining

$$\gamma_i(A_i)=\theta_i(u)$$

for some $u\in A_0^i$. This is well defined, since by assumption each A^i is connected, and $d^0\theta=0$, and thus we have that for each $e=[u,v]\in A_1^i$, $\theta_i(u)-\theta_i(v)=0$.

For any $1\le i<j\le s$, C_α^{ij} a connected component of A^{ij}, and $u\in C_{\alpha,0}^{ij}$, we have that,

$$\begin{aligned} F^1(z)_{ij}(u) &= z_{ij}(C_\alpha^{ij}) \\ &= \theta_i(u)-\theta_j(u). \end{aligned}$$

It is now easy to check that $\delta_0\gamma = z$, proving that z is a coboundary in L^1. This proves the injectivity of $\mathrm{H}^1(F)$. □

We now define a homomorphism of complexes, $G^\bullet\colon N^\bullet \to M^\bullet$, as follows.

First observe that for $1 \le i < j < \ell \le s$, there are natural restriction homomorphisms,

$$r_i^\bullet\colon C^\bullet(A) \to C^\bullet(A^i),$$

For $\phi \in C^0(A)$,

$$G^0(\phi) = \bigoplus_{1 \le i \le s} \gamma_i^0(\phi_i).$$

For $\psi \in C^1(A)$,

$$G^1(\phi) = \bigoplus_{1 \le i \le s} \gamma_i^1(\psi).$$

For $\nu \in C^2(A)$,

$$G^2(\nu) = \bigoplus_{1 \le i \le s} \gamma_i^2(\nu).$$

We now prove that,

Proposition 6.8. *The induced homomorphism,*

$$\mathrm{H}^1(G^\bullet)\colon \mathrm{H}^1(N^\bullet) \to \mathrm{H}^1(M^\bullet)$$

is an isomorphism.

Proof: We first prove that $\mathrm{H}^1(G^\bullet)$ is surjective.

Let $z = \phi \oplus \psi \in M^1$ be a cocycle, where

$$\phi \in \bigoplus_{1 \le i \le s} \mathrm{C}^1(A^i),\ \psi \in \bigoplus_{1 \le i < j \le s} \mathrm{C}^0(A^{ij}).$$

Since z is a cocycle, that is $D_0(z) = 0$, we have from the definition of D_0 that,

$$\begin{aligned} d^1\phi &= 0, \\ \delta_1^0\psi &= 0, \\ \delta_0^1\phi + d^0\psi &= 0. \end{aligned}$$

For $1 \le i < j < \ell \le s$, and $u \in A_0^{ij\ell}$,

$$\psi_{j\ell}(u) - \psi_{i\ell}(u) + \psi_{ij}(u) = 0. \tag{6.4}$$

We now define $\theta \in \bigoplus_{1 \le i \le s} \mathrm{C}^0(A^i)$ such that, $\delta_0(\theta) = \psi$.

For $1 \le i \le s$ and $u \in A_0^i$ we define,

$$\theta_i(u) = \frac{1}{n_u} \sum_{\substack{1 \leqslant j \leqslant s \\ j \ne i,\, u \in A_0^j}} (-1)^{i-j}\psi_{ij}(u),$$

where $n_u = \#\{j \mid u \in A_0^j\}$.

It is easy to check using (6.4) that $\delta_0(\theta) = \psi$. Now define, $z' = (\phi - d^0\theta) \oplus 0$. Now, $z' \in G^1(N^1)$, since for $1 \le i < j \le s$, and $e = [u, v] \in A_1^{ij}$,

$$\begin{aligned}(\phi - d^0\theta)_i(e) - (\phi - d^0\theta)_j(e) &= \phi_i(e) - \phi_j(e) - (\theta_i - \theta_j)(u - v) \\ &= (\phi_i(e) - \phi_j(e)) - \psi_{ij}(u) + \psi_{ij}(v) \\ &= (\delta_0^1\phi - d^0\psi)_{ij}(e) \\ &= 0.\end{aligned}$$

Also, $z - z' = d^0\theta \oplus \psi = (d^0 + \delta_1^0)\theta$ is a coboundary. This show that $\mathrm{H}^1(G^\bullet)$ is surjective.

Finally, since G^1 is obviously injective, it is clear that if the image of $z \in N^1$ is a coboundary in M^1, then it must also be a coboundary in N^1, which shows that $\mathrm{H}^1(G^\bullet)$ is injective as well. □

We are now in a position to prove,

Theorem 6.9. *Let A be a simplicial complex and $A^1, \ldots, A^k$ sub-complexes of A such that, each A_i is connected and*

$$\begin{aligned} A &= A^1 \cup \cdots \cup A^s, \\ \mathrm{H}^1(A^i) &= 0, 1 \le i \le s, \end{aligned}$$

and let $L^\bullet$ be the complex defined above. Then,

$$\mathrm{H}^1(A) \cong \mathrm{H}^1(L^\bullet).$$

Proof: The theorem follows directly from Proposition 6.7 and Proposition 6.8 proved above. □

6.1.4 The Mayer-Vietoris Theorem

In the next chapter, we will use heavily certain relations between the homology groups of two semi-algebraic sets and those of their unions and intersections. We start by indicating similar relations between the homology groups of the unions and intersections of sub-complexes of a simplicial complex. It turns out to be convenient to formulate these relations in terms of exact sequences.

A sequence of vector space homomorphisms,

$$\cdots \xrightarrow{\phi_{i-2}} F_{i-1} \xrightarrow{\phi_{i-1}} F_i \xrightarrow{\phi_i} F_{i+1} \xrightarrow{\phi_{i+1}} \cdots$$

is **exact** if and only if $\mathrm{Im}(\phi_i) = \mathrm{Ker}(\phi_{i+1})$ for each i.

Let $\mathrm{C}_\bullet, \mathrm{C}'_\bullet, \mathrm{C}''_\bullet$ be chain complexes, and let $\phi_\bullet\colon \mathrm{C}_\bullet \to \mathrm{C}'_\bullet$, $\psi_\bullet\colon \mathrm{C}'_\bullet \to \mathrm{C}''_\bullet$ be chain homomorphisms. We say that the sequence $0 \to \mathrm{C}_\bullet \xrightarrow{\phi_\bullet} \mathrm{C}'_\bullet \xrightarrow{\psi_\bullet} \mathrm{C}''_\bullet \to 0$ is a **short exact sequence of chain complexes** if in each dimension p the sequence $0 \to \mathrm{C}_p \xrightarrow{\phi_p} \mathrm{C}'_p \xrightarrow{\psi_p} \mathrm{C}''_p \to 0$ is an exact sequence of vector spaces.

We need the following lemma from homological algebra.

Lemma 6.10. [Zigzag Lemma] *Let* $0 \to C_\bullet \xrightarrow{\phi} C'_\bullet \xrightarrow{\psi} C''_\bullet \to 0$ *be a short exact sequence of chain complexes. Then, there exist connecting homomorphisms,* $H_p(\partial)$, *making the following sequence exact*

$$\cdots \xrightarrow{H_p(\phi_\bullet)} H_p(C'_\bullet) \xrightarrow{H_p(\psi_\bullet)} H_p(C''_\bullet) \xrightarrow{H_p(\partial)} H_{p-1}(C_\bullet) \xrightarrow{H_{p-1}(\phi_\bullet)} H_{p-1}(C'_\bullet) \cdots$$

Proof: The proof is by "chasing" the following diagram:

$$\begin{array}{ccccccccc}
0 & \to & C_{p+1} & \xrightarrow{\phi_{p+1}} & C'_{p+1} & \xrightarrow{\psi_{p+1}} & C''_{p+1} & \to & 0 \\
 & & \downarrow \partial_{p+1} & & \downarrow \partial'_{p+1} & & \downarrow \partial''_{p+1} & & \\
0 & \to & C_p & \xrightarrow{\phi_p} & C'_p & \xrightarrow{\psi_p} & C''_p & \to & 0 \\
 & & \downarrow \partial_p & & \downarrow \partial'_p & & \downarrow \partial''_p & & \\
0 & \to & C_{p-1} & \xrightarrow{\phi_{p-1}} & C'_{p-1} & \xrightarrow{\psi_{p-1}} & C''_{p-1} & \to & 0.
\end{array}$$

We have already seen in the proof of Lemma 6.1 that the chain homomorphisms $\phi_\bullet$ and $\psi_\bullet$ actually take boundaries to boundaries and hence the homomorphisms ϕ_p (resp. ψ_p) descend to homomorphisms on the homology vector spaces $H_p(C_\bullet) \to H_p(C'_\bullet)$ (resp., $H_p(C'_\bullet) \to H_p(C''_\bullet)$). We denote these homomorphisms by $H_p(\phi_\bullet)$ (resp. $H_p(\psi_\bullet)$).

We now define the homomorphism $H_p(\partial)\colon H_p(C''_\bullet) \to H_{p-1}(C_\bullet)$. Let α'' be a cycle in C''_p. By the exactness of the second row of the diagram we know that ψ_p is surjective and thus there exists $\alpha' \in C'_p$ such that $\psi_p(\alpha') = \alpha''$. Let $\beta' = \partial'_p(\alpha')$.

We show that $\beta' \in \mathrm{Ker}(\psi_{p-1})$. Using the commutativity of the diagram, we have that

$$\psi_{p-1}(\beta') = \psi_{p-1}(\partial'_p(\alpha')) = \partial''_p(\psi_p(\alpha')) = \partial''_p(\alpha'') = 0,$$

the last equality by virtue of the fact that α'' is a cycle.

By the exactness of the third row of the diagram, we have that $\mathrm{Im}(\phi_{p-1}) = \mathrm{Ker}(\psi_{p-1})$ and hence ϕ_{p-1} is injective. Thus, there exists a unique $\beta \in C_{p-1}$ such that $\beta' = \phi_{p-1}(\beta)$. Moreover, β is a cycle in C_{p-1}. To see this, observe that

$$\phi_{p-2}(\partial_{p-1}(\beta)) = \partial'_{p-1}(\phi_{p-1}(\beta)) = \partial'_{p-1}(\beta') = \partial'_{p-1}(\partial'_p(\beta')) = 0.$$

Since ϕ_{p-2} is injective, it follows that $\partial_{p-1}(\beta) = 0$, whence β is a cycle. Define $H_p(\partial)(\overline{\alpha''}) = \overline{\beta}$, where $\overline{\beta}$ represents the homology class of the cycle β.

We now check that $\mathrm{H}_p(\partial)$ is a well-defined homomorphism and that the long sequence of homology is exact as claimed.

We first prove that the map defined above indeed is a well-defined homomorphism $\mathrm{H}_p(\mathrm{C}''_\bullet) \to \mathrm{H}_{p-1}(\mathrm{C}_\bullet)$. We first check that the homology class $\overline{\beta}$ does not depend on the choice of $\alpha' \in \mathrm{C}'_p$ used in its definition. Let $\alpha'_1 \in \mathrm{C}'_p$ be such that $\psi_p(\alpha'_1) = \alpha''$. Let $\beta'_1 = \partial'_p(\alpha'_1)$. Now, β'_1 is also in $\mathrm{Ker}(\psi_{p-1})$ and by the exactness of the third row of the diagram, there exists a unique cycle $\beta_1 \in \mathrm{C}_{p-1}$ such that $\beta'_1 = \phi_{p-1}(\beta_1)$.

Now, $\alpha' - \alpha'_1 \in \mathrm{Ker}(\psi_p)$. Hence, there exists $\alpha_0 \in \mathrm{C}_p$ such that $\phi_p(\alpha_0) = \alpha' - \alpha'_1$, and using the commutativity of the diagram and the fact that ϕ_{p-1} is injective, we have that $\partial_p(\alpha_0) = \beta - \beta'_1$, whence $\beta_1 - \beta = 0$ in $\mathrm{H}_{p-1}(\mathrm{C}_\bullet)$. This shows that $\overline{\beta}$ is indeed independent of the choice of α'.

We now show that the $\mathrm{Im}(\mathrm{H}_p(\partial)) = \mathrm{Ker}(\mathrm{H}_{p-1}(\phi_\bullet))$. Exactness at the other terms is easy to verify and is left as an exercise.

Let $\beta \in \mathrm{C}_{p-1}$ be a cycle such that $\overline{\beta} \in \mathrm{H}_{p-1}(\mathrm{C}_\bullet)$ is in the image of $\mathrm{H}_p(\partial)$. Let $\alpha'' \in \mathrm{C}''_p$ be such that $\mathrm{H}_p(\partial)(\overline{\alpha''_p}) = \overline{\beta}$ and let $\alpha' \in \mathrm{C}'_p$, $\beta' \in \mathrm{C}'_{p-1}$ be as above. Then, $\beta' = \phi_{p-1}(\beta) = \partial'_p(\alpha') \in B_{p-1}(\mathrm{C}'_\bullet)$. Descending to homology, this shows that $\mathrm{H}_{p-1}(\phi_\bullet)(\overline{\beta}) = 0$, and $\beta \in \mathrm{Ker}(\mathrm{H}_{p-1}(\phi_\bullet))$.

Now, let $\beta \in \mathrm{C}_{p-1}$, such that $\overline{\beta} \in \mathrm{Ker}(\mathrm{H}_{p-1}(\phi_\bullet))$. This implies that $\phi_{p-1}(\beta) \in \mathrm{Im}(\partial'_p)$. Hence, there exists $\alpha' \in \mathrm{C}'_p$ such that $\partial'_p(\alpha') = \phi_{p-1}(\beta)$. Let $\alpha'' = \psi_p(\alpha')$. Since, $\psi_{p-1}(\partial'_p(\alpha')) = \psi_{p-1}(\phi_{p-1}(\beta)) = 0$ by commutativity of the diagram, we have that $\partial''_p(\alpha) = 0$. Hence, α is a cycle and it is easy to verify that $\mathrm{H}_p(\partial)(\overline{\alpha}) = \overline{\beta}$ and hence $\overline{\beta} \in \mathrm{Im}(\mathrm{H}_p(\partial))$. □

Another tool from homological algebra is the following Five Lemma.

Lemma 6.11. [Five Lemma] *Let*

$$\begin{array}{ccccccccc}
\mathrm{C}_1 & \xrightarrow{\phi_1} & \mathrm{C}_2 & \xrightarrow{\phi_2} & \mathrm{C}_3 & \xrightarrow{\phi_3} & \mathrm{C}_4 & \xrightarrow{\phi_4} & \mathrm{C}_5 \\
\downarrow a & & \downarrow b & & \downarrow c & & \downarrow d & & \downarrow e \\
\mathrm{D}_1 & \xrightarrow{\psi_1} & \mathrm{D}_2 & \xrightarrow{\psi_2} & \mathrm{D}_3 & \xrightarrow{\psi_3} & \mathrm{D}_4 & \xrightarrow{\psi_4} & \mathrm{D}_5
\end{array}$$

be a commutative diagram such that each row is exact. Then if a, b, d, e are isomorphisms, so is c.

Proof: We first show that c is injective. Let $c(x_3) = 0$ for some $x_3 \in \mathrm{C}_3$. Then $d \circ \phi_3(x_3) = 0 \Rightarrow \phi_3(x_3) = 0$, because d is an isomorphism. Hence, $x_3 \in \ker(\phi_3) = \mathrm{Im}(\phi_2)$. Let $x_2 \in \mathrm{C}_2$ be such that $x_3 = \phi_2(x_2)$. But then, $\psi_2 \circ b(x_2) = 0 \Rightarrow b(x_2) \in \ker(\psi_2) = \mathrm{Im}(\psi_1)$. Let $y_1 \in \mathrm{D}_1$ be such that $\psi_1(y_1) = b(x_2)$. Since a is an isomorphism there exists $x_1 \in \mathrm{C}_1$ such that $y_1 = a(x_1)$ and $\psi_1 \circ a(x_1) = b(x_2) = b \circ \phi_1(x_1)$. Since b is an isomorphism this implies that $x_2 = \phi_1(x_1)$, and thus $x_3 = \phi_2 \circ \phi_1(x_1) = 0$.

Next we show that c is surjective. Let $y_3 \in D_3$. Since d is surjective there exists $x_4 \in C_4$ such that $\psi_3(y_3) = d(x_4)$. Now, $\psi_4 \circ \psi_3(y_3) = 0 = e \circ \phi_4(x_4)$. Since e is injective this implies that $x_4 \in \ker(\phi_4) = \mathrm{Im}(\phi_3)$. Let $x_3 \in C_3$ be such that $x_4 = \phi_3(x_3)$. Then, $d \circ \phi_3(x_3) = \psi_3 \circ c(x_3) = \psi_3(y_3)$. Hence, $c(x_3) - y_3 \in \ker(\psi_3) = \mathrm{Im}(\psi_2)$. Let $y_2 \in D_2$ be such that $\psi_2(y_2) = c(x_3) - y_3$. There exists $x_2 \in C_2$ such that $\psi_2 \circ b(x_2) = c(x_3) - y_3 = c \circ \phi_2(x_2)$. But then, $c(x_3 - \phi_2(x_2)) = y_3$ showing that c is surjective. □

We use Lemma 6.10 to prove the existence of the so called Mayer-Vietoris sequence.

Theorem 6.12. [Mayer-Vietoris] *Let K be a simplicial complex and let K_1, K_2 be sub-complexes of K. Then there is an exact sequence*

$$\cdots \to \mathrm{H}_p(K_1) \oplus \mathrm{H}_p(K_2) \to \mathrm{H}_p(K_1 \cup K_2) \to \mathrm{H}_{p-1}(K_1 \cap K_2) \to \cdots.$$

Proof: We define homomorphisms $\phi_\bullet$, $\psi_\bullet$ so that the sequence

$$0 \to \mathrm{C}_\bullet(K_1 \cap K_2) \xrightarrow{\phi_\bullet} \mathrm{C}_\bullet(K_1) \oplus \mathrm{C}_\bullet(K_2) \xrightarrow{\psi_\bullet} \mathrm{C}_\bullet(K_1 \cup K_2) \to 0$$

is exact.

There are natural inclusion homomorphisms $i_1: \mathrm{C}_\bullet(K_1 \cap K_2) \to \mathrm{C}_\bullet(K_1)$ and i_2: $\mathrm{C}_\bullet(K_1 \cap K_2) \to \mathrm{C}_\bullet(K_2)$, as well as $j_1: \mathrm{C}_\bullet(K_1) \to \mathrm{C}_\bullet(K_1 \cup K_2)$ and $j_2: \mathrm{C}_\bullet(K_2) \to \mathrm{C}_\bullet(K_1 \cup K_2)$.

For $c \in \mathrm{C}_\bullet(K_1 \cap K_2)$, we define $\phi(c) = (i_1(c), -i_2(c))$.

For $(d, e) \in \mathrm{C}_\bullet(K_1) \oplus \mathrm{C}_\bullet(K_2)$, we define $\psi(d, e) = j_1(d) + j_2(e)$.

It is an exercise to check that, with these choices of $\phi_\bullet$ and $\psi_\bullet$, the sequence

$$0 \to \mathrm{C}_\bullet(K_1 \cap K_2) \xrightarrow{\phi_\bullet} \mathrm{C}_\bullet(K_1) \oplus \mathrm{C}_\bullet(K_2) \xrightarrow{\psi_\bullet} \mathrm{C}_\bullet(K_1 \cup K_2) \to 0$$

is exact. Now, apply Lemma 6.10 to complete the proof. □

6.1.5 Chain Homotopy

We identify a property that guarantees that two chain homomorphisms induce identical homomorphisms in homology. The property is that they are chain homotopic.

Given two chain complexes, $\mathrm{C}_\bullet = (\mathrm{C}_p, \partial_p)$ and $\mathrm{C}'_\bullet = (\mathrm{C}'_p, \partial'_p)$, two chain homomorphisms $\phi_\bullet, \psi_\bullet: \mathrm{C}_\bullet \to \mathrm{C}'_\bullet$ are **chain homotopic** (denoted $\phi_\bullet \sim \psi_\bullet$) if there exists a sequence of homomorphisms, $\gamma_p: \mathrm{C}_p \to \mathrm{C}'_{p+1}$ such that

$$\partial'_{p+1} \circ \gamma_p + \gamma_{p-1} \circ \partial_p = \phi_p - \psi_p \tag{6.5}$$

for all p. The collection $\gamma_\bullet$ of the homomorphisms γ_p is called a **chain homotopy** between $\mathrm{C}_\bullet$ and $\mathrm{C}'_\bullet$.

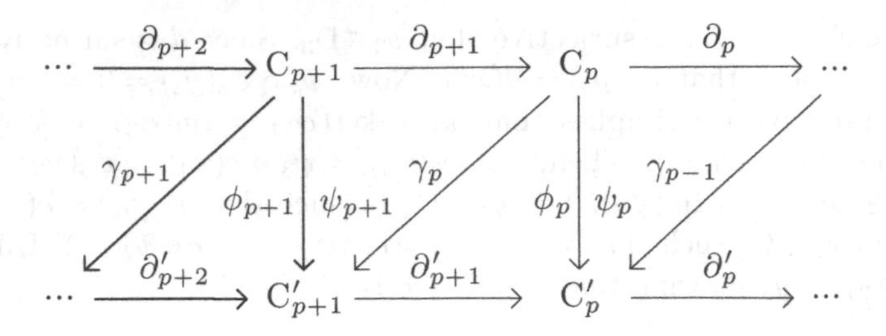

Lemma 6.13. *Chain homotopy is an equivalence relation among chain homomorphisms from* $C_\bullet$ *to* $C'_\bullet$.

Proof: Clearly every chain homomorphism $\phi_\bullet: C_\bullet \to C'_\bullet$ is chain homotopic to itself (choose $\gamma_p = 0$).

Also, if $\gamma_p: C_p \to C'_{p+1}$ gives a chain homotopy between chain homomorphisms $\phi_\bullet$ and $\psi_\bullet$, then $-\gamma_\bullet$ gives a chain homotopy between $\psi_\bullet$ and $\phi_\bullet$.

Finally, let $\gamma_p: C_p \to C'_{p+1}$ be a chain homotopy between the chain homomorphisms $\phi_\bullet$ and $\psi_\bullet$ and let $\lambda_p: C_p \to C'_{p+1}$ be a chain homotopy between the chain homomorphisms $\psi_\bullet$ and $\eta_\bullet$.

Then, the homomorphisms $\gamma_p + \lambda_p$ give a chain homotopy between $\phi_\bullet$ and $\eta_\bullet$. This is because

$$\begin{aligned} & \partial'_{p+1} \circ (\gamma_p + \lambda_p) + (\gamma_{p-1} + \lambda_{p-1}) \circ \partial_p \\ = \; & \partial'_{p+1} \circ \gamma_p + \gamma_{p-1} \circ \partial_p + \partial'_{p+1} \circ \lambda_p + \gamma_{p-1} \circ \lambda_p \\ = \; & \phi_p - \psi_p + \psi_p - \eta_p \\ = \; & \phi_p - \eta_p. \end{aligned}$$

□

Proposition 6.14. *If* $\phi_\bullet \sim \psi_\bullet: C_\bullet \to C'_\bullet$, *then*

$$H_\star(\phi_\bullet) = H_\star(\psi_\bullet): H_\star(C_\bullet) \to H_\star(C'_\bullet).$$

Proof: Let c be a p-cycle in $C_\bullet$, that is $c \in \mathrm{Ker}(\partial_p)$. Since $\phi_\bullet$ and $\psi_\bullet$ are chain homotopic, there exists a sequence of homomorphisms $\gamma_p: C_p \to C'_{p+1}$ satisfying equation (6.1).

Thus,

$$(\partial'_{p+1} \circ \gamma_p + \gamma_{p-1} \circ \partial_p)(c) = (\phi_p - \psi_p)(c).$$

Now, since c is a cycle, $\partial_p(c) = 0$, and moreover, $\partial'_{p+1}(\gamma_p(c))$ is a boundary. Thus, $(\phi_p - \psi_p)(c) = 0$ in $H_p(C'_\bullet)$. □

Example 6.15. As an example of chain homotopy, consider the simplicial complex K whose simplices are all the faces of a single simplex $[a_0, a_1, \ldots, a_k] \subset \mathrm{R}^k$. Consider the chain homomorphisms $C_\bullet(\phi)$ and $C_\bullet(\psi)$ induced by the simplicial maps $\phi = \mathrm{Id}_K$ and ψ such that $\psi(a_i) = a_0, 1 \le i \le k$.

Then, $C_\bullet(\phi)$ and $C_\bullet(\psi)$ are chain homotopic by the chain homotopy γ defined by $\gamma_p([a_{i_0}, \ldots, a_{i_p}]) = [a_0, a_{i_0}, \ldots, a_{i_p}]$ for $p \geq 0$ and $\gamma_p = 0$ otherwise.

Clearly for $p > 0$,

$$\begin{aligned}&(\partial_{p+1} \circ \gamma_p + \gamma_{p-1} \circ \partial_p)([a_{i_0}, \ldots, a_{i_p}])\\ &= [a_{i_0}, \ldots, a_{i_p}] - [a_0, \partial_p([a_{i_0}, \ldots, a_{i_p}])] + [a_0, \partial_p([a_{i_0}, \ldots, a_{i_p}])\\ &= [a_{i_0}, \ldots, a_{i_p}]\\ &= (\phi_p - \psi_p)([a_{i_0}, \ldots, a_{i_p}]).\end{aligned}$$

For $p = 0$,

$$\begin{aligned}(\partial_1 \circ \gamma_0 + \gamma_{-1} \circ \partial_0)([a_{i_0}]) &= [a_{i_0}] - [a_0]\\ &= (\phi_0 - \psi_0)([a_{i_0}]).\end{aligned}$$

It is now easy to deduce that $H_0(K) = \mathbb{Q}$ and $H_i(K) = 0$ for all $i > 0$. □

A simplicial complex K is **acyclic** if $H_0(K) = \mathbb{Q}$, $H_i(K) = 0$ for all $i > 0$.

Lemma 6.16. *Let K be the simplicial complex whose simplices are all the faces of a simplex $s = [a_0, a_1, \ldots, a_k] \subset R^k$. Then, K is acyclic, and its barycentric subdivision* $ba(K)$ *is acyclic.*

Proof: Recall from Section 5.6 that for every ascending sequence of simplices in K,

$$s_0 \prec s_1 \prec \cdots \prec s_j,$$

the simplex $[ba(s_0), \ldots, ba(s_j)]$ is included in $ba(K)$.

Consider the chain homomorphisms, ϕ, ψ: $C_\bullet(ba(K)) \to C_\bullet(ba(K))$, induced by the simplicial maps Id and ψ defined by $\psi(ba(s_i)) = ba(s)$, for each $s_i \in K$.

Then, ϕ and ψ are chain homotopic by the chain homotopy γ defined by $\gamma_p([ba(s_0), \ldots, ba(s_p)]) = [ba(s), ba(s_0), \ldots, ba(s_p)]$ for $p \geq 0$ and $\gamma_p = 0$ otherwise.

Clearly for $p > 0$,

$$\begin{aligned}&(\partial_{p+1} \circ \gamma_p + \gamma_{p-1} \circ \partial_p)([ba(s_0), \ldots, ba(s_p)])\\ &= [ba(s_0), \ldots, ba(s_p)] - [ba(s), \partial_p([ba(s_0), \ldots, ba(s_p)])]\\ &\quad + [ba(s), \partial_p([ba(s_0), \ldots, ba(s_p)])]\\ &= [ba(s_0), \ldots, ba(s_p)]\\ &= (\phi_p - \psi_p)([ba(s_0), \ldots, ba(s_p)]).\end{aligned}$$

For $p = 0$,

$$\begin{aligned}(\partial_1 \circ \gamma_0 + \gamma_{-1} \circ \partial_0)([ba(s_i)]) &= [ba(s_i)] - [ba(s)]\\ &= (\phi_0 - \psi_0)([ba(s_i)]).\end{aligned}$$

It is now easy to deduce that $\mathrm{H}_0(\mathrm{ba}(K)) = \mathbb{Q}$ and $\mathrm{H}_i(\mathrm{ba}(K)) = 0$ for all $i > 0$. □

We now identify a criterion that is sufficient to show that two homomorphisms are chain homotopic in the special case of chain complexes coming from simplicial complexes. The key notion is that of an **acyclic carrier function**.

Let K, L be two complexes. A function ξ which maps every simplex $s \in K$ to a sub-complex $\xi(s)$ of L is called a **carrier function** provided

$$s' \prec s \Rightarrow \xi(s') \subset \xi(s)$$

for all $s, s' \in K$. Moreover, if $\xi(s)$ is acyclic for all $s \in K$, ξ is called an **acyclic carrier function.** A chain homomorphism $\phi_\bullet: \mathrm{C}_\bullet(K) \to \mathrm{C}_\bullet(L)$ is **carried by a carrier function** ξ if for all p and each $s \in K_p$, $\phi_p(s)$ is a chain in $\xi(s)$.

The most important property of a carrier function is the following.

Lemma 6.17. *If $\phi_\bullet$, $\psi_\bullet: \mathrm{C}_\bullet(K) \to \mathrm{C}_\bullet(L))$ are chain homomorphisms carried by the same acyclic carrier ξ, then $\phi_\bullet \sim \psi_\bullet$.*

Proof: Let ∂ (resp. ∂') be the boundary maps of $\mathrm{C}_\bullet(K)$ (resp. $\mathrm{C}_\bullet(L)$). We construct a chain homotopy γ dimension by dimension.

For $s_0 \in \mathrm{C}_0(K)$, $\phi_0(s_0) - \psi_0(s_0)$ is a chain in $\xi(s_0)$ which is acyclic. Since $\xi(s_0)$ is acyclic, $\phi_0(s_0) - \psi_0(s_0)$ must also be a boundary. Thus, there exists a chain $t \in \mathrm{C}_1(L)$ such that $\partial_1'(t) = \phi_0(s_0) - \psi_0(s_0)$, and we let $\gamma_0(s_0) = t$.

Now, assume that for all $q < p$ we have constructed γ_q such that $(\phi - \psi)_q = \partial_{q+1}' \circ \gamma_q + \gamma_{q-1} \circ \partial_q$ and $\gamma_q(s) \subset \xi(s)$ for all q-simplices s.

We define $\gamma_p(s)$ for p-simplices s and extend it linearly to p-chains. Notice first that $(\phi - \psi)_p(s) \subset \xi(s)$ by hypothesis and that $\gamma_{p-1} \circ \partial_p(s)$ is a chain in $\xi(s)$ by the induction hypothesis. Hence $((\phi - \psi)_p - \gamma_{p-1} \circ \partial_p)(s)$ is a chain in $\xi(s)$ and let this chain be t. Then,

$$\begin{aligned}
\partial_p'(t) &= \partial_p'((\phi - \psi)_p - \gamma_{p-1} \circ \partial_p)(s) \\
&= (\partial_p' \circ (\phi - \psi)_p - \partial_p' \circ \gamma_{p-1} \circ \partial_p)(s) \\
&= ((\phi - \psi)_{p-1} \circ \partial_p)(s) - (\partial_p' \circ \gamma_{p-1} \circ \partial_p)(s) \\
&= ((\phi - \psi)_{p-1} - \partial_p' \circ \gamma_{p-1})(\partial_p(s)) \\
&= \gamma_{p-2} \circ \partial_{p-1} \circ \partial_p(s) \\
&= 0
\end{aligned}$$

so that t is a cycle.

But, since $t = ((\phi - \psi)_p - \gamma_{p-1} \circ \partial_p)(s)$ is a chain in $\xi(s)$ and $\xi(s)$ is acyclic, t must be a boundary as well. Thus, there is a chain, t', such that $t = \partial_{p+1}(t')$ and we define $\gamma_p(s) = t'$. It is straightforward to check that this satisfies all the conditions. □

Two simplicial maps ϕ, ψ: $K \to L$ are **contiguous** if $\phi(s)$ and $\psi(s)$ are faces of the same simplex in L for every simplex $s \in K$.

Two simplicial maps ϕ, ψ: $K \to L$ belong to the same contiguity class if there is a sequence of simplicial maps $\phi_i, i = 0, \ldots, n$, such that $\phi_0 = \phi, \phi_n = \psi$, and ϕ_i and ϕ_{i+1} are contiguous for $0 \leq i < n$.

Proposition 6.18. *If the chain homomorphisms*

$$\mathrm{C}_\bullet(\phi), \mathrm{C}_\bullet(\psi): \mathrm{C}_\bullet(K) \to \mathrm{C}_\bullet(L)$$

are induced by simplicial maps that belong to the same contiguity class, then $\mathrm{H}_\star(\phi) = \mathrm{H}_\star(\psi)$.

Proof: We show that two contiguous simplicial maps induce chain homotopic chain homomorphisms, which will prove the proposition. In order to show this, we construct an acyclic carrier, ξ, for both ϕ and ψ. For a simplex $s \in K$, let t be the smallest dimensional simplex of L such that $\phi(s)$ and $\psi(s)$ are both faces of t (in fact any such t will do). Let $\xi(s)$ be the sub-complex of L consisting of all faces of t. Clearly, ξ is an acyclic carrier of both ϕ and ψ, which implies that they are chain homotopic. □

6.1.6 The Simplicial Homology Groups Are Invariant Under Homeomorphism

We shall show that if K and L are two simplicial complexes that are homeomorphic then the homology groups of K and L are isomorphic.

6.1.6.1 Homology and Barycentric Subdivision

The first step is to show that the homology groups of a simplicial complex K are isomorphic to those of its barycentric subdivision (see Definition page 182).

Theorem 6.19. *Let K be a simplicial complex and* $\mathrm{ba}(K)$ *its barycentric subdivision. Then,* $\mathrm{H}_\star(K) \cong \mathrm{H}_\star(\mathrm{ba}(K))$.

As a consequence, we can iterate the operation of barycentric subdivision by setting $K^{(1)} = \mathrm{ba}(K)$ and, in general, $K^{(n)} = \mathrm{ba}(K^{(n-1)})$, thus obtaining finer and finer subdivisions of the complex K.

Corollary 6.20. *Let K be a simplicial complex. Then,* $\mathrm{H}_\star(K) \cong \mathrm{H}_\star(K^{(n)})$ *for all* $n > 0$.

In order to prove Theorem 6.19, we define some simplicial maps between the barycentric subdivision of a simplicial complex and the simplicial complex itself that will allow us to relate their homology groups.

Given a simplicial complex K and its barycentric subdivision $\mathrm{ba}(K)$, a **Sperner map** is a map ω: $\mathrm{ba}(K)_0 \to K_0$ such that $\omega(\mathrm{ba}(s))$ is one of the vertices of s for each simplex $s \in K$.

Lemma 6.21. *Any Sperner map can be linearly extended to a simplicial map.*

Proof: Let ω: $\mathrm{ba}(K)_0 \to K_0$ be a Sperner map. Then an oriented simplex $[\mathrm{ba}(s_0), ..., \mathrm{ba}(s_i)]$ in $\mathrm{ba}(K)$ corresponds to $s_0 \prec \cdots \prec s_i$ in K, with $\omega(\mathrm{ba}(s_j)) \in s_i$, $0 \leq j \leq i$, and hence $[\omega(\mathrm{ba}(s_0)), ..., \omega(\mathrm{ba}(s_i))]$ is an oriented simplex in K. □

Given two simplicial complexes K, L and a simplicial map ϕ: $K \to L$, there is a natural way to define a simplicial map ϕ': $\mathrm{ba}(K) \to L'$ by setting $\phi'(\mathrm{ba}(s)) = \mathrm{ba}(\phi(s))$ for every $s \in K$ and extending it linearly to $\mathrm{ba}(K)$. One can check that ϕ' so defined is simplicial.

We define a new homomorphism

$$\alpha_\bullet: \mathrm{C}_\bullet(K) \to \mathrm{C}_\bullet(\mathrm{ba}(K)),$$

which will play the role of an inverse to any Sperner map ω, as follows:

It is defined on simplices recursively by,

$$\begin{aligned} \alpha_0(s) &= s, \\ \alpha_p(s) &= [\mathrm{ba}(s), \alpha_{p-1}(\partial_p(s))], p > 0, \end{aligned}$$

(see Definition 6.1.1.1) and is then extended linearly to $\mathrm{C}_\bullet(K)$. It is easy to verify that $\alpha_\bullet$ is also a chain homomorphism.

Lemma 6.22. *Given a simplicial complex K and a Sperner map ω,* $\mathrm{C}_\bullet(\omega) \circ \alpha_\bullet = \mathrm{Id}_{\mathrm{C}_\bullet(K)}$.

Proof: The proof is by induction on the dimension p. It is easily seen that the lemma holds when $p = 0$. Consider a simplex $s = [a_0, ..., a_p]$. Now,

$$\begin{aligned} (\mathrm{C}_p(\omega) \circ \alpha_p)(s) &= \mathrm{C}_p(\omega)([\mathrm{ba}(s), \alpha_{p-1}(\partial_p(s))] \\ &= [\mathrm{C}_0(\omega)(\mathrm{ba}(s)), \mathrm{C}_{p-1}(\omega) \circ \alpha_{p-1}(\partial_p(s))]. \end{aligned}$$

By induction hypothesis, $(\mathrm{C}_{p-1}(\omega) \circ \alpha_{p-1})(\partial_p(s)) = \partial_p(s)$. Since, $\mathrm{C}_0(\omega)(\mathrm{ba}(s))$ is a vertex of s it follows that $\mathrm{C}_p(\omega) \circ \alpha_p(s) = s$. This completes the induction. □

We now prove that $\alpha_\bullet \circ \mathrm{C}_\bullet(\omega) \sim \mathrm{Id}_{\mathrm{C}_\bullet(\mathrm{ba}(K))}$ for a Sperner map ω.

Lemma 6.23. *Let ω: $\mathrm{ba}(K) \to K$ be a Sperner map. Then*

$$\alpha_\bullet \circ \mathrm{C}_\bullet(\omega) \sim \mathrm{Id}_{\mathrm{C}_\bullet(\mathrm{ba}(K))}.$$

Proof: We construct an acyclic carrier carrying $\alpha_\bullet \circ \mathrm{C}_\bullet(\omega)$ and $\mathrm{Id}_{\mathrm{C}_\bullet(\mathrm{ba}(K))}$. Let $\mathrm{ba}(s)$ be a simplex of $\mathrm{ba}(K)$ and let $b = \mathrm{ba}(s)$ be the leading vertex of $\mathrm{ba}(s)$, where s is a simplex in K. Let $\xi(\mathrm{ba}(s))$ be the sub-complex of $\mathrm{ba}(K)$ consisting of all simplices in the barycentric subdivision of s.

Clearly, ξ carries both $\alpha_\bullet \circ \mathrm{C}_\bullet(\omega)$ and $\mathrm{Id}_{\mathrm{C}_\bullet(\mathrm{ba}(K))}$ and is acyclic by Lemma 6.16, and hence satisfies the conditions for being an acyclic carrier. □

Proof of Theorem 6.19: Follows immediately from the preceding lemmas. □

6.1.6.2 Homeomorphisms Preserve Homology

Our goal in this paragraph is to show that homeomorphic polyhedra in real affine space have isomorphic homology groups.

Theorem 6.24. *If two simplicial complexes* $K \subset \mathbb{R}^k, L \subset \mathbb{R}^\ell$ *are two simplicial complexes and* $f: |K| \to |L|$ *is a homeomorphism, then there exists an isomorphism* $\mathrm{H}_\star(f): \mathrm{H}_\star(K) \to \mathrm{H}_\star(L)$.

We will use the fact that our ground field is $\mathbb{R}$ in two ways. In the next lemma, we use the fact that $\mathbb{R}$ is (sequentially) compact in its metric topology in order to show the existence of a Lebesgue number for any finite open covering of a compact set in $\mathbb{R}^k$. Secondly, we will use the archimedean property of $\mathbb{R}$.

We first need a notation. For a vertex a of a simplicial complex K, its **star** $\mathrm{star}(a) \subset |K|$ is the union of the relative interiors of all simplices having a as a vertex, i.e. $\mathrm{star}(a) = \cup_{\{a\} \prec s}\, s^\circ$. If the simplicial complexes K and L have the same polyhedron and if to every vertex a of K there is a vertex b of L such that $\mathrm{star}(a) \subset \mathrm{star}(b)$, then we write $K < L$ and say K is **finer** than L.

It is clear that for any simplicial complex K, $K^{(n)} < K$. Also, if $K < L$ and $L < M$ then $K < M$.

In the next lemma, we show that given a family of open sets whose union contains a compact subset S of $\mathbb{R}^n$, any "sufficiently small" subset of S is contained in a single set of the family.

We define the **diameter** $\mathrm{diam}(S)$ of a set S as the smallest number d such that S is contained in a ball of radius $d/2$.

Lemma 6.25. *Let* $\mathcal{A}$ *be an open cover of a compact subset* S *of* $\mathbb{R}^n$. *Then, there exists* $\delta > 0$ *(called the Lebesgue number of the cover) such that for any subset* B *of* S *with* $\mathrm{diam}(B) < \delta$, *there exists an* $A \in \mathcal{A}$ *such that* $B \subset A$.

Proof: Assume not. Then there exists a sequence of numbers $\{\delta_n\}$ and sets $S_n \subset S$ such that $\delta_n \to 0$, $\mathrm{diam}(S_n) < \delta_n$, and $S_n \not\subset A$, for all $A \in \mathcal{A}$.

Choose a point p_n in each S_n. Since S is compact, the sequence $\{p_n\}$ has a convergent subsequence, and we pass to this subsequence and henceforth assume that the sequence $\{p_n\}$ is convergent and its limit point is p.

Now $p \in S$ since S is closed, and thus there exists a set A in the covering $\mathcal{A}$ such that $p \in A$. Also, because A is open, there exists an $\epsilon > 0$ such that the open ball $B(p, \epsilon) \subset A$.

Now choose n large enough so that $||p - p_n|| < \epsilon/2$ and $\delta_n < \epsilon/2$. We claim that $S_n \subset A$, which is a contradiction. To see this, observe that S_n contains a point p_n which is within $\epsilon/2$ of p, but S_n also has diameter less than $\epsilon/2$. Hence it must be contained inside the ball $B(p, \epsilon)$ and hence is contained in A. □

The **mesh** mesh(K) of a complex K is defined by

$$\mathrm{mesh}(K) = \max\{\mathrm{diam}(s) | s \in K\}.$$

The following lemma bounds the mesh of the barycentric subdivision of a simplicial complex in terms of the mesh of the simplicial complex itself.

Lemma 6.26. *Let K be a simplicial complex of dimension k. Then,*

$$\mathrm{mesh}(\mathrm{ba}(K)) \leq \frac{k}{k+1} \mathrm{mesh}(K).$$

Proof: First note that mesh(K) (resp. mesh(ba(K))) equals the length of the longest edge in K (resp. ba(K)). This follows from the fact that the diameter of a simplex equals the length of its longest edge.

Let $(\mathrm{ba}(s), \mathrm{ba}(s'))$ be an edge in ba(K), where $s \prec s'$ are simplices in K. Also, without loss of generality, let $s = [a_0, \ldots, a_p]$ and $s' = [a_0, \ldots, a_q]$.

Now,

$$\begin{aligned} \mathrm{ba}(s) - \mathrm{ba}(s') &= \frac{1}{p+1} \sum_{0 \leq i \leq p} a_i - \frac{1}{q+1} \sum_{0 \leq i \leq q} a_i \\ &= \left(\frac{1}{p+1} - \frac{1}{q+1}\right) \sum_{0 \leq i \leq p} a_i - \frac{1}{q+1} \sum_{p+1 \leq i \leq q} a_i \\ &= \frac{q-p}{q+1} \left(\frac{1}{p+1} \sum_{0 \leq i \leq p} a_i - \frac{1}{q-p} \sum_{p+1 \leq i \leq q} a_i\right). \end{aligned}$$

The points $1/(p+1) \sum_{0 \leq i \leq p} a_i$ and $1/(q-p) \sum_{p+1 \leq i \leq q} a_i$ are both in s'.

Hence, we have

$$||\mathrm{ba}(s) - \mathrm{ba}(s')|| \leq \frac{q-p}{q+1} \mathrm{mesh}(K) \leq \frac{q}{q+1} \mathrm{mesh}(K) \leq \frac{k}{k+1} \mathrm{mesh}(K). \quad \square$$

Lemma 6.27. *For any two simplicial complexes K and L such that $|K| = |L| \subset \mathbb{R}^k$, there exists $n > 0$ such that $K^{(n)} < L$.*

Proof: The sets star(a) for each vertex $a \in L_0$ give an open cover of the polyhedron $|L|$. Since the polyhedron is compact, for any such open cover there exists, by Lemma 6.25, a $\delta > 0$ such that any subset of the polyhedron of diameter $< \delta$ is contained in an element of the open cover, that is in star(a) for some $a \in L_0$. Using the fact that $\mathrm{mesh}(\mathrm{ba}(K)) \le k/(k+1)\,\mathrm{mesh}(K)$ and hence $\mathrm{mesh}(K^{(n)}) \le (k/(k+1))^n\,\mathrm{mesh}(K)$, we can choose n large enough so that for each $b \in K_0^{(n)}$, the set star(b) having diameter $< 2\,\mathrm{mesh}(K^{(n)})$ is contained in star(a) for some $a \in L_0$. □

Lemma 6.28. *Let K, L be two simplicial complexes with $K < L$, and such that $|K| = |L| \subset \mathbb{R}^k$. Then, there exists a well-defined isomorphism*

$$i(K, L)\colon \mathrm{H}_\star(K) \to \mathrm{H}_\star(L),$$

which respects composition. In other words, given another simplicial complex M with $|M| = |L|$ and $L < M$,

$$i(K, M) = i(L, M) \circ i(K, L).$$

Proof: Since $K < L$, for any vertex $a \in K_0$, there exists a vertex $b \in L_0$ such that $\mathrm{star}(a) \subset \mathrm{star}(b)$ since $K < L$. Consider a map $\phi\colon K_0 \to L_0$ that sends each vertex $a \in K_0$ to a vertex $b \in L_0$ satisfying $\mathrm{star}(a) \subset \mathrm{star}(b)$. Notice that this agrees with the definition of a Sperner map in the case where K is a barycentric subdivision of L. Clearly, such a map is simplicial. Note that even though the simplicial map ϕ is not uniquely defined, any other choice of the simplicial map satisfying the above condition is contiguous to ϕ and thus induces the same homomorphism between $\mathrm{H}_\star(K)$ and $\mathrm{H}_\star(L)$. Also, by Lemma 6.27, we can choose n such that $L^{(n)} < K$ and a simplicial map $\psi\colon L^{(n)} \to K$ that gives rise to a homomorphism

$$\mathrm{H}_\star(\psi)\colon \mathrm{H}_\star(L^{(n)}) \to \mathrm{H}_\star(K).$$

In addition, using Theorem 6.19, we have an isomorphism

$$\mathrm{H}_\star(\gamma)\colon \mathrm{H}_\star(L^{(n)}) \to \mathrm{H}_\star(L).$$

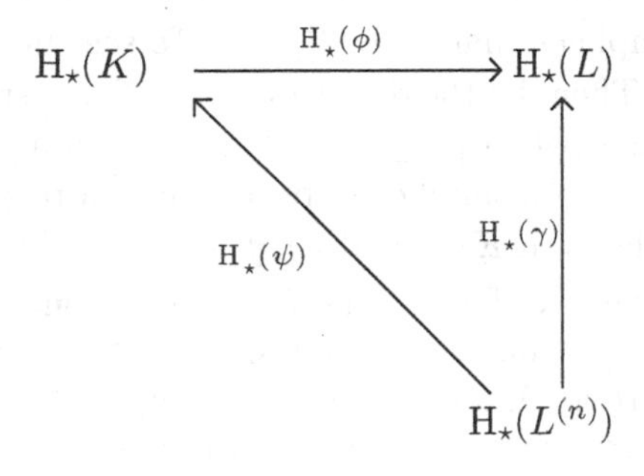

Again, note that the homomorphisms $\mathrm{H}_\star(\psi), \mathrm{H}_\star(\gamma)$ are well-defined, even though the simplicial maps from which they are induced are not. Moreover, $\mathrm{H}_\star(\gamma) = \mathrm{H}_\star(\phi) \circ \mathrm{H}_\star(\psi)$. To see this let $c \in L_0^{(n)}$, $a = \psi(c) \in K_0$, and $b = \phi(a) \in L_0$. Also, let $b' = \gamma(c) \in L_0$.

Then, $\mathrm{star}(c) \subset \mathrm{star}(a)$ and $\mathrm{star}(a) \subset \mathrm{star}(b)$, so that $\mathrm{star}(c) \subset \mathrm{star}(b)$. Also, $\mathrm{star}(c) \subset \mathrm{star}(b')$.

Let s be the simplex in L such that $b \in s$ and $c \in s^\circ$. Similarly, let t be the simplex in L such that $b' \in t$ and $c \in t^\circ$. But, this implies that $s^\circ \cap t^\circ \neq \emptyset$, implying that $s = t$. This proves that the simplicial maps $\phi \circ \psi$ and γ take a simplex s in $L^{(n)}$ to faces of the simplex in L containing s, and hence, $\phi \circ \psi$ and γ are contiguous, implying that $\mathrm{H}_\star(\gamma) = \mathrm{H}_\star(\phi) \circ \mathrm{H}_\star(\psi)$.

Now, since $\mathrm{H}_\star(\gamma)$ is surjective, so is $\mathrm{H}_\star(\phi)$. The same reasoning for the pair $L^{(n)} < K$ tells us that $\mathrm{H}_\star(\psi)$ is surjective. Now, since $\mathrm{H}_\star(\gamma)$ is injective and $\mathrm{H}_\star(\psi)$ is surjective, $\mathrm{H}_\star(\phi)$ is injective.

Define, $i(K, L) = \mathrm{H}_\star(\phi)$. Clearly, $i(K, L)$ is independent of the particular simplicial map ϕ chosen to define it. It also follows from the definition that the homomorphisms i respect composition. □

We next show that any continuous map between two polyhedrons can be suitably approximated by a simplicial map between some subdivisions of the two polyhedrons.

Given two simplicial complexes K, L and a continuous map $f: |K| \to |L|$, a simplicial map $\phi: K \to L$ is a **simplicial approximation** to f if $f(x) \in s^\circ$ implies $\phi(x) \in s$.

Proposition 6.29. *Given two simplicial complexes K, L and a continuous map f: $|K| \to |L|$, there exists an integer $n > 0$ and a simplicial map $\phi: K^{(n)} \to L$ that is a simplicial approximation to f.*

Proof: The family of open sets $\{\mathrm{star}(b) | b \in L_0\}$ is an open cover of L, and by continuity of f the family, $\{f^{-1}(\mathrm{star}(b)) | b \in L_0\}$ is an open cover of $|K|$. Let δ be the Lebesgue number of this cover of $|K|$ and choose n large enough so that $\mu(K^{(n)}) < \delta/2$. Thus, for every vertex a of $K_0^{(n)}$, $f(\mathrm{star}(a)) \subset \mathrm{star}(b)$ for some $b \in L_0$. It is easy to see that the map which sends a to such a b for every vertex $a \in K_0^{(n)}$ induces a simplicial map $\phi: K^{(n)} \to L$. To see this, let $s = [a_0, \ldots, a_m]$ be a simplex in $K^{(n)}$. Then, by the definition of ϕ, $\bigcap_{i=0}^m \mathrm{star}(\phi(a_i)) \neq \emptyset$ since it contains $f(s)$. Hence, $\{\phi(a_i) \mid 0 \le i \le m\}$ must span a simplex in L.

We now claim that ϕ is a simplicial approximation to f. Let $x \in |K|$ such that $x \in s^\circ$ for a simplex s in $K^{(n)}$, and let $f(x) \in t^\circ \subset |L|$.

Let $a \in K_0^{(n)}$ be a vertex of s, and let $b = \phi(a)$. From the definition of ϕ, we have that $f(\mathrm{star}(a)) \subset (\mathrm{star}(b))$, and since $x \in \mathrm{star}(a)$, $f(x) \in \mathrm{star}(b)$. Thus, $f(x) \in \cap_{a \in s} \mathrm{star}(\phi(a))$, and hence each $\phi(a)$, $a \in s$, is a vertex of the simplex t. Moreover, since $\phi(x)$ lies in the simplex spanned by $\{\phi(a) \mid a \in s\}$, it is clear that $\phi(x) \in t$. □

Proposition 6.30. *Given any two simplicial complexes* K and L, *as well as a continuous map* $f\colon |K| \to |L|$, *there exists a well-defined homomorphism* $\mathrm{H}_\star(f)\colon \mathrm{H}_\star(K) \to \mathrm{H}_\star(L)$ *such that if* N *is another simplicial complex and* $g\colon |L| \to |N|$ *is a continuous map, then*

$$\mathrm{H}_\star(g \circ f) = \mathrm{H}_\star(g) \circ \mathrm{H}_\star(f) \text{ and } \mathrm{H}_\star(\mathrm{Id}_{|K|}) = \mathrm{Id}_{\mathrm{H}_\star(K)}.$$

Proof: Choose n_1 large enough so that there is a simplicial approximation $\phi\colon K^{(n_1)} \to L$ to f. Define $\mathrm{H}_\star(f) = \mathrm{H}_\star(\phi) \circ i(K^{(n_1)}, K)^{-1}$. It is easy using Lemma 6.28 to see that $\mathrm{H}_\star(f)$ does not depend on the choice of n_1.

Now, suppose that we choose simplicial approximations $\psi\colon L^{(n_2)} \to N$ of g and $\phi\colon K^{(n_1)} \to L^{(n_2)}$ of f.

$$\begin{aligned} i(L^{(n_2)}, L) \circ \mathrm{H}_\star(\phi) \circ i(K^{(n_1)}, K)^{-1} &= \mathrm{H}_\star(f), \\ \mathrm{H}_\star(\psi) \circ i(L^{(n_2)}, L)^{-1} &= \mathrm{H}_\star(g), \\ \mathrm{H}_\star(g) \circ \mathrm{H}_\star(f) &= \mathrm{H}_\star(\psi) \circ \mathrm{H}_\star(\phi) \circ i(K^{(n_1)}, K)^{-1}. \end{aligned}$$

Note that $\psi \circ \phi\colon K^{(n_1)} \to N$ is a simplicial approximation of $g \circ f$. To see this, observe that for $x \in |K|$, $f(x) \in s$ implies that $\phi(x) \in s$, where s is a simplex in $L^{(n_2)}$. Since, ψ is a simplicial map $\psi(f(x)) \in t$ implies that $\psi(\phi(x)) \in t$ for any simplex t in N. This proves that $\psi \circ \phi$ is a simplicial approximation of $g \circ f$ and hence

$$\mathrm{H}_\star(\psi) \circ \mathrm{H}_\star(\phi) \circ i(K^{(n_1)}, K)^{-1} = \mathrm{H}_\star(g \circ f).$$

The remaining property that $\mathrm{H}_\star(\mathrm{Id}_{|K|}) = \mathrm{Id}_{\mathrm{H}_\star(K)}$ is now easy to check. □

Theorem 6.24 is now an immediate consequence of Proposition 6.30.

6.1.6.3 Semi-algebraic Homeomorphisms Preserve Homology

We next prove a result similar to Theorem 6.24 for semi-algebraic homeomorphisms between polyhedra defined over any real closed field.

Let K and L be two simplicial complexes contained in $\mathbb{R}^k$ whose vertices have rational coordinates. Since K and L have vertices with rational coordinates, they can be described by linear inequalities with rational coefficients and hence they are semi-algebraic subsets of $\mathbb{R}^k$. We denote by $\mathrm{Ext}(|K|, \mathrm{R})$ and $\mathrm{Ext}(|L|, \mathrm{R})$ the polyhedron defined by the same inequalities over R.

Theorem 6.31. *Let* K *and* L *be two simplicial complexes whose vertices have rational coordinates. If* $\mathrm{Ext}(|K|, \mathrm{R})$ *and* $\mathrm{Ext}(|L|, \mathrm{R})$ *are semi-algebraically homeomorphic for a real closed field* R, *then* $\mathrm{H}_\star(K) \cong \mathrm{H}_\star(L)$.

The theorem will follow from the transfer property stated in the next lemma.

Lemma 6.32. *Let K and L be two simplicial complexes whose vertices have rational coordinates. The following are equivalent*

- *There exists a semi-algebraic homeomorphism from* $\mathrm{Ext}(|K|, \mathbb{R}_{\mathrm{alg}})$ *to* $\mathrm{Ext}(|L|, \mathbb{R}_{\mathrm{alg}})$.
- *There exists a semi-algebraic homeomorphism from* $\mathrm{Ext}(|K|, \mathrm{R})$ *to* $\mathrm{Ext}(|L|, \mathrm{R})$ *for a real closed field* R.

Proof: It is clear that if $g\colon |K| = \mathrm{Ext}(|K|, \mathbb{R}_{\mathrm{alg}}) \to |L| = \mathrm{Ext}(|L|, \mathbb{R}_{\mathrm{alg}})$ is a semi-algebraic homeomorphism, then $\mathrm{Ext}(g, \mathrm{R})\colon \mathrm{Ext}(|K|, \mathrm{R}) \to \mathrm{Ext}(|L|, \mathrm{R})$ is a semi-algebraic homeomorphism, using the properties of the extension stated in Chapter 2 Exercise 2.16), since the property of a semi-algebraic function g of being a semi-algebraic homeomorphism can be described by a formula.

Conversely let R be a real closed field, and let $f\colon \mathrm{Ext}(|K|, \mathrm{R}) \to \mathrm{Ext}(|L|, \mathrm{R})$ be a semi-algebraic homeomorphism. Let $A = (a_1, \ldots, a_N) \in R^N$ be the vector of all the constants appearing in the definition of the semi-algebraic maps f.

Let $\Gamma_f \subset R^{2k}$ denote the graph of the semi-algebraic map f, and let $\phi_f(Z_1, \ldots, Z_{2k})$ denote the formula defining Γ. For $1 \leq i \leq N$, replace every appearance of the constant a_i in ϕ_f by a new variable Y_i to obtain a new formula ψ with $N + 2k$ variables $Y_1, \ldots, Y_N, Z_1, \ldots, Z_{2k}$. All constants appearing in ψ are now rational numbers.

For $b \in R^N$, let $\Gamma_f(b) \subset R^{2k}$ denote the set defined by $\psi(b, Z_1, \ldots, Z_{2k})$.

We claim that we can write a formula $\Phi_f(Y_1, \ldots, Y_N)$ such that, for every $b \in R^N$ satisfying Φ_f, the set $\Gamma_f(b) \subset \mathrm{R}^{2k}$ is the graph of a semi-algebraic homeomorphism from $\mathrm{Ext}(|K|, \mathrm{R})$ to $\mathrm{Ext}(|L|, \mathrm{R})$ (with the domain and range corresponding to the first k and last k coordinates respectively).

A semi-algebraic homeomorphism is a continuous, 1-1, and onto map, with a continuous inverse. Hence, in order to write such a formula, we first write formulas guaranteeing continuity, injectivity, surjectivity, and continuity of the inverse separately and then take their conjunction.

Thinking of $\bar{Y} = (Y_1, \ldots, Y_N)$ as parameters, let $\Phi_1(\bar{Y})$ be the first-order formula expressing that given $\bar{Y}$, for every open ball $B \subset R^k$, the set in R^k defined by

$$\{(Z_1, \ldots, Z_k) \mid \exists((Z_{k+1}, \ldots, Z_{2k}) \in B \wedge \psi(\bar{Y}, Z_1, \ldots, Z_{2k}))\}$$

is open in R^k. Since, we can clearly quantify over all open balls in R^k (quantify over all centers and radii), we can thus express the property of being open by a first-order formula, $\Phi_1(\bar{Y})$.

Similarly, it is an easy exercise to translate the properties of a semi-algebraic map being injective, surjective and having a continuous inverse, into formulas $\Phi_2(\bar{Y})$, $\Phi_3(\bar{Y})$, $\Phi_4(\bar{Y})$, respectively. Finally, to ensure that $\Gamma_f(b)$ is the graph of a map from $\mathrm{Ext}(|K|, \mathrm{R})$ to $\mathrm{Ext}(|L|, \mathrm{R})$, we recall that $\mathrm{Ext}(|K|, \mathrm{R})$ is defined by inequalities with rational coefficients and we can clearly write a formula $\Phi_5(\bar{Y})$ having the required property.

Now, take $\Phi_f = \Phi_1 \wedge \Phi_2 \wedge \Phi_3 \wedge \Phi_4 \wedge \Phi_5$. Since we know that $(a_1, \ldots, a_N) \in R^N$ satisfies Φ_f, and thus $\exists Y_1, \ldots, Y_N \; \Phi_f(Y_1, \ldots, Y_N)$ is true in R by the Tarski-Seidenberg principle (see Theorem 2.80), it is also true over $\mathbb{R}_{\mathrm{alg}}$. Hence, there exists $(b_1, \ldots, b_N) \in \mathbb{R}_{\mathrm{alg}}^N$ that satisfies Φ. By substituting $(b_1, \ldots, b_N)$ for $(a_1, \ldots, a_N)$ in the description of f, we obtain a description of a semi-algebraic homeomorphism

$$g: |K| = \mathrm{Ext}(|K|, \mathbb{R}_{\mathrm{alg}}) \to |L| = \mathrm{Ext}(|L|, \mathbb{R}_{\mathrm{alg}}). \qquad \square$$

Proof of Theorem 6.31: Let f: Ext($|K|$, R) $\to$ Ext($|L|$, R) be a semi-algebraic homeomorphism. Using Lemma 6.32, there exists a semi-algebraic homeomorphism g: $|K|$ = Ext($|K|$, $\mathbb{R}$) $\to$ $|L|$ = Ext($|L|$, $\mathbb{R}$). Hence, $\mathrm{H}_\star(K)$ and $\mathrm{H}_\star(L)$ are isomorphic using Theorem 6.24. $\square$

6.2 Simplicial Homology of Closed and Bounded Semi-algebraic Sets

6.2.1 Definitions and First Properties

We first define the simplicial homology groups of a closed and bounded semi-algebraic set S.

By Theorem 5.43, a closed, bounded semi-algebraic set S can be triangulated by a simplicial complex K with rational coordinates. Choose a semi-algebraic triangulation $f: |K| \to S$. **The homology groups $\mathbf{H}_p(S)$ are $\mathbf{H}_p(K)$**, $p \geq 0$. We denote by $\mathrm{H}_\star(S)$ the chain complex ($\mathrm{H}_p(S)$, 0) and call it the **homology of** S.

That the homology $\mathrm{H}_\star(S)$ does not depend on a particular triangulation up to isomorphism follows from the results of Section 6.1. Given any two triangulations, $f: |K| \to S, g: |L| \to S$, there exists a semi-algebraic homeomorphism, $\phi = g^{-1} f: |K| \to |L|$, and hence, using Theorem 6.31, $\mathrm{H}_\star(K)$ and $\mathrm{H}_\star(L)$ are isomorphic.

Note that two semi-algebraically homeomorphic closed and bounded semi-algebraic sets have isomorphic homology groups. Note too that the homology groups of S and those of its extension to a bigger real closed field are also isomorphic.

The homology groups of S are all finite dimensional vector spaces over $\mathbb{Q}$ (see Definition 6.2). The dimension of $\mathrm{H}_p(S)$ as a vector space over $\mathbb{Q}$ is called the ***p*-th Betti number** of S and denoted $\mathrm{b}_p(S)$.

$$\mathrm{b}(S) = \sum_i \mathrm{b}_i(S)$$

the sum of the Betti numbers of S. The **Euler-Poincaré characteristic** of S is

$$\chi(S) = \sum_i (-1)^i \mathrm{b}_i(S).$$

Note that $\chi(\emptyset) = 0$.

Using Proposition 6.3 and Theorem 5.43, we have the following result.

Proposition 6.33. *Let $S \subset \mathrm{R}^k$ be a closed and bounded semi-algebraic set, K be a simplicial complex in R^k and $h: |K| \to S$ be a semi-algebraic homeomorphism. Let $n_i(K)$ be the number of simplexes of dimension i of K. Then*

$$\chi(S) = \sum_i (-1)^i n_i(K).$$

In particular the Euler-Poincaré characteristic of a finite set of points is the cardinality of this set.

Proposition 6.34. *The number of connected components of a non-empty, closed, and bounded semi-algebraic set S is $\mathrm{b}_0(S)$.*

Proof: Let $f: |K| \to S$ be a triangulation of S. Hence,

$$\mathrm{H}_0(S) \cong \mathrm{H}_0(K).$$

Now apply Proposition 6.5. □

We now use Theorem 6.12 to relate the homology groups of the union and intersection of two closed and bounded semi-algebraic sets.

Theorem 6.35. [Semi-algebraic Mayer-Vietoris] *Let S_1, S_2 be two closed and bounded semi-algebraic sets. Then there is an exact sequence*

$$\cdots \to \mathrm{H}_p(S_1 \cap S_2) \to \mathrm{H}_p(S_1) \oplus \mathrm{H}_p(S_2) \to \mathrm{H}_p(S_1 \cup S_2) \to \mathrm{H}_{p-1}(S_1 \cap S_2) \to \cdots$$

Proof: We first obtain a triangulation of $S_1 \cup S_2$ that is simultaneously a triangulation of S_1, S_2, and $S_1 \cap S_2$ using Theorem 5.43 We then apply Theorem 6.12. □

From the exactness of the Mayer-Vietoris sequence, we have the following corollary.

Corollary 6.36. *Let S_1, S_2 be two closed and bounded semi-algebraic sets. Then,*

$$\begin{aligned}
\mathrm{b}_i(S_1) + \mathrm{b}_i(S_2) &\leq \mathrm{b}_i(S_1 \cup S_2) + \mathrm{b}_i(S_1 \cap S_2), \\
\mathrm{b}_i(S_1 \cap S_2) &\leq \mathrm{b}_i(S_1) + \mathrm{b}_i(S_2) + \mathrm{b}_{i+1}(S_1 \cup S_2), \\
\mathrm{b}_i(S_1 \cup S_2) &\leq \mathrm{b}_i(S_1) + \mathrm{b}_i(S_2) + \mathrm{b}_{i-1}(S_1 \cap S_2), \\
\chi(S_1 \cup S_2) &= \chi(S_1) + \chi(S_2) - \chi(S_1 \cap S_2).
\end{aligned}$$

Proof: Follows directly from Theorem 6.35. □

The Mayer-Vietoris sequence provides an easy way to compute the homology groups of some simple sets.

Proposition 6.37. *Consider the $(k-1)$-dimensional unit sphere $S^{k-1} \subset R^k$ for $k > 1$. If $k \geq 0$,*

$$\begin{aligned} \mathrm{H}_0(B_k) &\cong \mathbb{Q}, \\ \mathrm{H}_i(B_k) &\cong 0, \quad i > 0. \end{aligned}$$

If $k > 1$,

$$\begin{aligned} \mathrm{H}_0(S^{k-1}) &\cong \mathbb{Q}, \\ \mathrm{H}_i(S^{k-1}) &\cong 0, \quad 0 < i < k-1, \\ \mathrm{H}_{k-1}(S^{k-1}) &\cong \mathbb{Q}, \\ \mathrm{H}_i(S^{k-1}) &\cong 0, \quad i > k-1. \end{aligned}$$

Proof: We can decompose the unit sphere into two closed hemispheres, A, B, intersecting at the equator.

Each of the sets A, B is homeomorphic to the standard $(k-1)$-dimensional simplex, and $A \cap B$ is homeomorphic to the $(k-2)$-dimensional sphere S^{k-2}.

If $k = 1$, it is clear that $\mathrm{H}_0(S^1) \cong \mathrm{H}_1(S^1) \cong \mathbb{Q}$.

For $k \geq 2$, the statement is proved by induction on k.

Assume that the result holds for spheres of dimensions less than $k - 1 \geq 2$. The Mayer-Vietoris sequence for homology (Theorem 6.35) gives the exact sequence

$$\cdots \to \mathrm{H}_p(A \cap B) \to \mathrm{H}_p(A) \oplus \mathrm{H}_p(B) \to \mathrm{H}_p(A \cup B) \to \mathrm{H}_{p-1}(A \cap B) \to \cdots$$

Here, the homology groups of A and B are isomorphic to those of a $(k-1)$-dimensional closed ball, and thus $\mathrm{H}_0(A) \cong \mathrm{H}_0(B) \cong \mathbb{Q}$ and $\mathrm{H}_p(A) \cong \mathrm{H}_p(B) \cong 0$ for all $p > 0$. Moreover, the homology groups of $A \cap B$ are isomorphic to those of a $(k-2)$-dimensional sphere, and thus $\mathrm{H}_0(A \cap B) \cong \mathrm{H}_{k-2}(A \cap B) \cong \mathbb{Q}$ and $\mathrm{H}_p(A \cap B) \cong 0$, for $p \neq k-2, p \neq 0$. It now follows from the exactness of the above sequence that $\mathrm{H}_0(A \cup B) \cong \mathrm{H}_{k-1}(A \cup B) \cong \mathbb{Q}$, and $\mathrm{H}_p(A \cup B) \cong 0$, for $p \neq k-1, p \neq 0$

To see this, observe that the exactness of

$$\mathrm{H}_{k-1}(A) \oplus \mathrm{H}_{k-1}(B) \to \mathrm{H}_{k-1}(A \cup B) \to \mathrm{H}_{k-2}(A \cap B) \to \mathrm{H}_{k-2}(A) \oplus \mathrm{H}_{k-2}(B)$$

is equivalent to the following sequence being exact:

$$0 \to \mathrm{H}_{k-1}(A \cup B) \to \mathbb{Q} \to 0,$$

and this implies that the homomorphism $\mathrm{H}_{k-1}(A \cup B) \to \mathbb{Q}$ is an isomorphism. □

6.2.2 Homotopy

Let X, Y be two topological spaces. Two continuous functions f, g: $X \to Y$ are **homotopic** if there is a continuous function F: $X \times [0, 1] \to Y$ such that $F(x, 0) = f(x)$ and $F(x, 1) = g(x)$ for all $x \in X$. Clearly, homotopy is an equivalence relation among continuous maps from X to Y. It is denoted by $f \sim g$.

The sets X, Y are **homotopy equivalent** if there exist continuous functions $f: X \to Y$, $g: Y \to X$ such that $g \circ f \sim \mathrm{Id}_X$, $f \circ g \sim \mathrm{Id}_Y$. If two sets are homotopy equivalent, we also say that they have the same homotopy type.

Let X be a topological space and Y a closed subset of X. A **deformation retraction** from X to Y is a continuous function h: $X \times [0, 1] \to X$ such that $h(-,0) = \mathrm{Id}_X$ and such that $h(-,1)$ has its values in Y and such that for every $t \in [0,1]$ and every x in Y, $h(x,t) = x$. If there is a deformation retraction from X to Y, then X and Y are clearly homotopy equivalent.

Theorem 6.38. *Let K, L be simplicial complexes over $\mathbb{R}$ and f, g continuous homotopic maps from $|K|$ to $|L|$. Then*

$$\mathrm{H}_\star(f) = \mathrm{H}_\star(g): \mathrm{H}_\star(K) \to \mathrm{H}_\star(L).$$

The proposition follows directly from the following two lemmas.

Lemma 6.39. *Let K, L be simplicial complexes over $\mathbb{R}$ and δ the Lebesgue number of the cover $\{\mathrm{star}(b) | b \in L_0\}$. Let f, g: $|K| \to |L|$ be two continuous maps. If $\sup_{x \in |K|} |f(x) - g(x)| < \delta/3$, then f and g have a common simplicial approximation.*

Proof: For $b \in L_0$ let $B_b = \{x \in |L| \; |\mathrm{dist}(x, |L| - \mathrm{star}(b)) > \delta/3\}$. We first claim that $b \in B_b$ and hence, the family of sets $\{B_b | b \in L_0\}$ is an open covering of L. Consider the set $|L| \cap B(b, 2\,\delta/3)$. If $|L| \cap B(b, 2\,\delta/3) \subset \mathrm{star}(b)$, then clearly, $b \in B_b$. Otherwise, since $\mathrm{diam}(|L| \cap B(b, 2\,\delta/3)) < \delta$, there must exists a $b' \in L_0$ such that $|L| \cap B(b, 2\,\delta/3) \subset \mathrm{star}(b')$. But, then $b \in \mathrm{star}(b')$ implying that $b = b'$, which is a contradiction.

Let ϵ be the Lebesgue number of the open cover of $|K|$ given by $\{f^{-1}(B_b) | b \in L_0\}$. Then, there is an integer n such that $\mu(K^{(n)}) < \epsilon/2$. To every vertex $a \in K^{(n)}$, there is a vertex $b \in L_0$ such that $\mathrm{star}(a) \subset f^{-1}(B_b)$, and this induces a simplicial map, ϕ: $K^{(n)} \to L$, sending a to b. We now claim that ϕ is a simplicial approximation to both f and g.

Let $x \in |K|$ such that $x \in s^\circ$ for a simplex s in $K^{(n)}$, and let $f(x) \in t_1^\circ$ and $g(x) \in t_2^\circ$ for simplices t_1, $t_2 \in L$. Let $a \in K_0^{(n)}$ be a vertex of s, and let $b = \phi(a)$. From the definition of ϕ, we have that $f(\mathrm{star}(a)) \subset B_b \subset \mathrm{star}(b)$, and since $x \in \mathrm{star}(a)$, $f(x) \in \mathrm{star}(b)$. Moreover, since $|f(x) - g(x)| < \delta/3$ for all $x \in |K|$, $\mathrm{dist}(f(\mathrm{star}(a)), g(\mathrm{star}(a))) < \delta/3$, and hence $g(\mathrm{star}(a)) \subset \mathrm{star}(b)$ and $g(x) \in \mathrm{star}(b)$.

Thus, $f(x) \in \cap_{a \in s} \mathrm{star}(\phi(a))$, and hence each $\phi(a)$, $a \in s$ is a vertex of the simplex t_1. Moreover, since $\phi(x)$ lies in the simplex spanned by $\{\phi(a) | a \in s\}$, it is clear that $\phi(x) \in t_1$. Similarly, $\phi(x) \in t_2$, and hence ϕ is simultaneously a simplicial approximation to both f and g. □

Lemma 6.40. *Let K, L be simplicial complexes over $\mathbb{R}$ and suppose that f, g are homotopic maps from $|K| \to |L|$. Then, there is an integer n and simplicial maps ϕ, ψ: $K^{(n)} \to L$ that are in the same contiguity class and such that ϕ (resp. ψ) is a simplicial approximation of f (resp. g).*

Proof: Since $f \sim g$, there is a continuous map $F\colon |K| \times [0, 1] \to |L|$ such that $F(x, 0) = f(x)$ and $F(x, 1) = g(x)$. To a Lebesgue number δ of the cover $\mathrm{star}(b)$, $b \in L_0$, there exists a number ϵ such that $|t - t'| < \epsilon$ implies $\sup |F(x,t) - F(x,t')| < \delta/3$. This follows from the uniform continuity of F since $K \times [0, 1]$ is compact.

We now choose a sequence $t_0 = 0 < t_1 < t_2 < \cdots < t_n = 1$ such that $|t_{i+1} - t_i| < \epsilon$ and let $F(x, t_i) = f_i(x)$. By the previous lemma, f_i and f_{i+1} have a common simplicial approximation $\psi_i\colon K^{(n_i)} \to L$. Let $n = \max_i n_i$, and let $\phi_i\colon K^{(n)} \to L$ be the simplicial map induced by ψ_i. For each i, $0 \leq i < n$, ϕ_i and ϕ_{i+1} are contiguous and are simplicial approximations of f_i and f_{i+1} respectively. Moreover, ϕ_0 is a simplicial approximation of f and ϕ_n a simplicial approximation of g. Hence, they are in the same contiguity class. □

We will now transfer the previous results to semi-algebraic sets and maps over a general real closed field R. The method of transferring the results parallels those used at the end of Section 6.1.

Let X, Y be two closed and bounded semi-algebraic sets. Two semi-algebraic continuous functions $f, g\colon X \to Y$ are **semi-algebraically homotopic**, $f \sim_{sa} g$, if there is a continuous semi-algebraic function $F\colon X \times [0,1] \to Y$ such that $F(x,0) = f(x)$ and $F(x,1) = g(x)$ for all $x \in X$. Clearly, semi-algebraic homotopy is an equivalence relation among semi-algebraic continuous maps from X to Y.

The sets X, Y are **semi-algebraically homotopy equivalent** if there exist semi-algebraic continuous functions $f\colon X \to Y$, $g\colon Y \to X$ such that $g \circ f \sim_{sa} \mathrm{Id}_X$, $f \circ g \sim_{sa} \mathrm{Id}_Y$.

Let X be a closed and bounded semi-algebraic set and Y a closed semi-algebraic subset of X. A **semi-algebraic deformation retraction** from X to Y is a continuous semi algebraic function $h\colon X \times [0, 1] \to X$ such that $h(-,0) = \mathrm{Id}_X$ and such that $h(-,1)$ has its values in Y and such that for every $t \in [0,1]$ and every x in Y, $h(x,t) = x$. If there is a semi-algebraic deformation retraction from X to Y, then X and Y are clearly semi-algebraically homotopy equivalent.

Using the transfer principle and the same technique used in the proof of Theorem 6.31, it is possible to prove,

Proposition 6.41. *Let K, L be simplicial complexes with rational vertices, and let $f \sim_{sa} g$ be semi-algebraic continuous semi-algebraically homotopic maps from $\mathrm{Ext}(|K|, \mathrm{R})$ to $\mathrm{Ext}(|L|, \mathrm{R})$. Then*

$$\mathrm{H}_\star(f) = \mathrm{H}_\star(g)\colon \mathrm{H}_\star(K) \to \mathrm{H}_\star(L).$$

Finally, the following proposition holds in any real closed field.

Theorem 6.42. *Let R be a real closed field. Let X, Y be two closed, bounded semi-algebraic sets of R^k that are semi-algebraically homotopy equivalent. Then, $\mathrm{H}_\star(X) \cong \mathrm{H}_\star(Y)$.*

Proof: We first choose triangulations. Let $\phi\colon |K| \to X$ and $\psi\colon |L| \to Y$ be semi-algebraic triangulations of X and Y, respectively. Moreover, since X and Y are semi-algebraically homotopy equivalent, there exist semi-algebraic continuous functions $f\colon X \to Y$, $g\colon Y \to X$ such that $g \circ f \sim_{sa} \mathrm{Id}_X$, $f \circ g \sim_{sa} \mathrm{Id}_Y$.

Then, $f_1 = \psi^{-1} \circ f \circ \phi\colon |K| \to |L|$ and $g_1 = \phi^{-1} \circ g \circ \psi\colon |L| \to |K|$ give a semi-algebraic homotopy equivalence between $|K|$ and $L|$. These are defined over R. However, using the same method as in the proof of Lemma 6.32, we can show that in this case there exists $f_1'\colon |K| \to |L|$ and $g_1'\colon |L| \to |K|$ defined over $\mathbb{R}$ giving a homotopy equivalence between $|K|$ and $|L|$.

Now applying Proposition 6.41 and Proposition 6.30 we get

$$\mathrm{H}_\star(f_1' \circ g_1') = \mathrm{H}_\star(f_1') \circ \mathrm{H}_\star(g_1') = \mathrm{H}_\star(\mathrm{Id}_K) = \mathrm{Id}_{\mathrm{H}_\star(K)}$$

$$\mathrm{H}_\star(g_1' \circ f_1') = \mathrm{H}_\star(g_1') \circ \mathrm{H}_\star(f_1') = \mathrm{H}_\star(\mathrm{Id}_L) = \mathrm{Id}_{\mathrm{H}_\star(L)}.$$

This proves that $\mathrm{H}_\star(X) = \mathrm{H}_\star(K) \cong \mathrm{H}_\star(L) = \mathrm{H}_\star(Y)$. □

6.3 Homology of Certain Locally Closed Semi-Algebraic Sets

In Section 6.2 we have defined homology groups of closed and bounded semi-algebraic sets. Now, we consider more general semi-algebraic sets - namely, certain locally closed semi-algebraic sets. We first define homology groups for closed semi-algebraic sets, as well as for semi-algebraic sets which are realizations of sign conditions. These homology groups are homotopy invariant, but do not satisfy an addivity property useful in certain applications. In order to have a homology theory with the addivity property, we introduce the Borel-Moore homology groups and prove their basic properties.

6.3.1 Homology of Closed Semi-algebraic Sets and of Sign Conditions

We now define the homology groups for closed (but not necessarily bounded) semi-algebraic sets and for semi-algebraic sets defined by a single sign condition.

Let $S \subset \mathrm{R}^k$ be any closed semi-algebraic set. By Proposition 5.49 (conic structure at infinity), there exists $r \in \mathrm{R}$, $r > 0$, such that, for every $r' \geq r$, there exists a a semi-algebraic deformation retraction from S to $S_{r'} = S \cap B_k(0, r')$, and there exists a a semi-algebraic deformation retraction from S_r to $S_{r'}$. Thus the sets S_r and $S_{r'}$ are homotopy equivalent. So, by Theorem 6.42, $\mathrm{H}(S_r) = \mathrm{H}(S_{r'})$.

Notation 6.43. [Homology] We define $\mathrm{H}_\star(S) = \mathrm{H}_\star(S_r)$. □

We have the following useful result.

Proposition 6.44. *Let S_1, S_2 be two closed semi-algebraic sets. Then,*

$$\begin{aligned} \mathrm{b}_i(S_1) + \mathrm{b}_i(S_2) &\leq \mathrm{b}_i(S_1 \cup S_2) + \mathrm{b}_i(S_1 \cap S_2), \\ \mathrm{b}_i(S_1 \cap S_2) &\leq \mathrm{b}_i(S_1) + \mathrm{b}_i(S_2) + \mathrm{b}_{i+1}(S_1 \cup S_2), \\ \mathrm{b}_i(S_1 \cup S_2) &\leq \mathrm{b}_i(S_1) + \mathrm{b}_i(S_2) + \mathrm{b}_{i-1}(S_1 \cap S_2). \end{aligned}$$

Proof: Follows directly from Corollary 6.36 and the definition of the homology groups of a closed semi-algebraic set. □

We also define homology groups for semi-algebraic sets defined by a single sign condition.

Let $\mathcal{P} = \{P_1, \ldots, P_s\} \subset \mathrm{R}[X_1, \ldots, X_k]$ be a set of s polynomials, and let $\sigma \in \{0, 1, -1\}^{\mathcal{P}}$ be a realizable sign condition on $\mathcal{P}$. Without loss of generality, suppose

$$\begin{aligned} \sigma(P_i) &= 0 \ \text{ if } i = 1, \ldots, j, \\ \sigma(P_i) &= 1 \ \text{ if } i = j+1, \ldots, \ell, \\ \sigma(P_i) &= -1 \ \text{ if } i = \ell+1, \ldots, s. \end{aligned}$$

We denote by $\mathrm{Reali}(\sigma) \subset \mathrm{R}^k$ the realization of σ. Let $\delta > 0$ be a variable.

Consider the field $\mathrm{R}\langle\delta\rangle$ of algebraic Puiseux series in δ, in which δ is an infinitesimal. Let $\overline{\mathrm{Reali}}(\sigma) \subset \mathrm{R}\langle\delta\rangle^k$ be defined by

$$\sum_{1 \leq i \leq k} X_i^2 \leq 1/\delta \wedge P_1 = \cdots = P_j = 0$$

$$\wedge P_{j+1} \geq \delta \wedge \ldots \wedge P_\ell \geq \delta \wedge P_{\ell+1} \leq -\delta \wedge \ldots \wedge P_s \leq -\delta.$$

Proposition 6.45. *The set $\overline{\mathrm{Reali}}(\sigma)$ is a semi-algebraic deformation retract of the extension of $\mathrm{Reali}(\sigma)$ to $\mathrm{R}\langle\delta\rangle$.*

Proof: Consider the continuous semi-algebraic function f defined by

$$f(x) = \inf\left(1, \frac{1}{X_1^2 + \cdots + X_k^2}, \inf_{j+1 \leq i \leq s} (|P_i(x)|\right)$$

and note that

$$\overline{\mathrm{Reali}}(\sigma) = \{x \in \mathrm{Ext}(\mathrm{Reali}(\sigma), \mathrm{R}\langle\delta\rangle) \mid f(x) \geq \delta\}.$$

By Theorem 5.46 (Hardt's triviality), there exists $t_0 \in \mathrm{R}$ such that

$$\begin{gathered} \{x \in \mathrm{Reali}(\sigma) \mid t_0 \geq f(x) > 0\} \\ (\text{resp. } \{x \in \mathrm{Ext}(\mathrm{Reali}(\sigma), \mathrm{R}\langle\delta\rangle) \mid t_0 \geq f(x) \geq \delta\}) \end{gathered}$$

is homeomorphic to

$$\begin{gathered} \{x \in \mathrm{Reali}(\sigma) \mid f(x) = t_0\} \times (0, t_0] \\ (\text{resp. } \{x \in \mathrm{Ext}(\mathrm{Reali}(\sigma), \mathrm{R}\langle\delta\rangle) \mid f(x) = t_0\} \times [\delta, t_0]). \end{gathered}$$

Moreover, the corresponding homeomorphisms ϕ and ψ can be chosen such that $\phi|_{\{x\in\mathrm{Reali}(\sigma)|f(x)=t_0\}}$ and $\psi|_{\{x\in\mathrm{Ext}(\mathrm{Reali}(\sigma,\mathrm{R}\langle\delta\rangle))|f(x)=t_0\}}$ are identities. □

Notation 6.46. [Homology of a sign condition] We define

$$\mathrm{H}_\star(\mathrm{Reali}(\sigma)) = \mathrm{H}_\star(\overline{\mathrm{Reali}}(\sigma)).$$ □

Proposition 6.47. *Suppose that* $\mathrm{Reali}(\sigma)$ *and* $\mathrm{Reali}(\tau)$ *are semi-algebraically homotopy equivalent, then*

$$\mathrm{H}_\star(\mathrm{Reali}(\sigma)) = \mathrm{H}_\star(\mathrm{Reali}(\tau)).$$

Proof: By Proposition 6.45, $\overline{\mathrm{Reali}}(\sigma)$ and $\overline{\mathrm{Reali}}(\tau)$ are homotopy equivalent. Now apply Theorem 6.42. □

Exercise 6.1. Consider the unit disk minus a point which is the set D defined by

$$X^2+Y^2-1<0 \wedge X^2+Y^2>0.$$

Prove that

$$\begin{aligned} \mathrm{H}_0(D) &= \mathbb{Q}, \\ \mathrm{H}_1(D) &= \mathbb{Q}, \\ \mathrm{H}_2(D) &= 0. \end{aligned}$$

Remark 6.48. The homology groups we just defined agree with the singular homology groups [150] in the case when $\mathrm{R} = \mathbb{R}$: it is a consequence of Proposition 6.45 and the fact that the singular homology groups are homotopy invariants [150]. □

6.3.2 Homology of a Pair

We now define the simplicial homology groups of pairs of closed and bounded semi-algebraic sets.

Let K be a simplicial complex and A a sub-complex of K. Then, there is a natural inclusion homomorphism, $i\colon \mathrm{C}_p(A) \to \mathrm{C}_p(K)$, between the corresponding chain groups. Defining the group $\mathrm{C}_p(K, A) = \mathrm{C}_p(K)/i(\mathrm{C}_p(A))$, it is easy to see that the boundary maps $\partial_p\colon \mathrm{C}_p(K) \to \mathrm{C}_{p-1}(K)$ descend to maps $\partial_p\colon \mathrm{C}_p(K,A) \to \mathrm{C}_{p-1}(K,A)$, so that we have a short exact sequence of complexes,

$$0 \to \mathrm{C}_\bullet(A) \to \mathrm{C}_\bullet(K) \to \mathrm{C}_\bullet(K,A) \to 0.$$

Given a pair (K,A), where A is a sub-complex of K, the group

$$\mathrm{H}_p(K,A) = \mathrm{H}_p(\mathrm{C}_\bullet(K,A))$$

is the **p-th simplicial homology group** of the pair (K, A).

It is clear from the definition that $\mathrm{H}_p(K, A)$ is a finite dimensional $\mathbb{Q}$-vector space. The dimension of $\mathrm{H}_p(K, A)$ as a $\mathbb{Q}$-vector space is called the **p-th Betti number of the pair** (K, A) and denoted $\mathrm{b}_p(K, A)$. The **Euler-Poincaré characteristic** of the pair (K, A) is

$$\chi(K, A) = \sum_i (-1)^i \mathrm{b}_i(K, A).$$

We now define the simplicial homology groups of a pair of closed and bounded semi-algebraic sets $T \subset S \subset \mathrm{R}^k$. By Theorem 5.43, such a pair of closed, bounded semi-algebraic sets can be triangulated using a pair of simplicial complexes (K, A) with rational coordinates, where A is a sub-complex of K. The **p-th simplicial homology group** of the pair (S, T), $\mathrm{H}_p(S, T)$, is $\mathrm{H}_p(K, A)$. The dimension of $\mathrm{H}_p(S, T)$ as a $\mathbb{Q}$-vector space is called the **p-th Betti number of the pair** (S, T) and denoted $\mathrm{b}_p(S, T)$. The **Euler-Poincaré characteristic** of the pair (S, T) is

$$\chi(S, T) = \sum_i (-1)^i \mathrm{b}_i(S, T).$$

Exercise 6.2. Consider the pair (S, T) where S is the closed unit disk defined by $X^2 + Y^2 - 1 \leqslant 0$ and T is the union of the origin and the circle of radius one defined by $X^2 + Y^2 - 1 = 0 \vee X^2 + Y^2 = 0$.

Prove that

$$\begin{aligned} \mathrm{H}_0(S, T) &= \mathbb{Q}, \\ \mathrm{H}_1(S, T) &= \mathbb{Q}, \\ \mathrm{H}_2(S, T) &= \mathbb{Q}. \end{aligned}$$

Proposition 6.49. *Let $T \subset S \subset \mathrm{R}^k$ be a pair of closed and bounded semi-algebraic set, (K, A) be a pair of simplicial complexes in R^k, with A a sub-complex of K and let $h: |K| \to S$ be a semi-algebraic homeomorphism such that the image of $|K|$ is T. Then*

$$\begin{aligned} \chi(S, T) &= \chi(K, A) \\ &= \chi(K) - \chi(A) \\ &= \chi(S) - \chi(T). \end{aligned}$$

Proof: From the short exact sequence of chain complexes,

$$0 \to \mathrm{C}_\bullet(A) \to \mathrm{C}_\bullet(K) \to \mathrm{C}_\bullet(K, A) \to 0,$$

applying Lemma 6.10, we obtain the following long exact sequence of homology groups:

$$\cdots \to \mathrm{H}_p(A) \to \mathrm{H}_p(K) \to \mathrm{H}_p(K, A) \to \mathrm{H}_{p-1}(A) \to \mathrm{H}_{p-1}(K) \to \cdots$$

and

$$\cdots \to \mathrm{H}_p(T) \to \mathrm{H}_p(S) \to \mathrm{H}_p(S,T) \to \mathrm{H}_{p-1}(S) \to \mathrm{H}_{p-1}(T) \to \cdots \qquad (6.6)$$

The claim follows. □

Proposition 6.50. *Let $T \subset S \subset \mathrm{R}^k$ be a pair of closed and bounded semi-algebraic set, (K, A) be a pair of simplicial complexes in R^k, with A a sub-complex of K and let $h: |K| \to S$ be a semi-algebraic homeomorphism such that the image of $|K|$ is T. Let $n_i(K)$ be the number of simplexes of dimension i of K, and let $m_i(A)$ be the number of simplexes of dimension i of A. Then*

$$\begin{aligned} \chi(S,T) &= \chi(K,A) \\ &= \sum_i (-1)^i n_i(K) - \sum_i (-1)^i m_i(A). \end{aligned}$$

Proof: By Proposition 6.49, $\chi(K,A) = \chi(K) - \chi(A)$. The proposition is now a consequence of of Proposition 6.3. □

Let (X, A), (Y, B) be two pairs of semi-algebraic sets. The pairs are (X, A), (Y, B) are **semi-algebraically homotopy equivalent** if there exist continuous semi-algebraic functions $f: X \to Y$, $g: Y \to X$ such that, $\mathrm{Im}(f|_A) \subset B$, $\mathrm{Im}(g|_B) \subset A$ and such that $g \circ f \sim \mathrm{Id}_X$, $g|_B \circ f|_A \sim \mathrm{Id}_A$, $f \circ g \sim \mathrm{Id}_Y$, and $f|_A \circ g|_B \sim \mathrm{Id}_B$. If two pairs are semi-algebraically homotopy equivalent, we also say that they have the same homotopy type.

We have the following proposition which is a generalization of Proposition 6.35 to pairs of closed, bounded semi-algebraic sets.

Proposition 6.51. *Let R be a real closed field. Let (X, A), (Y, B) be two pairs of closed, bounded semi-algebraic sets of R^k that are semi-algebraically homotopy equivalent. Then, $\mathrm{H}_*(X,A) \cong \mathrm{H}_*(Y,B)$.*

Proof: Since (X, A) and (Y, B) are semi-algebraically homotopy equivalent, there exist continuous semi-algebraic functions $f: X \to Y$, $g: Y \to X$ such that, $\mathrm{Im}(f|_A) \subset B$, $\mathrm{Im}(g|_B) \subset A$ and such that $g \circ f \sim \mathrm{Id}_X$, $g|_B \circ f|_A \sim \mathrm{Id}_A$, $f \circ g \sim \mathrm{Id}_Y$, and $f|_A \circ g_B \sim \mathrm{Id}_B$.

After choosing triangulations of X and Y (respecting the subsets A and B, respectively) and using the same construction as in the proof of Proposition 6.35, we see that f induces isomorphisms, $\mathrm{H}_*(f)$: $\mathrm{H}_*(X) \to \mathrm{H}_*(Y)$, $\mathrm{H}_*(f)$: $\mathrm{H}_*(A) \to \mathrm{H}_*(B)$, as well an homomorphism $\mathrm{H}_*(f): \mathrm{H}_*(X,A) \to \mathrm{H}_*(Y,B)$ such that the following diagram commutes.

$$\begin{array}{ccccccccc} \mathrm{H}_i(A) & \longrightarrow & \mathrm{H}_i(X) & \longrightarrow & \mathrm{H}_i(X,A) & \longrightarrow & \mathrm{H}_{i-1}(A) & \longrightarrow & \mathrm{H}_{i-1}(X) \\ \downarrow \mathrm{H}_i(f) & & \downarrow \mathrm{H}_i(f) & & \downarrow \mathrm{H}_i(f) & & \downarrow \mathrm{H}_{i-1}(f) & & \downarrow \mathrm{H}_{i-1}(f) \\ \mathrm{H}_i(B) & \longrightarrow & \mathrm{H}_i(Y) & \longrightarrow & \mathrm{H}_i(Y,B) & \longrightarrow & \mathrm{H}_{i-1}(B) & \longrightarrow & \mathrm{H}_{i-1}(Y) \end{array}$$

The rows correspond to the long exact sequence of the pairs (X,A) and (Y,B) (see (6.6)) and the vertical homorphisms are those induced by f.

Now applying Lemma 6.11 (Five Lemma) we see that $\mathrm{H}_*(f): \mathrm{H}_*(X,A) \to \mathrm{H}_*(Y,B)$ is also an isomorphism. □

6.3.3 Borel-Moore Homology

In this section we will consider **basic locally closed semi-algebraic sets** which are, by definition, intersections of closed semi-algebraic sets with basic open ones. Let $S \subset \mathrm{R}^k$ be a basic locally closed semi-algebraic set and let $S_r = S \cap B_k(0,r)$. The **p-th Borel-Moore homology group** of S, denoted by $\mathrm{H}_p^{\mathrm{BM}}(S)$, is defined to be the p-th simplicial homology group of the pair $(\overline{S_r}, \overline{S_r} \setminus S_r)$ for large enough $r > 0$. Its dimension is the p-th Borel-Moore Betti number and is denoted by $\mathrm{b}_p^{\mathrm{BM}}(S)$. We denote by $\mathrm{H}_\star^{\mathrm{BM}}(S)$ the chain complex $(\mathrm{H}_p^{\mathrm{BM}}(S), 0)$ and call it the **Borel-Moore homology of S**.

Note that, for a basic locally closed semi-algebraic set S, both $\overline{S_r}$ and $\overline{S_r} \setminus S_r$ are closed and bounded and hence $\mathrm{H}_i(\overline{S_r}, \overline{S_r} \setminus S_r)$ is well defined. It follows clearly from the definition that for a closed and bounded semi-algebraic set, the Borel-Moore homology groups coincide with the simplicial homology groups.

Exercise 6.3. Let D be the plane minus the origin.
Prove that

$$\begin{aligned} \mathrm{H}_0^{\mathrm{BM}}(D) &= \mathbb{Q}, \\ \mathrm{H}_1^{\mathrm{BM}}(D) &= \mathbb{Q}, \\ \mathrm{H}_2^{\mathrm{BM}}(D) &= \mathbb{Q}. \end{aligned}$$

We will show that the Borel-Moore homology is invariant under semi-algebraic homeomorphisms by proving that the Borel-Moore homology coincides with the simplicial homology of the Alexandrov compactification which we introduce below.

Suppose that $X \subset \mathrm{R}^k$ is a basic locally closed semi-algebraic set, which is not simultaneously closed and bounded, and that X is the intersection of a closed semi-algebraic set, V, with the open semi-algebraic set defined by strict inequalities, $P_1 > 0, \ldots, P_m > 0$. We will assume that $X \neq \mathrm{R}^k$. Otherwise, we embed X in R^{k+1}.

We now define the Alexandrov compactification of X, denoted by $\dot{X}$, having the following properties:

- $\dot{X}$ is closed and bounded,
- there exists a semi-algebraic continuous map, $\eta: X \to \dot{X}$, which is a homeomorphism onto its image,
- $\dot{X} \setminus X$ is a single point.

Let $T_1, \ldots, T_m$ be new variables and π: $\mathrm{R}^{k+m} \to \mathrm{R}^k$ be the projection map forgetting the new coordinates. Consider the closed semi-algebraic set $Y \subset \mathrm{R}^{k+m}$ defined as the intersection of $\pi^{-1}(V)$ with the set defined by

$$T_1^2 P_1 - 1 = \cdots = T_m^2 P_m - 1 = 0, \ T_1 \geq 0, \ldots, T_m \geq 0.$$

Clearly, Y is homeomorphic to X. After making an affine change of coordinates we can assume that Y does not contain the origin.

Let ϕ: $\mathrm{R}^{k+m} \setminus \{0\} \to \mathrm{R}^{k+m}$ be the continuous map defined by

$$\phi(x) = \frac{x}{|x|^2}.$$

We define the **Alexandroff compactification** of X by

$$\dot{X} = \phi(Y) \cup \{0\},$$

and

$$\eta = \phi \circ \pi|_Y^{-1}.$$

In case X is closed and bounded, we define $\dot{X} = X$.

We now prove,

Lemma 6.52.

a) $\eta(X)$ *is semi-algebraically homeomorphic to* X*,*

b) $\dot{X}$ *is a closed and bounded semi-algebraic set.*

Proof: We follow the notations introduced in the definition of $\dot{X}$. It is easy to verify that ϕ is a homeomorphism and since and $\pi|_Y^{-1}$ is also a homeomorphism, it follows that η is a homeomorphism onto its image.

We now prove that $\dot{X}$ is closed and bounded. It is clear from the definition of Y, that Y is a closed and unbounded subset of R^{k+m}. Since $0 \notin Y$, $\phi(Y)$ is bounded. Moreover, if $x \in \overline{\phi(Y)}$, but $x \notin \phi(Y)$, then $x = 0$. Otherwise, if $x \neq 0$, then $\phi^{-1}(x)$ must belong to the closure of Y and hence to Y since Y is closed. But this would imply that $x \in \phi(Y)$. This shows that $\dot{X} = \phi(Y) \cup \{0\}$ is closed. □

We call $\dot{X}$ to be the Alexandrov compactification of X. We now show that the Alexandrov compactification is unique up to semi-algebraic homeomorphisms.

Theorem 6.53. *Suppose that* X *is as above and* Z *is a closed and bounded semi-algebraic set such that,*

a) There exists a semi-algebraic continuous map, ϕ: $X \to Z$*, which gives a homeomorphism between* X *and* $\phi(X)$*,*

b) $Z \setminus \phi(X)$ *is a single point.*

Then, Z is semi-algebraically homeomorphic to $\dot{X}$.

Proof: Let $Z = \phi(X) \cup \{z\}$. We have that $\dot{X} = \eta(X) \cup \{0\}$. Since $\eta(X)$ and $\phi(X)$ are each homeomorphic to X, there is an induced homeomorphism, ψ: $\eta(X) \to \phi(X)$. Extend ψ to $\dot{X}$ by defining $\psi(0) = z$. It is easy to check that this extension is continuous and thus gives a homeomorphism between $\dot{X}$ and Z. □

We finally prove that the Borel-Moore homology groups defined above is invariant under semi-algebraic homeomorphisms.

Theorem 6.54. *Let X be a basic locally closed semi-algebraic set. For any basic locally closed semi-algebraic set Y which is semi-algebraically homeomorphic to X, we have that $\mathrm{H}_\star^{\mathrm{BM}}(X) \cong \mathrm{H}_\star^{\mathrm{BM}}(Y)$.*

Proof: If X is closed and bounded there is nothing to prove since, $\dot{X} = X$ and $\mathrm{H}_\star^{\mathrm{BM}}(X) \cong \mathrm{H}_\star(X)$ by definition.

Otherwise, let X be the intersection of a closed semi-algebraic set, V, with the open semi-algebraic set defined by strict inequalities, $P_1 > 0, \ldots, P_m > 0$. We follow the notations used in the definition of the Alexandrov compactification above, as well as those used in the definition of Borel-Moore homology groups.

For $\varepsilon, \delta > 0$ we define, $X_{\varepsilon,\delta}$ to be the intersection of $V \cap B_k(0, \frac{1}{\delta})$ with the set defined by, $P_1 > \varepsilon, \ldots, P_m > \varepsilon$.

Let $\dot{X} \subset \mathrm{R}^{k+m}$ be the Alexandrov compactification of X defined previously, and let $B_{\varepsilon,\delta} = \overline{B_k(0,\delta)} \times \overline{B_m(0,\varepsilon)} \subset \mathrm{R}^{k+m}$. It follows from Theorem 5.48 that the pair $(\dot{X}, 0)$ is homotopy equivalent to $(\dot{X}, B_{\varepsilon,\delta})$ for all $0 < \varepsilon \ll \delta \ll 1$. Moreover, the pair $(\dot{X}, B_{\varepsilon,\delta})$ is homeomorphic to the pair, $(\overline{X_{\varepsilon,\delta}}, \overline{X_{\varepsilon,\delta}} \setminus X_{\varepsilon,\delta})$. It follows again from Theorem 5.48 that the pair, $(\overline{X_{\varepsilon,\delta}}, \overline{X_{\varepsilon,\delta}} \setminus X_{\varepsilon,\delta})$, is homotopy equivalent to $(\overline{X_{0,\delta}}, \overline{X_{0,\delta}} \setminus X_{0,\delta})$. However, by definition $\mathrm{H}_\star^{\mathrm{BM}}(X) \cong \mathrm{H}_\star(\overline{X_{0,\delta}}, \overline{X_{0,\delta}} \setminus X_{0,\delta})$ and hence we have shown that

$$\mathrm{H}_\star^{\mathrm{BM}}(X) \cong \mathrm{H}_\star(\dot{X}).$$

Since by Theorem 6.53, the Alexandrov compactification is unique up to semi-algebraic homeomorphisms, and by Theorem 6.31 the simplicial homology groups are also invariant under semi-algebraic homeomorphisms, this proves the theorem. □

We now prove the additivity of Borel-Moore homology groups. More precisely, we prove,

Theorem 6.55. *Let $A \subset X \subset \mathrm{R}^k$ be closed semi-algebraic sets. Then there exists an exact sequence,*

$$\cdots \to \mathrm{H}_i^{\mathrm{BM}}(A) \to \mathrm{H}_i^{\mathrm{BM}}(X) \to \mathrm{H}_i^{\mathrm{BM}}(X \setminus A) \to \cdots$$

Proof: Let $Y = X \setminus A$. By definition,

$\mathrm{H}_\star^{\mathrm{BM}}(Y) \cong \mathrm{H}_\star(\overline{Y_r}, \overline{Y_r} \setminus Y_r)$, where $Y_r = Y \cap B_k(0, r)$, with $r > 0$ and sufficiently large.

Similarly, let $X_r = X \cap B_k(0, r)$, and $A_r = X \cap B_k(0, r)$. Notice that, since X and A are closed, $\overline{Y_r} \setminus Y_r \subset \overline{A_r}$. Consider a semi-algebraic triangulation of h: $|K| \to \overline{X_r}$ which respects the subset $\overline{A_r}$. Let $K^1 \subset K^2$ denote the sub-complexes corresponding to $\overline{Y_r} \setminus Y_r \subset \overline{A_r}$. It is now clear from definition that, $C_\bullet(K, K^1) \cong C_\bullet(K, K^2)$ and hence $\mathrm{H}_\star(K, K^1) \cong \mathrm{H}_\star(K, K^2)$. But,

$$\mathrm{H}_\star(K, K^1) \cong \mathrm{H}_\star(\overline{Y_r}, \overline{Y_r} \setminus Y_r) \cong \mathrm{H}_\star^{\mathrm{BM}}(Y)$$

and $\mathrm{H}_*(K, K^2) \cong \mathrm{H}_*(\overline{X_r}, \overline{A_r})$.

This shows that

$$\mathrm{H}_\star^{\mathrm{BM}}(X \setminus A) \cong \mathrm{H}_\star(\overline{X_r}, \overline{A_r}).$$

The long exact sequence of homology for the pair $(\overline{X_r}, \overline{A_r})$ is

$$\cdots \to \mathrm{H}_i(\overline{A_r}) \to \mathrm{H}_i(\overline{X_r}) \to \mathrm{H}_i(\overline{X_r}, \overline{A_r}) \to \cdots$$

Using the isomorphisms proved above, and the fact that X (resp. A) and $\overline{X_r}$ (resp. $\overline{A_r}$) are homeomorphic, we get an exact sequence,

$$\cdots \to \mathrm{H}_i(\overline{A}) \to \mathrm{H}_i(\overline{X}) \to \mathrm{H}_i^{\mathrm{BM}}(X \setminus A) \to \cdots$$

□

6.3.4 Euler-Poincaré Characteristic

We define the Euler-Poincaré characteristic for basic locally closed semi-algebraic sets. This definition agrees with the previously defined Euler-Poincaré characteristic for closed and bounded semi-algebraic sets and is additive. The Euler-Poincaré characteristic is a discrete topological invariant of semi-algebraic sets which generalizes the cardinality of a finite set. Hence, additivity is a very natural property to require for Euler-Poincaré characteristic.

We define the **Euler-Poincaré characteristic** of a basic locally closed semi-algebraic set S by,

$$\chi(S) = \sum_i (-1)^i \mathrm{b}_i^{\mathrm{BM}}(S),$$

where $\mathrm{b}_i^{\mathrm{BM}}(S)$ is the dimension of $\mathrm{H}_i^{\mathrm{BM}}(S)$ as a $\mathbb{Q}$-vector space. In the special case of a closed and bounded semi-algebraic set, we recover the Euler-Poincaré characteristic already defined.

The Euler-Poincaré characteristic of a basic locally closed semi-algebraic set S, is related to the Euler-Poincaré characteristic of the closed and bounded semi-algebraic sets $\overline{S_r}$ and $\overline{S_r} \setminus S_r$ for all large enough $r > 0$, by the following lemma.

Lemma 6.56.

$$\chi(S) = \chi(\overline{S_r}) - \chi(\overline{S_r} \setminus S_r),$$

where $S_r = S \cap B_k(0, r)$ *and* $r > 0$ *and sufficiently large.*

Proof: Immediate consequence of the definition and Proposition 6.49 □

Proposition 6.57. [Additivity of Euler-Poincaré characteristic] *Let* S, S_1 *and* S_2 *be basic locally closed semi-algebraic sets such that* $S_1 \cup S_2 = S$, $S_1 \cap S_2 = \emptyset$. *Then*

$$\chi(S) = \chi(S_1) + \chi(S_2).$$

Proof: This is an immediate consequence of Theorem 6.55. □

Remark 6.58. Note that the additivity property of the Euler-Poincaré characteristic would not be satisfied if we had defined the Euler-Poincare characteristic in terms of the homology groups rather than in terms of the Borel-Moor homology groups.

For instance, the Euler-Poincaré of the line would be -1, that of a point would be 1, and that of the line minus the point is 2.

Using the definition of Euler-Poincaré characteristic through Borel-Moore homology, the Euler-Poincaré of the line is 0, that of a point is 1, and that of the line minus the point is -1. □

Remark 6.59. Notice that, unlike the ordinary homology (see Proposition 6.47), the Borel-Moore homology is not invariant under semi-algebraic homotopy. For instance, a line is semi-agebraically homotopy equivalent to a point, while their Euler-Poincaré characteristics differ as seen in Remark 6.58. □

Let $S \subset \mathrm{R}^k$ be a closed semi-algebraic set. Given $P \in \mathrm{R}[X_1, \ldots, X_k]$, we denote

$$\begin{aligned} \mathrm{Reali}(P = 0, S) &= \{x \in S \mid P(x) = 0\}, \\ \mathrm{Reali}(P > 0, S) &= \{x \in S \mid P(x) > 0\}, \\ \mathrm{Reali}(P < 0, S) &= \{x \in S \mid P(x) < 0\}, \end{aligned}$$

and $\chi(P = 0, S), \chi(P > 0, S), \chi(P < 0, S)$ the Euler-Poincaré characteristics of the corresponding sets.

The **Euler-Poincaré-query** of P for S is

$$\mathrm{EuQ}(P, S) = \chi(P > 0, S) - \chi(P < 0, S).$$

The following equality generalized the basic result of sign determination (Equation 2.6).

Proposition 6.60. *The following equality holds:*

$$\begin{bmatrix} 1 & 1 & 1 \\ 0 & 1 & -1 \\ 0 & 1 & 1 \end{bmatrix} \begin{bmatrix} \chi(P=0,S) \\ \chi(P>0,S) \\ \chi(P<0,S) \end{bmatrix} = \begin{bmatrix} \mathrm{EuQ}(1,S) \\ \mathrm{EuQ}(P,S) \\ \mathrm{EuQ}(P^2,Z) \end{bmatrix} \tag{6.7}$$

Proof: We need to prove

$$\begin{aligned} \chi(P=0,S)+\chi(P>0,S)+\chi(P<0,S) &= \mathrm{EuQ}(1,S), \\ \chi(P>0,S)-\chi(P<0,S) &= \mathrm{EuQ}(P,S), \\ \chi(P>0,S)+\chi(P<0,S) &= \mathrm{EuQ}(P^2,S). \end{aligned}$$

The claim is an immediate consequence of Proposition 6.57. □

6.4 Bibliographical Notes

Modern algebraic topology has its origins in the work of Poincaré [130]. The first proof of the independence of the simplicial homology groups from the triangulation of a polyhedron is due to Alexander [2]. The Mayer-Vietoris theorem first occurs in a paper by Vietoris [161]. The Borel-Moore homology groups first appear in [27].

7

Quantitative Semi-algebraic Geometry

In this chapter, we study various quantitative bounds on the number of connected components and Betti numbers of algebraic and semi-algebraic sets. The key method for this study is the critical point method, i.e. the consideration of the critical points of a well chosen projection. The critical point method also plays a key role for improving the complexity of algorithms in the last chapters of the book.

In Section 7.1, we explain a few basic results of Morse theory and use them to study the topology of a non-singular algebraic hypersurface in terms of the number of critical points of a well chosen projection. Bounding the number of these critical points by Bézout's theorem provides a bound on the sum of the Betti numbers of a non-singular bounded algebraic hypersurface in Section 7.2. Then we prove a similar bound on the sum of the Betti numbers of a general algebraic set.

In Section 7.3, we prove a bound on the sum of the i-th Betti numbers over all realizable sign conditions of a finite set of polynomials. In particular, the bound on the zero-th Betti numbers gives us a bound on the number of realizable sign conditions of a finite set of polynomials. We also explain why these bounds are reasonably tight.

In Section 7.4, we prove bounds on Betti numbers of closed semi-algebraic sets. In Section 7.5 we prove that any semi-algebraic set is semi-algebraically homotopic to a closed one and prove bounds on Betti numbers of general semi-algebraic sets.

7.1 Morse Theory

We first define the kind of hypersurfaces we are going to consider.

A **non-singular algebraic hypersurface** is the zero set $\mathrm{Zer}(Q,\mathrm{R}^k)$ of a polynomial $Q \in \mathrm{R}[X_1,\dots,X_k]$ such that the **gradient** of Q, i.e. the vector

$$\mathrm{Grad}(Q)(p) = \left(\frac{\partial Q}{\partial X_1}(p),\dots,\frac{\partial Q}{\partial X_k}(p)\right) \text{is never 0 for } p \in \mathrm{Zer}(Q,\mathrm{R}^k).$$

Exercise 7.1. Prove that a non-singular algebraic hypersurface is an $\mathcal{S}^\infty$ submanifold of dimension $k-1$. (Hint. Use the Semi-algebraic implicit function theorem (Theorem 3.25).)

Exercise 7.2. Let $\mathrm{Zer}(Q,\mathrm{R}^k)$ be a non-singular algebraic hypersurface. Prove that the gradient vector of Q at a point $p \in \mathrm{Zer}(Q,\mathrm{R}^k)$ is orthogonal to the tangent space $T_p(\mathrm{Zer}(Q,\mathrm{R}^k))$ to $\mathrm{Zer}(Q,\mathrm{R}^k)$ at p.

We denote by π the projection from R^k to the first coordinate sending $(x_1,\ldots,x_k)$ to x_1.

Notation 7.1. [Fiber] For $S \subset \mathrm{R}^k$, $X \subset \mathrm{R}$, let S_X denote $S \cap \pi^{-1}(X)$. We also use the abbreviations S_x, $S_{<x}$, and $S_{\le x}$ for $S_{\{x\}}$, $S_{(-\infty,x)}$, and $S_{(-\infty,x]}$. □

Let $\mathrm{Zer}(Q, \mathrm{R}^k)$ be a non-singular algebraic hypersurface and $p \in \mathrm{Zer}(Q,\mathrm{R}^k)$. Then, the derivative $\mathrm{d}\pi(p)$ of π on $\mathrm{Zer}(Q, \mathrm{R}^k)$ is a linear map from $T_p(\mathrm{Zer}(Q,\mathrm{R}^k))$ to R. Clearly, p is a critical point of π on $\mathrm{Zer}(Q,\mathrm{R}^k)$ if and only if

$$\frac{\partial Q}{\partial X_i}(p) = 0, 2 \le i \le k$$

(see Definition 5.55). In other words, p is a critical point of π on $\mathrm{Zer}(Q,\mathrm{R}^k)$ if and only if the gradient of Q is parallel to the X_1-axis, i.e. $T_p(\mathrm{Zer}(Q,\mathrm{R}^k))$ is orthogonal to the X_1 direction. A critical value of π on $\mathrm{Zer}(Q,\mathrm{R}^k)$ is the projection to the X_1-axis of a critical point of π on $\mathrm{Zer}(Q,\mathrm{R}^k)$.

Lemma 7.2. *Let* $\mathrm{Zer}(Q,\mathrm{R}^k)$ *be a bounded non-singular algebraic hypersurface. The set of values that are not critical for* π *is non-empty and open.*

Proof: The set of values that are not critical for π is clearly open, from the definition of a critical value. It is also non-empty by Theorem 5.56 (Sard's theorem) since the set of critical values is a finite subset of R. □

Also, as an immediate consequence of the Semi-algebraic implicit function theorem (Theorem 3.25), we have:

Proposition 7.3. *Let* $\mathrm{Zer}(Q,\mathrm{R}^k)$ *be a bounded non-singular algebraic hypersurface. If* x *is not a critical value of* π *on* $\mathrm{Zer}(Q, \mathrm{R}^k)$ *and* p *is a point of* $\mathrm{Zer}(Q,\mathrm{R}^k)_x$*, then for* ϵ *small enough* $\mathrm{Zer}(Q,\mathrm{R}^k) \cap B(p,\epsilon)_{<x}$ *is non-empty and semi-algebraically connected.*

We also have the following proposition.

Proposition 7.4. *Let* $\mathrm{Zer}(Q,\mathrm{R}^k)$ *be a bounded non-singular algebraic hypersurface. The set of critical points of* π *on* $\mathrm{Zer}(Q, \mathrm{R}^k)$ *meets every semi-algebraically connected component of* $\mathrm{Zer}(Q,\mathrm{R}^k)$.

Proof: Let C be a semi-algebraically connected component of $\mathrm{Zer}(Q, \mathrm{R}^k)$. Since C is semi-algebraic, closed, and bounded, its image by π is semi-algebraic, closed, and bounded, using Theorem 3.20. Thus $\pi(C)$ is a finite number of points and closed intervals and has a smallest element v. Using Proposition 7.3, it is clear that any $x \in C$ such that $\pi(x) = v$ is critical. □

We will now state and prove the first basic ingredient of Morse theory. In the remainder of the section, we assume $\mathrm{R} = \mathbb{R}$. We suppose that $\mathrm{Zer}(Q, \mathbb{R}^k)$ is a bounded algebraic non-singular hypersurface and denote by π the projection map sending $(x_1, \ldots, x_k)$ to x_1.

Consider the sets $\mathrm{Zer}(Q, \mathbb{R}^k)_{\leq x}$ as x varies from $-\infty$ to ∞. Thinking of X_1 as the horizontal axis, the set $\mathrm{Zer}(Q, \mathbb{R}^k)_{\leq x}$ is the part of $\mathrm{Zer}(Q, \mathbb{R}^k)$ to the left of the hyperplane defined by $X_1 = x$, and we study the changes in the homotopy type of this set as we sweep the hyperplane in the rightward direction. Theorem 7.5 states that there is no change in the homotopy type as x varies strictly between two critical values of π.

Theorem 7.5. [Morse lemma A] *Let $[a, b]$ be an interval containing no critical value of π. Then $\mathrm{Zer}(Q, \mathbb{R}^k)_{[a,b]}$ and $\mathrm{Zer}(Q, \mathbb{R}^k)_a \times [a, b]$ are homeomorphic, and $\mathrm{Zer}(Q, \mathbb{R}^k)_{\leq a}$ is homotopy equivalent to $\mathrm{Zer}(Q, \mathbb{R}^k)_{\leq b}$.*

Theorem 7.5 immediately implies:

Proposition 7.6. *Let $\mathrm{Zer}(Q, \mathbb{R}^k)$ be a non-singular bounded algebraic hypersurface, $[a, b]$ such that π has no critical value in $[a, b]$, and $d \in [a, b]$.*

- *The sets $\mathrm{Zer}(Q, \mathbb{R}^k)_{[a,b]}$ and $\mathrm{Zer}(Q, \mathbb{R}^k)_d$ have the same number of semi-algebraically connected components.*
- *Let S be a semi-algebraically connected component of $\mathrm{Zer}(Q, \mathbb{R}^k)_{[a,b]}$. Then, for every $d \in [a, b]$, S_d is semi-algebraically connected.*

The proof of Theorem 7.5 is based on local existence and uniqueness of solutions to systems of differential equations. Let U be an open subset of $\mathbb{R}^k$. A **vector field** Γ on U is a C^∞ map from an open set U of $\mathbb{R}^k$ to $\mathbb{R}^k$. To a vector field is associated a system of differential equations

$$\frac{\mathrm{d}x_i}{\mathrm{d}T} = \Gamma_i(x_1, \ldots, x_k), 1 \leq i \leq k.$$

A **flow line** of the vector field Γ is a C^∞ map $\gamma: I \to \mathbb{R}^k$ defined on some interval I and satisfying

$$\frac{\mathrm{d}\gamma}{\mathrm{d}T}(t) = \Gamma(\gamma(t)), t \in I.$$

Theorem 7.7. *Let Γ be a vector field on an open subset V of $\mathbb{R}^k$ such that for every $x \in V$, $\Gamma(x) \neq 0$. For every $y \in V$, there exists a neighborhood U of y and $\epsilon > 0$, such that for every $x \in U$, there exists a unique flow line $\gamma_x: (-\epsilon, \epsilon) \to \mathbb{R}^k$ of Γ satisfying the initial condition $\gamma_x(0) = x$.*

Proof: Since Γ is C^∞, there exists a bounded neighborhood W of y and $L>0$ such that $|\Gamma(x_1)-\Gamma(x_2)|<L|x_1-x_2|$ for all $x_1,x_2\in W$. Let $A=\sup_{x\in W}|\Gamma(x)|$. Also, let $\epsilon>0$ be a small enough number such that the set

$$W'=\{x\in W\mid B_k(x,\epsilon A)\subset W\}$$

contains an open set U containing y.

Let $x\in U$. If γ_x: $[-\epsilon,\epsilon]\to\mathbb{R}^k$, with $\gamma_x(0)=x$, is a solution, then $\gamma_x([-\epsilon,\epsilon])\subset W'$. This is because $|\Gamma(x')|\le A$ for every $x'\in W$, and hence applying the Mean Value Theorem, $|x-\gamma_x(t)|\le|t|A$ for all $t\in[-\epsilon,\epsilon]$. Now, since $x\in U$, it follows that $\gamma_x([-\epsilon,\epsilon])\subset W'$.

We construct the solution γ_x: $[-\epsilon,\epsilon]\to W$ as follows. Let $\gamma_{x,0}(t)=x$ for all t and

$$\gamma_{x,n+1}(t)=x+\int_0^t\Gamma(\gamma_{x,n}(t))dt.$$

Note that $\gamma_{x,n}([-\epsilon,\epsilon])\subset W'$ for every $n\ge 0$. Now,

$$\begin{aligned}|\gamma_{x,n+1}(t)-\gamma_{x,n}(t)| &= \left|\int_0^t(\Gamma(\gamma_{x,n}(t))-\Gamma(\gamma_{x,n-1}(t)))dt\right|\\ &\le \left|\int_0^t|\Gamma(\gamma_{x,n}(t))-\Gamma(\gamma_{x,n-1}(t))|dt\right|\\ &\le \epsilon L|\gamma_{x,n}(t)-\gamma_{x,n-1}(t)|\end{aligned}$$

Choosing ϵ such that $\epsilon<1/L$, we see that for every fixed $t\in[-\epsilon,\epsilon]$, the sequence $\gamma_{x,n}(t)$ is a Cauchy sequence and converges to a limit $\gamma_x(t)$.

Moreover, it is easy to verify that $\gamma_x(t)$ satisfies the equation,

$$\gamma_x(t)=x+\int_0^t\Gamma(\gamma_x(t))dt.$$

Differentiating both sides, we see that $\gamma_x(t)$ is a flow line of the given vector field Γ, and clearly $\gamma_x(0)=x$.

The proof of uniqueness is left as an exercise. □

Given a C^∞ hypersurface $M\subset\mathbb{R}^k$, a C^∞ **vector field on** M, Γ, is a C^∞ map that associates to each $x\in M$ a tangent vector $\Gamma(x)\in T_x(M)$.

An important example of a vector field on a hypersurface is the gradient vector field. Let $\mathrm{Zer}(Q,\mathbb{R}^k)$ be a non-singular algebraic hypersurface and (a',b') such that π has no critical point on $\mathrm{Zer}(Q,\mathbb{R}^k)_{(a',b')}$. The **gradient vector field of** π on $\mathrm{Zer}(Q,\mathbb{R}^k)_{(a',b')}$ is the C^∞ vector field on $\mathrm{Zer}(Q,\mathbb{R}^k)_{(a',b')}$ that to every $p\in\mathrm{Zer}(Q,\mathbb{R}^k)_{(a',b')}$ associates $\Gamma(p)$ characterized by the following properties

- it belongs to $T_p(\mathrm{Zer}(Q,\mathbb{R}^k))$,

- it belongs to the plane generated by the gradient $\mathrm{Grad}(Q)(p)$, and the unit vector of the X_1-axis,
- its projection on the X_1-axis is the negative of the unit vector.

The flow lines of the gradient vector field correspond to curves on the hypersurface along which the X_1 coordinate decreases maximally. A straightforward computation shows that, for $p \in \mathrm{Zer}(Q, \mathbb{R}^k)$,

$$\Gamma(p) = -\frac{G(p)}{\sum_{2 \leq i \leq k} \left(\frac{\partial Q}{\partial X_i}(p)\right)^2},$$

where

$$G(p) = \left(\sum_{2 \leq i \leq k} \left(\frac{\partial Q}{\partial X_i}(p)\right)^2, -\frac{\partial Q}{\partial X_1}\frac{\partial Q}{\partial X_2}(p), \ldots, -\frac{\partial Q}{\partial X_1}\frac{\partial Q}{\partial X_k}(p)\right).$$

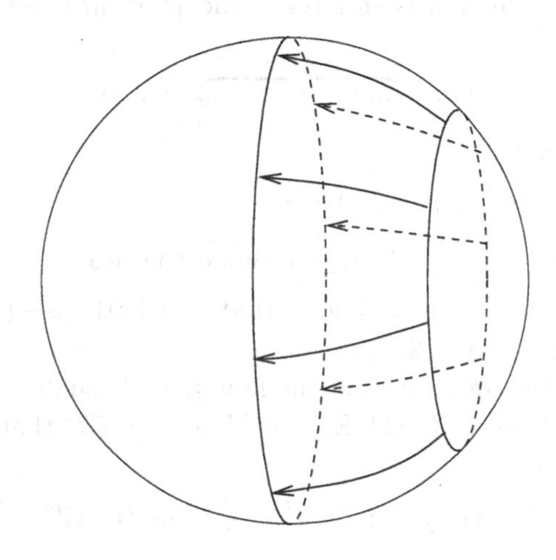

Fig. 7.1. Flow of the gradient vector field on the 2-sphere

Proof of Theorem 7.5: By Lemma 7.2 we can chose $a' < a$, $b' > b$ such that π has no critical point on $\mathrm{Zer}(Q, \mathbb{R}^k)_{(a',b')}$. Consider the gradient vector field of π on $\mathrm{Zer}(Q, \mathbb{R}^k)_{(a',b')}$.

By Corollary 5.51, the set $\mathrm{Zer}(Q, \mathbb{R}^k)_{(a',b')}$ can be covered by a finite number of open sets such that for each open set U' in the cover, there is an open U of $\mathbb{R}^{k-1}$ and a diffeomorphism $\Phi: U \to U'$.

Using the linear maps $d\Phi_x^{-1}: T_x(M) \to T_{\Phi^{-1}(x)}\mathbb{R}^{k-1}$, we associate to the gradient vector field of π on $U' \subset \mathrm{Zer}(Q, \mathbb{R}^k)_{(a',b')}$ a C^∞ vector field on U.

By Theorem 7.7, for each point $x \in \mathrm{Zer}(Q, \mathbb{R}^k)_{[a,b]}$, there exists a neighborhood W of $\Phi^{-1}(x)$ and an $\epsilon > 0$ such that the induced vector field in $\mathbb{R}^{k-1}$ has a solution $\gamma_x(t)$ for $t \in (-\epsilon, \epsilon)$ and such that $\gamma_x(0) = \Phi^{-1}(x)$. We consider its image $\Phi(\gamma_x)$ on $\mathrm{Zer}(Q, \mathbb{R}^k)_{(a',b')}$. Thus, for each point $x \in \mathrm{Zer}(Q, \mathbb{R}^k)_{[a,b]}$, we have a neighborhood U_x, a number $\epsilon_x > 0$, and a curve

$$\Phi \circ \gamma_x: (-\epsilon_x, \epsilon_x) \to \mathrm{Zer}(Q, \mathbb{R}^k)_{(a',b')},$$

such that $\Phi \circ \gamma_x(0) = x$, and $\mathrm{d}\Phi \circ \gamma_x \mathrm{d}t = \Gamma(\Phi \circ \gamma_x(t))$ for all $t \in (-\epsilon_x, \epsilon_x)$.

Since $\mathrm{Zer}(Q, \mathbb{R}^k)_{[a,b]}$ is compact, we can cover it using a finite number of the neighborhoods U_x and let $\epsilon_0 > 0$ be the least among the corresponding ϵ_x's. For $t \in [0, b-a]$, we define a one-parameter family of smooth maps

$$\alpha_t: \mathrm{Zer}(Q, \mathbb{R}^k)_b \to \mathrm{Zer}(Q, \mathbb{R}^k)_{\leq b}$$

as follows:

Let $x \in \mathrm{Zer}(Q, \mathbb{R}^k)_b$. If $|t| \leq \epsilon_0/2$, we let $\alpha_t(x) = \gamma_x(t)$. If $|t| > \epsilon_0/2$, we write $t = n\,\epsilon_0/2 + \delta$, where n is an integer and $|\delta| < \epsilon_0/2$. We let

$$\alpha_t(x) = \overbrace{\alpha_{\epsilon_0/2} \circ \cdots \circ \alpha_{\epsilon_0/2}}^{n \text{ times}} \circ \alpha_\delta(x).$$

Observe the following.

- For every $x \in \mathrm{Zer}(Q, \mathbb{R}^k)_b$, $\alpha_0(x) = x$.
- By construction, $\dfrac{\mathrm{d}\alpha_t(x)}{\mathrm{d}t} = \Gamma(\alpha_t(x))$. Since the projection on the X_1 axis of $\Gamma(\alpha_t(x)) = (-1, 0, \ldots, 0)$, it follows that $\pi(\alpha_t(x)) = b - t$.
- $\alpha_t(\mathrm{Zer}(Q, \mathbb{R}^k)_b) = \mathrm{Zer}(Q, \mathbb{R}^k)_{b-t}$.
- It follows from the uniqueness of the flowlines through every point of the gradient vector field on $\mathrm{Zer}(Q, \mathbb{R}^k)_{[a,b]}$ (Theorem 7.7) that each α_t defined above is injective.

We now claim that the map $f: \mathrm{Zer}(Q, \mathbb{R}^k)_{[a,b]} \to \mathrm{Zer}(Q, \mathbb{R}^k)_a \times [a, b]$ defined by

$$f(x) = (\alpha_{b-a}(\alpha_{b-\pi(x)}^{-1}(x)), \pi(x))$$

is a homeomorphism. This is an immediate consequence of the properties of α_t listed above.

Next, consider the map $F(x, t)$: $\mathrm{Zer}(Q, \mathbb{R}^k)_{\leq b} \times [0, 1] \to \mathrm{Zer}(Q, \mathbb{R}^k)_{\leq b}$ defined as follows:

$$\begin{aligned} F(x, s) &= x, && \text{if } \pi(x) \leq b - s\,(b-a) \\ &= \alpha_{s(b-a)}(\alpha_{b-\pi(x)}^{-1}(x)), && \text{otherwise.} \end{aligned}$$

Clearly, F is a deformation retraction from $\mathrm{Zer}(Q, \mathbb{R}^k)_{\leq b}$ to $\mathrm{Zer}(Q, \mathbb{R}^k)_{\leq a}$, so that $\mathrm{Zer}(Q, \mathbb{R}^k)_{\leq b}$ is homotopy equivalent to $\mathrm{Zer}(Q, \mathbb{R}^k)_{\leq a}$.

This completes the proof. □

Theorem 7.5 states that there is no change in homotopy type on intervals containing no critical values. The remainder of the section is devoted to studying the changes in homotopy type that occur at the critical values. In this case, we will not be able to use the gradient vector field of π to get a flow as the gradient becomes zero at a critical point. We will, however, show how to modify the gradient vector field in a neighborhood of a critical point so as to get a new vector field that agrees with the gradient vector field outside a small neighborhood. The flow corresponding to this new vector field will give us a homotopy equivalence between $\mathrm{Zer}(Q,\mathbb{R}^k)_{\leq c+\epsilon}$ and $\mathrm{Zer}(Q,\mathbb{R}^k)_{\leq c-\epsilon}\cup B$, where c is a critical value of π, $\epsilon>0$ is sufficiently small, and B a topological ball attached to $\mathrm{Zer}(Q,\mathbb{R}^k)_{\leq c-\epsilon}$ by its boundary. The key notion necessary to work this idea out is that of a Morse function.

Definition 7.8. [Morse function] Let $\mathrm{Zer}(Q, \mathbb{R}^k)$ be a bounded non-singular algebraic hypersurface and π the projection on the X_1-axis sending $x=(x_1,\ldots,x_k)\in\mathbb{R}^k$ to $x_1 \in \mathbb{R}$. Let $p \in \mathrm{Zer}(Q, \mathbb{R}^k)$ be a critical point of π. The tangent space $T_p(\mathrm{Zer}(Q, \mathbb{R}^k))$ is the $(k-1)$-dimensional space spanned by the $X_2, \ldots, X_k$ coordinates with origin p. By virtue of the Implicit Function Theorem (Theorem 3.25), we can choose $(X_2, \ldots, X_k)$ to be a local system of coordinates in a sufficiently small neighborhood of p. More precisely, we have an open neighborhood $U\subset\mathbb{R}^{k-1}$ of $p'=(p_2,\ldots,p_k)$ and a mapping $\phi: U\to\mathbb{R}$, such that, with $x'=(x_2,\ldots,x_k)$, and

$$\Phi(x')=(\phi(x'),x')\in\mathrm{Zer}(Q,\mathbb{R}^k), \tag{7.1}$$

the mapping Φ is a diffeomorphism from U to $\Phi(U)$.

The critical point p is **non-degenerate** if the $(k-1)\times(k-1)$ Hessian matrix

$$\mathrm{Hes}_\pi(p')=\left[\frac{\partial^2\phi}{\partial X_i\partial X_j}(p')\right],\ 2\leq i,j\leq k, \tag{7.2}$$

is invertible. Note that $\mathrm{Hes}_\pi(p')$ is a real symmetric matrix and hence all its eigenvalues are real (Theorem 4.42). Moreover, if p is a non-degenerate critical point, then all eigenvalues are non-zero. The number of positive eigenvalues of $\mathrm{Hes}_\pi(p')$ is the **index** of the critical point p.

The function π is a **Morse function** if all its critical points are non-degenerate and there is at most one critical point of π above each $x\in\mathrm{R}$. □

We next show that to require π to be a Morse function is not a big loss of generality, since an orthogonal change of coordinates can make the projection map π a Morse function on $\mathrm{Zer}(Q,\mathbb{R}^k)$.

Proposition 7.9. *Up to an orthogonal change of coordinates, the projection π to the X_1-axis is a Morse function.*

The proof of Proposition 7.9 requires some preliminary work.

We start by proving:

Proposition 7.10. *Let d be the degree of Q. Suppose that the projection π on the X_1-axis has only non-degenerate critical points. The number of critical points of π is finite and bounded by $d(d-1)^{k-1}$.*

Proof: The critical points of π can be characterized as the real solutions of the system of k polynomial equations in k variables

$$Q = \frac{\partial Q}{\partial X_2} = 0, \ldots, \frac{\partial Q}{\partial X_k} = 0 .$$

We claim that every real solution p of this system is non-singular, i.e. the Jacobian matrix

$$\begin{bmatrix} \frac{\partial Q}{\partial X_1}(p) & \frac{\partial^2 Q}{\partial X_2 \partial X_1}(p) & \cdots & \frac{\partial^2 Q}{\partial X_k \partial X_1}(p) \\ \vdots & \vdots & & \vdots \\ \frac{\partial Q}{\partial X_k}(p) & \frac{\partial^2 Q}{\partial X_2 \partial X_k}(p) & \cdots & \frac{\partial^2 Q}{\partial X_k \partial X_k}(p) \end{bmatrix}$$

is non-singular. Differentiating the identity (7.3) and evaluating at p, we obtain for $2 \le i, j \le k$, with $p' = (p_2, \ldots, p_k)$,

$$\frac{\partial^2 Q}{\partial X_j \partial X_i}(p) = -\frac{\partial Q}{\partial X_1}(p) \frac{\partial^2 \phi}{\partial X_j \partial X_i}(p') .$$

Since $\frac{\partial Q}{\partial X_1}(p) \neq 0$ and $\frac{\partial Q}{\partial X_i}(p) = 0$, for $2 \le i \le k$, the claim follows.By Theorem 4.106 (Bézout's theorem), the number of critical points of π is less than or equal to the product

$$\deg(Q) \deg\left(\frac{\partial Q}{\partial X_2}\right) \cdots \deg\left(\frac{\partial Q}{\partial X_k}\right) = d\,(d-1)^{k-1} .$$ □

We interpret geometrically the notion of non-degenerate critical point.

Proposition 7.11. *Let $p \in \mathrm{Zer}(Q, \mathrm{R}^k)$ be a critical point of π. Let $g\colon \mathrm{Zer}(Q, \mathrm{R}^k) \to S^{k-1}(0,1)$ be the Gauss map defined by*

$$g(x) = \frac{\mathrm{Grad}(Q(x))}{\|\mathrm{Grad}(Q(x))\|}.$$

The Gauss map is an $\mathcal{S}^\infty$-diffeomorphism in a neighborhood of p if and only if p is a non-degenerate critical point.

Proof: Since p is a critical point of π, $g(p) = (\pm 1, 0, \ldots, 0)$. Using Notation 7.8, for $x' \in U$, $x = \Phi(x') = (\phi(x'), x')$, and applying the chain rule,

$$\frac{\partial Q}{\partial X_i}(x) + \frac{\partial Q}{\partial X_1}(x) \frac{\partial \phi}{\partial X_i}(x') = 0, \ 2 \le i \le k. \tag{7.3}$$

Thus

$$g(x) = \pm \frac{1}{\sqrt{1 + \sum_{i=2}^{k} \left(\frac{\partial \phi}{\partial X_i}(x')\right)^2}} \left(-1, \frac{\partial \phi}{\partial X_2}(x'), \dots, \frac{\partial \phi}{\partial X_k}(x')\right).$$

Taking the partial derivative with respect to X_i of the j-th coordinate g_j of g, for $2 \le i, j \le k$, and evaluating at p, we obtain

$$\frac{\partial g_j}{\partial X_i}(p) = \pm \frac{\partial^2 \phi}{\partial X_j \, \partial X_i}(p'),\ 2 \le i, j \le k.$$

The matrix $[\partial g_i / \partial X_i(p)], 2 \le i,\ j \le k,$ is invertible if and only if p is a non-degenerate critical point of ϕ by (7.2). □

Proposition 7.12. *Up to an orthogonal change of coordinates, the projection π to the X_1-axis has only non-degenerate critical points.*

Proof: Consider again the Gauss map $g: \mathrm{Zer}(Q, \mathbb{R}^k) \to S^{k-1}(0,1)$, defined by

$$g(x) = \frac{\mathrm{Grad}(Q(x))}{\|\mathrm{Grad}(Q(x))\|}.$$

According to Sard's theorem (Theorem 5.56) the dimension of the set of critical values of g is at most $k-2$. We prove now that there are two antipodal points of $S^{k-1}(0,\ 1)$ such that neither is a critical value of g. Assume the contrary and argue by contradiction. Since the dimension of the set of critical values is at most $k-2$, there exists a non-empty open set U of regular values in $S^{k-1}(0,1)$. The set of points that are antipodes to points in U is non-empty, open in $S^{k-1}(0,1)$ and all critical, contradicting the fact that the critical set has dimension at most $k-2$.

After rotating the coordinate system, we may assume that (1, 0, ..., 0) and $(-1, 0, \dots, 0)$ are not critical values of g. The claim follows from Proposition 7.11. □

It remains to prove that it is possible to ensure, changing the coordinates if necessary, that there is at most one critical point of π above each $x \in \mathrm{R}$.

Suppose that the projection π on the X_1-axis has only non-degenerate critical points. These critical points are finite in number according to Proposition 7.10. We can suppose without loss of generality that all the critical points have distinct X_2 coordinates, making if necessary an orthogonal change of coordinates in the variables $X_2, \dots, X_k$ only.

Lemma 7.13. *Let δ be a new variable and consider the field $\mathbb{R}\langle\delta\rangle$ of algebraic Puiseux series in δ. The set S of points $\overline{p} = (\overline{p_1}, \dots, \overline{p_k}) \in \mathrm{Zer}(Q, \mathbb{R}\langle\delta\rangle^k)$ with gradient vector $Grad(Q)(\overline{p})$ proportional to $(1, \delta, 0, \dots, 0)$ is finite. Its number of elements is equal to the number of critical points of π. Moreover there is a point $\overline{p}$ of S infinitesimally close to every critical point p of π and the signature of the Hessian at p and $\overline{p}$ coincide.*

Proof: Note that, modulo the orthogonal change of variable

$$X_1' = X_1 + \delta X_2, X_2' = X_2 - \delta X_1, X_i' = X_i, i \geq 3,$$

a point $\overline{p}$ such that $\mathrm{Grad}(Q)(\overline{p})$ is proportional to $(1, \delta, 0, \ldots, 0)$ is a critical point of the projection π' on the X_1'-axis, and the corresponding critical value of π' is $\overline{p_1} + \delta \overline{p_2}$.

Since $\mathrm{Zer}(Q, \mathbb{R}^k)$ is bounded, a point $\overline{p} \in \mathrm{Zer}(Q, \mathbb{R}\langle\delta\rangle^k)$ always has an image by $\lim_\delta$. If $\overline{p}$ is such that $\mathrm{Grad}(Q)(\overline{p})$ is proportional to $(1, \delta, 0, \ldots, 0)$, then $\mathrm{Grad}(Q)(\lim_\delta (\overline{p}))$ is proportional to $(1, 0, \ldots, 0, 0)$, and thus $p = \lim_\delta (p)$ is a critical point of π. Suppose without loss of generality that $\mathrm{Grad}(Q)(p) = (1, 0, \ldots, 0, 0)$. Since p is a non-degenerate critical point of π, Proposition 7.11 implies that there is a semi-algebraic neighborhood U of $p' = (p_2, \ldots, p_k)$ such that $g \circ \Phi$ is a diffeomorphism from U to a semi-algebraic neighborhood of $(1, 0, \ldots, 0, 0) \in S^{k-1}(0, 1)$. Denoting by g' the inverse of the restriction of g to $\Phi(U)$ and considering

$$\mathrm{Ext}(g', \mathbb{R}\langle\delta\rangle) \colon \mathrm{Ext}(g(\Phi(U)), \mathbb{R}\langle\delta\rangle) \to \mathrm{Ext}(\Phi(U), \mathbb{R}\langle\delta\rangle),$$

there is a unique $\overline{p} \in \mathrm{Ext}(\Phi(U), \mathbb{R}\langle\delta\rangle)$ such that $\mathrm{Grad}(Q)(\overline{p})$ is proportional to $(1, \delta, 0, \ldots, 0)$. Moreover, denoting by J the Jacobian of $\mathrm{Ext}(g', \mathbb{R}\langle\delta\rangle)$, the value $J(1, 0, 0, \ldots, 0) = t$ is a non-zero real number. Thus the signature of the Hessian at p and $\overline{p}$ coincide. □

Proof of Proposition 7.9: Since J is the Jacobian of $\mathrm{Ext}(g', \mathbb{R}\langle\delta\rangle)$, the value $J(1, 0, 0, \ldots, 0) = t$ is a non-zero real number, $\lim_\delta (J(y)) = t$ for every $y \in \mathrm{Ext}(S^{k-1}(0, 1), \mathbb{R}\langle\delta\rangle)$ infinitesimally close to $(1, 0, 0, \ldots, 0)$. Using the mean value theorem (Corollary 2.23)

$$o(|\overline{p} - p|) = o\left(\left|\frac{1}{\sqrt{1 + \delta^2}}(1, \delta, 0, \ldots, 0) - (1, 0, 0, \ldots, 0)\right|\right) = 1.$$

Thus $o(\overline{p_i} - p_i) \geq 1, i \geq 1$.

Let $b_{i,j} = \dfrac{\partial^2 \phi}{\partial X_i \partial X_j}(p)$, $2 \leq i \leq k, 2 \leq j \leq k$. Taylor's formula at p for ϕ gives

$$\overline{p_1} = p_1 + \sum_{2 \leq i \leq k, 2 \leq j \leq k} b_{i,j} (\overline{p_i} - p_i)(\overline{p_j} - p_j) + c,$$

with $o(c) \geq 2$. Thus $o(\overline{p_1} - p_1) \geq 2$.

It follows that the critical value of π' at $\overline{p}$ is $\overline{p_1} + \delta \overline{p_2} = p_1 + \delta p_2 + w$, with $o(w) > 1$.

Thus, all the critical values of π' on $\mathrm{Zer}(Q, \mathbb{R}\langle\delta\rangle^k$ are distinct since all values of p_2 are. Using Proposition 3.17, we can replace δ by $d \in \mathbb{R}$, and we have proved that there exists an orthogonal change of variable such that π is a Morse function. □

We are now ready to state the second basic ingredient of Morse theory, which is describing precisely the change in the homotopy type that occurs in $\mathrm{Zer}(Q, \mathbb{R}^k)_{\leq x}$ as x crosses a critical value when π is a Morse function.

Theorem 7.14. [Morse lemma B] *Let* $\mathrm{Zer}(Q, \mathbb{R}^k)$ *be a non-singular bounded algebraic hypersurface such that the projection* π *to the* X_1*-axis is a Morse function. Let* p *be a non-degenerate critical point of* π *of index* λ *and such that* $\pi(p) = c$.

Then, for all sufficiently small $\epsilon > 0$*, the set* $\mathrm{Zer}(Q, \mathbb{R}^k)_{\leq c+\epsilon}$ *has the homotopy type of the union of* $\mathrm{Zer}(Q, \mathbb{R}^k)_{\leq c-\epsilon}$ *with a ball of dimension* $k-1-\lambda$*, attached along its boundary.*

We first prove a lemma that will allow us to restrict to the case where $x = 0$ and where Q is a quadratic polynomial of a very simple form.

Let $\mathrm{Zer}(Q, \mathbb{R}^k), U, \phi, \Phi$ be as above (see page 243).

Lemma 7.15. *Let* $\mathrm{Zer}(Q, \mathbb{R}^k)$ *be a non-singular bounded algebraic hypersurface such that the projection* π *to the* X_1*-axis is a Morse function. Let* $p \in \mathrm{Zer}(Q, \mathbb{R}^k)$ *be a non-degenerate critical point of the map* π *with index* λ*. Then there exists an open neighborhood* V *of the origin in* $\mathbb{R}^{k-1}$ *and a diffeomorphism* Ψ *from* U *to* V *such that, denoting by* Y_i *the* i*-th coordinate of* $\Psi(X_2, \ldots, X_k)$*,*

$$\phi(Y_2, \ldots, Y_k) = \sum_{2 \leq i \leq \lambda+1} Y_i^2 - \sum_{\lambda+2 \leq i \leq k} Y_i^2.$$

Proof: We assume without loss of generality that p is the origin. Also, by Theorem 4.42, we assume that the matrix

$$\mathrm{Hes}(0) = \left[\frac{\partial^2 \phi}{\partial X_i \partial X_j}(0)\right],\ 2 \leq i, j \leq k,$$

is diagonal with its first λ entries $+1$ and the remaining -1.

Let us prove that there exists a C^∞ map M from U to the space of symmetric $(k-1) \times (k-1)$ matrices, $X \mapsto M(X) = (m_{ij}(X))$, such that

$$\phi(X_2, \ldots, X_k) = \sum_{2 \leq i, j \leq k} m_{ij}(X) X_i X_j.$$

Using the fundamental theorem of calculus twice, we obtain

$$\begin{aligned}
\phi(X_2, \ldots, X_k) &= \sum_{2 \leq j \leq k} X_j \int_0^1 \frac{\partial \phi}{\partial X_j}(t X_2, \ldots, t X_k) \mathrm{d}t \\
&= \sum_{2 \leq i \leq k} \sum_{2 \leq j \leq k} X_i X_j \int_0^1 \int_0^1 \frac{\partial^2 \phi}{\partial X_i \partial X_j}(s t X_2, \ldots, s t X_k) \mathrm{d}t\, \mathrm{d}s.
\end{aligned}$$

Take

$$m_{ij}(X_2, \ldots, X_k) = \int_0^1 \int_0^1 \frac{\partial^2 \phi}{\partial X_i \partial X_j}(s t X_2, \ldots, s t X_k) \mathrm{d}t\, \mathrm{d}s.$$

Note that the matrix $M(X_2, \ldots, X_k)$ obtained above clearly satisfies $M(0) = H(0)$, and $M(x_2, \ldots, x_k)$ is close to $H(x_2, \ldots, x_k)$ for $(x_2, \ldots, x_k)$ in a sufficiently small neighborhood of the origin.

Using Theorem 4.42 again, there exists a C^∞ map N from a sufficiently small neighborhood V of 0 in R^{k-1} to the space of $(k-1)\times(k-1)$ real invertible matrices such that

$$\forall x \in V, N(x)^t M(x) N(x) = H(0).$$

Let $Y = N(X)^{-1}X$. Since $N(X)$ is invertible, the map sending X to Y maps V diffeomorphically into its image. Also,

$$\begin{aligned} X^t M(X) X &= Y^t N(X)^t M(X) N(X) Y \\ &= Y^t H(0) Y \\ &= \sum_{2\le i\le \lambda+1} Y_i^2 - \sum_{\lambda+2\le i\le k} Y_i^2. \end{aligned}$$

□

Using Lemma 7.15, we observe that in a small enough neighborhood of a critical point, a hypersurface behaves like one defined by a quadratic equation. So it suffices to analyze the change in the homotopy type of $\mathrm{Zer}(Q, \mathbb{R}^k)_{\le x}$ as x crosses 0 and the hypersurface defined by a quadratic polynomial of a very simple form. The change in the homotopy type consists in "attaching a handle along its boundary", which is the process we describe now.

A **j-ball** is an embedding of $\overline{B_j}(0,1)$, the closed j-dimensional ball with radius 1, in $\mathrm{Zer}(Q, \mathbb{R}^k)$. It is a homeomorphic image of $\overline{B_j}(0,1)$ in $\mathrm{Zer}(Q, \mathbb{R}^k)$.

Let

$$P = X_1 - \sum_{2\le i\le \lambda+1} X_i^2 + \sum_{\lambda+2\le i\le k} X_i^2,$$

and π the projection onto the X_1 axis restricted to $\mathrm{Zer}(P, \mathbb{R}^k)$.

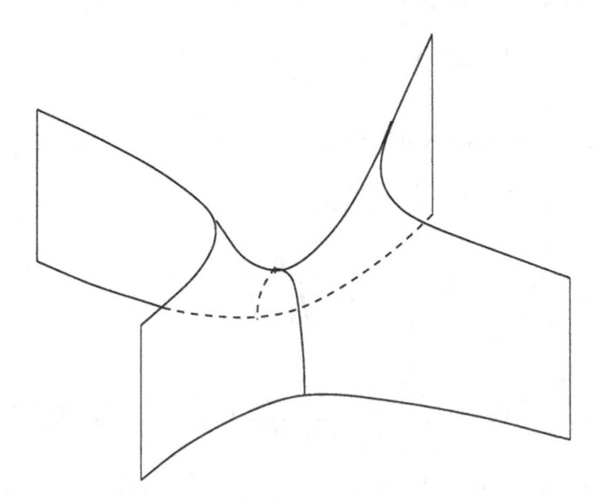

Fig. 7.2. The surface $\mathrm{Zer}(X_1 - X_2^2 + X_3^2, \mathbb{R}^3)$ near the origin

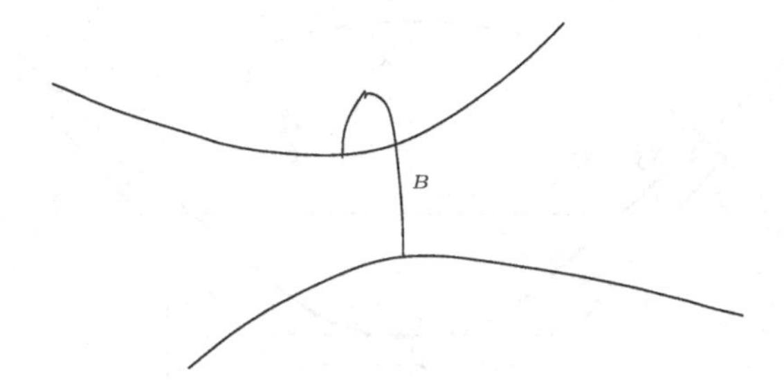

Fig. 7.3. The retract of $\mathrm{Zer}(X_1 - X_2^2 + X_3^2, \mathbb{R}^3)$ near the origin

Let B be the set defined by

$$X_2 = \cdots = X_{\lambda+1} = 0, X_1 = -\sum_{\lambda+2 \le i \le k} X_i^2, -\epsilon \le X_1 \le 0.$$

Note that B is a $(k - \lambda - 1)$-ball and $B \cap \mathrm{Zer}(P, \mathbb{R}^k)_{\le -\epsilon}$ is the set defined by

$$X_2 = \cdots = X_{\lambda+1} = 0, X_1 = -\epsilon, \sum_{\lambda+2 \le i \le k} X_i^2 = \epsilon,$$

which is also the boundary of B.

Lemma 7.16. *For all sufficiently small $\epsilon > 0$, and $r > 2\sqrt{\epsilon}$, there exists a vector field Γ' on $\mathrm{Zer}(P, \mathbb{R}^k)_{[-\epsilon,\epsilon]} \setminus B$, having the following properties:*

1. *Outside the ball $B_k(r)$, $2\epsilon\Gamma'$ equals the gradient vector field, Γ, of π on $\mathrm{Zer}(P, \mathbb{R}^k)_{[-\epsilon,\epsilon]}$.*
2. *Associated to Γ' there is an one parameter continuous family of smooth maps α_t: $\mathrm{Zer}(P, \mathbb{R}^k)_\epsilon \to \mathrm{Zer}(P, \mathbb{R}^k)_{[-\epsilon,\epsilon]}$, $t \in [0, 1)$, such that for $x \in \mathrm{Zer}(P, \mathbb{R}^k)_\epsilon, t \in [0,1)$,*
   a) *Each α_t is injective,*
   b) $\dfrac{\mathrm{d}\alpha_t(x)}{\mathrm{d}t} = \Gamma'(\alpha_t(x))$,
   c) $\alpha_0(x) = x$,
   d) $\lim_{t\to 1} \alpha_t(x) \in \mathrm{Zer}(P, \mathbb{R}^k)_{-\epsilon} \cup B$,
   e) *for every $y \in \mathrm{Zer}(P, \mathbb{R}^k)_{[-\epsilon,\epsilon]} \setminus B$ there exists a unique $z \in \mathrm{Zer}(P, \mathbb{R}^k)_\epsilon$ and $t \in [0,1)$ such that $\alpha_t(z) = y$.*

Proof of Lemma 7.16: In the following, we consider $\mathbb{R}^{k-1}$ as a product of the coordinate subspaces spanned by $X_2, \ldots, X_{\lambda+1}$ and $X_{\lambda+2}, \ldots, X_k$, respectively, and denote by Y(resp. Z) the vector of variables $(X_2, \ldots, X_{\lambda+1})$ (resp. $(X_{\lambda+2}, \ldots, X_k)$). We denote by $\phi: \mathbb{R}^k \to \mathbb{R}^{k-1}$ the projection map onto the hyperplane $X_1 = 0$. Let $S = \phi(\overline{B}_k(r))$.

We depict the flow lines of the flow we are going to construct (projected onto the hyperplane defined by $X_1 = 0$) in the case when $k = 3$ and $\lambda = 1$ in Figure 7.4.

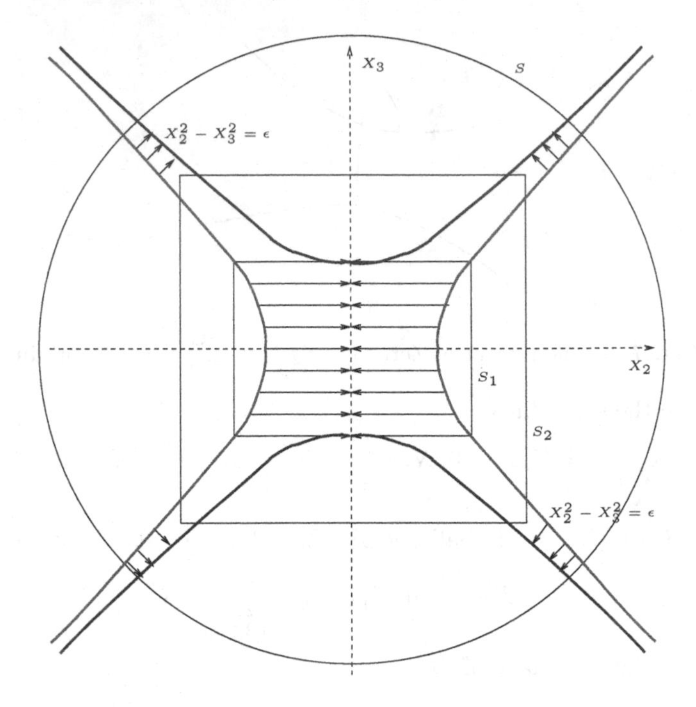

Fig. 7.4. S_1 and S_2

Consider the following two subsets of S.

$$S_1 = \overline{B}_\lambda(\sqrt{2\,\epsilon}) \times \overline{B}_{k-1-\lambda}(\sqrt{\epsilon})$$

and

$$S_2 = \overline{B}_\lambda(2\,\sqrt{\epsilon}) \times \overline{B}_{k-1-\lambda}(2\,\sqrt{\epsilon}).$$

In $\mathrm{Zer}(P,\ \mathbb{R}^k)_{[-\epsilon,\epsilon]} \cap \phi^{-1}(S_1)$, consider the flow lines whose projection onto the hyperplane $X_1 = 0$ are straight segments joining the points $(y_2, \ldots, y_k) \in \phi(\mathrm{Zer}(P, \mathbb{R}^k)_\epsilon)$ to $(0, \ldots 0, y_{\lambda+2}, \ldots, y_k)$.

These correspond to the vector field on $\mathrm{Zer}(P,\ \mathbb{R}^k)_{[-\epsilon,\epsilon]} \cap \phi^{-1}(S_1) \setminus B$ defined by

$$\Gamma_1 = \left(-\frac{1}{|Z|^2 + \epsilon}, \frac{-Y}{2|Y|^2(|Z|^2 + \epsilon)}, 0\right).$$

Let $p = (\epsilon, y, z) \in \mathrm{Zer}(P, \mathbb{R}^k)_\epsilon \cap \phi^{-1}(S_1)$ and q the point in $\mathrm{Zer}(P, \mathbb{R}^k)$ having the same Z coordinates but having $Y = 0$. Then, $\pi(q) = |z|^2 + \epsilon$. Thus, the decreases uniformly from ϵ to $-|z|^2$ along the flow lines of the vector field Γ_1. For a point $p = (x_1, y, z) \in \mathrm{Zer}(Q, \mathbb{R}^k)_{[-\epsilon,\epsilon]} \cap \phi^{-1}(S_1) \setminus B$, we denote by $g(p)$ the limiting point on the flow line through p of the vector field Γ_1 as it approaches $Y = 0$.

In $\mathrm{Zer}(P, \mathbb{R}^k)_{[-\epsilon,\epsilon]} \cap \phi^{-1}(S \setminus S_2)$, consider the flow lines of the vector field

$$\Gamma_2 = \left(-\frac{1}{2\,\epsilon}, -\frac{Y}{4\,\epsilon(|Y|^2 + |Z|^2)}, \frac{Z}{4\,\epsilon(|Y|^2 + |Z|^2)} \right).$$

Notice that Γ_2 is $\frac{1}{2\epsilon}$ times the gradient vector field on

$$\mathrm{Zer}(P, \mathbb{R}^k)_{[-\epsilon,\epsilon]} \cap \phi^{-1}(S \setminus S_2).$$

For a point $p = (x_1, y, z) \in \mathrm{Zer}(P, \mathbb{R}^k)_{[-\epsilon,\epsilon]} \cap \phi^{-1}(S \setminus S_2)$, we denote by $g(p)$ the point on the flow line through p of the vector field Γ_2 such that $\pi(g(p)) = -\epsilon$.

We patch these vector fields together in

$$\mathrm{Zer}(P, \mathbb{R}^k)_{[-\epsilon,\epsilon]} \cap \phi^{-1}(S_2 \setminus S_1)$$

using a C^∞ function that is 0 in S_1 and 1 outside S_2. Such a function $\mu: \mathbb{R}^{k-1} \to \mathbb{R}$ can be constructed as follows. Define

$$\lambda(x) = \begin{cases} 0 & \text{if } x \leq 0, \\ 1 - 2^{-\frac{1}{4x^2}} & \text{if } 0 < x \leq \frac{1}{2}, \\ 2^{-\frac{1}{4(1-x)^2}} & \text{if } \frac{1}{2} < x \leq 1, \\ 1 & \text{if } x \geq 1. \end{cases}$$

Take

$$\mu(y, z) = \lambda\left(\frac{|y| - \sqrt{2\,\epsilon}}{\sqrt{2\epsilon}(\sqrt{2} - 1)} \right) \lambda\left(\frac{|z| - \sqrt{\epsilon}}{\sqrt{\epsilon}} \right).$$

Then, on $\mathrm{Zer}(P, \mathbb{R}^k)_{[-\epsilon,\epsilon]} \cap \phi^{-1}(S_2 \setminus S_1)$ we consider the vector field

$$\Gamma'(p) = \mu(\phi(p))\Gamma_2(p) + (1 - \mu(\phi(p)))\Gamma_1(p).$$

Notice that it agrees with the vector fields defined on

$$\mathrm{Zer}(P, \mathbb{R}^k)_{[-\epsilon,\epsilon]} \cap \phi^{-1}(S \setminus S_2),\ \mathrm{Zer}(P, \mathbb{R}^k)_{[-\epsilon,\epsilon]} \cap \phi^{-1}(S_1).$$

For a point $p = (x_1, y, z) \in \mathrm{Zer}(Q, \mathbb{R}^k)_{[-\epsilon,\epsilon]} \cap \phi^{-1}(S_2 \setminus S_1)$, we denote by $g(p)$ the point on the flow line through p of the vector field Γ_2 such that $\pi(g(p)) = -\epsilon$.

Denote the flow through a point $p \in \mathrm{Zer}(P, \mathbb{R}^k)_\epsilon \cap \phi^{-1}(S)$ of the vector field Γ' by $\gamma_p: [0, 1] \to \mathrm{Zer}(P, \mathbb{R}^k)_{[-\epsilon,\epsilon]}$, with $\gamma_p(0) = p$.

For $x \in \mathrm{Zer}(P, \mathbb{R}^k)_\epsilon$ and $t \in [0, 1]$, define $\alpha_t(x) = \gamma_x(t)$. By construction of the vector field Γ, α_t has the required properties. □

Before proving Theorem 7.14 it is instructive to consider an example.

Example 7.17. Consider a smooth torus in R^3 (see Figure 7.5). There are four critical points p_1, p_2, p_3 and p_4 with critical values v_1, v_2, v_3 and v_4 and indices $2, 1, 1$ and 0 respectively, for the projection map to the X_1 coordinate.

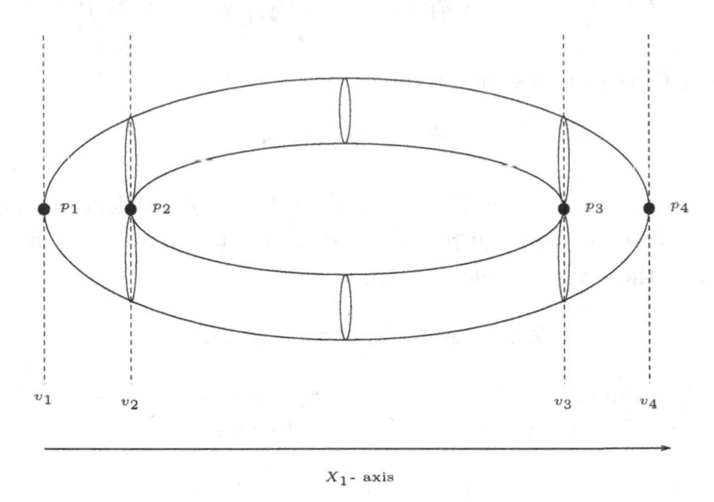

Fig. 7.5. Changes in the homotopy type of the smooth torus in $\mathbb{R}^3$ at the critical values

The changes in homotopy type at the corresponding critical values are described as follows: At the critical value v_1 we add a 0-dimensional ball. At the critical values v_2 and v_3 we add 1-dimensional balls and finally at v_4 we add a 2-dimensional ball. □

Proof of Theorem 7.14: We construct a vector field Γ' on $\mathrm{Zer}(Q,\mathbb{R}^k)_{[c-\epsilon,c+\epsilon]}$ that agrees with the gradient vector field Γ everywhere except in a small neighborhood of the critical point p. At the critical point p, we use Lemma 7.15 to reduce to the quadratic case and then use Lemma 7.16 to construct a vector field in a neighborhood of the critical point that agrees with Γ outside the neighborhood. We now use this vector field, as in the proof of Theorem 7.5, to obtain the required homotopy equivalence. □

We also need to analyze the topological changes that occur to sets bounded by non-singular algebraic hypersurfaces.

We are also going to prove the following versions of Theorem 7.5 (Morse Lemma A) and Theorem 7.14 (Morse Lemma B).

Proposition 7.18. *Let S be a bounded set defined by $Q \geq 0$, bounded by the non-singular algebraic hypersurface* $\mathrm{Zer}(Q,\mathbb{R}^k)$. *Let $[a,b]$ be an interval containing no critical value of π on* $\mathrm{Zer}(Q,\mathbb{R}^k)$. *Then $S_{[a,b]}$ is homeomorphic to $S_a \times [a,b]$ and $S_{\leq a}$ is homotopy equivalent to $S_{\leq b}$.*

Proposition 7.19. *Let S be a bounded set defined by $Q \geq 0$, bounded by the non-singular algebraic hypersurface* $\mathrm{Zer}(Q, \mathbb{R}^k)$*. Suppose that the projection π to the X_1-axis is a Morse function. Let p be the non-degenerate critical point of π on ∂W of index λ such that $\pi(p) = c$. For all sufficiently small $\epsilon > 0$, the set $S_{\leq c+\epsilon}$ has*

- *the homotopy type of $S_{\leq c-\epsilon}$ if $(\partial Q/\partial X_1)(p) < 0$,*
- *the homotopy type of the union of $S_{\leq c-\epsilon}$ with a ball of dimension $k-1-\lambda$ attached along its boundary, if $(\partial Q/\partial X_1)(p) > 0$.*

Example 7.20. Consider the set in $\mathbb{R}^3$ bounded by the smooth torus. Suppose that this set is defined by the single inequality $Q \geq 0$. In other words, Q is positive in the interior of the torus and negative outside. Referring back to Figure 7.5, we see that at the critical points p_2 and p_4, $(\partial Q/\partial X_1)(p) < 0$ and hence according to Proposition 7.19 there is no change in the homotopy type at the two corresponding critical values v_2 and v_4. However, $(\partial Q/\partial X_1)(p) > 0$ at p_1 and p_3 and hence we add a 0-dimensional and an 1-dimensional balls at the two critical values v_1 and v_3 respectively. □

Proof of Proposition 7.18: Suppose that S, defined by $Q \geq 0$, is bounded by the non-singular algebraic hypersurface $\mathrm{Zer}(Q, \mathbb{R}^k)$. We introduce a new variable, X_{k+1}, and consider the polynomial $Q_+ = Q - X_{k+1}^2$ and the corresponding algebraic set $\mathrm{Zer}(Q_+, \mathbb{R}^{k+1})$. Let ϕ: $\mathbb{R}^{k+1} \to \mathbb{R}^k$ be the projection map to the first k coordinates.

Topologically, $\mathrm{Zer}(Q_+, \mathbb{R}^{k+1})$ consists of two copies of S glued along $\mathrm{Zer}(Q, \mathbb{R}^k)$. Moreover, denoting by π' the projection from $\mathbb{R}^{k+1}$ to $\mathbb{R}$ forgetting the last k coordinates, $\mathrm{Zer}(Q_+, \mathbb{R}^{k+1})$ is non-singular and the critical points of π' on $\mathrm{Zer}(Q_-, \mathbb{R}^{k+1})$ are the critical points of π on $\mathrm{Zer}(Q, \mathbb{R}^k)$ (considering $\mathrm{Zer}(Q, \mathbb{R}^k)$ as a subset of the hyperplane defined by the equation $X_{k+1} = 0$). We denote by Γ_+ the gradient vector field on $\mathrm{Zer}(Q_+, \mathbb{R}^{k+1})$.

Since Q_+ is a polynomial in X_1, ..., X_k and X_{k+1}^2, the gradient vector field Γ_+ on $\mathrm{Zer}(Q_+, \mathbb{R}^{k+1})$ is symmetric with respect to the reflection changing X_{k+1} to $-X_{k+1}$. Hence, we can project Γ_+ and its associated flowlines down to the hyperplane defined by $X_{k+1} = 0$ and get a vector field as well as its flowlines in S.

Now, the proof is exactly the same as the proof of Theorem 7.5 above, using the vector field Γ_+ instead of Γ, and projecting the associated vector field down to $\mathbb{R}^k$, noting that the critical values of the projection map onto the first coordinate restricted to $\mathrm{Zer}(Q_+, \mathbb{R}^{k+1})$ are the same as those of $\mathrm{Zer}(Q, \mathbb{R}^k)$. □

For the proof of Proposition 7.19, we first study the quadratic case.

Let π the projection onto the X_1 axis and

$$P = X_1 - \sum_{2 \leq i \leq \lambda+1} X_i^2 + \sum_{\lambda+2 \leq i \leq k} X_i^2.$$

Let B_+ be the set defined by

$$X_2 = \cdots = X_{\lambda+1} = 0, X_1 = -\sum_{\lambda+2 \leq i \leq k} X_i^2, -\epsilon \leq X_1 \leq 0,$$

and let B_- be the set defined by

$$X_2 = \cdots = X_{\lambda+1} = 0, X_1 \leq -\sum_{\lambda+2 \leq i \leq k} X_i^2, -\epsilon \leq X_1 \leq 0.$$

Note that, B_+ is a $(k-\lambda-1)$-ball and $B_- \cap \mathrm{Zer}(P, \mathbb{R}^k)_{\leq -\epsilon}$ is the set defined by

$$X_2 = \cdots = X_{\lambda+1} = 0, X_1 = -\epsilon, \sum_{\lambda+2 \leq i \leq k} X_i^2 \leq \epsilon,$$

which is also the boundary of B_+.

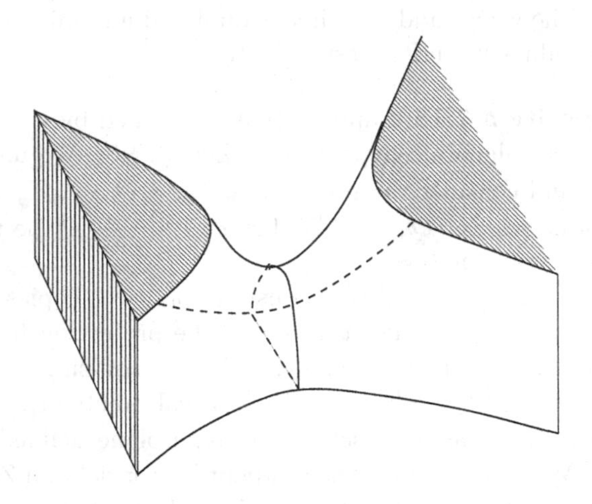

Fig. 7.6. Set defined by $X_1 - X_2^2 + X_3^2 \leq 0$ near the origin

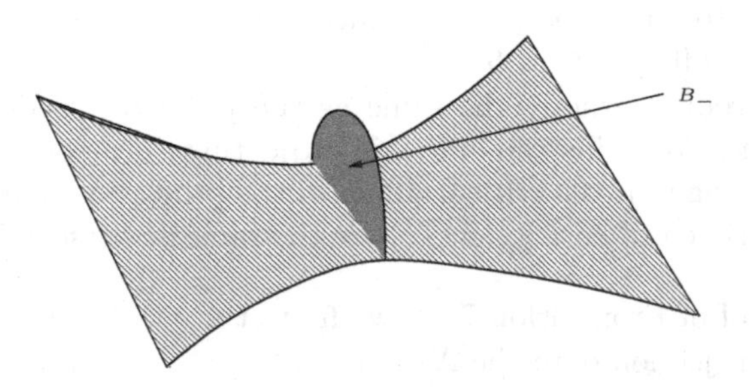

Fig. 7.7. Retract of the set $X_1 - X_2^2 + X_3^2 \leq 0$

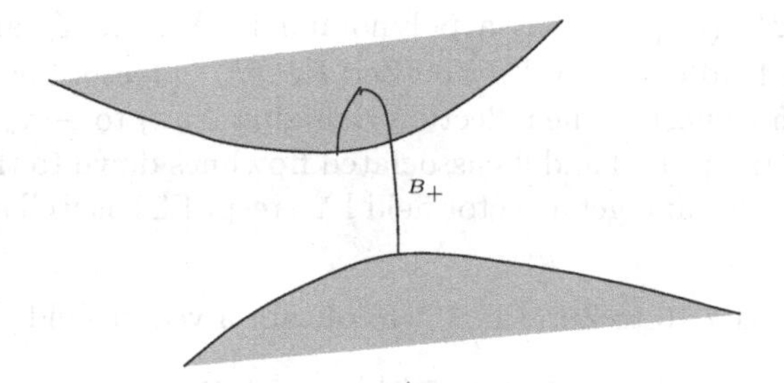

Fig. 7.8. Retract of the set $X_1 - X_2^2 + X_3^2 \geq 0$

Lemma 7.21. *Let* $P_+ = P - X_{k+1}^2, P_- = = P + X_{k+1}^2$.

1. *Let S' be the set defined by $P \geq 0$. Then, for all sufficiently small $\epsilon > 0$ and $r > 2\sqrt{(\epsilon)}$, there exists a vector field Γ'_+ on $S'_{[-\epsilon,\epsilon]} \setminus B_+$, having the following properties:*
   a) *Outside the ball $B_k(r)$, $2\epsilon\Gamma'_+$ equals the projection on $\mathbb{R}^k$ of the gradient vector field, Γ_+, of π on $\mathrm{Zer}(P_+, \mathbb{R}^{k+1})_{[-\epsilon,\epsilon]}$.*
   b) *Associated to Γ'_+, there is a one parameter family of smooth maps $\alpha_t^+ : S'_\epsilon \to S'_{[-\epsilon,\epsilon]}, t \in [0, 1)$, such that for $x \in S'_\epsilon, t \in [0, 1)$,*
      i. *Each α_t^+ is injective,*
      ii.
      $$\frac{\mathrm{d}\alpha_t^+(x)}{\mathrm{d}t} = \Gamma'_+(\alpha_t^+(x)),$$
      iii. $\alpha_0^+(x) = x$,
      iv. $\lim_{t \to 1} \alpha_t^+(x) \in S'_{-\epsilon} \cup B_+$ *and,*
      v. *for every $y \in S_{[-\epsilon,\epsilon]} \setminus B_+$ there exists a unique $z \in S_\epsilon$ and $t \in [0, 1)$ such that $\alpha_t(z) = y$.*
2. *Similarly, let T' be the set defined by $P \leq 0$. Then, for all sufficiently small $\epsilon > 0$ and $r > 2\sqrt{(\epsilon)}$, there exists a vector field Γ'_- on $T'_{[-\epsilon,\epsilon]} \setminus B_+$ having the following properties:*
   a) *Outside the ball $B_k(r)$, $2\epsilon\Gamma'_-$ the projection on $\mathbb{R}^k$ of the gradient vector field, Γ_-, of π on $\mathrm{Zer}(P_-, \mathbb{R}^{k+1})_{[-\epsilon,\epsilon]}$.*
   b) *Associated to Γ'_-, there is a one parameter continuous family of smooth maps $\alpha_t^- : T_\epsilon \to T_{[-\epsilon,\epsilon]}, t \in [0, 1)$, such that for $x \in T_\epsilon, t \in [0, 1)$*
      i. *Each α_t^- is injective,*
      ii.
      $$\frac{\mathrm{d}\alpha_t^-(x)}{\mathrm{d}t} = \Gamma'_-(\alpha_t^-(x)),$$
      iii. $\alpha_0^-(x) = x$,
      iv. $\lim_{t \to 1} \alpha_t^-(x) \in T'_{-\epsilon} \cup B_-$ *and,*
      v. *for every $y \in T'_{[-\epsilon,\epsilon]} \setminus B_-$, there exists a unique $z \in T'_\epsilon$ and $t \in [0, 1)$ such that $\alpha_t(z) = y$.*

Proof: Since P_+ (resp. P_-) is a polynomial in X_1, ..., X_k and X_{k+1}^2, the gradient vector field Γ_+ (resp. Γ_-) on $\mathrm{Zer}(P_+, \mathbb{R}^{k+1})$ (resp. $\mathrm{Zer}(P_-, \mathbb{R}^{k+1})$) is symmetric with respect to the reflection changing X_{k+1} to $-X_{k+1}$. Hence, we can project Γ_+ (resp. Γ_-) and its associated flowlines down to the hyperplane defined by $X_{k+1}=0$ and get a vector field $\Gamma_+^\star$ (resp. $\Gamma_-^\star$) as well as its flowlines in S' (resp. T').

1. Apply Lemma 7.16 to $\mathrm{Zer}(P_+, \mathbb{R}^k)$ to obtain a vector field Γ'_+ on

$$\mathrm{Zer}(P_+, \mathbb{R}^{k+1})_{[-\epsilon,\epsilon]} \setminus B_+$$

   coinciding with $\Gamma_+^\star$. Figure 7.8 illustrates the situation in the case $k=3$ and $\lambda=1$.
2. Apply Lemma 7.16 to $\mathrm{Zer}(Q_-, \mathbb{R}^k)$ to obtain a vector field Γ'_- on

$$\mathrm{Zer}(Q_-, \mathbb{R}^{k+1})_{[-\epsilon,\epsilon]} \setminus \phi^{-1}(B_-)$$

   coinciding with $\Gamma_-^\star$. Figures 7.8 and 7.8 illustrate the situation in the case $k=3$ and $\lambda=1$. □

We are now in a position to prove Proposition 7.19.

Proof of Proposition 7.19: First, use Lemma 7.15 to reduce to the quadratic case, and then use Lemma 7.21, noting that the sign of $\partial Q/\partial X_1\}(p)$ determines which case we are in. □

7.2 Sum of the Betti Numbers of Real Algebraic Sets

For a closed semi-algebraic set S, let $\mathrm{b}(S)$ denote the sum of the Betti numbers of the simplicial homology groups of S. It follows from the definitions of Chapter 6 that $\mathrm{b}(S)$ is finite (see page 198).

According to Theorem 5.47, there are a finite number of algebraic subsets of R^k defined by polynomials of degree at most d, say $V_1, \ldots, V_p$, such that any algebraic subset V of R^k so defined is semi-algebraically homeomorphic to one of the V_i. It follows immediately that any algebraic subset of R^k defined by polynomials of degree at most d is such that $\mathrm{b}(V) \le \max\{\mathrm{b}(V_1), \ldots, \mathrm{b}(V_p)\}$. Let $\mathrm{b}(k,d)$ be the smallest integer which bounds the sum of the Betti numbers of any algebraic set defined by polynomials of degree d in R^k. The goal of this section is to bound the Betti numbers of a bounded non-singular algebraic hypersurface in terms of the number of critical values of a function defined on it and to obtain explicit bounds for $\mathrm{b}(k,d)$.

Remark 7.22. Note that $\mathrm{b}(k, d) \ge d^k$ since the solutions to the system of equations,

$$(X_1-1)(X_1-2)\cdots(X_1-d)=\cdots=(X_k-1)(X_k-2)\cdots(X_k-d)=0$$

consist of d^k isolated points and the only non-zero Betti number of this set is $\mathrm{b}_0 = d^k$. (Recall that for a closed and bounded semi-algebraic set S, $\mathrm{b}_0(S)$ is the number of semi-algebraically connected components of S by Proposition 6.34.) □

We are going to prove the following theorem.

Theorem 7.23. [Oleinik-Petrovski/Thom/Milnor bound]

$$\mathrm{b}(k,d) \le d\,(2\,d-1)^{k-1}.$$

The method for proving Theorem 7.23 will be to use Theorems 7.5 and 7.14, which give enough information about the homotopy type of $\mathrm{Zer}(Q,\, \mathbb{R}^k)$ to enable us to bound $\mathrm{b}(\mathrm{Zer}(Q,\, \mathbb{R}^k))$ in terms of the number of critical points of π.

A first consequence of Theorems 7.5 and 7.14 is the following result.

Theorem 7.24. *Let $\mathrm{Zer}(Q,\mathbb{R}^k)$ be a non-singular bounded algebraic hypersurface such that the projection π on the X_1-axis is a Morse function. For $0 \le i \le k-1$, let c_i be the number of critical points of π restricted to $\mathrm{Zer}(Q,\mathbb{R}^k)$, of index i. Then,*

$$\begin{aligned} \mathrm{b}(\mathrm{Zer}(Q,\mathbb{R}^k)) &\le \sum_{i=0}^{k-1} c_i, \\ \chi(\mathrm{Zer}(Q,\mathbb{R}^k)) &= \sum_{i=0}^{k-1} (-1)^{k-1-i}\, c_i. \end{aligned}$$

In particular, $\mathrm{b}(\mathrm{Zer}(Q,\mathbb{R}^k))$ is bounded by the number of critical points of π restricted to $\mathrm{Zer}(Q,\mathbb{R}^k)$.

Proof: Let $v_1 < v_2 < \cdots < v_\ell$ be the critical values of π on $\mathrm{Zer}(Q,\mathbb{R}^k)$ and p_i the corresponding critical points, such that $\pi(p_i) = v_i$. Let λ_i be the index of the critical point p_i. We first prove that $\mathrm{b}(\mathrm{Zer}(Q,\mathbb{R}^k)_{\le v_i}) \le i$.

First note that $\mathrm{Zer}(Q,\mathbb{R}^k)_{\le v_1}$ is $\{p_1\}$ and hence

$$\mathrm{b}(\mathrm{Zer}(Q,\mathbb{R}^k)_{\le v_1}) = \mathrm{b}_0(\mathrm{Zer}(Q,\mathbb{R}^k)_{\le v_1}) = 1.$$

By Theorem 7.5, the set $\mathrm{Zer}(Q,\, \mathbb{R}^k)_{\le v_{i+1}-\epsilon}$ is homotopy equivalent to the set $\mathrm{Zer}(Q,\mathbb{R}^k)_{\le v_i+\epsilon}$ for any small enough $\epsilon > 0$, and thus

$$\mathrm{b}(\mathrm{Zer}(Q,\mathbb{R}^k)_{\le v_{i+1}-\epsilon}) = \mathrm{b}(\mathrm{Zer}(Q,\mathbb{R}^k)_{\le v_i+\epsilon}).$$

By Theorem 7.14, the homotopy type of $\mathrm{Zer}(Q,\mathbb{R}^k)_{\le v_i+\epsilon}$ is that of the union of $\mathrm{Zer}(Q,\mathbb{R}^k)_{\le v_i-\epsilon}$ with a topological ball. Recall from Proposition 6.44 that if S_1, S_2 are two closed semi-algebraic sets with non-empty intersection, then

$$\mathrm{b}_i(S_1 \cup S_2) \le \mathrm{b}_i(S_1) + \mathrm{b}_i(S_2) + \mathrm{b}_{i-1}(S_1 \cap S_2), 0 \le i \le k-1.$$

Recall also from Proposition 6.34 that for a closed and bounded semi-algebraic set S, $b_0(S)$ equals the number of connected components of S. Since, $S_1 \cap S_2 \neq \emptyset$, for $i=0$ we have the stronger inequality,

$$b_0(S_1 \cup S_2) \leq b_0(S_1) + b_0(S_2) - 1.$$

By Proposition 6.37, for $\lambda > 1$ we have that

$$\begin{aligned} b_0(B_\lambda) &= b_0(S^{\lambda-1}) \\ &= b_{\lambda-1}(S^{\lambda-1}) \\ &= 1, \\ b_i(B_\lambda) &= 0, i>0, \\ b_i(S^{\lambda-1}) &= 0, 0<i<\lambda-1. \end{aligned}$$

It follows that, for $\lambda > 1$, attaching a λ-ball can increase b_λ by at most one, and none of the other Betti numbers can increase.

For $\lambda = 1$, $b_{\lambda-1}(S^{\lambda-1}) = b_0(S^0) = 2$. It is an exercise to show that in this case, b_1 can increase by at most one and no other Betti numbers can increase. (Hint. The number of cycles in a graph can increase by at most one on addition of an edge.)

It thus follows that

$$b(\mathrm{Zer}(Q, \mathrm{R}^k)_{\leq v_i+\epsilon}) \leq b(\mathrm{Zer}(Q, \mathrm{R}^k)_{\leq v_i-\epsilon}) + 1.$$

This proves the first part of the lemma.

We next prove that for $1 < i \leq \ell$ and small enough $\epsilon > 0$,

$$\chi(\mathrm{Zer}(Q, \mathrm{R}^k)_{\leq v_i+\epsilon}) = \chi(\mathrm{Zer}(Q, \mathrm{R}^k)_{\leq v_{i-1}+\epsilon}) + (-1)^{k-1-\lambda_i}.$$

By Theorem 7.5, the set $\mathrm{Zer}(Q, \mathrm{R}^k)_{\leq v_i-\epsilon}$ is homotopy equivalent to the set $\mathrm{Zer}(Q, \mathrm{R}^k)_{\leq v_{i-1}+\epsilon}$ for any small enough $\epsilon > 0$, and thus

$$\chi(\mathrm{Zer}(Q, \mathrm{R}^k)_{\leq v_i-\epsilon}) = \chi(\mathrm{Zer}(Q, \mathrm{R}^k)_{\leq v_{i-1}+\epsilon}).$$

By Theorem 7.14, the homotopy type of $\mathrm{Zer}(Q, \mathrm{R}^k)_{\leq v_i+\epsilon}$ is that of the union of $\mathrm{Zer}(Q, \mathrm{R}^k)_{\leq v_i-\epsilon}$ with a topological ball of dimension $k-1-\lambda_i$. Recall from Corollary 6.36 (Equation 6.36) that if S_1, S_2 are two closed and bounded semi-algebraic sets with non-empty intersection, then

$$\chi(S_1 \cup S_2) = \chi(S_1) + \chi(S_2) - \chi(S_1 \cap S_2).$$

Hence,

$$\begin{aligned} \chi(\mathrm{Zer}(Q, \mathrm{R}^k)_{\leq v_i+\epsilon}) &= \chi(\mathrm{Zer}(Q, \mathrm{R}^k)_{\leq v_{i-1}+\epsilon}) \\ &= \chi(\overline{B}_{k-1-\lambda_i}) \\ &- \chi(S^{k-2-\lambda_i}). \end{aligned}$$

Now, it follows from Proposition 6.37 and the definition of Euler-Poincaré characteristic, that $\chi(\overline{B}_{k-1-\lambda_i}) = 1$ and $\chi(S^{k-2-\lambda_i}) = 1 + (-1)^{k-2-\lambda_i}$.

Substituting in the equation above we obtain that

$$\chi(\mathrm{Zer}(Q,\mathbb{R}^k)_{\leq v_i+\epsilon}) = \chi(\mathrm{Zer}(Q,\mathbb{R}^k)_{\leq v_{i-1}+\epsilon}) + (-1)^{k-1-\lambda_i}.$$

The second part of the theorem is now an easy consequence. □

We shall need the slightly more general result.

Proposition 7.25. *Let* $\mathrm{Zer}(Q,\mathbb{R}^k)$ *be a non-singular bounded algebraic hypersurface such that the projection* π *on the* X_1*-axis has non-degenerate critical points on* $\mathrm{Zer}(Q,\mathbb{R}^k)$*. For* $0 \leq i \leq k-1$*, let* c_i *be the number of critical points of* π *restricted to* $\mathrm{Zer}(Q,\mathbb{R}^k)$*, of index* i*. Then,*

$$\begin{aligned} \mathrm{b}(\mathrm{Zer}(Q,\mathbb{R}^k)) &\leq \sum_{i=0}^{k-1} c_i, \\ \chi(\mathrm{Zer}(Q,\mathbb{R}^k)) &= \sum_{i=0}^{k-1} (-1)^{k-1-i} c_i. \end{aligned}$$

In particular, $\mathrm{b}(\mathrm{Zer}(Q,\mathbb{R}^k))$ *is bounded by the number of critical points of* π *restricted to* $\mathrm{Zer}(Q,\mathbb{R}^k)$*.*

Proof: Use Lemma7.13 and Theorem 7.24. □

Using Theorem 7.24, we can estimate the sum of the Betti numbers in the bounded case.

Proposition 7.26. *Let* $\mathrm{Zer}(Q,\mathbb{R}^k)$ *be a bounded non-singular algebraic hypersurface with* Q *a polynomial of degree* d*. Then*

$$\mathrm{b}(\mathrm{Zer}(Q,\mathbb{R}^k)) \leq d\,(d-1)^{k-1}.$$

Proof: Using Proposition 7.9, we can suppose that π is a Morse function. Applying Theorem 7.24 to the function $\pi\colon \mathrm{Zer}(Q,\mathbb{R}^k) \to \mathbb{R}$, it follows that the sum of the Betti numbers of $\mathrm{Zer}(Q,\mathbb{R}^k)$ is less than or equal to the number of critical points of π. Now apply Proposition 7.10. □

In order to obtain Theorem 7.23, we will need the following Proposition.

Proposition 7.27. *Let* S *be a bounded set defined by* $Q \geq 0$*, bounded by the non-singular algebraic hypersurface* $\mathrm{Zer}(Q,\mathbb{R}^k)$*. Let the projection map* π *be a Morse function on* $\mathrm{Zer}(Q,\mathbb{R}^k)$*. Then, the sum of the Betti numbers of* S *is bounded by half the number of critical points of* π *on* $\mathrm{Zer}(Q,\mathbb{R}^k)$*.*

Proof: We use the notation of the proof of Proposition 7.18. Let $v_1 < v_2 < \cdots < v_\ell$ be the critical values of π on $\mathrm{Zer}(Q, \mathbb{R}^k)$ and $p_1, \ldots, p_\ell$ the corresponding critical points, such that $\pi(p_i) = v_i$. We denote by J the subset of $\{1, \ldots, \ell\}$ such that the direction of $\mathrm{Grad}(Q)(p)$ belongs to S (see Proposition 7.18).

We are going to prove that

$$\mathrm{b}(S_{\leq v_i}) \leq \#(j \in J, j \leq i).$$

First note that $S_{\leq v_1}$ is $\{p_1\}$ and hence $\mathrm{b}(S_{\leq v_1}) = 1$. By Proposition 7.18 $S_{\leq v_{i+1}-\epsilon}$ is homotopic to $S_{\leq v_i+\epsilon}$ for any small enough $\epsilon > 0$, and thus

$$\mathrm{b}(S_{\leq v_{i+1}-\epsilon}) = \mathrm{b}(S_{\leq v_i+\epsilon}).$$

By Theorem 7.14, the homotopy type of $S_{\leq v_i+\epsilon}$ is that of $S_{\leq v_i-\epsilon}$ if $i \notin J$ and that of the union of $S_{\leq v_i-\epsilon}$ with a topological ball if $i \in J$.

It follows that

$$\begin{cases} \mathrm{b}(S_{\leq v_i+\epsilon}) = \mathrm{b}(S_{\leq v_i-\epsilon}) & \text{if } i \notin J \\ \mathrm{b}(S_{\leq v_i+\epsilon}) \leq \mathrm{b}(S_{\leq v_i-\epsilon}) + 1 & \text{if } i \in J. \end{cases}$$

By switching the direction of the X_1 axis if necessary, we can always ensure that $\#(J)$ is at most half of the critical points. □

Proposition 7.28. *If* $\mathrm{R} = \mathbb{R}$,

$$\mathrm{b}(k, d) \leq d\,(2\,d - 1)^{k-1}.$$

Proof: Let $V = \mathrm{Zer}(\{P_1, \ldots, P_\ell\}, \mathbb{R}^k)$ with the the degrees of the P_i's bounded by d. By remark on page 226, it suffices to estimate the sum of the Betti numbers of $V \cap \overline{B}_k(0, r)$. Let

$$F(X) = \frac{P_1^2 + \cdots + P_\ell^2}{r^2 - \|X\|^2}.$$

By Sard's theorem (Theorem 5.56), the set of critical values of F is finite. Hence, there is a positive $a \in \mathbb{R}$ so that no $b \in (0, a)$ is a critical value of F and thus the set $W_b = \{x \in \mathbb{R}^k \mid P(x, b) = 0\}$, where

$$P(X, b) = P_1^2 + \cdots + P_\ell^2 + b\,(\|X\|^2 - r^2))$$

is a non-singular hypersurface in $\mathbb{R}^k$. To see this observe that, for $x \in \mathbb{R}^k$

$$P(x, b) = \partial P / \partial X_1(x, b) = \cdots = \partial P / \partial X_k(x, b) = 0$$

implies that $F(x) = b$ and $\partial F / \partial X_1(x) = \cdots = \partial F / \partial X_k(x) = 0$ implying that b is a critical value of F which is a contradiction.

Moreover, W_b is the boundary of the closed and bounded set

$$K_b = \{x \in \mathbb{R}^k \mid P(x, b) \leq 0\}.$$

By Proposition 7.26, the sum of the Betti numbers of W_b is less than or equal to $2\,d\,(2\,d - 1)^{k-1}$.

Also, using Proposition 7.27, the sum of the Betti numbers of K_b is at most half that of W_b.

We now claim that $V\cap\overline{B}_k(0,r)$ is homotopy equivalent to K_b for all small enough $b>0$. We replace b in the definition of the set K_b by a new variable T, and consider the set $K\subset\mathbb{R}^{k+1}$ defined by $\{(x,t)\in\mathbb{R}^{k+1}|P(x,t)\le 0\}$. Let π_X (resp. π_T) denote the projection map onto the X (resp. T) coordinates.

Clearly, $V\cap\overline{B}_k(0,r)\subset K_b$. By Theorem 5.46 (Semi-algebraic triviality), for all small enough $b>0$, there exists a semi-algebraic homeomorphism,

$$\phi\colon K_b\times(0,b]\to K\cap\pi_T^{-1}((0,b]),$$

such that $\pi_T(\phi(x,\ s)) = s$ and ϕ is a semi-algebraic homeomorphism from $V\cap B_k(0,r)\times(0,b]$ to itself.

Let $G\colon K_b\times[0,\ b]\to K_b$ be the map defined by $G(x,\ s)=\pi_X(\phi(x,\ s))$ for $s>0$ and $G(x,0)=\lim_{s\to 0+}\pi_X(\phi(x,\ s))$. Let $g\colon K_b\to V\cap\overline{B}_k(0,\ r)$ be the map $G(x,0)$ and $i\colon V\cap\overline{B}_k(0,\ r)\to K_b$ the inclusion map. Using the homotopy G, we see that $i\circ g\sim \mathrm{Id}_{K_b}$, and $g\circ i\sim\mathrm{Id}_{V\cap B_k(0,r)}$, which shows that $V\cap B_k(0,r)$ is homotopy equivalent to K_b as claimed.

Hence,

$$\mathrm{b}(V\cap\overline{B}_k(0,r))=\mathrm{b}(K_b)\le 1/2\,\mathrm{b}(W_b)\le d\,(2\,d-1)^{k-1}. \qquad\square$$

Proof of Theorem 7.23: It only remains to prove that Proposition 7.28 is valid for any real closed field R. We first work over the field of real algebraic numbers $\mathbb{R}_{\mathrm{alg}}$. We identify a system of ℓ polynomials $(P_1,\ldots,P_\ell)$ in k variables of degree less than or equal to d with the point of $\mathbb{R}_{\mathrm{alg}}^N$, $N=\ell\binom{k+d-1}{d}$, whose coordinates are the coefficients of $P_1,\ldots,P_\ell$. Let

$$Z=\{(P_1,\ldots,P_\ell,x)\in\mathbb{R}_{\mathrm{alg}}^N\times\mathbb{R}_{\mathrm{alg}}^k\mid P_1(x)=\cdots=P_\ell(x)=0\},$$

and let $\Pi\colon Z\to\mathbb{R}_{\mathrm{alg}}^N$ be the canonical projection. By Theorem 5.46 (Semi-algebraic Triviality), there exists a finite partition of $\mathbb{R}_{\mathrm{alg}}^N$ into semi-algebraic sets $A_1,\ldots,A_m$, semi-algebraic sets $F_1,\ldots,F_m$ contained in $\mathbb{R}_{\mathrm{alg}}^k$, and semi-algebraic homeomorphisms $\theta_i\colon\Pi^{-1}(A_i)\to A_i\times F_i$, for $i=1,\ldots,m$, such that the composition of θ_i with the projection $A_i\times F_i\to A_i$ is $\Pi|_{\Pi^{-1}(A_i)}$. The F_i are algebraic subsets of $\mathbb{R}_{\mathrm{alg}}^k$ defined by ℓ equations of degree less than or equal to d. The sum of the Betti numbers of $\mathrm{Ext}(F_i,\mathbb{R})$ is less than or equal to $d\,(2\,d-1)^{k-1}$. So, by invariance of the homology groups under extension of real closed field (Section 6.2), the same bound holds for the sum of the Betti numbers of F_i. Now, let $V\subset\mathrm{R}^k$ be defined by k equations $P_1=\cdots=P_\ell=0$ of degree less than or equal to d with coefficients in R. We have

$$\mathrm{Ext}(\Pi^{-1},\mathrm{R})(P_1,\ldots,P_\ell)=\{(P_1,\ldots,P_\ell)\}\times V.$$

The point $(P_1,\ldots,P_\ell)\in\mathrm{R}^N$ belongs to some $\mathrm{Ext}(A_i,\mathrm{R})$, and the semi-algebraic homeomorphism $\mathrm{Ext}(\theta_i,\mathrm{R})$ induces a semi-algebraic homeomorphism from V onto $\mathrm{Ext}(F_i,\ \mathrm{R})$. Again, the sum of the Betti numbers of $\mathrm{Ext}(F_i,\ \mathrm{R})$ is less than or equal to $d\,(2\,d-1)^{k-1}$, and the same bound holds for the sum of the Betti numbers of V. $\square$

7.3 Bounding the Betti Numbers of Realizations of Sign Conditions

Throughout this section, let $\mathcal{Q}$ and $\mathcal{P} \neq \emptyset$ be finite subsets of $\mathrm{R}[X_1, \ldots, X_k]$, let $Z = \mathrm{Zer}(\mathcal{Q}, \mathrm{R}^k)$, and let k' be the dimension of $Z = \mathrm{Zer}(\mathcal{Q}, \mathrm{R}^k)$.

Notation 7.29. [Realizable sign conditions] We denote by

$$\mathrm{SIGN}(\mathcal{P}) \subset \{0, 1, -1\}^{\mathcal{P}}$$

the set of all realizable sign conditions for $\mathcal{P}$ over R^k, and by

$$\mathrm{SIGN}(\mathcal{P}, \mathcal{Q}) \subset \{0, 1, -1\}^{\mathcal{P}}$$

the set of all realizable sign conditions for $\mathcal{P}$ over $\mathrm{Zer}(\mathcal{Q}, \mathrm{R}^k)$. □

For $\sigma \in \mathrm{SIGN}(\mathcal{P}, \mathcal{Q})$, let $\mathrm{b}_i(\sigma)$ denote the i-th Betti number of

$$\mathrm{Reali}(\sigma, Z) = \{x \in \mathrm{R}^k \mid \bigwedge_{Q \in \mathcal{Q}} Q(x) = 0, \bigwedge_{P \in \mathcal{P}} \mathrm{sign}(P(x)) = \sigma(P)\}.$$

Let $\mathrm{b}_i(\mathcal{Q}, \mathcal{P}) = \sum_\sigma \mathrm{b}_i(\sigma)$. Note that $\mathrm{b}_0(\mathcal{Q}, \mathcal{P})$ is the number of semi-algebraically connected components of basic semi-algebraic sets defined by $\mathcal{P}$ over $\mathrm{Zer}(\mathcal{Q}, \mathrm{R}^k)$.

We denote by $\deg(\mathcal{Q})$ the maximum of the degrees of the polynomials in $\mathcal{Q}$ and write $\mathrm{b}_i(d, k, k', s)$ for the maximum of $\mathrm{b}_i(\mathcal{Q}, \mathcal{P})$ over all $\mathcal{Q}, \mathcal{P}$, where $\mathcal{Q}$ and $\mathcal{P}$ are finite subsets of $\mathrm{R}[X_1, \ldots, X_k]$, $\deg(\mathcal{Q}, \mathcal{P}) \leqslant d$ whose elements have degree at most d, $\#(\mathcal{P}) = s$ (i.e. $\mathcal{P}$ has s elements), and the algebraic set $\mathrm{Zer}(\mathcal{Q}, \mathrm{R}^k)$ has dimension k'.

Theorem 7.30.

$$\mathrm{b}_i(d, k, k', s) \leq \sum_{1 \leq j \leq k' - i} \binom{s}{j} 4^j d (2d - 1)^{k-1}.$$

So we get, in particular a bound on the total number of semi-algebraically connected components of realizable sign conditions.

Proposition 7.31.

$$\mathrm{b}_0(d, k, k', s) \leq \sum_{1 \leq j \leq k'} \binom{s}{j} 4^j d (2d - 1)^{k-1}.$$

Remark 7.32. When $d = 1$, i.e. when all equations are linear, it is easy to find directly a bound on the number of non-empty sign conditions. The number of non-empty sign conditions $f(k', s)$ defined by s linear equations on a flat of dimension k' satisfies the recurrence relation

$$f(k', s+1) \leq f(k', s) + 2 f(k'-1, s),$$

since a flat L of dimension $k'-1$ meets at most $f(k'-1,s)$ non-empty sign condition defined by s polynomials on a flat of dimension k', and each such non-empty sign condition is divided in at most three pieces by L.

In Figure 7.9 we depict the situation with four lines in R^2 defined by four linear polynomials. The number of realizable sign conditions in this case is easily seen to be 33.

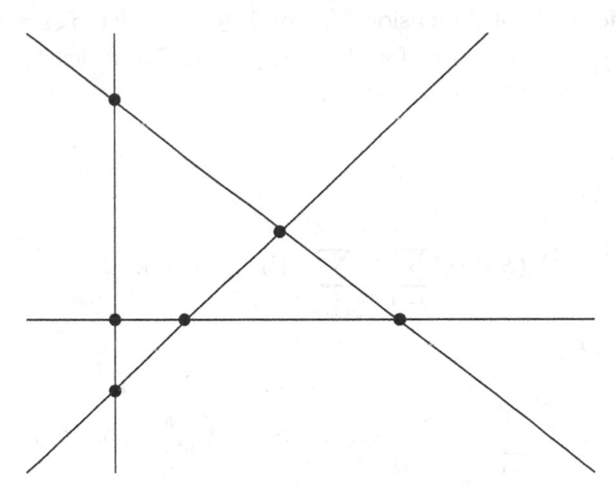

Fig. 7.9. Four lines in R^2

Moreover, when the linear equations are in general position,

$$f(k',s+1)=f(k',s)+2\,f(k'-1,s). \tag{7.4}$$

Since $f(k',0)=1$, the solution to Equation (7.4) is given by

$$f(k',s)=\sum_{i=0}^{k'}\sum_{j=0}^{k'-i}\binom{s}{i}\binom{s-i}{j}. \tag{7.5}$$

Since all the realizations are convex and hence contractible, this bound on the number of non-empty sign conditions is also a bound on

$$\mathrm{b}_0(1,k,k',s)=\mathrm{b}(1,k',k',s)$$

We note that

$$f(k',s)\leq\sum_{1\leq j\leq k'}\binom{s}{j}4^j,$$

the right hand side being the bound appearing in Proposition 7.31 with $d=1$.
□

The following proposition, Proposition 7.33, plays a key role in the proofs of these theorems. Part (a) of the proposition bounds the Betti numbers of a union of s semi-algebraic sets in R^k in terms of the Betti numbers of the intersections of the sets taken at most k at a time.

Part (b) of the proposition is a dual version of Part (a) with unions being replaced by intersections and vice-versa, with an additional complication arising from the fact that the empty intersection, corresponding to the base case of the induction, is an arbitrary real algebraic variety of dimension k', and is generally not acyclic.

Let $S_1, \ldots, S_s \subset \mathrm{R}^k$, $s \geq 1$, be closed semi-algebraic sets contained in a closed semi-algebraic set T of dimension k'. For $1 \leq t \leq s$, let $S_{\leq t} = \bigcap_{1 \leq j \leq t} S_j$, and $S^{\leq t} = \bigcup_{1 \leq j \leq t} S_j$. Also, for $J \subset \{1, \ldots, s\}$, $J \neq \emptyset$, let $S_J = \bigcap_{j \in J} S_j$, and $S^J = \bigcup_{j \in J} S_j$. Finally, let $S^{\emptyset} = T$.

Proposition 7.33.

a) For $0 \leq i \leq k'$,

$$\mathrm{b}_i(S^{\leq s}) \leq \sum_{j=1}^{i+1} \sum_{\substack{J \subset \{1, \ldots, s\} \\ \#(J) = j}} \mathrm{b}_{i-j+1}(S_J). \tag{7.6}$$

b) For $0 \leq i \leq k'$,

$$\mathrm{b}_i(S_{\leq s}) \leq \sum_{j=1}^{k'-i} \sum_{\substack{J \subset \{1, \ldots, s\} \\ \#(J) = j}} \mathrm{b}_{i+j-1}(S^J) + \binom{s}{k'-i} \mathrm{b}_{k'}(S^{\emptyset}). \tag{7.7}$$

Proof : a)We prove the claim by induction on s. The statement is clearly true for $s = 1$, since $\mathrm{b}_i(S_1)$ appears on the right hand side for $j = 1$ and $J = \{1\}$.

Using Proposition 6.44 (6.44), we have that

$$\mathrm{b}_i(S^{\leq s}) \leq \mathrm{b}_i(S^{\leq s-1}) + \mathrm{b}_i(S_s) + \mathrm{b}_{i-1}(S^{\leq s-1} \cap S_s). \tag{7.8}$$

Applying the induction hypothesis to the set $S^{\leq s-1}$, we deduce that

$$\mathrm{b}_i(S^{\leq s-1}) \leq \sum_{j=1}^{i+1} \sum_{\substack{J \subset \{1, \ldots, s-1\} \\ \#(J) = j}} \mathrm{b}_{i-j+1}(S_J). \tag{7.9}$$

Next, we apply the induction hypothesis to the set

$$S^{\leq s-1} \cap S_s = \bigcup_{1 \leq j \leq s-1} (S_j \cap S_s)$$

to get that

$$\mathrm{b}_{i-1}(S^{\leq s-1} \cap S_s) \leq \sum_{j=1}^{i} \sum_{\substack{J \subset \{1, \ldots, s-1\} \\ \#(J) = j}} \mathrm{b}_{i-j}(S_{J \cup \{s\}}). \tag{7.10}$$

Adding the inequalities (7.9) and (7.10), we get

$$\mathrm{b}_i(S^{\leq s-1}) + \mathrm{b}_i(S_s) + \mathrm{b}_{i-1}(S^{\leq s-1} \cap S_s) \leq \sum_{j=1}^{i+1} \sum_{\substack{J \subset \{1, \ldots, s\} \\ \#(J) = j}} \mathrm{b}_{i-j+1}(S_J).$$

We conclude using (7.8).

b) We first prove the claim when $s=1$. If $0 \leq i \leq k'-1$, the claim is

$$\mathrm{b}_i(S_1) \leq \mathrm{b}_{k'}(S^{\emptyset}) + \left(\mathrm{b}_i(S_1) + \mathrm{b}_{k'}(S^{\emptyset})\right),$$

which is clear. If $i=k'$, the claim is $\mathrm{b}_{k'}(S_1) \leq \mathrm{b}_{k'}(S^{\emptyset})$. If the dimension of S_1 is k', consider the closure V of the complement of S_1 in T. The intersection W of V with S_1, which is the boundary of S_1, has dimension strictly smaller than k' by Theorem 5.42 thus $\mathrm{b}_{k'}(W)=0$. Using Proposition 6.44

$$\mathrm{b}_{k'}(S_1) + \mathrm{b}_{k'}(V) \leq \mathrm{b}_{k'}(S^{\emptyset}) + \mathrm{b}_{k'}(W),$$

and the claim follows. On the other hand, if the dimension of S_1 is strictly smaller than k', $\mathrm{b}_{k'}(S_1)$=0.

The claim is now proved by induction on s. Assume that the induction hypothesis (7.7) holds for $s-1$ and for all $0 \leq i \leq k'$.

From Proposition 6.44 (6.44), we have

$$\mathrm{b}_i(S_{\leq s}) \leq \mathrm{b}_i(S_{\leq s-1}) + \mathrm{b}_i(S_s) + \mathrm{b}_{i+1}(S_{\leq s-1} \cup S_s). \tag{7.11}$$

Applying the induction hypothesis to the set $S_{\leq s-1}$, we deduce that

$$\begin{aligned}\mathrm{b}_i(S_{\leq s-1}) \;\leq\; & \sum_{j=1}^{k'-i} \sum_{\substack{J \subset \{1,\ldots,s-1\} \\ \#(J)=j}} \mathrm{b}_{i+j-1}(S^J) \\ & + \binom{s-1}{k'-i} \mathrm{b}_{k'}(S^{\emptyset}).\end{aligned}$$

Next, applying the induction hypothesis to the set,

$$S_{\leq s-1} \cup S_s = \bigcap_{1 \leq j \leq s-1} (S_j \cup S_s),$$

we get that

$$\begin{aligned}\mathrm{b}_{i+1}(S_{\leq s-1} \cup S_s) \;\leq\; & \sum_{j=1}^{k'-i-1} \sum_{\substack{J \subset \{1,\ldots,s-1\} \\ \#(J)=j}} \mathrm{b}_{i+j}(S^{J \cup \{s\}}) \\ & + \binom{s-1}{k'-i-1} \mathrm{b}_{k'}(S^{\emptyset}).\end{aligned} \tag{7.12}$$

Adding the inequalities (7.11) and (7.11), we get

$$\mathrm{b}_i(S_{\leq s}) \leq \sum_{j=1}^{k'-i} \sum_{\substack{J \subset \{1,\ldots,s\} \\ \#(J)=j}} \mathrm{b}_{i+j-1}(S^J) + \binom{s}{k'-i} \mathrm{b}_{k'}(S^{\emptyset}).$$

We conclude using (7.11). □

Let $\mathcal{P} = \{P_1, \ldots, P_s\}$, and let δ be a new variable. We will consider the field $\mathrm{R}\langle\delta\rangle$ of algebraic Puiseux series in δ, in which δ is an infinitesimal.

Let $S_i = \mathrm{Reali}(P_i^2(P_i^2 - \delta^2) \geq 0, \mathrm{Ext}(Z, \mathrm{R}\langle\delta\rangle))$, $1 \leq i \leq s$, and let S be the intersection of the S_i with the closed ball in $\mathrm{R}\langle\delta\rangle^k$ defined by

$$\delta^2 \left(\sum_{1 \leq i \leq k} X_i^2 \right) \leq 1.$$

In order to estimate $\mathrm{b}_i(S)$, we prove that $\mathrm{b}_i(\mathcal{P}, \mathcal{Q})$ and $\mathrm{b}_i(S)$ are equal and we estimate $\mathrm{b}_i(S)$.

Proposition 7.34.

$$\mathrm{b}_i(\mathcal{P}, \mathcal{Q}) = \mathrm{b}_i(S).$$

Proof: Consider a sign condition σ on $\mathcal{P}$ such that, without loss of generality,

$$\begin{aligned} \sigma(P_i) &= 0 \quad \text{if} i \in I \\ \sigma(P_j) &= 1 \quad \text{if} j \in J \\ \sigma(P_\ell) &= -1 \quad \text{if} \ell \in \{1, \ldots, s\} \setminus (I \cup J), \end{aligned}$$

and denote by $\overline{\mathrm{Reali}}(\sigma)$ the subset of $\mathrm{Ext}(Z, \mathrm{R}\langle\delta\rangle)$ defined by

$$\delta^2 \left(\sum_{1 \leq i \leq k} X_i^2 \right) \leq 1, P_i = 0, i \in I,$$

$$P_j \geq \delta, j \in J, P_\ell \leq -\delta, \ell \in \{1, \ldots, s\} \setminus (I \cup J).$$

Note that S is the disjoint union of the $\overline{\mathrm{Reali}}(\sigma)$ for all realizable sign conditions σ.

Moreover, by definition of the homology groups of sign conditions (Notation 6.46) $\mathrm{b}_i(\sigma) = \mathrm{b}_i(\overline{\mathrm{Reali}}(\sigma))$, so that

$$\mathrm{b}_i(\mathcal{P}, \mathcal{Q}) = \sum_{\sigma} \mathrm{b}_i(\sigma) = \mathrm{b}_i(S).$$ □

Proposition 7.35.

$$\mathrm{b}_i(S) \leq \sum_{j=1}^{k'-i} \binom{s}{j} 4^j d (2d-1)^{k-1}.$$

Before estimating $\mathrm{b}_i(S)$, we estimate the Betti numbers of the following sets.

Let $j \geq 1$,

$$V_j = \mathrm{Reali}\left(\bigvee_{1 \leq i \leq j} P_i^2(P_i^2 - \delta^2) = 0, \mathrm{Ext}(Z, \mathrm{R}\langle\delta\rangle) \right),$$

and

$$W_j = \mathrm{Reali}\left(\bigvee_{1 \leq i \leq j} P_i^2(P_i^2 - \delta^2) \geq 0, \mathrm{Ext}(Z, \mathrm{R}\langle\delta\rangle) \right).$$

Note that W_j is the union of $S_1, \ldots, S_j$.

Lemma 7.36.

$$\mathrm{b}_i(V_j) \leq (4^j - 1)\, d\, (2\, d - 1)^{k-1}.$$

Proof: Each of the sets

$$\mathrm{Reali}\big(P_i^2(P_i^2 - \delta^2)) = 0, \mathrm{Ext}(Z, \mathrm{R}\langle\delta\rangle)\big)$$

is the disjoint union of three algebraic sets, namely

$$\mathrm{Reali}(P_i = 0, \mathrm{Ext}(Z, \mathrm{R}\langle\delta\rangle)),$$
$$\mathrm{Reali}(P_i = \delta, \mathrm{Ext}(Z, \mathrm{R}\langle\delta\rangle)),$$
$$\mathrm{Reali}(P_i = -\delta, \mathrm{Ext}(Z, \mathrm{R}\langle\delta\rangle)).$$

Moreover, each Betti number of their union is bounded by the sum of the Betti numbers of all possible non-empty sets that can be obtained by taking, for $1 \leq \ell \leq j$, ℓ-ary intersections of these algebraic sets, using part (a) of Proposition 7.33. The number of possible ℓ-ary intersection is $\binom{j}{\ell}$. Each such intersection is a disjoint union of 3^ℓ algebraic sets. The sum of the Betti numbers of each of these algebraic sets is bounded by $d\,(2\,d - 1)^{k-1}$ by using Theorem 7.23.

Thus,

$$\mathrm{b}_i(V_j) \leq \sum_{\ell=1}^{j} \binom{j}{\ell} 3^\ell d\, (2\, d - 1)^{k-1} = (4^j - 1)\, d\, (2\, d - 1)^{k-1}. \qquad \square$$

Lemma 7.37.

$$\mathrm{b}_i(W_j) \leq (4^j - 1)\, d\, (2\, d - 1)^{k-1} + \mathrm{b}_i(Z).$$

Proof: Let $Q_i = P_i^2(P_i^2 - \delta^2)$ and

$$F = \mathrm{Reali}\left(\bigwedge_{1 \leq i \leq j} (Q_i \leq 0) \vee \bigvee_{1 \leq i \leq j} (Q_i = 0), \mathrm{Ext}(Z, \mathrm{R}\langle\delta\rangle) \right).$$

Apply inequality (6.44), noting that

$$W_j \cup F = \mathrm{Ext}(Z, \mathrm{R}\langle\delta\rangle),\ W_j \cap F = W_{j,0}.$$

Since $\mathrm{b}_i(Z) = \mathrm{b}_i(\mathrm{Ext}(Z, \mathrm{R}\langle\delta\rangle))$, we get that

$$\mathrm{b}_i(W_j) \leq \mathrm{b}_i(W_j \cap F) + \mathrm{b}_i(W_j \cup F) = \mathrm{b}_i(V_j) + \mathrm{b}_i(Z).$$

We conclude using Lemma 7.36. $\square$

Proof of Proposition 7.35: Using part b) of Proposition 7.33 and Lemma 7.37, we get

$$\begin{aligned} \mathrm{b}_i(S) \;&\leq\; \sum_{j=1}^{k'-i} \binom{s}{j} ((4^j-1)\, d\, (2\,d-1)^{k-1} + \mathrm{b}_i(Z)) \\ &+ \binom{s}{k'-i} \mathrm{b}_{k'}(Z). \end{aligned}$$

By Theorem 7.23, for all $i < k'$,

$$\mathrm{b}_i(Z) + \mathrm{b}_{k'}(Z) \leq d\,(2\,d-1)^{k-1}.$$

Thus, we have

$$\mathrm{b}_i(S) \leq \sum_{j=1}^{k'-i} \binom{s}{j} 4^j d\,(2\,d-1)^{k-1}.$$

□

Theorem 7.30 follows clearly from Proposition 7.34 and Proposition 7.35.

7.4 Sum of the Betti Numbers of Closed Semi-algebraic Sets

Let $\mathcal{P}$ and $\mathcal{Q}$ be finite subsets of $\mathrm{R}[X_1, \ldots, X_k]$.

A $(\mathcal{Q}, \mathcal{P})$-**closed formula** is a formula constructed as follows:

- For each $P \in \mathcal{P}$,

$$\bigwedge_{Q \in \mathcal{Q}} Q = 0 \wedge P = 0, \quad \bigwedge_{Q \in \mathcal{Q}} Q = 0 \wedge P \geq 0, \quad \bigwedge_{Q \in \mathcal{Q}} Q = 0 \wedge P \leq 0,$$

- If Φ_1 and Φ_2 are $(\mathcal{Q}, \mathcal{P})$-closed formulas, $\Phi_1 \wedge \Phi_2$ and $\Phi_1 \vee \Phi_2$ are $(\mathcal{Q}, \mathcal{P})$-closed formulas.

Clearly, $\mathrm{Reali}(\Phi)$, the realization of a $(\mathcal{Q}, \mathcal{P})$-closed formula Φ, is a closed semi-algebraic set. We denote by $\mathrm{b}(\Phi)$ the sum of its Betti numbers.

We write $\overline{\mathrm{b}}(d, k, k', s)$ for the maximum of $\mathrm{b}(\Phi)$, where Φ is a $(\mathcal{Q}, \mathcal{P})$-closed formula, $\mathcal{Q}$ and $\mathcal{P}$ are finite subsets of $\mathrm{R}[X_1, \ldots, X_k]$ $\deg(\mathcal{Q}, \mathcal{P}) \leqslant d$, $\#(\mathcal{P}) = s$, and the algebraic set $\mathrm{Zer}(\mathcal{Q}, \mathrm{R}^k)$ has dimension k'.

Our aim in this section is to prove the following result.

Theorem 7.38.

$$\overline{\mathrm{b}}(d, k, k', s) \leq \sum_{i=0}^{k'} \sum_{j=1}^{k'-i} \binom{s}{j} 6^j d\,(2\,d-1)^{k-1}.$$

For the proof of Theorem 7.38, we are going to introduce several infinitesimal quantities. Given a list of polynomials $\mathcal{P} = \{P_1, \ldots, P_s\}$ with coefficients in R, we introduce s new variables $\delta_1, \cdots, \delta_s$ and inductively define

$$\mathrm{R}\langle \delta_1, \ldots, \delta_{i+1} \rangle = \mathrm{R}\langle \delta_1, \ldots, \delta_i \rangle \langle \delta_{i+1} \rangle.$$

Note that δ_{i+1} is infinitesimal with respect to δ_i, which is denoted by

$$\delta_1 \gg \ldots \gg \delta_s.$$

We define $\mathcal{P}_{>i} = \{P_{i+1}, \ldots, P_s\}$ and

$$\begin{aligned}\Sigma_i &= \{P_i = 0, P_i = \delta_i, P_i = -\delta_i, P_i \geq 2\,\delta_i, P_i \leq -2\,\delta_i\},\\ \Sigma_{\leq i} &= \{\Psi \mid \Psi = \bigwedge_{j=1,\ldots,i} \Psi_i, \Psi_i \in \Sigma_i\}.\end{aligned}$$

If Φ is a $(\mathcal{Q}, \mathcal{P})$-closed formula, we denote by $\mathrm{Reali}_i(\Phi)$ the extension of $\mathrm{Reali}(\Phi)$ to $\mathrm{R}\langle\delta_1, \ldots, \delta_i\rangle^k$. For $\Psi \in \Sigma_{\leq i}$, we denote by $\mathrm{Reali}_i(\Phi \wedge \Psi)$ the intersection of the realization of Ψ with $\mathrm{Reali}_i(\Phi)$ and by $\mathrm{b}(\Phi \wedge \Psi)$ the sum of the Betti numbers of $\mathrm{Reali}_i(\Phi \wedge \Psi)$.

Proposition 7.39. *For every $(\mathcal{Q}, \mathcal{P})$-closed formula Φ,*

$$\mathrm{b}(\Phi) \leq \sum_{\substack{\Psi \in \Sigma_{\leq s}\\ \mathrm{Reali}_s(\Psi) \subset \mathrm{Reali}_s(\Phi)}} \mathrm{b}(\Psi).$$

The main ingredient of the proof of the proposition is the following lemma.

Lemma 7.40. *For every $(\mathcal{Q}, \mathcal{P})$-closed formula Φ and every $\Psi \in \Sigma_{\leq i}$,*

$$\mathrm{b}(\Phi \wedge \Psi) \leq \sum_{\psi \in \Sigma_{i+1}} \mathrm{b}(\Phi \wedge \Psi \wedge \psi).$$

Proof: Consider the formulas

$$\begin{aligned}\Phi_1 &= \Phi \wedge \Psi \wedge (P_{i+1}^2 - \delta_{i+1}^2) \geq 0,\\ \Phi_2 &= \Phi \wedge \Psi \wedge (0 \leq P_{i+1}^2 \leq \delta_{i+1}^2).\end{aligned}$$

Clearly, $\mathrm{Reali}_{i+1}(\Phi \wedge \Psi) = \mathrm{Reali}_{i+1}(\Phi_1 \vee \Phi_2)$. Using Proposition 6.44, we have that,

$$\mathrm{b}(\Phi \wedge \Psi) \leq \mathrm{b}(\Phi_1) + \mathrm{b}(\Phi_2) + \mathrm{b}(\Phi_1 \wedge \Phi_2).$$

Now, since $\mathrm{Reali}_{i+1}(\Phi_1 \wedge \Phi_2)$ is the disjoint union of

$$\mathrm{Reali}_{i+1}(\Phi \wedge \Psi \wedge (P_{i+1} = \delta_{i+1})),\ \mathrm{Reali}_{i+1}(\Phi \wedge \Psi \wedge (P_{i+1} = -\delta_{i+1})),$$

$$\mathrm{b}(\Phi_1 \wedge \Phi_2) = \mathrm{b}(\Phi \wedge \Psi \wedge (P_{i+1} = \delta_{i+1})) + \mathrm{b}(\Phi \wedge \Psi \wedge (P_{i+1} = -\delta_{i+1})).$$

Moreover,

$$\begin{aligned}\mathrm{b}(\Phi_1) &= \mathrm{b}(\Phi \wedge \Psi \wedge (P_{i+1}\\ &\quad \geq 2\,\delta_{i+1})) + \mathrm{b}(\Phi \wedge \Psi \wedge (P_{i+1} \leq -2\,\delta_{i+1})),\\ \mathrm{b}(\Phi_2) &= \mathrm{b}(\Phi \wedge \Psi \wedge (P_{i+1} = 0)).\end{aligned}$$

Indeed, by Theorem 5.46 (Hardt's triviality), denoting

$$F_t = \{x \in \mathrm{Reali}_i(\Phi \wedge \Psi) \mid P_{i+1}(x) = t\},$$

there exists $t_0 \in \mathrm{R}\langle\delta_1, \ldots, \delta_i\rangle$ such that

$$F_{[-t_0,0)\cup(0,t_0]} = \{x \in \mathrm{Reali}_i(\Phi \wedge \Psi) \mid t_0^2 \geq P_{i+1}(x) > 0\}$$

and

$$([-t_0, 0) \times F_{-t_0}) \cup ((0, t_0] \times F_{t_0})$$

are homeomorphic. This implies clearly that

$$F_{[\delta_{i+1},t_0]} = \{x \in \mathrm{Reali}_{i+1}(\Phi \wedge \Psi) \mid t_0 > P_{i+1}(x) \geq \delta_{i+1}\}$$

and

$$F_{[2\delta_{i+1},t_0]} = \{x \in \mathrm{Reali}_{i+1}(\Phi \wedge \Psi) \mid t_0 \geq P_{i+1}(x) \geq 2\,\delta_{i+1}\}$$

are homeomorphic, and moreover the homeomorphism can be chosen such that it is the identity on the fibers F_{-t_0} and F_{t_0}.

Hence,

$$\mathrm{b}(\Phi_1) = \mathrm{b}(\Phi \wedge \Psi \wedge (P_{i+1} \geq 2\,\delta_{i+1})) + \mathrm{b}(\Phi \wedge \Psi \wedge (P_{i+1} \leq -2\delta_{i+1})).$$

Note that $F_0 = \mathrm{Reali}_{i+1}(\Phi \wedge \Psi \wedge (P_{i+1} = 0))$ and $F_{[-\delta_{i+1},\delta_{i+1}]} = \mathrm{Reali}_{i+1}(\Phi_2)$.

Thus, it remains to prove that $\mathrm{b}(F_{[-\delta_{i+1},\delta_{i+1}]}) = \mathrm{b}(F_0)$. By Theorem 5.46 (Hardt's triviality), for every $0 < u < 1$, there is a fiber preserving semi-algebraic homeomorphism

$$\phi_u\colon F_{[-\delta_{i+1},-u\delta_{i+1}]} \to [-\delta_{i+1}, -u\delta_{i+1}] \times F_{-u\delta_{i+1}}$$

and a semi-algebraic homeomorphism

$$\psi_u\colon F_{[u\delta_{i+1},\delta_{i+1}]} \to [u\,\delta_{i+1}, \delta_{i+1}] \times F_{u\delta_{i+1}}.$$

We define a continuous semi-algebraic homotopy g from the identity of $F_{[-\delta_{i+1},\delta_{i+1}]}$ to $\lim_{\delta_{i+1}}$ (from $F_{[-\delta_{i+1},\delta_{i+1}]}$ to F_0) as follows:

- $g(0, -)$ is $\lim_{\delta_{i+1}}$,
- for $0 < u \leq 1$, $g(u, -)$ is the identity on $F_{[-u\delta_{i+1},u\delta_{i+1}]}$ and sends $F_{[-\delta_{i+1},-u\delta_{i+1}]}$ (resp. $F_{[u\delta_{i+1},\delta_{i+1}]}$) to $F_{-u\delta_{i+1}}$ (resp. $F_{u\delta_{i+1}}$) by ϕ_u (resp. ψ_u) followed by the projection to $F_{u\delta_{i+1}}$ (resp. $F_{-u\delta_{i+1}}$).

Thus,

$$\mathrm{b}(F_{[-\delta_{i+1},\delta_{i+1}]}) = \mathrm{b}(F_0).$$

Finally,

$$\mathrm{b}(\Phi \wedge \Psi) \leq \sum_{\psi \in \Sigma_{i+1}} \mathrm{b}(\Phi \wedge \Psi \wedge \psi).$$

□

Proof of Proposition 7.39: Starting from the formula Φ, apply Lemma 7.40 with Ψ the empty formula. Now, repeatedly apply Lemma 7.40 to the terms appearing on the right-hand side of the inequality obtained, noting that for any $\Psi \in \Sigma_{\leq s}$,

- either $\mathrm{Reali}_s(\Phi \wedge \Psi) = \mathrm{Reali}_s(\Psi)$ and $\mathrm{Reali}_s(\Psi) \subset \mathrm{Reali}_s(\Phi)$,
- or $\mathrm{Reali}_s(\Phi \wedge \Psi) = \emptyset$. □

Using an argument analogous to that used in the proof of Theorem 7.30, we prove the following proposition.

Proposition 7.41. *For* $0 \le i \le k'$,

$$\sum_{\Psi \in \Sigma_{\le s}} \mathrm{b}_i(\Psi) \le \sum_{j=1}^{k'-i} \binom{s}{j} 6^j d\,(2\,d-1)^{k-1}.$$

We first prove the following Lemma 7.42 and Lemma 7.43.

Let $\mathcal{P} = \{P_1, \ldots, P_j\} \subset R[X_1, \ldots, X_k]$, and let $Q_i = P_i^2 (P_i^2 - \delta_i^2)^2 (P_i^2 - 4\,\delta_i^2)$. Let $j \ge 1$,

$$\begin{aligned} V_j' &= \mathrm{Reali}\left(\bigvee_{1 \le i \le j} Q_i = 0, \mathrm{Ext}(Z, \mathrm{R}\langle \delta_1, \ldots, \delta_j \rangle) \right), \\ W_j' &= \mathrm{Reali}\left(\bigvee_{1 \le i \le j} Q_i \ge 0, \mathrm{Ext}(Z, \mathrm{R}\langle \delta_1, \ldots, \delta_j \rangle) \right). \end{aligned}$$

Lemma 7.42.

$$\mathrm{b}_i(V_j') \le (6^j - 1)\, d\,(2\,d-1)^{k-1}.$$

Proof: The set $\mathrm{Reali}((P_i^2 (P_i^2 - \delta_i^2)^2 (P_i^2 - 4\,\delta_i^2) = 0), Z)$ is the disjoint union of

$$\begin{gathered} \mathrm{Reali}(P_i = 0, \mathrm{Ext}(Z, \mathrm{R}\langle \delta_1, \ldots, \delta_j \rangle)), \\ \mathrm{Reali}(P_i = \delta_i, \mathrm{Ext}(Z, \mathrm{R}\langle \delta_1, \ldots, \delta_j \rangle)), \\ \mathrm{Reali}(P_i = -\delta_i, \mathrm{Ext}(Z, \mathrm{R}\langle \delta_1, \ldots, \delta_j \rangle)), \\ \mathrm{Reali}(P_i = 2\,\delta_i, \mathrm{Ext}(Z, \mathrm{R}\langle \delta_1, \ldots, \delta_j \rangle)), \\ \mathrm{Reali}(P_i = -2\,\delta_i, \mathrm{Ext}(Z, \mathrm{R}\langle \delta_1, \ldots, \delta_j \rangle)). \end{gathered}$$

Moreover, the i-th Betti number of their union V_j' is bounded by the sum of the Betti numbers of all possible non-empty sets that can be obtained by taking intersections of these sets using part (a) of Proposition 7.33.

The number of possible ℓ-ary intersection is $\binom{j}{\ell}$. Each such intersection is a disjoint union of 5^ℓ algebraic sets. The i-th Betti number of each of these algebraic sets is bounded by $d\,(2\,d-1)^{k-1}$ by Theorem 7.23.

Thus,

$$\mathrm{b}_i(V_j') \le \sum_{\ell=1}^{j} \binom{j}{\ell} 5^\ell d\,(2\,d-1)^{k-1} = (6^j - 1)\, d\,(2\,d-1)^{k-1}.$$ □

Lemma 7.43.

$$\mathrm{b}_i(W_j') \leq (6^j - 1)\, d\, (2\, d - 1)^{k-1} + \mathrm{b}_i(Z).$$

Proof: Let

$$F = \mathrm{Reali}\left(\bigwedge_{1 \leq i \leq j} Q_i \leq 0 \vee \bigvee_{1 \leq i \leq j} Q_i = 0, \mathrm{Ext}(Z, \mathrm{R}\langle \delta_1, \ldots, \delta_i \rangle) \right).$$

Now,

$$W_j' \cup F = Z, W_j' \cap F = V_j'.$$

Using inequality (6.44) we get that

$$\mathrm{b}_i(W_j') \leq \mathrm{b}_i(W_j' \cap F) + \mathrm{b}_i(W_j' \cup F) = \mathrm{b}_i(V_j') + \mathrm{b}_i(Z)$$

since $\mathrm{b}_i(Z) = \mathrm{b}_i(\mathrm{Ext}(Z, \mathrm{R}\langle \delta_1, \ldots, \delta_i \rangle))$. We conclude using Lemma 7.42. □

Now, let

$$S_i = \mathrm{Reali}\big(P_i^2 (P_i^2 - \delta_i^2)^2 (P_i^2 - 4 \delta_i^2) \geq 0, \mathrm{Ext}(Z, \mathrm{R}\langle \delta_1, \ldots, \delta_s \rangle)\big),\ 1 \leq i \leq s,$$

and let S be the intersection of the S_i with the closed ball in $\mathrm{R}\langle \delta_1, \ldots, \delta_s, \delta \rangle^k$ defined by $\delta^2 \left(\sum_{1 \leq i \leq k} X_i^2 \right) \leq 1$. Then, it is clear that

$$\sum_{\Psi \in \Sigma_{\leq s}} \mathrm{b}_i(\Psi) = \mathrm{b}_i(S).$$

Proof of Proposition 7.41: Since, for all $i < k'$,

$$\mathrm{b}_i(Z) + \mathrm{b}_{k'}(Z) \leq d\, (2\, d - 1)^{k-1}$$

by Theorem 7.23 we get that,

$$\sum_{\Psi \in \Sigma_{\leq s}} \mathrm{b}_i(\Psi) = \mathrm{b}_i(S) \leq \sum_{j=1}^{k'-i} \binom{s}{j} (6^j - 1) d\, (2\, d - 1)^{k-1} + \binom{s}{k' - i} \mathrm{b}_{k'}(Z)$$

using part (b) of Proposition 7.33 and Lemma 7.43.

Thus, we have that

$$\sum_{\Psi \in \Sigma_{\leq s}} \mathrm{b}_i(\Psi) \leq \sum_{j=1}^{k'-i} \binom{s}{j} 6^j\, d\, (2\, d - 1)^{k-1}.$$ □

Proof of Theorem 7.38: Theorem 7.38 now follows from Proposition 7.39 and Proposition 7.41. □

7.5 Sum of the Betti Numbers of Semi-algebraic Sets

We first describe a construction for replacing any given semi-algebraic subset of a bounded semi-algebraic set by a closed bounded semi-algebraic subset and prove that the new set has the same homotopy type as the original one. Moreover, the polynomials defining the bounded closed semi-algebraic subset are closely related (by infinitesimal perturbations) to the polynomials defining the original subset. In particular, their degrees do not increase, while the number of polynomials used in the definition of the new set is at most twice the square of the number used in the definition of the original set. This construction will be useful later in Chapter 16.

Definition 7.44. Let $\mathcal{P} \subset \mathrm{R}[X_1, \ldots, X_k]$ be a finite set of polynomials with t elements, and let S be a bounded $\mathcal{P}$-closed set. We denote by $\mathrm{SIGN}(S)$ the set of realizable sign conditions of $\mathcal{P}$ whose realizations are contained in S.

Recall that, for $\sigma \in \mathrm{SIGN}(\mathcal{P})$ we define the level of σ as $\#\{P \in \mathcal{P} | \sigma(P) = 0\}$. Let, $\varepsilon_{2t} \gg \varepsilon_{2t-1} \gg \cdots \gg \varepsilon_2 \gg \varepsilon_1 > 0$ be infinitesimals, and denote by R_i the field $\mathrm{R}\langle\varepsilon_{2t}\rangle \cdots \langle\varepsilon_i\rangle$. For $i > 2t$, $\mathrm{R}_i = \mathrm{R}$ and for $i \leqslant 0, \mathrm{R}_i = \mathrm{R}_1$.

We now describe the construction. For each level m, $0 \leq m \leq t$, we denote by $\mathrm{SIGN}_m(S)$ the subset of $\mathrm{SIGN}(S)$ of elements of level m.

Given $\sigma \in \mathrm{SIGN}_m(\mathcal{P}, S)$, let $\mathrm{Reali}(\sigma_+^{\mathrm{c}})$ be the intersection of $\mathrm{Ext}(S, \mathrm{R}_{2m})$ with the closed semi-algebraic set defined by the conjunction of the inequalities,

$$\begin{array}{ll} -\varepsilon_{2m} \leq P \leq \varepsilon_{2m} & \text{for each } P \in \mathcal{A} \text{ such that } \sigma(P) = 0, \\ P \geq 0 & \text{for each } P \in \mathcal{A} \text{ such that } \sigma(P) = 1, \\ P \leq 0 & \text{for each } P \in \mathcal{A} \text{ such that } \sigma(P) = -1. \end{array}$$

and let $\mathrm{Reali}(\sigma_+^{\mathrm{o}})$ be the intersection of $\mathrm{Ext}(S, \mathrm{R}_{2m-1})$ with the open semi-algebraic set defined by the conjunction of the inequalities,

$$\begin{array}{ll} -\varepsilon_{2m-1} < P < \varepsilon_{2m-1} & \text{for each } P \in \mathcal{A} \text{ such that } \sigma(P) = 0, \\ P > 0 & \text{for each } P \in \mathcal{A} \text{ such that } \sigma(P) = 1, \\ P < 0 & \text{for each } P \in \mathcal{A} \text{ such that } \sigma(P) = -1. \end{array}$$

Notice that, denoting $\mathrm{Reali}(\sigma)_i = \mathrm{Ext}(\mathrm{Reali}(\sigma), \mathrm{R}_i)$,

$$\begin{aligned} \mathrm{Reali}(\sigma)_{2m} &\subset \mathrm{Reali}(\sigma_+^{\mathrm{c}}), \\ \mathrm{Reali}(\sigma)_{2m-1} &\subset \mathrm{Reali}(\sigma_+^{\mathrm{o}}). \end{aligned}$$

Let $X \subset S$ be a $\mathcal{P}$-semi-algebraic set such that

$$X = \bigcup_{\sigma \in \Sigma} \mathrm{Reali}(\sigma)$$

with $\Sigma \subset \mathrm{SIGN}(S)$. We denote $\Sigma_m = \Sigma \cap \mathrm{SIGN}_m(S)$ and define a sequence of sets, $X^m \subset \mathrm{R}'^k$, $0 \leq m \leq t$ inductively by

- $X^0 = \mathrm{Ext}(X, \mathrm{R}_1)$.

– For $0 \le m \le t$,

$$X^{m+1} = \left(X^m \cup \bigcup_{\sigma \in \Sigma_m} \mathrm{Reali}(\sigma_+^c)_1 \right) \setminus \bigcup_{\sigma \in \mathrm{SIGN}_m(S) \setminus \Sigma_m} \mathrm{Reali}(\sigma_+^o)_1 ,$$

with $\mathrm{Reali}(\sigma_+^c)_i = \mathrm{Ext}(\mathrm{Reali}(\sigma_+^c), \mathrm{R}_i)$, $\mathrm{Reali}(\sigma_+^o)_i = \mathrm{Ext}(\mathrm{Reali}(\sigma_+^o), \mathrm{R}_i)$.

We denote by X' the set X^{t+1}. □

Theorem 7.45. *The sets* $\mathrm{Ext}(X, R_1)$ *and* X' *are semi-algebraically homotopy equivalent. In particular,*

$$\mathrm{H}_*(X) \cong \mathrm{H}_*(X').$$

For the purpose of the proof we introduce several new families of sets defined inductively.

For each p, $0 \le p \le t+1$ we define sets, $Y_p \subset \mathrm{R}_{2p}^k$, $Z_p \subset \mathrm{R}_{2p-1}^k$ as follows.

– We define

$$\begin{aligned} Y_p^p &= \mathrm{Ext}(X, \mathrm{R}_{2p}) \cup \bigcup_{\sigma \in \Sigma_p} \mathrm{Reali}(\sigma_+^c)_{2p} \\ Z_p^p &= \mathrm{Ext}(Y_p^p, \mathrm{R}_{2p-1}) \setminus \bigcup_{\sigma \in \mathrm{SIGN}_p(S) \setminus \Sigma_p} \mathrm{Reali}(\sigma_+^o)_{2p-1}. \end{aligned}$$

– For $p \le m \le t$, we define

$$\begin{aligned} Y_p^{m+1} &= \left(Y_p^m \cup \bigcup_{\sigma \in \Sigma_m} \mathrm{Reali}(\sigma_+^c)_{2p} \right) \setminus \bigcup_{\sigma \in \mathrm{SIGN}_m(S) \setminus \Sigma_m} \mathrm{Reali}(\sigma_+^o)_{2p} \\ Z_p^{m+1} &= \left(Z_p^m \cup \bigcup_{\sigma \in \Sigma_m} \mathrm{Reali}(\sigma_+^c)_{2p-1} \right) \setminus \bigcup_{\sigma \in \mathrm{SIGN}_m(S) \setminus \Sigma_m} \mathrm{Reali}(\sigma_+^o)_{2p-1}. \end{aligned}$$

We denote by $Y_p \subset \mathrm{R}_{2p}^k$ the set Y_p^{t+1} and by $Z_p \subset \mathrm{R}_{2p-1}^k$ the set Z_p^{t+1}.

Note that

– $X = Y_{t+1} = Z_{t+1}$,
– $Z_0 = X'$.

Notice also that for each $p, 0 \le p \le t$,

– $\mathrm{Ext}(Z_{p+1}^{p+1}, \mathrm{R}_{2p}) \subset Y_p^p$,
– $Z_p^p \subset \mathrm{Ext}(Y_p^p, \mathrm{R}_{2p-1})$

The following inclusions follow directly from the definitions of Y_p and Z_p.

Lemma 7.46. *For each* $p, 0 \le p \le t$,

– $\mathrm{Ext}(Z_{p+1}, \mathrm{R}_{2p}) \subset Y_p$,
– $Z_p \subset \mathrm{Ext}(Y_p, \mathrm{R}_{2p-1})$.

We now prove that in both the inclusions in Lemma 7.46 above, the pairs of sets are in fact semi-algebraically homotopy equivalent. These suffice to prove Theorem 7.45.

Lemma 7.47. *For* $1 \le p \le t$, Y_p *is semi-algebraically homotopy equivalent to* $\mathrm{Ext}(Z_{p+1}, \mathrm{R}_{2p})$.

Proof: Let $Y_p(u) \subset \mathrm{R}_{2p+1}^k$ denote the set obtained by replacing the infinitesimal ε_{2p} in the definition of Y_p by u, and for $u_0 > 0$, we will denote by

$$Y_p((0, u_0]) = \{(x, u) | x \in Y_p(u), u \in (0, u_0]\} \subset \mathrm{R}_{2p+1}^{k+1}.$$

By Hardt's triviality theorem there exist $u_0 \in \mathrm{R}_{2p+1}$, $u_0 > 0$ and a homeomorphism,

$$\psi: Y_p(u_0) \times (0, u_0] \to Y_p((0, u_0]),$$

such that

- $\pi_{k+1}(\phi(x, u)) = u$,
- $\psi(x, u_0) = (x, u_0)$ for $x \in Y_p(u_0)$,
- for all $u \in (0, u_0]$, and for every sign condition σ on

$$\cup_{P \in \mathcal{P}} \{P, P \pm \varepsilon_{2t}, \ldots, P \pm \varepsilon_{2p+1}\},$$

  $\psi(\cdot, u)$ defines a homeomorphism of $\mathrm{Reali}(\sigma, Y_p(u_0))$ to $\mathrm{Reali}(\sigma, Y_p(u))$.

Now, we specialize u_0 to ε_{2p} and denote the map corresponding to ψ by ϕ. For $\sigma \in \Sigma_p$, we define, $\mathrm{Reali}(\sigma_{++}^o)$ to be the set defined by

$$\begin{array}{ll} -2\varepsilon_{2p} < P < 2\varepsilon_{2p} & \text{for each } P \in \mathcal{A} \text{ such that } \sigma(P) = 0, \\ P > -\varepsilon_{2p} & \text{for each } P \in \mathcal{A} \text{ such that } \sigma(P) = 1, \\ P < \varepsilon_{2p} & \text{for each } P \in \mathcal{A} \text{ such that } \sigma(P) = -1. \end{array}$$

Let $\lambda: Y_p \to \mathrm{R}_{2p}$ be a semi-algebraic continuous function such that,

$$\begin{array}{ll} \lambda(x) = 1 & \text{on } Y_p \cap \cup_{\sigma \in \Sigma_p} \mathrm{Reali}(\sigma_+^c), \\ \lambda(x) = 0 & \text{on } Y_p \setminus \cup_{\sigma \in \Sigma_p} \mathrm{Reali}(\sigma_{++}^o), \\ 0 < \lambda(x) < 1 & \text{else.} \end{array}$$

We now construct a semi-algebraic homotopy,

$$h: Y_p \times [0, \varepsilon_{2p}] \to Y_p,$$

by defining,

$$\begin{array}{lll} h(x, t) & = \pi_{1\ldots k} \circ \phi(x, \lambda(x)t + (1 - \lambda(x))\varepsilon_{2p}) & \text{for } 0 < t \le \varepsilon_{2p} \\ h(x, 0) & \lim_{t \to 0^+} h(x, t), & \text{else.} \end{array}$$

Note that the last limit exists since S is closed and bounded. We now show that, $h(x, 0) \in \mathrm{Ext}(Z_{p+1}, \mathrm{R}_{2p})$ for all $x \in Y_p$.

Let $x \in Y_p$ and $y = h(x, 0)$.

There are two cases to consider.

- $\lambda(x) < 1$. In this case, $x \in \mathrm{Ext}(Z_{p+1}, \mathrm{R}_{2p})$ and by property (3) of ϕ and the fact that $\lambda(x) < 1$, $y \in \mathrm{Ext}(Z_{p+1}, \mathrm{R}_{2p})$.

- $\lambda(x) \geqslant 1$. Let σ_y be the sign condition of $\mathcal{P}$ at y and suppose that $y \notin \mathrm{Ext}(Z_{p+1}, \mathrm{R}_{2p})$. There are two cases to consider.
  - $\sigma_y \in \Sigma$. In this case, $y \in X$ and hence there must exist
$$\tau \in \mathrm{SIGN}_m(S) \setminus \Sigma_m,$$
    with $m > p$ such that $y \in \mathrm{Reali}(\tau_+^o)$.
  - $\sigma_y \notin \Sigma$. In this case, taking $\tau = \sigma_y$, $\mathrm{level}(\tau) > p$ and $y \in \mathrm{Reali}(\tau_+^o)$. It follows from the definition of y, and property (3) of ϕ, that for any $m > p$, and every $\rho \in \mathrm{SIGN}_m(S)$,
    - $y \in \mathrm{Reali}(\rho_+^o)$ implies that $x \in \mathrm{Reali}(\rho_+^o)$,
    - $x \in \mathrm{Reali}(\rho_+^c)$ implies that $y \in \mathrm{Reali}(\rho_+^c)$.

    Thus, $x \notin Y_p$ which is a contradiction.

  It follows that,
  - $h(\cdot, \varepsilon_{2p})\colon Y_p \to Y_p$ is the identity map,
  - $h(Y_p, 0) = \mathrm{Ext}(Z_{p+1}, \mathrm{R}_{2p})$,
  - $h(\cdot, t)$ restricted to $\mathrm{Ext}(Z_{p+1}, \mathrm{R}_{2p})$ gives a semi-algebraic homotopy between
$$h(\cdot, \varepsilon_{2p})|_{\mathrm{Ext}(Z_{p+1}, \mathrm{R}_{2p})} = \mathrm{id}_{\mathrm{Ext}(Z_{p+1}, \mathrm{R}_{2p})}$$
    and
$$h(\cdot, 0)|_{\mathrm{Ext}(Z_{p+1}, \mathrm{R}_{2p})}.$$

  Thus, Y_p is semi-algebraically homotopy equivalent to $\mathrm{Ext}(Z_{p+1}, \mathrm{R}_{2p})$.

□

Lemma 7.48. *For each* p, $0 \leq p \leq t$, Z_p *is semi-algebraically homotopy equivalent to* $\mathrm{Ext}(Y_p, \mathrm{R}_{2p-1})$.

Proof: For the purpose of the proof we define the following new sets for $u \in \mathrm{R}_{2p}$.

- Let $Z_p'(u) \subset \mathrm{R}_{2p}^k$ be the set obtained by replacing in the definition of Z_p, ε_{2j} by $\varepsilon_{2j} - u$ and ε_{2j-1} by $\varepsilon_{2j-1} + u$ for all $j > p$, and ε_{2p} by $\varepsilon_{2p} - u$, and ε_{2p-1} by u. For $u_0 > 0$ we will denote
$$Z_p'((0, u_0]) = \{(x, u) \mid x \in Z_p'(u), u \in (0, u_0]\}.$$
  $Z_p'((0, u_0])$ the set $\{(x, u) \mid x \in Z_p'(u), u \in (0, u_0]\}$.
- Let $Y_p'(u) \subset \mathrm{R}_{2p}^k$ be the set obtained by replacing in the definition of Y_p, ε_{2j} by $\varepsilon_{2j} - u$ and ε_{2j-1} by $\varepsilon_{2j-1} + u$ for all $j > p$ and ε_{2p} by by $\varepsilon_{2p} - u$.
- For $\sigma \in \mathrm{Sign}_m(S)$, with $m \geq p$, let $\mathrm{Reali}(\sigma_+^c)(u) \subset \mathrm{R}_{2p}^k$ denote the set obtained by replacing ε_{2m} by $\varepsilon_{2m} - u$ in the definition of $\mathrm{Reali}(\sigma_+^c)$.
- For $\sigma \in \mathrm{Sign}_m(S)$, with $m > p$, let $\mathrm{Reali}(\sigma_+^o)(u) \subset \mathrm{R}_{2p}^k$ denote the set obtained by replacing ε_{2m-1} by $\varepsilon_{2m-1} + u$ in the definition of $\mathrm{Reali}(\sigma_+^o)$.
- Finally, for $\sigma \in \mathrm{Sign}_p(S)$ let $\mathrm{Reali}(\sigma_+^o)(u) \subset \mathrm{R}_{2p-1}^k$ denote the set obtained by replacing in the definition of $\mathrm{Reali}(\sigma_o^c)$, ε_{2p-1} by u.

Notice that by definition, for any $u, v \in \mathrm{R}_{2p}$ with $0 < u \leq v$, $Z'_p(u) \subset Y'_p(u)$, $Z'_p(v) \subset Z'_p(u)$, $Y'_p(v) \subset Y'_p(u)$, and

$$\bigcup_{0<s\leq u} Y'_p(s) = \bigcup_{0<s\leq u} Z'_p(s).$$

We denote by Z'_p (respectively, Y'_p) the set $Z'_p(\varepsilon_{2p-1})$ (respectively, $Y'_p(\varepsilon_{2p-1})$). It is easy to see that Y'_p is semi-algebraically homotopy equivalent to $\mathrm{Ext}(Y_p, \mathrm{R}_{2p-1})$, and Z'_p is semi-algebraically homotopy equivalent to Z_p. We now prove that, Y'_p is semi-algebraically homotopy equivalent to Z'_p, which suffices to prove the lemma.

Let $\mu: Y'_p \to \mathrm{R}_{2p-1}$ be the semi-algebraic map defined by

$$\mu(x) = \sup_{u \in (0, \varepsilon_{2p-1}]} \{u \mid x \in Z'_p(u)\}.$$

We prove separately (Lemma 7.49 below) that μ is continuous. Note that the definition of the set $Z'_p(u)$ (as well as the set $Y'_p(u)$) is more complicated than the more natural one consisting of just replacing ε_{2p-1} in the definition of Z_p by u, is due to the fact that with the latter definition the map μ defined below is not necessarily continuous.

We now construct a continuous semi-algebraic map,

$$h: Y'_p \times [0, \varepsilon_{2p-1}] \to Y'_p$$

as follows.

By Hardt's triviality theorem there exist $u_0 \in \mathrm{R}_{2p}$, with $u_0 > 0$ and a semi-algebraic homeomorphism,

$$\psi: Z'_p(u_0) \times (0, u_0] \to Z'_p((0, u_0]),$$

such that

- $\pi_{k+1}(\psi(x, u)) = u$,
- $\psi(x, u_0) = (x, u_0)$ for $x \in Z'_p(u_0)$,
- for all $u \in (0, u_0]$ and for every sign condition σ of the family,

$$\bigcup_{P \in \mathcal{P}} \{P, P \pm \varepsilon_{2t}, \ldots, P \pm \varepsilon_{2p+1}\},$$

  the map $\psi(\cdot, u)$ restricts to a homeomorphism of $\mathrm{Reali}(\sigma, Z'_p(u_0))$ to $\mathrm{Reali}(\sigma, Z'_p(u))$.

We now specialize u_0 to ε_{2p-1} and denote by ϕ the corresponding map,

$$\phi: Z'_p \times (0, \varepsilon_{2p-1}] \to Z'_p((0, \varepsilon_{2p-1}]).$$

Note, that for every u, $0 < u \leq \varepsilon_{2p-1}$, ϕ gives a homeomorphism,

$$\phi_u: Z'_p(u) \to Z'_p.$$

Hence, for every pair, u, u', $0 < u \leq u' \leq \varepsilon_{2p-1}$, we have a homeomorphism, $\theta_{u,u'}: Z'_p(u) \to \mathbb{Z}'_p(u')$ obtained by composing ϕ_u with $\phi_{u'}^{-1}$.

For $0 \leq u' < u \leq \varepsilon_{2p-1}$, we let $\theta_{u,u'}$ be the identity map. It is clear that $\theta_{u,u'}$ varies continuously with u and u'.

For $x \in Y'_p, t \in [0, \varepsilon_{2p-1}]$ we now define,

$$h(x,t) = \theta_{\mu(x),t}(x).$$

It is easy to verify from the definition of h and the properties of ϕ listed above that, h is continuous and satisfies the following.

- $h(\cdot, 0): Y'_p \to Y'_p$ is the identity map,
- $h(Y'_p, \varepsilon_{2p-1}) = Z'_p$,
- $h(\cdot, t)$ restricts to a homeomorphism $Z'_p \times t \to Z'_p$ for every $t \in [0, \varepsilon_{2p-1}]$.

This proves the required homotopy equivalence. □

We now prove that the function μ used in the proof above is continuous.

Lemma 7.49. *The semi-algebraic map* $\mu: Y'_p \to \mathrm{R}_{2p-1}$ *defined by*

$$\mu(x) = \sup_{u \in (0, \varepsilon_{2p-1}]} \{u \mid x \in Z'_p(u)\}$$

is continuous.

Proof : Let $0 < \delta \ll \varepsilon_{2p-1}$ be a new infinitesimal. In order to prove the continuity of μ (which is a semi-algebraic function defined over R_{2p-1}), it suffices, by Proposition 3.5 to show that

$$\lim_\delta \mathrm{Ext}(\mu, \mathrm{R}_{2p-1}\langle\delta\rangle)(x') = \lim_\delta \mathrm{Ext}(\mu, \mathrm{R}_{2p-1}\langle\delta\rangle)(x)$$

for every pair of points $x, x' \in \mathrm{Ext}(Y'_p, \mathrm{R}_{2p-1}\langle\delta\rangle)$ such that $\lim_\delta x = \lim_\delta x'$.

Consider such a pair of points $x, x' \in \mathrm{Ext}(Y'_p, \mathrm{R}_{2p-1}\langle\delta\rangle)$. Let $u \in (0, \varepsilon_{2p-1}]$ be such that $x \in Z'_p(u)$. We show below that this implies $x' \in Z'_p(u')$ for some u' satisfying $\lim_\delta u' = \lim_\delta u$.

Let m be the largest integer such that there exists $\sigma \in \Sigma_m$ with $x \in \mathrm{Reali}(\sigma_+^{\mathrm{c}})(u)$. Since $x \in Z'_p(u)$ such an m must exist.

We have two cases:

- $m > p$: Let $\sigma \in \Sigma_m$ with $x \in \mathrm{Reali}(\sigma_+^{\mathrm{c}})(u)$. Then, by the maximality of m, we have that for each $P \in \mathcal{P}$, $\sigma(P) \neq 0$ implies that $\lim_\delta P(x) \neq 0$. As a result, we have that $x' \in \mathrm{Reali}(\sigma_+^{\mathrm{c}})(u')$ for all

$$u' < u - \max_{P \in \mathcal{P}, \sigma(P) = 0} |P(x) - P(x')|,$$

  and hence we can choose u' such that $x' \in \mathrm{Reali}(\sigma_+^{\mathrm{c}})(u')$ and $\lim_\delta u' = \lim_\delta u$.
- $m \leq p$: If $x' \notin Z'_p(u)$ then since $x' \in Y'_p \subset Y'_p(u)$,

$$x' \in \cup_{\sigma \in \mathrm{SIGN}_p(\mathcal{P}, S) \setminus \Sigma_p} \mathrm{Reali}(\sigma_+^{\mathrm{o}})(u).$$

Let $\sigma \in \mathrm{SIGN}_p(S) \setminus \Sigma_p$ be such that $x' \in \mathrm{Reali}(\sigma_+^o)(u)$. We prove by contradiction that $\lim_\delta \max_{P\in\mathcal{P},\sigma(P)=0} |P(x')| = u$.

Assume that

$$\lim_\delta \max_{P\in\mathcal{P},\sigma(P)=0} |P(x')| \neq u.$$

Since, $x \notin \mathrm{Reali}(\sigma_+^o)(u)$ by assumption, and $\lim_\delta x' = \lim_\delta x$, there must exist $P \in \mathcal{P}$, $\sigma(P) \neq 0$, and $\lim_\delta P(x) = 0$. Letting τ denote the sign condition defined by $\tau(P) = 0$ if $\lim_\delta P(x) = 0$ and $\tau(P) = \sigma(P)$ else, we have that $\mathrm{level}(\tau) > p$ and x belongs to both $\mathrm{Reali}(\tau_+^o)(u)$ as well as $\mathrm{Reali}(\tau_+^c)(u)$.

Now there are two cases to consider depending on whether τ is in Σ or not. If $\tau \in \Sigma$, then the fact that $x \in \mathrm{Reali}(\tau_+^c)(u)$ contradicts the choice of m, since $m \leq p$ and $\mathrm{level}(\tau) > p$. If $\tau \notin \Sigma$ then x gets removed at the level of τ in the construction of $Z_p'(u)$, and hence $x \in \mathrm{Reali}(\rho_+^c)(u)$ for some $\rho \in \Sigma$ with $\mathrm{level}(\rho) > \mathrm{level}(\tau) > p$. This again contradicts the choice of m. Thus, $\lim_\delta \max_{P\in\mathcal{P},\sigma(P)=0} |P(x')| = u$ and since $x' \notin \cup_{\sigma\in\mathrm{SIGN}_p(\mathcal{C},S)\setminus\Sigma_p} \mathrm{Reali}(\sigma_+^o)(u')$ for all $u' < \max_{P\in\mathcal{P},\sigma(P)=0} |P(x')|$, we can choose u' such that $\lim_\delta u' = \lim_\delta u$, and $x' \notin \cup_{\sigma\in\mathrm{SIGN}_p(\mathcal{P},S)\setminus\Sigma_p} \mathrm{Reali}(\sigma_+^o)(u')$.

In both cases we have that $x' \in Z_p'(u')$ for some u' satisfying $\lim_\delta u' = \lim_\delta u$, showing that $\lim_\delta \mu(x') \geq \lim_\delta \mu(x)$. The reverse inequality follows by exchanging the roles of x and x' in the previous argument. Hence,

$$\lim_\delta \mu(x') = \lim_\delta \mu(x),$$

proving the continuity of μ. □

Proof of Theorem 7.45: The theorem follows immediately from Lemmas 7.47 and 7.48. □

We now define the **Betti numbers** of a general $\mathcal{P}$-semi-algebraic set and bound them. Given a $\mathcal{P}$-semi-algebraic set $Y \subset \mathrm{R}^k$, we replace it by

$$X = \mathrm{Ext}(Y, \mathrm{R}\langle\varepsilon\rangle) \cap \overline{B}_k(0, 1/\varepsilon).$$

Taking $S = \overline{B}_k(0, 1/\varepsilon)$, we know by Theorem 7.45 that there is a closed and bounded semi-algebraic set $X' \subset \mathrm{R}\langle\varepsilon\rangle_1^k$ such that $\mathrm{Ext}(X, \mathrm{R}\langle\varepsilon\rangle_1)$ and X' are semi-algebraically homotopy equivalent.We define the Betti numbers $\mathrm{b}_i(Y) := \mathrm{b}_i(X')$. Note that this definition is clearly homotopy invariant since Y and X' has are semi-algebraically homotopy equivalent. We denote by $\mathrm{b}(Y) = \mathrm{b}(X')$ the sum of the Betti numbers of Y.

Theorem 7.50. *Let Y be a $\mathcal{P}$-semi-algebraic set where $\mathcal{P}$ is a family of at most s polynomials of degree d in k variables. Then*

$$\mathrm{b}(Y) \leq \sum_{i=0}^{k} \sum_{j=1}^{k-i} \binom{2s^2+1}{j} 6^j d (2d-1)^{k-1}.$$

Proof: Take $S = \overline{B}_k(0, 1/\varepsilon)$ and $X = \mathrm{Ext}(Y, \mathrm{R}\langle\varepsilon\rangle) \cap \overline{B}_k(0, 1/\varepsilon)$. Defining X' according to Definition 7.44, apply Theorem 7.38 to X', noting that the number of polynomials defining X' is $2\,s^2+1$, and their degrees are bounded by d. □

7.6 Bibliographical Notes

The inequalities relating the number of critical points of a Morse function (Theorem 7.24) with the Betti numbers of a compact manifold was proved in its full generality by Morse [121], building on prior work by Birkhoff ([25], page 42). Their generalization "with boundary" (Propositions 7.18 and 7.18) can be found in [81, 82]. Using these inequalities, Thom [157], Milnor[118], Oleinik and Petrovsky [124, 125] proved the bound on the sum of the Betti numbers of algebraic sets presented here. Subsequently, using these bounds Warren [165] proved a bound of $(4e\,s\,d/k)^k$ on the number of connected components of the realizations of strict sign conditions of a family of polynomials of s polynomials in k variables of degree at most d and Alon [3] derived a bound of $(8e\ s\ d/k)^k$ on the number of all realizable sign conditions (not connected components). All these bounds are in terms of the product $s\,d$.

The first result in which the dependence on s (the combinatorial part) was studied separately from the dependence on d (algebraic) appeared in [14], where a bound of $\binom{O(s)}{k'} O(d)^k$ was proved on the number of connected components of all realizable sign conditions of a family of polynomials restricted to variety of dimension k'. The generalization of the Thom-Milnor bound on the sum of the Betti numbers of basic semi-algebraic sets to more general closed semi-algebraic sets restricted to a variety was first done in [11] and later improved in [17]. The first result bounding the individual Betti numbers of a basic semi-algebraic set appears in [10] and that bounding the individual Betti numbers over all realizable sign conditions in [17]. The construction for replacing any given semi-algebraic subset of a bounded semi-algebraic set by a closed bounded semi-algebraic subset in Section 7.5, as well as the bound on the sum of the Betti numbers of a general semi-algebraic set, appears in [62].

8

Complexity of Basic Algorithms

In Section 8.1, we discuss a few notions needed to analyze the complexity of algorithms and illustrate them by several simple examples. In Section 8.2, we study basic algorithms for linear algebra, including computations of determinants and characteristic polynomials of matrices, and signatures of quadratic forms. In Section 8.3, we compute remainder sequences and the related subresultant polynomials. The algorithms in this chapter are very basic and will be used throughout the other chapters of the book.

8.1 Definition of Complexity

An **algorithm** is a computational procedure that takes an input and after performing a finite number of allowed operations produces an output.

A typical input to an algorithm in this book will be a set of polynomials with coefficients in a ring A or a matrix with coefficients in A or a formula involving certain polynomials with coefficients in A.

Each of our algorithms will depend on a specified **structure**. The specified structure determines which operations are allowed in the algorithm. We list the following structures that will be used most:

- **Ring structure:** the only operations allowed are addition, subtraction, multiplication between elements of a given ring, and deciding whether an element of the ring is zero.
- **Ordered ring structure:** in addition to the ring structure operations, we can also **compare** two elements in a given ordered ring. That is, given a, b in the ordered ring we can decide whether $a = b$, $a > b$, or $a < b$.
- **Ring with integer division structure:** in addition to the ring structure operations, it is also possible to do **exact division** by an element of $\mathbb{Z}$ which can be performed when we know in advance that the result of the division belongs to the ring. In such a ring $n \cdot 1 \neq 0$ when $n \in \mathbb{Z}, n \neq 0$.

- **Integral domain structure:** in addition to the ring structure operations, it is also possible to do **exact division** by an element of a given integral domain which can be performed when we know in advance that the result of a division belongs to the integral domain.
- **Field structure:** in addition to the ring structure operations, we can also perform **division** by any element of a given field, which can be performed only by a non-zero element.
- **Ordered integral domain structure:** in addition to the integral domain structure operations, we can also **compare** two elements of a given ordered integral domain. That is, given a, b in the ordered integral domain, we can decide whether $a = b$, $a > b$, or $a < b$,
- **Ordered field structure:** in addition to the field structure operations, we can also **compare** two elements of a given ordered field.

Which structure is associated to the algorithm will be systematically indicated in the description of the algorithm.

The **size of the input** is always a vector of integers. Typical parameters we use to describe the size of the input are the dimensions of a matrix, the number of polynomials, their degrees, and their number of variables.

The **complexity** of an algorithm in a structure is a function associating to a vector of integers v describing the size of the input a bound on the number of operations performed by the algorithm in the structure when it runs over all possible inputs of size v.

Remark 8.1. In this definition of complexity, there are many manipulations that are cost free. For example, given a matrix, we can access an element for free. Also the cost of reading the input or writing the output is not taken into account. □

The same computation has a different complexity depending on the structure which is specified. In a ring A, the complexity of a single addition or multiplication is 1. However, if the ring A is D$[X]$, where D is a ring, then the cost of adding two polynomials is one in D$[X]$, while the cost of the same operation in D clearly depends on the degree of the two polynomials.

To illustrate the discussion, we consider first a few basic examples used later in the book.

We consider first arithmetic operations on univariate polynomials.

Algorithm 8.1. **[Addition of Univariate Polynomials]**

- **Structure:** a ring A.
- **Input:** two univariate polynomials in A$[X]$:

$$\begin{aligned} P &= a_p X^p + a_{p-1} X^{p-1} + a_{p-2} X^{p-2} + \cdots + a_0, \\ Q &= b_q X^q + b_{q-1} X^{q-1} + \cdots + b_0. \end{aligned}$$

- **Output:** the sum $P + Q$.

- **Complexity:** $p+1$,where p is a bound on the degree of P and Q.
- **Procedure:** For every $k \leq p$, compute the coefficient c_k of X^k in $P+Q$,

$$c_k := a_k + b_k.$$

Here, the size of the input is one natural number p, a bound on the degree of the two polynomials. The computation takes place in the ring A.

Algorithm 8.2. **[Multiplication of Univariate Polynomials]**

- **Structure:** a ring A.
- **Input:** two univariate polynomials

$$\begin{aligned} P &= a_p X^p + a_{p-1} X^{p-1} + a_{p-2} X^{p-2} + \cdots + a_0, \\ Q &= b_q X^q + b_{q-1} X^{q-1} + \cdots + b_0. \end{aligned}$$

  in $\mathrm{A}[X]$, with $p \geq q$.
- **Output:** the product PQ.
- **Complexity:** $(p+1)(q+1)+pq$,where p is a bound on the degree of P and q a bound on the degree of Q.
- **Procedure:** For each $k \leq p+q$, compute the coefficient c_k of X^k in PQ,

$$c_k := \begin{cases} \sum_{i=0}^{k} a_{k-i} b_i, & \text{if } 0 \leq k \leq q, \\ \sum_{i=0}^{q} a_{k-i} b_i, & \text{if } q < k < p, \\ \sum_{i=0}^{k-p} a_{k-i} b_i. & \text{if } p \leq k \leq p+q. \end{cases}$$

Here the size of the input is two natural numbers, a bound on the degree of each of the two polynomials. The computation takes place in the ring A.

Complexity analysis: For every k, $0 \leq k \leq q$, there are k additions and $k+1$ multiplications in A, i.e. $2k+1$ arithmetic operations. For every k, such that $q < k < p$, there are q additions and $q+1$ multiplications in A, i.e. $2q+1$ arithmetic operations. For every k, $p \leq k < p+q$, there are $p+q-k$ additions and $p+q-k+1$ multiplications in A, i.e. $2(p+q-k)+1$ arithmetic operations. Since $\sum_{k=0}^{q} k = (q+1)q/2$,

$$\sum_{k=0}^{q} (2k+1) = \sum_{k=p}^{p+q} (2(p+q-k)+1) = (q+1)^2.$$

So there are all together

$$2(q+1)^2 + (p-q-1)(2q+1) = (p+1)(q+1) + pq$$

arithmetic operations performed by the algorithm. □

From now on, our estimates on the complexity of an algorithm will often use the notation O.

Notation 8.2. [Big O] Let f and g be mappings from $\mathbb{N}^\ell$ to $\mathbb{R}$ and h be a function from $\mathbb{R}$ to $\mathbb{R}$. The expression "$f(v)$ is $h(O(g(v)))$" means that there exists a natural number b such that for all $v \in \mathbb{N}^\ell$, $f(v) \leq h(b\, g(v))$. The expression "$f(v)$ is $h(\tilde{O}(g(v)))$" means that there exist natural number a such that for all $v \in \mathbb{N}^\ell$, $f(v) \leq h(g(v) \log_2(g(v))^a)$. □

For example, the complexity of the algorithms presented for the addition and multiplication of polynomials are $O(p)$ and $O(p\,q)$.

Remark 8.3. The complexity of computing the product of two univariate polynomials depends on the algorithm used. The complexity of the multiplication of two univariate polynomials of degree at most d is $O(d^2)$ when the multiplication is done naively, as in Algorithm 8.2, $O(d^{\log_2(3)})$ when Karatsuba's method is used, $O(d \log_2(d)) = \tilde{O}(d)$ using the Fast Fourier Transform (FFT). We decided not to enter into these developments and refer the interested reader to [64]. □

Algorithm 8.3. **[Euclidean Division]**

- **Structure:** a field K.
- **Input:** two univariate polynomials

$$\begin{aligned} P &= a_p X^p + a_{p-1} X^{p-1} + a_{p-2} X^{p-2} + \dots + a_0, \\ Q &= b_q X^q + b_{q-1} X^{q-1} + \dots + b_0. \end{aligned}$$

  in $\mathrm{K}[X]$, with $b_q \neq 0$.
- **Output:** $\mathrm{Quo}(P, Q)$ and $\mathrm{Rem}(P, Q)$, the quotient and remainder in the Euclidean division of P by Q.
- **Complexity:** $(p - q + 1)\,(2\,q + 3)$,where p is a bound on the degree of P and q a bound on the degree of Q.
- **Procedure:**
  - Initialization: $C := 0$, $R := P$.
  - For every j from p to q,

$$\begin{aligned} C &:= C + \frac{\mathrm{cof}_j(R)}{b_q} X^{j-q}, \\ R &:= R - \frac{\mathrm{cof}_j(R)}{b_q} X^{j-q} Q. \end{aligned}$$

  - Output C, R.

Here the size of the input is two natural numbers, a bound on the degree of one polynomial and the degree of the other. The computation takes place in the field K.

Complexity analysis: There are $p - q + 1$ values of j to consider. For each value of j, there is one division, $q + 1$ multiplications and $q + 1$ subtractions. Thus, the complexity is bounded by $(p - q + 1)\,(2\,q + 3)$. □

The complexity of an algorithm defined in terms of arithmetic operations often does not give a realistic estimate of the actual computation time when the algorithm is implemented. The reason behind this is the intermediate growth of coefficients during the computation. This is why, in the case of integer entries, we also take into account the bitsizes of the integers which occur in the input. The **bitsize** of a non-zero integer is the number of bits in its binary representation. More precisely, the bitsize of n is τ if and only if $2^{\tau-1} \leq |n| < 2^{\tau}$. The bitsize of a rational number is the sum of the bitsizes of its numerators and denominators.

Adding n integers of bitsizes bounded by τ gives an integer of bitsize bounded by $\tau + \nu$ where ν is the bitsize of n: indeed, if for every $1 \leq i \leq n$, we have $m_i < 2^{\tau}$, then $m_1 + \cdots + m_n < n\, 2^{\tau} < 2^{\tau+\nu}$.

Multiplying n integers of bitsizes bounded by τ gives an integer of size bounded by $n\,\tau$: indeed, if for every $1 \leq i \leq n$, $m_i < 2^{\tau}$, then $m_1 \cdots m_n < 2^{n\tau}$.

When the input of the algorithms belongs to $\mathbb{Z}$, it is thus natural to discuss the **binary complexity** of the algorithms, i.e. to estimate the number of bit operations.

Most of the time, the binary complexity of our algorithms is obtained in two steps. First we compute the number of arithmetic operations performed, second we estimate the bitsize of the integers on which these operations are performed. These bitsize estimates do not follow in general from an analysis of the steps of the algorithm itself, but are consequences of bounds coming from the mathematical nature of the objects considered. For example, when all the intermediate results of a computation are determinants of matrices with integer entries, we can make use Hadamard's bound (see Proposition 8.10).

Remark 8.4. The binary complexity of an addition of two integers of bitsize τ is $O(\tau)$. The binary cost of a multiplication of two integers of bitsize τ depends strongly of the algorithm used: $O(\tau^2)$ when the multiplication is done naively, $O(\tau^{\log_2(3)})$ when Karatsuba's method, is used, $O(\tau \log_2(\tau) \log_2(\log_2(\tau))) = \tilde{O}(d\, \tau)$ using FFT. These developments are not included in the book. We refer the interested reader to [64]. □

Now we describe arithmetic operations on multivariate polynomials.

Algorithm 8.4. **[Addition of Multivariate Polynomials]**

- **Structure:** a ring A.
- **Input:** two multivariate polynomials P and Q in $\mathrm{A}[X_1, \ldots, X_k]$ whose degrees are bounded by d.
- **Output:** the sum $P + Q$.
- **Complexity:** $\binom{d+k}{k} \leq (d+1)^k$.
- **Procedure:** For every monomial m of degree $\leq d$ in k variables, denoting by a_m, b_m, and c_m the coefficients of m in P, Q, and $P + Q$, compute

$$c_m := a_m + b_m.$$

Studying the complexity of this algorithm requires the following lemma.

Lemma 8.5. *The number of monomials of degree* $\le d$ *in* k *variables is* $\binom{d+k}{k} \le (d+1)^k$.

Proof: By induction on k and d. The result is true for $k=1$ and every d since there are $d+1$ monomials of degree less than or equal to d. Since either a monomial does not depend on X_k or is a multiple of X_k, the number of monomials of degree $\le d$ in k variables is the sum of the number of monomials of degree $\le d$ in $k-1$ variables and the number of monomials of degree $\le d-1$ in k variables. Finally, note that $\binom{d+k-1}{k-1} + \binom{d-1+k}{k} = \binom{d+k}{k}$.

The estimate $\binom{d+k}{k} \le (d+1)^k$ is also proved by induction on k and d. The estimate is true for $k=1$ and every d, and also for $d=1$ and every $k \ge 0$. Suppose by induction hypothesis that $\binom{d+k-1}{k-1} \le (d+1)^{k-1}$ and $\binom{d-1+k}{k} \le d^k$. Then

$$\binom{d+k}{k} \le (d+1)^{k-1} + d^k \le (d+1)^{k-1} + d\,(d+1)^{k-1} = (d+1)^k. \qquad \square$$

Complexity analysis of Algorithm 8.4:

The complexity is $\binom{d+k}{k} \le (d+1)^k$,using Lemma 8.5, since there is one addition to perform for each m.

If $\mathrm{A} = \mathbb{Z}$, and the bitsizes of the coefficients of P and Q are bounded by τ, the bitsizes of the coefficients of their sum are bounded by $\tau + 1$. $\square$

Algorithm 8.5. **[Multiplication of Multivariate Polynomials]**

- **Structure:** a ring A.
- **Input:** two multivariate polynomials P and Q in $\mathrm{A}[X_1, \ldots, X_k]$ whose degrees are bounded by p and q.
- **Output:** the product PQ.
- **Complexity:** $\le 2\binom{p+k}{k}\binom{q+k}{k} \le 2\,(p+1)^k\,(q+1)^k$.
- **Procedure:** For every monomial m (resp. n, resp. u) of degree $\le p$ (resp. $\le q$, resp. $\le p+q$) in k variables, denoting by a_m, b_n, and c_u the coefficients of m in P (resp. Q, resp. $P \cdot Q$), compute

$$c_u := \sum_{n+m=u} a_n\, b_m.$$

Complexity analysis: Given that there are at most $\binom{p+k}{k}$ monomials of degree $\le p$ and $\binom{q+k}{k}$ monomials of degree $\le q$, there are at most $\binom{p+k}{k}\binom{q+k}{k}$ multiplications and $\binom{p+k}{k}\binom{q+k}{k}$ additions to perform. The complexity is at most $2\binom{p+k}{k}\binom{q+k}{k} \le 2(p+1)^k\,(q+1)^k$.

If $\mathrm{A} = \mathbb{Z}$, and the bitsizes of the coefficients of P and Q are bounded by τ and σ, the bitsizes of the coefficients of their product are bounded by $\tau + \sigma + k\,\nu$ where ν is the bitsize of $p + q + 1$, since there are at most $(p+q+1)^k$ monomials of degree $p+q$ in k variables. $\square$

Algorithm 8.6. **[Exact Division of Multivariate Polynomials]**

- **Structure:** a field K.
- **Input:** two multivariate polynomials P and Q in $\mathrm{K}[X_1, \ldots, X_k]$ whose degrees are bounded by p and $q \leq p$ and such that Q divides P in $\mathrm{K}[X_1, \ldots, X_k]$.
- **Output:** the polynomial C such that $P = CQ$.
- **Complexity:** $\leq \binom{p+k}{k}(2\binom{q+k}{k}+1) \leq (2(p+1)^k+1)(q+1)^k$.
- **Procedure:**
  - Initialization: $C := 0$, $R := P$.
  - While $R \neq 0$, order using the graded lexicographical ordering the monomials of P and Q and denote by m and n the leading monomial of P and Q so obtained. Since Q divides P, it is clear that n divides m. Denoting by a_m and b_n the coefficient of m and n in P and Q,
$$\begin{aligned} C &:= C + \frac{a_m m}{b_n n} \\ R &:= R - \frac{a_m m}{b_n n} Q. \end{aligned}$$
  - Output C.

Proof of correctness: The equality $P = CQ + R$ is maintained throughout the algorithm. Moreover, since Q divides P, Q divides R. The algorithm terminates with $R = 0$, since the leading monomial of R decreases strictly for the graded lexicographical ordering in each call to the loop. □

Complexity analysis: There are at most $\binom{p+k}{k}$ monomials to consider before the loop terminates, and there are for each call to the loop at most one division, $\binom{q+k}{k}$ multiplications and $\binom{q+k}{k}$ additions to perform. The complexity is
$$\binom{p+k}{k}\left[2\binom{q+k}{k}+1\right] \leq (2(p+1)^k+1)(q+1)^k.$$

Note that the choice of the leading monomial for the graded lexicographical ordering is cost free in our model of complexity. □

We consider now how to evaluate a univariate polynomial P at a value b.

Notation 8.6. [Horner] Let $P = a_p X^p + \cdots + a_0 \in \mathrm{A}[X]$, where A is a ring. The evaluation process uses the **Horner polynomials associated to** P, which are defined inductively by

$$\begin{aligned} \mathrm{Hor}_0(P, X) &= a_p, \\ &\vdots \\ \mathrm{Hor}_i(P, X) &= X\,\mathrm{Hor}_{i-1}(P, X) + a_{p-i}. \end{aligned}$$

for $0 \leq i \leq p$, so that

$$\mathrm{Hor}_i(P, X) = a_p X^i + a_{p-1} X^{i-1} + \cdots + a_{p-i}. \tag{8.1}$$

Note that $\mathrm{Hor}_p(P,X)=P(X)$. □

Algorithm 8.7. **[Evaluation of a Univariate Polynomial]**

- **Structure:** a ring A.
- **Input:** $P=a_p X^p+\cdots+a_0\in \mathrm{A}[X]$ and $b\in \mathrm{A}$.
- **Output:** the value $P(b)$.
- **Complexity:** $2p$.
- **Procedure:**
  - Initialize $\mathrm{Hor}_0(P,b):=a_p$.
  - For i from 1 to p,

$$\mathrm{Hor}_i(P,b):=b\,\mathrm{Hor}_{i-1}(P,b)+a_{p-i}.$$

  - Output $\mathrm{Hor}_p(P,b)=P(b)$.

Here the size of the input is a number, a bound on the degree of P. The computation takes place in the ring A.

Complexity analysis: The number of arithmetic operations is $2\,p$: p additions and p multiplications. □

When the polynomial has coefficients in $\mathbb{Z}$, we have the following variant.

Algorithm 8.8. **[Special Evaluation of a Univariate Polynomial]**

- **Structure:** the ring $\mathbb{Z}$.
- **Input:** $P=a_p X^p+\cdots+a_0\in \mathbb{Z}[X]$ and $b/c\in\mathbb{Q}$ with $b\in\mathbb{Z},c\in\mathbb{Z}$.
- **Output:** the value $c^p\,P(b/c)$.
- **Complexity:** $4\,p$.
- **Procedure:**
  - Initialize $\bar{H}_0(P,b):=a_p,\ d:=1$.
  - For i from 1 to p,

$$\begin{aligned} d &:= c\,d \\ \overline{\mathrm{Hor}}_i(P,b) &:= b\,\bar{H}_{i-1}(P,b)+d\,a_{p-i}. \end{aligned}$$

  - Output $\overline{\mathrm{Hor}}_p(P,b)=c^p\,P(b/c)$.

Complexity analysis: The number of arithmetic operations is $4\,p$: p additions and $3\,p$ multiplications. If τ is a bound on the bitsizes of the coefficients of P and τ' is a bound on the bitsizes of b and c, the bitsize of $\overline{\mathrm{Hor}}_i(P,b)$ is $\tau+i\tau'+\nu$, where ν is the bitsize of $p+1$, since the bitsize of the product of an integer of bitsize τ with i-times the product of an integer of bitsize τ' is $\tau+i\tau'$, and the bitsize of the sum of $i+1$ numbers of size λ is bounded by $\lambda+\nu$. □

The Horner process can also be used for computing the translate of a polynomial.

Algorithm 8.9. **[Translation]**

- **Structure:** a ring A.
- **Input:** $P(X) = a_p X^p + \cdots + a_0$ in $\mathrm{A}[X]$ and an element $c \in \mathrm{A}$.
- **Output:** the polynomial $T = P(X - c)$.
- **Complexity:** $p\,(p+1)$.
- **Procedure:**
  - Initialization: $T := a_p$.
  - For i from 1 to p,
  $$T := (X - c)\,T + a_{p-i}.$$
  - Output T.

Proof of correctness: It is immediate to verify that after step i,

$$T = a_p\,(X - c)^i + \cdots + a_{p-i}.$$

So after step p, $T = P(X - c)$. □

Complexity analysis: In step i, the computation of $(X - c)\,T$ takes i multiplications by c and i additions (multiplications by X are not counted). The complexity is the sum of the $p\,(p+1)/2$ multiplications by c and $p\,(p+1)/2$ additions and is bounded by $p\,(p+1)$. □

When the polynomial is with coefficients in $\mathbb{Z}$, we have the following variant.

Algorithm 8.10. **[Special Translation]**

- **Structure:** the ring $\mathbb{Z}$.
- **Input:** $P(X) = a_p X^p + \cdots + a_0$ in $\mathbb{Z}[X]$ and $b/c \in \mathbb{Q}$, with $b \in \mathbb{Z}, c \in \mathbb{Z}$.
- **Output:** the polynomial $c^p\,P(X - b/c)$.
- **Complexity:** $3\,p\,(p+3)/2$.
- **Procedure:**
  - Initialization: $\bar{T}_0 := a_p$, $d := 1$.
  - For i from 1 to p,
  $$\begin{aligned} d &:= c\,d \\ \bar{T}_i &:= (c\,X - b)\,\bar{T}_{i-1} + d \cdot a_{p-i}. \end{aligned}$$
  - Output $\bar{T}_p$.

Proof of correctness: It is immediate to verify that after step i,

$$\bar{T}_i = c^i\,(a_p\,(X - b/c)^i + \cdots + a_{p-i}).$$

So after step p, $\bar{T}_p = c^p\,P(X - b/c)$. □

Complexity analysis: In step i, the computation of $\bar{T}$ takes $2\,i + 2$ multiplications and i additions. The complexity is the sum of the $p(p+3)$ multiplications and $p\,(p+1)/2$ additions and is bounded by $3\,p\,(p+3)/2$.

Let τ be a bound on the bitsizes of the coefficients of P, τ' a bound on the bitsizes of b and c, and ν is the bitsize of $p+1$. Since

$$\sum_{k=0}^{i} a_{p-k}\,(b\,X-c)^{i-k} = \sum a_{p-k}\binom{j}{i-k}\,b^j\,(-c)^{i-k-j}X^j,$$

the bitsizes of the coefficients of $\bar{T}_i$ is $\tau+i\,(1+\tau')+\nu$: the bitsize of a binomial coefficient $\binom{i-k}{j}$ is at most i, the bitsize of the product of an integer of bitsize τ with the product of $i-k$ integers of bitsize τ' is bounded by $\tau+i\,\tau'$, and the bitsize of the sum of $i+1$ numbers of size λ is bounded by $\lambda+\nu$. □

Remark 8.7. Using fast arithmetic, a translation by 1 in a polynomial of degree d and bit size τ can be computed with binary complexity $\tilde{O}(d\,\tau)$ [64]. □

We give an algorithm computing the coefficients of a polynomial knowing its Newton sums.

Algorithm 8.11. **[Newton Sums]**

- **Structure:** a ring D with division in $\mathbb{Z}$.
- **Input:** the Newton sums N_i, $i=0,\ldots,p$, of a monic polynomial

$$P = X^p + a_{p-1}\,X^{p-1} + \cdots + a_0$$

  in D$[X]$.
- **Output:** the list of coefficients $1, a_{p-1},\ldots,a_0$ of P.
- **Complexity:** $p\,(p+1)$.
- **Procedure:**
  - $a_p := 1$.
  - For i from 1 to p,

$$a_{p-i} := \frac{-1}{i}\left(\sum_{j=1}^{i} a_{p-i+j}\,N_j\right).$$

Proof of correctness: Follows from Equation (4.1). Note that we have to know in advance that $P\in$ D$[X]$. □

Complexity analysis: The computation of each a_{p-i} takes $2\,i+1$ arithmetic operations in D. Since the complexity is bounded by

$$\sum_{i=1}^{p} (2\,i+1) = 2\,\frac{p\,(p-1)}{2} + p = p\,(p+1).$$ □

Note also that the Newton formulas (Equation (4.1)) could also be used to compute the Newton sums from the coefficients.

We end this list of examples with arithmetic operations on matrices.

Algorithm 8.12. **[Addition of Matrices]**

- **Structure:** a ring A.

- **Input:** two $n \times m$ matrices $M = [m_{i,j}]$ and $N = [n_{i,j}]$ with entries in A.
- **Output:** the sum $S = [s_{i,j}]$ of M and N.
- **Complexity:** $n\,m$.
- **Procedure:** For every i, j, $i \leq n$, $j \leq m$,

$$s_{i,j} := m_{i,j} + n_{i,j}.$$

Here the size of the input is two natural numbers n, m. The computation takes place in the ring A.

Complexity analysis: The complexity is $n\,m$ in A since there are $n\,m$ entries to compute and each of them is computed by one single addition.

If $\mathrm{A} = \mathbb{Z}$, and the bitsizes of the entries of M and N are bounded by τ, the bitsizes of the entries of their sum are bounded by $\tau + 1$.

If $\mathrm{A} = \mathbb{Z}[Y]$, $Y = Y_1, \ldots, Y_t$, and the degrees in Y of the entries of M and N are bounded by c, while the bitsizes of the entries of M and N are bounded by τ, the degrees in Y of the entries of their sum is bounded by c, and the bitsizes of the coefficients of the entries of their sum are bounded by $\tau + 1$. □

Algorithm 8.13. **[Multiplication of Matrices]**

- **Structure:** a ring A.
- **Input:** two matrices $M = [m_{i,j}]$ and $N = [n_{j,k}]$ of size $n \times m$ and $m \times \ell$ with entries in A.
- **Output:** the product $P = [p_{i,k}]$ of M and N.
- **Complexity:** $n\,\ell\,(2\,m - 1)$.
- **Procedure:** For each i, k, $i \leq n, k \leq \ell$,

$$p_{i,k} = \sum_{j=1}^{m} m_{i,j}\, n_{j,k}.$$

Complexity analysis: For each i, k there are m multiplications and $m - 1$ additions. The complexity is $n\ell(2m - 1)$.

If $\mathrm{A} = \mathbb{Z}$, and the bitsizes of the entries of M and N are bounded by τ and σ, the bitsizes of the entries of their product are bounded by $\tau + \sigma + \mu$, where μ is the bitsize of m.

If $\mathrm{A} = \mathbb{Z}[Y]$, $Y = Y_1, \ldots, Y_k$, and the degrees in Y of the entries of M and N are bounded by p and q, while the bitsizes of the entries of M and N are bounded by τ and σ, the degrees in Y of the entries of their product are bounded by $p + q$, and the bitsizes of the coefficients of the entries of their product are bounded by $\tau + \sigma + k\nu + \mu$ where μ is the bitsize of m and ν is the bitsize of $p + q + 1$, since the number of monomials of degree $p + q$ in k variables is bounded by $(p + q + 1)^k$. □

Algorithm 8.14. **[Multiplication of Several Matrices]**

- **Structure:** a ring A.
- **Input:** m matrices $M_1 \ldots M_m$ of size $n \times n$, with entries in A.
- **Output:** the product P of $M_1, \ldots, M_m$.
- **Complexity:** $(m-1)\, n^2\, (2\, n - 1)$.
- **Procedure:** Initialize $N_1 := M_1$. For i from 2 to m define $N_i = N_{i-1}\, M_i$.

Complexity analysis: For each i from 2 to m, and j, k from 1 to n, there are n multiplications and $n-1$ additions. The complexity is $(m-1)\, n^2\, (n-1)$.

If $\mathrm{A} = \mathbb{Z}$, and the bitsizes of the entries of the M_i are bounded by τ, the bitsizes of the entries of their product are bounded by $m\,(\tau + \mu)$ where μ is the bitsize of n.

If $\mathrm{A} = \mathbb{Z}[Y]$, $Y = Y_1, \ldots, Y_k$, and the degrees in Y of the entries of the M_i are bounded by p, while the bitsizes of the entries of the M_i are bounded by τ, the degrees in Y of the entries of their product are bounded by $m\, p$, and the bitsizes of the coefficients of the entries of their product are bounded by $m\,(\tau + \mu) + k\nu$ where μ is the bitsize of n and ν is the bitsize of $k p + 1$. □

Remark 8.8. The complexity of computing the product of two matrices depends on the algorithm used. The complexity of the multiplication of two square matrices of size n is $O(n^3)$ when the multiplication is done naively, as in Algorithm 8.13, $O(n^{\log_2(7)})$ when Strassen's method is used. Even more efficient algorithms are known but we have decided not to include this topic in this book. The interested reader is referred to [64].

Similar remarks were made earlier for the multiplications of polynomials and of integers, and apply also to the euclidean remainder sequence and to most of the algorithms dealing with univariate polynomials and linear algebra presented in Chapters 8 and 9. Explaining sophisticated algorithms would have required a lot of effort and many more pages. In order to prove the complexity estimates we present in Chapters 10 to 15, complexities of $n^{O(1)}$ for algorithms concerning univariate polynomials and linear algebra (where n is a bound on the degrees or on the size of the matrices) are sufficient. □

8.2 Linear Algebra

8.2.1 Size of Determinants

Proposition 8.9. [Hadamard] *Let M be an $n \times n$ matrix with integer entries. Then the determinant of M is bounded by the product of the euclidean norms of the columns of M.*

Proof: If $\det(M) = 0$, the result is certainly true. Otherwise, the column vectors of M, $v_1, \ldots, v_n$, span R^n. We denote by $u \cdot v$ the inner product of u and v. Using the Gram-Schmidt orthogonalization process (Proposition 4.40), there are vectors $w_1, \ldots, w_n$ with the following properties

- $w_i - v_i$ belong to the vector space spanned by $w_1, \ldots, w_{i-1}$,
- $\forall i \, \forall j \; j \neq i, w_i \cdot w_j = 0$.

Moreover, denoting $u_i = w_i - v_i$,

$$\begin{aligned} \|w_i\|^2 + \|u_i\|^2 &= \|v_i\|^2, \\ \|w_i\| &\leq \|v_i\|. \end{aligned}$$

Then it is clear that

$$|\det(M)| = \prod_{i=1}^{n} \|w_i\| \leq \prod_{i=1}^{n} \|v_i\|. \qquad \square$$

Corollary 8.10. *Let M be an $n \times n$ matrix with integer entries of bitsizes at most τ. Then the bitsize of the determinant of M is bounded by $n(\tau + \nu/2)$, where ν is the bitsize of n.*

Proof: If $n < 2^\nu$ and $|m_{i,j}| < 2^\tau$ then $\sqrt{\sum_{i=1}^{n} m_{i,j}^2} < \sqrt{n} 2^\tau < 2^{\tau + \nu/2}$.
Thus $|\det(M)| < 2^{n(\tau + \nu/2)}$, using Lemma 8.9. $\square$

The same kind of behavior is observed when we consider degrees of polynomials rather than bitsize. Things are even simpler, since there is no carry to take into account in the degree estimates.

Proposition 8.11. *Let M be an $n \times n$ matrix with entries that are polynomials in $Y_1, \ldots, Y_k$ of degrees d. Then the determinant considered as a polynomial in $Y_1, \ldots, Y_k$ has degree in $Y_1, \ldots, Y_k$ bounded by $d\,n$.*

Proof: This follows from $\det(M) = \sum_{\sigma \in \mathcal{S}_n} (-1)^{\varepsilon(\sigma)} \prod_{i=1}^{n} m_{\sigma(i),i}$, where $\varepsilon(\sigma)$ is the signature of σ. $\square$

Moreover we have

Proposition 8.12. *Let M be an $n \times n$ matrix with entries that are polynomials in $Y_1, \ldots, Y_k$ of degrees d in $Y_1, \ldots, Y_k$ and coefficients in $\mathbb{Z}$ of bitsize τ. Then the determinant considered as a polynomial in $Y_1, \ldots, Y_k$ has degrees in $Y_1, \ldots, Y_k$ bounded by $d\,n$, and coefficients of bitsize $(\tau + \nu)\,n + k\,\mu$ where ν is the bitsize of n and μ is the bitsize of $n\,d + 1$.*

Proof: The only thing which remains to prove is the result on the bitsize. Performing the multiplication of n monomials appearing in the entries of the matrix produces integers of bitsize $\tau\,n$ Since the number of monomials of a polynomial of degree $n\,d$ in k variables is bounded by $(n\,d + 1)^k$ by Lemma 8.5, the bitsizes of the coefficients of the products of n entries of the matrix are bounded by $(\tau + \nu)\,n + k\,\mu$. Since there are $n!$ terms in the determinant, and the bitsize of $n!$ is bounded by $n\,\nu$ the final bound is $(\tau + \nu)\,n + k\,\mu$. $\square$

8.2.2 Evaluation of Determinants

The following method, which is the standard row reduction technique, can be used to compute the determinant of a square matrix with coefficients in a field.

Algorithm 8.15. **[Gauss]**

- **Structure:** a field K.
- **Input:** an $n \times n$ matrix $M = [m_{i,j}]$ with coefficients in K.
- **Output:** the determinant of M.
- **Complexity:** $O(n^3)$.
- **Procedure:**
  - Initialization: $k := 0$ and $g_{i,j}^{(0)} := m_{i,j}$.
  - For k from 0 to $n-2$,
    - If for every $j = k+1, \ldots, n$, $g_{k+1,j}^{(k)} = 0$, output $\det(M) = 0$.
    - Otherwise, exchanging columns if needed, suppose $g_{k+1,k+1}^{(k)} \neq 0$.
    - For i from $k+2$ to n,
$$g_{i,k+1}^{(k+1)} := 0,$$
      - For j from $k+2$ to n,
$$g_{i,j}^{(k+1)} := g_{i,j}^{(k)} - \frac{g_{i,k+1}^{(k)}}{g_{k+1,k+1}^{(k)}} \, g_{k+1,j}^{(k)}. \tag{8.2}$$
  - Output
$$\det(M) = (-1)^s \, g_{1,1}^{(0)} \cdots g_{n,n}^{(n-1)} \tag{8.3}$$
    (where s is the number of exchanges of columns in the intermediate computations).

Example 8.13. Consider the following matrix

$$M := \begin{bmatrix} a_1 & b_1 & c_1 \\ a_2 & b_2 & c_2 \\ a_3 & b_3 & c_3 \end{bmatrix},$$

and suppose $a_1 \neq 0$ and $b_2 \, a_1 - b_1 \, a_2 \neq 0$. Performing the first step of Algorithm 8.15 (Gauss), we get

$$\begin{aligned} g_{22}^{(1)} &= \frac{a_1 b_2 - b_1 a_2}{a_1} \\ g_{23}^{(1)} &= \frac{a_1 c_2 - c_1 a_2}{a_1} \\ g_{32}^{(1)} &= \frac{a_1 b_3 - b_1 a_3}{a_1} \\ g_{33}^{(1)} &= \frac{a_1 c_3 - c_1 a_3}{a_1} \end{aligned}$$

After the first step of reduction we have obtained the matrix

$$M_1 = \begin{bmatrix} a_1 & b_1 & c_1 \\ 0 & g_{22}^{(1)} & g_{23}^{(1)} \\ 0 & g_{32}^{(1)} & g_{33}^{(1)} \end{bmatrix}.$$

Note that the determinant of M_1 is the same as the determinant of M since M_1 is obtained from M by adding a multiple of the first row to the second and third row.

Performing the second step of Algorithm 8.15 (Gauss), we get

$$g_{33}^{(2)} = \frac{c_3\, a_1\, b_2 - c_3\, b_1\, a_2 - c_1\, a_3\, b_2 - c_2\, a_1\, b_3 + c_2\, b_1\, a_3 + c_1\, a_2\, b_3}{b_2\, a_1 - b_1\, a_2}$$

After the second step of reduction we have obtained the triangular matrix

$$M' = \begin{bmatrix} a_1 & b_1 & c_1 \\ 0 & g_{22}^{(1)} & g_{23}^{(1)} \\ 0 & 0 & g_{33}^{(2)} \end{bmatrix}.$$

Note that the determinant of M' is the same as the determinant of M since M' is obtained from M_1 by adding a multiple of the second row to the third row.

Finally, since $g_{11}^{(0)} = a_1$,

$$\det(M) = \det(M') = g_{11}^{(0)}\, g_{22}^{(1)}\, g_{33}^{(2)}.$$

□

Proof of correctness: The determinant of the $n \times n$ matrix $M' = [g_{i,j}^{(i-1)}]$ obtained at the end of the algorithm is equal to the determinant of M since the determinant does not change when a multiple of another row is added to a row. Thus, taking into account exchanges of rows,

$$\det(M) = \det(M') = (-1)^s\, g_{11}^{(0)} \cdots g_{nn}^{(n-1)}.$$

□

Complexity analysis: The number of calls to the main loop are at most $n-1$, the number of elements computed in each call to the loop is at most $(n-i)^2$, and the computation of an element is done by 3 arithmetic operations. So, the complexity is bounded by

$$3\left(\sum_{i=1}^{n-1} (n-i)^2\right) = \frac{2\, n^3 - 3\, n^2 + n}{2} = O(n^3).$$

Note that if we are interested only in the bound $O(n^3)$, we can estimate the number of elements computed in each call to the loop by n^2 since being more precise changes the constant in front of n^3 but not the fact that the complexity is bounded by $O(n^3)$.

□

Remark 8.14. It is possible, using other methods, to compute determinants of matrices of size n in parallel complexity $O(\log_2(n)^2)$ using $n^{O(1)}$ processors [127]. As a consequence it is possible to compute them in complexity $n^{O(1)}$, using only $\log_2(n)^{O(1)}$ space at a given time [127]. □

As we can see in Example 8.13, it is annoying to see denominators arising in a determinant computation, since the determinant belongs to the ring generated by the entries of the matrix. This is fixed in what follows.

Notation 8.15. [Bareiss] Let $M_{i,j}^{(k)}$ be the $(k+1)\times(k+1)$ matrix obtained by taking

$$\begin{cases} m_{i',j'}^{(k)} = m_{i',j'} & \text{for } i'=1,\dots,k,\ j'=1,\dots,k,\\ m_{k+1,j'}^{(k)} = m_{i,j'} & \text{for } j'=1,\dots,k,\\ m_{i',k+1}^{(k)} = m_{i',j} & \text{for } i'=1,\dots,k,\\ m_{k+1,k+1}^{(k)} = m_{i,j}. \end{cases}$$

and define $b_{i,j}^{(k)} = \det(M_{i,j}^{(k)})$. Then $b_{k,k}^{(k-1)}$ is the **principal k-th minor** of M, i.e. the determinant of the submatrix extracted from M on the k first rows and columns. It follows from the definition of the $b_{i,j}^{(k)}$ that if M has entries in an integral domain D then $b_{i,j}^{(k)} \in \mathrm{D}$. □

In the following discussion, we always suppose without loss of generality that if $b_{k+1,k+1}^{(k)} = 0$ then $b_{k+1,j}^{(k)} = 0$ for $j = k+2,\dots,n$, since this condition is fulfilled after a permutation of columns.

Note that by (8.3), if $i, j \geq k+1$,

$$b_{i,j}^{(k)} = g_{1,1}^{(0)} \cdots g_{k,k}^{(k-1)}\, g_{i,j}^{(k)}. \tag{8.4}$$

Indeed, denoting by ${g'_{i,j}}^{(k)}$ the output of Gauss's method applied to $M_{i,j}^{(k)}$, it is easy to check that ${g'_{i,i}}^{(i-1)} = g_{i,i}^{(i-1)}$ for $i = 1,\dots,k$, and ${g'_{k+1,k+1}}^{(k)} = g_{i,j}^{(k)}$.

Proposition 8.16.

$$b_{i,j}^{(k+1)} = \frac{b_{k+1,k+1}^{(k)}\, b_{i,j}^{(k)} - b_{i,k+1}^{(k)}\, b_{k+1,j}^{(k)}}{b_{k,k}^{(k-1)}}.$$

Proof: The result follows easily from the recurrence (8.2) and equation (8.4). Indeed (8.4) implies

$$\begin{aligned} \frac{b_{k+1,k+1}^{(k)}\, b_{i,j}^{(k)} - b_{i,k+1}^{(k)}\, b_{k+1,j}^{(k)}}{b_{k,k}^{(k-1)}} &= \frac{(g_{1,1}^{(0)} \cdots g_{k,k}^{(k-1)})^2\, (g_{k+1,k+1}^{(k)}\, g_{i,j}^{(k)} - g_{i,k+1}^{(k)}\, g_{k+1,j}^{(k)})}{g_{1,1}^{(0)} \cdots g_{k,k}^{(k-1)}} \\ &= g_{1,1}^{(0)} \cdots g_{k,k}^{(k-1)}\, (g_{k+1,k+1}^{(k)}\, g_{i,j}^{(k)} - g_{i,k+1}^{(k)}\, g_{k+1,j}^{(k)}). \end{aligned}$$

On the other hand, (8.2) implies that

$$g_{k+1,k+1}^{(k)} g_{i,j}^{(k)} - g_{i,k+1}^{(k)} g_{k+1,j}^{(k)} = g_{k+1,k+1}^{(k)} g_{i,j}^{(k+1)}. \tag{8.5}$$

So

$$\frac{b_{k+1,k+1}^{(k)} b_{i,j}^{(k)} - b_{i,k+1}^{(k)} b_{k+1,j}^{(k)}}{b_{k,k}^{(k-1)}} = g_{1,1}^{(0)} \cdots g_{k,k}^{(k-1)} g_{k+1,k+1}^{(k)} g_{i,j}^{(k+1)}.$$

Using again (8.4),

$$g_{1,1}^{(0)} \cdots g_{k,k}^{(k-1)} g_{k+1,k+1}^{(k)} g_{i,j}^{(k+1)} = b_{i,j}^{(k+1)}, \tag{8.6}$$

and the result follows. □

Note that (8.6) implies that, if $b_{k+1,k+1}^{(k)} \neq 0$,

$$g_{i,j}^{(k+1)} = \frac{b_{i,j}^{(k+1)}}{b_{k+1,k+1}^{(k)}}. \tag{8.7}$$

A new algorithm for computing the determinant follows from Proposition 8.16.

Algorithm 8.16. **[Dogdson-Jordan-Bareiss]**

- **Structure:** a domain D.
- **Input:** an $n \times n$ matrix $M = [m_{i,j}]$ with coefficients in D.
- **Output:** the determinant of M.
- **Complexity:** $O(n^3)$.
- **Procedure:**
  - Initialization: $k := 0$ and $b_{i,j}^{(0)} := m_{i,j}$, $b_{0,0}^{(-1)} := 1$.
  - For k from 0 to $n-2$,
    - If for every $j = k+1, \ldots, n$, $b_{k+1,j}^{(k)} = 0$, output $\det(M) = 0$.
    - Otherwise, exchanging columns if needed, suppose that $b_{k+1,k+1}^{(k)} \neq 0$.
    - For i from $k+2$ to n,

$$b_{i,k+1}^{(k+1)} := 0,$$

      - For j from $k+2$ to n,

$$b_{i,j}^{(k+1)} := \frac{b_{k+1,k+1}^{(k)} b_{i,j}^{(k)} - b_{i,k+1}^{(k)} b_{k+1,j}^{(k)}}{b_{k,k}^{(k-1)}}. \tag{8.8}$$

  - Output

$$\det(M) = (-1)^s b_{n,n}^{(n-1)} \tag{8.9}$$

(where s is the number of exchanges of columns in the intermediate computation

Proof of correctness: The correctness follows from Proposition 8.16. Note that although divisions are performed, they are always exact divisions, since we know from Proposition 8.16 that all the intermediate computations obtained by a division in the algorithm are determinants extracted from M and hence belong to D. □

Complexity analysis: The number of calls to the main loop are at most $n-1$, the number of elements computed in each call to the loop is at most $(n-i)^2$, and the computation of an element is done by 4 arithmetic operations. So, the complexity is bounded by

$$4\left(\sum_{i=1}^{n-1} (n-i)^2\right) = \frac{4\,n^3 - 6\,n^2 + 2\,n}{3} = O(n^3).$$

If M is a matrix with integer coefficients having bitsize at most τ, the arithmetic operations in the algorithm are performed on integers of bitsize $n\,(\tau + \nu)$, where ν is the bitsize of n, using Hadamard's bound (Corollary 8.10). □

Example 8.17. Consider again

$$M := \begin{bmatrix} a_1 & b_1 & c_1 \\ a_2 & b_2 & c_2 \\ a_3 & b_3 & c_3 \end{bmatrix}.$$

Performing the first step of Algorithm 8.16 (Dogdson-Jordan-Bareiss), we get

$$\begin{aligned} b_{22}^{(1)} &= a_1\,b_2 - b_1\,a_2, \\ b_{23}^{(1)} &= a_1\,c_2 - c_1\,a_2, \\ b_{32}^{(1)} &= a_1\,b_3 - b_1\,a_3, \\ b_{33}^{(1)} &= a_1\,c_3 - c_1\,a_3. \end{aligned}$$

which are determinants extracted from M.

Performing the second step of Algorithm 8.16 (Dogdson-Jordan-Bareiss), we get

$$\begin{aligned} b_{33}^{(2)} &= \frac{(a_1\,b_2 - b_1\,a_2)\,(a_1\,c_3 - c_1\,a_3) - (a_1\,c_2 - c_1\,a_2)\,(a_1\,b_3 - b_1\,a_3)}{a_1} \\ &= c_3\,a_1\,b_2 - c_3\,b_1\,a_2 - c_1\,a_3\,b_2 - c_2\,a_1\,b_3 + c_2\,b_1\,a_3 + c_1\,a_2\,b_3. \end{aligned}$$

Finally,

$$\det(M) = b_{33}^{(2)}.$$

□

Remark 8.18. It is easy to see than either Algorithm 8.15 (Gauss) or Algorithm 8.16 (Dogdson-Jordan-Bareiss) can be adapted as well to compute the rank of the matrix with the same complexity. □

Exercise 8.1. Describe algorithms for computing the rank of a matrix by adapting Algorithm 8.15 (Gauss) and Algorithm 8.16 (Dogdson-Jordan-Bareiss).

8.2.3 Characteristic Polynomial

Let A be a ring and M be a matrix $M = (m_{ij}) \in \mathrm{A}^{n \times n}$. The first idea of the method we present to compute the characteristic polynomial is to compute the traces of the powers of M, and to use Algorithm 8.11 (Newton Sums) to recover the characteristic polynomial. Indeed the trace of M^i is the i-th Newton sum of the characteristic polynomial of M. The second idea is to notice that, in order to compute the trace of a product of two matrices M and N, it is not necessary to compute the product $M\,N$, since $\mathrm{Tr}(MN) = \sum_{k,\ell} m_{k,\ell} n_{\ell,k}$. So rather than computing all the powers M^i of M, $i = 2, \ldots, n$ then all the corresponding traces, it is enough, defining r as the smallest integer $> \sqrt{n}$, to compute the powers M^i for $i = 2, \ldots, r-1$, the powers M^{jr} for $j = 2, \ldots, r-1$ and then $\mathrm{Tr}(M^{rj+i}) = \mathrm{Tr}(M^i M^{jr})$.

Algorithm 8.17. **[Characteristic Polynomial]**

- **Structure:** a ring with integer division A.
- **Input:** an $n \times n$ matrix $M = [m_{i,j}]$, with coefficients in A.
- **Output:** $\mathrm{CharPol}(M) = \det(X\,\mathrm{Id}_n - M)$, the characteristic polynomial of M.
- **Complexity:** $O(n^{3.5})$.
- **Procedure:**
  - Define r as the smallest integer $> \sqrt{n}$.
  - Computation of powers M^i for $i < r$ and their traces.
    - $B_0 := \mathrm{Id}_n$, $N_0 := n$.
    - For i from 0 to $r-2$
      $B_{i+1} := M B_i$, $N_{i+1} := \mathrm{Tr}(B_{i+1})$.
  - Computation of powers M^{rj} for $j < r$ and their traces.
    - $C_1 := M B_{r-1}$, $N_r = \mathrm{Tr}(C_1)$.
    - For j from 1 to $r-2$
      $C_{j+1} = C_1 C_j$, $N_{(j+1)r} = \mathrm{Tr}(C_{j+1})$.
  - Computation of traces of M^k for $k = jr + i, i = 1, \ldots, r-1, j = 1, \ldots, r-1$.
    - For i from 1 to $r-1$
      - For j from 1 to $r-1$
        $N_{jr+i} = \mathrm{Tr}(B_i C_j)$.
  - Computation of the coefficients of $\det(X\,\mathrm{Id}_n - M)$: use Algorithm 8.11 (Newton Sums) taking as $i-\mathrm{th}$ Newton sum $N_i, i = 0, \ldots, n$.

Proof of correctness: Since a square matrix with coefficients in a field K can be triangulated over C, the fact that the trace of M^i is the Newton sums of the eigenvalues is clear in the case of an integral domain. For a general ring with integer division, it is sufficient to specialize the preceding algebraic identity expressed in the ring $\mathbb{Z}[U_{i,j}, i = 1, \ldots, n, j = 1, \ldots n]$ by replacing $U_{i,j}$ with the entries $m_{i,j}$ of the matrix. □

Complexity analysis: The first step and second step take $O(r\,n^3)=O(n^{3.5})$ arithmetic operations. The third step take $O(n^3)$ arithmetic operations, and the fourth step $O(n^2)$.

If the entries of M are elements of $\mathbb{Z}$ of bitsize at most τ, and the bitsize of n is ν, the bitsizes of the intermediate computations are bounded by $O((\tau+\nu)\,n)$ using the complexity analysis of Algorithm 8.14 (Multiplication of Several Matrices) The arithmetic operations performed are multiplications between integers of bitsizes bounded by $(\tau\,+\,\nu)\,\sqrt{n}$ and integers of bitsizes bounded by $(\tau+\nu)\,n$.

If the entries of M are elements of $\mathbb{Z}[Y]$, $Y=Y_1,\ldots,Y_k$ of degrees at most d and of bitsizes at most τ, the degrees in Y and bitsizes of the intermediate computations are $d\,n$ and $(\tau\,+\,2\,\nu)\,n$ where n u is the bitsize of $n\,d+1$ using the complexity analysis of Algorithm 8.14 (Multiplication of Several Matrices). The arithmetic operations performed are multiplications between integers of bitsizes bounded by $(\tau+2\,\nu)\,\sqrt{n}$ and integers of bitsizes bounded by $(\tau+2\,\nu)\,n$. □

Remark 8.19. a) In the case of a field of characteristic zero, the rank of M is easily computed from its characteristic polynomial $\mathrm{CharPol}(M)$: it is the degree of the monomial of least degree in $\mathrm{CharPol}(M)$.

b) Algorithm 8.17 (Characteristic polynomial) provides the determinant of M in $O(n^{3.5})$ arithmetic operations in an arbitrary ring with integer division, substituting 0 to X in $\mathrm{CharPol}(M)$. □

8.2.4 Signature of Quadratic Forms

A general method for computing the signature of quadratic form using the characteristic polynomial is based on the following result.

Proposition 8.20. *If Φ is a quadratic form with associated symmetric matrix M of size n, with entries in a real closed field* R *and*

$$\mathrm{CharPol}(M)=\det(X\,\mathrm{Id}_n-M)=X^n+a_{n-1}X^{n-1}+\cdots+a_0$$

is the characteristic polynomial of M, then

$$\mathrm{Sign}(M)=\mathrm{Var}(1,a_{n-1},\ldots,a_0)-\mathrm{Var}((-1)^n,(-1)^{n-1}a_{n-1},\ldots,a_0),$$

(see Notation 2.32).

Proof: All the roots of the characteristic polynomial of a symmetric matrix belong to R by Theorem 4.42 and we can apply Proposition 2.33 (Descartes' law of signs) and Remark 2.38. □

Algorithm 8.18. **[Signature Through Descartes]**

- **Structure:** an ordered integral domain D.
- **Input:** an $n \times n$ symmetric matrix $M = [m_{i,j}]$, with coefficients in D.
- **Output:** the signature of the quadratic form associated to M.
- **Complexity:** $O(n^{3,5})$.
- **Procedure:** Compute the characteristic polynomial of M

$$\mathrm{CharPol}(M) = \det(X \,\mathrm{Id}_n - M) = X^n + a_{n-1} X^{n-1} + \cdots + a_0$$

using Algorithm 8.17 (Characteristic polynomial) and output

$$\mathrm{Var}(1, a_{n-1}, \ldots, a_0) - \mathrm{Var}((-1)^n, (-1)^{n-1} a_{n-1}, \ldots, a_0).$$

Complexity analysis: The complexity is bounded by $O(n^{3.5})$, according to the complexity analysis of Algorithm 8.17 (Characteristic polynomial). Moreover, if the entries of A are elements of $\mathbb{Z}$ of bitsize at most τ, the arithmetic operations performed are multiplications between integers of bitsizes bounded by τ and integers of bitsizes bounded by $(\tau + 2\,\nu)\, n + \nu + 2$ where ν is the bitsize of n. □

8.3 Remainder Sequences and Subresultants

8.3.1 Remainder Sequences

We now present some results concerning the computation of the signed remainder sequence that was defined in Chapter 1 (Definition 1.2).

The following algorithm follows immediately from the definition.

Algorithm 8.19. **[Signed Remainder Sequence]**

- **Structure:** a field K.
- **Input:** two univariate polynomials P and Q with coefficients K.
- **Output:** the signed remainder sequence of P and Q.
- **Complexity:** $O(p\,q)$, where p is the degree of P and q the degree of Q.
- **Procedure:**
  - Initialization: $i := 1$, $\mathrm{SRemS}_0(P, Q) := P$, $\mathrm{SRemS}_1(P, Q) := Q$.
  - While $\mathrm{SRemS}_i(P, Q) \neq 0$
    - $\mathrm{SRemS}_{i+1}(P, Q) = -\mathrm{Rem}(\mathrm{SRemS}_{i-1}(P, Q), \mathrm{SRemS}_i(P, Q))$,
    - $i := i + 1$.

Complexity analysis: Let P and Q have degree p and q. The number of steps in the algorithm is at most $q + 1$. Denoting by $d_i = \deg(\mathrm{SRemS}_i(P, Q))$, the complexity of computing $\mathrm{SRemS}_{i+1}(P, Q)$ knowing $\mathrm{SRemS}_{i-1}(P, Q)$ and $\mathrm{SRemS}_i(P, Q)$ is bounded by $(d_{i-1} - d_i + 1)\,(2\,d_i + 3)$ by Algorithm 8.3. Summing over all i and bounding d_i by q, we get the bound $(p + q + 1)\,(2\,q + 3)$, which is $O(p\,q)$. □

An important variant of Signed Euclidean Division is the following Extended Signed Euclidean Division computing the extended signed remainder sequence (Definition 1.10).

Algorithm 8.20. **[Extended Signed Remainder Sequence]**

- **Structure:** a field K.
- **Input:** two univariate polynomials P and Q with coefficients in K.
- **Output:** the extended signed remainder sequence $\mathrm{Ex}(P, Q)$.
- **Complexity:** $O(pq)$, where p is the degree of P and q the degree of Q.
- **Procedure:**
  - Initialization: $i := 1$,

$$\begin{aligned} \mathrm{SRemS}_0(P,Q) &:= P,\\ \mathrm{SRemS}_1(P,Q) &:= Q,\\ \mathrm{SRemU}_0(P,Q) = \mathrm{SRemV}_1(P,Q) &:= 1,\\ \mathrm{SRemV}_0(P,Q) = \mathrm{SRemU}_1(P,Q) &:= 0. \end{aligned}$$

  - While $\mathrm{SRemS}_i(P, Q) \neq 0$
    - Compute

$$\begin{aligned} A_{i+1} &= \mathrm{Quo}(\mathrm{SRemS}_{i-1}(P,Q), \mathrm{SRemS}_i(P,Q)),\\ \mathrm{SRemS}_{i+1}(P,Q) &= -\mathrm{SRemS}_{i-1}(P,Q) + A_{i+1}\,\mathrm{SRemS}_i(P,Q),\\ \mathrm{SRemU}_{i+1}(P,Q) &= -\mathrm{SRemU}_{i-1}(P,Q) + A_{i+1}\,\mathrm{SRemU}_i(P,Q),\\ \mathrm{SRemV}_{i+1}(P,Q) &= -\mathrm{SRemV}_{i-1}(P,Q) + A_{i+1}\,\mathrm{SRemV}_i(P,Q). \end{aligned}$$

    - $\mathrm{Ex}_i(P,Q) = (\mathrm{SRemS}_i(P,Q), \mathrm{SRemU}_i(P,Q), \mathrm{SRemV}_i(P,Q))$
    - $i := i + 1$.

Proof of correctness: Immediate by Proposition 1.9. □

Complexity analysis: Suppose that P and Q have respective degrees p and q. It is immediate to check that the complexity is $O(pq)$, as in Algorithm 8.19 (Signed Remainder Sequence). □

If we also take into consideration the growth of the bitsizes of the coefficients in the signed remainder sequence, an exponential behavior of the preceding algorithms is a priori possible. If the coefficients are integers of bitsize τ, the bitsizes of the coefficients in the signed remainder sequence of P and Q could be exponential in the degrees of the polynomials P and Q since the bitsize of the coefficients could be doubled at each computation of a remainder in the euclidean remainder sequence.

The bitsizes of the coefficients in the signed remainder sequence can indeed increase dramatically as we see in the next example.

Example 8.21. Consider the following numerical example:

$$\begin{aligned} P := \; & 9X^{13} - 18X^{11} - 33X^{10} + 102X^8 + 7X^7 - 36X^6\\ & - 122X^5 + 49X^4 + 93X^3 - 42X^2 - 18X + 9. \end{aligned}$$

The greatest common divisor of P and P' is of degree 5. The leading coefficients of the signed remainder sequence of P and P' are:

$$\begin{array}{c} \frac{36}{13}, \\ -\frac{10989}{16}, \\ -\frac{2228672}{165649}, \\ -\frac{900202097355}{4850565316}, \\ -\frac{3841677139249510908}{543561530761725025}, \\ -\frac{6648854900739944448789496725}{676140352527579535315696712}, \\ -\frac{200117670554781699308164692478544184}{1807309302290980501324553958871415645}. \end{array}$$

□

8.3.2 Signed Subresultant Polynomials

Now we define and study the subresultant polynomials. Their coefficients are determinants extracted from the Sylvester matrix, and they are closely related to the remainder sequence. Their coefficients of highest degree are the subresultant coefficients introduced in Chapter 4 and used to study the geometry of semi-algebraic sets in Chapter 5. We are going to use them in this chapter to estimate the bitsizes of the coefficients in the signed remainder sequence. They will be also used for real root counting with a good control on the size of the intermediate computations.

8.3.2.1 Polynomial Determinants

We first study polynomial determinants, which will be useful in the study of subresultants.

Let K be a field of characteristic 0. Consider the K-vector space $\mathcal{F}_n$, consisting of polynomials whose degrees are less than n, equipped with the basis

$$\mathcal{B} = X^{n-1}, \ldots, X, 1.$$

We associate to a list of polynomials $\mathcal{P} = P_1, \ldots, P_m$, with $m \leq n$ a matrix $\mathrm{Mat}(\mathcal{P})$ whose rows are the coordinates of the P_i's in the basis $\mathcal{B}$. Note that $\mathrm{Mat}(\mathcal{B})$ is the identity matrix of size n.

Let $0 < m \leq n$. A mapping Φ from $(\mathcal{F}_n)^m$ to $\mathcal{F}_{n-m+1}$ is **multilinear** if for $\lambda \in \mathrm{K}, \mu \in \mathrm{K}$

$$\Phi(\ldots, \lambda A_i + \mu B_i, \ldots) = \lambda \Phi(\ldots, A_i, \ldots) + \mu \Phi(\ldots, B_i, \ldots).$$

A mapping Φ from $(\mathcal{F}_n)^m$ to $\mathcal{F}_{n-m+1}$ is **alternating** if

$$\Phi(\ldots, A, \ldots, A, \ldots) = 0.$$

A mapping Φ from $(\mathcal{F}_n)^m$ to $\mathcal{F}_{n-m+1}$ is **antisymmetric** if

$$\Phi(..., A, ..., B, ...) = -\Phi(..., B, ..., A, ...).$$

Lemma 8.22. *A mapping from* $(\mathcal{F}_n)^m$ *to* $\mathcal{F}_{n-m+1}$ *which is multilinear and alternating is antisymmetric.*

Proof: Since Φ is alternating,

$$\begin{aligned}\Phi(..., A+B, ..., A+B, ...) &= \Phi(..., A, ..., A, ...)\\ &= \Phi(..., B, ..., B, ...)\\ &= 0.\end{aligned}$$

Using multilinearity, we get easily

$$\Phi(..., A, ..., B, ...) + \Phi(..., B, ..., A, ...) = 0.$$ □

Proposition 8.23. *There exists a unique multilinear alternating mapping* Φ *from* $(\mathcal{F}_n)^m$ *to* $\mathcal{F}_{n-m+1}$ *satisfying, for every* $n > i_1 > ... > i_{m-1} > i$

$$\begin{cases}\Phi(X^{i_1}, ..., X^{i_{m-1}}, X^i) = X^i & \text{if for every } j < m \ i_j = n - j.\\ \Phi(X^{i_1}, ..., X^{i_{m-1}}, X^i) = 0 & \text{otherwise.}\end{cases}$$

Proof: Decomposing each P_i in the basis $\mathcal{B}$ of monomials and using multilinearity and antisymmetry, it is clear that a multilinear and alternating mapping Φ from $\mathcal{F}_n^m$ to $\mathcal{F}_{n-m+1}$ depends only on the values $\Phi(X^{i_1}, ..., X^{i_{m-1}}, X^i)$ for $n > i_1 > ... > i_{m-1} > n$. This proves the uniqueness.

In order to prove existence, let m_i, $i \le n$, be the $m \times m$ minor of $\mathrm{Mat}(\mathcal{P})$ based on the columns $1, ..., m-1, n-i$, then

$$\Phi(\mathcal{P}) = \sum_{i \le n-m} m_i X^i \tag{8.10}$$

satisfies all the properties required. □

The (m,n)-**polynomial determinant** mapping, denoted $\mathrm{pdet}_{m,n}$, is the unique multilinear alternating mapping from $\mathcal{F}_n^m$ to $\mathcal{F}_{n-m+1}$ satisfying the properties of Proposition 8.23.

When $n = m$, it is clear that $\mathrm{pdet}_{n,n}(\mathcal{P}) = \det(\mathrm{Mat}(\mathcal{P}))$, since det is known to be the unique multilinear alternating map sending the identity matrix to 1.

On the other hand, when $m = 1$, $\mathrm{pdet}(P)_{1,n}(X^i) = X^i$ and, by linearity, $\mathrm{pdet}_{1,n}(P) = P$.

If follows immediately from the definition that

Lemma 8.24. *Let* $\mathcal{P} = P_1, ..., P_m$.

If $\mathcal{Q} = Q_1, ..., Q_m$ *is such that* $Q_i = P_i$, $i \ne j$, $Q_j = P_j + \sum_{j \ne i} \lambda_j P_j$, *then* $\mathrm{pdet}_{m,n}(\mathcal{Q}) = \mathrm{pdet}_{m,n}(\mathcal{P})$.

If $\mathcal{Q} = P_m, \ldots, P_1$, *then* $\operatorname{pdet}_{n,m}(\mathcal{Q}) = \varepsilon_m \operatorname{pdet}_{m,n}(\mathcal{P})$, *where* $\varepsilon_m = (-1)^{m(m-1)/2}$ *(see Notation 4.26).*

We consider now a sequence $\mathcal{P}$ of polynomials with coefficients in a ring D. Equation (8.10) provides a definition of the (m, n)-polynomial determinant $\operatorname{pdet}_{m,n}(\mathcal{P})$ of $\mathcal{P}$. Note that $\operatorname{pdet}_{m,n}(\mathcal{P}) \in \mathrm{D}[X]$.

We can express the polynomial determinant as the classical determinant of a matrix whose last column has polynomial entries in the following way:

If $\mathcal{P} = P_1, \ldots, P_m$ we let $\operatorname{Mat}(\mathcal{P})^*$ be the $m \times m$ matrix whose first $m-1$ columns are the first $m-1$ columns of $\operatorname{Mat}(\mathcal{P})$ and such that the elements of the last column are the polynomials $P_1, \ldots, P_m$.

With this notation, we have

Lemma 8.25.

$$\operatorname{pdet}_{m,n}(\mathcal{P}) = \det(\operatorname{Mat}(\mathcal{P})^*).$$

Proof: Using the linearity of det $(\operatorname{Mat}(\mathcal{P})^*)$ as a function of its last column, it is clear that det $(\operatorname{Mat}(\mathcal{P})^*) = \sum_{i \leq n} m_i X^i$, using the notation of Proposition 8.23. For $i > n - m$, $m_i = 0$ since it is the determinant of a matrix with two equal columns. □

Remark 8.26. Expanding det $(\operatorname{Mat}(\mathcal{P})^*)$ by its last column we observe that $\operatorname{pdet}_{m,n}(\mathcal{P})$ is a linear combination of the P_i with coefficients equal (up to sign) $(m-1) \times (m-1)$ to minors extracted on the $m-1$ first columns of $\mathcal{P}$. It is thus a linear combination with coefficients in D of the P_i's. □

The following immediate consequences of Lemma 8.25 will be useful.

Lemma 8.27. *Let* $\mathcal{P} = P_1, \ldots, P_\ell, P_{\ell+1}, \ldots, P_m$ *be such that*

$$\deg(P_i) = n - i, i \leq \ell, \deg(P_i) < n - 1 - \ell, \ell < i \leq m,$$

with

$$\begin{aligned} P_i &= p_{i,n-i} X^{n-i} + \cdots + p_{i,0}, i \leqslant \ell, \\ P_i &= p_{i,n-1-\ell} X^{n-1-\ell} + \cdots + p_{i,0}, \ell < i \leq m. \end{aligned}$$

Then

$$\operatorname{pdet}_{m,n}(\mathcal{P}) = \prod_{i=1}^{\ell} p_{i,n-i} \operatorname{pdet}_{m-\ell,n-\ell}(\mathcal{Q}),$$

where $\mathcal{Q} = P_{\ell+1}, \ldots, P_m$.

Proof: Let

$$\begin{aligned} P_i &= p_{i,n-i} X^{n-i} + \cdots + p_{i,0}, i \leqslant \ell, \\ P_i &= p_{i,n-1-\ell} X^{n-1-\ell} + \cdots + p_{i,0}, \ell < i \leq m. \end{aligned}$$

The shape of the matrix $\mathrm{Mat}(\mathcal{P})$ is as follows

$$\begin{bmatrix} p_{1,n-1} & \cdots & \cdots & \cdots & & \cdots & \cdots & p_{1,0} \\ 0 & \ddots & & & & & & \\ \vdots & & \ddots & & & & & \\ \vdots & & & \ddots & p_{\ell,n-\ell} & \cdots & \cdots & p_{\ell,0} \\ \vdots & & & & 0 & p_{\ell+1,n-\ell-1} & \cdots & p_{\ell+1,0} \\ \vdots & & & & \vdots & \vdots & & \vdots \\ 0 & \cdots & \cdots & & 0 & p_{m,n-\ell-1} & \cdots & p_{m,0} \end{bmatrix}$$

Using Lemma 8.25, develop the determinant det $(\mathrm{Mat}(\mathcal{P})^*)$ by its first ℓ columns. □

Lemma 8.28. *Let* $\mathcal{P} = P_1, , \ldots, P_m$ *be such that for every* i, $1 \le i \le m$, *we have* $\deg(P_i) < n - 1$. *Then*

$$\mathrm{pdet}_{m,n}(\mathcal{P}) = 0.$$

Proof: Using Lemma 8.25, develop the determinant $\det(\mathrm{Mat}(\mathcal{P})^*)$ by its first column which is zero. □

8.3.2.2 Definition of Signed Subresultants

For the remainder of this chapter, let P and Q be two non-zero polynomials of degrees p and q, with $q < p$, with coefficients in an integral domain D. The fraction field of D is denoted by K. Let

$$\begin{aligned} P &= a_p X^p + a_{p-1} X^{p-1} + a_{p-2} X^{p-2} + \cdots + a_0, \\ Q &= b_q X^q + b_{q-1} X^{q-1} + \cdots + b_0. \end{aligned}$$

We define the signed subresultants of P and Q and some related notions.

Notation 8.29. [Signed subresultant] For $0 \le j \le q$, the **j-th signed subresultant of P and Q**, denoted $\mathrm{sResP}_j(P, Q)$, is the polynomial determinant of the sequence of polynomials

$$X^{q-j-1} P, \ldots, P, Q, \ldots, X^{p-j-1} Q,$$

with associated matrix the Sylvester-Habicht matrix $\mathrm{SyHa}_j(P, Q)$ (Notation 4.21). Note that $\mathrm{SyHa}_j(P, Q)$ has $p + q - 2j$ rows and $p + q - j$ columns. Clearly, $\deg(\mathrm{sResP}_j(P, Q)) \le j$. By convention, we extend these definitions for $q < j \le p$ by

$$\begin{aligned} \mathrm{sResP}_p(P, Q) &= P, \\ \mathrm{sResP}_{p-1}(P, Q) &= Q, \\ \mathrm{sResP}_j(P, Q) &= 0, \ q < j < p - 1. \end{aligned}$$

Also by convention $\mathrm{sResP}_{-1}(P, Q) = 0$. Note that

$$\mathrm{sResP}_q(P, Q) = \varepsilon_{p-q} b_q^{p-q-1} Q.$$

□

The j-th signed subresultant coefficient of P and Q $\mathrm{sRes}_j(P, Q)$, (Notation 4.21) is the coefficient of X^j in $\mathrm{sResP}_j(P, Q)$, $j < p$.

If $\deg(\mathrm{sResP}_j(P, Q)) = j$ (equivalently if $\mathrm{sRes}_j(P, Q) \neq 0$) we say that $\mathrm{sResP}_j(P, Q)$ is **non-defective**. If $\deg(\mathrm{sResP}_j(P, Q)) = k < j$ we say that $\mathrm{sResP}_j(P, Q)$ is **defective** of degree k.

8.3.3 Structure Theorem for Signed Subresultants

We are going to see that the non-zero signed subresultants are proportional to the polynomials in the signed remainder sequence. Moreover, the signed subresultant polynomials present the gap structure, graphically displayed by the following diagram: when sResP_{j-1} is defective of degree k, sResP_{j-1} and sResP_k are proportional, $\mathrm{sResP}_{j-2}, \ldots, \mathrm{sResP}_{k+1}$ are zero.

The structure theorem for signed subresultants describes precisely this situation. We write s_j for $\mathrm{sRes}_j(P, Q)$ and t_j for $\mathrm{lcof}(\mathrm{sResP}_j(P, Q))$. Note that if $\deg(\mathrm{sResP}_j(P, Q)) = j$, $t_j = s_j$. In particular $t_p = s_p = \mathrm{sign}(a_p)$.

Theorem 8.30. [Structure theorem for subresultants] *Let* $0 \leq j < i \leq p+1$. *Suppose that* $\mathrm{sResP}_{i-1}(P, Q)$ *is non-zero and of degree* j.

- *If* $\mathrm{sResP}_{j-1}(P, Q)$ *is zero, then* $\mathrm{sResP}_{i-1}(P, Q) = \gcd(P, Q)$, *and for* $\ell \leq j-1$, $\mathrm{sResP}_\ell(P, Q)$ *is zero.*
- *If* $\mathrm{sResP}_{j-1}(P, Q) \neq 0$ *has degree* k *then*

$$\begin{aligned} & s_j t_{i-1} \mathrm{sResP}_{k-1}(P, Q) \\ = & -\mathrm{Rem}(s_k t_{j-1} \mathrm{sResP}_{i-1}(P, Q), \mathrm{sResP}_{j-1}(P, Q)). \end{aligned}$$

If $j \leq q$, $k < j-1$, $\mathrm{sResP}_k(P, Q)$ *is proportional to* $\mathrm{sResP}_{j-1}(P, Q)$. *More precisely*

$$\begin{aligned} \mathrm{sResP}_\ell(P, Q) &= 0, j-1 > \ell > k \\ s_k &= \varepsilon_{j-k} \frac{t_{j-1}^{j-k}}{s_j^{j-k-1}} \\ t_{j-1} \mathrm{sResP}_k(P, Q) &= s_k \mathrm{sResP}_{j-1}(P, Q). \end{aligned}$$

(where $\varepsilon_i = (-1)^{i(i-1)/2}$*)*

Note that Theorem 8.30 implies that sResP_{i-1} and sResP_j are proportional. The following corollary of Theorem 8.30 will be used later in this chapter.

Corollary 8.31. *If* $\mathrm{sResP}_{j-1}(P, Q)$ *is of degree* k,

$$s_j^2 \mathrm{sResP}_{k-1}(P, Q) = -\mathrm{Rem}(s_k t_{j-1} \mathrm{sResP}_j(P, Q), \mathrm{sResP}_{j-1}(P, Q)).$$

Proof: Immediate from Theorem 8.30, using

$$s_j t_{i-1} \mathrm{sResP}_{k-1}(P,Q) = -\mathrm{Rem}(s_k t_{j-1} \mathrm{sResP}_{i-1}(P,Q), \mathrm{sResP}_{j-1}(P,Q))$$

and the proportionality between sResP_{i-1} and sResP_j. □

Note that we have seen in Chapter 4 (Proposition 4.24) that $\deg(\gcd(P,Q))$ is the smallest j such that $\mathrm{sRes}_j(P, Q) \neq 0$. The Structure Theorem 8.30 makes this statement more precise:

Corollary 8.32. *The last non-zero signed subresultant of P and Q is non-defective and a greatest common divisor of P and Q.*

Proof: Suppose that $\mathrm{sResP}_j(P, Q) \neq 0$, and $\forall \ell < k\ \mathrm{sResP}_\ell(P,Q) = 0$. By Theorem 8.30 there exists i such that $\deg(\mathrm{sResP}_{i-1})(P, Q) = j$, and $\mathrm{sResP}_{i-1}(P, Q)$ and $\mathrm{sResP}_j(P, Q)$ are proportional. So $\mathrm{sResP}_j(P, Q)$ is non-defective and $\mathrm{sResP}_{i-1}(P,Q)$ is a greatest common divisor of P and Q, again by Theorem 8.30. □

Moreover, a consequence of the Structure Theorem 8.30 is that signed subresultant polynomials are closely related to the polynomials in the signed remainder sequence.

In the non-defective case, we have:

Corollary 8.33. *When all* $\mathrm{sResP}_j(P, Q)$ *are non-defective,* $j = p, \ldots, 0$, *the signed subresultant polynomials are proportional up to a square to the polynomials in the signed remainder sequence.*

Proof: We consider the signed remainder sequence

$$\begin{aligned}
\mathrm{SRemS}_0(P,Q) &= P,\\
\mathrm{SRemS}_1(P,Q) &= Q,\\
&\ \vdots\\
\mathrm{SRemS}_{\ell+1}(P,Q) &= -\mathrm{Rem}(\mathrm{SRemS}_{\ell-1}(P,Q), \mathrm{SRemS}_\ell(P,Q)),\\
&\ \vdots\\
\mathrm{SRemS}_p(P,Q) &= -\mathrm{Rem}(\mathrm{SRemS}_{p-2}(P,Q), \mathrm{SRemS}_{p-1}(P,Q)),\\
\mathrm{SRemS}_{p+1}(P,Q) &= 0,
\end{aligned}$$

and prove by induction on ℓ that $\mathrm{sResP}_{p-\ell}(P, Q)$ is proportional to $\mathrm{SRemS}_\ell(P,Q)$.

The claim is true for $\ell = 0$ and $\ell = 1$ by definition of $\mathrm{sResP}_{p(P,Q)}$ and $\mathrm{sResP}_{p-1}(P,Q)$.

Suppose that the claim is true up to ℓ. In the non-defective case, the Structure Theorem 8.30 b) implies

$$\begin{aligned}
&s_{p-\ell+1}^2 \mathrm{sResP}_{p-\ell-1}(P,Q)\\
= &-\mathrm{Rem}(s_{p-\ell}^2 \mathrm{sResP}_{p-\ell+1}(P,Q), \mathrm{sResP}_{p-\ell}(P,Q)).
\end{aligned}$$

By induction hypothesis, $\mathrm{sResP}_{p-\ell+1}(P, Q)$ and $\mathrm{sResP}_{p-\ell}(P, Q)$ are proportional to $\mathrm{SRemS}_{\ell-1}(P, Q)$ and $\mathrm{SRemS}_{\ell}(P, Q)$. Thus, by definition of the signed remainder sequence and by equation (8.12) $\mathrm{sResP}_{p-\ell-1}(P, Q)$, is proportional to $\mathrm{SRemS}_{\ell+1}(P, Q)$. □

More generally, the signed subresultants are either proportional to polynomials in the signed remainder sequence or zero.

Let us illustrate this property by an example in the defective case. Let

$$\begin{aligned} P &= X^{11} - X^{10} + 1, \\ P' &= 11X^{10} - 10X^9, \end{aligned}$$

The signed remainder sequence is

$$\begin{aligned} \mathrm{SRemS}_0(P, P') &= X^{11} - X^{10} + 1, \\ \mathrm{SRemS}_1(P, P') &= 11\,X^{10} - 10X^9 \\ \mathrm{SRemS}_2(P, P') &= \frac{10\,X^9}{121} - 1, \\ \mathrm{SRemS}_3(P, P') &= -\frac{1331X}{10} + 121, \\ \mathrm{SRemS}_4(P, P') &= \frac{275311670611}{285311670611}. \end{aligned}$$

The non-zero signed subresultant polynomials are the following:

$$\begin{aligned} \mathrm{sResP}_{11}(P, P') &= X^{11} - X^{10} + 1, \\ \mathrm{sResP}_{10}(P, P') &= 11\,X^{10} - 10\,X^9, \\ \mathrm{sResP}_9(P, P') &= 10\mathrm{X}^9\text{-}121, \\ \mathrm{sResP}_8(P, P') &= -110X + 100, \\ \mathrm{sResP}_1(P, P') &= 2143588810\mathrm{X}\text{-}1948717100, \\ \mathrm{sResP}_0(P, P') &:= \text{-}275311670611. \end{aligned}$$

It is easy to check that $\mathrm{sResP}_8(P, P')$ and $\mathrm{sResP}_1(P, P')$ are proportional.

Corollary 8.34. *If* $\mathrm{SRemS}_{\ell-1}(P, Q)$ *and* $\mathrm{SRemS}_{\ell}(P, Q)$ *are two successive polynomials in the signed remainder sequence of* P *and* Q, *of degrees* $d(\ell-1)$ *and* $d(\ell)$, *then* $\mathrm{sResP}_{d(\ell-1)-1}(P, Q)$ *and* $\mathrm{sResP}_{d(\ell)}(P, Q)$ *are proportional to* $\mathrm{SRemS}_{\ell}(P, Q)$.

Proof: The proof if by induction on ℓ. Note first that $P = \mathrm{SRemS}_0$ is proportional to sResP_p. The claim is true for $\ell = 1$ by definition of $\mathrm{sResP}_p(P, Q)$, $\mathrm{sResP}_{p-1}(P, Q)$, and $\mathrm{sResP}_q(P, Q)$. Suppose that the claim is true up to ℓ. The Structure Theorem 8.30 b) implies (with $i = d(\ell-2)$, $j = d(\ell-1)$, $k = d(\ell)$) that $\mathrm{sResP}_{d(\ell)-1}(P, Q)$ is proportional to

$$\mathrm{Rem}(\mathrm{sResP}_{d(\ell-2)-1}, \mathrm{sResP}_{d(\ell-1)-1})(P, Q).$$

By the induction hypothesis, $\mathrm{sResP}_{d(\ell-2)-1}(P,Q)$ and $\mathrm{sResP}_{d(\ell-1)-1}(P,Q)$ are proportional to $\mathrm{SRemS}_{\ell-1}(p, Q)$ and $\mathrm{SRemS}_{\ell}(P, Q)$. It follows that $\mathrm{sResP}_{d(\ell)-1}(P, Q)$ is proportional to $\mathrm{SRemS}_{\ell+1}(P, Q)$. Moreover $\mathrm{sResP}_{d(\ell)-1}(P, Q)$ and $\mathrm{sResP}_{d(\ell+1)}(P, Q)$ are proportional by the Structure Theorem 8.30. □

The proof of the structure theorem relies on the following proposition relating the signed subresultants of P and Q and of Q and $-R$, with $R=\mathrm{Rem}(P,Q)$.

We recall that P and Q are two non-zero polynomials of degrees p and q, with $q<p$, with coefficients in an integral domain D, with

$$\begin{aligned} P &= a_p X^p + a_{p-1} X^{p-1} + a_{p-2} X^{p-2} + \cdots + a_0, \\ Q &= b_q X^q + b_{q-1} X^{q-1} + \cdots + b_0. \end{aligned}$$

The following proposition generalizes Proposition 4.36.

Proposition 8.35. *Let r be the degree of $R=\mathrm{Rem}(P,Q)$.*

$$\mathrm{sResP}_j(P,Q) \;=\; \varepsilon_{p-q} b_q^{p-r}\,\mathrm{sResP}_j(Q,-R) \;\; \text{if } j<q-1,$$

where $\varepsilon_i=(-1)^{i(i-1)/2}$.

Proof: Replacing the polynomials $X^{q-j-1}P$, ..., P by the polynomials $X^{q-j-1}R$, ..., R in $\mathrm{SyHaPol}_j(P, Q)$ does not modify the polynomial determinant. Indeed,

$$R=P-\sum_{i=0}^{p-q} c_i\,(X^i Q),$$

where $C=\sum_{i=0}^{p-q} c_i X^i$ is the quotient of P in the euclidean division of P by Q, and adding to a polynomial of a sequence a multiple of another polynomial of the sequence does not change the polynomial determinant, by Lemma 8.24.

Reversing the order of the polynomials multiplies the polynomial determinant by ε_{p+q-2j} using again Lemma 8.24. Replacing R by -R multiplies the polynomial determinant by $(-1)^{q-j}$, by Lemma 8.24, and $(-1)^{q-j}\varepsilon_{p+q-2j}=\varepsilon_{p-q}$ (see Notation 4.26). So, defining

$$A_j=\mathrm{pdet}_{p+q-2j,p+q-j}(X^{p-j-1}Q,\ldots,Q,-R\ldots,-X^{q-j-1}R),$$

we have

$$\mathrm{sResP}_j(P,Q)=\varepsilon_{p-q}A_j.$$

If $j\le r$,

$$\begin{aligned} A_j &= b_q^{p-r}\mathrm{pdet}_{q+r-2j,q+r-j}(X^{r-j-1}Q,\ldots,Q,-R\ldots,-X^{q-j-1}R) \\ &= b_q^{p-r}\mathrm{sResP}_j(Q,-R), \end{aligned}$$

using Lemma 8.27.

If $r<j<q-1$,

$$\mathrm{pdet}_{p+q-2j,p+q-j}(X^{p-j-1}Q,\ldots,Q,-R\ldots,-X^{q-j-1}R)=0,$$

using Lemma 8.27 and Lemma 8.28, since $\deg(-X^{q-j-1}R)<q-1$. □

Proof of Theorem 8.30: For $q<j\le p$, the only thing to check is that

$$\mathrm{sign}(a_p)^2\,\mathrm{sResP}_{q-1}(P,Q)=-\mathrm{Rem}(s_q t_{p-1}\mathrm{sResP}_p(P,Q),\mathrm{sResP}_{p-1}(P,Q)),$$

since $s_p=\mathrm{sign}(a_p)$ (Notation 4.21) Indeed

$$\begin{aligned}\mathrm{sResP}_{q-1}(P,Q) &= \varepsilon_{p-q}b_q^{p-q+1}R\\ &= -\varepsilon_{p-q+2}b_q^{p-q+1}R\\ &= -\mathrm{Rem}(\varepsilon_{p-q}b_q^{p-q+1}P,Q)\\ &= -\mathrm{Rem}(s_q t_{p-1}P,Q)\end{aligned}$$

since $s_q=\varepsilon_{p-q}b_q^{p-q}, t_{p-1}=b_q$, and $\mathrm{sResP}_p(P,Q)=P, \mathrm{sResP}_{p-1}(P,Q)=Q$.

The remainder of the proof is by induction on the length of the remainder sequence of P and Q.

Suppose that the theorem is true for $Q,-R$. The fact that the theorem holds for P, Q for $j\le r$ is clear by Proposition 8.35, since $\mathrm{sResP}_j(P,Q)$ and $\mathrm{sResP}_j(Q,-R)$, $j\le r$, are proportional, with the same factor of proportionality $\varepsilon_{p-q}b_q^{p-r}$.

For $r<j\le q$, the only thing to check is that

$$s_q t_{p-1}\mathrm{sResP}_{r-1}(P,Q)=-\mathrm{Rem}(s_r t_{q-1}\mathrm{sResP}_{p-1}(P,Q),\mathrm{sResP}_{q-1}(P,Q)),$$

which follows from the induction hypotheses

$$s_q'^2\,\mathrm{sResP}_{r-1}(Q,-R)=-\mathrm{Rem}(s_r' t_{q-1}'\mathrm{sResP}_q(Q,-R),\mathrm{sResP}_{q-1}(Q,-R)),$$

where we write s_j' for $\mathrm{sRes}_j(Q,-R)$ and t_j' for $\mathrm{lcof}(\mathrm{sResP}_j(Q,-R))$, noting that $t_q'=s_q'=\mathrm{sign}(b-q)$, since

$$\begin{aligned}&t_{p-1}s_q\,\mathrm{sResP}_{r-1}(P,Q)\\ =\;& b_q(\varepsilon_{p-q}b_q^{p-q})\,\varepsilon_{p-q}b_q^{p-r}\mathrm{sResP}_{r-1}(Q,-R)\\ =\;& -b_q^{2p-q-r+1}\mathrm{Rem}(s_r' t_{q-1}'\mathrm{sResP}_q(Q,-R),\mathrm{sResP}_{q-1}(Q,-R))\\ =\;& -\mathrm{Rem}(s_r t_{q-1}\mathrm{sResP}_{p-1}(P,Q),\mathrm{sResP}_{q-1}(P,Q)),\end{aligned}$$

by Proposition 8.35, noting that $t_{q-1}=\varepsilon_{p-q}b_q^{p-q+1}t_{q-1}'$, $s_r=\varepsilon_{p-q}b_q^{p-r}s_r'$ and using that $\mathrm{sResP}_{q-1}(P,Q)$ is proportional to $\mathrm{sResP}_{q-1}(Q,-R)$. □

The following proposition gives a useful precision to Theorem 8.30.

Proposition 8.36. *Using the notation of Theorem 8.30 and defining C_{k-1} as the quotient of* $s_k t_{j-1}\mathrm{sResP}_{i-1}(P,Q)$ by $\mathrm{sResP}_{j-1}(P,Q)$, *we have* $C_{k-1}\in \mathrm{D}[X]$.

Before proving it, we need an analogue of Proposition 1.9 for subresultants.

Notation 8.37. [Subresultant cofactors] Define $\mathrm{sResU}_j(P, Q)$ (resp. $\mathrm{sResV}_j(P, Q)$) as $\det(M_i)$ (resp. $\det(N_i)$), where M_i (resp. N_i) is the square matrix obtained by taking the first $p+q-2j-1$ columns of $\mathrm{SyHa}_j(P, Q)$ and with last column equal to $(X^{q-1-j}, \ldots, X, 1, 0, \ldots, 0)^t$ (resp. $(0, \ldots, 0, 1, X, \ldots, X^{p-1-j})^t$).

Note that if $P, Q \in \mathrm{D}[X]$, then $\mathrm{sResU}_j(P, Q), \mathrm{sResV}_j(P, Q) \in \mathrm{D}[X]$. □

Proposition 8.38. *Let $j \le q$. Then,*

a) $\deg(\mathrm{sResU}_{j-1}(P, Q)) \le q - j, \deg(\mathrm{sResV}_{j-1}(P, Q)) \le p - j,$

$$\mathrm{sResP}_j(P, Q) = \mathrm{sResU}_j(P, Q)P + \mathrm{sResV}_j(P, Q)Q.$$

b) If $\mathrm{sResP}_j(P, Q)$ is not 0 and if U and V are such that

$$UP + VQ = \mathrm{sResP}_j(P, Q),$$

$\deg(U) \le q - j - 1$, *and* $\deg(V) \le p - j - 1$, *then* $U = \mathrm{sResU}_j(P, Q)$ *and* $V = \mathrm{sResV}_j(P, Q)$.

c) If $\mathrm{sResP}_j(P, Q)$ is non-defective, then

$$\deg(\mathrm{sResU}_{j-1}(P, Q)) = q - j, \deg(\mathrm{sResV}_{j-1}(P, Q)) = p - j,$$

and $\mathrm{lcof}(\mathrm{sResV}_{j-1}(P, Q)) = a_p \mathrm{sRes}_j(P, Q)$.

Proof: a) The conditions

$$\deg(\mathrm{sResU}_{j-1}(P, Q)) \le q - j, \deg(\mathrm{sResV}_{j-1}(P, Q)) \le p - j$$

follow from the definitions of $\mathrm{sResU}_{j-1}(P, Q)$ and $\mathrm{sResV}_{j-1}(P, Q)$. By Lemma 8.25, $\mathrm{sResP}_j(P, Q) = \det(\mathrm{SyHa}_j(P, Q)^*)$, where $\mathrm{SyHa}_j(P, Q)^*$ is the square matrix obtained by taking the first $p+q-2j-1$ columns of $\mathrm{SyHa}_j(P, Q)$ and with last column equal to

$$(X^{q-1-j}P, \ldots, XP, P, Q, \ldots, X^{p-j-1}Q)^t.$$

Expanding the determinant by its last column, we obtain the claimed identity.

b) Suppose $\deg(U) \le q - j - 1$, $\deg(V) \le p - j - 1$, and

$$\mathrm{sResP}_j(P, Q) = UP + VQ$$

so that

$$(\mathrm{sResU}_j(P, Q) - U)P + (\mathrm{sResV}_j(P, Q) - V)Q = 0.$$

If $\mathrm{sResU}_j(P, Q) - U$ is not 0, then $\mathrm{sResV}_j(P, Q) - V$ cannot be 0, and it follows from Proposition 1.5 that $\deg(\gcd(P, Q)) > j$. But this is impossible since $\mathrm{sResP}_j(P, Q)$ is a non-zero polynomial of degree $\le j$ belonging to the ideal generated by P and Q.

c) Since $\mathrm{sResP}_j(P,Q)$ is non-defective, it follows that $\mathrm{sRes}_j(P,Q)\neq 0$. By considering the determinant of the matrix $\mathrm{SyHa}_{j-1}(P,Q)^*$, it is clear that the coefficient of X^{p-j} in $\mathrm{sResV}_{j-1}(P,Q)$ is $a_p\mathrm{sRes}_j(P,Q)$. Moreover,

$$\deg(\mathrm{sResV}_{j-1})=p-j, \deg(\mathrm{sResU}_{j-1}(P,Q))=q-j.$$

□

We omit P and Q in the notation in the next paragraphs. For sResP_{i-1} non-zero of degree j, we define

$$B_{j,i}=\begin{bmatrix} \mathrm{sResU}_{i-1} & \mathrm{sResV}_{i-1} \\ \mathrm{sResU}_{j-1} & \mathrm{sResV}_{j-1} \end{bmatrix},$$

where $\mathrm{sResU}_{i-1}, \mathrm{sResV}_{i-1}, \mathrm{sResU}_{j-1}, \mathrm{sResV}_{j-1}\in \mathrm{D}[X]$ are the polynomials of the $(i-1)$-th and $(j-1)$-th relations of Proposition 8.38, whence

$$\begin{bmatrix} \mathrm{sResP}_{i-1} \\ \mathrm{sResP}_{j-1} \end{bmatrix}=B_{j,i}\cdot\begin{bmatrix} P \\ Q \end{bmatrix}. \tag{8.11}$$

Lemma 8.39. *If* sResP_{i-1} *is non-zero of degree* j*, then*

$$\det(B_{j,i})=s_j t_{i-1}.$$

Proof: Eliminating Q from the system (8.11), we have

$$\begin{aligned} &(\mathrm{sResU}_{i-1}\mathrm{sResV}_{j-1}-\mathrm{sResU}_{j-1}\mathrm{sResV}_{i-1})\,P \\ =\;& \mathrm{sResV}_{j-1}\mathrm{sResP}_{i-1}-\mathrm{sResV}_{i-1}\mathrm{sResP}_{j-1}. \end{aligned}$$

Since $\deg(S\,R_{i-1})=j$, $\deg(S\,R_j)=j$ by the Structure Theorem 8.30, and $\deg(\mathrm{sResV}_{j-1})=p-j$. Using $\deg(SR_{j-1})\leq j-1$ and $\deg(\mathrm{sResV}_{i-1})\leq p-i< p-j$, we see that the right hand side of equation (8.14) has degree p. The leading coefficient of sResV_{j-1} is $a_p s_j$ by Proposition 8.38. Hence

$$\mathrm{sResU}_{i-1}\mathrm{sResV}_{j-1}-\mathrm{sResU}_{j-1}\mathrm{sResV}_{i-1}=s_j t_{i-1}\neq 0.$$ □

Corollary 8.40. *If* sResP_{i-1} *is non-zero of degree* j*, then*

$$B_{j,i}^{-1}=\frac{1}{s_j t_{i-1}}\begin{bmatrix} \mathrm{sResV}_{j-1} & -\mathrm{sResV}_{i-1} \\ -\mathrm{sResU}_{j-1} & \mathrm{sResU}_{i-1} \end{bmatrix}, \text{and } s_j t_{i-1}B_{j,i}^{-1}\in \mathrm{D}[X].$$

Now we study the transition between two consecutive couples of signed subresultant polynomials sResP_{i-1}, sResP_{j-1} and sResP_{j-1}, sResP_{k-1}, where sResP_{i-1} is of degree j, sResP_{j-1} is of degree k, and $0\leq k<j\leq p$.

The **signed subresultant transition matrix** is

$$T_j=\begin{bmatrix} 0 & 1 \\ -\dfrac{s_k t_{j-1}}{s_j t_{i-1}} & \dfrac{C_{k-1}}{s_j t_{i-1}} \end{bmatrix}\in \mathrm{K}[X]^{2\times 2},$$

so that

$$\mathrm{sResP}_{k-1} = -\frac{s_k t_{j-1}}{s_j t_{i-1}} \mathrm{sResP}_{i-1} + \frac{C_{k-1}}{s_j t_{i-1}} \mathrm{sResP}_{j-1} \tag{8.12}$$

and

$$\begin{bmatrix} \mathrm{sResP}_{j-1} \\ \mathrm{sResP}_{k-1} \end{bmatrix} = T_j \begin{bmatrix} \mathrm{sResP}_{i-1} \\ \mathrm{sResP}_{j-1} \end{bmatrix} \tag{8.13}$$

by the Structure Theorem 8.30.

Lemma 8.41. *If* sResP_{i-1} *is non-zero of degree* j *and* sResP_{j-1} *is non-zero of degree* k*, then*

$$B_{k,j} = T_j B_{j,i}.$$

Proof: Let

$$T_j B_{j,i} = \begin{bmatrix} A & B \\ C & D \end{bmatrix}.$$

A simple degree calculation shows that deg $(A) \le q - j$, $\deg(B) \le p - j$, and $\deg(C) = q - k$, and $\deg(D) = p - k$. From equations (8.13) and (8.11) we see that

$$\begin{aligned} \mathrm{sResP}_{j-1} &= A P + B Q \\ \mathrm{sResP}_{k-1} &= C P + D Q. \end{aligned}$$

The conclusion follows from the uniqueness asserted in Proposition 8.38 b). □

Proof of Proposition 8.36: From Lemma 8.41, we see that $T_j = B_{k,j} B_{j,i}^{-1}$, which together with the definition of $B_{k,j}$ and Corollary 8.40 shows that

$$\frac{C_{k-1}}{s_j t_{i-1}} = \frac{1}{s_j t_{i-1}} (-\mathrm{sResU}_{k-1} \mathrm{sResV}_{i-1} + \mathrm{sResV}_{k-1} \mathrm{sResU}_{i-1}),$$

whence $C_{k-1} = \mathrm{sResU}_{k-1} \mathrm{sResV}_{i-1} - \mathrm{sResV}_{k-1} \mathrm{sResU}_{i-1} \in \mathrm{D}[X]$. □

Proposition 8.42. *Let* $j \le q$, $\deg(\mathrm{sResP}_j) = j$ $\deg(\mathrm{sResP}_{j-1}) = k \le j - 1$,

$$\begin{aligned} S_{j-1} &= \mathrm{sResP}_{j-1}, \\ S_{j-1-\delta} &= \frac{(-1)^\delta t_{j-1} S_{j-\delta}}{s_j}, \textit{for } \delta = 1, \dots, j - k - 1. \end{aligned}$$

Then all of these polynomials are in $\mathrm{D}[X]$ *and* $\mathrm{sResP}_k = S_k$.

Proof: Add the $j - k - 1 - \delta$ polynomials $X^{k+\delta+1}, \dots, X^j$ to $\mathrm{SyHaPol}_{j-1}$ to obtain $M_{j-1-\delta}$. It is easy to see that the polynomial determinant of $M_{j-1-\delta}$ is $S_{j-1-\delta}$. □

8.3.4 Size of Remainders and Subresultants

Observe, comparing the following example with Example 8.21, that the bit-sizes of coefficients in the signed subresultant sequence can be much smaller than in the signed remainder sequence.

Example 8.43. We consider, as in Example 8.21,

$$P := 9X^{13} - 18X^{11} - 33X^{10} + 102X^8 + 7X^7 - 36X^6 - 122X^5 + 49X^4 + 93X^3 - 42X^2 - 18X + 9.$$

The subresultant coefficients of P and P' for j from 11 to 5 are:

$$\begin{array}{c} 37908 \\ -72098829 \\ -666229317948 \\ -1663522740400320 \\ -2181968897553243072 \\ -151645911413926622112 \\ -165117711302736225120, \end{array}$$

the remaining subresultants being 0. □

The difference in bitsizes of coefficients between signed remainder and signed subresultant sequences observed in Example 8.21 and Example 8.43 is a general fact.

First, let us see that the size of subresultants is well controlled. Indeed, using Proposition 8.10 we obtain the following:

Proposition 8.44. [Size of signed subresultants] *If P and Q have degrees p and q and have coefficients in $\mathbb{Z}$ which have bitsizes at most τ, then the bitsizes of the coefficients of* $\mathrm{sResP}_j(P, Q)$ *and of* sResU_j *and* sResV_j *are at most $(\tau + \nu_j)(p + q - 2j)$, where ν_j is the bitsize of $p + q - 2j$.*

We also have, using Proposition 8.11,

Proposition 8.45. [Degree of signed subresultants] *If P and Q have degrees p and q and have coefficients in* $\mathrm{R}[Y_1, \ldots, Y_k]$ *which have degrees d in $Y_1, \ldots, Y_k$ then the degree of* $\mathrm{sResP}_j(P, Q)$ *in $Y_1, \ldots, Y_k$ is at most $d(p + q - 2j)$.*

We finally have, using Proposition 8.12,

Proposition 8.46. *If P and Q have degrees p and q and have coefficients in $\mathbb{Z}[Y_1, \ldots, Y_k]$ which have degrees d in $Y_1, \ldots, Y_k$ of bitsizes τ, then the degree of* $\mathrm{sResP}_j(P, Q)$ *in $Y_1, \ldots, Y_k$ is at most $d(p + q - 2j)$, and the bitsizes of the coefficients of* $\mathrm{sResP}_j(P, Q)$ *are at most $(\tau + \nu)(p + q - 2j) + k\mu$ where ν is the bitsize of $p + q$ and μ is the bitsize of $(p + q)d + 1$.*

The relationship between the signed subresultants and remainders provides a bound for the bitsizes of the coefficients of the polynomials appearing in the signed remainder sequence.

Theorem 8.47. [Size of signed remainders] *If $P \in \mathbb{Z}[X]$ and $Q \in \mathbb{Z}[X]$ have degrees p and $q < p$ and have coefficients of bitsizes at most τ, then the numerators and denominators of the coefficients of the polynomials in the signed remainder sequence of P, Q have bitsizes at most $(p+q)(q+1)(\tau+\nu)+\tau$, where ν is the bitsize of $p+q$.*

Proof: Denote by

$$P = S_0, Q = S_1, S_2, \ldots, S_k$$

the polynomials in the signed remainder sequence of P and Q. Let $d_j = \deg(S_j)$. According to Theorem 8.30 S_ℓ is proportional to $\mathrm{sResP}_{d_{\ell-1}-1}$, which defines $\beta_\ell \in \mathbb{Q}$ such that

$$S_\ell = \beta_\ell \mathrm{sResP}_{d_{\ell-1}-1}.$$

Consider successive signed remainders of respective degrees $i = d_{\ell-3}, j = d_{\ell-2}$, and $k = d_{\ell-1}$. According to Theorem 8.30,

$$s_j t_{i-1} \mathrm{sResP}_{k-1} = -\mathrm{Rem}(s_k t_{j-1} \mathrm{sResP}_{i-1}, \mathrm{sResP}_{j-1}),$$

which implies that

$$\beta_\ell = \frac{s_j t_{i-1}}{s_k t_{j-1}} \beta_{\ell-2}$$

since

$$S_\ell = -\mathrm{Rem}(S_{\ell-2}, S_{\ell-1}).$$

Denoting by D_ℓ and N_ℓ the bitsizes of the numerator and denominator of β_ℓ, and using Proposition 8.44, we get the estimates

$$\begin{aligned} N_\ell &\leq 2(p+q)(\tau+\nu) + N_{\ell-2}, \\ D_\ell &\leq 2(p+q)(\tau+\nu) + D_{\ell-2}. \end{aligned}$$

Since N_0, D_0, N_1, and D_1 are bounded by τ and $\ell \leq q+1$, the claim follows. □

This quadratic behavior of the bitsizes of the coefficients of the signed remainder sequence is often observed in practice (see Example 8.21).

8.3.5 Specialization Properties of Subresultants

Since the signed subresultant is defined as a polynomial determinant which is a multilinear form with respect to its rows, and given the convention for sResP_p (see Notation 8.29), we immediately have the following:

Let $f: \mathrm{D} \to \mathrm{D}'$ be a ring homomorphism, and let f also denote the induced homomorphism from $f: \mathrm{D}[X] \to \mathrm{D}'[X]$.

Proposition 8.48. *Suppose that* $\deg(f(P)) = \deg(P)$, $\deg(f(Q)) = \deg(Q)$. *Then for all $j \leq p$,*

$$\mathrm{sResP}_j(f(P), f(Q)) = f(\mathrm{sResP}_j(P, Q)).$$

Applying this to the ring homomorphism from $\mathbb{Z}[Y][X]$ to $\mathrm{R}[X]$ obtained by assigning values $(y_1, \ldots, y_\ell) \in \mathrm{R}^\ell$ to the variables $(Y_1, \ldots, Y_\ell)$, we see that the signed subresultants after specialization are obtained by specializing the coefficients of the signed subresultants.

Example 8.49. Consider, for example, the general polynomial of degree 4:

$$P = X^4 + a X^2 + b X + c.$$

The signed subresultant sequence of P and P' is formed by the polynomials (belonging to $\mathbb{Z}[a, b, c][X]$)

$$\begin{aligned}
\mathrm{sResP}_4(P, P') &= X^4 + a X^2 + b X + c \\
\mathrm{sResP}_3(P, P') &= 4 X^3 + 2 a X + b \\
\mathrm{sResP}_2(P, P') &= -4 (2 a X^2 + 3 b X + 4 c) \\
\mathrm{sResP}_1(P, P') &= 4 ((8 a c - 9 b^2 - 2 a^3) X - a^2 b - 12 b c) \\
\mathrm{sResP}_0(P, P') &= 256 c^3 - 128 a^2 c^2 + 144 a b^2 c + 16 a^4 c - 27 b^4 - 4 a^3 b^2,
\end{aligned}$$

which agree, up to squares in $\mathbb{Q}(a, b, c)$, with the signed remainder sequence for P and P' when there is a polynomial of each degree in the signed remainder sequence (see example 1.15). If $a = 0$, the subresultant sequence of the polynomial $P = X^4 + b X + c$ and P' is

$$\begin{aligned}
\mathrm{sResP}_4(P, P') &= X^4 + b X + c \\
\mathrm{sResP}_3(P, P') &= 4 X^3 + b \\
\mathrm{sResP}_2(P, P') &= -4(3 b X + 4 c) \\
\mathrm{sResP}_1(P, P') &= -12 b (3 b X + 4 c) \\
\mathrm{sResP}_0(P, P') &= -27 b^4 + 256 c^3,
\end{aligned}$$

which is the specialization of the signed subresultant sequence of P with $a = 0$. Comparing this with Example 1.15, we observe that the polynomials in the signed subresultant sequence are multiples of the polynomials in the signed remainder sequence obtained when $a = 0$. We also observe the proportionality of sResP_2 and sResP_1, which is a consequence of the Structure Theorem 8.30. □

Note that if f: $\mathrm{D} \rightarrow \mathrm{D}'$ is a ring homomorphism such that $\deg(f(P)) = \deg(P)$, $\deg(f(Q)) < \deg(Q)$, then for all $j \leq \deg(f(Q))$

$$f(\mathrm{sResP}_j(P, Q)) = \mathrm{lcof}(f(P))^{\deg(Q) - \deg(f(Q))} \mathrm{sResP}_j(f(P), f(Q)),$$

using Lemma 8.27.

8.3.6 Subresultant Computation

We now describe an algorithm for computing the subresultant sequence, based upon the preceding results.

Let P and Q be polynomials in $\mathrm{D}[X]$ with $\deg(P) = p$, $\deg(Q) = q < p$. The **signed subresultant sequence** is the sequence

$$\mathrm{sResP}(P, Q) = \mathrm{sResP}_p(P, Q), \ldots, \mathrm{sResP}_0(P, Q).$$

Algorithm 8.21. **[Signed Subresultant]**

- **Structure:** an ordered integral domain D.
- **Input:** two univariate polynomials

$$\begin{aligned} P &= a_p X^p + \cdots + a_0 \\ Q &= b_q X^q + \ldots + b_0 \end{aligned}$$

  with coefficients D of respective degrees p and q, $p > q$.
- **Output:** the sequence of signed subresultant polynomials and signed subresultant coefficients.
- **Complexity:** $O(p q)$, where p is the degree of P and q the degree of Q.
- **Procedure:**
  - Initialize:

$$\begin{aligned} \mathrm{sResP}_p &:= P, \\ s_p = t_p &:= \mathrm{sign}(a_p), \\ \mathrm{sResP}_{p-1} &:= Q, \\ t_{p-1} &:= b_q, \\ \mathrm{sResP}_q &:= \varepsilon_{p-q} b_q^{p-q-1} Q, \\ s_q &:= \varepsilon_{p-q} b_q^{p-q}, \\ \mathrm{sResP}_\ell = s_\ell &:= 0 \quad \textit{for } \ell \text{ from } q+1 \text{ to } p-2 \end{aligned}$$

    $i := p+1$, $j := p$.
  - While $\mathrm{sResP}_{j-1} \neq 0$,
    - $k := \deg(\mathrm{sResP}_{j-1})$,
      - If $k = j-1$,
        - $s_{j-1} := t_{j-1}$.
        - $\mathrm{sResP}_{k-1} := -\mathrm{Rem}(s_{j-1}^2 \mathrm{sResP}_{i-1}, \mathrm{sResP}_{j-1}) / (s_j t_{i-1})$.
      - If $k < j-1$,
        - $s_{j-1} := 0$.
        - Compute s_k and sResP_k: for δ from 1 to $j-k-1$:

$$\begin{aligned} t_{j-\delta-1} &:= (-1)^\delta (t_{j-1} t_{j-\delta}) / s_j, \\ s_k &:= t_k \\ \mathrm{sResP}_k &:= s_k \mathrm{sResP}_{j-1} / t_{j-1}. \end{aligned}$$

        - Compute s_ℓ and sResP_ℓ for ℓ from $j-2$ to $k+1$:

$$\mathrm{sResP}_\ell = s_\ell := 0.$$

        - Compute sResP_{k-1}:

$$\mathrm{sResP}_{k-1} := -\mathrm{Rem}(t_{j-1} s_k \mathrm{sResP}_{i-1}, \mathrm{sResP}_{j-1}) / (s_j t_{i-1}).$$

  - $t_{k-1} := \mathrm{lcof}(\mathrm{sResP}_{k-1})$.
  - $i := j, j := k$.
- For $\ell = 0$ to $j - 2$

$$\mathrm{sResP}_\ell = s_\ell := 0.$$

- Output $\mathrm{sResP} := \mathrm{sResP}_p, \ldots, \mathrm{sResP}_0$, $\mathrm{sRes} := s_p, \ldots, s_0$.

Proof of correctness: The correctness of the algorithm follows from Theorem 8.30. □

Complexity analysis: All the intermediate results in the computation belong to $\mathrm{D}[X]$ by the definition of the signed subresultants as polynomial determinants (Notation 8.29) and Proposition 8.42.

The computation of sResP_{k-1} takes $j + 2$ multiplications to compute $s_k\, t_{j-1}\, \mathrm{sResP}_{i-1}$, $(j - k + 1)\,(2\,k + 3)$ arithmetic operations to perform the euclidean division of $s_k\, t_{j-1}\, \mathrm{sResP}_{i-1}$ by sResP_{j-1}, one multiplication and k divisions to obtain the result. The computation of s_k takes $j - k - 1$ multiplications and $j - k - 1$ exact divisions. The computation of sResP_k takes $k + 1$ multiplications and $k + 1$ exact divisions. So computing sResP_{k-1} and sResP_k takes $O((j - k)\, k)$ arithmetic operations.

Finally the complexity of computing the signed subresultant sequence is $O(p\ q)$, similarly to the computation of the signed remainder sequence when $q < p$ (Algorithm 8.19).

When P and Q are in $\mathbb{Z}[X]$, with coefficients of bitsizes bounded by τ, the bitsizes of the integers in the operations performed by the algorithm are bounded by $(\tau + \nu)\,(p + q)$ where ν is the bitsize of $p + q$ according to Proposition 8.44. □

Remark 8.50. Note that initializing $s_p = t_p$: $= 1$ Algorithm 8.21 (Signed Subresultant) is also valid in a domain, and computes correct values of

$$\mathrm{sResP}_{p-1}, \ldots, \mathrm{sResP}_0,\ \mathrm{sRes}_{p-1}, \ldots, \mathrm{sRes}_0.$$ □

Note that Algorithm 8.21 (Signed Subresultant) provides an algorithm for computing the resultant of two polynomial of degree p and q, $q < p$, with complexity $O(pq)$, since $\mathrm{sRes}_0(P, Q)$ is up to a sign equal to the resultant of P and Q, while a naive computation of the resultant as a determinant would have complexity $O(p^3)$. This improvement is due to the special structure of the Sylvester-Habicht matrix, which is taken into account in the subresultant algorithm. Algorithm 8.21 (Signed Subresultant) can be used to compute the resultant with complexity $O(p\,q)$ in the special case $p = q$ as well.

Exercise 8.2. Describe an algorithm computing the resultant of P and Q with complexity $O(p^2)$ when $\deg(P) = \deg(Q) = p$. Hint: consider $Q_1 = a_p\, Q - b_p\, P$ and prove that $a_p^{p-1}\, \mathrm{Res}(P, Q) = \mathrm{Res}(P, Q_1)$.

The signed subresultant coefficients are also computed in time $O(p\,q)$ using Algorithm 8.21 (Signed Subresultant), while computing them from their definition as determinants using Algorithm 8.16 (Dogdson-Jordan-Bareiss) would cost $O(p^4)$, since there are $O(p)$ determinants of matrices of size $O(p)$ to compute.

Algorithm 8.22. **[Extended Signed Subresultant]**

- **Structure:** an ordered integral domain D.
- **Input:** two univariate polynomials

$$\begin{aligned} P &= a_p X^p + \cdots + a_0 \\ Q &= b_q X^q + \cdots + b_0 \end{aligned}$$

  with coefficients D of respective degrees p and q, $p > q$.
- **Output:** the sequence of signed subresultant polynomials and the corresponding sResU and sResV.
- **Complexity:** $O(p\,q)$, where p is the degree of P and q the degree of Q.
- **Procedure:**
  - Initialize:

$$\begin{aligned} \mathrm{sResP}_p &:= P, \\ s_p = t_p &:= \mathrm{sign}(a_p), \\ \mathrm{sResP}_{p-1} &:= Q, \\ t_{p-1} &:= b_q, \\ \mathrm{sResU}_p = \mathrm{sResV}_{p-1} &:= 1, \\ \mathrm{sResV}_p = \mathrm{sResU}_{p-1} &:= 0, \\ \mathrm{sResP}_q &:= \varepsilon_{p-q} b_q^{p-q-1} Q, \\ s_q &:= \varepsilon_{p-q} b_q^{p-q}, \\ \mathrm{sResU}_q &:= 0, \\ \mathrm{sResV}_q &:= \varepsilon_{p-q} b_q^{p-q}, \\ \mathrm{sResP}_\ell = s_\ell = \mathrm{sResU}_\ell = \mathrm{sResV}_\ell &:= 0 \quad \textit{for } \ell \text{ from } q+1 \text{ to } p-2 \end{aligned}$$

    $i := p+1$, $j := p$.
  - While $\mathrm{sResP}_{j-1} \neq 0$
    - $k := \deg(\mathrm{sResP}_{j-1})$
      - If $k = j-1$,

$$\begin{aligned} s_{j-1} &:= t_{j-1}, \\ C_{k-1} &:= \mathrm{Quo}(s_{j-1}^2 \mathrm{sResP}_{i-1}, \mathrm{sResP}_{j-1}), \\ \mathrm{sResP}_{k-1} &:= (-s_{j-1}^2 \mathrm{sResP}_{i-1} + C_{k-1} \mathrm{sResP}_{j-1})/(s_j t_{i-1}), \\ \mathrm{sResU}_{k-1} &:= (-s_{j-1}^2 \mathrm{sResU}_{i-1} + C_{k-1} \mathrm{sResU}_{j-1})/(s_j t_{i-1}), \\ \mathrm{sResV}_{k-1} &:= (-s_{j-1}^2 \mathrm{sResV}_{i-1} + C_{k-1} \mathrm{sResV}_{j-1})/(s_j t_{i-1}). \end{aligned}$$

– If $k < j-1$,

$$s_{j-1} := 0.$$

Compute $\mathrm{sResP}_k, \mathrm{sResU}_k, \mathrm{sResV}_k$: for δ from 1 to $j-k-1$

$$\begin{aligned} t_{j-\delta-1} &:= (-1)^{\delta}(t_{j-1}t_{j-\delta})/s_j, \\ s_k &:= t_k, \\ \mathrm{sResP}_k &:= s_k \mathrm{sResP}_{j-1}/t_{j-1}, \\ \mathrm{sResU}_k &:= s_k \mathrm{sResU}_{j-1}/t_{j-1}, \\ \mathrm{sResV}_k &:= s_k \mathrm{sResV}_{j-1}/t_{j-1}. \end{aligned}$$

Compute s_ℓ, $\mathrm{sResP}_\ell, \mathrm{sResU}_\ell, \mathrm{sResV}_\ell$: for ℓ from $j-2$ to $k+1$:

$$\mathrm{sResP}_\ell = s_\ell = \mathrm{sResU}_\ell = \mathrm{sResV}_\ell := 0.$$

Compute $\mathrm{sResP}_{k-1}, \mathrm{sResU}_{k-1}, \mathrm{sResV}_{k-1}$:

$$\begin{aligned} C_{k-1} &:= \mathrm{Quo}(s_k t_{j-1} \mathrm{sResP}_{i-1}, \mathrm{sResP}_{j-1}), \\ \mathrm{sResP}_{k-1} &:= (-s_k t_{j-1} \mathrm{sResP}_{i-1} + C_{k-1} \mathrm{sResP}_{j-1})/(s_j t_{i-1}), \\ \mathrm{sResU}_{k-1} &:= (-s_k t_{j-1} \mathrm{sResU}_{i-1} + C_{k-1} \mathrm{sResU}_{j-1})/(s_j t_{i-1}), \\ \mathrm{sResV}_{k-1} &:= (-s_k t_{j-1} \mathrm{sResV}_{i-1} + C_{k-1} \mathrm{sResV}_{j-1})/(s_j s_{i-1}). \end{aligned}$$

– $t_{k-1} := \mathrm{lcof}(\mathrm{sResP}_{k-1})$.
– $i := j, j := k$.

– For $\ell = j-2$ to 0:

$$\mathrm{sResP}_\ell = s_\ell = \mathrm{sResU}_\ell = \mathrm{sResV}_\ell := 0.$$

– Output $\mathrm{sResP} := \mathrm{sResP}_p, \ldots, \mathrm{sResP}_0$, $\mathrm{sResU} := \mathrm{sResU}_p, \ldots, \mathrm{sResU}_0$, $\mathrm{sResV} := \mathrm{sResV}_p, \ldots, \mathrm{sResV}_0$.

Proof of correctness: The correctness of the algorithm follows from Theorem 8.30 and Proposition 8.38 b) since it is immediate to verify that, with sResU and sResV computed in the algorithm above,

$$\begin{aligned} \mathrm{sResP}_{i-1} &= \mathrm{sResU}_{i-1} P + \mathrm{sResV}_{i-1} Q, \\ \mathrm{sResP}_{j-1} &= \mathrm{sResU}_{j-1} P + \mathrm{sResV}_{j-1} Q. \end{aligned}$$

This implies that $\mathrm{sResP}_{k-1} = \mathrm{sResU}_{i-1} P + \mathrm{sResV}_{i-1} Q$. □

Complexity analysis: The complexity is clearly $O(pq)$ as in Algorithm 8.21 (Signed Subresultant).

When P and Q are in $\mathbb{Z}[X]$, with coefficients of bitsizes bounded by τ, the bitsizes of the integers in the operations performed by the algorithm are bounded by $(\tau + \nu)(p + q)$, where ν is the bitsize of $p + q$ according to Proposition 8.44. □

Remark 8.51. Algorithm 8.21 (Signed Subresultant) and Algorithm 8.22 (Extended Signed Subresultant) use exact divisions and are valid only in an integral domain, and not in a general ring. In a ring with division by integers, the algorithm computing determinants indicated in Remark 8.19 can always be used for computing the signed subresultant coefficients. The complexity obtained is $(p\,q)^{O(1)}$ arithmetic operations in the ring D of coefficients of P and Q, which is sufficient for the complexity estimates obtained in later chapters. □

8.4 Bibliographical Notes

Bounds on determinants are due to Hadamard [80]. A variant of Dogdson-Jordan-Bareiss's algorithm appears in [54] (see also [9]). Note that Dogdson is better known as Lewis Carrol.

The idea of using the traces of the powers of the matrix for computing its characteristic polynomial is a classical method due to Leverrier [106]. The improvement we present is due to Preparata and Sarwate [132]. It is an instance of the "baby step -giant step" method.

Subresultant polynomials and their connection with remainders were already known to Euler [56] and have been studied by Habicht [79]. Subresultants appear in computer algebra with Collins [44], and they have been studied extensively since then.

There are much more sophisticated algorithms than the ones presented in this book, and with much better complexity, for polynomial and matrix multiplication (see von zur Gathen and Gerhard's Modern Computer Algebra [64]).

9

Cauchy Index and Applications

In Section 9.1, several real root and Cauchy index counting methods are described. Section 9.2 deals with the closely related topic of Hankel matrices and quadratic forms. In Section 9.3 an important application of Cauchy index to counting complex roots with positive real part is described. The only ingredient used in later chapters of the book coming from Chapter 9 is the computation of the Tarski-query.

9.1 Cauchy Index

9.1.1 Computing the Cauchy Index

A first algorithm for computing the Cauchy index follows from Algorithm 8.19 (Signed Remainder Sequence), using Theorem 2.58 (Sturm).

Algorithm 9.1. **[Sturm Cauchy Index]**

- **Structure:** an ordered field K.
- **Input:** $P \in \mathrm{K}[X] \setminus \{0\}$, $Q \in \mathrm{K}[X]$.
- **Output:** the Cauchy index $\mathrm{Ind}(Q/P)$.
- **Complexity:** $O(p\,q)$, where p is the degree of P and q the degree of Q.
- **Procedure:** Compute the signed remainder sequence of P and Q, using Algorithm 8.19, then compute the difference in sign variations at $-\infty$ and $+\infty$ from the degrees and signs of leading coefficients of the polynomials in this sequence.

Proof of correctness: The correctness follows from Theorem 2.58 (Sturm). □

Complexity analysis: The complexity of the algorithm is $O(p\,q)$ according to the complexity analysis of the Algorithm 8.19 (Signed Remainder Sequence). Indeed, there are only $O(p)$ extra sign determinations tests to perform. □

Remark 9.1. Note that a much more sophisticated method for computing the Cauchy index is based on ideas from [145, 119]. In this approach, the sign variations in the polynomials of the signed remainder sequence evaluated at $-\infty$ and ∞ are computed from the quotients and the gcd, with complexity $O((p+q)\log(p+q)^2) = \tilde{O}(p+q)$ using the fact that the quotients can be computed from the leading terms of the polynomials in the remainder sequence. The same remark applies for Algorithm 9.2. □

This algorithm gives the following method for computing a Tarski-query. Recall that the Tarski-query of Q for P is the number

$$\mathrm{TaQ}(Q,P) = \sum_{x \in \mathrm{R}, P(x)=0} \mathrm{sign}(Q(x)).$$

Algorithm 9.2. **[Remainder Univariate Tarski-query]**

- **Structure:** an ordered field K.
- **Input:** $P \in \mathrm{K}[X] \setminus \{0\}$, $Q \in \mathrm{K}[X]$.
- **Output:** the Tarski-query $\mathrm{TaQ}(Q,P)$.
- **Complexity:** $O(p\,(p+q))$, where p is the degree of P and q the degree of Q.
- **Procedure:** Call Algorithm 9.1 (Sylvester Cauchy index) with input P and $P'Q$.

Proof of correctness: The correctness follows from Theorem 2.61 (Sylvester's theorem). □

Complexity analysis: Suppose that P and Q have respective degree p and q. The complexity of the algorithm is $O(p\,(p+q))$ according to the complexity analysis of Algorithm 9.1 (Sylvester Cauchy index). □

Exercise 9.1. Design an algorithm computing $\mathrm{TaQ}(Q, P; a, b)$ with complexity $O((p+q)^2)$, where $\mathrm{TaQ}(Q,P;a,b)$.

Another algorithm for computing the Cauchy index using the subresultant polynomials is based on Theorem 4.31. Its main advantage is that the bitsize of intermediate computations are much better controlled.

We first compute generalized permanences minus variations (see Notation 4.30).

Algorithm 9.3. **[Generalized Permancences minus Variations]**

- **Structure:** an ordered integral domain D.
- **Input:** $s = s_p, \ldots, s_0$ be a finite list of elements in D such that $s_p \neq 0$.
- **Output:** $\mathrm{PmV}(s)$.
- **Complexity:** $O(p)$.
- **Procedure:**
  - Initialize n to 0.

- Compute the number ℓ of non-zero elements of s and define the list $(s'(1), m(1)), \ldots, (s'(\ell), m(\ell)) = (s_p, p), (s_q, q), \ldots,$ of non-zero elements of s with their index.
- For every i from 1 to $\ell - 1$, if $m(i) - m(i+1)$ is odd

$$n := n + (-1)^{(m(i)-m(i+1))(m(i)-m(i+1)-1)/2} \operatorname{sign}(s'(i)\, s'(i+1)).$$

Algorithm 9.4. **[Cauchy Index]**

- **Structure:** an ordered integral domain D.
- **Input:** $P \subset \mathrm{D}[X] \setminus \{0\}$, $Q \in \mathrm{D}[X]$.
- **Output:** the Cauchy index $\mathrm{Ind}((Q/P)$.
- **Complexity:** $O(p q)$, where p is the degree of P and q the degree of Q.
- **Procedure:** If $q \geqslant p$, replace Q by the signed pseudo-remainder of Q and P. Using Algorithm 8.21 (Signed Subresultant), compute the sequence sRes of principal signed subresultant coefficient of P and Q, and then compute $\mathrm{PmV}(\mathrm{sRes}(P, Q))$ (Notation 4.30).

Proof of correctness: The correctness follows from Theorem 4.31. □

Complexity analysis: The complexity of the algorithm is $O(p q)$ according to the complexity analysis of Algorithm 8.21 (Signed Subresultant), since there are only $O(p)$ extra sign evaluations to perform.

When P and Q are in $\mathbb{Z}[X]$, with coefficients of bitsizes bounded by τ, the bitsizes of the integers in the operations performed by the algorithm are bounded by $(\tau + \nu)\,(p + q)$, where ν is the bitsize of $p + q$. This follows from Proposition 8.44. □

Remark 9.2. Similar ideas to that of [145, 119] (see Remark 9.1) can be used for computing the Cauchy index with complexity $\tilde{O}(q\, \tau)$ and binary complexity $\tilde{O}((p + q)^2 \tau)$ [107]. The same remark applies for Algorithm 9.5.□

This algorithm gives the following method for computing Tarski-queries.

Algorithm 9.5. **[Univariate Tarski-query]**

- **Structure:** an ordered integral domain D.
- **Input:** $P \in \mathrm{D}[X] \setminus \{0\}$, $Q \in \mathrm{D}[X]$.
- **Output:** the Tarski-query $\mathrm{TaQ}(Q, P)$.
- **Complexity:** $O((p + q)\, q)$, where p is the degree of P and q the degree of Q.
- **Procedure:**
  - If $\deg(Q) = 0$, $Q = b_0$, compute the sequence $\mathrm{sRes}(P, P')$ of signed subresultant coefficient of P and P' using Algorithm 8.21 (Signed Subresultant), and compute $\mathrm{PmV}(\mathrm{sRes}(P, P'))$ (Definition 4.30). Output

$$\begin{cases} \mathrm{PmV}(\mathrm{sRes}(P, P')) & \text{if } b_0 > 0, \\ -\mathrm{PmV}(\mathrm{sRes}(P, P')) & \text{if } b_0 < 0. \end{cases}$$

- If $\deg(Q) = 1$, $Q = b_1 X + b_0$, compute $R := P' Q - p\, b_1 P$, the sequence $\mathrm{sRes}(P, R)$ of signed subresultant coefficient of P and R, using Algorithm 8.21 (Signed Subresultant), and compute $\mathrm{PmV}(\mathrm{sRes}(P, R))$ (Definition 4.30).
- If $\deg(Q) > 1$ use Algorithm 8.21 (Signed Subresultant) to compute the sequence $\mathrm{sRes}(-P' Q, P)$ of signed subresultant coefficient of $-P' Q$ and Q, and compute $\mathrm{PmV}(\mathrm{sRes}(-P' Q, P))$ (Definition 4.30). Output

$$\begin{cases} \mathrm{PmV}(\mathrm{sRes}(-P' Q, P)) + \mathrm{sign}(b_q) & \text{if } q-1 \text{ is odd,} \\ \mathrm{PmV}(\mathrm{sRes}(-P' Q, P)) & \text{otherwise.} \end{cases}$$

Proof of correctness: The correctness follows from Corollary 9.6 and Lemma 9.5. □

Complexity analysis: The complexity of the algorithm is $O((p + q)\, p)$, according to the complexity analysis of the Algorithm 8.21 (Signed Subresultant).

Suppose P and Q in $\mathbb{Z}[X]$ with coefficients of bitsizes bounded by τ, and denote by ν the bitsize of $2p + q - 1$.

When $q > 1$, the bitsizes of the coefficients of $P'Q$ are bounded by $2\tau + \nu$. When $q = 1$, the bitsizes of the coefficients of $\bar{R}$ are bounded by $2\tau + 2\nu$. When $q = 0$, the bitsizes of the coefficients of P' are bounded by $\tau + \nu$.

Thus the bitsizes of the integers in the operations performed by the algorithm are bounded by $(2\tau + 2\nu)(2p + q - 1)$, according to Proposition 8.44. □

9.1.2 Bezoutian and Cauchy Index

We give in this section yet another way of obtaining the Cauchy index. Let P and Q be two polynomials with:

$$\begin{aligned} P &= a_p X^p + a_{p-1} X^{p-1} + \cdots + a_0 \\ Q &= b_{p-1} X^{p-1} + \cdots + b_0, \end{aligned}$$

with $\deg(P) = p$, $\deg(Q) = q \leq p - 1$.

Notation 9.3. [Bezoutian] The **Bezoutian** of P and Q is

$$\mathrm{Bez}(P, Q) = \frac{Q(Y)\, P(X) - Q(X)\, P(Y)}{X - Y}.$$

If $\mathcal{B} = b_1(X), \ldots b_p(X)$ is a basis of $\mathrm{K}[X]/(P(X))$, $\mathrm{Bez}(P, Q)$ can be uniquely written

$$\mathrm{Bez}(P, Q) = \sum_{i,j=1}^{p} c_{i,j}\, b_i(X)\, b_j(Y).$$

The matrix of $\mathrm{Bez}(P,Q)$ in the basis $\mathcal{B}$ is the symmetric matrix with i,j-th entry the coefficient $c_{i,j}$ of $b_i(X)\,b_j(Y)$ in $\mathrm{Bez}(P,Q)$. Note that the signature of the matrix of $\mathrm{Bez}(P,Q)$ in the basis $\mathcal{B}$ does not depend of $\mathcal{B}$ by Sylvester's inertia law (Theorem 4.38). □

Theorem 9.4. *The following equalities hold*

$$\begin{aligned}\mathrm{Rank}(\mathrm{Bez}(P,Q)) &= \deg(P)-\deg(\gcd(P,Q))\\ \mathrm{Sign}(\mathrm{Bez}(P,Q)) &= \mathrm{Ind}(Q/P).\end{aligned}$$

The proof of the Theorem will use the following results.

Lemma 9.5. *Suppose $s=s_0,\ldots,s_{c-2},s_{c-1},\ldots,s_{2n-2}$, with $2n-1\geq c\geq n$, and $s_0=\ldots,=s_{c-2}=0$, $s_{c-1}\neq 0$, and let H be the $n\times n$ matrix defined by $h_{i,j}=s_{i+j-2}$. Then*

$$\begin{aligned}\mathrm{Rank}(H) &= 2n-c,\\ \mathrm{Sign}(H) &= \begin{cases}\mathrm{sign}(s_{c-1}) & \text{if } c \text{ is odd},\\ 0 & \text{if } c \text{ is even}.\end{cases}\end{aligned}$$

The proof of the lemma is based on the following proposition.

Proposition 9.6. *Let H be a semi-algebraic continuous mapping from an interval I of R into the set of symmetric matrix of dimension n. If, for every $t\in I$, the rank of $H(t)$ is always equal to the same value, then, for every $t\in I$, the signature of $H(t)$ is always equal to the same value.*

Proof: Let r be the rank of $H(t)$, for every $t\in I$. The number of zero eigenvalues of $H(t)$ is $n-r$ for every $t\in I$, by Corollary 4.44. The number of positive and negative eigenvalues of $H(t)$ is thus also constant, since roots vary continuously (see Theorem 5.12 (Continuity of roots)). Thus, by Corollary 4.44, for every $t\in I$, the signature of $H(t)$ is always equal to the same value. □

Proof of Lemma 9.5: Let $s_c=\cdots=s_{2n-2}=0$. In this special case, the rank of H is obviously $2n-c$. Since the associated quadratic form is

$$\Phi=\sum_{i=c+1-n}^{n} s_{c-1}\,f_i\,f_{c+1-i}$$

and, if $c+1-i\neq i$,

$$4f_i\,f_{c+1-i}=(f_i+f_{c+1-i})^2-(f_i-f_{c+1-i})^2,$$

it is easy to see that the signature of Φ is 0 if c is even, and is 1 (resp. -1) if $s_{c-1}>0$ (resp. $s_{c-1}<0$).

Defining, for $t \in [0, 1]$, $s_t = 0, \ldots, 0, s_{c-1}, t\, s_c, \ldots, t\, s_{2n-2}$ the quadratic form with associated matrix H_t defined by $h_{t,i,j} = s_{t,i+j-2}$ is of rank $2\,n - c$ for every $t \in [0, 1]$, since the $c - n$ first columns of H_t are zero and its $2\,n - c$ last columns are clearly independent. Thus the rank of H_t is constant as t varies. Thus by Proposition 9.6 the signature of H_t is constant as t varies. This proves the claim, taking $t = 1$. □

We denote by R and C the remainder and quotient of the euclidean division of P by Q.

Lemma 9.7. *The following equalities hold.*

$$\begin{aligned}\mathrm{Rank}(\mathrm{Bez}(P,Q)) &= \mathrm{Rank}(\mathrm{Bez}(Q,-R)) + p - q\\ \mathrm{Sign}(\mathrm{Bez}(P,Q)) &= \begin{cases}\mathrm{Sign}(\mathrm{Bez}(Q,-R)) + \mathrm{sign}(a_p\, b_q) & \text{if } p - q \text{ is odd,}\\ \mathrm{Sign}(\mathrm{Bez}(Q,-R)) & \text{if } p - q \text{ is even.}\end{cases}\end{aligned}$$

Proof: We consider the matrix $M(P, Q)$ of coefficients of $\mathrm{Bez}(P, Q)$ in the canonical basis

$$X^{p-1}, \ldots, 1,$$

and the matrix $M'(P, Q)$ of coefficients of $\mathrm{Bez}(P, Q)$ in the basis

$$X^{p-q-1}\, Q(X), \ldots, X\, Q(X), Q(X), X^{q-1}, \ldots, 1.$$

Let $c = p - q$, $C = u_c X^c + \cdots + u_0$, $s = \overbrace{0, \ldots, 0}^{c-1 \text{ times}}, u_c \ldots, u_1$ of length $2\,c - 1$ and H the $c \times c$ matrix defined by $h_{i,j} = s_{i+j-2}$. Since $P = C\,Q + R$, and $\deg(R) < q$,

$$\mathrm{Bez}(P, Q) = \frac{C(X) - C(Y)}{X - Y}\, Q(Y)\, Q(X) + \mathrm{Bez}(Q, -R),$$

the matrix $M'(P, Q)$ is the block matrix

$$\begin{bmatrix} H & 0\\ 0 & M(Q,-R)\end{bmatrix}.$$

The claim follows from Lemma 9.5 since the leading coefficient of C is a_p / b_q. □

Proof of Theorem 9.4: The proof of the Theorem proceeds by induction on the number n of elements with distinct degrees in the signed subresultant sequence.

If $n = 2$, Q divides P and $R = 0$. We have

$$\mathrm{Rank}(\mathrm{Bez}(P, Q)) = \deg(P) - \deg(Q).$$

We also have

$$\mathrm{Ind}(Q/P) = \begin{cases}\mathrm{sign}(a_p\, b_q) & \text{if } p - q \text{ is odd,}\\ 0 & \text{if } p - q \text{ is even.}\end{cases}$$

by Lemma 4.34 and

$$\mathrm{Sign}(\mathrm{Bez}(P,Q)) = \begin{cases} \mathrm{sign}(a_p b_q) & \text{if } p-q \text{ is odd,} \\ 0 & \text{if } p-q \text{ is even.} \end{cases}$$

by Lemma 9.7.

Let us suppose that the Theorem holds for $n-1$ and consider P and Q such that their signed subresultant sequence has n elements with distinct degrees. The signed subresultant sequence of Q and $-R$ has $n-1$ elements with distinct degrees and by induction hypothesis,

$$\begin{aligned} \mathrm{Rank}(\mathrm{Bez}(Q,-R)) &= \deg(Q) - \deg(\gcd(Q,-R)) \\ \mathrm{Sign}(\mathrm{Bez}(Q,-R) &= \mathrm{Ind}(-R/Q) \end{aligned}$$

By Lemma 9.4 and Lemma 9.7, since $\gcd(P,Q) = \gcd(Q,-R)$.

$$\begin{aligned} \mathrm{Rank}(\mathrm{Bez}(P,Q)) &= \deg(P) - \deg(\gcd(P,Q)) \\ \mathrm{Sign}(\mathrm{Bez}(Q,-R)) &= \mathrm{Ind}(Q/P) \end{aligned}$$

□

Theorem 9.4 has the following corollaries.

Corollary 9.8. *Let P and Q be polynomials in* $\mathrm{D}[X]$ *and R the remainder of $P'Q$ divided by P. Then* $\mathrm{Sign}(\mathrm{Bez}(P,R)) = \mathrm{TaQ}(Q,P)$.

Proof: Apply Theorem 9.4 and Proposition 2.57, noticing that

$$\mathrm{Ind}(P'Q/P) = \mathrm{Ind}(R/P).$$

□

Corollary 9.9. *Let P be a polynomial in* $\mathrm{D}[X]$. *Then* $\mathrm{Sign}(\mathrm{Bez}(P, P'))$ *is the number of roots of P in* R.

Is follows immediately from Theorem 9.4 that the determinant of the matrix of $\mathrm{Bez}(P, Q)$ in the canonical basis $X^{p-1}, \ldots, 1$, is 0 if and only if $\deg(\gcd(P,Q)) > 0$. On the other hand, by Proposition 4.15 $\mathrm{Res}(P, Q) = 0$ if and only if $\deg(\gcd(P, Q)) > 0$. This suggests a close connection between the determinant of the matrix of $\mathrm{Bez}(P,Q)$ in the canonical basis $X^{p-1}, \ldots, 1$, and $\mathrm{Res}(P,Q)$.

Proposition 9.10. *Let $M(P,Q)$ be the matrix of coefficients of* $\mathrm{Bez}(P,Q)$ *in the canonical basis $X^{p-1}, \ldots, 1$. Then*

$$\det(M(P,Q)) = \varepsilon_p a_p^{p-q} \mathrm{Res}(P,Q),$$

with $\varepsilon_i = (-1)^{i(i-1)/2}$.

Proof: If $\deg(\gcd(P,Q)) > 0$, both quantities are 0, so the claim is true.

Suppose now that $\gcd(P, Q)$ is a constant. The proof is by induction on the number n of elements in the signed remainder sequence.

If $n=2$, and $Q=b, q=0$,

$$\begin{aligned}\operatorname{Res}(P,Q) &= b^{p-q},\\ \det(M(P,Q)) &= \varepsilon_p a_p^p b^p\end{aligned}$$

and the claim holds.

Suppose that the claim holds for $n-1$. According to Proposition 8.35 and Equation (4.3) in Notation 4.26

$$\varepsilon_p \operatorname{Res}(P,Q) = \varepsilon_q \varepsilon_{p-q} b_q^{p-r} \operatorname{Res}(Q,-R). \tag{9.1}$$

On the other hand, by the proof of Lemma 9.7, and using its notation, the matrix $M'(P,Q)$ of coefficients of $\operatorname{Bez}(P,Q)$ in the basis

$$X^{p-q-1}Q, \ldots, XQ, Q, X^{q-1}, \ldots, 1$$

is the matrix

$$\begin{bmatrix} H & 0 \\ 0 & M(Q,-R) \end{bmatrix}$$

with, for $1 \le i \le p-q$, $h_{p-q+1-i,i} = a_p/b_q$. Thus, using the fact that the leading coefficient of Q is b_q,

$$\det(\operatorname{Bez}(P,Q)) = \varepsilon_{p-q} a_p^{p-q} b_q^{p-q} \det(\operatorname{Bez}(Q,-R)). \tag{9.2}$$

The induction hypothesis

$$\det(M(Q,-R)) = \varepsilon_q b_q^{q-r} \operatorname{Res}(Q,-R),$$

Equation (9.1) and Equation (9.2), imply the claim for P, Q. □

9.1.3 Signed Subresultant Sequence and Cauchy Index on an Interval

We show that the Cauchy index on an interval can be expressed in terms of appropriately counted sign variations in the signed subresultant sequence. The next definitions introduce the sign counting function to be used.

Notation 9.11. Let $s = s_n, 0\ldots, 0, s'$, be a finite sequence of elements in an ordered field K such that $s_n \neq 0$, $s' = \emptyset$ or $s' = s_m, \ldots, s_0$, $s_m \neq 0$. The **modified number of sign variations** in s is defined inductively as follows

$$\operatorname{MVar}(s) = \begin{cases} 0 & \text{if } s' = \emptyset, \\ \operatorname{MVar}(s') + 1 & \text{if } s_n s_m < 0, \\ \operatorname{MVar}(s') + 2 & \text{if } s_n s_m > 0 \text{ and } n - m = 3 \\ \operatorname{MVar}(s') & \text{if } s_n s_m > 0 \text{ and } n - m \neq 3, \end{cases}$$

In other words, we modify the usual definition of the number of sign variations by counting 2 sign variations for the groups: $+, 0, 0, +$ and $-, 0, 0, -$.

Let $\mathcal{P} = P_0, P_1, \ldots, P_d$ be a sequence of polynomials in $\mathrm{D}[X]$ and a be an element of $\mathrm{R} \cup \{-\infty, +\infty\}$ which is not a root of $\gcd(\mathcal{P})$. Then $\mathrm{MVar}(\mathcal{P}; a)$, the **modified number of sign variations** of $\mathcal{P}$ at a, is the number defined as follows:

- Delete from $\mathcal{P}$ those polynomials that are identically 0 to obtain the sequence of polynomials $\mathcal{Q} = Q_0, \cdots, Q_s$ in $\mathrm{D}[X]$,
- Define $\mathrm{MVar}(\mathcal{P}; a)$ as $\mathrm{MVar}(Q_0(a), \cdots, Q_s(a))$.

Let a and b be elements of $\mathrm{R} \cup \{-\infty, +\infty\}$ which are not roots of $\gcd(\mathcal{P})$. The difference between the number of modified sign variations in $\mathcal{P}$ at a and b is denoted by

$$\mathrm{MVar}(\mathcal{P}; a, b) = \mathrm{MVar}(\mathcal{P}; a) - \mathrm{MVar}(\mathcal{P}; b).$$ □

For example, if $\mathcal{P} = X^5, X^2 - 1, 0, X^2 - 1, X + 2, 1,$ the modified number of sign variations of $\mathcal{P}$ at 1 is 2 while the number of signs variations of $\mathcal{P}$ at 1 is 0.

Theorem 9.12.

$$\mathrm{MVar}(\mathrm{sResP}(P, Q); a, b) = \mathrm{Ind}(Q/P; a, b).$$

Note that when polynomials of all possible degrees $\leq p$ appear in the remainder sequence, Theorem 9.12 is an immediate consequence of Theorem 2.58, since the signed remainder sequence and the signed subresultant sequence are proportional up to squares by Corollary 8.33.

The proof of Theorem 9.12 uses the following lemma.

Lemma 9.13. *Let Let $R = \mathrm{Rem}(P, Q)$ and let $\sigma(a)$ be the sign of PQ at a and $\sigma(b)$ be the sign of PQ at b. Then*

$$\begin{aligned} &\mathrm{MVar}(\mathrm{sResP}(P, Q); a, b) \\ = &\begin{cases} \mathrm{MVar}(\mathrm{sResP}(Q, -R); a, b) + \sigma(b) & \text{if } \sigma(a)\,\sigma(b) = -1, \\ \mathrm{MVar}(\mathrm{sResP}(Q, -R); a, b) & \text{if } \sigma(a)\,\sigma(b) = 1. \end{cases} \end{aligned}$$

Proof: We denote $L = \mathrm{sResP}(P, Q)$ and $L' = \mathrm{sResP}(Q, -R)$.

Suppose that x be not a root of P, Q, or R. According to Proposition 8.35, and the conventions in Notation 8.29,

$$\begin{aligned} \mathrm{sResP}_p(P, Q) &= P, \\ \mathrm{sResP}_{p-1}(P, Q) &= Q, \\ \mathrm{sResP}_q(P, Q) &= \varepsilon_{p-q}\, b_q^{p-q-1}\, Q, \\ \mathrm{sResP}_q(Q, -R) &= Q, \\ \mathrm{sResP}_{q-1}(P, Q) &= -\varepsilon_{p-q}\, b_q^{p-q+1}\, R, \\ \mathrm{sResP}_{q-1}(Q, -R) &= -R, \\ \mathrm{sResP}_j(P, Q) &= \varepsilon_{p-q}\, b_q^{p-r}\, \mathrm{sResP}_j(Q, -R),\ j \leq r. \end{aligned}$$

Hence for every x which is not a root of P, Q and $-R$, in particular for a and b, the following holds, denoting by c_r the leading coefficient of $-R$.

If $P(x)\,Q(x)>0$

– if $\varepsilon_{q-r}\,c_r^{q-r-1}>0$

$$\mathrm{MVar}(L;x)=\begin{cases}\mathrm{MVar}(L';x)+2 & \text{if } \varepsilon_{p-q}\,b_q^{p-q-1}<0 \text{ and } b_q^{q-r-1}<0,\\ \mathrm{MVar}(L';x)+1 & \text{if } \varepsilon_{p-q}\,b_q^{p-r}<0,\\ \mathrm{MVar}(L';x) & \text{if } \varepsilon_{p-q}\,b_q^{p-q-1}>0 \text{ and } b_q^{q-r-1}>0,\end{cases}$$

– if $\varepsilon_{q-r}\,c_r^{q-r-1}<0$

$$\mathrm{MVar}(L;x)=\begin{cases}\mathrm{MVar}(L';x)+1 & \text{if } \varepsilon_{p-q}\,b_q^{p-q-1}<0 \text{ and } b_q^{q-r-1}>0,\\ \mathrm{MVar}(L';x) & \text{if } \varepsilon_{p-q}\,b_q^{p-r}>0,\\ \mathrm{MVar}(L';x)-1 & \text{if } \varepsilon_{p-q}\,b_q^{p-q-1}>0 \text{ and } b_q^{q-r-1}<0.\end{cases}$$

If $P(x)\,Q(x)<0$,

– if $\varepsilon_{q-r}\,c_r^{q-r-1}>0$

$$\mathrm{MVar}(L;x)=\begin{cases}\mathrm{MVar}(L';x)+3 & \text{if } \varepsilon_{p-q}\,b_q^{p-q-1}<0 \text{ and } b_q^{q-r-1}<0,\\ \mathrm{MVar}(L';x)+2 & \text{if } \varepsilon_{p-q}\,b_q^{p-r}<0,\\ \mathrm{MVar}(L';x)+1 & \text{if } \varepsilon_{p-q}\,b_q^{p-q-1}>0 \text{ and } b_q^{q-r-1}>0,\end{cases}$$

– if $\varepsilon_{q-r}\,c_r^{q-r-1}<0$

$$\mathrm{MVar}(L;x)=\begin{cases}\mathrm{MVar}(L';x)+2 & \text{if } \varepsilon_{p-q}\,b_q^{p-q-1}<0 \text{ and } b_q^{q-r-1}>0,\\ \mathrm{MVar}(L';x)+1 & \text{if } \varepsilon_{p-q}\,b_q^{p-r}>0,\\ \mathrm{MVar}(L';x) & \text{if } \varepsilon_{p-q}\,b_q^{p-q-1}>0 \text{ and } b_q^{q-r-1}<0\end{cases}$$

The lemma follows easily. □

Proof of Theorem 9.12: We can assume without loss of generality that a and b are not roots of a non-zero polynomial in the signed subresultant sequence. Indeed if $a<a'<b'<b$ with $(a,a']$ and $[b',b)$ containing no root of the polynomials in the signed subresultant sequence,

$$\mathrm{Ind}(Q/P;a,b)=\mathrm{Ind}(Q/P;a',b').$$

We also have

$$\mathrm{MVar}(\mathrm{sResP}(P,Q);a,b)=\mathrm{MVar}(\mathrm{sResP}(P,Q);a',b').$$

Indeed if a is a root of of sResP_{j-1},

– when sResP_{j-1} is non-defective, we have

$$\begin{aligned}\mathrm{MVar}(\mathrm{sResP}_{j-2},\mathrm{sResP}_{j-1},\mathrm{sResP}_j;a) &= 1,\\ \mathrm{MVar}(\mathrm{sResP}_{j-2},\mathrm{sResP}_{j-1},\mathrm{sResP}_j;a') &= 1,\end{aligned}$$

– when $\mathrm{sResP}_{j-1}(P,Q)$ is defective of degree k, we have

$$\begin{aligned}&\mathrm{MVar}(\mathrm{sResP}_j,\mathrm{sResP}_{j-1},\mathrm{sResP}_k,\mathrm{sResP}_{k-1};a')\\ =\ &\mathrm{MVar}(\mathrm{sResP}_j,\mathrm{sResP}_{j-1},\mathrm{sResP}_k,\mathrm{sResP}_{k-1};a)\\ =\ &\begin{cases}2 \text{ if } \mathrm{sResP}_j(a)\mathrm{sResP}_{k-1}(a)>0\\ 1 \text{ if } \mathrm{sResP}_j(a)\mathrm{sResP}_{k-1}(a)<0.\end{cases}\end{aligned}$$

The proof of the theorem proceeds by induction on the number n of elements with distinct degrees in the signed subresultant sequence. The base case $n=2$ corresponds to $\deg(Q)=0$, $R=0$ and follows from Lemma 9.13 and Lemma 2.60. Let us suppose that the Theorem holds for $n-1$ and consider P and Q such that their signed subresultant sequence has n elements with distinct degrees. The signed subresultant sequence of Q and $-R$ has $n-1$ elements with distinct degrees and by the induction hypothesis,

$$\mathrm{MVar}(\mathrm{sResP}(Q,-R);a,b)=\mathrm{Ind}(-R/Q;a,b).$$

So, by Lemma 9.13 and Lemma 2.60,

$$\mathrm{MVar}(\mathrm{sResP}(P,Q);a,b)=\mathrm{Ind}(Q/P;a,b).$$

□

Corollary 9.14. *Let P, $Q\in \mathrm{D}[X]$. Let R be the remainder of $P'Q$ and P. If $a<b$ are elements of $\mathrm{R}\cup\{-\infty,+\infty\}$ that are not roots of P, then*

$$\mathrm{MVar}(\mathrm{sResP}(P,R);a,b)=\mathrm{TaQ}(Q,P;a,b).$$

Proof: Apply Theorem 9.12 and Proposition 2.57, since

$$\mathrm{Ind}(P'Q/P;a,b)=\mathrm{Ind}(R/P;a,b),$$

by Remark 2.55. □

Corollary 9.15. *Let P be a polynomial in $\mathrm{D}[X]$. If $a<b$ are elements of $\mathrm{R}\cup\{-\infty,+\infty\}$ which are not roots of P, then $\mathrm{MVar}(\mathrm{sResP}(P,P');a,b)$ is the number of roots of P in (a,b).*

Exercise 9.2. Using Algorithm 8.21 (Signed Subresultant), design an algorithm computing $\mathrm{TaQ}(Q,P;a,b)$ with complexity $O((p+q)\,p)$. If $P\in\mathbb{Z}[X]$, $Q\in\mathbb{Z}[X]$, a, b are rational numbers, and τ is a bound on the bitsize of the coefficients of P and Q and on a and b, estimate the bitsize of the rationals computed by this algorithm.

9.2 Hankel Matrices

Hankel matrices are important because of their relation with rational functions and sequences satisfying linear recurrence relations. We define Hankel matrices and quadratic forms and indicate how to compute the corresponding signature.

9.2.1 Hankel Matrices and Rational Functions

Hankel matrices are symmetric matrix with equal entries on the anti-diagonals. More precisely **Hankel matrices** of size p are matrices with entries $a_{i+1,j+1}$ (i from 0 to $p-1$, j from 0 to $p-1$) such that $a_{i+1,j+1}=a_{i'+1,j'+1}$ whenever $i+j=i'+j'$.

A typical Hankel matrix is the matrix $\mathrm{Newt}_k(P)$ considered in Chapter 4.

Notation 9.16. [Hankel] Let $\bar{s}=s_0,\ldots,s_n,\ldots$ be an infinite sequence. We denote $\bar{s}_n=s_0,\ldots,s_{2n-2}$, and $\mathrm{Han}(\bar{s}_n)$ the Hankel matrix whose $i+1,j+1$ entry is s_{i+j} for $0\le i,j\le n-1$, and by $\mathrm{han}(\bar{s}_n)$ the determinant of $\mathrm{Han}(\bar{s}_n)$. □

Theorem 9.17. *Let* K *be a field. Let $\bar{s}=s_0,\ldots,s_n,\ldots$ be an infinite sequence of elements of* K *and $p\in\mathbb{N}$. The following conditions are equivalent:*

a) *The elements $s_0,\ldots,s_n,\ldots$ satisfy a linear recurrence relation of order p with coefficients in* K

$$a_p\, s_n=-a_{p-1}\, s_{n-1}-\ldots-a_0\, s_{n-p}, \tag{9.3}$$

$a_p\neq 0,\ n\ge p.$

b) *There exists a polynomial $P\in\mathrm{K}[X]$ of degree p and a linear form λ on $\mathrm{K}[X]/(P)$ such that $\lambda(X^i)=s_i$ for every $i\ge 0$.*

c) *There exist polynomials $P,Q\in\mathrm{K}[X]$ with $\deg(Q)<\deg(P)=p$ such that*

$$Q/P=\sum_{j=0}^{\infty} s_j/X^{j+1} \tag{9.4}$$

d) *There exists an $r\le p$ such that the ranks of all the Hankel matrices $\mathrm{Han}(\bar{s}_r),\mathrm{Han}(\bar{s}_{r+1}),\mathrm{Han}(\bar{s}_{r+2}),\ldots$ are equal to r.*

e) *There exists an $r\le p$ such that*

$$\mathrm{han}(\bar{s}_r)\neq 0,\quad \forall\, n>r\ \mathrm{han}(\bar{s}_n)=0.$$

A sequence satisfying the equivalent properties of Theorem 9.17 is a **linear recurrent sequence of order** p.

The proof of the theorem uses the following definitions and results. Let

$$P=a_p X^p+a_{p-1}X^{p-1}+\cdots+a_1 X+a_0$$

be a polynomial of degree p. The Horner polynomials associated to P are defined inductively by

$$\begin{aligned}\mathrm{Hor}_0(P,X) &= a_p,\\ &\vdots\\ \mathrm{Hor}_i(P,X) &= X\,\mathrm{Hor}_{i-1}(P,X)+a_{p-i},\\ &\vdots\end{aligned}$$

for $i=0,\ldots,p-1$ (see Notation 8.6).

The Horner polynomials

$$\mathrm{Hor}_0(P,X),\ldots,\mathrm{Hor}_{p-1}(P,X)$$

are obviously a basis of the quotient ring $\mathrm{K}[X]/(P)$.

The **Kronecker form** is the linear form ℓ_P defined on $\mathrm{K}[X]/(P(X))$, by

$$\ell_P(1)=\cdots=\ell_P(X^{p-2})=0, \ell_P(X^{p-1})=1/a_p.$$

If $Q\in\mathrm{K}[X]$, $\mathrm{Rem}(Q,P)$ is the canonical representative of its equivalence class in $\mathrm{K}[X]/(P(X))$, and $\ell_P(Q)$ denotes $\ell_P(\mathrm{Rem}(Q,P))$.

Proposition 9.18. *For* $0\le i\le p-1, 0\le j\le p-1$,

$$\ell_P(X^j\,\mathrm{Hor}_{p-1-i}(P,X))=\begin{cases}1, & j=i,\\ 0, & j\ne i.\end{cases}$$

Proof: The claim is clear from the definitions if $j\le i$.

If $i<j\le p-1$, since

$$a_pX^p+\cdots+a_{i+1}X^{i+1}=-(a_iX^i+\cdots+a_0 \bmod P(X),$$

we have

$$\begin{aligned}X^{i+1}\,\mathrm{Hor}_{p-1-i}(P,X) &= -(a_iX^i+\cdots+a_0) \bmod P(X)\\ X^j\,\mathrm{Hor}_{p-1-i}(P,X) &= -X^{j-i-1}(a_iX^i+\cdots+a_0) \bmod P(X),\end{aligned}$$

and, by definition of ℓ_P

$$\ell_P(X^j\,\mathrm{Hor}_{p-1-i}(P,X))=-\ell_P(X^{j-i-1}(a_iX^i+\cdots+a_0))=0.$$ □

Corollary 9.19. *For every* $Q\in\mathrm{K}[X]$,

$$Q=\sum_{i=0}^{p-1}\ell_P(Q\,X^i)\,\mathrm{Hor}_{p-1-i}(P,X) \bmod P(X). \tag{9.5}$$

Proof: By (9.18),

$$\mathrm{Hor}_{p-1-j}(P,X)=\sum_{i=0}^{p-1}\ell_P(\mathrm{Hor}_{p-1-j}(P,X)\,X^i)\,\mathrm{Hor}_{p-1-i}(P,X) \bmod P(X).$$

The claim follows by the linearity of ℓ_P after expressing $Q_1=\mathrm{Rem}(Q,P)$ in the Horner basis. □

Proof of Theorem 9.17: $a)\Rightarrow c)$: Take

$$\begin{aligned}P &= a_pX^p+a_{p-1}X^{p-1}+\cdots+a_0,\\ Q &= s_0\,\mathrm{Hor}_{p-1}(P,X)+\cdots+s_{p-1}\,\mathrm{Hor}_0(P,X).\end{aligned}$$

Note that if $Q = b_{p-1} X^{p-1} + \cdots + b_0$, then

$$b_{p-n-1} = a_p\, s_n + \cdots + a_{p-n}\, s_0 \tag{9.6}$$

for $0 \le n \le p-1$, identifying the coefficients of X^{p-n-1} on both sides of (9.5). Let t_n be the infinite sequence defined by the development of the rational fraction Q/P as a series in $1/X$:

$$Q/P = \left(\sum_{n=0}^{\infty} t_n / X^{n+1}\right). \tag{9.7}$$

Thus,

$$Q = \left(\sum_{n=0}^{\infty} t_n / X^{n+1}\right) P. \tag{9.8}$$

Identifying the coefficients of X^{p-n-1} on both sides of (9.8) proves that for $0 \le n < p$

$$b_{p-n-1} = a_p\, t_n + \cdots + a_{p-n}\, t_0,$$

and for $n \ge p$

$$a_p\, t_n + a_{p-1}\, t_{n-1} + \cdots + a_0\, t_{n-p} = 0.$$

Since $a_p \neq 0$, the sequences s_n and t_n have the same p initial values and satisfy the same recurrence relation. Hence they coincide.

c) $\Rightarrow$ *b)*: For $i = 0, \ldots, p-1$, take $\lambda(X^i) = \ell_P(Q\, X^i)$, where ℓ_P is the Kronecker form. Since $Q = \sum_{k=0}^{p-1} s_k \operatorname{Hor}_{p-1-k}(P, X)$, using (9.5),

$$\lambda(X^j) = \ell_P(Q\, X^j) = \sum_{k=0}^{p-1} s_k\, \ell_P(X^j \operatorname{Hor}_{p-1-k}(P, X)) = s_j. \tag{9.9}$$

b) $\Rightarrow$ *a)* is clear, taking for a_i the coefficient of X^i in P and noticing that

$$\begin{aligned} a_p\, s_n &= \lambda(a_p\, X^n) \\ &= -\lambda(a_{p-1}\, X^{n-1} + \cdots + a_0\, X^{n-p}) \\ &= -a_{p-1}\, \lambda(X^{n-1}) + \cdots + a_0\, \lambda(X^{n-p}) \\ &= -a_{p-1}\, s_{n-1} + \cdots + a_0\, s_{n-p}. \end{aligned}$$

a) $\Rightarrow$ *d)*: For $(n, m) \in \mathbb{N}^2$, define $v_{m,n}$ as the vector $(s_m, \ldots, s_{m+n})$. The recurrence relation (9.3) proves that for $m \ge p$,

$$a_p\, v_{m,n} = -a_{p-1}\, v_{m-1,n} - \ldots - a_0\, v_{m-p,n}.$$

It is easy to prove by induction on n that the vector space generated by $v_{0,n}, \ldots, v_{n,n}$ is of dimension $\le p$, which proves the claim.

d) $\Rightarrow$ *e)*: is clear.

e) $\Rightarrow$ *a)*: Let $r \le p$ be such that

$$\operatorname{han}(\bar{s}_r) \neq 0, \forall\, n > r \operatorname{han}(\bar{s}_n) = 0.$$

Then the vector $v_{n-r,r}$, $n \geq r$, is a linear combination of $v_{0,r}$, ..., $v_{r-1,r}$. Developing the determinant of the square matrix with columns

$$v_{0,r}, \ldots, v_{r-1,r}, v_{n-r,r}$$

on the last columns gives

$$\mu_r\, s_n + \mu_{r-1}\, s_{n-1} + \cdots + \mu_0\, s_{n-r} = 0,$$

with μ_i the cofactor of the $i-1$-th element of the last column. Since $\mu_r \neq 0$, take $a_i = \mu_{p-r+i}$. □

9.2.2 Signature of Hankel Quadratic Forms

Hankel quadratic forms are quadratic forms associated to Hankel matrices. We design an algorithm for computing the signature of a general Hankel form. Note that we have already seen a special case where the signature of a Hankel quadratic form is particularly easy to determine in Lemma 9.5.

Given $\bar{s}_n = s_0, \ldots, s_{2n-2}$, we write

$$\overline{\mathrm{Han}}(\bar{s}_n) = \sum_{i=0}^{n-1} \sum_{j=0}^{n-1} s_{i+j}\, f_i\, f_j.$$

Note that the Hermite quadratic form seen in Chapter 4 is a Hankel form. Let

$$Q/P = \sum_{j=0}^{\infty} s_j / X^{j+1} \in \mathrm{K}[[1/X]],$$

and $\mathrm{Hor}_{p-1}(P, X)$, ..., $\mathrm{Hor}_0(P, X)$ the Horner basis of P. We indicate now the relationship between the Hankel quadratic form

$$\overline{\mathrm{Han}}(\bar{s}_p) = \sum_{i=0}^{p-1} \sum_{j=0}^{p-1} s_{i+j}\, f_i\, f_j.$$

and the quadratic form $\mathrm{Bez}(P, Q)$ (see Notation 9.3).

Proposition 9.20. *The matrix of coefficients of* $\mathrm{Bez}(P, Q)$ *in the Horner basis* $\mathrm{Hor}_{p-1}(P, X), \ldots, \mathrm{Hor}_0(P, X)$ *is* $\overline{\mathrm{Han}}(\bar{s}_p)$, *i.e.*

$$\mathrm{Bez}(P, Q) = \sum_{i,j=0}^{p-1} s_{i+j}\, \mathrm{Hor}_{p-1-i}(P, X)\, \mathrm{Hor}_{p-1-j}(P, Y). \tag{9.10}$$

Proof: Indeed,

$$\mathrm{Bez}(P, Q) = \frac{Q(Y) - Q(X)}{X - Y}\, P(X) + Q(X)\, \frac{P(X) - P(Y)}{X - Y} \bmod P(X),$$

which implies

$$\mathrm{Bez}(P, Q) = Q(X)\, \frac{P(X) - P(Y)}{X - Y} \bmod P(X),$$

noting that

$$\frac{P(X)-P(Y)}{X-Y}=\sum_{i=0}^{p-1} X^i \operatorname{Hor}_{p-1-i}(P,Y),$$

using Corollary 9.19 and Equation (9.9),

$$\begin{aligned} & Q(X)\,\frac{P(X)-P(Y)}{X-Y} \\ = & \sum_{i=0}^{p-1} Q(X)\,X^i \operatorname{Hor}_{p-1-i}(P,Y) \\ = & \sum_{i=0}^{p-1}\sum_{j=0}^{p-1} \ell_P(Q(X)\,X^{i+j}) \operatorname{Hor}_{p-1-i}(P,X) \operatorname{Hor}_{p-1-j}(P,Y) \bmod P(X) \\ = & \sum_{i=0}^{p-1}\sum_{j=0}^{p-1} s_{i+j} \operatorname{Hor}_{p-1-i}(P,X) \operatorname{Hor}_{p-1-i}(P,Y) \bmod P(X). \end{aligned}$$

This proves (9.10). □

Remark 9.21. So, by Proposition 4.55, the Hermite quadratic form $\operatorname{Her}(P,Q)$ is nothing but $\operatorname{Bez}(P,\ R)$ expressed in the Horner basis of P, with R the remainder of $P'Q$ by P. This proves that Theorem 9.4 is a generalization of Theorem 4.57. □

Let $\bar{s}_n = 0, \ldots, 0, s_{c-1}, \ldots, s_{2n-2}$, $s_{c-1} \neq 0$, $c < n$, and define the series S in $1/X$ by

$$S=\sum_{j=0}^{2n-2} s_j/X^{j+1}.$$

Consider the inverse S^{-1} of S, which is a Laurent series in $1/X$ and define $C \in \mathrm{K}[X], T \in \mathrm{K}[[1/X]]$ by $S^{-1}=C+T$, i.e. $(C+T)\,S=1$. Since S starts with s_{c-1}/X^c, it is clear that $\deg(C)=c$. Let

$$C=u_c X^c+\cdots+u_0,$$

and $\bar{u} = \overbrace{0,\ldots,0}^{c-1 \text{ times}}, u_c \ldots, u_1$ of length $2c-1$, and $T=\sum_{j=0}^{\infty} t_j/X^{j+1}$. Note that $u_c=1/s_{c-1}$. Let $\bar{t}_{n-c}=t_0,\ldots,t_{2n-2c-2}$.

Lemma 9.22.

$$\operatorname{Sign}(\overline{\operatorname{Han}}(\bar{s}_n))=\begin{cases} \operatorname{Sign}(\operatorname{Han}(\bar{t}_{n-c}))+\operatorname{sign}(s_{c-1}) & \text{if } c \text{ is odd,} \\ \operatorname{Sign}(\operatorname{Han}(\bar{t}_{n-c})) & \text{if } c \text{ is even.} \end{cases}$$

Lemma 9.22 is a consequence of the following Lemma, which uses Toeplitz matrices, i.e. matrices with equal entries on the parallels to the diagonal. More precisely a **Toeplitz matrix** of size n is a matrix with entries $a_{i,j}$ (i from 1 to n, j from 1 to n) such that $a_{i,j}=a_{i',j'}$ whenever $j-i=j'-i'$.

Notation 9.23. [Toeplitz] Let $\bar{v} = v_0, v_1, \ldots, v_{n-1}$. We denote by $\mathrm{To}(\bar{v})$ the triangular Toeplitz matrix of size n whose i,j-th entry is v_{j-i} for $0 \le i, j \le n$, with $j - i \ge 0$, 0 otherwise. □

Lemma 9.24.

$$\mathrm{Han}(\bar{s}_n) = \mathrm{To}(\bar{v})^t \begin{bmatrix} \mathrm{Han}(\bar{u}) & 0 \\ 0 & \mathrm{Han}(\bar{t}_{n-c}) \end{bmatrix} \mathrm{To}(\bar{v}),$$

with $\bar{v} = s_{c-1}, \ldots, s_{n+c-2}$.

We first explain how Lemma 9.22 is a consequence of Lemma 9.24 before proving the lemma itself.

Proof of Lemma 9.22: Using Lemma 9.24, the quadratic forms associated to $\mathrm{Han}(\bar{s}_n)$ and

$$\begin{bmatrix} \mathrm{Han}(\bar{u}) & 0 \\ 0 & \mathrm{Han}(\bar{t}_{n-c}) \end{bmatrix}$$

have the same signature, and

$$\mathrm{Sign}(\overline{\mathrm{Han}}(\bar{s}_n)) = \mathrm{Sign}(\overline{\mathrm{Han}}(\bar{t}_{n-c})) + \mathrm{Sign}(\overline{\mathrm{Han}}(\bar{u})).$$

The claim follows from Lemma 9.5, since, as noted above, $u_c = 1/s_{c-1}$. □

The proof of Lemma 9.24 requires some preliminary work.

Let $P = a_p X^p + \cdots + a_0$ and $Q = b_q X^q + \cdots + b_0$, $q = \deg(Q) < p = \deg(P)$, such that

$$Q/P = \sum_{j=0}^{\infty} s_j / X^{j+1} \in \mathrm{K}[[1/X]].$$

Then, $s_{p-q-1} = b_q/a_p \neq 0$. If $p - q \le n$, let $C \in \mathrm{K}[X], T \in \mathrm{K}[[1/X]]$ be defined by $S(C + T) = 1$. It is clear that $\deg(C) = p - q$, $P = CQ + R$, with $\deg(R) < q$ and

$$T = -R/Q = \sum_{j=0}^{\infty} t_j / X^{j+1}.$$

By Proposition 9.20, the matrix of coefficients of $\mathrm{Bez}(P, Q)$ in the Horner basis $\mathrm{Hor}_{p-1}(P, X), \ldots, \mathrm{Hor}_0(P, X)$ is $\overline{\mathrm{Han}}(\bar{s}_p)$.

We consider now the matrix of coefficients of $\mathrm{Bez}(P, Q)$ in the basis

$$X^{p-q-1} Q(X), \ldots, X Q(X), Q(X), \mathrm{Hor}_{q-1}(Q, X), \ldots, \mathrm{Hor}_0(Q, X).$$

Since $P = CQ + R$, $\deg(R) < q$,

$$\mathrm{Bez}(P, Q) = \frac{C(X) - C(Y)}{X - Y} Q(Y) Q(X) + \mathrm{Bez}(Q, -R),$$

the matrix of coefficients of $\mathrm{Bez}(P, Q)$ in the basis

$$X^{p-q-1} Q(X), \ldots, X Q(X), Q(X), \mathrm{Hor}_{q-1}(Q, X), \ldots, \mathrm{Hor}_0(Q, X),$$

is the block Hankel matrix

$$\begin{bmatrix} \mathrm{Han}(\bar{u}) & 0 \\ 0 & \mathrm{Han}(\bar{t}_q) \end{bmatrix}.$$

Proof of Lemma 9.24: Take

$$P = X^{2n-1}, Q = s_{c-1}X^{2n-c-1} + \cdots + s_{2n-2}.$$

Note that

$$Q/P = \sum_{j=c-1}^{\infty} s_j/X^{j+1} \in \mathrm{K}[[1/X]]$$

with $s_j = 0$ for $j > 2n-2$. The Horner basis of P is $X^{n-2}, \ldots, X, 1$. The change of basis matrix from the Horner basis of P to

$$X^{c-1}Q(X), \ldots, X\,Q(X), Q(X), \mathrm{Hor}_{q-1}(Q, X), \ldots, \mathrm{Hor}_0(Q, X)$$

is $\mathrm{To}(\bar{w})$, with $\bar{w} = s_{c-1}, s_c, \ldots, s_{2n+c-3}$. Thus, according to the preceding discussion,

$$\mathrm{Han}(\bar{s}_{2n-1}) = \mathrm{To}(\bar{w})^t \begin{bmatrix} \mathrm{Han}(\bar{u}) & 0 \\ 0 & \mathrm{Han}(\bar{t}_{2n-c-1}) \end{bmatrix} \mathrm{To}(\bar{w}). \tag{9.11}$$

Lemma 9.24 follows easily from Equation (9.11), suppressing the last $n-1$ lines and columns in each matrix. □

The following result gives a method to compute the signature of a Hankel form.

Proposition 9.25. *Let*

$$\begin{aligned} P &= a_p X^p + \cdots + a_0 \\ Q &= b_q X^q + \cdots + b_0 \end{aligned}$$

be coprime, $q = \deg(Q) < p = \deg(P)$*, such that*

$$Q/P = \sum_{j=0}^{\infty} s_j/X^{j+1} \in \mathrm{K}[[1/X]].$$

Let $\mathrm{han}(\bar{s}_{[0..n]}) = 1, \mathrm{han}(\bar{s}_1), \cdots, \mathrm{han}(\bar{s}_n)$.

a) Suppose $p \le n$*. Then*

$$\mathrm{Sign}(\overline{\mathrm{Han}}(\bar{s}_n)) = \mathrm{PmV}(\mathrm{han}(\bar{s}_{[0..n]})) = \mathrm{PmV}(\mathrm{sRes}(P, Q)) = \mathrm{Ind}(Q/P).$$

b) Suppose $p > n$*. Denoting by* j *the biggest natural number* $\le p-n$ *such that the subresultant* sResP_j *is non-defective and by* $\mathrm{sRes}(P, Q)_{[p..j]}$ *the sequence of* $\mathrm{sRes}_i(P, Q)$*,* $i = p, \ldots, j$*, we have*

$$\mathrm{Sign}(\overline{\mathrm{Han}}(\bar{s}_n)) = \mathrm{PmV}(\mathrm{han}(\bar{s}_{[0..p-j]})) = \mathrm{PmV}(\mathrm{sRes}(P, Q)_{[p..j]}).$$

The proof of the proposition uses the following lemma

Lemma 9.26. *For all $k \in \{1, \cdots, p\}$, we have:*

$$\mathrm{sRes}_{p-k}(P, Q) = a_p^{2k+2-p+q}\, \mathrm{han}(\bar{s}_k).$$

Proof: Let

$$\Delta = \begin{bmatrix} a_p & a_{p-1} & & & & \ddots & \ddots & a_{p-2k+2} \\ 0 & a_p & a_{p-1} & & & & \ddots & a_{p-2k+3} \\ \vdots & \ddots & \ddots & \ddots & & & \ddots & \\ \vdots & & \ddots & a_p & a_{p-1} & & & a_{p-k} \\ \vdots & & & 0 & b_{p-1} & & & b_{p-k} \\ \vdots & & \iddots & \iddots & & & \iddots & \\ \vdots & \iddots & \iddots & & & \iddots & & \\ 0 & b_{p-1} & & & & \iddots & & b_{p-2k+2} \\ b_{p-1} & & & & \iddots & & \iddots & b_{p-2k+1} \end{bmatrix},$$

(the first coefficients of Q may be zero).

We have $\det(\Delta) = a_p^{p-q-1}\, \mathrm{sRes}_{p-k}(P, Q)$. From the relations (9.6), we deduce $a_p^{p-q-1}\, \mathrm{sRes}_{p-k}(P, Q) = \det(D) \det(D')$ with

$$D = \begin{bmatrix} 1 & 0 & \cdots & \cdots & \cdots & \cdots & \cdots & \cdots & \cdots & 0 \\ 0 & 1 & 0 & & & & & & & \vdots \\ \vdots & \ddots & \ddots & \ddots & & & & & & \vdots \\ \vdots & & \ddots & 1 & 0 & \cdots & \cdots & \cdots & \cdots & 0 \\ \vdots & & & 0 & s_0 & s_1 & \cdots & \cdots & \cdots & s_{k-1} \\ \vdots & & \iddots & \iddots & \iddots & & & & \iddots & \vdots \\ \vdots & \iddots & \iddots & \iddots & & & & \iddots & & \vdots \\ 0 & s_0 & s_1 & & & & \iddots & & & s_{2k-3} \\ s_0 & s_1 & \cdots & \cdots & \cdots & s_{k-1} & & & \iddots & s_{2k-2} \end{bmatrix},$$

$$D' = \begin{bmatrix} a_p & \cdots & \cdots & \cdots & \cdots & \cdots & \cdots & \cdots & a_{p-2k+2} \\ 0 & a_p & \ddots & & & & & & \vdots \\ \vdots & \ddots & \ddots & \ddots & & & & & \vdots \\ 0 & & 0 & a_p & \ddots & & & & \vdots \\ 0 & & & 0 & a_p & \ddots & & & \vdots \\ \vdots & & & & \ddots & \ddots & \ddots & & \vdots \\ \vdots & & & & & \ddots & \ddots & \ddots & \vdots \\ \vdots & & & & & & \ddots & \ddots & \vdots \\ 0 & \cdots & \cdots & \cdots & \cdots & \cdots & \cdots & 0 & a_p \end{bmatrix}.$$

This implies that

$$a_p^{p-q-1}\, \mathrm{sRes}_{p-k}(P, Q) = a_p^{2k+1}\, \mathrm{han}(\bar{s}_k).$$

□

Proof of Proposition 9.25: a) It follows from Lemma 9.26 and Theorem 4.31 that

$$\mathrm{PmV}(\mathrm{han}(\bar{s}_{[0..n]})) = \mathrm{PmV}(\mathrm{sRes}(P,Q)) = \mathrm{Ind}(Q/P).$$

So it remains to prove that

$$\mathrm{Sign}(\overline{\mathrm{Han}}(\bar{s}_n)) = \mathrm{Ind}(Q/P).$$

The proof is by induction on the number of elements m of the euclidean remainder sequence of P and Q.

If $m=2$, then $Q=b$ is a constant, and the equality

$$\mathrm{Ind}(b/P) = \begin{cases} \mathrm{sign}(a_p b) & \text{if } p \text{ is odd,} \\ 0 & \text{if } p \text{ is even,} \end{cases}$$

is part of the proof of Theorem 4.31. The equality

$$\mathrm{Sign}(\overline{\mathrm{Han}}(s_n)) = \begin{cases} \mathrm{sign}(a_p b) & \text{if } p \text{ is odd,} \\ 0 & \text{if } p \text{ is even,} \end{cases}$$

follows from Lemma 9.22, since here $p \le n, C = P/b, c = p, T = 0, s_{p-1} = b/a_p$. Thus the theorem is true when $m=2$.

If $m > 2$, the theorem follows by induction from Lemma 9.4 and Lemma 9.22, since $\deg(C) = p - q$, $s_{p-q-1} = b_q a_p \neq 0$,

$$T = -R/Q = \sum_{j=0}^{\infty} t_j / X^{j+1},$$

and the signed remainder sequence of Q and $-R$ has $m-1$ elements.

b) It follows from Lemma 9.26 that

$$\mathrm{PmV}(\mathrm{han}(\bar{s}_{[0..p-j]}) = \mathrm{PmV}(\mathrm{sRes}(P,Q)_{[p..j]}).$$

So it remains to prove that

$$\mathrm{Sign}(\overline{\mathrm{Han}}(\bar{s}_n)) = \mathrm{PmV}(\mathrm{sRes}(P,Q)_{[p..j]}).$$

The proof is by induction on the number of elements m of non-zero elements in $\mathrm{sRes}(P,Q)_{[p..j]}$.

If $m=2$, then $n < p - q$,

$$\mathrm{PmV}(\mathrm{sRes}_p(P,Q), 0, \ldots, 0, \mathrm{sRes}_q(P,Q)) = \begin{cases} \mathrm{sign}(a_p b_q) & \text{if } p-q \text{ is odd,} \\ 0 & \text{if } p-q \text{ is even,} \end{cases}$$

according to the definitions of $\mathrm{sRes}_p(P,Q)$, $\mathrm{sRes}_q(P,Q)$ (see Notation 8.29), and D (see Notation 9.1). The equality

$$\mathrm{Sign}(\overline{\mathrm{Han}}(\bar{s}_n)) = \begin{cases} \mathrm{sign}(a_p b_q) & \text{if } p-q \text{ is odd,} \\ 0 & \text{if } p-q \text{ is even,} \end{cases}$$

is a particular case of Lemma 9.5 since $s_{p-q-1}=b_q/a_p\neq 0$. Thus the theorem is true when $n=2$.

If $m>2$, the theorem follows by induction from Lemma 9.4 and Lemma 9.22, since $\mathrm{sRes}(Q,-R)_{[q..j]}$ has $m-1$ non-zero elements by Proposition 8.35. □

Remark 9.27. a) Proposition 9.25 is a generalization of Theorem 4.57, taking for Q the remainder of $P'Q$ divided by P.

b) Note than given any

$$\bar{s}_n = s_0, \ldots, s_{m-1}, 0, \ldots, 0$$

such that $s_{m-1}\neq 0$, $P=T^m$ and $Q=s_0 T^{m-1}+\cdots+s_{m-1}$ satisfy the hypotheses of Proposition 9.25. Thus Proposition 9.25 provides a general method for computing the signature of a Hankel form.

c) Note also that when the Hankel matrix $\mathrm{Han}(\bar{s}_n)$ is invertible, $p=n$ and $\mathrm{Sign}(\overline{\mathrm{Han}}(\bar{s}_n))=\mathrm{PmV}(\mathrm{han}(\bar{s}_{[0..n]}))$. The signature of a quadratic form associated to an invertible Hankel matrix is thus determined by the signs of the principal minors of its associated matrix. This generalizes Remark 4.59.□

We are now ready to describe an algorithm computing the signature of a Hankel quadratic form. The complexity of this algorithm will turn out to be better than the complexity of the algorithm computing the signature of a general quadratic form (Algorithm 8.18 (Signature through Descartes)), because the special structure of a Hankel matrix is taken into account in the computation.

Algorithm 9.6. **[Signature of Hankel Form]**

- **Structure:** an ordered integral domain D.
- **Input:** $2n-1$ numbers $\bar{s}_n=s_0,\ldots,s_{2n-2}$ in D.
- **Output:** the signature of the Hankel quadratic form $\overline{\mathrm{Han}}(\bar{s}_n)$.
- **Complexity:** $O(n^2)$.
- **Procedure:**
  - If $s_i=0$ for every $i=0,\ldots,2n-2$, output 0.
  - If $s_i=0$ for every $i=0,\ldots,c-2$, $s_{c-1}\neq 0$, $c\geq n$, output

$$\begin{cases} \mathrm{sign}(s_{c-1}) & \text{if } c \text{ is odd,} \\ 0 & \text{if } c \text{ is even.} \end{cases}$$

  - Otherwise, let m, $0<m\leq 2n-2$, be such that $s_{m-1}\neq 0$, $s_i=0, i\geq m$.
    - If $m\leq n$, output

$$\begin{cases} \mathrm{sign}(s_{m-1}) & \text{if } m \text{ is odd,} \\ 0 & \text{if } m \text{ is even.} \end{cases}$$

    - If $m>n$, take $P:=T^m$, $Q:=s_0\,T^{m-1}+\cdots+s_{m-1}$ and apply a modified version of Algorithm 8.21 (Signed Subresultant) to P and Q stopping at the first non-defective $\mathrm{sResP}_j(P,Q)$ such that $j\leq m-n$. Compute $\mathrm{PmV}(\mathrm{sRes}(P,Q)_{[m..j]})$.

Proof of correctness: Use Lemma 9.5, Proposition 9.25 together with Remark 9.27 a). □

Complexity analysis: The complexity of this algorithms is $O(n^2)$, by the complexity analysis of Algorithm 8.21 (Signed Subresultant).

When s_0, ..., s_{2n-2} are in $\mathbb{Z}$, of bitsizes bounded by τ, the bitsizes of the integers in the operations performed by the algorithm are bounded by $O((\tau+\nu)\, n)$ where ν is the bitsize of n according to Proposition 8.44. □

Remark 9.28. It is possible to compute the signature of a Hankel matrix with complexity $\tilde{O}(n\,\tau)$ and binary complexity $\tilde{O}(n^2\,\tau)$ using Remark 9.2. □

9.3 Number of Complex Roots with Negative Real Part

So far Cauchy index was used only for the computation of Tarski-queries. We describe an important application of Cauchy index to the determination of the number of complex roots with negative real part of a polynomial with real coefficients.

Let $P(X) = a_p X^p + \cdots + a_0 \in \mathrm{R}[X]$, $a_p \neq 0$, where R is real closed, and $\mathrm{C} = \mathrm{R}[i]$ is algebraically closed. Define $F(X), G(X)$ by

$$P(X) = F(X^2) + X\,G(X^2). \tag{9.12}$$

Note that if $p = 2m$ is even

$$\begin{aligned} F &= a_{2m} X^m + a_{2m-2} X^{m-1} + \cdots, \\ G &= a_{2m-1} X^{m-1} + a_{2m-3} X^{m-2} + \cdots, \end{aligned}$$

and if $p = 2m+1$ is odd

$$\begin{aligned} F &= a_{2m} X^m + a_{2m-2} X^{m-1} + \cdots, \\ G &= a_{2m+1} X^m + a_{2m-1} X^{m-1} + \cdots. \end{aligned}$$

We are going to prove the following result.

Theorem 9.29. *Let $n(P)$ be the difference between the number of roots of P with positive real parts and the number of roots of P with negative real parts.*

$$n(P) = \begin{cases} -\mathrm{Ind}(G/F) + \mathrm{Ind}(X\,G/F) & \text{if } p \text{ is even,} \\ -\mathrm{Ind}(F/X\,G) + \mathrm{Ind}(F/G) & \text{if } p \text{ is odd.} \end{cases}$$

This result has useful consequences in control theory. When considering a linear differential equation with coefficients depending on parameters a_i,

$$a_p\, y^{(p)}(t) + a_{p-1}\, y^{(p-1)}(t) + \cdots + a_0\, y(t) = 0, a_p \neq 0, \tag{9.13}$$

it is important to determine for which values of the parameters all the roots of the characteristic equation

$$P = a_p X^p + a_{p-1} X^{p-1} + \cdots + a_0 = 0, a_p \neq 0, \tag{9.14}$$

have negative real parts. Indeed if $x_1, \ldots, x_r$ are the complex roots of P with respective multiplicities $\mu_1, \ldots, \mu_r$, the functions

$$e^{x_i t}, \ldots, t^{\mu_i - 1} e^{x_i t}, i = 1, \ldots, r$$

form a basis of solutions of Equation (9.13) and when all the x_i have negative real parts, all the solutions of Equation (9.13) tend to 0 as t tends to $+\infty$, for every possible initial value. This is the reason why the set of polynomials of degree p which have all their complex roots with negative real part is called the **domain of stability** of degree p.

We shall prove the following result, as a corollary of Theorem 9.29.

Theorem 9.30. [Liénard/Chipart] *The polynomial*

$$P = a_p X^p + \cdots + a_0, a_p > 0,$$

belongs to the domain of stability of degree p if and only if $a_i > 0$, $i = 0, \ldots, p$, and

$$\begin{cases} \mathrm{sRes}_m(F, G) > 0, \ldots, \mathrm{sRes}_0(F, G) > 0 & \text{if } p = 2m \text{ is even,} \\ \mathrm{sRes}_{m+1}(X G, F) > 0, \ldots, \mathrm{sRes}_0(X G, F) > 0 & \text{if } p = 2m + 1 \text{ is odd.} \end{cases}$$

As a consequence, we can decide whether or not P belongs to the domain of stability by testing the signs of some polynomial conditions in the a_i, without having to approximate the roots of P.

Exercise 9.3. Determine the conditions on a, b, c, d characterizing the polynomials $P = a X^3 + b X^2 + c X + d$, belonging to the domain of stability of degree 3.

The end of the section is devoted to the proof of Theorem 9.29 and Theorem 9.30.

Define $A(X)$, $B(X)$, $\deg(A) = p$, $\deg(B) < p$, as the real and imaginary parts of $(-i)^p P(i X)$. In other words,

$$\begin{aligned} A &= a_p X^p - a_{p-2} X^{p-2} + \cdots, \\ B &= -a_{p-1} X^{p-1} + a_{p-3} X^{p-3} + \cdots, \end{aligned}$$

so that when p is even A is even and B is odd (resp. when p is odd A is odd and B is even). Note that, using the definition of F and G (see (9.12)),

- if $p = 2m$,

$$A = (-1)^m F(-X^2), B = (-1)^m X G(-X^2). \tag{9.15}$$

– if $p=2m+1$,

$$A=(-1)^m X G(-X^2), B=-(-1)^m F(-X^2). \tag{9.16}$$

We are going to first prove the following result.

Theorem 9.31. [Cauchy] *Let $n(P)$ be the difference between the number of roots of P with positive real part and the number of roots of P with negative real part. Then,*

$$n(P)=\mathrm{Ind}(B/A).$$

A preliminary result on Cauchy index is useful.

Lemma 9.32. *Denote by $t \mapsto (A_t, B_t)$ a semi-algebraic and continuous map from $[0,1]$ to the set of pairs of polynomials (A,B) of $\mathrm{R}[X]$ with A monic of degree p, $\deg(B)<p$ (identifying pairs of polynomials with their coefficients). Suppose that A_0 has a root x of multiplicity μ in (a, b) and no other root in $[a,b]$, and B_0 has no root in $[a,b]$. Then, for t small enough,*

$$\mathrm{Ind}((B_0/A_0;a,b)=\mathrm{Ind}((B_t/A_t;a,b).$$

Proof: Using Theorem 5.12 (Continuity of roots), there are two cases to consider:

- If μ is odd, the number n of roots of A_t in $[a,b]$ with odd multiplicity is odd, and thus the sign of A_t changes n times while the sign of B_t is fixed, and hence for t small enough,

$$\mathrm{Ind}(B_0/A_0;a,b)=\mathrm{Ind}(B_t/A_t;a,b)=\mathrm{sign}(A_0^{(\mu)}(0)\, B_0(x)).$$

- If μ is even, the number of roots of A_t in $[a, b]$ with odd multiplicity is even, and thus for t small enough,

$$\mathrm{Ind}(B_0/A_0;a,b)=\mathrm{Ind}(B_t/A_t;a,b)=0.$$

□

Proof of Theorem 9.31: We can suppose without loss of generality that $P(0)\neq 0$.

If A and B have a common root $a+i\,b$, $a\in \mathrm{R}, b\in \mathrm{R}$, $b-i\,a$ is a root of P.

- If $b=0$, $i\,a$ and $-i\,a$ are roots of P, and $P=(X^2+a^2)Q$. Denoting

$$(-i)^{p-2}Q(i\,X)=C(X)+i\,D(X),\ \ C\in\mathrm{R}[X], D\in\mathrm{R}[X],$$

we have

$$\begin{aligned} A &= (X^2-a^2)\,C,\\ B &= (X^2-a^2)\,D. \end{aligned}$$

- If $b \neq 0$, $b + i\,a$, $b - i\,a$, $-b + i\,a$, $-b - i\,a$, are roots of P and

$$P = (X^4 + 2\,(a^2 - b^2)\,X^2 + (a^2 + b^2)^2)\,Q.$$

- Denoting

$$(-i)^{p-4}\,Q(i\,X) = C(X) + i\,D(X),\ \ C \in \mathrm{R}[X], D \in \mathrm{R}[X],$$

  we have

$$\begin{aligned} A &= (X^4 - 2\,(a^2 - b^2)\,X^2 + (a^2 + b^2)^2)\,C, \\ B &= (X^4 - 2\,(a^2 - b^2)\,X^2 + (a^2 + b^2)^2)\,D. \end{aligned}$$

In both cases $n(P) = n(Q)$, $\mathrm{Ind}(B/A) = \mathrm{Ind}(D/C)$.

So we can suppose without loss of generality that P has no two roots on the imaginary axis and no two roots with opposite real part, and A and B are coprime.

Let $x_1 = a_1 + i\,b_1$, ..., $x_r = a_r + i\,b_r$, be the roots of P with multiplicities μ_1, ..., μ_r, c be a positive number smaller than the difference between two distinct absolute values of a_i, M a positive number bigger than twice the absolute value of the b_i. Consider for $t \in [0,1]$, and $i = 1, \ldots, r$,

$$x_{i,t} = (1 - t)\,x_i + t(a_i + c/M\,b_i),$$

and the polynomial

$$P_t(X) = (X - x_{1,t})^{\mu_1} \cdots (X - x_{r,t})^{\mu_r}.$$

Note that $P_0 = P$, P_1 has only real roots, and for every $t \in [0,1]$ no two roots with opposite real parts, and hence for every $t \in [0,1]$, defining

$$(-i)^p\,P_t(i\,X) = A_t(X) + i\,B_t(X), A_t \in \mathrm{R}[X], B_t \in \mathrm{R}[X],$$

the polynomial A_t and B_t are coprime. Thus $\mathrm{Res}(A_t, B_t) \neq 0$ and by Proposition 9.10, denoting by $M(A_t, B_t)$ the matrix of coefficients of $\mathrm{Bez}(A_t, B_t)$ in the canonical basis X^{p-1}, ..., 1, $\det(M(A_t, B_t)) \neq 0$. Thus the rank of $M(A_t, B_t)$ is constantly p as t varies in $[0,1]$. Hence by Proposition 9.6 the signature of $M(A_t, B_t)$ is constant as t varies in $[0,1]$. We have proved that, for every $t \in [0,1]$, $\mathrm{Ind}(B_t/A_t) = \mathrm{Ind}(B/A)$.

So, we need only to prove the claim for a polynomial P with all roots real and no opposite roots. This is done by induction on the degree of P.

The claim is obvious for a polynomial of degree 1 since if $P = X - a$, $A = X$, and $B = a$, $\mathrm{Ind}(a/X)$ is equal to 1 when $a > 0$ and -1 when $a < 0$.

Suppose that the claim is true for every polynomial of degree $< p$ and consider P of degree p. Let a be the root of P with minimum absolute value among the roots of P and $P = (X - a)Q$.

If $a > 0$, we are going to prove, denoting

$$(-i)^{p-1}\,Q(i\,X) = C(X) + i\,D(X), C \in \mathrm{R}[X], D \in \mathrm{R}[X],$$

that

$$\mathrm{Ind}(B/A) = \mathrm{Ind}(D/C) + 1. \tag{9.17}$$

We define $P_t = (X - t)Q$, $t \in (0, a]$ and denote

$$(-i)^p P_t(i X) = A_t(X) + i B_t(X), A_t \in \mathrm{R}[X], B_t \in \mathrm{R}[X].$$

Note that $P_a = P$, and for every $t \in (0, a]$, P_t has only real roots, no opposite roots, and A_t and B_t are coprime. Thus $\mathrm{Res}(A_t, B_t) \neq 0$ and by Proposition 9.10, denoting by $M(A_t, B_t)$ be the matrix of coefficients of $\mathrm{Bez}(A_t, B_t)$ in the canonical basis $X^{p-1}, \ldots, 1$, $\det(M(A_t, B_t)) \neq 0$. Thus the rank of $M(A_t, B_t)$ is constantly p as t varies in $(0, a]$. Thus by Proposition 9.6 the signature of $M(A_t, B_t)$ is constant as t varies in $(0, a]$. We have proved that, for every $t \in (0, a]$,

$$\mathrm{Ind}(B_t/A_t) = \mathrm{Ind}(B/A). \tag{9.18}$$

We now prove that

$$\mathrm{Ind}(B_t/A_t) = \mathrm{Ind}(D/C) + 1, \tag{9.19}$$

if t is small enough. Note that

$$\begin{aligned} A_t(X) + i B_t(X) &= (-i)^p P_t(i X) \\ &= (X + i t)\,(-i)^{p-1} Q(i X) \\ &= (X + i t)(C(X) + i D(X)), \\ A_t(X) &= X C(X) - t D(X), \\ B_t(X) &= X D(X) + t C(X). \end{aligned}$$

For t small enough, A_t is close to $X C(X)$ and B_t close to $X D(X)$.

- If p is odd, $C(0) \neq 0$, $D(0) = 0$ since D is odd and C and D have no common root. For t small enough, using Theorem 5.12 (Continuity of roots), A_t has a simple root y close to 0. The sign of $B_t(y)$ is the sign of $t C(0)$. Hence for $[a, b]$ small enough containing 0, and t sufficiently small,

$$\mathrm{Ind}(D/C; a, b) = 0, \mathrm{Ind}(B_t/A_t; a, b) = 1.$$

- If p is even, $C(0) = 0$, $D(0) \neq 0$ since C is odd and C and D have no common root.
  - If $C'(0)\ D(0) > 0$, there is a jump from $-\infty$ to $+\infty$ in D/C at 0, and $A_t(0)$ has two roots close to 0, one positive and one negative. Hence for $[a, b]$ small enough containing 0, and t sufficiently small,

$$\mathrm{Ind}(D/C; a, b) = 1, \mathrm{Ind}(B_t/A_t; a, b) = 2.$$

  - If $C'(0)\ D(0) < 0$, there is a jump from $+\infty$ to $-\infty$ in D/C at 0 and $A_t(0)$ has no root close to 0. Hence for $[a, b]$ small enough containing 0, and t sufficiently small,

$$\mathrm{Ind}(D C; a, b) = -1, \mathrm{Ind}(B_t/A_t; a, b) = 0.$$

Using Lemma 9.32 at the neighborhood of non-zero roots of C, Equation (9.19) follows. Equation (9.17) follows from Equation (9.18) and Equation (9.19).

If $a<0$, a similar analysis, left to the reader, proves that

$$\mathrm{Ind}(B/A)=\mathrm{Ind}(D/C)-1.$$

□

Proof of Theorem 9.29:

– If $p=2m$, let

$$\varepsilon=\begin{cases}\mathrm{sign}_{x<0,x\to 0}(G(x)/F(x)) & \text{if } \lim_{x<0,x\to 0}|G(x)/F(x)|=\infty,\\ 0 & \text{otherwise.}\end{cases}$$

Then, since by (9.15) $A=(-1)^m F(-X^2), B=(-1)^m X G(-X^2)$,

$$\begin{aligned}\mathrm{Ind}(B/A) &= \mathrm{Ind}\big(XG(-X^2)/F(-X^2)\big)\\ &= \mathrm{Ind}\big(XG(-X^2)/F(-X^2);-\infty,0\big)\\ &\quad+\mathrm{Ind}\big(XG(-X^2)/F(-X^2);0,+\infty\big)+\varepsilon\\ &= 2\,\mathrm{Ind}\big(XG(-X^2)/F(-X^2);-\infty,0\big)+\varepsilon\\ &= -2\,\mathrm{Ind}\big(G(-X^2)/F(-X^2);-\infty,0\big)+\varepsilon\\ &= -2\,\mathrm{Ind}(G(X)/F(X);-\infty,0)-\varepsilon\\ &= -(G(X)/F(X);-\infty,0)+\mathrm{Ind}(XG(X)/F(X);-\infty,0)+\varepsilon\\ &= -\mathrm{Ind}(G/F)+\mathrm{Ind}(XG/F).\end{aligned}$$

– If $p=2m+1$, let

$$\varepsilon=\begin{cases}\mathrm{sign}_{x<0,x\to 0}(F(x)/G(x)) & \text{if } \lim_{x<0,x\to 0}(F(x)/G(x))\neq 0,\\ 0 & \text{otherwise.}\end{cases}$$

Then, since by (9.16) $A=(-1)^m XG(-X^2), B=-(-1)^m F(-X^2)$,

$$\begin{aligned}\mathrm{Ind}(B/A) &= -\mathrm{Ind}\big(F(-X^2)/XG(-X^2)\big)\\ &= -\mathrm{Ind}\big(F(-X^2)/XG(-X^2);-\infty,0\big)\\ &\quad-\mathrm{Ind}\big(F(-X^2)/XG(-X^2);0,+\infty\big)-\varepsilon\\ &= -2\,\mathrm{Ind}\big(F(-X^2)/XG(-X^2);-\infty,0\big)-\varepsilon\\ &= -2\,\mathrm{Ind}(F(X)/XG(X);-\infty,0)-\varepsilon\\ &= -\mathrm{Ind}(F(X)/XG(X);-\infty,0)\\ &\quad+\mathrm{Ind}(F(X)/G(X);-\infty,0)-\varepsilon\\ &= -\mathrm{Ind}(F/XG)+\mathrm{Ind}(F/G).\end{aligned}$$

This proves the theorem, using Theorem 9.31. □

Proof of Theorem 9.30: If

$$P=a_pX^p+\cdots+a_0,\ a_p>0$$

belongs to the domain of stability of degree p, it is the product of a_p, polynomials $X+u$ with $u>0\in \mathrm{R}$, and X^2+sX+t with $s>0\in \mathrm{R}, t>0\in \mathrm{R}$, and hence all the a_i, $i=0, \ldots, p$, are strictly positive. Thus, F and G have no positive real root, and $\mathrm{sign}(F(0)G(0))=\mathrm{sign}(a_0 a_1)=1$. Hence,

– If $p=2m$ is even,

$$\begin{aligned} \mathrm{Ind}(G/F) &= -\mathrm{Ind}(XG/F), \\ -p &= -\mathrm{Ind}(G/F)+\mathrm{Ind}(XG/F) \\ m &= \mathrm{Ind}(G/F), \end{aligned}$$

– If $p=2m+1$ is odd,

$$\begin{aligned} \mathrm{Ind}(F/XG) &= -\mathrm{Ind}(F/G)+1, \\ -p &= -\mathrm{Ind}(F/XG)+\mathrm{Ind}(F/G) \\ m+1 &= \mathrm{Ind}(F/XG). \end{aligned}$$

The proof of the theorem follows, using Theorem 4.31. □

9.4 Bibliographical Notes

The use of quadratic forms for studying the Cauchy index appears in [90]. Bezoutians were considered by Sylvester [153]. The signature of Hankel forms has been studied by Frobenius [61].

A survey of some results appearing in this chapter can be found in [99, 63]. However it seems that the link between the various approaches for computing the Cauchy index, using subresultants, is recent [70, 140].

The domain of stability has attracted much attention, notably by Routh [139], Hurwitz [93], and Liénart/Chipart [108].

10

Real Roots

In Section 10.1 we describe classical bounds on the roots of polynomials. In Section 10.2 we study real roots of univariate polynomials by a method based on Descartes's law of sign and Bernstein polynomials. These roots are characterized by intervals with rational endpoints. The method presented works only for archimedean real closed fields. In the second part of the chapter we study exact methods working in general real closed fields. Section 10.3 is devoted to exact sign determination in a real closed field and Section 10.4 to characterization of roots in a real closed field.

Besides their aesthetic interest, the specific methods of Section 10.2 are important in practical computations. This is the reason why we describe them, though they are less general than the methods of the second part of the chapter.

10.1 Bounds on Roots

We have already used a bound on the roots of a univariate polynomial in Chapter 5 (see Proposition 5.9). The following classical bound will also be useful.

In this section, we consider a polynomial

$$P = a_p X^p + \cdots + a_q X^q, p > q, a_q a_p \neq 0,$$

with coefficients in an ordered field K, a real closed field R containing K, and $C = R[i]$.

Notation 10.1. [Cauchy bound] We denote

$$\begin{aligned} C(P) &= \sum_{q \leq i \leq p} \left|\frac{a_i}{a_p}\right|, \\ c(P) &= \left(\sum_{q \leq i \leq p} \left|\frac{a_i}{a_q}\right|\right)^{-1}. \end{aligned}$$

□

Lemma 10.2. [Cauchy] *The absolute value of any root of P in* R *is smaller than $C(P)$.*

Proof: Let $x \in \mathrm{R}$ be a root of $P = a_p X^p + \cdots + a_q X^q, p > q$. Then

$$a_p x = - \sum_{q \leq i \leq p-1} a_i x^{i-p+1}.$$

If $|x| \geq 1$ this gives

$$\begin{aligned} |a_p| \, |x| &\leq \sum_{q \leq i \leq p-1} |a_i| \, |x|^{i-p+1} \\ &\leq \sum_{q \leq i \leq p-1} |a_i|. \end{aligned}$$

Thus it is clear that $|x| \leq C(P)$.

If $|x| \leq 1$, we have $|x| \leq 1 \leq C(P)$, since $C(P) \geq 1$. □

Similarly, we have

Lemma 10.3. *The absolute value of any non-zero root of P in* R *is bigger than $c(P)$.*

Proof: This follows from Lemma 10.2 by taking the reciprocal polynomial $X^p P(1/X)$. □

Corollary 10.4. *If $P \in \mathbb{Z}[X]$ had degree at most p, coefficients of bit length at most τ, and p has bitsize at most ν, the absolute values of the roots of P in* R *are bounded by $2^{\tau+\nu}$.*

Proof: Follows immediately from Lemma 10.2, since $p + 1 \leqslant 2^{\nu}$. □

The following proposition will be convenient when the polynomials we consider depend on parameters.

Notation 10.5. [Modified Cauchy bound] We denote

$$\begin{aligned} C'(P) &= (p+1) \cdot \sum_{q \leq i \leq p} \frac{a_i^2}{a_p^2}, \\ c'(P) &= \left((p+1) \cdot \sum_{q \leq i \leq p} \frac{a_i^2}{a_q^2} \right)^{-1}. \end{aligned}$$

□

Lemma 10.6. *The absolute value of any root of P in* R *is smaller than $C'(P)$.*

Proof: Let $x \in \mathrm{R}$ be a root of $P = a_p X^p + \cdots + a_q X^q, p > q$. Then

$$a_p x = - \sum_{q \leq i \leq p-1} a_i x^{i-p+1}.$$

If $|x| \geq 1$, this gives

$$\begin{aligned}(a_p x)^2 &= \left(\sum_{q \leq i \leq p-1} a_i x^{i-p-1}\right)^2 \\ &\leq (p+1)\left(\sum_{q \leq i \leq p-1} a_i^2\right).\end{aligned}$$

Thus $|x| \leq C'(P)$. If $|x| \leq 1$, we have $|x| \leq 1 \leq C'(P)$, since $C(P) \geq 1$. □

Lemma 10.7. *The absolute value of any non-zero root of P in* R *is bigger than $c'(P)$.*

Proof: This follows from Lemma 10.6 by taking the reciprocal polynomial $X^p P(1/X)$. □

Our next aim is to give a bound on the divisors of a polynomial with integer coefficients.

We are going to use the following notions. If

$$P = a_p X^p + \cdots + a_0 \in \mathrm{C}[X], a_p \neq 0,$$

the **norm of** P is

$$\|P\| = \sqrt{|a_p|^2 + \cdots + |a_0|^2}.$$

The **length of** P is

$$\mathrm{Len}(P) = |a_p| + \cdots + |a_0|.$$

If $z_1, \ldots, z_p$ are the roots of P in C counted with multiplicity so that

$$P = a_p \prod_{i=1}^{p} (X - z_i), \tag{10.1}$$

the **measure** of P is

$$\mathrm{Mea}(P) = |a_p| \prod_{i=1}^{p} \max(1, |z_i|).$$

These three quantities are related as follows

Proposition 10.8.

$$\mathrm{Len}(P) \leq 2^p \, \mathrm{Mea}(P).$$

Proposition 10.9.

$$\mathrm{Mea}(P) \leq \|P\|.$$

Proof of Proposition 10.8: By Lemma 2.12,

$$a_{p-k} = (-1)^k \left(\sum_{1 \leq i_1 < \cdots < i_k \leq p} z_{i_1} \ldots z_{i_k}\right) a_p.$$

Thus $|a_{p-k}| \leq \binom{p}{k} \operatorname{Mea}(P)$, and

$$\operatorname{Len}(P) \leq \sum_{k=0}^{p} \binom{p}{k} \operatorname{Mea}(P) = 2^p \operatorname{Mea}(P). \qquad \square$$

The proof of Proposition 10.9 relies on the following lemma.

Lemma 10.10. *If $P \in \mathrm{C}[X]$ and $\alpha \in \mathrm{C}$, then*

$$\|(X-\alpha)\, P(X)\| = \|(\bar{\alpha} X - 1)\, P(X)\|.$$

Proof: We have

$$\begin{aligned} \|(X-\alpha)\, P(X)\|^2 &= \sum_{j=0}^{p+1} (a_{j-1} - \alpha\, a_j)\, (\bar{a}_{j-1} - \bar{\alpha}\bar{a}_j) \\ &= (1 + |\alpha|^2)\, \|P\|^2 - \sum_{j=0}^{p} (\alpha a_j \bar{a}_{j-1} + \bar{\alpha}\, \bar{a}_j\, a_{j-1}), \end{aligned}$$

where $a_{-1} = a_{p+1} = 0$, since

$$(a_{j-1} - \alpha a_j)\, (\bar{a}_{j-1} - \bar{\alpha}\bar{a}_j) = |a_{j-1}|^2 + |\alpha|^2\, |a_j|^2 - (\alpha a_j \bar{a}_{j-1} + \bar{\alpha}\bar{a}_j\, a_{j-1}).$$

Similarly

$$\begin{aligned} \|(\bar{\alpha} X - 1)\, P(X)\|^2 &= \sum_{j=0}^{p+1} (\bar{\alpha} a_{j-1} - a_j)\, (\alpha \bar{a}_{j-1} - \bar{a}_j) \\ &= (1 + |\alpha|^2)\, \|P\|^2 - \sum_{j=0}^{p} (\alpha a_j \bar{a}_{j-1} + \bar{\alpha}\, \bar{a}_j a_{j-1}). \end{aligned}$$

$\square$

Proof of Proposition 10.9: Let $z_1, \ldots, z_k$ be the roots of P outside of the unit disk. Then, by definition, $M(P) = |a_p| \prod_{i=1}^{k} |z_i|$. We consider the polynomial

$$R = a_p \prod_{i=1}^{k} (\bar{z}_i X - 1) \prod_{i=k+1}^{n} (X - z_i) = b_p X^p + \cdots + b_0.$$

Noting that $|b_p| = \operatorname{Mea}(P)$, and applying Lemma 10.10 k times, we obtain $\|P\| = \|R\|$. Since $\|P\|^2 = \|R\|^2 \geq |b_p|^2 = \operatorname{Mea}(P)^2$, the claim is proved. $\square$

Proposition 10.11. *If $P \in \mathbb{Z}[X]$ and $Q \in \mathbb{Z}[X]$, $\deg(Q) = q$, Q divides P, then*

$$\begin{aligned} \operatorname{Mea}(Q) &\leq \operatorname{Mea}(P), \\ \operatorname{Len}(Q) &\leq 2^q\, \|P\|. \end{aligned}$$

Proof: Since the leading coefficient of Q divides the leading coefficient of P and every root of Q is a root of P, it is clear that $\mathrm{Mea}(Q) \le \mathrm{Mea}(P)$. The other part of the claim follows using Proposition 10.9 and Proposition 10.8. □

Corollary 10.12. *If $P \in \mathbb{Z}[X]$ and $Q \in \mathbb{Z}[X]$ divides P in $\mathbb{Z}[X]$, then the bitsize of any coefficient of Q is bounded by $q + \tau + \nu$, where τ is a bound on the bitsizes of the coefficients of P and ν is the bitsize of $p + 1$.*

Proof: Notice that $\|P\| < (p + 1)\, 2^{\tau}$, $2^{\tau-1} \le \mathrm{Len}(Q)$, and apply Proposition 10.11. □

The preceding bound can be used to estimate the bitsizes of the coefficients of the separable part of a polynomial. The **separable part of** P is a separable polynomial with the same set of roots as P in C. The separable part of P is unique up to a multiplicative constant.

Lemma 10.13. *The polynomial $P/\gcd(P, P')$ is the separable part of P.*

Proof: Decompose P as a product of linear factors over C:

$$P = (X - z_1)^{\mu_1} \cdots (X - z_r)^{\mu_r},$$

with $z_1, \ldots, z_r$ distinct. Then since $z_1, \ldots, z_r$ are roots of P' of multiplicities $\mu_1 - 1, \cdots, \mu_r - 1$,

$$\begin{aligned} \gcd(P, P') &= (X - z_1)^{\mu_1 - 1} \cdots (X - z_r)^{\mu_r - 1}, \\ P/\gcd(P, P') &= (X - z_1) \cdots (X - z_r). \end{aligned}$$

is separable. □

More generally, it is convenient to consider the **gcd-free part of** P with respect to Q, which is the divisor D of P such that $DQ = \mathrm{lcm}(P, Q)$. It is clear that $D = P/\gcd(P, Q)$. The gcd-free part of P with respect to Q is unique up to a multiplicative constant.

The greatest common divisor of P and Q and the gcd-free part of P with respect to Q can be computed using Algorithm 8.22 (Extended Signed Subresultant).

Proposition 10.14. *If $\deg(\gcd(P, Q)) = j$, then $\mathrm{sResP}_j(P, Q)$ is the greatest common divisor of P and Q and $\mathrm{sResV}_{j-1}(P, Q)$ is the gcd-free part of P with respect to Q.*

Proof: Let j be the degree of $\gcd(P,\ Q)$. According to Theorem 8.30, $\mathrm{sResP}_j(P,\ Q)$ is a greatest common divisor of P and Q. Moreover, Theorem 8.30 implies that $\mathrm{sResP}_{j-1}(P, Q) = 0$. Since, by Proposition 8.38,

$$\mathrm{sResU}_{j-1}(P, Q)\, P + \mathrm{sResV}_{j-1}(P, Q)\, Q = \mathrm{sResP}_{j-1}(P, Q) = 0,$$

then $\mathrm{sResU}_{j-1}(P, Q)\ P = -\mathrm{sResV}_{j-1}(P, Q)\ Q$ is a multiple of the least common multiple of P and Q and is of degree $\geq p+q-j$. On the other hand by Proposition 8.38 a),

$$\deg(\mathrm{sResU}_{j-1}(P,Q)) \leq q-j,$$
$$\deg(\mathrm{sResV}_{j-1}(P,Q)) \leq p-j.$$

It follows immediately that $\mathrm{sResU}_{j-1}(P,Q)$ is proportional to $Q/\gcd(P,Q)$ and $\mathrm{sResV}_{j-1}(P,Q)$ is proportional to $P/\gcd(P,Q)$. □

Corollary 10.15. *If* $\deg(\gcd(P,\ P')) = j$, $\mathrm{sResP}_j(P,\ P')$ *is the greatest common divisor of* P *and* P' *and* $\mathrm{sResV}_{j-1}(P,P')$ *is the separable part of* P.

According to the preceding results, we are going to compute the gcd and gcd-free part using Algorithm 8.22 (Extended Signed Subresultants). In the case of integer coefficients, it will be possible to improve slightly the algorithm, using the following definitions and results.

If $P \in \mathbb{Z}[X]$, denote by $\mathrm{cont}(P)$ the **content** of P, which is the greatest common divisor of the coefficients of P.

Lemma 10.16. *Let* $P_1 \in \mathbb{Z}[X]$, $P_2 \in \mathbb{Z}[X]$. *If* $\mathrm{cont}(P_1) = \mathrm{cont}(P_2) = 1$, *then* $\mathrm{cont}(P_1\, P_2) = 1$.

Proof: Consider a prime number p. Reducing the coefficients of P_1 and P_2 modulo p, notice that if P_1 and P_2 are not zero modulo p, $P_1 P_2 = P$ is also not zero modulo p. Thus $\mathrm{cont}(P)$ is divisible by no prime number p, and hence is equal to 1. □

Lemma 10.17. *If* $P \in \mathbb{Z}[X], P = P_1\, P_2, P_1 \in \mathbb{Q}[X]$ *and* $P_2 \in \mathbb{Q}[X]$ *there exist* $\bar{P}_1$ *and* $\bar{P}_2$ *in* $\mathbb{Z}[X]$, *proportional to* P_1 *and* P_2 *resp., such that* $\bar{P}_1\, \bar{P}_2 = P$.

Proof: We can easily find $\bar{P}_1 \in \mathbb{Z}[X]$ proportional to P_1 such that $\mathrm{cont}(\bar{P}_1) = 1$. Let c be the least common multiple of the denominators of the coefficients of $\bar{P}_2 = P/\bar{P}_1$. Then $c\,\bar{P}_2 \in \mathbb{Z}[X]$ and $\mathrm{cont}(\bar{P}_2) = d$ is prime to c. Consider $\bar{P}_1$ and $c\,\bar{P}_2/d$, which belong to $\mathbb{Z}[X]$. Both of these polynomials have content equal to 1 and hence $\mathrm{cont}(c\,P/d) = 1$, by Lemma 10.16. Since c and d are coprime, $c = 1, \mathrm{cont}(P) = d$. Hence $\bar{P}_2 \in \mathbb{Z}[X]$. □

Algorithm 10.1. **[Gcd and Gcd-free Part]**

- **Structure:** an integral domain D.
- **Input:** two univariate polynomials $P = a_p\ X^p + \cdots + a_0$ and $Q = b_q X^q + \cdots + b_0$ with coefficients D and of respective degrees p and q, $p > q$.
- **Output:** the greatest common divisor of P and Q and the gcd-free part of P with respect to Q.
- **Complexity:** $O(p^2)$.

- **Procedure:**
  - Run Algorithm 8.22 (Extended Signed Subresultants) with inputs P and Q. Let $j = \deg(\gcd(P, Q))$.
  - If $\mathrm{D} = \mathbb{Z}$, output $a_p \mathrm{sResP}_j / \mathrm{sRes}_j$, $a_p \mathrm{sResV}_{j-1} \mathrm{lcof}(\mathrm{sResV}_{j-1})$.
  - Otherwise, output sResP_j, sResV_{j-1}.

Proof of correctness of Algorithm 10.1: The correctness of the algorithm when $\mathrm{D} \neq \mathbb{Z}$ follows from the correctness of Algorithm 8.22 (Extended Signed Subresultants) and Corollary 10.15.

When $\mathrm{D} = \mathbb{Z}$, Lemma 10.17 implies that there exists a in $\mathbb{Z}$ with a dividing a_p such that $a\, \mathrm{sResP}_j / \mathrm{sRes}_j \in \mathbb{Z}[X]$ and there exists b in $\mathbb{Z}$ with b dividing a_p such that $b\, S V_{j-1} / \mathrm{lcof}(\mathrm{sResV}_{j-1}) \in \mathbb{Z}[X]$. Thus

$$a_p \mathrm{sResP}_j / \mathrm{sRes}_j \in \mathbb{Z}[X],$$
$$a_p \mathrm{sResV}_{j-1} / \mathrm{lcof}(\mathrm{sResV}_{j-1}) \in \mathbb{Z}[X]$$

□

Complexity analysis: The complexity is $O(p^2)$, using the complexity analysis of Algorithm 8.22 (Extended Signed Subresultants).

When $P \in \mathbb{Z}[X]$, with the bitsizes of its coefficients bounded by τ, the bitsizes of the integers in the operations performed by the algorithm are bounded by $O(\tau\, p)$ according to Proposition 8.44. Moreover using Corollary 10.12 the bitsize of the output is $j + \tau + \nu$ and $p - j + \tau + \nu$ with ν the bitsize of $p + 1$. Note that the bitsize produced by the subresultant algorithm would be $(p + q - 2j)\, (\tau + \mu)$ with μ the bitsize of $p + q$, so the normalization step at the end of the algorithm when $\mathrm{D} = \mathbb{Z}$ improves the bounds on the bitsizes of the result. □

Remark 10.18. Algorithm 10.1 (Gcd and Gcd-Free part) is based on the Algorithm 8.22 (Extended Signed Subresultants) which uses exact divisions and is valid only in an integral domain, and not in a general ring. In a ring, the algorithm for computing determinants indicated in Remark 8.19 can always be used for computing the signed subresultant coefficients, and hence the separable and the gcd-free part. The complexity is $(p q)^{O(1)}$ arithmetic operations in the ring D of coefficients of P and Q, which is sufficient for the complexity estimates proved in later chapters. □

Remark 10.19. The computation of the gcd and gcd free-part can be performed with complexity $\tilde{O}(d)$ using [145, 119], and with binary complexity $\tilde{O}(d^2\, \tau)$, using [107]. □

Now we study the minimal distance between roots of a polynomial P.

If $P = a_p \prod_{i=1}^{p} (X - z_i) \in \mathrm{C}[Z]$, the **minimal distance between the roots of** P is

$$\mathrm{sep}(P) = \min \{|z_i - z_j|, z_i \neq z_j\}.$$

We denote by $\mathrm{Disc}(P)$ the discriminant of P (see Notation 4.1).

Proposition 10.20.

$$\begin{aligned}\mathrm{sep}(P) &\geq (p/\sqrt{3})^{-1}\, p^{-p/2}|\mathrm{Disc}(P)|^{1/2}\,\mathrm{Mea}(P)^{1-p}\\ &\geq (p/\sqrt{3})^{-1}\, p^{-p/2}|\mathrm{Disc}(P)|^{1/2}\|P\|^{1-p}.\end{aligned}$$

Proof: Consider the Vandermonde matrix

$$V(z_1, \ldots, z_p) = \begin{bmatrix} 1 & 1 & \cdots & 1 \\ z_1 & z_2 & \cdots & z_p \\ \vdots & \vdots & & \vdots \\ z_1^{p-1} & z_2^{p-1} & \cdots & z_p^{p-1} \end{bmatrix}.$$

We know by Equation (4.4) that $\mathrm{Disc}(P) = a_p^{2p-2} \det(V(z_1, \ldots, z_p))^2$. We suppose without loss of generality that $|z_1 - z_2| = \mathrm{sep}(P)$ and $|z_1| \geq |z_2|$. Using Hadamard's inequality (Proposition 8.9) on

$$V' = \begin{bmatrix} 0 & 1 & \cdots & 1 \\ z_1 - z_2 & z_2 & \cdots & z_p \\ \vdots & \vdots & & \vdots \\ z_1^{p-1} - z_2^{p-1} & z_2^{p-1} & \cdots & z_p^{p-1} \end{bmatrix},$$

and noticing that $\det(V(z_1, \ldots, z_p)) = \det(V')$, we get

$$|\mathrm{Disc}(P)|^{1/2} \leq |a_p|^{p-1} \left(\sum_{j=0}^{p} |z_1^j - z_2^j|^2 \right)^{1/2} \prod_{i \neq 1} (1 + |z_i|^2 + \cdots + |z_i|^{2(p-1)})^{1/2}.$$

Now,

$$\begin{aligned}\prod_{i \neq 1} (1 + |z_i|^2 + \cdots + |z_i|^{2(p-1)})^{1/2} &\leq \prod_{i \neq 1} (p \max(1, |z_i|)^{2(p-1)})^{1/2}\\ &\leq p^{(p-1)/2} \left(\frac{\mathrm{Mea}(P)}{|a_p| \max(1, |z_1|)} \right)^{p-1}.\end{aligned}$$

On the other hand since it is clear that $\sum_{j=0}^{p-1} j^2 \leq p^3/3$,

$$\begin{aligned}|z_1^j - z_2^j| &\leq j\, |z_1 - z_2| \max(1, |z_1|)^{p-1},\\ \sum_{j=0}^{p-1} |z_1^j - z_2^j|^2 &\leq \left(\sum_{j=0}^{p-1} j^2 \right) |z_1 - z_2|^2 \max(1, |z_1|)^{2p-2},\\ &\leq (p^3/3) |z_1 - z_2|^2 \max(1, |z_1|)^{2p-2},\\ \left(\sum_{j=0}^{p-1} |z_1^j - z_2^j|^2 \right)^{1/2} &\leq (p^{3/2}/\sqrt{3}) |z_1 - z_2| \max(1, |z_1|)^{p-1}.\end{aligned}$$

Finally

$$|\mathrm{Disc}(P)|^{1/2} \leq \mathrm{sep}(P)\, (p/\sqrt{3}) p^{p/2}\, \mathrm{Mea}(P)^{p-1}.$$

□

Proposition 10.21. *If* $P\in\mathbb{Z}[X]$,

$$\mathrm{sep}(P)\geq(p/\sqrt{3})^{-1}\,p^{-p/2}\mathrm{Mea}(P)^{1-p}\geq(p/\sqrt{3})^{-1}\,p^{-p/2}\|P\|^{1-p}.$$

Proof: If P is separable, $\mathrm{Disc}(P)$ is a non-zero integer, by Proposition 4.2 and Remark 4.28. Hence $|\mathrm{Disc}(P)|\geq 1$ and the claim follows by Proposition 10.20.

If P is not separable, its separable part Q divides P and belongs to $\mathbb{Z}[X]$. Thus by Proposition 10.11, $\mathrm{Mea}(Q)\leq\mathrm{Mea}(P)$. The conclusion follows, using Proposition 10.20 for Q and $|\mathrm{Disc}(Q)|\geq 1$. □

Corollary 10.22. *If P is of degree at most p with coefficients in $\mathbb{Z}$ of bitsize bounded by τ*

$$\mathrm{sep}(P)\geq(p/\sqrt{3})^{-1}\,p^{-p/2}(p+1)^{(1-p)/2}2^{\tau(1-p)}.$$

Proof: It is clear that $\|P\|\leq(p+1)^{1/2}2^{\tau}$. □

Proposition 10.20 can be generalized as follows.

Proposition 10.23. *Let $Z=\mathrm{Zer}(P,\mathrm{C})$ and $G=(Z,E)$ a directed acyclic graph such that*

a) for all $(z_i,z_j)\in E$, $|z_i|\leqslant|z_j|$,
b) the in-degree of G is at most 1, i.e. for every $z_j\in Z$, there is at most one element z_i of Z such that $(z_i,z_j)\in E$.

Then, denoting by m the number of elements of E,

$$\begin{aligned}\prod_{(z_i,z_j)\in E}|z_i-z_j| &\geq (p/\sqrt{3})^{-m}\,p^{-p/2}|\mathrm{Disc}(P)|^{1/2}\mathrm{Mea}(P)^{1-p}\\ &\geq (p/\sqrt{3})^{-m}p^{-p/2}|\mathrm{Disc}(P)|^{1/2}\|P\|^{1-p}.\end{aligned}$$

Proof: We can suppose without loss of generality that $(z_i,z_j)\in E$ implies $j<i$. Consider, as in the proof of Proposition 10.20, the Vandermonde matrix $V(z_1,\ldots,z_p)$.

Denote by A the subset of $\{1,\ldots,p\}$ consisting of elements j such that there exists i (necessarily greater than j) such that $(z_i,z_j)\in E$, and note that by condition (b), the number of elements of A is m. For $j\in A$ in increasing order, replace the j-th column of $V(z_1,\ldots,z_p)$ by the difference between the j-th and i-th column and denote by V_G the corresponding matrix. Because of condition (b), $\det(V(z_1,\ldots,z_p))=\det(V_G)$.

Using Hadamard's inequality (Proposition 8.9) on V_G and using the fact that, by Equation (4.4), $\mathrm{Disc}(P)=a_p^{2p-2}\det(V(z_1,\ldots,z_p))^2$

$$\begin{aligned}&\mathrm{Disc}(P)|^{1/2}\\ \leq\ &|a_p|^{p-1}\prod_{(z_i,z_j)\in E}\left(\sum_{k=0}^{p}|z_i^k-z_j^k|^2\right)^{1/2}\prod_{j\notin A}(1+|z_j|^2+\cdots+|z_j|^{2(p-1)})^{1/2}.\end{aligned}$$

Now,

$$\begin{aligned}\prod_{j\notin A}(1+|z_j|^2+\cdots+|z_j|^{2(p-1)})^{1/2} &\le \prod_{j\notin A}(p\max(1,|z_j|)^{2(p-1)})^{1/2}\\ &\le p^{(p-m)/2}\left(\frac{\mathrm{Mea}(P)}{|a_p|\prod_{j\in A}\max(1,|z_j|)}\right)^{p-1}.\end{aligned}$$

On the other hand since it is clear that $\sum_{k=0}^{p-1} k^2\le p^3/3$,

$$\begin{aligned}|z_i^k-z_j^k| &\le j\,|z_i-z_j|\max(1,|z_j|)^{p-1},\\ \sum_{k=0}^{p-1}|z_i^k-z_j^k|^2 &\le \left(\sum_{k=0}^{p-1}k^2\right)|z_i-z_j|^2\max(1,|z_j|)^{2p-2},\\ &\le (p^3/3)|z_i-z_j|^2\max(1,|z_j|)^{2p-2},\\ \left(\sum_{k=0}^{p-1}|z_i^k-z_j^k|^2\right)^{1/2} &\le (p^{3/2}/\sqrt{3})\,|z_i-z_j|\max(1,|z_j|)^{p-1},\\ \prod_{(z_i,z_j)\in E}\left(\sum_{k=0}^{p-1}|z_i^k-z_j^k|^2\right)^{1/2} &\le (p^{3/2}/\sqrt{3})^m\,B,\end{aligned}$$

with $B=\prod_{(z_i,z_j)\in E}|z_i-z_j|\prod_{j\in A}\max(1,|z_j|)^{p-1}$.

Finally

$$|\mathrm{Disc}(P)|^{1/2}\le\prod_{(z_i,z_j)\in E}|z_i-z_j|\,(p/\sqrt{3})^m p^{p/2}\,\mathrm{Mea}(P)^{p-1}.$$ □

Corollary 10.24. *Let P be of degree at most p with coefficients in $\mathbb{Z}$ of bitsize bounded by τ. Let Z=Zer(P,C) and $G=(Z,E)$ a directed acyclic graph such that*

a) for all $(z_i,z_j)\in E$, $|z_i|\leqslant|z_j|$,
b) the in-degree of G is at most 1.

Then, denoting by m the number of elements of E,

$$\prod_{(z_i,z_j)\in E}|z_i-z_j|\ge(p/\sqrt{3})^{-m}\,p^{-p/2}\,(p+1)^{(1-p)/2}\,2^{\tau(1-p)}.$$

Proof: It is clear that $\|P\|\le(p+1)^{1/2}\,2^{\tau}$. □

10.2 Isolating Real Roots

Throughout this section, R is an archimedean real closed field. Let P be a polynomial of degree p in R[X]. We are going to explain how to perform exact computations for determining several properties of the roots of P in R: characterization of a root, sign of another polynomial at a root, and comparisons between roots of two polynomials.

The characterization of the roots of P in R will be performed by finding intervals with rational end points. Our method will be based on Descartes's law of signs (Theorem 2.33) and the properties of the Bernstein basis defined below.

Notation 10.25. [Bernstein polynomials] The **Bernstein polynomials** of degree p for ℓ, r are

$$\mathrm{Bern}_{p,i}(\ell,r)=\binom{p}{i}\frac{(X-\ell)^i\,(r-X)^{p-i}}{(r-\ell)^p},$$

for $i=0,\ldots,p$. □

Remark 10.26. Note that $\mathrm{Bern}_{p,i}(\ell,r)=(-1)^p\,\mathrm{Bern}_{p,p-i}(r,\ell)$ and that

$$\begin{aligned}\mathrm{Bern}_{p,i}(\ell,r) &= \frac{(X-\ell)}{r-\ell}\frac{p}{i}\mathrm{Bern}_{p-1,i-1}(\ell,r)\\ &= \frac{(r-X)}{r-\ell}\frac{p}{p-i}\mathrm{Bern}_{p-1,i}(\ell,r).\end{aligned}$$

□

In order to prove that the Bernstein polynomials form a basis of polynomials of degree $\le p$, we are going to need three simple transformations of P.

- **Reciprocal polynomial in degree p:**

$$\mathrm{Rec}_p(P(X))=X^p\,P(1/X).$$

The non-zero roots of P are the inverses of the non-zero roots of $\mathrm{Rec}(P)$.

- **Contraction by ratio λ:** for every non-zero λ,

$$\mathrm{Co}_\lambda(P(X))=P(\lambda X).$$

The roots of $\mathrm{Co}_\lambda(P)$ are of the form x/λ, where x is a root of P.

- **Translation by c:** for every c,

$$\mathrm{T}_c(P(X))=P(X-c).$$

The roots of $\mathrm{T}_c(P(X))$ are of the form $x+c$ where x is a root of P.

These three transformations clearly define bijections from the set of polynomials of degree at most p into itself.

Proposition 10.27. *Let $P=\sum_{i=0}^{p} b_i\,\mathrm{Bern}_{p,i}(\ell,r)\in \mathrm{R}[X]$ be of degree $\le p$. Let $\mathrm{T}_{-1}(\mathrm{Rec}_p(\mathrm{Co}_{r-\ell}(\mathrm{T}_{-\ell}(P))))=\sum_{i=0}^{p} c_i X^i$. Then $\binom{p}{i} b_i=c_{p-i}$.*

Proof: Performing the contraction of ratio $r-\ell$ after translating by $-\ell$ transforms $\binom{p}{i}$ $(X-\ell)^i$ $(r-X)^{p-i}/(r-\ell)^p$ into $\binom{p}{i}$ X^i $(1-X)^{p-i}$. Translating by -1 after taking the reciprocal polynomial in degree p transforms $\binom{p}{i} X^i(1-X)^{p-i}$ into $\binom{p}{i} X^{p-i}$. □

Remark 10.28. Proposition 10.27 immediately provides an algorithm for converting a polynomial from the monomial basis to the Bernstein basis for ℓ, r.□

Corollary 10.29. *The Bernstein polynomials for c, d form a basis of the vector space of polynomials of degree* $\leq p$.

Corollary 10.30. *Let* $P \in \mathbb{Z}[X]$ *be of degree* $\leq p$. *If the bitsizes of the coefficients of P are bounded by* τ *in the monomial basis* $1, X, \ldots, X^{p-1}$ *and the bitsizes of the rational numbers r and* ℓ *are bounded by* τ', *then there exists* $\lambda(P, \ell, r) \in \mathbb{Z}$ *such that the bitsizes of the coefficients of* $\lambda(P, \ell, r)\, P$ *in the Bernstein basis for* (r, ℓ) *are integers of bitsize bounded by* $O(\tau + p\,\tau' + p\log_2(p))$.

Proof: Let $\ell = a/b, r - \ell = a'/b'$, with a, b, a', b' in $\mathbb{Z}$. Consider

$$p!\,\mathrm{T}_{-1}(\mathrm{Rec}_p(b'^p\,\mathrm{Co}_{a'/b'}(b^p\,\mathrm{T}_{-c}(P)))) = \sum_{i=0}^{p} d_i X^i.$$

It is clear that d_i is an integer multiple of $p!$. Thus the quotient $\bar{b_i}$ of d_i by $\binom{p}{i}$ is an integer, and we obtained $\lambda(P)\,P$ with integer coefficients in the Bernstein basis of (c, d). The claim follows immediately from Proposition 10.27 and the complexity analysis of Algorithm 8.10 (Special translation), noting that the bitsize of $p!$ is $O(p\log_2(p))$ by Stirling's formula. □

Remark 10.31. The list $b = b_0, \ldots, b_p$ of coefficients of P in the Bernstein basis of r, ℓ gives the value of P at ℓ (resp. r), which is equal to b_0 (resp. b_p). Moreover, the sign of P at the right of ℓ (resp. left of r) is given by the first non-zero element (resp. last non-zero element) of the list b. □

We denote as usual by $\mathrm{Var}(b)$ the number of sign variations in a list b.

Note that, if $\mathrm{Var}(b) = 0$, where $b = b_0, \ldots, b_p$ is the list of coefficients of P in the Bernstein basis of ℓ, r, the sign of P on (c, d) is the sign of any non zero element of b, since the Bernstein polynomials for ℓ, r are positive on (ℓ, r), thus P has no root in (ℓ, r). More generally, we have:

Proposition 10.32. *Let P be of degree p. We denote by* $b = b_0, \ldots, b_p$ *the coefficients of P in the Bernstein basis of* ℓ, r. *Let* $\mathrm{num}(P; (\ell, r))$ *be the number of roots of P in* (ℓ, r) *counted with multiplicities. Then*

- $\mathrm{Var}(b) \geq \mathrm{num}(P; (\ell, r))$,
- $\mathrm{Var}(b) - \mathrm{num}(P; (\ell, r))$ *is even.*

Proof: The claim follows immediately from Descartes's law of signs (Theorem 2.33), using Proposition 10.27. Indeed, the image of (ℓ, r) under translation by $-\ell$ followed by contraction of ratio $r - \ell$ is $(0, 1)$. The image of $(0, 1)$ under the inversion $z \mapsto 1/z$ is $(1, +\infty)$. Finally, translating by -1 gives $(0, +\infty)$. □

The coefficients $b = b_0, \ldots, b_p$ of P in the Bernstein basis of ℓ, r give a rough idea of the shape of the polynomial P on the interval $[\ell, r]$. The **control line of P on** $[\ell, r]$ is the union of the segments $[M_i, M_{i+1}]$ for $i = 0, \ldots, p-1$, with

$$M_i = \left(\frac{i\,r + (p-i)\,\ell}{p}, b_i \right).$$

It is clear from the definitions that the graph of P goes through M_0 and M_p and that the line M_0, M_1 (resp M_{p-1}, M_p) is tangent to the graph of P at M_0 (resp. M_p).

Example 10.33. We take $p = 3$, and consider the polynomial

$$P = -33\,X^3 + 69\,X^2 - 30\,X + 4$$

with coefficients $(4, -6, 7, 10)$ in the Bernstein basis for $0, 1$

$$(1-X)^3, 3\,X\,(1-X)^2, 3\,X^2\,(1-X), X^3.$$

In Figure 10.1 we depict the graph of P on $[0, 1]$, the control line, and the X-axis.

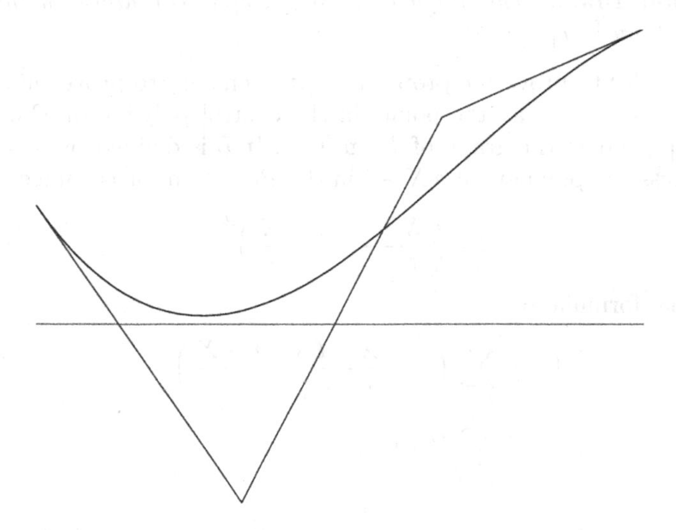

Fig. 10.1. Graph of P and control line of P on $[0, 1]$

□

The **control polygon of P on** $[\ell, r]$ is the convex hull of the points M_i for $i = 0, \ldots, p$.

Example 10.34. Continuing Example 10.33, we draw the graph of P on $[0, 1]$ and the control polygon (see Figure 10.2).

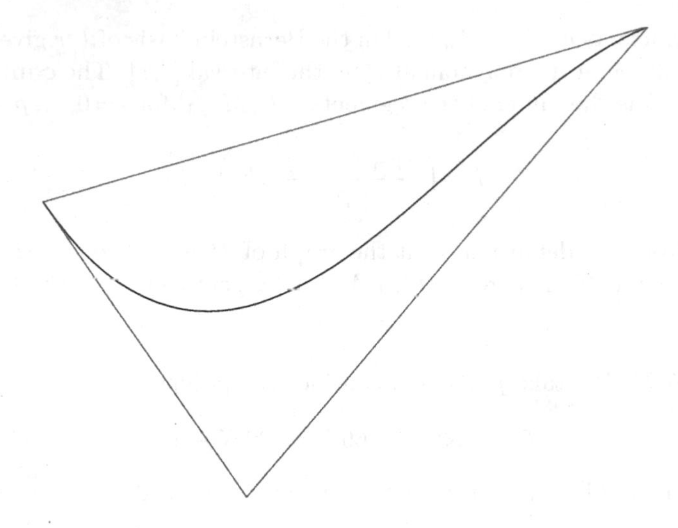

Fig. 10.2. Graph of P on $[0,1]$ and the control polygon

□

Proposition 10.35. *The graph of P on $[\ell, r]$ is contained in the control polygon of P on $[\ell, r]$.*

Proof: In order to prove the proposition, it is enough to prove that any line L above (resp. under) all the points in the control polygon of P on $[\ell, r]$ is above (resp. under) the graph of P on $[\ell, r]$. If L is defined by $Y = a X + b$, let us express the polynomial $a X + b$ in the Bernstein basis. Since

$$1 = \left(\frac{X-\ell}{r-\ell} + \frac{r-X}{r-\ell}\right)^p,$$

the binomial formula gives

$$\begin{aligned} 1 &= \sum_{i=0}^{p} \binom{p}{i} \left(\frac{X-\ell}{r-\ell}\right)^i \left(\frac{r-X}{r-\ell}\right)^{p-i} \\ &= \sum_{i=0}^{p} \mathrm{Bern}_{p,i}(\ell, r). \end{aligned}$$

Since

$$X = \left(r\left(\frac{X-\ell}{r-\ell}\right) + \ell\left(\frac{r-X}{r-\ell}\right)\right)\left(\frac{X-\ell}{r-\ell} + \frac{r-X}{r-\ell}\right)^{p-1},$$

the binomial formula together with Remark 10.31 gives

$$\begin{aligned} X &= \sum_{i=0}^{p-1} \left(r\left(\frac{X-\ell}{r-\ell}\right) + \ell\left(\frac{r-X}{r-\ell}\right)\right) \mathrm{Bern}_{p-1,i}(\ell, r), \\ &= \sum_{i=0}^{p} \left(\frac{i\,r + (p-i)\,\ell}{p}\right) \mathrm{Bern}_{p,i}(\ell, r). \end{aligned}$$

Thus,

$$a\,X+b=\sum_{i=0}^{p}\left(a\left(\frac{i\,r+(p-i)\,\ell}{p}\right)+b\right)\mathrm{Bern}_{p,i}(\ell,r).$$

It follows immediately that if L is above every M_i, i.e. if

$$a\left(\frac{i\,r+(p-i)\,\ell}{p}\right)+b\geq b_i$$

for every i, then L is above the graph of P on $[\ell, r]$, since $P=\sum_{i=0}^{p} b_i\,\mathrm{Bern}_{p,i}(\ell,r)$ and the Bernstein basis of ℓ, r is non-negative on $[\ell,r]$. A similar argument holds for L under every M_i. □

The following algorithm computes the coefficients of P in the Bernstein bases of ℓ,m and m,r from the coefficients of P in the Bernstein basis of ℓ,r.

Algorithm 10.2. **[Bernstein Coefficients]**

- **Structure:** an archimedean real closed field R.
- **Input:** a list $b=b_0,\ldots,b_p$ representing a polynomial P of degree $\leq p$ in the Bernstein basis of ℓ,r, and a number $m\in\mathrm{R}$.
- **Output:** the list b' representing P in the Bernstein basis of ℓ,m, the list b'' representing P in the Bernstein basis of m,r.
- **Complexity:** $O(p^2)$.
- **Procedure:**
  - Define $\alpha=\dfrac{r-m}{r-\ell}$, $\beta=\dfrac{m-\ell}{r-\ell}$.
  - Initialization: $b_j^{(0)}:=b_j$, $j=0,\ldots,p$.
  - For $i=1,\ldots,p$,
    - For $j=0,\ldots,p-i$, compute
      $b_j^{(i)}:=\alpha\,b_j^{(i-1)}+\beta\,b_{j+1}^{(i-1)}$
  - Output

$$\begin{aligned} b' &= b_0^{(0)},\ldots,b_0^{(j)},\ldots,b_0^{(p)},\\ b'' &= b_0^{(p)},\ldots,b_j^{(p-j)},\ldots,b_p^{(0)}. \end{aligned}$$

Algorithm 10.2 (Bernstein Coefficients) can be visualized with the following triangle.

$$\begin{array}{ccccccccc}
b_0^{(0)} & & \ldots & \ldots & \ldots & \ldots & & b_p^{(0)} \\
& b_0^{(1)} & & \ldots & \ldots & \ldots & b_{p-1}^{(1)} & \\
& \ddots & & \ldots & \ldots & & \therefore & \\
& & b_0^{(i)} & & \ldots & b_{p-i}^{(i)} & & \\
& & & b_0^{(p-1)} & b_1^{(p-1)} & & & \\
& & & & b_0^{(p)} & & &
\end{array}$$

with $b_j^{(i)}:=\alpha\,b_j^{(i-1)}+\beta\,b_{j+1}^{(i-1)}$, $\alpha=(r-m)/(r-\ell)$, $\beta=(m-\ell)/(r-\ell)$.

The coefficients of P in the Bernstein basis of ℓ, r appear in the top side of the triangle and the coefficients of P in the Bernstein basis of ℓ, m and m, r appear in the two other sides of the triangle. Note that $b_0^{(p)} = P(m)$.

Notation 10.36. [Reverted list] We denote by $\tilde{a}$ the list obtained by reversing the list a. □

Proof of correctness: It is enough to prove the part of the claim concerning ℓ, m. Indeed, by Remark 10.31, $\tilde{b}$ represents $(-1)^p P$ in the Bernstein basis of r, ℓ, and the claim is obtained by applying Algorithm 10.2 (Bernstein Coefficients) to $\tilde{b}$ at m. The output is $\tilde{b}''$ and $\tilde{b}$ and the conclusion follows using again Remark 10.31.

Let $\delta_{p,i}$ be the list of length $p + 1$ consisting all zeroes except a 1 at the $i + 1$-th place. Note that $\delta_{p,i}$ is the list of coefficients of $\mathrm{Bern}_{p,i}(\ell, t)$ in the Bernstein basis of ℓ, r. We will prove that the coefficients of $\mathrm{Bern}_{p,i}(\ell, m)$ in the Bernstein basis of ℓ, m coincide with the result of Algorithm 10.2 (Bernstein Coefficients) performed with input $\delta_{p,i}$. The correctness of Algorithm 10.2 (Bernstein Coefficients) for ℓ, m then follows by linearity.

First notice that, since $\alpha = (r - m)/(r - \ell)$, $\beta = (m - \ell)/(r - \ell)$,

$$\begin{aligned} \frac{X-\ell}{r-\ell} &= \beta\frac{X-\ell}{m-\ell}, \\ \frac{r-X}{r-\ell} &= \alpha\frac{X-\ell}{m-\ell} + \frac{m-X}{m-\ell}. \end{aligned}$$

Thus

$$\begin{aligned} \left(\frac{X-\ell}{r-\ell}\right)^i &= \beta^i\left(\frac{X-\ell}{m-\ell}\right)^i \\ \left(\frac{r-X}{r-\ell}\right)^{p-i} &= \sum_{k=0}^{p-i}\binom{p-i}{k}\alpha^k\left(\frac{X-\ell}{m-\ell}\right)^k\left(\frac{m-X}{m-\ell}\right)^{p-i-k}. \end{aligned}$$

It follows that

$$\mathrm{Bern}_{p,i}(\ell, r) = \binom{p}{i}\sum_{j=i}^{p}\binom{p-i}{j-i}\alpha^{j-i}\beta^i\left(\frac{X-\ell}{m-\ell}\right)^j\left(\frac{m-X}{m-\ell}\right)^{p-j}. \tag{10.2}$$

Since

$$\begin{aligned} \binom{p}{i}\binom{p-i}{j-i} &= \binom{j}{i}\binom{p}{j}, \\ \mathrm{Bern}_{p,i}(\ell, r) &= \sum_{j=i}^{p}\binom{j}{i}\alpha^{j-i}\beta^i\binom{p}{j}\left(\frac{X-\ell}{m-\ell}\right)^j\left(\frac{m-X}{m-\ell}\right)^{p-j}. \end{aligned}$$

Finally,

$$\mathrm{Bern}_{p,i}(\ell, r) = \sum_{j=i}^{p}\binom{j}{i}\alpha^{j-i}\beta^i\,\mathrm{Bern}_{p,j}(\ell, m).$$

On the other hand, we prove by induction on p that Algorithm 10.2 (Bernstein Coefficients) with input $\delta_{p,i}$ outputs the list $\delta'_{p,i}$ starting with i zeroes and with $(j+1)$-th element $\binom{j}{i}\alpha^{j-i}\beta^i$ for $j=i,\ldots,p$.

The result is clear for $p=i=0$. If Algorithm 10.2 (Bernstein Coefficients) applied to $\delta_{p-1,i-1}$ outputs $\delta'_{p-1,i-1}$, the equality

$$\binom{j}{i}\alpha^{j-i}\beta^i=\alpha\binom{j-1}{i}\alpha^{j-i-1}\beta^i+\beta\binom{j-1}{i-1}\alpha^{j-i}\beta^{i-1}$$

proves by induction on j that Algorithm 10.2 (Bernstein Coefficients) applied to $\delta_{p,i}$ outputs $\delta'_{p,i}$. So the coefficients of $\mathrm{Bern}_{p,i}(\ell,r)$ in the Bernstein basis of ℓ,m coincide with the output of Algorithm 10.2 (Bernstein Coefficients) with input $\delta_{p,i}$. □

Corollary 10.37. *Let b, b' and b'' be the lists of coefficients of P in the Bernstein basis of ℓ,r, ℓ,m, and m,r respectively.*

- *Algorithm 10.2 (Bernstein Coefficients) outputs b' and b'' when applied to b with weights*

$$\alpha=\frac{r-m}{r-\ell},\beta=\frac{m-\ell}{r-\ell}.$$

- *Algorithm 10.2 (Bernstein Coefficients) outputs b and $\tilde{b}''$ when applied to b' with weights*

$$\alpha'=\frac{m-r}{m-\ell},\beta'=\frac{r-\ell}{m-\ell}.$$

- *Algorithm 10.2 (Bernstein Coefficients) outputs $\tilde{b}'$ and b when applied to b'' with weights*

$$\alpha''=\frac{r-\ell}{r-m},\beta''=\frac{\ell-m}{r-m}.$$

Complexity analysis of Algorithm 10.2: The number of multiplications in the algorithm is $2\,p\,(p+1)/2$, the number of additions is $p\,(p+1)/2$. □

The following variant of Algorithm 10.2 (Bernstein Coefficients)is be useful in the case of a polynomial with integer coefficients in the Bernstein basis of ℓ,r since it avoids denominators.

Algorithm 10.3. **[Special Bernstein Coefficients]**

- **Structure:** an archimedean real closed field R.
- **Input:** a list $b=b_0,\ldots,b_p$ representing a polynomial P of degree $\le p$ in the Bernstein basis of ℓ,r.
- **Output:** the list b' representing $2^p P$ in the Bernstein basis of $\ell,(\ell+r)/2$ and the list b'' representing $2^p P$ in the Bernstein basis of $(\ell+r)/2,r$.
- **Complexity:** $O(p^2)$.

- **Procedure:**
  - Initialization: $b_j^{(0)} := b_j$, for $j = 0, \ldots, p$.
  - For $i = 1, \ldots, p$,
    - For $j = 0, \ldots, p - i$, compute
    $$b_j^{(i)} := b_j^{(i-1)} + b_{j+1}^{(i-1)}.$$
    - Output
    $$\begin{aligned} b' &= 2^p b_0^{(0)}, \ldots, 2^{p-j} b_0^{(j)}, \ldots, b_0^{(p)}, \\ b'' &= b_0^{(p)}, \ldots, 2^{p-j} b_j^{(p-j)}, \ldots, 2^p b_p^{(0)}. \end{aligned}$$

Complexity analysis: The number of additions in the algorithm is $p(p+1)/2$. The number of multiplications by 2 is $p\ (p + 1)$. Note that if $b \in \mathbb{Z}^{p+1}$, then $b' \in \mathbb{Z}^{p+1}$ and $b'' \in \mathbb{Z}^{p+1}$. If the bitsize of the b_i is bounded by τ, the bitsizes of the b_i' and b_i'' is bounded by $p + \tau$. □

Proposition 10.38. *Let P be a univariate polynomial of degree p. Let b be the Bernstein coefficients of P on (ℓ, r) and b' the Bernstein coefficients of P on (ℓ', r'). Denoting by $c_i = \binom{p}{i} b_i$, $Q = \sum_{i=0}^{p} c_i X^i$, $c_i' = \binom{p}{i} b_i'$, and $Q' = \sum_{i=0}^{p} c_i' X^i$, we have*

$$Q' = \mathrm{T}_{-1}(\mathrm{Rec}_p(\mathrm{Co}_{r'-\ell'}(\mathrm{T}_{\ell-\ell'}(\mathrm{Co}_{1/(r-\ell)}(\mathrm{Rec}_p(\mathrm{T}_1(Q))))))).$$

Proof: Apply Proposition 10.27. □

Remark 10.39. It is possible to output b' (and also b'') with arithmetic complexity $\tilde{O}(d)$ and binary complexity $\tilde{O}(d(\tau + d))$ using Proposition 10.38 with $\ell' = \ell, r' = (\ell + r)/2$, and Remark 8.7. □

Algorithm 10.2 (Bernstein Coefficients) gives a geometric construction of the control polygon of P on $[\ell, m]$ and on $[m, r]$ from the control polygon of P on $[\ell, r]$. The points of the new control polygons are constructed by taking iterated barycenters with weights α and β. The construction is illustrated in Figure 10.3, where we show how the control line of P on $[0, 1/2]$ is obtained from the control line of P on $[0, 1]$.

Example 10.40. Continuing Example 10.34, the Special Bernstein Coefficients Algorithm computes the following triangle.

$$\begin{array}{ccccccc} 4 & & -6 & & 7 & & 10 \\ & -2 & & 1 & & 17 & \\ & & -1 & & 18 & & \\ & & & 17 & & & \end{array}$$

The Bernstein coefficients of $8P$ on (0,1/2) are $32, -8, -2, 17$, the Bernstein coefficients of $8P$ on (1/2,1) are $17, 36, 68, 80$.

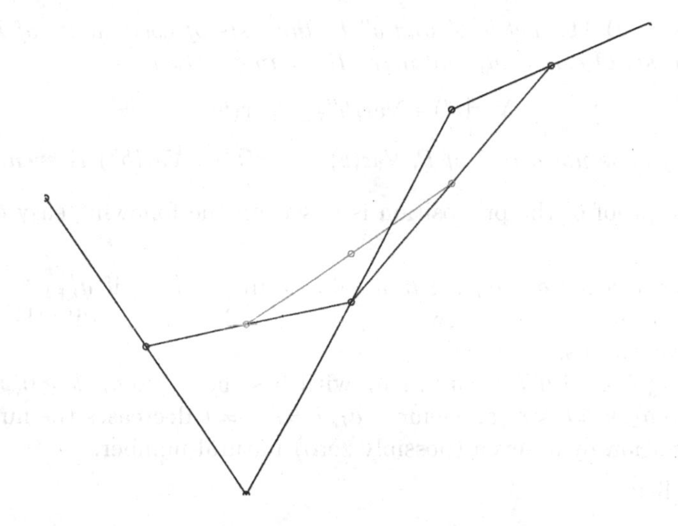

Fig. 10.3. Construction of the control line of P on [0, 1/2] by Bernstein Coefficients Algorithm

In Figure 10.4 we show the graph of P on [0, 1] and the control line on $[0,1/2]$.

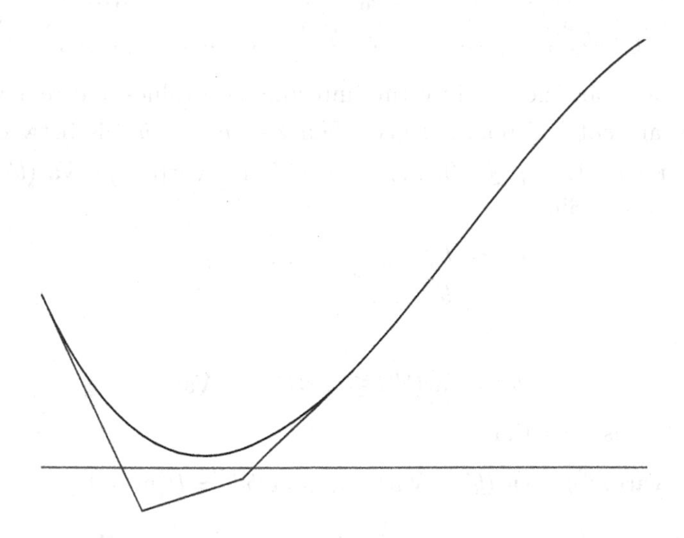

Fig. 10.4. Graph of P on $[0,1]$ and control line of P on $[0,1/2]$

□

We denote as usual by $\mathrm{Var}(b)$ the number of sign variations in a list b.

Proposition 10.41. *Let b, b' and b'' be the lists of coefficients of P in the Bernstein basis of ℓ, r, ℓ, m, and m, r. If $\ell < m < r$, then*

$$\mathrm{Var}(b') + \mathrm{Var}(b'') \leq \mathrm{Var}(b).$$

Moreover, if m is not a root of P, $\mathrm{Var}(b) - \mathrm{Var}(b') - \mathrm{Var}(b'')$ is even.

Proof: The proof of the proposition is based on the following easy observations:

- Inserting in a list $a = a_0, \ldots, a_n$ a value x in $[a_i, a_{i+1}]$ if $a_{i+1} \geq a_i$ (resp. in $[a_{i+1}, a_i]$ if $a_{i+1} < a_i$) between a_i and a_{i+1} does not modify the number of sign variations.
- Removing from a list $a = a_0, \ldots, a_n$ with first non-zero a_k, $k \geq 0$, and last non-zero a_ℓ, $k \leq \ell \leq n$, an element a_i, $i \neq k, i \neq \ell$ decreases the number of sign variation by an even (possibly zero) natural number.

Indeed the lists

$$\begin{aligned}
b &= b_0^{(0)}, \ldots, \ldots, \ldots, \ldots, \ldots, b_p^{(0)} \\
b^{(1)} &= b_0^{(0)}, b_0^{(1)}, \ldots, \ldots, \ldots, \ldots, b_{p-1}^{(1)}, b_p^{(0)} \\
&\ldots \\
b^{(i)} &= b_0^{(0)}, \ldots, \ldots, b_0^{(i)}, \ldots, \ldots, b_{p-i}^{(i)}, \ldots, \ldots, b_p^{(0)} \\
&\ldots \\
b^{(p-1)} &= b_0^{(0)}, \ldots, \ldots, \ldots, \ldots, b_0^{(p-1)}, b_1^{(p-1)}, \ldots, \ldots, \ldots, \ldots, b_p^{(0)} \\
b^{(p)} &= b_0^{(0)}, \ldots, \ldots, \ldots, \ldots, \quad b_0^{(p)} \quad, \ldots, \ldots, \ldots, \ldots, b_p^{(0)}
\end{aligned}$$

are successively obtained by inserting intermediate values and removing elements that are not end points, since when $\ell < m < r$, $b_j^{(i)}$ is between $b_j^{(i-1)}$ and $b_{j+1}^{(i-1)}$, for $i = 1, \ldots, p, j = 0, \ldots, p - i - 1$. Thus $\mathrm{Var}(b^{(p)}) \leq \mathrm{Var}(b)$ and the difference is even. Since

$$\begin{aligned}
b' &= b_0^{(0)}, \ldots, \ldots, \ldots, \ldots, \ldots, b_0^{(p)}, \\
b'' &= b_0^{(p)}, \ldots, \ldots, \ldots, \ldots, \ldots, b_p^{(0)},
\end{aligned}$$

it is clear that

$$\mathrm{Var}(b') + \mathrm{Var}(b'') \leq \mathrm{Var}(b^{(p)}) \leq \mathrm{Var}(b).$$

If $P(m) \neq 0$, it is clear that

$$\mathrm{Var}(b^{(p)}) = \mathrm{Var}(b') + \mathrm{Var}(b''), \text{since } b_0^{(p)} = P(m) \neq 0. \qquad \square$$

Example 10.42. Continuing Example 10.40, we observe, denoting by b, b' and b'', the lists of coefficients of P in the Bernstein basis of $0, 1$, $0, 1/2$, and $1/2, 1$, that $\mathrm{Var}(b) = 2$. This is visible in the figure: the control line for $[0, 1]$ cuts twice the X-axis. Similarly, $\mathrm{Var}(b') = 2$. This is visible in the figure: the control line for $[0, 1/2]$ also cuts twice the X-axis. Similarly, it is easy to check that $\mathrm{Var}(b'') = 0$.

We cannot decide from this information whether P has two roots in $(0,1/2)$ or no root in $(0,1/2)$. □

Let $b(P,(\ell,r))$ be the list of coefficients of P in the Bernstein basis of ℓ,r, $\ell<r$. The interval (ℓ,r) is **active** if $\mathrm{Var}(b(P,(\ell,r)))>0$.

Remark 10.43. It is clear from Proposition 10.41 that if $a_0<\cdots<a_N$, the number of active intervals among (a_i,a_{i+1}) is at most p. □

Let $P\in \mathrm{R}[X]$ and let $b(P,(\ell,r))$ be the list of coefficients of P in the Bernstein basis of ℓ,r. We describe cases where the number $\mathrm{Var}(b(P,(\ell,r)))$ coincides with the number of roots of P on (ℓ,r). Denote by $\mathcal{C}(\ell,r)$ the closed disk with $[\ell,r]$ as a diameter, by $\mathcal{C}_1(\ell,r)$ the closed disk whose boundary circumscribes the equilateral triangle T_1 based on $[\ell,r]$ (see Figure 10.5), and by $\mathcal{C}_2(\ell,r)$ the closed disk symmetric to $\mathcal{C}_1(\ell,r)$ with respect to the X-axis (see Figure 10.5)

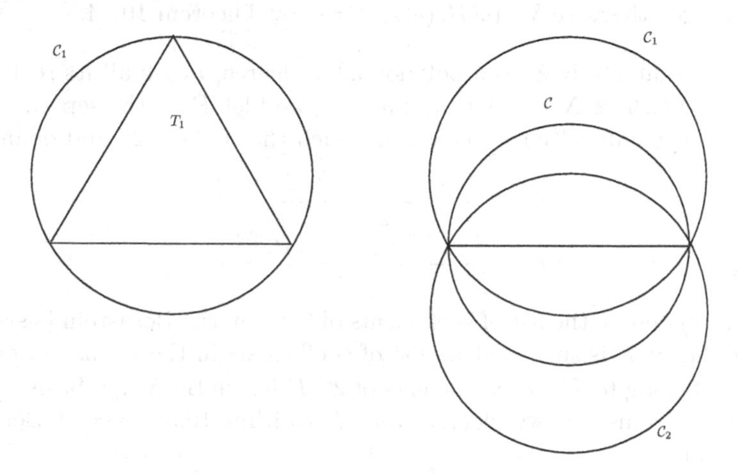

Fig. 10.5. $\mathcal{C}(\ell,r)$, $\mathcal{C}_1(\ell,r)$ and $\mathcal{C}_2(\ell,r)$

Theorem 10.44. [Theorem of three circles]
Let P be a separable polynomial of $\mathrm{R}[X]$.
If P has no root in $\mathcal{C}(\ell,r)$, then $\mathrm{Var}(b(P,(\ell,r)))=0$.
If P has exactly one root in $\mathcal{C}_1(\ell,r)\cup\mathcal{C}_2(\ell,r)$, then $\mathrm{Var}(b(P,(\ell,r)))=1$.

Proof: We identify R^2 with $\mathrm{C}=\mathrm{R}[i]$. The image of the complement of $\mathcal{C}(\ell,r)$ (resp. $\mathcal{C}_1(\ell,r)\cup\mathcal{C}_2(\ell,r)$) under translation by $-\ell$ followed by contraction by ratio $r-\ell$ is the complement of $\mathcal{C}(0,1)$ (resp. $\mathcal{C}_1(0,1)\cup\mathcal{C}_2(0,1)$). The image of the complement of $\mathcal{C}(0,1)$ under the inversion $z\mapsto 1/z$ is the half plane of complex numbers with real part less than 1. Translating by -1, we get the half plane of complex numbers with non-positive real part.

The image of the complement of $\mathcal{C}_1(0,1) \cup \mathcal{C}_2(0,1)$, under the inversion $z \mapsto 1/z$ is the sector

$$\{(x + i\, y) \in \mathrm{R}[i] \mid |\, y\, | \leqslant \sqrt{3}(1 - x)\}.$$

Translating this region by -1, we get the cone $\mathcal{B}$ defined in Proposition 2.40.

The statement follows from Proposition 2.39, Proposition 2.44 and Proposition 10.27. □

Corollary 10.45. *If P is separable,* $\mathrm{Var}(b(P, (\ell, r))) \geqslant 2$ *implies that P has at least two roots in $\mathcal{C}(\ell, r)$ or the interval (ℓ, r) contains exactly one real root and $\mathcal{C}_1(\ell, r) \cup \mathcal{C}_2(\ell, r)$ contains a pair of conjugate roots.*

Proof: If P has no root in $\mathcal{C}(\ell, r)$, then $\mathrm{Var}(b(P, (\ell, r))) = 0$, by Theorem 10.44. Thus, P has at least one root in $\mathcal{C}(\ell, r)$. If this is the only root in $\mathcal{C}(\ell, r)$,the root is in (ℓ, r) and $\mathcal{C}_1(\ell, r) \cup \mathcal{C}_2(\ell, r)$ must contain a pair of conjugate roots because otherwise $\mathrm{Var}(b(P, (\ell, r))) = 1$, by Theorem 10.44. □

Suppose that $P \in \mathrm{R}[X]$ is a polynomial of degree p with all its real zeroes in $(-2^N, 2^N)$ (where N is a natural number) and let $\bar{P}$ be the separable part of P. Consider natural numbers k and c such that $0 \leq c \leq 2^k$ and define

$$\begin{aligned} \ell &= \frac{-2^{N+k} + c\, 2^{N+1}}{2^k} \\ r &= \frac{-2^{N+k} + (c+1)\, 2^{N+1}}{2^k} \end{aligned}$$

Let $b(\bar{P}, \ell, r)$ denote the list of coefficients of $2^{kp}\bar{P}$ in the Bernstein basis of (ℓ, r). Note that if $\bar{P}$ is such that its list of coefficients in the Bernstein basis of $(-2^N, 2^N)$ belong to $\mathbb{Z}$, the coefficients of $2^{kp}\bar{P}$ in the Bernstein basis of (ℓ, r) belong to $\mathbb{Z}$. This follows clearly from Algorithm 10.3 (Special Bernstein Coefficients).

Remark 10.46. Let sep be the minimal distance between two roots of P in C, and let N be such that the real roots of P belong to $(-2^N, 2^N)$, and $k \geq -\log_2(\mathrm{sep}) + N + 1$. Since the circle of center $(\ell + r)/2, 0)$ and radius $r - \ell$ contains $\mathcal{C}_1(\ell, r) \cup \mathcal{C}_2(\ell, r)$, and two points inside this circle have distance at most $2(r - \ell)$, it is clear that the polynomial $\bar{P}$ has at most one root in (ℓ, r) and has no other complex root in $\mathcal{C}_1(\ell, r) \cup \mathcal{C}_2(\ell, r)$. So, $\mathrm{Var}(b(\bar{P}, \ell, r))$ is zero or one, using Theorem 10.44.

Thus, it is possible to decide, whether $\bar{P}$ has exactly one root in (ℓ, r) or has no root on (ℓ, r), by testing whether $\mathrm{Var}(b(\bar{P}, \ell, r))$ is zero or one. □

Example 10.47. Continuing Example 10.42, let us study the roots of P on $[0, 1]$, as a preparation to a more formal description of Algorithm 10.4 (Real Root Isolation).

The Bernstein coefficients of P for $0,1$ are $4,-6,7,10$. There maybe roots of P on $(0,1)$ as there are sign variations in these Bernstein coefficients.

As already seen in Example 10.42,a first application of Algorithm 10.3 (Special Bernstein Coefficients) gives

$$\begin{array}{ccccccc} 4 & & -6 & & 7 & & 10 \\ & -2 & & 1 & & 17 & \\ & & -1 & & 18 & & \\ & & & 17 & & & \end{array}$$

There maybe roots of P on $(0,1/2)$ as there are sign variations in the Bernstein coefficients of $8\,P$ on $(0,1/2)$ which are $32,-8,-2,17$. There are no roots of P on $(1/2,1)$ as there are no sign variations in the Bernstein coefficients of $8\,P$ on $(1/2,1)$ which are $17,36,68,80$.

Let us apply once more Algorithm 10.3 (Special Bernstein Coefficients):

$$\begin{array}{ccccccc} 32 & & -8 & & -2 & & 17 \\ & 24 & & -10 & & 15 & \\ & & 14 & & 5 & & \\ & & & 19 & & & \end{array}$$

The Bernstein coefficients of 64 P on $(0,1/4)$ are 256, 96, 28, 19, and the Bernstein coefficients of $64\,P$ on $(1/4,1/2)$ are $19,10,60,136$.There are no sign variations on the sides of the triangle so there are no roots of P on $(0,1/4)$ and on $(1/4,1/2)$.

Finally there are no roots of P on $[0,1]$. □

Definition 10.48. [Isolating list] Let Z be a finite subset of R. An **isolating list for Z** is a finite list L of rational points and open intervals with rational end points of R, such that each element of L contains exactly one element of Z, every element of Z belongs to an element of L and two elements of L have an empty intersection. □

Algorithm 10.4. **[Real Root Isolation]**

- **Structure:** the ring $\mathbb{Z}$.
- **Input:** a non-zero polynomial $P\in\mathbb{Z}[X]$.
- **Output:** a list isolating for the zeroes of P in R.
- **Binary complexity:** $O(p^5(\tau+\log_2(p))^2))$, where p is a bound on the degree of P, and τ is a bound on the bitsize of the coefficients of P.
- **Procedure:**
  - Compute $N\in\mathbb{N}$, $N\geqslant\log_2(C(P))$ (Notation 10.1) such that $(-2^N,2^N)$ contains the roots of P in R.
  - Compute $\bar{P}$, the separable part of P using Algorithm 10.1 (Gcd and Gcd-Free part). Replace $\overline{P}$ by $\lambda(\bar{P},-2^N,2^N)\,\bar{P}$, using Corollary 10.30 and its notation. Compute $b(\bar{P},-2^N,2^N)$, the Bernstein coefficients of $\bar{P}$ on $(-2^N,2^N)$, using Remark 10.28.

- Initialization: $\text{Pos} := \{b(\bar{P}, -2^N, 2^N)\}$ and $L(P)$ is the empty list.
- While Pos is non-empty,
  - Remove an element $b(\bar{P}, \ell, r)$ from Pos.
  - If $\text{Var}(b(\bar{P}, \ell, r)) = 1$, add (ℓ, r) to $L(P)$.
  - If $\text{Var}(b(\bar{P}, \ell, r)) > 1$,
    - Compute $b(\bar{P}, \ell, m)$ and $b(\bar{P}, m, r)$, with $m = (\ell + r)/2$, using Algorithm 10.3 (Special Bernstein Coefficients) and add them to Pos.
    - If the sign of $\bar{P}(m)$ is 0, see Remark 10.31 add $\{m\}$ to $L(P)$.

Proof of correctness: The algorithm terminates since R is archimedean, using Remark 10.46. Its correctness follows from Theorem 10.44. Note that, since there is only one root of $\bar{P}$ on each interval $[a, b]$ of $L(P)$, we have $\bar{P}(a)\,\bar{P}(b) < 0$. □

The complexity analysis requires some preliminary work.

Remark 10.49. Note that by Corollary 10.45, the binary tree T produced by Algorithm 10.4 enjoys the following properties:

- the interval labeling the root of the tree T contains all roots of P in R,
- at every leaf node labelled by (ℓ, r) of T, the interval (ℓ, r) contains either no root or one single root of P,
- at every node labelled by (ℓ, r) of T which is not a leaf, either P has at least two roots in $\mathcal{C}(\ell, r)$, or the interval (ℓ, r) contains exactly one real root and the union of the two circles $\mathcal{C}_1(\ell, r) \cup \mathcal{C}_2(\ell, r)$ contains two conjugate roots. □

So, we consider binary trees labeled by open intervals with rational endpoints, such that if a node of the tree is labeled by (ℓ, r), its children are labeled either by (ℓ, m) or by (m, r), with $m = (\ell + r)/2$. Such a tree T is an **isolating tree** for P if the following properties holds:

- the interval labeling the root of the tree T contains all the roots of P in R,
- at every leaf node labelled by (ℓ, r) of T, the interval (ℓ, r) contains either no root or one single root of P,
- at every node labelled by (ℓ, r) of T which is not a leaf, either P has at least two roots in $\mathcal{C}(\ell, r)$, or the interval (ℓ, r) contains exactly one root of P and the union of the two circles $\mathcal{C}_1(\ell, r) \cup \mathcal{C}_2(\ell, r)$ contains two conjugate roots.

As noted above, the binary tree produced by Algorithm 10.4 (Real Root Isolation) is an isolating tree for P.

Let $P \in \mathbb{Z}[X]$ be of degree p, and let τ be a bound on the bitsize of the coefficients of P and ν a bound on the bitsize of p. By Corollary 10.4 all the roots of P belong to the interval

$$u_0 = (-2^{\tau+\nu}, 2^{\tau+\nu}). \tag{10.3}$$

Proposition 10.50. *Let T be an isolating tree for P with root u_0 and L its set of leaves. Given a leaf $u \in L$, denote by h_u its depth. Then,*

$$\sum_{u \in L} h_u \leqslant 2\,(2\,\tau + 3\,\nu + 3)\,p.$$

Before proving Proposition 10.50 we need to study in some properties of T in more detail. Note that a node of T is labeled by an interval (ℓ, r). Note also that a leaf of T is

- either a leaf of type 1, when P has a root on (ℓ, r),
- or a leaf of type 0, when P has no root on (ℓ, r).

In order to bound the number of nodes of T we introduce a subtree T' of T defined by pruning certain leaves from T:

- If a leaf u has a sibling that is not a leaf, we prune u.
- If u and v are both leaves and siblings of each other, then we prune exactly one of them; the only constraint is that a leaf of type 0 is pruned preferably to a leaf of type 1.

We denote by L' the set of leaves of T'.

Clearly,

$$\sum_{u \in L} h_u \leqslant 2 \sum_{u \in L'} h_u \tag{10.4}$$

So in order to bound $\sum_{u \in L} h_u$ it suffices to bound $\sum_{u \in L'} h_u$.

If $u = (\ell, r)$ is an interval, we denote by $w(u) = r - \ell$ the **width** of the interval u. We define $w_0 = w(u_0)$ where u_0 is the interval labeling the root of the tree T. The number of nodes along the path from any $u \in L'$ to the root of T' is exactly $\log_2(w_0/w(u))$. Thus

$$\sum_{u \in L'} h_u \leqslant \sum_{u \in U} \log_2\left(\frac{w_0}{w(u)}\right). \tag{10.5}$$

Let $u \in L'$ be a root of T'. We are going to define two roots of P, α_u and β_u of P such that

$$|\,\alpha_u - \beta_u\,| < 4\,w(u).$$

Furthermore we will show that if u, u' have the same type (both type 0, or both type 1) then $\{\alpha_u, \beta_u\}$ and $\{\alpha_{u'}, \beta_{u'}\}$ are disjoint.

Let v be the parent of the leaf u.

a) If u is of type 1, then u contains a root α_u, and the union of the two circles $\mathcal{C}_1(v) \cup \mathcal{C}_2(v)$ contains two conjugate roots, and we denote by β_v one of these. Then

$$|\,\alpha_u - \beta_u\,| < (2/\sqrt{3})w(v) = (4/\sqrt{3})w(u) < 4\,w(u). \tag{10.6}$$

Let u' be another leaf of type 1 and v' its parent. Clearly, $\alpha_u \neq \alpha_{u'}$. We claim that it is possible to choose β_u and $\beta_{u'}$ such that $\beta_u \neq \beta_{u'}$. Consider the case when v and v' are siblings. Moreover, assume that β_u and $\overline{\beta_u}$ are the only non-real roots in $\mathcal{C}_1(v) \cup \mathcal{C}_2(v)$ and $\mathcal{C}_1(v') \cup \mathcal{C}_2(v')$. Then it must be the case that either $\beta_u \in \mathcal{C}_1(v) \cap \mathcal{C}_1(v')$, or $\beta_u \in \mathcal{C}_2(v) \cap \mathcal{C}_2(v')$. In either case, we can choose $\beta_{u'} = \overline{\beta_u}$, distinct from β_u. Thus $\{\alpha_u, \beta_u\}$ and $\{\alpha_{u'}, \beta_{u'}\}$ are disjoint.

b) If u is of type 0, P has one root α_u in $\mathcal{C}(v)$. Clearly, α_u is non-real, otherwise u would either have a non-leaf or a leaf of type 1 as a sibling in T, and would have been pruned from T. Thus $\mathcal{C}(v)$ contains $\overline{\alpha_u} \neq \alpha_u$ and we define $\beta_u = \overline{\alpha_u}$. Then,

$$|\alpha_u - \beta_u| < 2\,w(u) < 4\,w(u). \tag{10.7}$$

If u' is another leaf of type 1, then $\{\alpha_u, \overline{\alpha_u}\}$and $\{\alpha_{u'}, \overline{\alpha_{u'}}\}$ are disjoint, since $\mathcal{C}(v)$ and $\mathcal{C}(v')$ are disjoint.

Taking logarithms and substituting (10.6) and (10.7) in (10.5) we get

$$\sum_{u \in L'} h_u \leqslant \sum_{u \in U} \log_2\left(\frac{4\,w_0}{|\alpha_u - \beta_u|}\right) \tag{10.8}$$

Lemma 10.51. *We have $\#(L') \leqslant p$. More precisely denoting by L_0 the leaves of type 0 of T' and by L_1 the leaves of type 1 of T',*

a) $\#(L_0)$ is at most half the number of non-real roots of P.
b) $\#(L_1)$ is at most the number of real roots of P.

Proof: As shown above, to every $u \in L_0$ we can associate a unique pair of non-real roots (α_u, β_u), $\beta_u = \overline{\alpha_u}$. Since the non-real roots come in pair, the upper bound of U_0 follows.

Again by the arguments above, to each $u \in L_1$ we can associate a unique real root α_u and the claim on L_1 follows.

Finally $\#(L') \leqslant p$. □

Proof of Proposition 10.50:

From Corollary 10.4, we know that $\log_2(w_0) \leqslant \tau + \nu + 1$, where ν is a bound on the bitsize of p, and from Lemma 10.51, we have $\#(L') \leqslant p$. So, from (10.8), we obtain

$$\sum_{u \in L'} h_u \leqslant (\tau + \nu + 3)\,p - \sum_{u \in L'} \log_2(|\alpha_u - \beta_u|). \tag{10.9}$$

It remains to lower bound $\sum_{u \in L'} \log_2(|\alpha_u - \beta_u|)$. This will be done using Corollary 10.24. Consider the graph G whose edge set is $E_0 \cup E_1$ where

$$\begin{aligned} E_0 &= \{(\alpha_u, \beta_u) \mid u \in L_0\}, \\ E_1 &= \{(\alpha_u, \beta_u) \mid u \in L_1\}. \end{aligned}$$

We want to show that G satisfies the hypotheses of Corollary 10.24. First of all, for any $u \in L'$, we can reorder the pair (α_u, β_u) such that $|\alpha_u| \leqslant |\beta_u|$, without affecting (10.9).

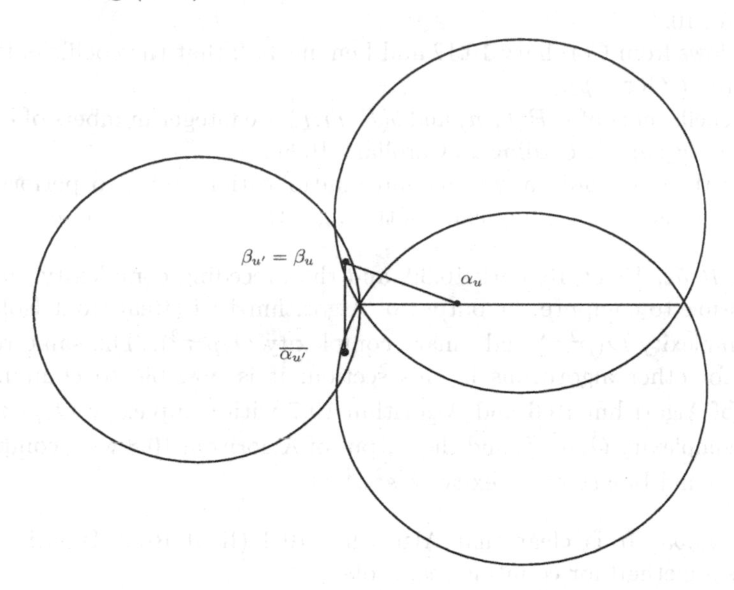

Fig. 10.6. A type 0 leaf and a type 1 leaf sharing the same root

Now we show that the in-degree of G is at most 1. Clearly the edge sets E_0 and E_1 have in-degree at most 1. However in $E_0 \cup E_1$, a case like that illustrated in Figure 10.6 can happen. That is, a $u \in L_1$ and a $u' \in L_0$ such that $\beta_u = \beta_{u'}$. But in such a case we can always reorder the edge $(\alpha_{u'}, \beta_{u'})$ to $(\beta_{u'}, \alpha_{u'})$ since $\beta_{u'} = \overline{\alpha_{u'}}$, and thus reduce the in-degree to 1.

Now we may apply Corollary 10.24 to G, and get

$$\prod_{u \in L'} |\alpha_u - \beta_u| \geq (p/\sqrt{3})^{-\#(U)}\, p^{-p/2}\, (p+1)^{(1-p)/2}\, 2^{\tau(1-p)}.$$

Taking logarithms of both sides gives

$$-\sum_{u \in L'} \log_2(|\alpha_u - \beta_u|) \leqslant (\tau + 2\nu)\, p, \tag{10.10}$$

since $\#(L') \leqslant p$ by Lemma 10.51, and $\log_2(p) < \log_2(p+1) \leqslant \nu$.

Substituting in (10.9), we obtain

$$\sum_{u \in L'} h_u \leqslant (\tau + \nu + 3)\, p + (\tau + 2\nu)\, p = (2\tau + 3\nu + 3)\, p.$$

The proposition is proved, since $\sum_{u \in L} h_u \leqslant 2 \sum_{u \in L'} h_u$. □

Complexity analysis of Algorithm 10.4: The computation of $\bar{P}$ takes $O(p^2)$ arithmetic operations using Algorithm 10.1.

Using Proposition 10.50 and Remark 10.49, the number of nodes produced by the algorithm is at most $O(p\,(\tau+\log_2(d)))$.

At each node the computation takes $O(p^2)$ arithmetic operations using Algorithm 10.3.

It follows from Corollary 10.12 and Lemma 10.2 that the coefficients of $\bar{P}$ are of bitsize $O(\tau+p)$.

The coefficients of $b(\bar{P},\ell,m)$ and $b(\bar{P},m,r)$ are integer numbers of bitsizes $O(p^2\,(\tau+\log_2(p)))$ according to Corollary 10.30.

Since there are only additions and multiplications by 2 to perform, the estimate for the binary complexity of the algorithm is $O(p^5\,(\tau+log_2(p))^2)$. □

Remark 10.52. Using Remark 10.39, and the preceding complexity analysis, it is possible to compute the output of Algorithm 10.4 (Real Root Isolation) with complexity $\tilde{O}(p^2\,\tau)$ and binary complexity $\tilde{O}(p^4\tau^2)$. The same remark applies for other algorithms in this section: it is possible to compute the output of Algorithm 10.6 and Algorithm 10.7 with complexity $\tilde{O}(p^2\,\tau)$ and binary complexity $\tilde{O}(p^4\tau^2)$ and the output of Algorithm 10.8 with complexity $\tilde{O}(s^2\,p^2\,\tau)$ and binary complexity $\tilde{O}(s^2\,p^4\tau^2)$. □

Remark 10.53. It is clear that Algorithm 10.4 (Real Root Isolation) also provides a method for counting real roots.

Note that if $(\ell,\ r)$ is an open interval isolating a root x of P, and $m=(\ell+r)/2$ it is easy to decide whether $x=m$, $x\in(\ell,m)$, or $x\in(m,r)$ by applying once more Algorithm 10.3 (Special Bernstein Coefficients), since the sign of $P(m)$ is part of the output. □

We now give a variant of Algorithm 10.4 (Real Root Isolation) where the computations take place in the basis of monomials.

Notation 10.54. Let $P\in\mathbb{Z}[X]$ be a polynomial of degree p having all its real roots between $(-2^N,2^N)$. We define for $r-\ell=a\,2^k, a\in\mathbb{Z}, k\in\mathbb{Z}, a$ odd,

$$\begin{aligned} P[\ell,r] &= P((r-\ell)X+\ell) \qquad \text{if } k\geqslant 0\\ &= 2^{kp}\,P((r-\ell)X+\ell) \ \text{ if } k<0 \end{aligned}$$

In other words, the roots of $P[\ell,r]$ in $(0,1)$ are in 1-1 correspondence with the roots of P on (ℓ,r), and $P[\ell,r]\in\mathbb{Z}[X]$. Note that

$$\begin{aligned} P[-2^N,2^N] &= \mathrm{Co}_{2^{N+1}}(\mathrm{T}_{2^N}(P))\\ &= \mathrm{Co}_2(\mathrm{T}_{-1}(\mathrm{Co}_{2^N}(P))) \end{aligned} \tag{10.11}$$

and, if $m=(\ell+r)/2$.

$$\begin{aligned} P[\ell,m] &= 2^q\,\mathrm{Co}_{1/2}(P[\ell,r]) \ \text{ if } m\notin\mathbb{Z},\\ &= \mathrm{Co}_{1/2}(P[\ell,r]) \qquad \text{if } m\in\mathbb{Z},\\ P[m,r] &= \mathrm{T}_{-1}(P[\ell,m]). \end{aligned} \tag{10.12}$$

□

The following proposition explains how to recover the sign variations in the Bernstein coefficients of P for (ℓ, r), from $P[\ell, r]$.

Proposition 10.55.

$$\mathrm{Var}(b(P,(\ell,r)))=\mathrm{Var}(\mathrm{T}_{-1}(\mathrm{Rec}_p(P[\ell,r]))).$$

Proof: Follows immediately from Proposition 10.27. □

Algorithm 10.5. **[Descartes' Real Root Isolation]**

- **Structure:** the ring $\mathbb{Z}$.
- **Input:** a non-zero polynomial $P \in \mathbb{Z}[X]$.
- **Output:** a list isolating for the zeroes of P in R.
- **Binary complexity:** $O(p^5(\tau + \log_2(p))^2))$, where p is a bound on the degree of P, and τ is a bound on the bitsize of the coefficients of P.
- **Procedure:**
  - Compute $N \in \mathbb{N}$, $N \geqslant \log_2(C(P))$ (Notation 10.1) such that $(-2^N, 2^N)$ contains the roots of P in R.
  - Compute $\bar{P}$, the separable part of P using Algorithm 10.1 (Gcd and Gcd-Free part) and denote by q its degree.
  - Compute $\overline{P}[-2^N, 2^N] = \mathrm{Co}_2(\mathrm{T}_{-1}(\mathrm{Co}_{2^N}(\bar{P})))$, using Algorithm 8.10.
  - Initialization: $\mathrm{Pos} := \{\overline{P}[-2^N, 2^N]\}$ and $L(P)$ is the empty list.
  - While Pos is non-empty,
    - Remove an element $\overline{P}[\ell, r]$ from Pos.
    - If $\mathrm{Var}(\mathrm{T}_{-1}(\mathrm{Rec}_q(\overline{P}[\ell, r]))) = 1$, add (ℓ, r) to $L(P)$.
    - If $\mathrm{Var}(\mathrm{T}_{-1}(\mathrm{Rec}_q(\overline{P}[\ell, r]))) > 1$,
      - Let $m = (\ell + r)/2$. Compute,

$$\begin{aligned}\overline{P}[\ell,m] &= 2^q\,\mathrm{Co}_{1/2}(\overline{P}[\ell,r]) \text{ if } m \notin \mathbb{Z},\\ &= \mathrm{Co}_{1/2}(\overline{P}[\ell,r]) \quad \text{if } m \in \mathbb{Z},\\ \overline{P}[m,r] &= \mathrm{T}_{-1}(\overline{P}[\ell,m]),\end{aligned}$$

using Algorithm 8.10, add

$$\overline{P}[\ell,m] \text{ and } \overline{P}[m,r]$$

to Pos.

    - If the sign of $\bar{P}(m)$ is 0, add $\{m\}$ to $L(P)$.

Proof of correctness: The algorithm terminates since R is archimedean. The correctness of the algorithm follows from the correctness of Algorithm 10.4 (Real Root Isolation), using Proposition 10.55. □

Complexity analysis: The computation of $\bar{P}$ takes $O(p^2)$ arithmetic operations using Algorithm 10.1. Using Proposition 10.50 and Remark 10.49, the number of nodes produced by the algorithm is at most $O(p\,(\tau+\log_2(d)))$.At each node the computation takes $O(p^2)$ arithmetic operations. It follows from Corollary 10.12 and Lemma 10.2 that the coefficients of $\bar{P}$ are of bitsize $O(\tau+p)$. Finally the coefficients of $\overline{P}[\ell, r]$ and $\mathrm{T}_{-1}(\mathrm{Rec}(\overline{P}[\ell, r]))$ are bounded by $O(p^2\,(\tau+\log_2(p)))$ and only multiplication by 2 and additions are performed. □

To evaluate the sign of another polynomial Q at the root of a polynomial characterized by an isolating interval, it may be necessary to refine the isolating intervals further. We need the following definition.

Definition 10.56. [Isolating list with signs] Let Z be a finite subset of R and a finite list $\mathcal{Q}$ of polynomial of $\mathrm{R}[X]$. An **isolating list with signs for Z and $\mathcal{Q}$** is a finite list L of couples (I,σ) such that I is a rational point or an open interval with rational end points, and σ is an element of $\{-1,0,1\}^{\mathcal{Q}}$. Every element of Z belongs to some I with (I,σ) in L and for every (I,σ) in L, there exists one and only one element x in I and σ is the sign condition realized by the family $\mathcal{Q}$ at x. □

Algorithm 10.6. **[Sign at a Real Root]**

- **Structure:** the ring $\mathbb{Z}$.
- **Input:** a polynomial $P\in\mathbb{Z}[X]$, a list $L(P)$ isolating for the zeroes of P in R and a polynomial $Q\in\mathbb{Z}[X]$.
- **Output:** an isolating list with signs for the zeroes of P in R and $\{Q\}$.
- **Binary complexity:** $O(p^5\,(\tau+\log_2(p))^2))$, where p is a bound on the degree of P, and τ is a bound on the bitsize of the coefficients of P.
- **Procedure:**
  - First step: Identify the common roots of P and Q as follows. This is done as follows:
  - Compute the greatest common divisor G of $\bar{P}$ and Q. Note that G is separable. If the structure is $\mathbb{Z}$, replace G by $\lambda(G,-2^N,2^N)\,G$ using Corollary 10.30 and its notation.
  - Initialization:
    - Set $N(P):=\emptyset$, $\mathrm{NCom}(P,Q):=\emptyset$ ($\mathrm{NCom}(P,Q)$ will contain points or intervals corresponding to roots of P which are not roots of Q). For every $\{a\}\in L(P)$, add $(\{a\},\mathrm{sign}(Q(a)))$ to $N(P)$.
    - Compute $b(G,\ell,r)$, the Bernstein coefficients of G, for the intervals $(\ell,r)\in L(P)$ using Proposition 10.27. Set

$$\mathrm{Pos}:=\{(b(\bar{P},\ell,r),b(G,\ell,r))\}\text{ for the intervals }(\ell,r)\in L(P).$$

  - While Pos is non-empty,
    - Remove an element $b(\bar{P},\ell,r),b(G,\ell,r)$ from Pos.
    - If $\mathrm{Var}(b(G,\ell,r))=1$, add $((\ell,r),0)$ to $N(P)$.

- If $\mathrm{Var}(b(G,\ell,r))=0$, add $(b(\bar{P},(\ell,r))$ to $\mathrm{NCom}(P,Q)$.
- If $\mathrm{Var}(b(G, \ell, r)) > 1$, compute $(b(\bar{P},\ell,m), b(G,\ell,m))$ and $(b(\bar{P},m,r), b(G,m,r))$ with $m=(\ell+r)/2$ using Algorithm 10.3 (Special Bernstein Coefficients).
  - If $\bar{P}(m)=0$, add $(\{m\},\mathrm{sign}(Q(m)))$ to $N(P)$.
  - If the signs of $\bar{P}$ at the right of ℓ and at m coincide, add $(b(\bar{P},m,r), b(G,m,r))$ to Pos.
  - If the signs of $\bar{P}$ at the right of ℓ and at m differ, add $(b(\bar{P},\ell,m), b(G,\ell,m))$ to Pos.
- Second step: Find the sign of Q at the roots of P where Q is non-zero. This is done as follows:
- Initialization: Pos := $\mathrm{NCom}(P, Q)$. If the structure is $\mathbb{Z}$, replace Q by $\lambda(Q,-2^N,2^N)\,Q$ using Corollary 10.30 and its notation.
- While Pos is non-empty,
  - Remove an element $b(\bar{P},\ell,r)$ from Pos. Compute $b(Q,\ell,r)$ the Bernstein coefficients of Q on (ℓ,r) using Proposition 10.27.
  - If $\mathrm{Var}(b(Q,\ell,r)=0$, add $((\ell,r),\tau)$ to $N(P)$, where τ is the sign of any element of $b(Q,\ell,r)$.
  - If $\mathrm{Var}(b(Q,\ell,r)) \neq 0$, compute $b(\bar{P},\ell,m)$ and $b(\bar{P},\ell,r)$ using Algorithm 10.3 (Special Bernstein Coefficients).
    - If $\bar{P}(m)=0$, add $(\{m\},\mathrm{sign}(Q(m))$ to $N(P)$.
    - If the signs of $\bar{P}$ at the right of ℓ and at m coincide, add $(b(\bar{P},m,r)$ to Pos.
    - If the signs of $\bar{P}$ at the right of ℓ and at m differ, add $(b(\bar{P},\ell,m)$ to Pos.

Proof of correctness: The algorithm terminates since R is archimedean. Its correctness follows from Theorem 10.37. Note that on any interval output, denoting by x the root of P in the interval, either $Q(x)=0$ or the sign of Q on the interval is everywhere equal to the sign of $Q(x)$. □

Complexity analysis:

Note first that the binary tree produced by Algorithm 10.6 is isolating for the polynomial $P\,Q$. Thus its number of nodes is at most $O(p\,(\tau+\log_2(d))$, by Proposition 10.50.

The computation of G takes $O(p^2)$ arithmetic operations, as well as the computation of of $b(G,\ell,m)$, $b(\bar{P},\ell,m)$, and $b(\bar{P},m,r)$.

We skip the details on the bit length as they are very similar to the ones in the binary complexity analysis of Algorithm 10.4 (Real Root Isolation).

Finally, the estimate for the binary complexity of the algorithm is $O(p^5(\tau+\log_2(p))^2)$. □

Remark 10.57. Note that it is easy to design variants to Algorithm 10.6, Algorithm 10.7, Algorithm 10.8 using Descartes' isolation technique rather than Casteljau's, with the same complexity. □

Remark 10.58. Similarly to Remark 10.52, using Remark 8.7 , it is possible to compute the output of Algorithm 10.5 (Descartes' Real Root Isolation) with complexity $\tilde{O}(p^2\tau)$ and binary complexity $\tilde{O}(p^4\tau^2)$. The same remark applies for the Descartes' variants of the other algorithms in this section: it is possible to compute the output of Algorithm 10.6 and Algorithm 10.7 with complexity $\tilde{O}(p^2\,\tau)$ and binary complexity $\tilde{O}(p^4\tau^2)$ and the output of Algorithm 10.8 with complexity $\tilde{O}(s^2\,p^2\,\tau)$ and binary complexity $\tilde{O}(s^2\,p^4\tau^2)$ using Descartes' variant (see Remark 10.57). □

We indicate now how to compare the roots of two polynomials in R.

Algorithm 10.7. **[Comparison of Real Roots]**

- **Structure:** the ring $\mathbb{Z}$.
- **Input:** a polynomial P and a polynomial Q in $\mathbb{Z}[X]$.
- **Output:** a isolating list for the zeroes of $\{P, Q\}$ in R and $\{P, Q\}$.
- **Binary complexity:** $O(p^5\,(\tau + \log_2(p))^2))$, where p is a bound on the degree of P, and τ is a bound on the bitsize of the coefficients of P.
- **Procedure:**
  - Compute ℓ such that $(-2^N, 2^N)$ contains the roots of P and Q using Lemma 10.2.
  - Isolate the roots of P (resp. Q) using Algorithm 10.4 and perform the sign determination for Q (resp. P) at these roots using Algorithm 10.6. Merge these two lists by taking the point or the interval of smallest length in case of non-empty intersection.

Proof of correctness: The algorithm terminates since R is archimedean. Its correctness follows from Theorem 10.37. Note that because of the dichotomy process, any two elements of $L(P)$ and $L(Q)$ are either disjoint or one is included in the other. □

Complexity analysis: Follows from the binary complexity analysis of Algorithm 10.6 (Sign at a Real Root). □

Finally, we are able, given a finite set of univariate polynomials, to describe the real roots of these polynomials as well as points in the intervals they define.

Algorithm 10.8. **[Real Univariate Sample Points]**

- **Structure:** the ring $\mathbb{Z}$.
- **Input:** a finite set of univariate polynomials $\mathcal{P}$ with coefficients in $\mathbb{Z}$.
- **Output:** an isolating list with signs for the roots of $\mathcal{P}$ in R and $\mathcal{P}$, an element between each two consecutive roots of elements of $\mathcal{P}$, an element of R smaller than all these roots, and an element of R greater than all these roots.
- **Binary complexity:** $O(s^2\,p^5\,(\tau + \log_2(p))^2)$, where p is a bound on the degree of P, and τ is a bound on the bitsize of the coefficients of P.

- **Procedure:**
  - For every pair P, Q of elements of $\mathcal{P}$ perform Algorithm 10.7.
  - Compute a rational point in between two consecutive roots using the isolating sets.
  - Compute a rational point smaller than all these roots and a rational point greater than all the roots of polynomials in $\mathcal{P}$ using Lemma 10.2.

Proof of correctness: The algorithm terminates since R is archimedean. Its correctness follows from Theorem 10.37. □

Complexity analysis: Follows from the binary complexity analysis of Algorithm 10.7 (Comparison of Real Roots). □

10.3 Sign Determination

We consider now a general real closed field R, not necessarily archimedean. Note that the approximation of the elements of R by rational numbers cannot be performed anymore. Our aim is to give a method for determining the sign conditions realized by a family of polynomials on a finite set Z of points in R^k.

This general method will be applied in two special cases: the zero set of a univariate polynomial in R in this chapter and the zero set of a zero-dimensional polynomial system in R^k later in the book.

Let Z be a finite subset of R^k. We denote

$$\begin{aligned} \mathrm{Reali}(P=0, Z) &= \{x \in Z \mid P(x)=0\}, \\ \mathrm{Reali}(P>0, Z) &= \{x \in Z \mid P(x)>0\}, \\ \mathrm{Reali}(P<0, Z) &= \{x \in Z \mid P(x)<0\}, \end{aligned}$$

and $c(P=0, Z)$, $c(P>0, Z)$, $c(P<0, Z)$ the corresponding numbers of elements. The Tarski-query of P for Z is

$$\mathrm{TaQ}(P, Z) = \sum_{x \in Z} \mathrm{sign}(Q(x)) = c(P>0, Z) - c(P<0, Z).$$

We consider the computation of $\mathrm{TaQ}(P, Z)$ as a basic black box. We have already seen several algorithms for computing it when $Q \in \mathrm{R}[X]$, and $Z = \mathrm{Zer}(Q, \mathrm{R})$ (Algorithms 9.2 and 9.5). Later in the book, we shall see other algorithms for the multivariate case.

Consider $\mathcal{P} = P_1, \ldots, P_s$, a finite list of polynomials in $\mathrm{R}[X_1, \ldots, X_k]$.

Let σ be a sign condition on $\mathcal{P}$, i.e. an element of $\{0, 1, -1\}^{\mathcal{P}}$. The **realization of the sign condition σ on Z** is

$$\mathrm{Reali}(\sigma, Z) = \{x \in Z \mid \bigwedge_{P \in \mathcal{P}} \mathrm{sign}(P(x)) = \sigma(P)\}.$$

The cardinality of $\mathrm{Reali}(\sigma, Z)$ is denoted $c(\sigma, Z)$.

We write $\mathrm{SIGN}(\mathcal{P}, Z)$ for the list of sign conditions realized by $\mathcal{P}$ on Z, i.e. the list of $\sigma \in \{0, 1, -1\}^{\mathcal{P}}$ such that $\mathrm{Reali}(\sigma, Z)$ is non-empty, and $c(\mathcal{P}, Z)$ for the corresponding list of cardinals $c(\sigma, Z) = \#(\mathrm{Reali}(\sigma, Z))$ for $\sigma \in \mathrm{SIGN}(\mathcal{P}, Z)$.

Our aim is to determine $\mathrm{SIGN}(\mathcal{P}, Z)$, and, more precisely, to compute the numbers $c(\mathcal{P}, Z)$. The only information we are going to use to compute $\mathrm{SIGN}(\mathcal{P}, Z)$ is the Tarski-query of products of elements of $\mathcal{P}$.

A method for sign determination in the univariate case was already presented in Chapter 2 (Section 2.3). This method can be immediately generalized to the multivariate case, as we see now.

Given $\alpha \in \{0, 1, 2\}^{\mathcal{P}}$, we write σ^α for $\prod_{P \in \mathcal{P}} \sigma(P)^{\alpha(P)}$, and $\mathcal{P}^\alpha$ for $\prod_{P \in \mathcal{P}} P^{\alpha(P)}$, with $\sigma \in \{0, 1, -1\}^{\mathcal{P}}$. When $\mathrm{Reali}(\sigma, Z) \neq \emptyset$, the sign of $\mathcal{P}^\alpha$ is fixed on $\mathrm{Reali}(\sigma, Z)$ and is equal to σ^α with the understanding that $0^0 = 1$.

We order the elements of $\mathcal{P}$ so that $\mathcal{P} = \{P_1, \ldots, P_s\}$. As in Chapter 2, we order $\{0, 1, 2\}^{\mathcal{P}}$ lexicographically. We also order $\{0, 1, -1\}^{\mathcal{P}}$ lexicographically (with $0 \prec 1 \prec -1$).

Given $A = \alpha_1, \ldots, \alpha_m$, a list of elements of $\{0, 1, 2\}^{\mathcal{P}}$ with $\alpha_1 <_{\mathrm{lex}} \ldots <_{\mathrm{lex}} \alpha_m$, we define

$$\begin{aligned} \mathcal{P}^A &= \mathcal{P}^{\alpha_1}, \ldots, \mathcal{P}^{\alpha_m}, \\ \mathrm{TaQ}(\mathcal{P}^A, Z) &= \mathrm{TaQ}(\mathcal{P}^{\alpha_1}, Z), \ldots, \mathrm{TaQ}(\mathcal{P}^{\alpha_m}, Z). \end{aligned}$$

Given $\Sigma = \sigma_1, \ldots, \sigma_n$, a list of elements of $\{0, 1, -1\}^{\mathcal{P}}$, with $\sigma_1 <_{\mathrm{lex}} \ldots <_{\mathrm{lex}} \sigma_n$, we define

$$\begin{aligned} \mathrm{Reali}(\Sigma, Z) &= \mathrm{Reali}(\sigma_1, Z), \ldots, \mathrm{Reali}(\sigma_n, Z), \\ c(\Sigma, Z) &= c(\sigma_1, Z), \ldots, c(\sigma_n, Z). \end{aligned}$$

The **matrix of signs of $\mathcal{P}^A$ on Σ** is the $m \times n$ matrix $\mathrm{Mat}(A, \Sigma)$ whose i, j-th entry is $\sigma_j^{\alpha_i}$.

Proposition 10.59. *If* $\cup_{\sigma \in \Sigma} \mathrm{Reali}(\sigma, Z) = Z$, *then*

$$\mathrm{Mat}(A, \Sigma) \cdot c(\Sigma, Z) = \mathrm{TaQ}(\mathcal{P}^A, Z).$$

Proof: This is obvious since the $(i, j) -$ th entry of $\mathrm{Mat}(\mathcal{P}^A, \Sigma)$ is $\sigma_j^{\alpha_i}$. □

When the matrix $\mathrm{Mat}(A, \Sigma)$ is invertible, we can compute $c(\Sigma, Z)$ from $\mathrm{TaQ}(\mathcal{P}^A, Z)$.

Note also that when $\mathcal{P} = \{P\}$, $A = \{0, 1, 2\}^{\{P\}}$, and $\Sigma = \{0, 1, -1\}^{\{P\}}$, the conclusion of Proposition 10.59 is

$$\begin{bmatrix} 1 & 1 & 1 \\ 0 & 1 & -1 \\ 0 & 1 & 1 \end{bmatrix} \cdot \begin{bmatrix} c(P = 0, Z) \\ c(P > 0, Z) \\ c(P < 0, Z) \end{bmatrix} = \begin{bmatrix} \mathrm{TaQ}(1, Z) \\ \mathrm{TaQ}(P, Z) \\ \mathrm{TaQ}(P^2, Z) \end{bmatrix}. \quad (10.13)$$

This is a generalization to Z of Equation (2.6) which had been stated for the set of zeroes of a univariate polynomial.

In order to compute each $c(\sigma, Z)$ knowing all $\mathrm{TaQ}(\mathcal{P}^\alpha, Z)$, we take $A = \{0,1,2\}^{\mathcal{P}}$ and $\Sigma = \{0,1,-1\}^{\mathcal{P}}$.

As in Chapter 2, Notation 2.71 (Total matrix of signs), let M_s be the $3^s \times 3^s$ matrix defined inductively by

$$M_1 = \begin{bmatrix} 1 & 1 & 1 \\ 0 & 1 & -1 \\ 0 & 1 & 1 \end{bmatrix}$$

and

$$M_{t+1} = M_t \otimes M_1.$$

We generalize Proposition 2.72 and obtain

Proposition 10.60. *Let $\mathcal{P}$ be a set of polynomials with s elements, $A = \{0,1,2\}^{\mathcal{P}}$, and $\Sigma = \{0,1,-1\}^{\mathcal{P}}$ ordered lexicographically. Then,*

$$\mathrm{Mat}(A, \Sigma) = M_s.$$

Proof: The proof is by induction on s. If s=1, the claim is Equation (10.13). If the claim holds for s, it holds also for $s+1$ given the definitions of M_{s+1} and $\mathrm{Mat}(\mathcal{P}^A, \Sigma)$, and the orderings on $A = \{0,1,2\}^{\mathcal{P}}$ and $\Sigma = \{0,1,-1\}^{\mathcal{P}}$. □

As a consequence:

Corollary 10.61.

$$M_s \cdot c(\Sigma, Z) = \mathrm{TaQ}(\mathcal{P}^A, Z).$$

The preceding results give the following algorithm for sign determination, by using repeatedly the Tarski-query black box.

Algorithm 10.9. **[Naive Sign Determination]**

- **Input:** a finite subset $Z \subset \mathrm{R}^k$ with r elements and a finite list $\mathcal{P} = P_1, \ldots, P_s$ of polynomials in $\mathrm{R}[X_1, \ldots, X_k]$.
- **Output:** the list of sign conditions realized by $\mathcal{P}$ on Z, $\mathrm{SIGN}(\mathcal{P}, Z)$.
- **Blackbox:** For a polynomial P, the Tarski-query $\mathrm{TaQ}(P, Z)$.
- **Complexity:** 3^s calls to the Tarski-query black box.
- **Procedure:**
  - Define $A = \{0,1,2\}^{\mathcal{P}}$ and $\Sigma = \{0,1,-1\}^{\mathcal{P}}$, ordered lexicographically.
  - Call the Tarski-query black box 3^s times with input the elements of $\mathcal{P}^A$ to obtain $\mathrm{TaQ}(\mathcal{P}^A, Z)$. Solve the $3^s \times 3^s$ system

    $$M_s \cdot c(\Sigma, Z) = \mathrm{TaQ}(\mathcal{P}^A, Z)$$

    to obtain the vector $c(\Sigma, Z)$of length 3^s. Output the set of sign conditions σ with $c(\sigma, Z) \neq 0$.

Complexity analysis: The number of calls to the Tarski-query black box is 3^s. The calls to the Tarski-query black box are done for polynomials which are products of at most s polynomials of the form P or P^2, $P\in\mathcal{P}$. □

To avoid the exponential number of calls to the Tarski-query black box in Algorithm 10.9 (Naive Sign Determination), notice that $\#(\mathrm{SIGN}(\mathcal{P},Z))\leq\#(Z)$, so that the number of realizable sign conditions does not exceed $\#(Z)$. We are now going to determine the non-empty sign conditions inductively getting rid of the empty sign conditions at each step of the computation, in order to control the size of the data we manipulate.

Let $Z\subset\mathrm{R}^k$ be a finite set, and let $\mathcal{P}$ be a finite list of polynomials in $\mathrm{R}[X_1, \ldots, X_k]$. A list A of elements in $\{0, 1, 2\}^{\mathcal{P}}$ is **adapted to sign determination for $\mathcal{P}$ on Z** if the matrix of signs of $\mathcal{P}^A$ over $\mathrm{SIGN}(\mathcal{P},Z)$ is invertible.

Example 10.62. Consider the set of polynomials $\{P\}$. In this case, $\{0,1,2\}^{\{P\}}$ can be identified with $\{0,1,2\}$. Note that when Z is non-empty, $\mathrm{SIGN}(\{P\},Z)$ is also non-empty.

- If $\mathrm{SIGN}(\{P\}, Z) = \{0, 1, -1\}$, 0, 1, 2 is adapted to sign determination for $\{P\}$ on Z, since $\{P\}^{0,1,2}=1,P,P^2$, and the matrix of signs of $1,P,P^2$ over $0,1,-1$ is
$$\begin{bmatrix} 1 & 1 & 1 \\ 0 & 1 & -1 \\ 0 & 1 & 1 \end{bmatrix},$$
which is invertible.
- If $\mathrm{SIGN}(\{P\},Z)=\{1,-1\}$ (resp. $\{0,1\}$, resp. $\{0,-1\}$), $0,1$ is adapted to sign determination for $\{P\}$ on Z, since $\{P\}^{0,1}=1,P$ and the matrix of signs of $1,P$ over $1,-1$ (resp. $0,1$, resp. $0,-1$) is
$$\begin{bmatrix} 1 & 1 \\ 1 & -1 \end{bmatrix} \quad (\text{resp. } \begin{bmatrix} 1 & 1 \\ 0 & 1 \end{bmatrix}, \text{resp. } \begin{bmatrix} 1 & 1 \\ 0 & -1 \end{bmatrix}),$$
which is invertible.
- If $\mathrm{SIGN}(\{P\}, Z) = \{0\}$ (resp. $\{1\}$, resp. $\{-1\}$), 0 is adapted to sign determination for $\{P_i\}$ on Z, since $\{P\}^0=1$ and the matrix of signs of 1 over 0 (resp. 1, resp. -1) is $\begin{bmatrix} 1 \end{bmatrix}$, which is invertible.

□

Let $Z\subset\mathrm{R}^k$ be a finite set, $\mathcal{P}$ be a finite list of polynomials in $\mathrm{R}[X_1,\ldots,X_k]$. We now describe a method for determining a list of elements in $\{0,1,2\}^{\mathcal{P}}$ adapted to sign determination for $\mathcal{P}$ on Z from the set $\mathrm{SIGN}(\mathcal{P},Z)$. The definition of this list $\mathrm{Ada}(\mathcal{P}, Z)$ is by induction on the number of elements of $\mathcal{P}$.

Before defining $\mathrm{Ada}(\mathcal{P},Z)$ we need the following definition.

Definition 10.63. [Extension of a sign condition] A sign condition $\tau \in \text{SIGN}(\{P\} \cup \mathcal{P}, Z)$ **extends** $\sigma \in \text{SIGN}(\mathcal{P}, Z)$ if $\sigma(Q) = \tau(Q), Q \in \mathcal{P}$. □

We now define $\text{Ada}(\mathcal{P}, Z)$.

Definition 10.64. [Adapted family]

- If $\mathcal{P} = \{P\}$,
  - if $\#(\text{SIGN}(\{P\}, Z)) = 3$, define $\text{Ada}(\{P\}, Z) = 0, 1, 2$,
  - if $\#(\text{SIGN}(\{P\}, Z)) = 2$, define $\text{Ada}(\{P\}, Z) = 0, 1$,
  - if $\#(\text{SIGN}(\{P\}, Z)) = 1$, define $\text{Ada}(\{P\}, Z) = 0$.
- If $\mathcal{P} = \{P\} \cup \mathcal{Q}$, let $\text{SIGN}(\mathcal{Q}, Z)_2$ be the subset of $\text{SIGN}(\mathcal{Q}, Z)$ of sign conditions σ such that there are at least two distinct sign conditions of $\text{SIGN}(\mathcal{P}, Z)$ extending σ, and $\text{SIGN}(\mathcal{Q}, Z)_3$) be the subset of $\text{SIGN}(\mathcal{Q}, Z)$ of sign conditions σ such that there are three distinct sign conditions of $\text{SIGN}(\mathcal{P}, Z)$ extending σ. Let

$$\begin{aligned} Z_2 &= \bigcup_{\sigma \in \text{SIGN}(\mathcal{Q}, Z)_2} \text{Reali}(\sigma, Z), \\ Z_3 &= \bigcup_{\sigma \in \text{SIGN}(\mathcal{Q}, Z)_3} \text{Reali}(\sigma, Z). \end{aligned}$$

  Note that

$$\begin{aligned} \text{SIGN}(\mathcal{Q}, Z_2) &= \text{SIGN}(\mathcal{Q}, Z)_2, \\ \text{SIGN}(\mathcal{Q}, Z_3) &= \text{SIGN}(\mathcal{Q}, Z)_3. \end{aligned}$$

  For $\alpha \in \{0, 1, 2\}$ and $\beta \in \{0, 1, 2\}^{\mathcal{Q}}$, we define $\alpha \times \beta \in \{0, 1, 2\}^{\mathcal{P}}$ by

$$\begin{cases} (\alpha \times \beta)(P) = \alpha(P), \\ (\alpha \times \beta)(Q) = \beta(Q) \text{ if } Q \in \mathcal{Q}. \end{cases}$$

  Define

$$\text{Ada}(\mathcal{P}, Z) = 0 \times \text{Ada}(\mathcal{Q}, Z), 1 \times \text{Ada}(\mathcal{Q}, Z_2), 2 \times \text{Ada}(\mathcal{P}, Z_3).$$

□

Proposition 10.65. *The list* $\text{Ada}(\mathcal{P}, Z)$ *is adapted to sign determination for* $\mathcal{P}$ *on* Z.

Proof: The proof proceeds by induction on the number of elements of $\mathcal{P}$. The claim is true for $\mathcal{P} = \{P\}$, as seen in Example 10.62.

If $\mathcal{P} = \{P\} \cup \mathcal{Q}$, we want to prove that

$$\text{Mat}(\text{Ada}(\mathcal{P}, Z), \text{SIGN}(\mathcal{P}, Z))$$

is invertible. Denoting by C_τ its column indexed by τ, consider a zero linear combination of its columns:

$$\sum_{\tau \in \text{SIGN}(\mathcal{P}, Z)} \lambda_\tau C_\tau = 0.$$

We want to prove that all λ_τ are zero.

If $\sigma \in \mathrm{SIGN}(\mathcal{Q}, Z)_3$, we denote by $\sigma_1 <_{\mathrm{lex}} \sigma_2 <_{\mathrm{lex}} \sigma_3$ the sign conditions of $\mathrm{SIGN}(\mathcal{P}, Z)$ extending σ.

Similarly, if $\sigma \in \mathrm{SIGN}(\mathcal{Q}, Z)_2 \setminus \mathrm{SIGN}(\mathcal{Q}, Z)_3$, we denote by

$$\sigma_1 <_{\mathrm{lex}} \sigma_2$$

the sign conditions of $\mathrm{SIGN}(\mathcal{P}, Z)$ extending σ.

Finally if $\sigma \in \mathrm{SIGN}(\mathcal{Q}, Z) \setminus \mathrm{SIGN}(\mathcal{Q}, Z)_2$, we denote by σ_1 the sign condition of $\mathrm{SIGN}(\mathcal{P}, Z)$ extending σ.

Since by induction hypothesis $\mathrm{Mat}(\mathrm{Ada}(\mathcal{Q}, Z), \mathrm{SIGN}(\mathcal{Q}, Z))$ is invertible,

$$\begin{aligned} \lambda_{\sigma_1} &= 0, & &\text{for every } \sigma \in \mathrm{SIGN}(\mathcal{Q}, Z) \setminus \mathrm{SIGN}(\mathcal{Q}, Z)_2, \\ \lambda_{\sigma_1} + \lambda_{\sigma_2} &= 0, & &\text{for every } \sigma \in \mathrm{SIGN}(\mathcal{Q}, Z)_2 \setminus \mathrm{SIGN}(\mathcal{Q}, Z)_3, \\ \lambda_{\sigma_1} + \lambda_{\sigma_2} + \lambda_{\sigma_3} &= 0, & &\text{for every } \sigma \in \mathrm{SIGN}(\mathcal{Q}, Z)_3. \end{aligned}$$

By induction hypothesis, the matrix $\mathrm{Mat}(\mathrm{Ada}(\mathcal{Q}, Z_2), \mathrm{SIGN}(\mathcal{Q}, Z_2))$ is invertible, then $\sigma_1(P)\,\lambda_{\sigma_1} - \sigma_2(P)\,\lambda_{\sigma_2} = 0$, for every $\sigma \in \mathrm{SIGN}(\mathcal{Q}, Z)_2 \setminus \mathrm{SIGN}(\mathcal{Q}, Z)_3$ and $\lambda_{\sigma_2} - \lambda_{\sigma_3} = 0$, for every $\mathrm{SIGN}(\mathcal{Q}, Z)_3$, Thus $\lambda_{\sigma_1} = \lambda_{\sigma_2} = 0$, for every $\sigma \in \mathrm{SIGN}(\mathcal{Q}, Z)_2 \setminus \mathrm{SIGN}(\mathcal{Q}, Z)_3$. Finally, using again the induction hypothesis, $\mathrm{Mat}(\mathrm{Ada}(\mathcal{Q}, Z_3), \mathrm{SIGN}(\mathcal{Q}, Z_3))$ is invertible, then $\lambda_{\sigma_2} + \lambda_{\sigma_3} = 0$ for every $\sigma \in \mathrm{SIGN}(\mathcal{Q}, Z)_3$. Thus $\lambda_{\sigma_1} = \lambda_{\sigma_2} = \lambda_{\sigma_3} = 0$ for every $\sigma \in \mathrm{SIGN}(\mathcal{Q}, Z)_3$.

This proves that the matrix

$$\mathrm{Mat}(\mathrm{Ada}(\mathcal{P}, Z), \mathrm{SIGN}(\mathcal{P}, Z))$$

is invertible. □

Lemma 10.66. *Let* $Z' \subset Z$, $r = \#(\mathrm{SIGN}(\mathcal{P}, Z))$, $r' = \#(\mathrm{SIGN}(\mathcal{P}, Z'))$. *The matrix* $\mathrm{Mat}(\mathrm{Ada}(\mathcal{P}, Z'), \mathrm{SIGN}(\mathcal{P}, Z'))$ *coincides with the matrix obtained by extracting from* $\mathrm{Mat}(\mathrm{Ada}(\mathcal{P}, Z), \mathrm{SIGN}(\mathcal{P}, Z'))$ *its* r' *first linearly independent rows.*

Proof: The proof is by induction on the number of elements of $\mathcal{P}$.

The claim is clearly true is $\mathcal{P} = \{P\}$.

Suppose now that $\mathcal{P} = \{P\} \cup \mathcal{Q}$ and the claim holds for $\mathcal{Q}$.

Note that by Definition 10.64, $\mathrm{Ada}(\mathcal{P}, Z')$ is a sublist of $\mathrm{Ada}(\mathcal{P}, Z)$, so that the rank of $\mathrm{Mat}(\mathrm{Ada}(\mathcal{P}, Z), \mathrm{SIGN}(\mathcal{P}, Z'))$ is r'.

Similarly,

$$\begin{aligned} r_1' &= \#(\mathrm{SIGN}(\mathcal{Q}, Z')) = \mathrm{Rank}(\mathrm{Mat}(\mathrm{Ada}(\mathcal{Q}, Z), \mathrm{SIGN}(\mathcal{Q}, Z'))), \\ r_2' &= \#(\mathrm{SIGN}(\mathcal{Q}, Z')_2) = \mathrm{Rank}(\mathrm{Mat}(\mathrm{Ada}(\mathcal{Q}, Z_2), \mathrm{SIGN}(\mathcal{Q}, Z'))), \\ r_3' &= \#(\mathrm{SIGN}(\mathcal{Q}, Z')_3) = \mathrm{Rank}(\mathrm{Mat}(\mathrm{Ada}(\mathcal{Q}, Z_3), \mathrm{SIGN}(\mathcal{Q}, Z')). \end{aligned}$$

It follows immediately that,

$$\begin{aligned} \mathrm{Rank}(\mathrm{Mat}(0 \times \mathrm{Ada}(\mathcal{Q}, Z), \mathrm{SIGN}(\mathcal{P}, Z'))) &\le r_1', \\ \mathrm{Rank}(\mathrm{Mat}(1 \times \mathrm{Ada}(\mathcal{Q}, Z_2, \mathrm{SIGN}(\mathcal{P}, Z'))) &\le r_2', \\ \mathrm{Rank}(\mathrm{Mat}(2 \times \mathrm{Ada}(\mathcal{Q}, Z_3), \mathrm{SIGN}(\mathcal{P}, Z'))) &\le r_3'. \end{aligned}$$

Finally, since $r'_1+r'_2+r'_3=r$,

$$\begin{aligned}\mathrm{Rank}(\mathrm{Mat}(0\times \mathrm{Ada}(\mathcal{Q},Z),\mathrm{SIGN}(\mathcal{P},Z'))) &= r'_1,\\ \mathrm{Rank}(\mathrm{Mat}(1\times \mathrm{Ada}(\mathcal{Q},Z_2),\mathrm{SIGN}(\mathcal{P},Z'))) &= r'_2,\\ \mathrm{Rank}(\mathrm{Mat}(2\times \mathrm{Ada}(\mathcal{Q},Z_3),\mathrm{SIGN}(\mathcal{P},Z'))) &= r'_3,\end{aligned}$$

and the first r' linearly independent rows of $\mathrm{Mat}(\mathrm{Ada}(\mathcal{P}, Z), \mathrm{SIGN}(\mathcal{P}, Z'))$ consist of r'_1 rows of $\mathrm{Mat}(0\times \mathrm{Ada}(\mathcal{Q},Z),\mathrm{SIGN}(\mathcal{P},Z'))$, r'_2 linearly independent rows of $\mathrm{Mat}(1\times \mathrm{Ada}(\mathcal{Q},Z_2),\mathrm{SIGN}(\mathcal{P},Z'))$ and r'_3 linearly independent rows of $\mathrm{Mat}(2\times \mathrm{Ada}(\mathcal{Q},Z_2),\mathrm{SIGN}(\mathcal{P},Z'))$. The corresponding r'_1 (resp. r'_2, resp. r'_3) rows of $\mathrm{Mat}(\mathrm{Ada}(\mathcal{Q}, Z), \mathrm{SIGN}(\mathcal{Q}, Z'))$ (resp. $\mathrm{Mat}(\mathrm{Ada}(\mathcal{Q}, Z_2)$, $\mathrm{SIGN}(\mathcal{Q}, Z'_2)$, resp. $\mathrm{Mat}(\mathrm{Ada}(\mathcal{Q}, Z_3), \mathrm{SIGN}(\mathcal{Q}, Z'_3))$ are linearly independent and are the rows indexed by $\mathrm{Ada}(\mathcal{Q},Z')$ (resp. $\mathrm{Ada}(\mathcal{Q},Z'_2)$, resp. $\mathrm{Ada}(\mathcal{Q},Z'_3)$) by the induction hypothesis. The claim follows from Definition 10.64.

□

Algorithm 10.10. **[Adapted Family]**

- **Input:** the set $\mathrm{SIGN}(\{P\}\cup\mathcal{Q},Z)$, the list $\mathrm{Ada}(\mathcal{Q},Z)$.
- **Output:** the list $\mathrm{Ada}(\{P\}\cup\mathcal{Q},Z)$.
- **Procedure:**
  - If $\mathcal{Q}=\emptyset$,
  - if $\#(\mathrm{SIGN}(\{P\},Z))=3$, define $\mathrm{Ada}(\{P\},Z)=0,1,2$,
  - if $\#(\mathrm{SIGN}(\{P\},Z))=2$, define $\mathrm{Ada}(\{P\},Z)=0,1$,
  - if $\#(\mathrm{SIGN}(\{P\},Z))=1$, define $\mathrm{Ada}(\{P\},Z)=0$.
  - Using the notation in Definition 10.64, let

$$\begin{aligned} r_1 &= \#(\mathrm{SIGN}(\mathcal{Q},Z)),\\ r_2 &= \#(\mathrm{SIGN}(\mathcal{Q},Z)_2),\\ r_3 &= \#(\mathrm{SIGN}(\mathcal{Q},Z)_3).\end{aligned}$$

  Then $\#(\mathrm{SIGN}(\{P\}\cup\mathcal{Q},Z))=r_1+r_2+r_3$.

  Consider the matrix $\mathrm{Mat}(\mathrm{Ada}(\mathcal{Q}, Z), \mathrm{SIGN}(\mathcal{Q}, Z)_2)$ and extract from it the first r_2 linearly independent rows, which correspond to a sublist A_2 of $\mathrm{Ada}(\mathcal{Q},Z)$.

  Similarly, consider the matrix $\mathrm{Mat}(\mathrm{Ada}(\mathcal{Q},Z),\mathrm{SIGN}(\mathcal{Q},Z)_3)$ and extract from it the first r_3 linearly independent rows which correspond to a sublist A_3 of $\mathrm{Ada}(\mathcal{Q},Z)$.

  Output

$$\mathrm{Ada}(\{P\}\cup\mathcal{Q},Z)=0\times\mathrm{Ada}(\mathcal{Q},Z),1\times A_2,2\times A_3.$$

Correctness of Algorithm 10.10 : Follows immediately from Definition 10.64 and Lemma 10.66.

□

Notation 10.67. [Sign determination] Let $\mathcal{P} = P_1, \ldots, P_s$. For $1 \le i \le s$ we define $\mathcal{P}_i = P_i, \ldots, P_s$. For $\sigma \in \{0, 1, -1\}$ and $\tau \in \{0, 1, -1\}^{\mathcal{P}_{i+1}}$, we define $\sigma \wedge \tau \in \{0, 1, -1\}^{\mathcal{P}_i}$ by

$$\begin{cases} (\sigma \wedge \tau)(P_i) = \sigma(P_i) \\ (\sigma \wedge \tau)(P) = \tau(P) \quad \text{if } P \in \mathcal{P}_{i+1}, \end{cases}$$

If $\Sigma = \sigma_1, \ldots, \sigma_m$ is a list of elements of $\{0, 1, -1\}$ with $\sigma_1 <_{\text{lex}} \ldots <_{\text{lex}} \sigma_m$ and $T = \tau_1, \ldots, \tau_n$ is a list of element of $\{0, 1, -1\}^{\mathcal{P}_i}$ with $\tau_1 <_{\text{lex}} \ldots <_{\text{lex}} \tau_n$, then $\Sigma \wedge T$ is the list

$$\sigma_1 \wedge \tau_1 <_{\text{lex}} \ldots <_{\text{lex}} \sigma_1 \wedge \tau_n <_{\text{lex}} \ldots <_{\text{lex}} \sigma_m \wedge \tau_1 <_{\text{lex}} \ldots <_{\text{lex}} \sigma_m \wedge \tau_n.$$

For $\alpha \in \{0, 1, 2\}$ and $\beta \in \{0, 1, 2\}^{\mathcal{P}_{i+1}}$, we define $\alpha \times \beta \in \{0, 1, 2\}^{\mathcal{P}_i}$ by

$$\begin{cases} (\alpha \times \beta)(P_i) = \alpha, \\ (\alpha \times \beta)(P) = \beta(P) \quad \text{if } P \in \mathcal{P}_{i+1}. \end{cases}$$

If $A = \alpha_1 <_{\text{lex}} \ldots <_{\text{lex}} \alpha_m$ and $B = \beta_1 <_{\text{lex}} \ldots <_{\text{lex}} \beta_n$ are lists of elements of $\{0, 1, 2\}$ and $\{0, 1, 2\}^{\mathcal{P}_{i+1}}$ we define $A \times B$ to be the list

$$\alpha_1 \times \beta_1 <_{\text{lex}} \ldots <_{\text{lex}} \alpha_1 \times \beta_n <_{\text{lex}} \ldots <_{\text{lex}} \alpha_m \times \beta_1 <_{\text{lex}} \ldots <_{\text{lex}} \alpha_m \times \beta_n$$

in $\{0, 1, 2\}^{\mathcal{P}_i}$.

The list $\mathcal{P}_i^{A \times B}$ is defined to be

$$P_i^{\alpha_1} \mathcal{P}_{i+1}^{\beta_1}, \ldots, P_i^{\alpha_1} \mathcal{P}_{i+1}^{\beta_n}, \ldots, P_i^{\alpha_m} \mathcal{P}_{i+1}^{\beta_1}, \ldots, P_i^{\alpha_m} \mathcal{P}_{i+1}^{\beta_n}.$$

□

Recall that the matrix of signs of $\mathcal{P}^B$ (of length m) on Σ (of length n) is the $m \times n$ matrix $\mathrm{Mat}(B, \Sigma)$ whose i, j-th entry is $\sigma_j^{\alpha_i}$, and that $\mathrm{TaQ}(\mathcal{P}^B, Z)$ is the vector $\mathrm{TaQ}(\mathcal{P}^{\beta_1}, Z), \ldots, \mathrm{TaQ}(\mathcal{P}^{\beta_m}, Z)$. Using Notation 2.69 (Tensor product) we have

Proposition 10.68. *If* $\cup_{\sigma \in \Sigma} \mathrm{Reali}(\sigma, Z) = Z$ $A = 0, 1, 2,$ *and* $T = \{0, 1, -1\}$, *let*

$$(\mathrm{Mat}(A, T) \otimes \mathrm{Mat}(B, \Sigma)) \cdot c(T \wedge \Sigma, Z) = \mathrm{TaQ}(\mathcal{P}_i^{A \times B}, Z).$$

Proof: Immediate from Proposition 10.59. □

We are now ready for the Sign Determination algorithm.

Algorithm 10.11. **[Sign Determination]**

- **Input:** a finite subset $Z \subset \mathrm{R}^k$ with r elements and a finite list $\mathcal{P} = P_1, \ldots, P_s$ of polynomials in $\mathrm{R}[X_1, \ldots, X_k]$.
- **Output:** the list of sign conditions realized by $\mathcal{P}$ on Z, $\mathrm{SIGN}(\mathcal{P}, Z)$.
- **Blackbox:** for a polynomial P, the Tarski-query $\mathrm{TaQ}(P, Z)$.
- **Complexity:** $1 + 2\, s\, r$ calls to the to the Tarski-query black box.

- **Procedure:**
  - Compute $r = \mathrm{TaQ}(1, Z)$ using the Tarski-query black box with input 1. If $r = 0$, output $\emptyset$.
  - Let $\mathcal{P}_i = P_i, \ldots, P_s$. We are going to determine iteratively, for i from s to 1, $\mathrm{SIGN}(\mathcal{P}_i, Z)$ the non-empty sign conditions for $\mathcal{P}_i$ on Z. More precisely, we are going to compute $\mathrm{SIGN}(\mathcal{P}_i, Z)$ and $\mathrm{Ada}(\mathcal{P}_i, Z)$, starting from $\mathrm{SIGN}(\mathcal{P}_{i+1}, Z)$ and $\mathrm{Ada}(\mathcal{P}_{i+1}, Z)$.
  - For i from s to 1,
    - Determine $\mathrm{SIGN}(P_i, Z)$, the list of sign conditions realized by P_i on Z, and a list B_i of elements in $\{0, 1, 2\}$ adapted to sign determination for P_i on Z as follows:
      - Use the Tarski-query black box with inputs P_i and P_i^2 to determine $\mathrm{TaQ}(P_i, Z)$ and $\mathrm{TaQ}(P_i^2, Z)$.
      - From these values, using the equality

$$\begin{bmatrix} 1 & 1 & 1 \\ 0 & 1 & -1 \\ 0 & 1 & 1 \end{bmatrix} \begin{bmatrix} c(P_i = 0, Z) \\ c(P_i > 0, Z) \\ c(P_i < 0, Z) \end{bmatrix} = \begin{bmatrix} \mathrm{TaQ}(1, Z) \\ \mathrm{TaQ}(P_i, Z) \\ \mathrm{TaQ}(P_i^2, Z) \end{bmatrix},$$

        compute $c(P_i = 0, Z)$, $c(P_i > 0, Z)$ and $c(P_i < 0, Z)$ and output $\mathrm{SIGN}(P_i, Z)$.
      - If $r(P_i) = \#(\mathrm{SIGN}(P_i, Z)) = 3$, let $B_i = 0, 1, 2$.
      - If $r(P_i) = \#(\mathrm{SIGN}(P_i, Z)) = 2$, let $B_i = 0, 1$.
      - If $r(P_i) = \#(\mathrm{SIGN}(P_i, Z)) = 1$, let $B_i = 0$.
      - Define $M_i = \mathrm{Mat}(B_i, \mathrm{SIGN}(P_i, Z))$.
    - If $i = s$, define $\mathrm{SIGN}(\mathcal{P}_s, Z) := \mathrm{SIGN}(P_s, Z)$, $\mathrm{Ada}(\mathcal{P}_s, Z) := B_s$.
    - If $i < s$, Compute $\mathrm{SIGN}(\mathcal{P}_i, Z)$, the list of sign conditions realized by $\mathcal{P}_i$ on Z, as follows:
      - Use the Tarski-query black box with input the elements of $\mathcal{P}_i^{B_i \times \mathrm{Ada}(\mathcal{P}_{i+1}, Z)}$ to determine $d' = \mathrm{TaQ}(\mathcal{P}_i^{B_i \times \mathrm{Ada}(\mathcal{P}_{i+1}, Z)}, Z)$.
      - Take the matrix

$$M_i' := \mathrm{Mat}(\mathrm{Ada}(\mathcal{P}_{i+1}, Z), \mathrm{SIGN}(\mathcal{P}_{i+1}, Z)) \otimes M_i.$$

        Compute the list $c' = c(\mathrm{SIGN}(P_i, Z) \wedge \mathrm{SIGN}(\mathcal{P}_{i+1}, Z))$ from the equality $M_i' \cdot c' = d'$ by inverting M_i'. Compute $\mathrm{SIGN}(\mathcal{P}_i, Z)$, removing from $\mathrm{SIGN}(P_i, Z) \wedge \mathrm{SIGN}(\mathcal{P}_{i+1}, Z)$ the sign conditions with empty realization, which correspond to the zeroes in c'.
      - Call Algorithm 10.10 (Adapted family) with input $\mathrm{SIGN}(\mathcal{P}_i, Z)$ and $\mathrm{Ada}(\mathcal{P}_{i+1}, Z)$, and compute $\mathrm{Ada}(\mathcal{P}_i, Z)$.
  - Output $\mathrm{SIGN}(\mathcal{P}, Z) = \mathrm{SIGN}(\mathcal{P}_1, Z)$.

Remark 10.69. We denote by $B(\mathrm{SIGN}(\mathcal{P}, Z)) \subset \{0, 1, 2\}^{\mathcal{P}}$ the set constructed inductively as follows:

$$\begin{aligned} B(\mathrm{SIGN}(\mathcal{P}_s, Z)) &= \{0, 1, 2\}_1 \\ B(\mathrm{SIGN}(\mathcal{P}_i, Z)) &= B(\mathrm{SIGN}(\mathcal{P}_{i+1}, Z)) \cup \{0, 1, 2\}_i \cup B_i \times \mathrm{Ada}(\mathcal{P}_{i+1}, Z), \end{aligned}$$

denoting by $\{0,1,2\}_i$ the subset of $\{0,1,2\}^{\mathcal{P}}$ with three elements defined by

$$\alpha \in \{0,1,2\}_i \Leftrightarrow \alpha(j) = 0 \; \forall j \neq i,$$

and identifying $\alpha \in \{0,1,2\}^{\mathcal{P}_i}$ with $\alpha' \in \{0,1,2\}^{\mathcal{P}}$ such that

$$\alpha'(P_j) = \alpha(P_j), j \geqslant i, \alpha'(P_j) = 0, j < i,$$

using the notation of Algorithm 10.11 (Sign Determination). It is easy to see that $B(\mathrm{SIGN}(\mathcal{P},\, Z))$ is nothing but the list of elements $\alpha \in \{0,\, 1,\, 2\}^{\mathcal{P}}$ such that the Tarski-query of P^{α} has been computed in Algorithm 10.11 (Sign Determination). Using Algorithm 10.10 (Adapted family), it is clear that $B(\mathrm{SIGN}(\mathcal{P}, Z))$ can be determined from $\mathrm{SIGN}(\mathcal{P}, Z)$. □

Proof of correctness of Algorithm 10.11: It follows from Corollary 10.68 and the correctness of Algorithm 10.10 (Adapted family). □

Before discussing the complexity of the Sign Determination Algorithm, we first give an example.

Example 10.70. Consider

$$\begin{aligned} P &= (X^3 - 1)(X^2 - 9), \\ Z &= \mathrm{Zer}(P, R), \\ P_1 &= X - 2, \\ P_2 &= X + 1, \\ P_3 &= X. \end{aligned}$$

The call to the Tarski-query black box with input 1 determines $\mathrm{TaQ}(1, Z) = 3$. So P has 3 real roots (which is not a real surprise).

The call to the Tarski-query black box with inputs P_3 and P_3^2 determines $\mathrm{TaQ}(P_3, Z) = 1$ and $\mathrm{TaQ}(P_3^2, Z) = 3$. Thus

$$\begin{bmatrix} 1 & 1 & 1 \\ 0 & 1 & -1 \\ 0 & 1 & 1 \end{bmatrix} \cdot \begin{bmatrix} c(P_3 = 0, Z) \\ c(P_3 > 0, Z) \\ c(P_3 < 0, Z) \end{bmatrix} = \begin{bmatrix} 3 \\ 1 \\ 3 \end{bmatrix},$$

which means, after solving the system, that P has

$$\begin{cases} 0 \text{ root with } P_3 = 0 \\ 2 \text{ roots with } P_3 > 0 \\ 1 \text{ root with } P_3 < 0 \end{cases}.$$

Hence $c(P_3 = 0, Z) = 0$. So we have $\mathrm{SIGN}(P_3, Z) = 1, -1$ and

$$\mathrm{Ada}(\mathcal{P}_3, Z) = B_3 = 0, 1.$$

The matrix $\mathrm{Mat}(\mathrm{Ada}(\mathcal{P}_3, Z), \mathrm{SIGN}(\mathcal{P}_3, Z))$ of signs of $1, P_3$ on $1, -1$ is

$$\begin{bmatrix} 1 & 1 \\ 1 & -1 \end{bmatrix}.$$

We now consider $\mathcal{P}_2 = P_2, P_3$.

The call to the Tarski-query black box with inputs P_2 and P_2^2 determines $\mathrm{TaQ}(P_2, Z) = 1, \mathrm{TaQ}(P_2^2, Z) = 3$. Hence,

$$\begin{bmatrix} 1 & 1 & 1 \\ 0 & 1 & -1 \\ 0 & 1 & 1 \end{bmatrix} \cdot \begin{bmatrix} c(P_2 = 0, Z) \\ c(P_2 > 0, Z) \\ c(P_2 < 0, Z) \end{bmatrix} = \begin{bmatrix} 3 \\ 1 \\ 3 \end{bmatrix},$$

which means, after solving the system, that P has

$$\begin{cases} 0 \text{ root with } P_2 = 0 \\ 2 \text{ roots with } P_2 > 0 \\ 1 \text{ root with } P_2 < 0 \end{cases}.$$

Hence $c(P_2 = 0, Z) = 0$. So we have $\mathrm{SIGN}(P_2, Z) = 1, -1$ and $B_2 = 0, 1$. The matrix M_2 of signs of $1, P_2$ on the sign conditions $1, -1$ is

$$\begin{bmatrix} 1 & 1 \\ 1 & -1 \end{bmatrix}.$$

The call to the Tarski-query black box with input $P_2 P_3$ yields $\mathrm{TaQ}(P_2 P_3, Z)$, which is equal to 3. Hence we have

$$M_2' = \mathrm{Mat}(\mathrm{Ada}(\mathcal{P}_3, Z), \mathrm{SIGN}(\mathcal{P}_3, Z)) \otimes M_2 = \begin{bmatrix} 1 & 1 & 1 & 1 \\ 1 & -1 & 1 & -1 \\ 1 & 1 & -1 & -1 \\ 1 & -1 & -1 & 1 \end{bmatrix},$$

$$\begin{bmatrix} 1 & 1 & 1 & 1 \\ 1 & -1 & 1 & -1 \\ 1 & 1 & -1 & -1 \\ 1 & -1 & -1 & 1 \end{bmatrix} \cdot \begin{bmatrix} c(P_2 > 0 \wedge P_3 > 0, Z) \\ c(P_2 > 0 \wedge P_3 < 0, Z) \\ c(P_2 < 0 \wedge P_3 > 0, Z) \\ c(P_2 < 0 \wedge P_3 < 0, Z) \end{bmatrix} = \begin{bmatrix} 3 \\ 1 \\ 1 \\ 3 \end{bmatrix}.$$

Solving the system we find that P has

$$\begin{cases} 2 \text{ roots with } P_2 > 0 \text{ and } P_3 > 0 \\ 0 \text{ roots with } P_2 > 0 \text{ and } P_3 < 0 \\ 0 \text{ roots with } P_2 < 0 \text{ and } P_3 > 0 \\ 1 \text{ root with } P_2 < 0 \text{ and } P_3 < 0 \end{cases}.$$

So we have $\mathrm{SIGN}(\mathcal{P}_2, Z) = (1, 1), (-1, -1)$. There is no sign condition on P_3 which is partitioned by sign conditions on P_2, so $\mathrm{Ada}(\mathcal{P}_2, Z) = (0,0), (1,0)$. The matrix $\mathrm{Mat}(\mathrm{Ada}(\mathcal{P}_2, Z), \mathrm{SIGN}(\mathcal{P}_2, Z))$ of signs of $1, P_3$ on the sign conditions $(1, 1), (-1, -1)$ is

$$\begin{bmatrix} 1 & 1 \\ 1 & -1 \end{bmatrix}.$$

Finally we consider $\mathcal{P} = P_1, P_2, P_3$.

The call to the Tarski-query black box with inputs P_1 and P_1^2 determines $\mathrm{TaQ}(P_1, Z) = -1, \mathrm{TaQ}(P_1^2, Z) = 3$. Hence $c(P_1 = 0, Z) = 0$. So,

$$\begin{bmatrix} 1 & 1 & 1 \\ 0 & 1 & -1 \\ 0 & 1 & 1 \end{bmatrix} \cdot \begin{bmatrix} c(P_1 = 0, Z) \\ c(P_1 > 0, Z) \\ c(P_1 < 0, Z) \end{bmatrix} = \begin{bmatrix} 3 \\ -1 \\ 3 \end{bmatrix},$$

which means, after solving the system, that P has

$$\begin{cases} 0 \text{ root with } P_1 = 0 \\ 1 \text{ root with } P_1 > 0 \\ 2 \text{ roots with } P_1 < 0 \end{cases}.$$

So we have $\mathrm{SIGN}(P_1, Z) = \{1, -1\}$, $B_1 = \{0, 1\}$. The matrix M_1 of signs of $1, P_1$ on $1, -1$ is

$$\begin{bmatrix} 1 & 1 \\ 1 & -1 \end{bmatrix}.$$

The call to the Tarski-query black box with input $P_1 P_3$ yields $\mathrm{TaQ}(P_1 P_3, Z)$ which is equal to 1. Hence we have

$$M_1' = \mathrm{Mat}(\mathrm{Ada}(\mathcal{P}_2, Z), \mathrm{SIGN}(\mathcal{P}_2, Z)) \otimes M_1 = \begin{bmatrix} 1 & 1 & 1 & 1 \\ 1 & -1 & 1 & -1 \\ 1 & 1 & -1 & -1 \\ 1 & -1 & -1 & 1 \end{bmatrix},$$

$$\begin{bmatrix} 1 & 1 & 1 & 1 \\ 1 & -1 & 1 & -1 \\ 1 & 1 & -1 & -1 \\ 1 & -1 & -1 & 1 \end{bmatrix} \cdot \begin{bmatrix} c(P_1 > 0, P_2 > 0, P_3 > 0, Z) \\ c(P_1 > 0, P_2 > 0, P_3 < 0, Z) \\ c(P_1 < 0, P_2 < 0, P_3 > 0, Z) \\ c(P_1 < 0, P_2 < 0, P_3 < 0, Z) \end{bmatrix} = \begin{bmatrix} 3 \\ -1 \\ 1 \\ 1 \end{bmatrix}.$$

Solving the system, we find that P has

$$\begin{cases} 1 \text{ root with } P_1 > 0 \text{ and } P_2 > 0 \text{ and } P_3 > 0 \\ 0 \text{ root with } P_1 > 0 \text{ and } P_2 < 0 \text{ and } P_3 < 0 \\ 1 \text{ root with } P_1 < 0 \text{ and } P_2 > 0 \text{ and } P_3 > 0 \\ 1 \text{ root with } P_1 < 0 \text{ and } P_2 < 0 \text{ and } P_3 < 0 \end{cases}.$$

So we have $\mathrm{SIGN}(\mathcal{P}) = \{(1, 1, 1), (-1, 1, 1), (-1, -1, -1)\}$. □

In order to study the complexity of the Algorithm 10.11 (Sign Determination) we need the following proposition.

Proposition 10.71. *Let Z be a finite subset of R^k and $r = \#(Z)$. For every $\alpha \in \mathrm{Ada}(\mathcal{P}, Z)$, the number $\#(\{P \in \mathcal{P} \mid \alpha(P) \neq 0\})$ is at most $\log_2(r)$.*

We need the following definition. Let α and β be elements of $\{0, 1, 2\}^{\mathcal{P}}$. We say that β **precedes** α if for every $P \in \mathcal{P}$, $\beta(P) \neq 0$ implies $\beta(P) = \alpha(P)$. Note that if β precedes α, then $\beta <_{\mathrm{lex}} \alpha$.

The proof of Proposition 10.71 is based on the following lemma.

Lemma 10.72. *If β precedes α and $\alpha \in \mathrm{Ada}(\mathcal{P}, Z)$ then $\beta \in \mathrm{Ada}(\mathcal{P}, Z)$.*

Proof: The proof is by induction on the number of elements of $\mathcal{P}$. The claim is obvious for $\mathcal{P} = \{P\}$.

If $\mathcal{P} = \{P\} \cup \mathcal{Q}$, $\alpha \in \{0,1,2\}^{\mathcal{P}}$ we denote by α' the element of $\{0,1,2\}^{\mathcal{Q}}$ such that $\alpha'(P) = \alpha(P)$, $P \in \mathcal{Q}$. Note that, by definition of $\mathrm{Ada}(\mathcal{P}, Z)$, if $\alpha' \notin \mathrm{Ada}(\mathcal{Q}, Z)$, $\alpha \notin \mathrm{Ada}(\mathcal{P}, Z)$.

Suppose that β precedes α and that $\beta \notin \mathrm{Ada}(\mathcal{P}, Z)$. There are several cases to consider:

- If $\alpha(P) = \beta(P) = 1$, $\beta' \notin \mathrm{Ada}(\mathcal{Q}, Z_2)$ by Definition 10.64. By induction hypothesis, $\alpha' \notin \mathrm{Ada}(\mathcal{Q}, Z_2)$ and $\alpha = 1 \times \alpha' \notin \mathrm{Ada}(\mathcal{P}, Z)$ again by Definition 10.64.
- If $\alpha(P) = \beta(P) = 2$, $\beta' \notin \mathrm{Ada}(\mathcal{Q}, Z_3)$ by Definition 10.64. By induction hypothesis, $\alpha' \notin \mathrm{Ada}(\mathcal{Q}, Z_3)$ and $\alpha = 2 \times \alpha' \notin \mathrm{Ada}(\mathcal{P}, Z)$ again by Definition 10.64.
- If $\beta(P) = 0$, $\beta' \notin \mathrm{Ada}(\mathcal{Q}, Z)$ by Definition 10.64, thus $\alpha' \notin \mathrm{Ada}(\mathcal{Q}, Z)$ by induction hypothesis, and $\alpha \notin \mathrm{Ada}(\mathcal{P}, Z)$ by Definition 10.64. □

Proof of Proposition 10.71:

Let α be such that $\#(\{P \in \mathcal{P} \mid \alpha(P) \neq 0\}) = k$. Since the number of elements β of $\{0,1,2\}^{\mathcal{P}}$ preceding α is 2^k, and the total number of polynomials in A_s is at most r, we have $2^k \le r$ and $k \le \log_2(r)$. So, the claim follows immediately from the Lemma 10.72. □

Complexity analysis: There are s steps in Algorithm 10.11 (Sign Determination). In each step, the number of calls to the Tarski-query black box is bounded by $2\,r$. Indeed, in Step i, there are at most $3\,r_{i-1}$ Tarski-queries to compute and r_{i-1} of these Tarski-queries have been determined in Step $i-1$. So, in Step i, there are at most $2\,r_{i-1}$ Tarski-queries to determine. The total number of calls to the to the Tarski-query black box is bounded by $1 + 2s\,r$. The calls to the Tarski-query black box are done for polynomials which are product of at most $\log_2(r)$ products of polynomials of the form P or P^2, $P \in \mathcal{P}$ by Proposition 10.71. □

Note that we did not count the complexity of performing the linear algebra involved in the algorithm. This is because when we consider particular ways of realization the Tarski-query black box later, we bound only the number of arithmetic operations in the ring. Since the complexity of linear algebra is polynomial in the size of the matrix, the maximum size of the matrices is $3r$, and their entries are $0, 1$ or -1, taking into account the linear algebra part of the algorithm would not change the linearity in s and the polynomial time in r character of the algorithm.

We finally describe how to get a family adapted to sign determination from $\mathrm{SIGN}(\mathcal{P}, Z)$ using Algorithm 10.10 (Adapted family).

Algorithm 10.12. **[Family adapted to Sign Determination]**

- **Input:** the set $\mathrm{SIGN}(\mathcal{P}, Z)$.
- **Output:** a list $\mathrm{Ada}(\mathcal{P}, Z)$ of elements in $\{0,1,2\}^{\mathcal{P}}$ adapted to sign determination for $\mathcal{P}$ on Z.
- **Procedure:**
  Let $\mathcal{P} = P_1, ..., P_s$, $\mathcal{P}_i = P_i, ..., P_s$, $i = 1, ..., s$. Note that $\mathrm{SIGN}(\mathcal{P}_i, Z)$ can be obtained from $\mathrm{SIGN}(\mathcal{P}, Z)$ by forgetting the $i-1$ first signs of elements of $\mathrm{SIGN}(\mathcal{P}, Z)$.
  - If $\#(\mathrm{SIGN}(\mathcal{P}_s, Z)) = 3$, define $\mathrm{Ada}(\mathcal{P}_s, Z) = 0, 1, 2$.
  - If $\#(\mathrm{SIGN}(\mathcal{P}_s, Z)) = 2$, define $\mathrm{Ada}(\mathcal{P}_s, Z) = 0, 1$.
  - If $\#(\mathrm{SIGN}(\mathcal{P}_s, Z)) = 1$, define $\mathrm{Ada}(\mathcal{P}_s, Z) = 0$.
  - For i from $s-1$ to 1, apply Algorithm 10.10 (Adapted family) to $\mathrm{SIGN}(\mathcal{P}_i, Z)$, $\mathrm{Ada}(\mathcal{P}_{i+1}, Z)$ to obtain $\mathrm{Ada}(\mathcal{P}_i, Z)$.
  - Output $\mathrm{Ada}(\mathcal{P}_1, Z)$.

We can now describe in a more specific way how the Tarski-query black box can be implemented in the univariate case.

Algorithm 10.13. **[Univariate Sign Determination]**

- **Structure:** an ordered integral domain D, contained in a real closed field R.
- **Input:** a non-zero univariate polynomial Q and a list $\mathcal{P}$ of univariate polynomials with coefficients in D. Let $Z = \mathrm{Zer}(Q, \mathrm{R})$.
- **Output:** the list of sign conditions realized by $\mathcal{P}$ on Z, $\mathrm{SIGN}(\mathcal{P}, Z)$, and a list A of elements in $\{0,1,2\}^{\mathcal{P}}$ adapted to sign determination for $\mathcal{P}$ on Z.
- **Complexity:** $O(s\,p^2\,(p + q\log_2(p)))$, where s is a bound on the number of polynomials in $\mathcal{P}$, p is a bound on the degree of Q and q is a bound on the degree of the polynomials in $\mathcal{P}$.
- **Procedure:** Perform Algorithm 10.11 (Sign Determination), using as Tarski-query black box Algorithm 9.5 (Univariate Tarski-query). Products of elements of $\mathcal{Q}$ are reduced modulo P each time a multiplication is performed.

Complexity analysis: According to the complexity of Algorithm 10.11 (Sign Determination), the number of calls to the Tarski-query black box is bounded by $1 + 2\,s\,p$, since $r \le p$. The calls to the Tarski-query black box are done for P and polynomials of degree at most q. The complexity is thus $O(s\,p^2\,(p + q\,))$, using the complexity of Algorithm 9.5 (Univariate Tarski-query).

When Q and $P \in \mathcal{P}$ are in $\mathbb{Z}[X]$ with coefficients of bitsize bounded by τ, the bitsize of the integers in the operations performed by the algorithm are bounded by $O((p + q\log_2(p))(\tau + \log_2(p + q\log_2(p))))$, according to Proposition 8.44. □

Remark 10.73. Using Remark 9.2, it is possible to compute the output of Algorithm 10.13 (Univariate Sign Determination) with complexity $\tilde{O}(s\,p\,(p+q))$ and binary complexity $\tilde{O}(s\,p\,(p+q)^2\,\tau)$. □

10.4 Roots in a Real Closed Field

We consider here too a general real closed field R, not necessarily archimedean. In such a field, it is not possible to perform real root isolation and to approximate roots by rational numbers. In order to characterize and compute the roots of a polynomial, in a sense made precise in this section, we are going to use Proposition 2.28 (Thom encoding) and the preceding sign determination method.

Let $P\in \mathrm{R}[X]$ and $\sigma\in\{0,1,-1\}^{\mathrm{Der}(P)}$, a sign condition on the set $\mathrm{Der}(P)$ of derivatives of P. By Definition 2.29, the sign condition σ is a Thom encoding of $x\in \mathrm{R}$ if $\sigma(P)=0$ and σ is the sign condition taken by the set $\mathrm{Der}(P)$ at x. We say that x is specified by σ. Given a Thom encoding σ, we denote by $x(\sigma)$ the root of P in R specified by σ.

The **ordered list of Thom encodings of** P is the ordered list $\sigma_1,\ldots,\sigma_r$ of Thom encodings of the roots $x(\sigma_1)<\cdots<x(\sigma_r)$ of P.

The ordered list of Thom encodings of a univariate polynomial can be obtained using sign determination as follows.

Algorithm 10.14. **[Thom Encoding]**

- **Structure:** an ordered integral domain D, contained in a real closed field R.
- **Input:** a non-zero polynomial $P\in \mathrm{D}[X]$ of degree p.
- **Output:** the ordered list of Thom encodings of the roots of P in R.
- **Complexity:** $O(p^4\log_2(p))$.
- **Procedure:** Apply Algorithm 10.13 (Univariate Sign Determination) to P and its derivatives $\mathrm{Der}(P')$. Order the Thom encodings using Proposition 2.28.

Complexity analysis: The complexity is $O(p^4\log_2(p))$ using the complexity of Algorithm 10.13 (Univariate Sign Determination), since Algorithm 10.13 is called with a family of at most p polynomials of degree at most p.

When $P\in\mathbb{Z}[X]$, with coefficients of bitsize bounded by τ, the bitsizes of the integers in the operations performed by the algorithm are bounded by $O(p\log_2(p)\,(\tau+\log_2(p)))$ according to Proposition 8.44. □

Remark 10.74. When arithmetic operations are performed naively, it follows from the preceding complexity analysis, using Remark 8.4, that the binary complexity of Algorithm 10.14 (Thom Encodings) is thus

$$O(p^6\log_2(p)^3\,(\tau+\log_2(p))^2).$$

Note that from a binary complexity point of view, Algorithms 10.4 (Real Root Isolation) is preferable to Algorithm 10.14 (Thom Encodings). It turns out that, in practice as well, Algorithm 10.4 is much better, as the number of nodes in the isolation tree of Algorithm 10.4 (Real Root Isolation) is much smaller in most cases than its theoretical value $O(p\,(\tau+\log_2(p)))$ given by Proposition 10.50. This is the reason why, even though it is less general than Algorithm 10.14 (Thom Encoding), Algorithm 10.4 (Real Root Isolation) is important. □

Remark 10.75. Using Remark 9.2, it is possible to compute the output of Algorithm 10.14 in complexity $\tilde{O}(p^3)$ and binary complexity $\tilde{O}(p^4\,\tau)$. Similarly the output of Algorithm 10.15 can be computed in complexity $\tilde{O}(p^2\,(p+q))$ and binary complexity $\tilde{O}(p^2\,(p+q)^2\,\tau)$, the output of Algorithm 10.16 and Algorithm 10.18 can be computed in complexity $\tilde{O}(p^3)$ and binary complexity $\tilde{O}(p^4\tau)$, and the output of Algorithm 10.17 and Algorithm 10.19 in complexity $\tilde{O}(s^2\,p^3)$ and binary complexity $\tilde{O}(s^2\,p^4\,\tau)$. □

Remark 10.76. The Thom Encoding algorithm is based on the Sign Determination algorithm which is in turn based on the Signed subresultant Algorithm. This algorithm uses exact divisions and is valid only in an integral domain, and not in a general ring. In a ring, the algorithm computing determinants indicated in Remark 8.19 can always be used for computing the signed subresultant coefficients, and hence the Thom encoding. The complexity obtained is $p^{O(1)}$ arithmetic operations in the ring D of coefficients of P, which is sufficient for the complexity estimates proved in later chapters. □

Algorithm 10.15. **[Sign at the Roots in a Real Closed Field]**

- **Structure:** an ordered integral domain D, contained in a real closed field R.
- **Input:** a polynomial $P\in \mathrm{D}[X]$ of degree p and a polynomial $Q\in \mathrm{D}[X]$ of degree q, the list $\mathrm{Thom}(P)$ of Thom encodings of the set Z of roots of P in R.
- **Output:** for every $\sigma\in \mathrm{Thom}(P)$ specifying the root x of P, the sign $Q(x)$.
- **Complexity:** $O(p^2\,(p\log_2(p)+q))$.
- **Procedure:**
  - Determine the non-empty sign conditions $\mathrm{SIGN}(Q,Z)$ for Q and the list $\mathrm{Ada}(Q)$ of elements in $\{0,1,2\}$ adapted to sign determination using Algorithm 10.11 (Sign Determination).
  - Construct from the list $\mathrm{Thom}(P)$ of Thom encodings of the roots of P the list $\mathrm{Ada}(\mathrm{Der}(P'))$ of elements in $\{0,1,2\}^{\mathrm{Der}(P')}$ adapted to sign determination using Algorithm (Family adapted for sign determination) 10.12
  - Determine the non-empty sign conditions for $\mathrm{Der}(P'), Q$ as follows:
  - Compute the list of Tarski-queries

$$d'=\mathrm{TaQ}((Q,\mathrm{Der}(P'))^{\mathrm{Ada}(Q)\times \mathrm{Ada}(\mathrm{Der}(P'))},Z).$$

- Let $M = \mathrm{Mat}(\mathrm{Der}(P'), \mathrm{SIGN}(\mathrm{Der}(P'), Z))$ and
$$M' = \mathrm{Mat}(\mathrm{Ada}(Q), \mathrm{SIGN}(Q, Z)) \otimes M.$$
Compute the list $c' = c(\mathrm{SIGN}(Q, Z) \wedge \mathrm{SIGN}(\mathrm{Der}(P'), Z))$ from the equality
$$M' \cdot c' = d'$$
by inverting M'. Output using the non-zero entries of c' the signs of $Q(x(\sigma))$, $\sigma \in \mathrm{SIGN}(\mathrm{Der}(P'), Z)$.

Proof of correctness: This is a consequence of Proposition 2.28 since the number of non-zero elements in c' is exactly $r = c(\mathrm{Der}(P))$. □

Complexity analysis: The complexity is $O(p^2 (p \log_2(p) + q))$ since there are at most $3\,p$ calls to Algorithm 9.5 (Univariate Tarski-query) for polynomials of degree p and $p \log_2(p) + q$.

When P and Q are in $\mathbb{Z}[X]$, and the bitsizes of the coefficients of P and Q are bounded by τ, the bitsizes of the intermediate computations and the output are bounded by $(\tau + \log_2(p + q))\, O(p \log_2(p + q) + q)$, using the complexity analysis of Algorithm 9.5 (Univariate Tarski-query). □

It is also possible to compare the roots of two polynomials in a real closed field by a similar method.

Let $\mathcal{P}$ be a finite subset of $\mathrm{R}[X]$. The **ordered list of Thom encodings of** $\mathcal{P}$ is the ordered list $\sigma_1, \ldots, \sigma_r$ of Thom encoding of elements of
$$Z = \{x \in \mathrm{R} \mid \bigvee_{P \in \mathcal{P}} P(x) = 0\} = \{x(\sigma_1) < \cdots < x(\sigma_r)\}.$$

Algorithm 10.16. **[Comparison of Roots in a Real Closed Field]**

- **Structure:** an ordered integral domain D, contained in a real closed field R.
- **Input:** two non-zero polynomials P and Q in $\mathrm{D}[X]$ of degree p.
- **Output:** the ordered list of the Thom encodings of $\{P, Q\}$.
- **Complexity:** $O(p^4 \log_2(p))$.
- **Procedure:** Apply Algorithm 10.13 (Univariate Sign Determination) to P and $\mathrm{Der}(P'), \mathrm{Der}(Q)$, then to Q and $\mathrm{Der}(Q'), \mathrm{Der}(P)$. Compare the roots using Proposition 2.28.

Complexity analysis: The complexity is $O(p^4 \log_2(p))$ since we call Algorithm 10.13 (Univariate Sign determination) twice, each time with a family of at most $2\,p$ polynomials of degree at most p.

When P and Q are in $\mathbb{Z}[X]$, with coefficients of bitsize bounded by τ, the bitsizes of the integers in the operations performed by the algorithm are bounded by $O(p \log_2(p)\, (\tau + \log_2(p)))$ according to Proposition 8.44. □

Finally, we are able, given a finite set of univariate polynomials, to describe the ordered list of real roots of these polynomials.

Algorithm 10.17. **[Partition of a Line]**

- **Structure:** an ordered integral domain D, contained in a real closed field R.
- **Input:** a finite family $\mathcal{P} \subset \mathrm{D}[X]$.
- **Output:** the ordered list of the roots of $\mathcal{P}$, described by Thom encodings.
- **Complexity:** $O(s^2\, p^4 \log_2(p))$, where p is a bound on the degree of the elements of $\mathcal{P}$, and s a bound on the number of elements of $\mathcal{P}$.
- **Procedure:** Characterize all the roots of the polynomials of $\mathcal{P}$ in R using Algorithm 10.14 (Thom Encoding). Using Algorithm 10.16, compare these roots for every couple of polynomials in $\mathcal{D}$. Output the ordered list of Thom encodings of $\mathcal{P}$.

Complexity analysis: Since there are $O(s^2)$ pairs of polynomials to consider, the complexity is clearly bounded by $O(s^2\, p^4 \log_2(p))$, using the complexity of Algorithms 10.16.

When $\mathcal{P} \subset \mathbb{Z}[X]$ and the coefficients of $P \in \mathcal{P}$ are of bitsize bounded by τ, the bitsizes of the integers in the operations performed by the algorithm are bounded by $O(p \log_2(p)\,(\tau + \log_2(p)))$ according to Proposition 8.44. □

It is also possible, using the same techniques, to find a point between two elements of R specified by Thom encodings.

Algorithm 10.18. **[Intermediate Points]**

- **Structure:** an ordered integral domain D, contained in a real closed field R.
- **Input:** two non-zero univariate polynomials P and Q in $\mathrm{R}[X]$ of degree bounded by p.
- **Output:** Thom encodings specifying values y in intervals between two consecutive roots of P and Q.
- **Complexity:** $O(p^4 \log_2(p))$.
- **Procedure:** Compute the Thom encodings of the roots of $(P\,Q)'$ in R using Algorithm 10.14 (Thom Encoding) and compare them to the roots of P and Q using Algorithm 10.16. Keep one intermediate point between two consecutive roots of PQ.

Proof of correctness: Let y be a root of P and z be a root of Q. Then there is a root of $(PQ)'$ in (y, z) by Rolle's theorem (Proposition 2.22). □

Complexity analysis: The complexity is clearly bounded by $O(p^4 \log_2(p))$ using the complexity analysis of Algorithms 10.14 and 10.16.

When P and Q are in $\mathbb{Z}[X]$, with coefficients of bitsize bounded by τ, the bitsize of the integers in the operations performed by the algorithm are bounded by $O(p \log_2(p)\,(\tau + \log_2(p)))$ according to Proposition 8.44. □

Remark 10.77. Note that Algorithm 10.18 (Intermediate Points) can also be used to produce intermediate points between zeros of one polynomial by setting $Q=1$. □

Finally we are able, given a finite set of univariate polynomials, to describe the real roots of these polynomials as well as points between consecutive roots.

Given a family $\mathcal{P}$ of univariate polynomials, an **ordered list of sample points for** $\mathcal{P}$ is an ordered list L of Thom encodings σ specifying the roots of the polynomials of $\mathcal{P}$ in R, an element between two such consecutive roots, an element of R smaller than all these roots, and an element of R greater than all these roots. Moreover σ, appears before τ in L if and only if $x(\sigma) \leq x(\tau)$. The sign of $Q(x(\sigma))$ is also output for every $Q \in \mathcal{P}, \sigma \in L$.

Algorithm 10.19. **[Univariate Sample Points]**

- **Structure:** an ordered integral domain D, contained in a real closed field R.
- **Input:** a finite subset $\mathcal{P} \subset \mathrm{D}[X]$.
- **Output:** an ordered list of sample points for $\mathcal{P}$.
- **Complexity:** $O(s^2\ p^4 \log_2(p))$, where s is a bound on the number of elements of $\mathcal{P}$ and p is a bound on the degree of the elements of $\mathcal{P}$.
- **Procedure:** Characterize all the roots of the polynomials in R using Algorithm 10.14 (Thom Encoding). Using Algorithm 10.16, compare these roots for every couple of polynomials in $\mathcal{P}$. Compute a itemize of a point in each interval between the roots by Algorithm 10.18 (Intermediate Points). Order all these Thom encodings and keep only one intermediate point in each open interval between roots of polynomials in $\mathcal{P}$. Use Proposition 10.1 to find a polynomial of degree 1 with coefficients in D whose root is smaller (resp. larger) than any root of any polynomial in $\mathcal{P}$.

Complexity analysis: Since there are $O(s^2)$ pairs of polynomials to consider, the complexity is clearly bounded by $O(s^2\ p^4 \log_2(p))$, using the complexity of Algorithms 10.16 and 10.18.

When $\mathcal{P} \subset \mathbb{Z}[X]$ and the coefficients of $P \in \mathcal{P}$ are of bitsize bounded by τ, the bitsizes of the integers in the operations performed by the algorithm are bounded by $O(p \log_2(p)\,(\tau + \log_2(p)))$ according to Proposition 8.44. □

10.5 Bibliographical Notes

The real root isolation method goes back to Vincent [162] and has been studied by Uspensky [158]. Bernstein's polynomials are important in Computer Aided Design [57], they have been used in real root isolation in [101]. The algorithm for computing the Bernstein coefficients described in this chapter was discovered by De Casteljau, an engineer. The complexity analysis of the real root isolation algorithm is due to [55].

The basic idea of the sign determination algorithm appears in [23]. The use of Thom encodings for characterizing real roots appears in [50].

11

Cylindrical Decomposition Algorithm

The cylindrical decomposition method described in Chapter 5 can be turned into algorithms for solving several important problems.

The first problem is the **general decision problem** for the theory of the reals. The general decision problem is to design a procedure to decide the truth or falsity of a sentence Φ of the form $(\mathrm{Qu}_1 X_1) \ldots (\mathrm{Qu}_k X_k) F(X_1, \ldots, X_k)$, where $\mathrm{Qu}_i \in \{\exists, \forall\}$ and $F(X_1, \ldots, X_k)$ is a quantifier free formula.

The second problem is the **quantifier elimination problem**. We are given a formula $\Phi(Y)$ of the form $(\mathrm{Qu}_1 X_1) \ldots (\mathrm{Qu}_k X_k) F(Y_1, \ldots, Y_\ell, X_1, \ldots, X_k)$, where $\mathrm{Qu}_i \in \{\exists, \forall\}$ and $F(Y_1, \ldots, Y_\ell, X_1, \ldots, X_k)$ is a quantifier free formula. The quantifier elimination problem is to output a quantifier free formula, $\Psi(Y)$, such that for any $y \in \mathrm{R}^\ell$, $\Phi(y)$ is true if and only if $\Psi(y)$ is true.

The general decision problem is a special case of the quantifier elimination problem, corresponding to $\ell = 0$.

In Chapter 2, we have already proved that every formula is equivalent to a quantifier free formula (Theorem 2.77). The method used in the proof can in fact be turned into an algorithm, but if we performed the complexity analysis of this algorithm, we would get a tower of exponents of height linear in the number of variables. We decided not to develop the complexity analysis of the method for quantifier elimination presented in Chapter 2, since the algorithms described in this chapter and in Chapter 14 have a much better complexity.

In Section 11.1, we describe the Cylindrical Decomposition Algorithm. The degrees of the polynomials output by this algorithm are doubly exponential in the number of variables. A general outline is included in the first part of the section, and technical details on the lifting phase are included in the second part. In Section 11.2 we use the Cylindrical Decomposition Algorithm to decide the truth of a sentence. In Section 11.3, a variant of the Cylindrical Decomposition Algorithm makes it possible to perform quantifier elimination. In Section 11.4, we prove that the complexity of quantifier elimination is intrinsically doubly exponential. In Section 11.5, another variant of the Cylindrical Decomposition Algorithm is used to compute a stratification. In the two variable case, the Cylindrical Decomposition Algorithm is particularly simple and is used for computing the topology of a real algebraic plane curve in Section 11.6. Finally in Section 11.7, a variant called Restricted Elimination is used to replace infinitesimal quantities by sufficiently small numbers.

11.1 Computing the Cylindrical Decomposition

11.1.1 Outline of the Method

We use the results in Section 5.1 of Chapter 5, in particular the definition of a cylindrical decomposition adapted to $\mathcal{P}$ (see Definitions 5.1 and 5.5) and the properties of the set $\mathrm{Elim}_{X_k}(\mathcal{P})$ (see Notation 5.15).

We denote, for $i = k-1, \ldots, 1$,

$$\mathrm{C}_i(\mathcal{P}) = \mathrm{Elim}_{X_{i+1}}(\mathrm{C}_{i+1}(\mathcal{P})),$$

with $\mathrm{C}_k(\mathcal{P}) = \mathcal{P}$, so that $\mathrm{C}_i(\mathcal{P}) \subset \mathrm{R}[X_1, \ldots, X_i]$. The family

$$\mathrm{C}(\mathcal{P}) = \bigcup_{i \le k} \mathrm{C}_i(\mathcal{P})$$

is the **cylindrifying family of polynomials associated to** $\mathcal{P}$. It follows from the proof of Theorem 5.6 that the semi-algebraically connected components of the sign conditions on $\mathrm{C}(\mathcal{P})$ are the cells of a cylindrical decomposition adapted to $\mathcal{P}$.

The Cylindrical Decomposition Algorithm consists of two phases: in the first phase the cylindrifying family of polynomials associated to $\mathcal{P}$ is computed and in the second phase the cells defined by these polynomials are used to define inductively, starting from $i = 1$, the cylindrical decomposition.

The computation of the cylindrifying family of polynomials associated to $\mathcal{P}$ is based on the following Elimination Algorithm, computing the family of polynomials $\mathrm{Elim}_{X_k}(\mathcal{P})$ defined in 5.15, using Notation 1.16.

Algorithm 11.1. **[Elimination]**

- **Structure:** an ordered integral domain D contained in a real closed field R.
- **Input:** a finite list of variables $X_1, \ldots, X_k$, a finite set $\mathcal{P} \subset \mathrm{D}[X_1, \ldots, X_k]$, and a variable X_k.
- **Output:** a finite set $\mathrm{Elim}_{X_k}(\mathcal{P}) \subset \mathrm{D}[X_1, \ldots, X_{k-1}]$. The set $\mathrm{Elim}_{X_k}(\mathcal{P})$ is such that the degree of $P \in \mathcal{P}$ with respect to X_k, the number of real roots of $P \in \mathcal{P}$, and the number of real roots common to $P \in \mathcal{P}$ and $Q \in \mathcal{P}$ is fixed on every semi-algebraically connected component of the realization of each sign condition on $\mathrm{Elim}_{X_k}(\mathcal{P})$.
- **Complexity:** $s^2\, d^{O(k)}$, where s is a bound on the number of elements of $\mathcal{P}$, and d is a bound on the degrees of the elements of $\mathcal{P}$.

- **Procedure:** Place in $\mathrm{Elim}_{X_k}(\mathcal{P})$ the following polynomials, computed by Algorithm 8.21 (Signed Subresultant), using Remark 8.50, when they are not in D:
  - For $P \in \mathcal{P}$, $\deg_{X_k}(P) = p \geq 2$, $R \in \mathrm{Tru}(P)$, $j = 0, \ldots, \deg_{X_k}(R) - 2$, $\mathrm{sRes}_j(R, \partial R/\partial X_k)$.
  - For $R \in \mathrm{Tru}(\mathcal{P})$, $S \in \mathrm{Tru}(\mathcal{P})$,
    - if $\deg_{X_k}(R) > \deg_{X_k}(S)$, $\mathrm{sRes}_j(R, S)$, $j = 0, \ldots, \deg_{X_k}(S) - 1$,
    - if $\deg_{X_k}(R) < \deg_{X_k}(S)$, $\mathrm{sRes}_j(S, R)$, $j = 0, \ldots, \deg_{X_k}(R) - 1$,
    - if $\deg_{X_k}(R) = \deg_{X_k}(S)$, $\mathrm{sRes}_j(S, \overline{R})$, with $\overline{R} = \mathrm{lcof}(S)\, R - \mathrm{lcof}(R)\, S$, $j = 0, \ldots, \deg_{X_k}(\overline{R}) - 1$.
  - For $R \in \mathrm{Tru}(\mathcal{P})$, $\mathrm{lcof}(R)$.

Proof of correctness: The correctness follows from Theorem 5.16. □

Complexity analysis of Algorithm 11.1: Consider

$$\mathrm{D}[X_1, \ldots, X_k] = \mathrm{D}[X_1, \ldots, X_{k-1}][X_k].$$

There are $O(s^2 d^2)$ subresultant sequences to compute, since there are $O(s^2)$ couples of polynomials in $\mathcal{P}$ and $O(d)$ truncations for each polynomial to consider. Each of these subresultant sequence takes $O(d^2)$ arithmetic operations in the integral domain $\mathrm{D}[X_1, \ldots, X_{k-1}]$ according to the complexity analysis of Algorithm 8.21 (Signed Subresultant). The complexity is thus $O(s^2 d^4)$ in the integral domain $\mathrm{D}[X_1, \ldots, X_{k-1}]$. There are $O(s^2 d^3)$ polynomials output.

The degree with respect to $X_1, \ldots, X_{k-1}$ of the polynomials throughout these computations is bounded by $2\, d^2$ by Proposition 8.45. Since each multiplication and exact division of polynomials of degree $2\, d^2$ in $k - 1$ variables costs $O(d)^{4(k-1)}$ (see Algorithms 8.5 and 8.6), the final complexity is $s^2\, O(d)^{4k} = s^2\, d^{O(k)}$.

When $\mathrm{D} = \mathbb{Z}$, and the bitsizes of the coefficients of $\mathcal{P}$ are bounded by τ, the bitsizes of the intermediate computations and output are bounded by $\tau d^{O(k)}$, using Proposition 8.46. □

Example 11.1. a) Let $P = X_1^2 + X_2^2 + X_3^2 - 1$. The output of Algorithm 11.1 (Elimination) with input the variable X_3 and the set $\mathcal{P} = \{P\}$ is (getting rid of irrelevant constant factors) the polynomial $\mathrm{sRes}_0(P, \partial P/\partial X_3) = X_1^2 + X_2^2 - 1$ (see Example 5.17).

b) Consider the two polynomials

$$P = X_2^2 - X_1\,(X_1 + 1)\,(X_1 - 2), Q = X_2^2 - (X_1 + 2)\,(X_1 - 1)\,(X_1 - 3).$$

The output of Algorithm 11.1 (Elimination) with input the variable Y and $\mathcal{P} = \{P, Q\}$ contains three polynomials: the discriminant of P with respect to X_2,

$$\mathrm{sRes}_0(P, \partial P/X_2) = 4\, X_1\,(X_1 + 1)\,(X_1 - 2),$$

the discriminant of Q with respect to Y,

$$\mathrm{sRes}_0(Q, \partial Q/\partial X_2) = 4\,(X_1+2)\,(X_1-1)\,(X_1-3),$$

and the resultant of P and Q with respect to Y,

$$\mathrm{sRes}_0(P, Q) = (-X_1^2 - 3\,X_1 + 6)^2,$$

since $\mathrm{sRes}_1(P, Q) = 0$ is a constant. □

Now we are ready to describe the two phases of the cylindrical decomposition method.

Let $\mathcal{S} = \mathcal{S}_1, \ldots, \mathcal{S}_k$ be a cylindrical decomposition of R^k. A **cylindrical set of sample points** of $\mathcal{S}$, $\mathcal{A} = \mathcal{A}_1, \ldots, \mathcal{A}_k$, is a list of k sets such that

- for every i, $1 \le i \le k$, $\mathcal{A}_i$ is a finite subset of R^i which intersects every $S \in \mathcal{S}_i$,
- for every i, $1 \le i \le k-1$, $\pi_i(\mathcal{A}_{i+1}) = \mathcal{A}_i$, where π_i is the projection from R^{i+1} to R^i forgetting the last coordinate.

Algorithm 11.2. **[Cylindrical Decomposition]**

- **Structure:** an ordered integral domain D contained in a real closed field R.
- **Input:** a finite ordered list of variables $X_1, \ldots, X_k$, and a finite set $\mathcal{P} \subset \mathrm{D}[X_1, \ldots, X_k]$.
- **Output:** a cylindrical set of sample points of a cylindrical decomposition $\mathcal{S}$ adapted to $\mathcal{P}$ and the sign of the elements of $\mathcal{P}$ on each cell of $\mathcal{S}_k$.
- **Complexity:** $(s\,d)^{O(1)^{k-1}}$, where s is a bound on the number of elements of $\mathcal{P}$, and d is a bound on the degrees of the elements of $\mathcal{P}$.
- **Procedure:**
  - Initialize $\mathrm{C}_k(\mathcal{P}) := \mathcal{P}$.
  - Elimination phase: Compute $\mathrm{C}_i(\mathcal{P}) = \mathrm{Elim}_{X_{i+1}}(\mathrm{C}_{i+1}(\mathcal{P}))$, for $i = k-1, \ldots, 1$, applying repeatedly $\mathrm{Elim}_{X_{i+1}}$ using Algorithm 11.1 (Elimination).
  - Lifting phase:
    - Compute the sample points of the cells in $\mathcal{S}_1$ by characterizing the roots of $\mathrm{C}_1(\mathcal{P})$ and choosing a point in each interval they determine.
    - For every $i = 2, \ldots, k$, compute the sample points of the cells of $\mathcal{S}_i$ from the sample points of the cells in $\mathcal{S}_{i-1}$ as follows: Consider, for every sample point x of a cell in $\mathcal{S}_{i-1}$, the list L of non-zero polynomials $P_i(x, X_i)$ with $P_i \in \mathrm{C}_i(\mathcal{P})$. Characterize the roots of L and choose a point in each interval they determine.
  - Output the sample points of the cells and the sign of $P \in \mathcal{P}$ on the corresponding cells of R^k.

We need to be more specific about how we describe and compute sample points. This will be explained fully in the next subsection.

Proof of correctness: The correctness of Algorithm 11.2 (Cylindrical Decomposition) follows from the proof of Theorem 5.6. □

Complexity analysis of the Elimination phase:. Using the complexity analysis of Algorithm 11.1 (Elimination), if the input polynomials have degree D, the degree of the output is $2\ (D^2)$ after one application of Algorithm 11.1 (Elimination). Thus, the degrees of the polynomials output after $k-1$ applications of Algorithm 11.1 (Elimination) are bounded by $f(d,k-1)$, where f satisfies the recurrence relation

$$f(d,i)=2\,f(d,i-1)^2, f(d,0)=d. \tag{11.1}$$

Solving the recurrence we get that $f(d,k)=2^{1+2+\ldots+2^{k-2}}d^{2^{k-1}}$, and hence the degrees of the polynomials in the intermediate computations and the output are bounded by $2^{1+2+\ldots+2^{k-2}}d^{2^{k-1}}=O(d)^{2^{k-1}}$, which is polynomial in d and doubly exponential in k. A similar analysis shows that the number of polynomials output is bounded by $(s\,d)^{3^{k-1}}$, which is polynomial in s and d and doubly exponential in k.

When $\mathrm{D}=\mathbb{Z}$, and the bitsizes of the coefficients of $\mathcal{P}$ are bounded by τ, the bitsizes of the the intermediate computations and the output are bounded by $\tau d^{O(1)^{k-1}}$, using the complexity analysis of Algorithm 11.1 (Elimination), which is performed $k-1$ times. □

Example 11.2. Let $P=X_1^2+X_2^2+X_3^2-1$. Continuing Example 5.17, we describe the output of the Cylindrical Decomposition Algorithm applied to $\mathcal{P}=\{P\}$.

We have

$$\begin{aligned} \mathrm{C}_3(\mathcal{P}) &= \{X_1^2+X_2^2+X_3^2-1\}, \\ \mathrm{C}_2(\mathcal{P}) &= \{X_1^2+X_2^2-1\}, \\ \mathrm{C}_1(\mathcal{P}) &= \{X_1^2-1\}. \end{aligned}$$

The sample points of $\mathbb{R}$ consists of five points, corresponding to the two roots of X^2-1 and one point in each of the three intervals they define: these are the semi-algebraically connected components of the realization of sign conditions defined by $\mathrm{C}_1(\mathcal{P})$. We choose a sample point in each cell and obtain

$$\{(S_1,-2),(S_2,-1),(S_3,0),(S_4,1),(S_5,2)\}.$$

The cells in $\mathbb{R}^2$ are obtained by taking the semi-algebraically connected components of the realization of sign conditions defined by $\mathrm{C}_1(\mathcal{P})\cup\mathrm{C}_2(\mathcal{P})$. There are thirteen such cells, listed in Example 5.4. The sample points in $\mathbb{R}^2$ consist of thirteen points, one in each cell. The projection of a sample point in a cell of $\mathbb{R}^2$ on its first coordinate is a point in a cell of $\mathbb{R}$. We choose a sample point in each cell and obtain

$$\begin{gathered} \{(S_{1,1},(-2,0)), \\ (S_{2,1},(-1,-1)),(S_{2,2},(-1,0)),(S_{2,3},(-1,1)), \\ (S_{3,1},(0,-2)),(S_{3,2},(0,-1)),(S_{3,3},(0,0)),(S_{3,4},(0,1)),(S_{3,5},(0,2)), \\ (S_{4,1},(1,-1)),(S_{4,2},(1,0)),(S_{4,3},(1,1)), \\ (S_{5,1},(2,0))\}. \end{gathered}$$

The cells in $\mathbb{R}^3$ are obtained by taking the semi-algebraically connected components of the realization of sign conditions defined by

$$\mathrm{C}_1(\mathcal{P}) \cup \mathrm{C}_2(\mathcal{P}) \cup \mathrm{C}_3(\mathcal{P}).$$

There are twenty five such cells, listed in Example 5.4. The sample points in $\mathbb{R}^3$ consist of twenty five points, one in each cell. The projection of a sample point in a cell of $\mathbb{R}^3$ is a point in a cell of $\mathbb{R}^2$. We choose the following sample points and obtain, indicating the cell, its sample point and the sign of P at this sample point:

$$\begin{gathered}
\{(S_{1,1,1},(-2,0,0),1),\\
(S_{2,1,1},(-1,-1,0),1),\\
(S_{2,2,1},(-1,0,-1),1),(S_{2,2,2},(-1,0,0),0),(S_{2,2,3},(-1,0,1),1),\\
(S_{2,3,1},(-1,1,0),1),\\
(S_{3,1,1},(0,-2,0),1),\\
(S_{3,2,1},(0,-1,-1),1),(S_{3,2,2},(0,-1,0),0),(S_{3,2,3},(0,-1,1),1),\\
(S_{3,3,1},(0,0,-2),1),(S_{3,3,2},(0,0,-1),0),\\
(S_{3,3,3},(0,0,0),-1),\\
(S_{3,3,4},(0,0,1),0),(S_{3,3,5},(0,0,2),1),\\
(S_{3,4,1},(0,1,-1),1),(S_{3,4,2},(0,1,0),0),(S_{3,4,3},(0,1,1),1),\\
(S_{3,5,1},(0,2,0),1),\\
(S_{4,1,1},(1,-1,0),1),\\
(S_{4,2,1},(1,0,-1),1),(S_{4,2,2},(1,0,0),0),(S_{4,2,3},(1,0,1),1),\\
(S_{4,3,1},(1,1,0),1),\\
(S_{5,1,1},(2,0,0),1)\}.
\end{gathered}$$

□

This example is particularly simple because we can choose all sample points with rational coordinates. This will not be the case in general: the coordinates of the sample points will be roots of univariate polynomials above sample points of cells of lower dimension, and the real roots techniques of Chapter 10 will have to be generalized to deal with the cylindrical situation.

11.1.2 Details of the Lifting Phase

In order to make precise the lifting phase of the Cylindrical Decomposition Algorithm, it is necessary to compute sample points on cells. In the archimedean case, this can be done using isolating intervals. For a general real closed field, Thom encodings will be used.

Since the degree bounds are already doubly exponential, and since we are going to give a much better algorithm in Chapter 14, we do not dwell on precise complexity analysis of the lifting phase.

The notion of a triangular system of equations is natural in the context of the lifting phase of the Cylindrical Decomposition Algorithm.

Definition 11.3. Let K be a field contained in an algebraically closed field C. A **triangular system of polynomials** with variables $X_1, \ldots, X_k$ is a list $\mathcal{T} = \mathcal{T}_1, \mathcal{T}_2, \ldots, \mathcal{T}_k$, where

$$\begin{aligned} \mathcal{T}_1 &\in \mathrm{K}[X_1], \\ \mathcal{T}_2 &\in \mathrm{K}[X_1, X_2], \\ &\vdots \\ \mathcal{T}_k &\in \mathrm{K}[X_1, \ldots, X_k], \end{aligned}$$

such that the polynomial system $\mathcal{T}$ is zero-dimensional, i.e. $\mathrm{Zer}(\mathcal{T}, \mathrm{C}^k)$ is finite. □

11.1.2.1 The Archimedean Case

In this case, we are going to use isolating intervals to characterize the sample points of the cells. We need a notion of parallelepiped isolating a point.

A **parallelepiped isolating** $z \in \mathbb{R}^k$ is a list $(\mathcal{T}_1, I_1), (\mathcal{T}_2, I_2), \ldots, (\mathcal{T}_k, I_k)$ where $\mathcal{T}_1 \in \mathbb{R}[X_1], \ldots, \mathcal{T}_k \in \mathbb{R}[X_1, \ldots, X_k]$, $\mathcal{T} = \mathcal{T}_1, \mathcal{T}_2, \ldots, \mathcal{T}_k$ is a triangular system, I_1 is an open interval with rational end points or a rational containing the root z_1 of $\mathcal{T}_1$ and no other root of $\mathcal{T}_1$ in $\mathbb{R}$, I_2 is an open interval with rational end points or a rational containing the root z_2 of $\mathcal{T}_2(z_1, X_2)$ and no other root of $\mathcal{T}_2(z_1, X_2)$ in $\mathbb{R}$, ..., I_k is an open interval with rational end points or a rational containing the root z_k of $\mathcal{T}_k(z_1, \ldots, z_{k-1}, X_k)$ and no other root of $\mathcal{T}_k(z_1, \ldots, z_{k-1}, X_k)$ in $\mathbb{R}$.

Given a parallelepiped isolating $z \in \mathbb{R}^k$ it is not difficult, using elementary properties of intervals, to give bounds on the value of $P(z)$ where P is a polynomial of $\mathbb{R}[X_1, \ldots, X_k]$. So if $Q \in \mathbb{R}[X_1, \ldots, X_k, X_{k+1}]$, it is not difficult to find a natural number N such that all the roots of $Q(z, X_k)$ belong to $(-2^N, 2^N)$ using Lemma 10.6.

As in the case of a univariate polynomial, the root isolation method can be used for characterizing real roots in the cylindrical situation. Note that testing equality to zero and deciding signs are necessary to evaluate the degrees of the polynomials and the numbers of sign variations in their coefficients. These testing equality to zero and deciding signs will be done through a recursive call to Algorithm 11.4 (Multivariate Sign at a Sample Point).

Let us first consider an example in order to illustrate this situation in the simple situation where $k = 2$.

Example 11.4. We want to isolate the real roots of the polynomials

$$\mathcal{T}_1 = 9X_1^2 - 1, \mathcal{T}_2 = (5X_2 - 1)^2 + 3X_1 - 1.$$

We first isolate the roots of $\mathcal{T}_1$ and get z_1 isolated by $(9X_1^2 - 1, [0, 1])$ and z_1' isolated by $(9X_1^2 - 1, [-1, 0])$.

We now want to isolate the roots of $\mathcal{T}_2(z_1, X_2) = (5\,X_2 - 1)^2 + 3\,z_1 - 1$, using Algorithm 10.4 (Real Root Isolation) in a recursive way. We first need to know the separable part of $\mathcal{T}_2(z_1, X_2)$. In order to compute it, it is necessary to decide whether $3\,z_1 - 1$ is 0. For this we call Algorithm 10.6 which computes the gcd of $3\,X_1 - 1$ and $9\,X_1^2 - 1$, which is $3\,X_1 - 1$, and checks whether $3\,X_1 - 1$ vanishes at z_1. This is the case since the sign of $3\,X_1 - 1$ changes between 0 and 1. So the separable part of $\mathcal{T}_2(z_1, X_2)$ is $5\,X_2 - 1$ and $\mathcal{T}_2(z_1, X_2)$ has a single double root above z_1.

We now isolate the roots of $\mathcal{T}_2(z_1', X_2) = (5\,X_2 - 1)^2 + 3\,z_1' - 1$. We follow again the method of Algorithm 10.4 (Real Root Isolation). We first need to know the separable part of $\mathcal{T}_2(z_1', X_2)$. In order it, it is necessary to decide whether $3z_1' - 1$ is 0. For this purpose we call Algorithm 10.6 which computes the gcd of $3\,X_1 - 1$ and $9\,X_1^2 - 1$, which is $3\,X_1 - 1$, and checks whether $3\,X_1 - 1$ vanishes at z_1'. This is not the case, since the sign of $3\,X_1 - 1$ does not changes between -1 and 0. In fact, $3\,z_1' - 1$ is negative. So $\mathcal{T}_2(z_1', X_2)$ is separable.

Continuing the isolation process, we finally find that $P_2(z_1', X_2)$ has two distinct real roots, one positive and one negative. □

Note that in this example it was not necessary to refine the intervals defining z_1 and z_1' to decide whether $\mathcal{T}_2(z_1, X_2)$ and $\mathcal{T}_2(z_1', X_2)$ were separable. However, in the general situation considered now, such refinements may be necessary, and are produced by the recursive calls.

Algorithm 11.3. **[Real Recursive Root Isolation]**

- **Structure:** the field of real numbers $\mathbb{R}$.
- **Input:** a parallelepiped isolating $z \in \mathbb{R}^k$, a polynomial $P \in \mathbb{R}[X_1, \ldots, X_{k+1}]$, and a natural number N such that all the roots of $P(z,\ X_k)$ belong to $(-2^N, 2^N)$.
- **Output:** a parallelepipeds isolating (z, y) for every y root of $P(z, X_{k+1})$.
- **Procedure:** Perform the computations in Algorithm 10.4 (Real Root Isolation), testing equality to zero and deciding signs necessary for the computation of the degrees and of the sign variations being done by recursive calls to Algorithm 11.4 (Real Recursive Sign at a Point) at level k.

So we need to find the sign of a polynomial at a point. This algorithm calls itself recursively.

Algorithm 11.4. **[Real Recursive Sign at a Point]**

- **Structure:** the field of real numbers $\mathbb{R}$.
- **Input:** a parallelepiped isolating $z \in \mathbb{R}^k$, a polynomial $Q(X_1, \ldots, X_k)$, and a natural number N such that all the roots of $Q(z_1, \ldots z_{k-1}, X_k)$ belong to $(-2^N, 2^N)$.
- **Output:** a parallelepiped isolating $z \in \mathbb{R}^k$, and the sign of $Q(z)$.
- **Procedure:**
  - If $k = 1$, perform Algorithm 10.6 (Sign at a Real Root).

- If $k > 1$, perform the computations of the Algorithm 10.6 (Sign at a Real Root), testing equality to zero and deciding signs necessary for the computation of the degrees and of the sign variations being done by recursive calls to Algorithm 11.4 (Multivariate Sign at a Point) with level $k - 1$.

We can compare the roots of two polynomials. Again Algorithm 11.4 is called to evaluate sign variations.

Algorithm 11.5. **[Recursive Comparison of Real Roots]**

- **Structure:** the field of real numbers $\mathbb{R}$.
- **Input:** a parallelepiped isolating $z \in \mathbb{R}^{k-1}$, a polynomial $P(X_1, \ldots, X_k)$, a polynomial $Q(X_1, \ldots, X_k)$, and a natural number N such that all the roots of $P(z, X_k), Q(z, X_k)$ belong to $(-2^N, 2^N)$.
- **Output:** a parallelepiped isolating (z, y) for every root y of $P(z, X_k)$ or $Q(z, X_k)$, and the signs of $Q(z, y)$ (resp. $P(z, y)$).
- **Procedure:** Perform the computations of the Algorithm 10.7 (Comparison of Real Roots), testing equality to zero and deciding signs necessary for the computation of the degrees and of the sign variations being done by recursive calls to Algorithm 11.4 (Real Recursive Sign at a Point) with level $k - 1$.

Finally, we can find sample points for a family of polynomials above a point.

Algorithm 11.6. **[Real Recursive Sample Points]**

- **Structure:** the field of real numbers $\mathbb{R}$.
- **Input:** a parallelepiped isolating $z \in \mathbb{R}^{k-1}$, a finite set of polynomials $\mathcal{P} \subset \mathbb{R}[X_1, \ldots, X_k]$, and a natural number N such that all the roots of $P(z, X_k)$ belong to $(-2^N, 2^N)$ for $P \in \mathcal{P}$.
- **Output:** a level k, parallelepipeds isolating the roots of the non-zero polynomials in $\mathcal{P}(z, X_k)$, an element between two consecutive roots of elements of $\mathcal{P}(z, X_k)$, an element of $\mathbb{R}$ smaller than all these roots and an element of $\mathbb{R}$ greater than all these roots. The sign of all $Q(z, y)$, $Q \in \mathcal{P}$ is also output for every root of an element of $\mathcal{P}(z, X_k)$.
- **Procedure:**
  - For every pair P, Q of elements of $\mathcal{P}$, perform Algorithm 11.5.
  - Compute a rational point in between two consecutive roots using the isolating intervals.
  - Compute a rational point smaller than all these roots and rational point greater than all the roots of polynomials in $\mathcal{P}(z, X_k)$ using Proposition 10.1.

The preceding algorithms make it possible to describe more precisely the Lifting Phase of the Cylindrical Decomposition Algorithm 11.2 in the real case.

Algorithm 11.7. **[Real Lifting Phase]**

- **Structure:** the field of real numbers $\mathbb{R}$.
- **Input:** $\mathrm{C}_i(\mathcal{P})$, for $i = k-1, \ldots, 1$.
- **Output:** a cylindrical set of sample points of a cylindrical decomposition $\mathcal{S}_1, \ldots, \mathcal{S}_k$ of $\mathbb{R}^k$ adapted to $\mathcal{P}$ and, for each sample point, the sign of the polynomials in $\mathcal{P}$ at this point.
- **Procedure:**
  - Run Algorithm 10.8 (Sample Points on a Line) with input $\mathrm{C}_1(\mathcal{P})$ to obtain the sample points of the cells in $\mathcal{S}_1$.
  - For every $i = 2, \ldots, k$, compute the sample points of the cells of $\mathcal{S}_i$ from the sample points of the cells in $\mathcal{S}_{i-1}$ as follows: Compute for every parallelepiped I specifying x a list denoted by L of non-zero polynomials $P_i(x, X_i)$ with $P_i \in \mathrm{C}_i(\mathcal{P})$ using Algorithm 11.4 (Real Recursive Sign at a Point). Run Algorithm 11.6 (Real Recursive Sample Points) with input L and x.

11.1.2.2 The case of a real closed field

In this case, we are going to use Thom encodings to characterize the sample points of the cells.

Definition 11.5. A triangular Thom encoding specifying

$$z = (z_1, \ldots, z_k) \in \mathrm{R}^k$$

is a pair $\mathcal{T}, \sigma$ where $\mathcal{T}$ is a triangular system of polynomials and $\sigma = \sigma_1, \ldots, \sigma_k$ is a list of Thom encodings such that

- σ_1 is the Thom encoding of a root z_1 of $\mathcal{T}_1$,
- σ_2 is the Thom encoding of a root z_2 of $\mathcal{T}_2(z_1, X_2)$,
- $\vdots$
- σ_k is the Thom encoding of a root z_k of $\mathcal{T}_k(z_1, \ldots, z_{k-1}, X_k)$.

In other words, denoting by $\mathrm{Der}(\mathcal{T})$ the set of derivatives of $\mathcal{T}_j$ with respect to X_j, $j = 1, \ldots, k$, σ is a sign condition on $\mathrm{Der}(\mathcal{T})$.

We denote by $z(\mathcal{T}, \sigma)$ the k-tuple specified by the Thom encoding $\mathcal{T}, \sigma$.□

The lifting phase of the Cylindrical Decomposition Algorithm in the case of a real closed field is based on the following recursive algorithms generalizing the corresponding univariate algorithms in Chapter 10.

Algorithm 11.8. **[Recursive Sign Determination]**

- **Structure:** an ordered integral domain D contained in a real closed field K.
- **Input:** a triangular system $\mathcal{T}$, and a list $\mathcal{Q}$ of elements of $\mathrm{D}[X_1, \ldots, X_k]$. Denote by $Z = \mathrm{Zer}(\mathcal{T}, \mathrm{R}^k)$.

- **Output:** the set $\mathrm{SIGN}(\mathcal{Q},\mathcal{T})$ of sign conditions realized by $\mathcal{Q}$ on Z.
- **Procedure:**
  - If $k=1$, perform Algorithm 10.13 (Univariate Sign Determination).
  - If $k>1$, perform the computations of the Algorithm 10.11(Sign Determination), deciding signs necessary for the determination of the necessary Tarski-queries by recursive calls to Algorithm 11.8 (Recursive Sign Determination) with level $k-1$.

Exercise 11.1. Prove that the complexity of Algorithm 11.8 (Recursive Sign Determination) is $s\, d^{O(k)}$, where s is a bound on the number of elements of $\mathcal{Q}$, d is a bound on the degrees on the elements of $\mathcal{T}$ and $\mathcal{Q}$.

Algorithm 11.9. **[Recursive Thom Encoding]**

- **Structure:** an ordered integral domain D contained in a real closed field R.
- **Input:** a triangular system $\mathcal{T}$.
- **Output:** the list $\mathrm{Thom}(\mathcal{T})$ of Thom encodings of the roots of $\mathcal{T}$.
- **Procedure:** Apply Algorithm 11.8 (Recursive Sign Determination) to $\mathcal{T}$ and $\mathrm{Der}(\mathcal{T})$.

Exercise 11.2. Prove that the complexity of Algorithm 11.9 (Recursive Thom Encoding) is $d^{O(k)}$, where d is a bound on the degrees on the elements of $\mathcal{T}$.

Let $\mathcal{T},\sigma$ be a triangular Thom encoding specifying a point $z=(z_1,\ldots,z_{k-1})$ of R^{k-1}.

Definition 11.6. A Thom encoding P,τ **above** $\mathcal{T},\sigma$ is

- a polynomial $P\in\mathrm{R}[X_1,\ldots,X_k]$,
- a sign condition τ on $\mathrm{Der}_{X_k}(P)$ such that σ, τ is the triangular Thom encoding of a root (z,a) of $\mathcal{T},P$. □

Algorithm 11.10. **[Recursive Comparison of Roots]**

- **Structure:** an ordered integral domain D, contained in a real closed field R.
- **Input:** a Thom encoding $\mathcal{T}$, σ specifying $z\in\mathrm{R}^{k-1}$, and two non-zero polynomials P and Q in $\mathrm{D}[X_1,\ldots,X_k]$.
- **Output:** the ordered list of the Thom encodings of the roots of P and Q over σ.
- **Procedure:** Apply Algorithm 11.8 (Recursive Sign Determination)

  $$\mathcal{T},P,\ \mathrm{Der}(\mathcal{T})\cup\mathrm{Der}(P)\cup\mathrm{Der}(Q),$$

  then to

  $$\mathcal{T},Q,\mathrm{Der}(\mathcal{T})\cup\mathrm{Der}(Q)\cup\mathrm{Der}(P).$$

Compare the roots using Proposition 2.28.

Exercise 11.3. Prove that the complexity of Algorithm 11.10 (Recursive Comparison of Roots) is $d^{O(k)}$, where d is a bound on the degrees on the elements of $\mathcal{T}$, P and Q.

We can also construct points between two consecutive roots.

Algorithm 11.11. **[Recursive Intermediate Points]**

- **Structure:** an ordered integral domain D contained in a real closed field R.
- **Input:** a Thom encoding $\mathcal{T}$, σ specifying $z \in \mathrm{R}^{k-1}$, and two non-zero polynomials P and Q in $\mathrm{D}[X_1, \ldots, X_k]$.
- **Output:** Thom encodings specifying values y intersecting intervals between two consecutive roots of $P(z, X_k)$ and $Q(z, X_k)$.
- **Procedure:** Compute the Thom encodings of the roots of the polynomial $\partial(P\,Q)/\partial X_k(z,\, X_k)$ above $\mathcal{T}$, σ using Algorithm 11.9 (Recursive Thom Encoding) and compare them to the roots of P and Q above σ using Algorithm 11.10 (Recursive Comparison of Roots). Keep one intermediate point between two consecutive roots of PQ.

Exercise 11.4. Prove that the complexity of Algorithm 11.11 (Recursive Intermediate Points) is $d^{O(k)}$, where d is a bound on the degrees on the elements of $\mathcal{T}$, P and Q.

Finally we can compute sample points on a line.

Algorithm 11.12. **[Recursive Sample Points]**

- **Structure:** an ordered integral domain D contained in a real closed field R.
- **Input:** a Thom encoding $\mathcal{T}$, σ specifying $z \in \mathrm{R}^{k-1}$, and a family of polynomials $\mathcal{P} \subset \mathrm{D}[X_1, \ldots, X_k]$.
- **Output:** an ordered list L of Thom encodings specifying the roots in R of the non-zero polynomials $P(z, X_k)$, $P \in \mathcal{P}$, an element between two such consecutive roots, an element of R smaller than all these roots, and an element of R greater than all these roots. Moreover (τ_1) appears before (τ_2) in L if and only if $x_k(\tau_1) \leq x_k(\tau_2)$. The sign of $Q(z, x_k(\tau))$ is also output for every $Q \in \mathcal{P}, \tau \in L$.
- **Procedure:** Characterize the roots of the polynomials in R using Algorithm 11.9 (Recursive Thom Encoding). Compare these roots using Algorithm 11.10 (Recursive Comparison of Roots) for every pair of polynomials in $\mathcal{P}$. Characterize a point in each interval between the roots by Algorithm 11.11 (Recursive Intermediate Points). Use Proposition 10.5 to find an element of R smaller and bigger than any root of any polynomial in $\mathcal{P}$ above z.

Exercise 11.5. Prove that the complexity of Algorithm 11.12 (Recursive Sample Points) is $s\,d^{O(k)}$, where s is a bound on the number of elements of $\mathcal{Q}$ and d is a bound on the degrees on the elements of $\mathcal{T}$ and $\mathcal{Q}$.

We are now ready to describe the lifting phase of the Cylindrical Decomposition Algorithm in the general case.

Algorithm 11.13. **[Lifting Phase]**

- **Structure:** an ordered integral domain D contained in a real closed field R.
- **Input:** $\mathrm{C}_i(\mathcal{P})$, for $i=k,\ldots,1$.
- **Output:** a set of sample points of a cylindrical decomposition $\mathcal{S}_1,\ldots,\mathcal{S}_k$, of R^k adapted to $\mathcal{P}$ and for each sample point the sign of the polynomials in $\mathcal{P}$ at this point.
- **Procedure:**
  - Run Algorithm 10.19 (Cylindrical Univariate Sample Points) with input $\mathrm{C}_1(\mathcal{P})$ to obtain the sample points of the cells in $\mathcal{S}_1$, (described by Thom encodings).
  - For every $i=1,\ldots,k-1$, compute the sample points of the cells of $\mathcal{S}_i$ from the sample points of the cells in $\mathcal{S}_{i-1}$, (described by triangular Thom encodings) as follows: Compute for every σ specifying a sample point x the list L of non-zero polynomials $P_i(x, X_i)$ with $P_i \in \mathrm{C}_i(\mathcal{P})$, using Algorithm 11.8 (Recursive Sign Determination). Run Algorithm 11.12 (Recursive Sample Points) with input L and σ.

Exercise 11.6. Prove that the complexity of Algorithm 11.13 (Lifting Phase) is $(s\,d)^{O(1)^k}$, where s is a bound on the number of elements of $\mathcal{P}$ and d is a bound on the degrees of the polynomials in $\mathcal{P}$.

11.2 Decision Problem

Now we explain how to decide the truth or falsity of a sentence using the Cylindrical Decomposition Algorithm applied to the family of polynomials used to build the sentence.

Let $\mathcal{P}$ be a finite subset of $\mathrm{R}[X_1,\ldots,X_k]$. A **$\mathcal{P}$-atom** is one of $P=0$, $P\neq 0$, $P>0$, $P<0$, where P is a polynomial in $\mathcal{P}$. A **$\mathcal{P}$-formula** is a formula (Definition page 58) written with $\mathcal{P}$-atoms. A **$\mathcal{P}$-sentence** is a sentence (Definition page 59) written with $\mathcal{P}$-atoms.

Notation 11.7. [Cylindrical realizable sign conditions]
For $z\in \mathrm{R}^k$, we denote by $\mathrm{sign}(\mathcal{P})(z)$ the sign condition on $\mathcal{P}$ mapping $P\in\mathcal{P}$ to $\mathrm{sign}(P)(z)\in\{0,1,-1\}$.

We are going to define inductively the tree of cylindrical realizable sign conditions, CSIGN($\mathcal{P}$), of $\mathcal{P}$. The importance of this notion is that the truth or falsity of any $\mathcal{P}$-sentence can be decided from CSIGN($\mathcal{P}$).

We denote by π_i the projection from R^{i+1} to R^i forgetting the last coordinate. By convention, $\mathrm{R}^0 = \{0\}$.

- For $z \in \mathrm{R}^k$, let $\mathrm{CSIGN}_k(\mathcal{P})(z) = \mathrm{sign}(\mathcal{P})(z)$.
- For i, $0 \le i < k$, and all $y \in \mathrm{R}^i$, we inductively define

$$\mathrm{CSIGN}_i(\mathcal{P})(y) = \{\mathrm{CSIGN}_{i+1}(\mathcal{P})(z) | z \in \mathrm{R}^{i+1}, \pi_i(z) = y\}.$$

Finally, we define the **tree of cylindrical realizable sign conditions of $\mathcal{P}$**, denoted CSIGN($\mathcal{P}$), by

$$\mathrm{CSIGN}(\mathcal{P}) = \mathrm{CSIGN}_0(\mathcal{P})(0). \qquad \square$$

Example 11.8. Consider two bivariate polynomials $P_1 = X_2, P_2 = X_1^2 + X_2^2 - 1$ and $\mathcal{P} = \{P_1, P_2\}$.

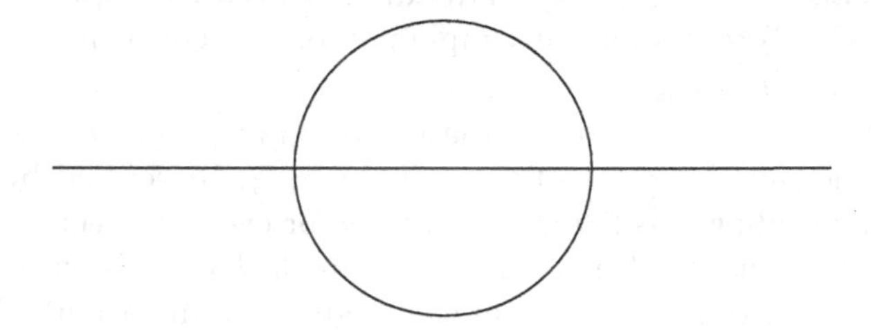

Fig. 11.1. $\mathrm{Zer}(P_1, \mathrm{R}^2)$ and $\mathrm{Zer}(P_2, \mathrm{R}^2)$

We order the set $\mathcal{P}$ with the order $P_1 < P_2$.

For $y \in \mathrm{R}^2$, $\mathrm{sign}(\mathcal{P})(y)$ is the mapping from $\mathcal{P}$ to $\{0, 1, -1\}$ sending (P_1, P_2) to $(\mathrm{sign}(P_1(y)),\ \mathrm{sign}(P_2(y)))$. Abusing notation, we denote the mapping $\mathrm{sign}(\mathcal{P})(y)$ by $(\mathrm{sign}(P_1(y)), \mathrm{sign}(P_2(y)))$.

For example if $y = (0, 0)$, $\mathrm{sign}(\mathcal{P})(0, 0) = (0, -1)$ since $\mathrm{sign}(P_1(0, 0)) = 0$ and $\mathrm{sign}(P_2(0, 0)) = -1$.

Fixing $x \in \mathrm{R}$, $\mathrm{CSIGN}_1(\mathcal{P})(x)$ is the set of all possible $\mathrm{sign}(\mathcal{P})(z)$ for $z \in \mathrm{R}^2$ such that $\pi_1(z) = x$. For example if $x = 0$, there are seven possibilities for $\mathrm{sign}(\mathcal{P})(z)$ as z varies in $\{0\} \times \mathrm{R}$:

$$(-1, 1), (-1, 0), (-1, -1), (0, -1), (1, -1), (1, 0), (1, 1).$$

So $\mathrm{CSIGN}_1(\mathcal{P})(0)$ is

$$\{(-1, 1), (-1, 0), (-1, -1), (0, -1), (1, -1), (1, 0), (1, 1)\}.$$

Similarly, if $x = 1$, there are three possibilities for $\mathrm{sign}(\mathcal{P})(z)$ as z varies in $\{1\} \times \mathrm{R}$:

$$(-1, 1), (0, 0), (1, 1).$$

So $\mathrm{CSIGN}_1(\mathcal{P})(1)$ is

$$\{(-1,1),(0,0),(1,1)\}.$$

If $x=2$, there are three possibilities for $\mathrm{sign}(\mathcal{P})(z)$ as z varies in $\{2\}\times\mathrm{R}$:

$$(-1,1),(0,1),(1,1).$$

So $\mathrm{CSIGN}_1(\mathcal{P})(2)$ is

$$\{(-1,1),(0,1),(1,1)\}.$$

Finally $\mathrm{CSIGN}(\mathcal{P})$ is the set of all possible $\mathrm{CSIGN}_1(\mathcal{P})(x)$ for $x\in\mathrm{R}$. It is easy to check that the three cases we have considered $(x=0,x=1,x=2)$ already give all possible $\mathrm{CSIGN}_1(\mathcal{P})(x)$ for $x\in\mathrm{R}$. So $\mathrm{CSIGN}(\mathcal{P})$ is the set with three elements

$$\begin{array}{c}\{\{(-1,1),(-1,0),(-1,-1),(0,-1),(1,-1),(1,0),(1,1)\},\\ \{(-1,1),(0,0),(1,1)\},\\ \{(-1,1),(0,1),(1,1)\}\}.\end{array}$$

□

We now explain how $\mathrm{CSIGN}(\mathcal{P})$ can be determined from a cylindrical set of sample points of a cylindrical decomposition adapted to $\mathcal{P}$ and the signs of $P\in\mathcal{P}$ at these points.

If $\mathcal{A}=\mathcal{A}_1,\ldots,\mathcal{A}_k$, $\mathcal{A}_i\subset\mathrm{R}^k$, $\pi_i(\mathcal{A}_{i+1})=\mathcal{A}_i$, where π_i is the projection from R^{i+1} to R^i forgetting the last coordinate, we define inductively the **tree of cylindrical realizable sign conditions** $\mathrm{CSIGN}(\mathcal{P},\mathcal{A})$ of $\mathcal{P}$ on $\mathcal{A}$.

- For $z\in\mathcal{A}_k$, let

$$\mathrm{CSIGN}_k(\mathcal{P},\mathcal{A})(z)=\mathrm{sign}(\mathcal{P})(z).$$

- For all i, $0\le i<k$, and all $y\in\mathcal{A}_i$, we inductively define

$$\mathrm{CSIGN}_i(\mathcal{P},\mathcal{A})(y)=\{\mathrm{CSIGN}_{i+1}(\mathcal{P},\mathcal{A})(z)|z\in\mathcal{A}_{i+1},\pi_i(z)=y\}.$$

Finally,

$$\mathrm{CSIGN}(\mathcal{P},\mathcal{A})=\mathrm{CSIGN}_0(\mathcal{P},\mathcal{A})(0).$$

Note that $\mathrm{CSIGN}(\mathcal{P})=\mathrm{CSIGN}(\mathcal{P},\mathrm{R}^k)$. Note also that $\mathrm{CSIGN}(\mathcal{P},\mathcal{A})$ is a subtree of $\mathrm{CSIGN}(\mathcal{P})$.

We are going to prove the following result.

Proposition 11.9. *Let $\mathcal{S}=\mathcal{S}_1,\ldots,\mathcal{S}_k$ be a cylindrical decomposition of R^k adapted to $\mathcal{P}$ and let $\mathcal{A}=\mathcal{A}_1,\ldots,\mathcal{A}_k$ be a cylindrical set of sample points for $\mathcal{S}$. Then*

$$\mathrm{CSIGN}(\mathcal{P},\mathcal{A})=\mathrm{CSIGN}(\mathcal{P}).$$

We first start by explaining how this works on an example.

Example 11.10. Let $P = X_1^2 + X_2^2 + X_3^2 - 1$ and $\mathcal{P} = \{P\}$. Since there is only one polynomial in $\mathcal{P}$, we identify $\{0, 1, -1\}^{\mathcal{P}}$ with $\{0, 1, -1\}$.

We use Example 11.2, where the cells and sample points of the cylindrical decomposition of $\{P = X_1^2 + X_2^2 + X_3^2 - 1\}$ were described.

The sign condition $\mathrm{sign}(\mathcal{P})(u)$ is fixed on each cell of R^3 by the sign of P at the sample point of the cell and thus

$$\mathrm{sign}(\mathcal{P})(z) = \begin{cases} -1 & \text{if } z \in S_{3,3,3} \\ 0 & \text{if } z \in S_{2,2,1} \cup S_{2,2,2} \cup S_{3,2,2} \\ & \quad \cup S_{3,3,2} \cup S_{3,3,4} \cup S_{3,4,2} \cup S_{4,2,2} \\ 1 & \text{otherwise.} \end{cases}$$

The set $\mathrm{CSIGN}_2(\mathcal{P})(y)$ is fixed on each cell of R^2 by its value at the sample point of the cell and thus

$$\mathrm{CSIGN}_2(\mathcal{P})(y) = \begin{cases} \{0, 1, -1\} & \text{if } y \in S_{3,3} \\ \{0, 1\} & \text{if } y \in S_{2,2} \cup S_{3,2} \cup S_{3,4} \cup S_{4,2} \\ \{1\} & \text{otherwise.} \end{cases}$$

The set $\mathrm{CSIGN}_1(\mathcal{P})(x)$ is fixed on each cell of R by its value at the sample point of the cell and thus

$$\mathrm{CSIGN}_1(\mathcal{P})(x) = \begin{cases} \{\{1\}, \{0, 1\}, \{0, 1, -1\}\} & \text{if } x \in S_3 \\ \{\{1\}, \{0, 1\}\} & \text{if } x \in S_2 \cup S_4 \\ \{\{1\}\} & \text{if } x \in S_1 \cup S_5. \end{cases}$$

Finally the set $\mathrm{CSIGN}(\mathcal{P})$ has three elements and

$$\mathrm{CSIGN}(\mathcal{P}) = \{\{\{1\}, \{0, 1\}, \{0, 1, -1\}\}, \{\{1\}, \{0, 1\}\}, \{\{1\}\}\}.$$

This means that there are three possible cases:

- there are values of $x_1 \in \mathrm{R}$ for which
  - for some value of $x_2 \in \mathrm{R}$, the only sign taken by $P(x_1, x_2, x_3)$ when x_3 varies in R is 1,
  - for some value of $x_2 \in \mathrm{R}$, the only signs taken by $P(x_1, x_2, x_3)$ when x_3 varies in R are 0 or 1,
  - for some value of $x_2 \in \mathrm{R}$, the signs taken by $P(x_1, x_2, x_3)$ when x_3 varies in R are 0, 1, or -1,
  - and these are the only possibilities,
- there are values of x_1 for which
  - for some value of $x_2 \in \mathrm{R}$, the only sign taken by $P(x_1, x_2, x_3)$ when x_3 varies in R is 1,
  - for some value of $x_2 \in \mathrm{R}$, the only signs taken by $P(x_1, x_2, x_3)$ when x_3 varies in R are 0 or 1,
  - and these are the only possibilities,
- there are values of x_1 for which
  - the only sign taken by $P(x_1, x_2, x_3)$ when (x_2, x_3) varies in R^2 is 1,

– and together these three cases exhaust all possible values of $x_1 \in \mathrm{R}$. □

Proposition 11.11. *Let $\mathcal{S} = \mathcal{S}_1, ..., \mathcal{S}_k$ be a cylindrical decomposition of* R^k *adapted to $\mathcal{P}$. For every $1 \leq i \leq k$ and every $S \in \mathcal{S}_i$,* $\mathrm{CSIGN}_i(y)$ *is constant as y varies in S.*

Proof: The proof is by induction on $k - i$.

If $i = k$, the claim is true since the sign of every $P \in \mathcal{P}$ is fixed on $S \in \mathcal{S}_k$.

Suppose that the claim is true for $i + 1$ and consider $S \in \mathcal{S}_i$. Let $T_1, ..., T_\ell$ be the cells of $\mathcal{S}_{i+1}$ such that $\pi_i(T_j) = S$. By induction hypothesis, $\mathrm{CSIGN}_{i+1}(\mathcal{P})(z)$ is constant as z varies in T_j. Since $\mathcal{S}$ is a cylindrical decomposition, $\bigcup_{j=1}^{\ell} T_j = S \times \mathrm{R}$. Thus

$$\mathrm{CSIGN}_i(\mathcal{P})(y) = \{\mathrm{CSIGN}_{i+1}(\mathcal{P})(z) | z \in \mathrm{R}^{i+1}, \pi_i(z) = y\}$$

is constant as y varies in S. □

Proof of Proposition 11.9: Let $\mathcal{A}_0 = \{0\}$. We are going to prove that for every $y \in \mathcal{A}_i$,

$$\mathrm{CSIGN}_i(\mathcal{P})(y) = \mathrm{CSIGN}_i(\mathcal{P}, \mathcal{A})(y).$$

The proof is by induction on $k - i$.

If $i = k$, the claim is true since $\mathcal{A}_k$ meets every cell of $\mathcal{S}_k$.

Suppose that the claim is true for $i + 1$ and consider $y \in \mathcal{A}_i$. Let $S \in \mathcal{S}_i$ be the cell containing y, and let $T_1, ..., T_\ell$ be the cells of $\mathcal{S}_{i+1}$ such that $\pi_i(T_j) = S$. Denote by z_j the unique point of $T_j \cap \mathcal{A}_{i+1}$ such that $\pi_i(z_j) = y$. By induction hypothesis,

$$\mathrm{CSIGN}_{i+1}(\mathcal{P})(z_j) = \mathrm{CSIGN}_{i+1}(\mathcal{P}, \mathcal{A})(z_j).$$

Since $\mathrm{CSIGN}_{i+1}(\mathcal{P})(z)$ is constant as z varies in T_j,

$$\begin{aligned}
\mathrm{CSIGN}_i(\mathcal{P})(y) &= \{\mathrm{CSIGN}_{i+1}(\mathcal{P})(z) | z \in \mathrm{R}^{i+1}, \pi_i(z) = y\} \\
&= \{\mathrm{CSIGN}_{i+1}(\mathcal{P}, \mathcal{A})(z) | z \in \mathcal{A}_{i+1}, \pi_i(z) = y\} \\
&= \mathrm{CSIGN}_i(\mathcal{P}, \mathcal{A})(y)
\end{aligned}$$

□

The Cylindrical Decision Algorithm is based on the following result. We are going to need a notation.

Notation 11.12. If $\mathcal{P} \subset \mathrm{K}[X_1, ..., X_k]$ is finite, $X = (X_1, ..., X_k)$, $F(X)$ is a $\mathcal{P}$-quantifier free formula, and $\sigma \in \mathcal{P}^{\{0,1,-1\}}$ is a sign condition on $\mathcal{P}$, we define $F^\star(\sigma) \in \{\mathrm{True}, \mathrm{False}\}$ as follows:

– If F is the atom $P = 0$, $P \in \mathcal{P}$, $F^\star(\sigma) = \mathrm{True}$ if $\sigma(P) = 0$, $F^\star(\sigma) = \mathrm{False}$ otherwise.

- If F is the atom $P>0$, $P \in \mathcal{P}$, $F^\star(\sigma) = \text{True}$ if $\sigma(P)=1$, $F^\star(\sigma) = \text{False}$ otherwise.
- If F is the atom $P<0$, $P \in \mathcal{P}$, $F^\star(\sigma) = \text{True}$ if $\sigma(P)=-1$, $F^\star(\sigma) = \text{False}$ otherwise.
- If $F = F_1 \wedge F_2$, $F^\star(\sigma) = F_1^\star(\sigma) \wedge F_2^\star(\sigma)$.
- If $F = F_1 \vee F_2$, $F^\star(\sigma) = F_1^\star(\sigma) \vee F_2^\star(\sigma)$.
- If $F = \neg(G)$, $F^\star(\sigma) = \neg(G^\star(\sigma))$. □

Example 11.13. If $F = X_1^2 + X_2^2 + X_3^2 - 1 > 0$, then

$$F^\star(\sigma) = \begin{cases} \text{True} & \text{if } \sigma = 1 \\ \text{False} & \text{if } \sigma = 0, -1 \end{cases}$$ □

Proposition 11.14. *The $\mathcal{P}$-sentence*

$$(\mathrm{Qu}_1 X_1)\ (\mathrm{Qu}_2 X_2) \ldots\ (\mathrm{Qu}_k X_k)\ F(X_1, \ldots, X_k),$$

where $F(X_1, \ldots, X_k)$ is quantifier free, $\mathrm{Qu}_i \in \{\exists, \forall\}$, is true if and only if

$$(\mathrm{Qu}_1 \sigma_1 \in \mathrm{CSIGN}(\mathcal{P}))\ (\mathrm{Qu}_2 \sigma_2 \in \sigma_1) \ldots\ (\mathrm{Qu}_k \sigma_k \in \sigma_{k-1}) F^\star(\sigma_k)$$

is true.

Example 11.15. We illustrate the statement of the proposition by an example. Consider again $\mathcal{P} = \{X_1^2 + X_2^2 + X_3^2 - 1\}$, and recall that

$$\mathrm{CSIGN}(\mathcal{P}) = \{\{\{1\}, \{0,1\}, \{0,1,-1\}\}, \{\{1\}, \{0,1\}\}, \{\{1\}\}\}$$

by Example 11.2.

The sentence $(\forall\, X_1)(\forall\, X_2)(\forall\, X_3)\ F$, with $F = X_1^2 + X_2^2 + X_3^2 - 1 > 0$ is false since taking $(x_1, x_2, x_3) = (0, 0, 0)$ we get $x_1^2 + x_2^2 + x_3^2 - 1 < 0$. It is also the case that

$$\forall \sigma_1 \in \mathrm{CSIGN}(\mathcal{P}) \quad \forall \sigma_2 \in \sigma_1 \quad \forall \sigma_3 \in \sigma_2 \quad F^\star(\sigma_3)$$

is false since taking $\sigma_1 = \{\{1\}, \{0,1\}, \{0,1,-1\}\}, \sigma_2 = \{0,1,-1\}, \sigma_3 = -1$, the value of $F^\star(\sigma_3)$ is false. □

Proof of Proposition 11.14: The proof is by induction on the number k of quantifiers, starting from the one outside.

Since $(\forall\, X)\ \Phi$ is equivalent to $\neg\ (\exists X)\ \neg\Phi$, we can suppose without loss of generality that Qu_1 is $\exists$.

The claim is certainly true when there is only one existential quantifier, by definition of $\mathrm{sign}(\mathcal{P})$.

Suppose that

$$(\exists X_1)\ (\mathrm{Qu}_2 X_2) \ldots\ (\mathrm{Qu}_k X_k)\ F(X_1, \ldots, X_k),$$

is true, and choose $a \in \mathrm{R}$ such that

$$(\mathrm{Qu}_2 X_2) \ldots (\mathrm{Qu}_k X_k)\ F(a, \ldots, X_k)$$

is true. Note that, if $\mathcal{P}_a$ is the set of polynomials obtained by substituting $a \in \mathrm{R}$ to X_1 in $\mathcal{P}$,

$$\mathrm{CSIGN}_1(\mathcal{P})(a) = \mathrm{CSIGN}(\mathcal{P}_a).$$

By induction hypothesis,

$$(\mathrm{Qu}_2 \sigma_2 \in \mathrm{CSIGN}(\mathcal{P}_a)) \ldots (\mathrm{Qu}_k \sigma_k \in \sigma_{k-1})\ F^\star(\sigma_k) \text{ is true.}$$

is true. So, taking $\sigma_1 = \mathrm{CSIGN}(\mathcal{P}_a) = \mathrm{CSIGN}(\mathcal{P})(a) \in \mathrm{CSIGN}(\mathcal{P})$,

$$\exists \sigma_1 \in \mathrm{CSIGN}(\mathcal{P}) \quad \mathrm{Qu}_2 \sigma_2 \in \sigma_1 \ldots \mathrm{Qu}_k \sigma_k \in \sigma_{k-1} F^\star(\sigma_k) \text{ is true.}$$

Conversely suppose

$$\exists \sigma_1 \in \mathrm{CSIGN}(\mathcal{P}) \quad \mathrm{Qu}_2 \sigma_2 \in \sigma_1 \ldots \mathrm{Qu}_k \sigma_k \in \sigma_{k-1} \quad F^\star(\sigma_k)$$

is true and choose $\sigma_1 \in \mathrm{CSIGN}(\mathcal{P})$ such that

$$\mathrm{Qu}_2 \sigma_2 \in \sigma_1 \ldots \mathrm{Qu}_k \sigma_k \in \sigma_{k-1} \quad F^\star(\sigma_k)$$

is true. By definition of $\mathrm{CSIGN}(\mathcal{P})$, $\sigma_1 = \mathrm{CSIGN}(\mathcal{P})(a)$ for some $a \in \mathrm{R}$, and hence

$$\mathrm{Qu}_2 \sigma_2 \in \mathrm{CSIGN}(\mathcal{P}_a) \ldots \mathrm{Qu}_k \sigma_k \in \sigma_{k-1} \quad F^\star(\sigma_k)$$

is true. By induction hypothesis,

$$(\mathrm{Qu}_2 X_2) \ldots (\mathrm{Qu}_k X_k)\ F(a, \ldots, X_k)$$

is true. Thus

$$(\exists X_1)\ (\mathrm{Qu}_2 X_2) \ldots (\mathrm{Qu}_k X_k)\ F(X_1, \ldots, X_k)$$

is true. □

Before giving a description of the Cylindrical Decision Algorithm, we explain how it works on the following example.

Example 11.16. We continue Example 11.10 to illustrate Proposition 11.14. We had determined

$$\mathrm{CSIGN}(\mathcal{P}) = \{\{\{\{1\}, \{0, 1\}, \{0, 1, -1\}\}, \{\{1\}, \{0, 1\}\}, \{\{1\}\}\}.$$

The formula

$$(\exists X_1)\ (\forall X_2)\ (\forall X_3)\ X_1^2 + X_2^2 + X_3^2 - 1 > 0$$

is certainly true since

$$\exists \sigma_1 \in \mathrm{CSIGN}(\mathcal{P}) \quad \forall\, \sigma_2 \in \sigma_1 \quad \forall \sigma_3 \in \sigma_2 \quad \sigma_3(P) = 1:$$

take $\sigma_1=\{\{1\}\}$. It is also the case that the formula

$$(\forall X_1)\ (\exists X_2)\ (\exists X_3)\ \ X_1^2+X_2^2+X_3^2-1>0$$

is true since

$$\forall \sigma_1 \in \mathrm{CSIGN}(\mathcal{P}) \quad \exists \sigma_2 \in \sigma_1 \quad \exists \sigma_3 \in \sigma_2 \quad \sigma_3(P)=1.$$

The formula

$$(\forall X_1)\ (\exists X_2)\ (\exists X_3)\ \ X_1^2+X_2^2+X_3^2-1=0$$

is false since it is not the case that

$$\forall \sigma_1 \in \mathrm{CSIGN}(\mathcal{P}) \quad \exists \sigma_2 \in \sigma_1 \quad \exists \sigma_3 \in \sigma_2 \quad \sigma_3(P)=0:$$

take $\sigma_1=\{\{1\}\}$ to obtain a counter-example. It is also easy to check that the formula

$$(\exists X_1)\ (\forall X_2)\ (\exists X_3)\ X_1^2+X_2^2+X_3^2-1=0$$

is false since it is not the case that

$$\exists \sigma_1 \in \mathrm{CSIGN}(\mathcal{P}) \quad \forall \sigma_2 \in \sigma_1 \quad \exists \sigma_3 \in \sigma_2 \quad \sigma_3(P)=0.$$

□

We are ready for the Decision Algorithm using cylindrical decomposition. We consider a finite set $\mathcal{P} \subset \mathrm{D}[X_1,\ldots,X_k]$, where D is an ordered integral domain.

Algorithm 11.14. **[Cylindrical Decision]**

- **Structure:** an ordered integral domain D contained in a real closed field R.
- **Input:** a finite set $\mathcal{P} \subset \mathrm{D}[X_1,\ldots,X_k]$, a $\mathcal{P}$-sentence

  $$\Phi=(\mathrm{Qu}_1 X_1)\ (\mathrm{Qu}_2 X_2)\ldots\ (\mathrm{Qu}_k X_k)\ F(X_1,\ldots,X_k),$$

  where $F(X_1,\ldots,X_k)$ is quantifier free, $\mathrm{Qu}_i \in \{\exists,\forall\}$.
- **Output:** True if Φ is true and False otherwise.
- **Procedure:**
  - Run Algorithm 11.2 (Cylindrical Decomposition) with input $X_1,\ldots,X_k$ and $\mathcal{P}$ using Algorithm 11.13 for the Lifting Phase.
  - Extract $\mathrm{CSIGN}(\mathcal{P})$ from the set of cylindrical sample points and the signs of the polynomials in $\mathcal{P}$ on the cells of R^k using Proposition 11.9.
  - Trying all possibilities, decide whether

    $$\mathrm{Qu}_1\sigma_1 \in \mathrm{CSIGN}(\mathcal{P}) \quad \mathrm{Qu}_2\sigma_2 \in \sigma_1 \ldots \mathrm{Qu}_k\sigma_k \in \sigma_{k-1} \quad F^\star(\sigma_k)=\mathrm{True},$$

    which is clearly a finite verification.

Proof of correctness: Follows from Proposition 11.14. Note that the two first steps of the computation depend only on $\mathcal{P}$ and not on Φ. As noted before $\mathrm{CSIGN}(\mathcal{P})$ allows us to decide the truth or falsity of every $\mathcal{P}$-sentence. □

Exercise 11.7. Prove that the complexity of Algorithm 11.14 (Cylindrical Decision) $(s\,d)^{O(1)^k}$, where s is a bound on the number of elements of $\mathcal{P}$ and d is a bound on the degrees of the polynomials in $\mathcal{P}$.

11.3 Quantifier Elimination

We start by explaining that the set of points of R^ℓ at which a $\mathcal{P}$-formula Φ with free variables $Y_1, \ldots, Y_\ell$ is true, is a union of cells in R^ℓ of a cylindrical decomposition adapted to $\mathcal{P}$.

Indeed, let $\mathcal{P} \subset \mathrm{R}[Y_1, \ldots, Y_\ell, X_1, \ldots, X_k]$ and let $\mathcal{S}_1, \ldots, \mathcal{S}_{\ell+k}$ a cylindrical decomposition of $\mathrm{R}^{k+\ell}$ adapted to $\mathcal{P}$. Let $S \in \mathcal{S}_i$. We denote $\mathrm{CSIGN}_i(\mathcal{P})(y)$ for $y \in S$ by $\mathrm{CSIGN}_i(\mathcal{P})(S)$, using Proposition 11.11.

Let $\Phi(Y) = (\mathrm{Qu}_1 X_1)\,(\mathrm{Qu}_2 X_2) \ldots (\mathrm{Qu}_k X_k) F(Y_1, \ldots, Y_\ell, X_1, \ldots, X_k)$, where $F(Y_1, \ldots, Y_\ell, X_1, \ldots, X_k)$ is quantifier free, $\mathrm{Qu}_i \in \{\exists, \forall\}$, be a $\mathcal{P}$-formula. Let $\mathcal{L}$ be the union of cells S of $\mathcal{S}_\ell$ such that

$$\mathrm{Qu}_1 \sigma_1 \in \mathrm{CSIGN}_\ell(\mathcal{P})(S)\; \mathrm{Qu}_2 \sigma_2 \in \sigma_1 \ldots \mathrm{Qu}_k\, \sigma_k \in \sigma_{k-1}\, F^\star(\sigma_k) = \mathrm{True}.$$

Then $\mathrm{Reali}(\Phi, \mathrm{R}^\ell) = \{y \in \mathrm{R}^\ell \mid \Phi(y)\} = \mathcal{L}$. So we are not far from quantifier elimination.

However, a union of cells of a cylindrical decomposition in R^ℓ is not necessarily the realization of a $\mathrm{C}_{\leq \ell}(\mathcal{P})$-quantifier free formulas, where $\mathrm{C}_{\leq \ell}(\mathcal{P}) = \bigcup_{i \leq \ell} \mathrm{C}_i(\mathcal{P})$. So a cylindrical decomposition does not always provide a $\mathrm{C}_{\leq \ell}(\mathcal{P})$-quantifier free formula equivalent to Φ. We give an example of this situation:

Example 11.17. Continuing Example 11.1 b), we consider $\mathcal{P} = \{P, Q\}$ with $P = X_2^2 - X_1(X_1+1)(X_1-2)$ and $Q = X_2^2 - (X_1+2)(X_1-1)(X_1-3)$.

We have seen in Example 11.1 b) that

$$\mathrm{C}_1(\mathcal{P}) = \{A, B, C\},$$

with

$$\begin{aligned} A(X_1) &= \mathrm{sRes}_0(P, \partial P/\partial X_2) \\ &= 4\,X_1\,(X_1+1)\,(X_1-2), \\ B(X_1) &= \mathrm{sRes}_0(Q, \partial Q/\partial X_2) \\ &= 4\,(X_1+2)\,(X_1-1)\,(X_1-3), \\ C(X_1) &= \mathrm{sRes}_0(P, Q) \\ &= (-X_1^2 - 3\,X_1 + 6)^2. \end{aligned}$$

The zero sets of P and Q in $\mathbb{R}^2$ are two cubic curves with no intersection (see Figure 11.2).

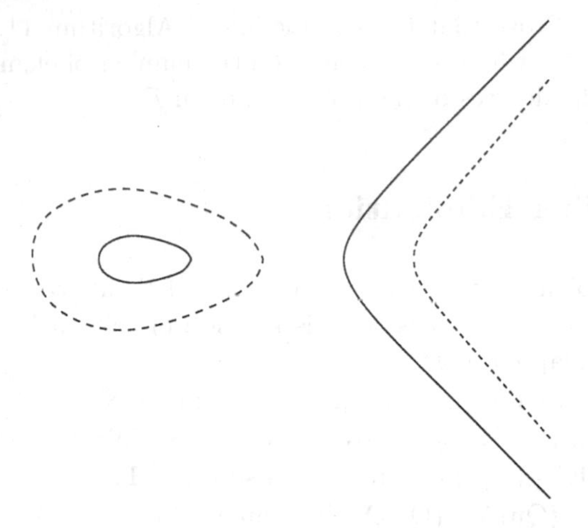

Fig. 11.2. $\mathrm{Zer}(P,\mathrm{R}^2)$ (solid) and $\mathrm{Zer}(Q,\mathrm{R}^2)$ (dotted)

This can be checked algebraically. The roots of $(-X_1^2-3X_1+6)^2$, which is the resultant of P and Q, are $a=-3/2+(1/2)\sqrt{33}\}$ and $b=3/2-(1/2)\sqrt{33}\}$. Substituting these values in P and Q gives polynomials of degree 2 without real roots.

The only subset of $\mathbb{R}$ defined by sign conditions on $\mathrm{C}_1(\mathcal{P})$ are

$$\begin{aligned}
\{-1,0\} &= \{x\in\mathbb{R} \mid A(x)=0\wedge B(x)>0\wedge C(x)>0\},\\
(-1,0)\cup(3,+\infty) &= \{x\in\mathbb{R} \mid A(x)>0\wedge B(x)>0\wedge C(x)>0\},\\
(-2,-1)\cup(0,1) &= \{x\in\mathbb{R} \mid A(x)<0\wedge B(x)>0\wedge C(x)>0\},\\
\{3\} &= \{x\in\mathbb{R} \mid A(x)>0\wedge B(x)=0\wedge C(x)>0\},\\
\{-2,1\} &= \{x\in\mathbb{R} \mid A(x)<0\wedge B(x)=0\wedge C(x)>0\},\\
\{2\} &= \{x\in\mathbb{R} \mid A(x)=0\wedge B(x)<0\wedge C(x)>0\},\\
(2,3) &= \{x\in\mathbb{R} \mid A(x)>0\wedge B(x)<0\wedge C(x)>0\},\\
(-\infty,-2)\cup(1,2)\setminus\{a,b\} &= \{x\in\mathbb{R} \mid A(x)<0\wedge B(x)<0\wedge C(x)>0\},\\
\{a,b\} &= \{x\in\mathbb{R} \mid A(x)<0\wedge B(x)<0\wedge C(x)=0\}.
\end{aligned}$$

The set $\{x\in\mathbb{R} \mid \exists y\in\mathbb{R}\ P(x,y)<0\wedge Q(x,y)>0\}=(2,+\infty)$ is the union of semi-algebraically connected components of semi-algebraic sets defined by sign conditions on $\mathrm{C}_1(\mathcal{P})$, but is not defined by any $\mathrm{C}_1(\mathcal{P})$-quantifier free formula. There are $\mathcal{P}$-formulas whose realization set cannot be described by $\mathrm{C}_1(\mathcal{P})$-quantifier free formulas. □

Fortunately, closing the set of polynomials under differentiation before each application of elimination of a variable provides an extended cylindrifying family whose realization of sign conditions are the cells of a cylindrical decomposition. This has been already proved in Theorem 5.34,

We denote by $\overline{\mathrm{C}}_k(\mathcal{P})$ the set of polynomials in $\mathcal{P}$ and all their derivatives with respect to X_k, and by $\overline{\mathrm{C}}_i(\mathcal{P})$ the set obtained by adding to the polynomials in $\mathrm{Elim}_{X_{i+1}}(\overline{\mathrm{C}}_{i+1}(\mathcal{P}))$, all their derivatives with respect to X_i, so that $\overline{\mathrm{C}}_i(\mathcal{P}) \subset \mathrm{R}[X_1, ..., X_i]$. According to Theorem 5.34, the realization of sign conditions on $\overline{\mathrm{C}}_{\leq i}(\mathcal{P}) = \bigcup_{j \leq i} \overline{\mathrm{C}}_j(\mathcal{P})$ are the sets of a cylindrical decomposition of R^i and the realization of sign conditions on $\overline{\mathrm{C}}(\mathcal{P}) = \overline{\mathrm{C}}_{\leq k}(\mathcal{P})$ are the sets of a cylindrical decomposition of R^k adapted to $\mathcal{P}$.

Algorithm 11.15. **[Improved Cylindrical Decomposition]**

- **Structure:** an ordered integral domain D contained in a real closed field R.
- **Input:** an ordered list of variables $X_1, ..., X_k$, a finite set $\mathcal{P} \subset \mathrm{D}[X_1, ..., X_k]$.
- **Output:** the finite set of polynomials $\overline{\mathrm{C}}_i(\mathcal{P}) \subset \mathrm{D}[X_1, ..., X_i]$ and the realizable sign conditions on $\overline{\mathrm{C}}_{\leq i}(\mathcal{P})$ for every $i = k, ..., 1$. The non-empty realizations of sign conditions on $\overline{\mathrm{C}}_{\leq i}(\mathcal{P})$, $i = 1, ..., k$ constitute a cylindrical decomposition of R^k adapted to $\mathcal{P}$.
- **Procedure:**
  - Add to the elements of $\mathcal{P}$ all their derivatives with respect to X_k, which defines $\overline{\mathrm{C}}_k(\mathcal{P})$,
  - Elimination phase: Compute $\overline{\mathrm{C}}_i(\mathcal{P})$ for $i = k - 1, ..., 1$, using Algorithm 11.1 (Elimination) and adding the derivatives with respect to X_i.
  - Lifting phase:
    - Compute the sample points of the cells in $\mathcal{S}_1$ by characterizing the roots of $\overline{\mathrm{C}}_1(\mathcal{P})$ and choosing a point in each interval they determine, using Algorithm 11.13 (Lifting Phase) or Algorithm 11.7 (Real Lifting Phase).
    - For every $i = 2, ..., k$, compute the sample points of the cells of $\mathcal{S}_i$ from the sample points of the cells in $\mathcal{S}_{i-1}$ as follows: Compute for every sample point x of the cells in $\mathcal{S}_{i-1}$, the list L of non-zero polynomials $P_i(x, X_i)$ with $P_i \in \overline{\mathrm{C}}_i(\mathcal{P})$, using Algorithm 11.4 (Real Recursive Sign at a Point) or Algorithm 11.8 (Recursive Sign Determination). Characterize the roots of L and choose a point in each interval they determine using Algorithm 11.13 (Lifting Phase) or Algorithm 11.7 (Real Lifting Phase).
  - Output the sample points of the cells with the sign condition on $\overline{\mathrm{C}}_{\leq i}(\mathcal{P})$ valid at the sample point of each cell of R^i.

Proof of correctness: Follows from Theorem 5.34. Note that the realization of sign conditions on $\overline{\mathrm{C}}_{\leq i}(\mathcal{P})$ are semi-algebraically connected subsets of R^i. □

Exercise 11.8. Prove that the complexity of Algorithm 11.15 (Improved Cylindrical Decision) $(s\,d)^{O(1)^k}$, where s is a bound on the number of elements of $\mathcal{P}$ and d is a bound on the degrees of the polynomials in $\mathcal{P}$.

We are going to see that the Improved Cylindrical Decomposition Algorithm with input $\mathcal{P}$ makes it possible to eliminate quantifiers of any $\mathcal{P}$-formula. We need the following notation:

For every non-empty sign condition σ on $\overline{\mathrm{C}}_{i\leq \ell}(\mathcal{P})$, $\mathrm{CSIGN}_\ell(\mathcal{P})(x)$ is constant as x varies in the realization of σ, by Proposition 11.11, and is denoted by $\mathrm{CSIGN}_\ell(\mathcal{P})(\sigma)$.

Algorithm 11.16. **[Cylindrical Quantifier Elimination]**

- **Structure:** an integral domain D contained in a real closed field R.
- **Input:** a finite set $\mathcal{P} \subset \mathrm{D}[Y_1, ..., Y_\ell][X_1, ..., X_k]$, a $\mathcal{P}$-formula

$$\Phi(Y) = (\mathrm{Qu}_1 X_1)\ (\mathrm{Qu}_2 X_2) ...\ (\mathrm{Qu}_k X_k)\ F(Y_1, ..., Y_\ell, X_1, ..., X_k).$$

  where $F(Y_1, ..., Y_\ell, X_1, ..., X_k)$ is quantifier free, $\mathrm{Qu}_i \in \{\exists, \forall\}$, with free variables $Y = Y_1, ..., Y_\ell$.
- **Output:** a quantifier free formula $\Psi(Y)$ equivalent to $\Phi(Y)$.
- **Procedure:**
  - Run Algorithm 11.15 (Improved Cylindrical Decomposition) with input $Y_1, ..., Y_\ell, X_1, ..., X_k$ and $\mathcal{P}$.
  - For every non-empty sign condition σ on $\overline{\mathrm{C}}_{\leq \ell}(\mathcal{P})$, extract $\mathrm{CSIGN}_\ell(\mathcal{P})(\sigma)$ from the sample points of the cells and the signs of the polynomials in $\mathcal{P}$ on the cells of R^k using Proposition 11.9.
  - Make the list $\mathcal{L}$ of the non-empty sign condition σ on $\overline{\mathrm{C}}_{\leq \ell}(\mathcal{P})$ for which

$$\mathrm{Qu}_1 \sigma_1 \in \mathrm{CSIGN}_\ell(\mathcal{P})(\sigma) \quad \mathrm{Qu}_2 \sigma_2 \in \sigma_1 ... \ \mathrm{Qu}_k \sigma_k \in \sigma_{k-1} \quad F^\star(\sigma_k) = \mathrm{True}.$$

  - Output

$$\Psi(Y) = \bigvee_{\sigma \in \mathcal{L}} \ \bigwedge_{P \in \mathrm{C}_{\leq \ell}(\mathcal{P})} \mathrm{sign}(P(Y_1, ..., Y_\ell)) = \sigma(P).$$

Proof of correctness: Follows from Theorem 5.34. □

Exercise 11.9. Prove that the complexity of Algorithm 11.16 (Cylindrical Quantifier Elimination) $(s\ d)^{O(1)^k}$, where s is a bound on the number of elements of $\mathcal{P}$ and d is a bound on the degrees of the polynomials in $\mathcal{P}$. When $\mathrm{D} = \mathbb{Z}$, and the bitsizes of the coefficients of $\mathcal{P}$ are bounded by τ, prove the bitsizes of the intermediate computations and the output are bounded by $\tau d^{O(1)^{k-1}}$.

11.4 Lower Bound for Quantifier Elimination

In this section, we prove that a doubly exponential complexity for the quantifier elimination problem is unavoidable. We first need a notion for the size of a formula, which we define as follows.

We define the size of atomic formulas $P>0, P<, P=0$ to be the number of coefficients needed to write the polynomial P in the dense form. Thus, if $P \in \mathrm{R}[X_1, \ldots, X_k]$ and $\deg(P)=d$,

$$\mathrm{size}(P>0)=\mathrm{size}(P=0)=\mathrm{size}(P<0):=\binom{d+k}{k}$$

by Lemma 8.5.

Next we define inductively, for formulas ϕ_1, ϕ_2,

$$\begin{aligned}
\mathrm{size}(\phi_1 \vee \phi_2)=\mathrm{size}(\phi_1 \wedge \phi_2) &:= \mathrm{size}(\phi_1)+\mathrm{size}(\phi_2)+1,\\
\mathrm{size}(\neg\phi_1) &:= \mathrm{size}(\phi_1)+1,\\
\mathrm{size}(\exists X\,\phi_1)=\mathrm{size}(\forall X\,\phi_1) &:= \mathrm{size}(\phi_1)+2.
\end{aligned}$$

We prove the following theorem.

Theorem 11.18. *There exist natural numbers $c, c'>0$ and a sequence of quantified formulas $\phi_n(X,Y)$ such that $\mathrm{size}(\phi_n) \le c' n$ and any quantifier-free formula equivalent to ϕ_n must have size at least $2^{c+2^{n-3}}$.*

Proof: We construct the formulas $\phi_n(X,Y)$ as follows. It is useful to consider $Z=X+iY$ as a complex variable. We now define a predicate, $\psi_n(W,Z)$ such that $\psi_n(W,Z)$ holds if and only if $W=Z^{2^{2^n}}$. Here, both W and Z should be thought of as complex variables. The predicate ψ_n is defined recursively as follows:

$$\begin{aligned}
\psi_0(W,Z) &:= (W-Z^2=0),\\
\psi_n(W,Z) &:= (\exists U)(\forall A\,\forall B)(((A=W \wedge B=U) \vee (A=U \wedge B=Z))\\
&\qquad \Rightarrow \psi_{n-1}(A,B)).
\end{aligned} \tag{11.2}$$

It is easy to check that formula (11.2) is equivalent to formula,

$$(\exists U)\,\psi_{n-1}(W,U) \wedge \psi_{n-1}(U,Z), \tag{11.3}$$

which is clearly equivalent to $W=Z^{2^{2^n}}$. Moreover the recursion in formula (11.2) implies that $\mathrm{size}(\psi_n(W,Z)) \leqslant c_1 n$, where c_1 is a natural number.

We now define $\phi_n(X,Y)$ to be the formula obtained by specializing W to 1 in the formula ψ_n, as well as writing the various complex variables appearing in the formula in terms of their real and imaginary parts. It is easy to check that $\mathrm{size}(\phi_n) \leqslant c' n$ where c' is a natural number, $c' \geqslant c_1$.

Now, let $\theta_n(X, Y)$ be a quantifier-free formula equivalent to $\phi_n(X, Y)$. Let $\mathcal{P}_n=\{P_1, \ldots, P_s\}$ denote the set of polynomials appearing in θ_n and let $\deg(P_i)=d_i$. From the definition of the size of a formula we have,

$$\mathrm{size}(\theta_n) \geq \sum_{i=1}^{s} d_i.$$

Clearly, the set $S \subset \mathrm{R}^2$ defined by θ_n has 2^{2^n} isolated points (corresponding to the different 2^{2^n}-th complex roots of unity). But S is a $\mathcal{P}_n$-semi-algebraic set. By Theorem 7.50, there exists a natural number C such that

$$\sum_{0 \leq i \leq 2} b_i(S) \leq C\,(s\,d)^4,$$

where $d = \sum_{1 \leq i \leq s} d_i$ is an upper bound on the degrees of the polynomials in $\mathcal{P}_n$. Moreover, $s \leq d$. Thus, we have that, $b_0(S) = 2^{2^n} \leq C\,(s\,d)^4 \leq C\,d^8$. Hence, $\mathrm{size}(\theta_n) \geq d \geq 2^{c+2^{n-3}}$. □

Notice however in the above proof the number of quantifier alternations in the formulas ϕ_n is linear in n. Later in Chapter 14, we will develop an algorithm for performing quantifier elimintation whose complexity is doubly exponential in the number of quantifier alternations, but whose complexity is only singly exponential if we fix the number of quantifier alternations allowed in the input formula.

11.5 Computation of Stratifying Families

When we want to decide the truth of a sentence, or eliminate quantifiers from a formula, the variables provided in the input play a special role in the problem considered and cannot be changed. However when we are only interested in computing the topology of a set, we are free to perform linear changes of coordinates.

We indicate now how to compute a cell stratification of R^k adapted to a finite set of polynomials $\mathcal{P}$ using Section 5.5 of Chapter 5.

Algorithm 11.17. **[Stratifying Cylindrical Decomposition]**

- **Structure:** an ordered integral domain D contained in a real closed field R.
- **Input:** a finite set $\mathcal{P} \subset \mathrm{D}[X_1, \ldots, X_k]$ of s polynomials of degree bounded by d.
- **Output:** a cell stratification of R^k adapted to $\mathcal{P}$. More precisely, a linear automorphism u such that the finite set of polynomials

$$\overline{\mathrm{C}}_i(u(\mathcal{P})) \subset \mathrm{D}[X_1, \ldots, X_i]$$

  are quasi-monic with respect to X_i and the realizable sign conditions on $\overline{\mathrm{C}}_{\leq i}(u(\mathcal{P}))$ for every $i = 1, \ldots, k$. The families $\mathcal{S}_i$, for $i = 1, \ldots, k$, consisting of all $\mathrm{Reali}(\sigma)$ with σ a realizable sign conditions on $\overline{\mathrm{C}}_{\leq i}(u(\mathcal{P}))$ constitute a cell stratification of R^k adapted to $u(\mathcal{P})$.

- **Procedure:**
  - Try successively
$$a_k = (a_{k,1}, \ldots, a_{k,k-1}) \in \{0, \ldots, s\,d\}^{k-1}$$
    and choose one such that after the linear change of variables u_k associating to
$$X_1, X_2 \ldots, X_{k-1}, X_k,$$
$$X_1 + a_{k,1} X_k,\ X_2 + a_{k,2} X_k, \ldots,\ X_{k-1} + a_{k,k-1} X_k, X_k$$
    the polynomials in $u_k(\mathcal{P})$ are monic in X_k.
  - Add to the elements of $u_k(\mathcal{P})$ all their derivatives with respect to X_k, which defines $\overline{\mathrm{C}}_k(u_k(\mathcal{P}))$.
  - **Elimination phase:**
    For $i = k-1, \ldots, 1$, denote by d_i and s_i a bound on the degree and number of the polynomials in $\overline{\mathrm{C}}_{i+1}(u_{i+1}(\mathcal{P}))$ and choose
$$a_i = (a_{i,1}, \ldots, a_{i,i-1}) \in \{0, \ldots, s_i d_i\}^{i-1}$$
    such that after the linear change of variables v_i associating to
$$X_1, X_2 \ldots, X_{i-1}, X_i, \ldots, X_k,$$
$$X_1 + a_{i,1} X_i,\ X_2 + a_{i,2} X_i, \ldots, X_{i-1} + a_{i,i-1} X_i, X_i, \ldots, X_k$$
    the polynomials in $\overline{\mathrm{C}}_{i+1}(u_i(\mathcal{P}))$ are monic in X_i, with $u_i = v_i \circ u_{i+1}$. Compute $\overline{\mathrm{C}}_i(u_i(\mathcal{P}))$ for $i = k-1, \ldots, 1$, using Algorithm 11.1 (Elimination) and adding the derivatives with respect to X_i. Define $u = u_1 = v_1 \circ \ldots \circ v_k$.
  - **Lifting phase:**
    - Compute the sample points of the cells in $\mathcal{S}_1$, by characterizing the roots of $\mathrm{C}_1(u(\mathcal{P}))$ and choosing a point in each interval they determine using Algorithm 11.13 (Lifting Phase) or Algorithm 11.7 (Real Lifting Phase).
    - For every $i = 2, \ldots, k$, compute the sample points of the cells of $\mathcal{S}_i$ from the sample points of the cells in $\mathcal{S}_{i-1}$, as follows: Compute, for every sample point x of a cell in $\mathcal{S}_{i-1}$, the list L of non-zero polynomials $Q \in \overline{\mathrm{C}}_i(u(\mathcal{P}))$ using Algorithm 11.4 (Real Recursive Sign at a Point) or Algorithm 11.8 (Recursive Sign Determination). Characterize the roots of L and chose a point in each interval they determine using Algorithm 11.13 (Lifting Phase) or Algorithm 11.7 (Real Lifting Phase).
  - Output the sample points of the cells with the sign condition on $\overline{\mathrm{C}}_{\leq i}(u(\mathcal{P}))$ valid at the sample point of each cell of R^j, $j \leq i$.

Proof of correctness: Follows from Lemma 4.73, Lemma 4.74, Section 5.5 of Chapter 5, Lemma 4.74, and the correctness of Algorithm 11.1 (Elimination). □

Exercise 11.10. Prove that the complexity of Algorithm 11.17 (Stratifying Cylindrical Decision) $(s\,d)^{O(1)^k}$, where s is a bound on the number of elements of $\mathcal{P}$ and d is a bound on the degrees of the polynomials in $\mathcal{P}$. When $\mathrm{D}=\mathbb{Z}$, and the bitsizes of the coefficients of $\mathcal{P}$ are bounded by τ, prove the bitsizes of the intermediate computations and the output are bounded by $\tau d^{O(1)^{k-1}}$.

Using cylindrical decomposition, it is possible to design algorithms for computing various topological informations.

Remark 11.19.

a) It is possible to compute a semi-algebraic description of the semi-algebraically connected components of a $\mathcal{P}$-semi-algebraic set with complexity $(s\ \ d)^{O(1)^k}$ as follows, using the proofs of Theorem 5.21 and Theorem 5.38 and the complexity analysis of Algorithm 11.17 (Stratifying Cylindrical Decomposition) (Exercise 11.10). Hint: Consider a $\mathcal{P}$-semi-algebraic set, compute a stratifying family adapted to $\mathcal{P}$ and a description of the cells associated to the corresponding Stratifying Cylindrical Decomposition. Determine the adjacency relations between cells from the list of non-empty sign conditions, using Proposition 5.39 (Generalized Thom's Lemma). Use these adjacencies to describe the connected components of S. We do not give details since we are going to see in Chapters 15 and 16 much better algorithms for the description of connected components of semi-algebraic sets.

b) A triangulation of a closed and bounded semi-algebraic set, as well as its homology groups, can be computed with complexity $(s\,d)^{O(1)^k}$, using the proof of Theorem 5.43, the definition of the homology groups of bounded and closed semi-algebraic sets in Chapter 6, and the complexity analysis of Algorithm 11.17 (Stratifying Cylindrical Decomposition) (Exercise 11.10). We do not give details either, even though we currently have no better algorithm for computing the homology groups.

c) A bound on the finite number of topological types of algebraic sets defined by polynomials of degree d follows from Algorithm 11.17 (Stratifying Cylindrical Decomposition). The bound is polynomial in d and doubly exponential in the number $\binom{d+k}{k}$ of monomials of degree d in k variables. Again, we do not give details for this quantitative version of Theorem 5.47. □

11.6 Topology of Curves

In this section, D is an ordered integral domain contained in a real closed field R and $\mathrm{C}=\mathrm{R}[i]$.

The simplest situation where the Cylindrical Decomposition Algorithm can be performed is the case of one single non-zero polynomial bivariate polynomial $P(X,Y)\in\mathrm{D}[X,Y]$.

The zero set of this polynomial is an algebraic subset $\mathrm{Zer}(P, \mathrm{R}^2) \subset \mathrm{R}^2$ contained in the plane R^2 and distinct from R^2.

Since an algebraic set is semi-algebraic, there are three possible cases:

- $\mathrm{Zer}(P, \mathrm{R}^2)$ is an algebraic set of dimension 1, which is called a **curve**.
- $\mathrm{Zer}(P, \mathrm{R}^2)$ is an algebraic set of dimension 0, i.e. a finite set of points.
- $\mathrm{Zer}(P, \mathrm{R}^2)$ is empty.

Typical examples of these three situations are the unit circle defined by $X^2+Y^2-1=0$, the origin defined by $X^2+Y^2=0$, and the empty set defined by $X^2+Y^2+1=0$.

Let us consider a polynomial $P(X,Y)$ in two variables, monic and separable, of degree d as a polynomial in Y. By separable we mean in this section that any gcd of P and $\partial P/\partial Y$ is an element of $\mathrm{R}(X)$. This is not a big loss of generality since it is always possible to make a polynomial monic by a change of variables of the form $X + a\,Y$, $a \in \mathbb{Z}$, by Lemma 4.73. Replacing P by a separable polynomial can be done using Algorithm 10.1 (Gcd and gcd-free part), and does not modify the zero set. The zero set of this polynomial is an algebraic set $\mathrm{Zer}(P, \mathrm{R}^2)$ contained in the plane.

The cylindrifying family of polynomials associated to P consists of the subdiscriminants of P with respect to Y (see page 102 and Proposition 4.27), which are polynomials in the variable X, denoted by $\mathrm{sDisc}_j(X)$ for j from 0 to $d-1$ (since P is monic in Y, sDisc_d and sDisc_{d-1} are constant). Denote by sDisc the list $\mathrm{sDisc}_d, \ldots \mathrm{sDisc}_0$. Note that $\mathrm{Disc}(X) = \mathrm{sDisc}_0(X)$ is the discriminant, and is not identically 0 since P is separable (see Equation (4.4) and Corollary 4.2). On intervals between the roots of $\mathrm{Disc}_0(X)$, the number of roots of $P(x, Y)$ in R is fixed (this is a special case of Theorem 5.16). The number of roots of $P(x,Y)$ can be determined by the signs of the other signed subresultant coefficients and is equal to PmV(sDisc) according to Theorem 4.33. Note that on an interval between two roots of $\mathrm{Disc}(X)$ in R, the signs of the subdiscriminants may change but PmV(sDisc) is fixed.

A cylindrical decomposition of the plane adapted to P can thus be obtained as follows: the cells of R are the roots of $\mathrm{Disc}(X)$ in R and the intervals they determine. Above each root of $\mathrm{Disc}(X)$ in R, $\mathrm{Zer}(P, \mathrm{R}^2)$ contains a finite number of points. These points and intervals between them are cells of R^2. Above each interval determined by the roots of $\mathrm{Disc}(X)$, there are 1 dimensional cells, called **curve segments**, which are graphs of functions from the interval to R, and 2 dimensional cells which are bands between these graphs. Finally, $\mathrm{Zer}(P, \mathrm{R}^2)$ is the union of the points and curve segments of this cylindrical decomposition and consists of a finite number of points projecting on the roots of $\mathrm{Disc}(X)$ and a finite number of curve segments homeomorphic to segments of R and projecting on intervals determined by the roots of $\mathrm{Disc}(X)$. If above every interval determined by the roots of $\mathrm{Disc}(X)$ there are no curve segments, $\mathrm{Zer}(P, \mathrm{R}^2)$ is at most a finite number of points. or empty. If moreover above every root of $\mathrm{Disc}(X)$ there are no points $\mathrm{Zer}(P, \mathrm{R}^2)$ is empty.

The purpose of the algorithm we are going to present is to compute exactly the topology of the curve $\mathrm{Zer}(P, \mathrm{R}^2)$, i.e. to determine a planar graph homeomorphic to $\mathrm{Zer}(P, \mathrm{R}^2)$. After performing, if necessary, a linear change of coordinates, this will be done by indicating adjacencies between points of the curve $\mathrm{Zer}(P, \mathrm{R}^2)$ above roots of $\mathrm{Disc}(X)$ and curve segments on intervals between these roots. In this study, the notions of critical points of the projection and of curves in generic position will be useful.

The **critical points** of the projection of $\mathrm{Zer}(P, \mathrm{C}^2)$ to the X-axis are the points $(x, y) \in \mathrm{Zer}(P, \mathrm{C}^2)$ such that y is a multiple root of $P(x, Y)$.

The critical points of the projection of $\mathrm{Zer}(P, \mathrm{C}^2)$ to the X-axis are of two kinds

- **singular points** of $\mathrm{Zer}(P, \mathrm{C}^2)$, i.e. points of $\mathrm{Zer}(P, \mathrm{C}^2)$ where

$$\partial P/\partial X(x, y) = \partial P/\partial Y(x, y) = 0,$$

- **ordinary critical points** points where the tangent to $\mathrm{Zer}(P, \mathrm{C}^2)$ is well defined and parallel to the Y-axis, i.e. points $(x, y) \in \mathrm{C}^2$ where

$$\partial P/\partial Y(x, y) = 0, \partial P/\partial X(x, y) \neq 0.$$

In both cases, the first coordinate of a critical point of the projection of $\mathrm{Zer}(P, \mathrm{C}^2)$ to the X-axis is a root of the discriminant $\mathrm{Disc}(X)$ of P considered as a polynomial in Y.

Computing the topology will be particularly easy for a curve in generic position, which is the notion we define now. Indeed, the critical points of the projection of $\mathrm{Zer}(P, \mathrm{C}^2)$ on the X_1-axis are easy to characterize in this case.

Let P be polynomial of degree d in $\mathrm{R}[X, Y]$ that is separable. The set $\mathrm{Zer}(P, \mathrm{C}^2)$ is in **generic position** if the following two conditions are satisfied:

- $\deg(P) = \deg_Y(P)$,
- for every $x \in \mathrm{C}$, $\gcd(P(x, Y), \partial P/\partial Y(x, Y))$ is either a constant or a polynomial of degree j with exactly one root of multiplicity j. In other words, there is at most one critical point (x, y) of the projection of $\mathrm{Zer}(P, \mathrm{C}^2)$ to the X_1-axis above any $x \in \mathrm{C}$.

Note that above an element x_1 of R which is a root of $\mathrm{Disc}(X)$, the unique root of $\gcd(P(x, Y), \partial P/\partial Y(x_{,Y}))$ is necessarily in R. So there is exactly a critical point with coordinates in R above each root of $\mathrm{Disc}(X)$ in R.

Example 11.20. If $P = (X^2 - Y + 1)(X^2 + Y^2 - 2Y)$, the set $\mathrm{Zer}(P, \mathrm{C}^2)$ is not in generic position since there are two critical points of the projection on X above the point 0, namely $(0, 0)$ and $(0, 1)$. □

The output of the algorithm computing the topology of a curve in generic position will be the following:

- the number r of roots of $\mathrm{Disc}(X)$ in R. We denote these roots by $x_1 < \cdots < x_r$, and by $x_0 = -\infty, x_{r+1} = +\infty$.

- the number m_i of roots of $P(x, Y)$ in R when x varies on (x_i, x_{i+1}), for $i = 0, \ldots, r$.
- the number n_i of roots of $P(x, Y)$ in R. We denote these roots by $y_{i,j}$, for $j = 1, \ldots, n_i$.
- a number $c_i \leq n_i$ such that if $C_i = (x_i, z_i)$ is the unique critical point of the projection of $\mathrm{Zer}(P, \mathrm{C}^2)$ on the X-axis above x_i, $z_i = y_{i,c_i}$.

More precisely, the output is

$$[m_0, [n_1, c_1], \ldots, m_{r-1}, [n_r, c_r], m_r]].$$

It is clear that a graph homeomorphic to $\mathrm{Zer}(P, \mathrm{R}^2) \subset \mathrm{R}^2$ can be drawn using the output since if $m_i \geq n_i$ (resp. $m_i \geq n_{i+1}$), C_i belong to the closure of $m_i - n_i$ (resp. $m_i - n_{i+1}$) curve segments above (x_i, x_{i+1}) getting glued at C_i (resp. C_{i+1}) and if $m_i = n_i - 1$ (resp. $m_i = n_{i+1} - 1$) the point C_i does not belong to the closure of a curve segment above (x_i, x_{i+1}).

The critical points of the projection of $\mathrm{Zer}(P, \mathrm{C}^2)$ on the X-axis in the generic position case are easy to determine.

We denote

$$\begin{aligned} \mathrm{sDiscP}_j(X, Y) &= \mathrm{sResP}_j(P(X, Y), \partial P / \partial Y(X, Y)) \\ \mathrm{sDisc}_j(X) &= \mathrm{sRes}_j(P(X, Y), \partial P / \partial Y(X, Y)) \end{aligned}$$

(P is considered as a polynomial in Y).

Proposition 11.21. *Let P be a polynomial of degree d in $\mathrm{R}[X, Y]$, separable and in generic position. If (x, y) is a critical point of the projection of $\mathrm{Zer}(P, \mathrm{C}^2)$ on the X-axis, and j is the multiplicity of y as a root of $P(x, Y)$ (considered as a polynomial in X_2), then*

$$\mathrm{Disc}(x) = \mathrm{sDisc}_1(x) = 0, \ldots, \mathrm{sDisc}_{j-1}(x) = 0, \mathrm{sDisc}_j(x) \neq 0,$$

and

$$y = -\frac{1}{j} \frac{\mathrm{sDisc}_{j,j-1}(x)}{\mathrm{sDisc}_j(x)},$$

where $sDisc_{j,j-1}(X)$ is the coefficient of Y^{j-1} in $\mathrm{sDiscP}_j(X, Y)$.

Proof: Since $(x, y) \in \mathrm{R}^2$ is a critical point of the projection of $\mathrm{Zer}(P, \mathrm{C}^2)$ on the X-axis, $j > 0$. Since P is in generic position, $\mathrm{sDiscP}_j(x, Y)$ is a degree j polynomial with only one root, y, which implies that

$$\mathrm{sDiscP}_j(x, Y) = \mathrm{sDisc}_j(x)(Y - y)^j, \tag{11.4}$$

and

$$y = -\frac{1}{j} \frac{\mathrm{sDisc}_{j,j-1}(x)}{\mathrm{sDisc}_j(x)},$$

identifying coefficients of degree $j - 1$ on both sides of (11.4). □

We can also count the number of points of the curve above the critical points using the following result.

Proposition 11.22. *Let $P(X,Y) \in \mathrm{D}[X,Y]$ be a degree d separable polynomial such that $\mathrm{lcof}_Y(P) \in \mathrm{D}$ and $x \in \mathrm{R}$. Let*

$$R(X,Y,Z) = (Y - Z)\,\partial P/\partial Y(X,Y) - d\,P(X,Y).$$

We denote

$$T_j(X,Z) = \mathrm{sRes}_j(P(X,Y), R(X,Y,Z))$$

(considered as polynomials in Y) and $T(X,Z)$ the list $T_j(X,Z)$, j from 0 to $d-1$. Let $r = \mathrm{PmV}(\mathrm{sDisc}(x))$.

a)

$$\#\{y \in \mathrm{R} \mid P(x,y) = 0\} = r$$

b) If $P(x,z) \neq 0$,

$$\#\{y \in \mathrm{R} \mid P(x,y) = 0 \wedge z < y\} = (r + \mathrm{PmV}(T(x,z)))/2.$$

c) If $P(x,z) = 0$,

$$\#\{y \in \mathrm{R} \mid P(x,y) = 0 \wedge z < y\} = (r + \mathrm{PmV}(T(x,z)) - 1)/2.$$

Proof: We denote

$$\begin{aligned}
Z_x &= \{y \in \mathrm{R} \mid P(x,y) = 0\}, \\
\mathrm{TaQ}(1, Z_x) &= \#(Z_x) \\
&= \#(\{y \in \mathrm{R} \mid P(x,y) = 0\}) \\
\mathrm{TaQ}(Y - z, Z_x) &= \#(\{y \in \mathrm{R} \mid P(x,y) = 0 \wedge y > z\}) \\
&- \#(\{y \in \mathrm{R} \mid P(x,y) = 0 \wedge y < z\}), \\
\mathrm{TaQ}((Y - z)^2, Z_x) &= \#\{y \in \mathrm{R} \mid P(x,y) = 0 \wedge y \neq z\}.
\end{aligned}$$

a) follows from Theorem 4.33. b) and c) follow from Equation (2.6) (see page 65) applied to $P(x,Y)$ and $Y - z$:

$$\begin{bmatrix} 1 & 1 & 1 \\ 0 & 1 & -1 \\ 0 & 1 & 1 \end{bmatrix} \cdot \begin{bmatrix} c(Y = z, P(x,Y) = 0) \\ c(Y > z, P(x,Y) = 0) \\ c(Y < z, P(x,Y) = 0) \end{bmatrix} = \begin{bmatrix} \mathrm{TaQ}(1, P(x,Y)) \\ \mathrm{TaQ}(Y - z, P(x,Y)) \\ \mathrm{TaQ}((Y - z)^2, P(x,Y)) \end{bmatrix}.$$

If $P(x,z) \neq 0$, we have $c(Y - z = 0, P(x,Y) = 0) = 0$. Thus

$$c(Y > z, P(x,Y) = 0) = (\mathrm{TaQ}(1, P(x,Y)) + \mathrm{TaQ}(Y - z, P(x,Y)))/2.$$

If $P(x,z) = 0$, we have $c(Y - z = 0, P(x,Y) = 0) = 1$. Thus

$$c(Y > z, P(x,Y)) = (\mathrm{TaQ}(1, P(x,Y)) + \mathrm{TaQ}(Y - z, P(x,Y)) - 1)/2. \quad \square$$

Next we show that it is always possible to perform a linear change of variables such that in the new coordinates the curve is in generic position. The idea is to maximize the number of distinct roots of the discriminant, which are the X-coordinates of the critical points.

Let P be a separable polynomial in $\mathrm{D}[X, Y]$, U a new variable and $Q(U,X,Y)$ the polynomial defined by:

$$Q(U,X,Y)=P(X+UY,Y).$$

If $a\in\mathbb{Z}$, let $\mathrm{Zer}(P_a,\mathrm{R}^2)$ denote the curve defined by the polynomial

$$P_a(X,Y)=Q(a,X,Y)=0.$$

We denote

$$\begin{aligned}\mathrm{sDiscP}_j(U,X,Y) &= \mathrm{sResP}_j(Q(U,X,Y),\partial Q/\partial Y(U,X,Y))\\ \mathrm{sDisc}_j(U,X) &= \mathrm{sRes}_j(Q(U,X,Y),\partial Q/\partial Y(U,X,Y))\end{aligned}$$

Note that $\mathrm{Disc}(U, X)$=$\mathrm{sDisc}_0(U, X)$ is the discriminant of $Q(U, X, Y)$ with respect to the variable Y. We denote by $\mathrm{sDisc}(U, X)$ the list of the subdiscriminants $\mathrm{sDisc}_j(U,X)$. Let $\Delta(U)$ be the non-zero subdiscriminant of smallest possible index of $\mathrm{Disc}(U,X)$ with respect to the variable X.

Proposition 11.23. *Let a be an integer number such that*

$$\deg_{X_2}(Q(a,X,Y))=\deg(Q(a,X,Y)) \quad \textit{and} \quad \Delta(a)\neq 0.$$

Then P_a is in generic position.

Proof: Suppose that $\deg_{X_2}(P_a(X, Y)) = \deg(P_a(X, Y))$ and P_a is not in generic position. Let δ be a new variable and consider the field $\mathrm{C}\langle\delta\rangle$ of algebraic Puiseux series in δ. We are going to prove the following property (P): the number of distinct roots of $\mathrm{Disc}(a+\delta,X)$ in $\mathrm{C}\langle\delta\rangle$, which is the discriminant of $P_{a+\delta}(X, Y)$ with respect to the variable Y, is bigger than the number of distinct roots of $\mathrm{Disc}(a, X)$ in C, which is the discriminant of $P_a(X,Y)$ with respect to the variable Y. Thus, by definition of Δ, $\Delta(a)=0$, and the statement is proved.

The end of the proof is devoted to proving property (P).

We first study the number of critical points of the projection of $\mathrm{Zer}(P_\delta,\mathrm{C}^2)$ on the X-axis close to a critical point of the projection of $\mathrm{Zer}(P,\mathrm{C}^2)$ to the X-axis.

- If (x,y) is a singular point of $\mathrm{Zer}(P_a,\mathrm{C}^2)$, $(x+\delta y,y)$ is a singular point of $\mathrm{Zer}(P_{a+\delta},\mathrm{C}^2)$.
- If (x,y) is an ordinary critical point of the projection of $\mathrm{Zer}(P_a,\mathrm{C}^2)$ on the X_1-axis, $(x+\delta y)$ is not a critical point of the projection of $\mathrm{Zer}(P_{a+\delta},\mathrm{C}^2)$ on the X_1-axis. However we are going to prove that there is an ordinary critical point of the projection of $\mathrm{Zer}(P_{a+\delta},\mathrm{C}^2)$ on the X_1-axis of the form $(x+\delta y+u,y+v)$, $o(u)>1, o(v)>0$.

If (x,y) is an ordinary critical point of the projection of $\mathrm{Zer}(P_a,\mathrm{C}^2)$ on the X-axis, there exists r such that

$$\begin{gathered}P_a(x,y)=0, \partial P_a/\partial X(x,y)=b\neq 0,\\ \partial P_a/\partial Y(x,y)=\cdots=\partial^{r-1}P_a/\partial Y^{r-1}(x,y)=0, \partial^r P_a/\partial Y^r(x,y)=r!\,c\neq 0.\end{gathered}$$

Writing Taylor's formula for P_a in the neighborhood of (x, y), we have

$$P_a(X,Y)=b(X-x)+c(Y-y)^r+A(X,Y),$$

with $A(X,Y)$ a linear combination of monomials multiple of $(X-y)$ or $(Y-y)^r$. We consider a new variable ε and a point of $\mathrm{Zer}(P_a, \mathrm{C}\langle\varepsilon\rangle^2)$ with coordinates $x+\xi$, $y+\varepsilon$, infinitesimally close to x, y. In other words we consider a solution ξ of

$$b\xi+c\varepsilon^r+A(x+\xi, y+\varepsilon)=0 \tag{11.5}$$

which is infinitesimal. Using the proof of Theorem 2.91, there is a solution of Equation (11.5) of the form $\xi=-(c/b)\varepsilon^r+w$, $0(w)>r$. Moreover

$$\begin{aligned}\partial P_a/\partial X(x+\xi, y+\varepsilon) &= b+w_1, o(w_1)>1,\\ \partial P_a/\partial Y(x+\xi, y+\varepsilon) &= c\,r\,\varepsilon^{r-1}+w_2, o(w_2)>r-1.\end{aligned}$$

Thus, if $(\partial P_a/\partial X(x+\xi', y+\varepsilon'), \partial P_a/\partial Y(x+\xi', y+\varepsilon'))\}$ is proportional to $(1, \delta)$, with ε' and ξ' infinitesimal in $\mathrm{C}\langle\delta\rangle$, we have

$$\begin{aligned}\varepsilon' &= d\delta^{1/(r-1)}+w_3\\ \xi' &= -(1/r)\,d\,\delta^{r/(r-1)}+w_4\end{aligned}$$

with $d^{r-1}=b/c\ r$, $o(w_3)>1/(r-1)$, $o(w_4)>r/(r-1)$. Thus there is a point $(x+\xi', y+\varepsilon')$ of $\mathrm{Zer}(P_a, \mathrm{C}\langle\delta\rangle^2)$ with gradient proportional to $(1, \delta)$, and $o(\varepsilon')>0, o(\xi')>1$. In other words, $(x+\delta y+\xi'+\delta\varepsilon', y+\varepsilon')$ is a critical point of the projection of $\mathrm{Zer}(P_{a+\delta}, \mathrm{C}\langle\delta\rangle^2)$ on the X-axis.

Suppose that (x, y_1) and (x, y_2) are two distinct critical point of the projection of $\mathrm{Zer}(P_a, \mathrm{C}^2)$ to the X_1-axis. According to the preceding discussion, there are two critical point of the projection of $\mathrm{Zer}(P_{a+\delta}, \mathrm{C}^2)$ to the X-axis, with first coordinates $x+\delta y_1+u_1$ and $x+\delta y_2+u_2$, with $o(u_1)>1, o(u_2)>1$.Note that $x+\delta y_1+u_1$ and $x+\delta y_2+u_2$ are distinct, since y_1-y_2 is not infinitesimal. We have proved that the number of distinct roots of $\mathrm{Disc}(a+\delta, X)$ in $\mathrm{C}\langle\delta\rangle$ is strictly bigger that the number of distinct roots of $\mathrm{Disc}(a, X)$ in C. □

The Topology of a Curve Algorithm can now be described. We perform the computation as if the curve was in generic position. If it is not, the algorithm detects it and then a new linear change of coordinates is performed.

Algorithm 11.18. **[Topology of a Curve]**

- **Structure:** an ordered integral domain D contained in a real closed field R (resp. the field of real numbers $\mathbb{R}$).
- **Input:** a separable polynomial P in $\mathrm{D}[X,Y]$ of degree d.
- **Output:** the topology of the curve $\mathrm{Zer}(P, \mathrm{R}^2)$, described by
  - An integer a such that $P_a(X,Y)=P(X+aY,Y)$ is in general position.
  - The number r of roots of $\mathrm{Disc}(a, X)$ in R. We denote these roots by $x_1<\cdots<x_r$, and by $x_0=-\infty, x_{r+1}=+\infty$.
  - The number m_i of roots of $P_a(x, Y)$ in R when x varies on (x_i, x_{i+1}), $i=0,\ldots,r$.

- The number n_i of roots of $P_a(x_i, Y)$ in R. We denote these roots by $y_{i,j}$, $j=1, \ldots, n_i$.
- An index $c_i \le n_i$ such that if $C_i = (x_i, z_i)$ is the unique critical point of the projection of $\mathrm{Zer}(P_a, \mathrm{C}^2)$ on the X_1-axis above x_i, $z_i = y_{i,c_i}$.

More precisely, the output is $[a, [m_0, [n_1, c_1], \ldots, m_{r-1}, [n_r, c_r], m_r]]$.

- **Complexity:** $O(d^{11} \log_2(d))$, where d is a bound on the degrees of P.
- **Procedure:**
  - Take $a := 0$.
  - ($\star$) Define $P_a(X, Y) := P(X + aX, Y)$.
  - If P_a is not quasi-monic with respect to Y, the curve is not in generic position. Take $a := a+1$, go to ($\star$).
  - Otherwise, compute the subdiscriminants $\mathrm{sDisc}(a, X)$ of $P_a(X, Y)$ (considered as a polynomial in Y). Characterize the roots $x_1, \ldots, x_r$ of $\mathrm{Disc}(a, X)$ using Algorithm 10.14 (Thom encoding) (resp. Algorithm 10.4 (Real Root Isolation)).
  - For every $1 \le i \le r$, determine $j(i)$ such that

$$\mathrm{Disc}_0(a, x_i) = 0, \ldots, \mathrm{sDisc}_{j(i)-1}(a, x_i) = 0, \mathrm{sDisc}_{j(i)}(a, x_i) \neq 0,$$

    and compute the sign ε_i of $\mathrm{sDisc}_{j(i)}(a, x_i)$ using Algorithm 10.15 (Sign at the Roots in a real closed field) (resp. Algorithm 10.6 (Sign at a Real Root)).
  - Define

$$z_i = -\frac{1}{j(i)} \frac{\mathrm{sDisc}_{j(i), j(i)-1}(a, x_i)}{\mathrm{sDisc}_{j(i)}(a, x_i)}.$$

  - Check whether, for every $i = 1, \ldots, r$, z_i is a root of multiplicity $j(i)$ of $\mathrm{sDisc}_{j(i)}(x_i, Y)$ using Algorithm 10.15 (Sign at the Roots in a real closed field) (resp. Algorithm 10.6 (Sign at a Real Root)).
  - If there exists i such that z_i is not a root of multiplicity $j(i)$ of $\mathrm{sDisc}_{j(i)}(a, x_i, Y)$, the curve is not in generic position. Take $a := a+1$, go to ($\star$).
  - If for every i, z_i is a root of multiplicity $j(i)$ of $\mathrm{sDisc}_{j(i)}(a, x_i, Y)$, the curve is in generic position.
  - For every $0 \le i \le r$ choose an intermediate point $t_i \in (x_i, x_{i+1})$ (with the convention $x_0 = -\infty, x_{r+1} = +\infty$) using Algorithm 10.18 (Intermediate Points) and Remark 10.77, and compute the number m_i of roots of $P_a(t_i, Y)$ evaluating the signs of the signed subresultant coefficients at t_i using Algorithm 10.11 (Sign Determination) (resp. choose an intermediate rational point $t_i \in (x_i, x_{i+1})$ using the isolating intervals for the x_i and compute the number m_i of roots of $P_a(t_i, Y)$ using Algorithm 10.4 (Real Root Isolation)).
  - Compute, for every $i \leqslant r$,

$$\begin{aligned} S &= \varepsilon_i \, (j(i) \, \mathrm{sDisc}_{j(i)}(a, X) \, Y + \mathrm{sDisc}_{j(i), j(i)-1}(X) \, (\partial P_a / \partial Y)) \\ R &= S - \varepsilon_i \, j(i) \, d \, \mathrm{sDisc}_{j(i)}(a, X) \, P_a \end{aligned}$$

and the list

$$T_i(X) = \text{sRes}(P_a, R).$$

- Evaluate the signs of the elements of $T_i(X)$ at x_i to determine the number of real roots of the polynomial $P_a(x_i, Y)$ which are strictly bigger than z_i using Proposition 11.22 and Algorithm 10.15 (Sign at the Roots in a real closed field) (resp. Algorithm 10.6 (Sign at a Real Root)).
- Decide whether $\text{Zer}(P, \mathrm{R}^2)$ is empty or a finite number of points.

Proof of correctness: By Lemma 4.74, there are only a finite number of values of a such that $\deg_{X_2}(Q(a, X, Y)) \neq \deg(Q(a, X, Y))$. Moreover there are only a finite number of zeros of $\Delta(X)$. So by Proposition 11.23, there are only a finite number of values of a such that the curve is not in generic position. The correctness of the algorithm follows from Proposition 11.21, and the correctness of Algorithm 10.14 (Thom encoding) (resp. Algorithm 10.4 (Real Root Isolation)), Algorithm 10.15 (Sign at the Roots in a real closed field) (resp. Algorithm 10.6 (Sign at a Real Root)), Algorithm 10.18 (Intermediate Points), and Algorithm 10.11 (Sign Determination). □

Complexity analysis: We estimate the complexity of the computation in a general real closed field.

Let d be the degree of $P(X, Y)$ a separable polynomial in $\mathrm{D}[X, Y]$.

The degree of $\text{Disc}(U, X)$ is $O(d^2)$, and the degree of δ is $O(d^4)$. So there are at most $O(d^4)$ values of a to try.

For each value of a, we compute the subdiscriminant sequence of $P_a(X, Y)$ (considered as polynomial in Y): this requires $O(d^2)$ arithmetic operations in $\mathrm{D}[X]$ by Algorithm 8.21 (Signed subresultant). The degrees of the polynomials in X produced in the intermediate computations of the algorithm are bounded by $2\ d^2$ by Proposition 8.45. The complexity in D for this computation is $O(d^6)$.

The Thom encoding of the roots of $\text{Disc}(a,\ X)$ takes $O(d^8\ \log_2(d))$ arithmetic operations using the complexity analysis of Algorithm 10.14 (Thom encoding). Checking whether $Q(X,\ Y)$ is in generic position takes also $O(d^8 \log_2(d))$ arithmetic operations using the complexity analysis of Algorithm 10.15 (Sign at the Roots in a real closed field), since there are at most $O(d^2)$ calls to this algorithm with polynomials of degree d^2.

Since there are at most $O(d^4)$ values of a to try, the complexity to reach generic position is $O(d^{11} \log_2(d))$.

The choice of the intermediate points, the determination of all m_i, the determination of $j(i)$ and ε_i takes again $O(d^8 \log_2(d))$ arithmetic operations using the complexity analysis of Algorithm 10.18 (Intermediate Points) and Algorithm 10.15 (Sign at the Roots in a real closed field).

The computation of one list T_i by Algorithm 8.21 (Signed subresultant) requires $O(d^2)$ arithmetic operations in $\mathrm{D}[X]$. The degree of the polynomials in X produced in the intermediate computations of the algorithm are bounded by $O(d^4)$ by Proposition 8.45. The complexity in D for the computation of the list $\bar{T}$ is $O(d^{10})$.

So the total complexity for computing the various T_i is $O(d^{11})$.

The only remaining computations are the determination of the signs taken by the polynomials of T_i evaluated at the roots of $\mathrm{Disc}(a, X)$. The degrees of the polynomials involved are $O(d^4)$ and and their number is $O(d^2)$. This takes $O(d^{10})$ operations using the complexity analysis of Algorithm 10.15 (Sign at the Roots in a real closed field).

So the total complexity for determining the topology of $\mathrm{Zer}(P, \mathrm{R}^2)$ is $O(d^{11}\log_2(d))$.

If $\mathrm{D} = \mathbb{Z}$ and the coefficients of P are of bitsizes bounded by τ, the coefficients of $P_a(X,Y)$ are $\tau + O(d)\,\nu$ where ν is the bitsize of d, since there are at most $O(d^4)$ values of a to try. For each value of a, the bitsizes of the integers produced in the intermediate computations of the subresultantnt sequence of $P_a(X,Y)$ are bounded by $O(d\,(\tau + d\nu))$ by the complexity analysis of Algorithm 8.21 (Signed subresultant).

The univariate sign determinations performed in the last steps of the algorithm produce integers of bitsizes $O(d^2\,(\tau + d\,\nu)\log_2(d)))$, according to the complexity analysis of Algorithm 10.14 (Thom encoding), Algorithm 10.15 (Sign at the Roots in a real closed field), Algorithm 10.18 (Intermediate Points). So the maximum bitsizes of the integers in the computation determining the topology of $\mathrm{Zer}(P, \mathrm{R}^2)$ is $O(d^2\,(\tau + d\,\nu)\log_2(d)))$. □

Example 11.24. Let us consider the real algebraic curve defined by the monic polynomial

$$P = 2\,Y^3 + (3\,X - 3)\,Y^2 + (3\,X^2 - 3\,X)\,Y + X^3.$$

The discriminant of P (considered as polynomial in Y) is

$$-27X^2(X^4 + 6X^2 - 3).$$

whose real roots are given by

$$x_1 = -\sqrt{-3 + 2\sqrt{3}}, \qquad x_2 = 0, \qquad x_3 = \sqrt{-3 + 2\sqrt{3}}.$$

Using the signed subresultant sequence, we determine that above each $x_{1,i}$, for $i = 1, \ldots, 3$, $P(x_i, Y)$ has two real roots, and only one of these roots is double. Thus the curve is already in generic position. The multiplicity of the unique critical point (x_i, z_i) as a root of $P(x_i, Y)$ is 2, and is given by the equation

$$z_i = -\frac{x_i(x_i^2 + 2x_i - 1)}{2\,x_i^2 - 2}.$$

We also determine that before x_1 and after x_3, $P(x, Y)$ has only one real root, while between x_1 and x_2, and between x_2 and x_3 $P(x, Y)$ has three real roots.

For $i = 1, 3$, there is one root of $P(x_i, Y)$ under z_i, while for $i = 2$ there is one root of $P(x_i, Y)$ above z_i.

Finally, the topology of the curve is given by the following graph

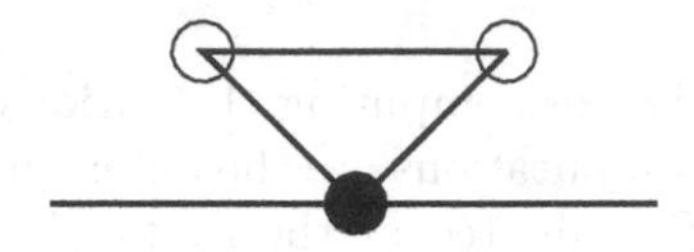

Fig. 11.3. Topology of the curve

where the points represent the critical points of the projection on the X-axis. The white points are critical points of the projection on the X-axis, while the black one is singular.

The topology can be described by 0 which is the value of a (the curve was given in generic position) and the list

$$[1, [2, 2], 3, [2, 1], 3, [2, 2], 1]$$

which can be read as follows: there are three roots of $\mathrm{Disc}(a, X)$, the number of branches above (x_0, x_1) is 1, the number of points above x_1 is 2 and the index of the critical point is 2, the number of branches above (x_1, x_2) is 3, the number of points above x_2 is 2 and the index of the critical point is 1, the number of branches above (x_2, x_3) is 3, the number of points above x_3 is 2, and the index of the critical point is 2, the number of branches above (x_3, x_4) is 1. □

11.7 Restricted Elimination

A variant of Algorithm 11.1 (Elimination) will be useful in the last chapters of the book. In this variant, we are interested at the signs of a family of polynomials at the roots of a polynomial, rather that at all the sign conditions of a family of polynomials.

We need some preliminary work.

We first prove a result similar to Theorem 5.12 and Proposition 5.13 which is adapted to the situation we are interested in studying.

Proposition 11.25. *Let P and Q be in $\mathrm{R}[X_1, \ldots X_k]$, and let S be a semi-algebraically connected semi-algebraic subset of R^{k-1} such that P is not identically 0 on S, and such that $\deg(P(x', X_k))$, the number of distinct roots of P in C, and $\deg(\gcd(P(x', X_k), Q(x', X_k)))$ are constant over S. Then there are ℓ continuous semi-algebraic functions $\xi_1 < \cdots < \xi_\ell$: $S \to \mathrm{R}$ such that, for every $x' \in S$, the set of real roots of $P(x', X_k)$ is exactly $\{\xi_1(x'), \ldots, \xi_\ell(x')\}$. Moreover, for $i = 1, \ldots, \ell$, the multiplicity of the root $\xi_i(x')$ is constant for $x' \in S$ and so is the sign of $Q(\xi_i(x'))$.*

Proof: Let $a' \in S$ and let $z_1, \ldots, z_j$ be the distinct roots of $P(a', X_k)$ in C. Let μ_i (resp. ν_i) be the multiplicity of z_i as a root of $P(a', X_k)$ (resp. $\gcd(P(a', X_k), Q(a', X_k))$). The degree of $P(a', X_k)$ is $\sum_{i=1}^{j} \mu_i$ and the degree of $\gcd(P(a', X_k), Q(a', X_k))$ is $\sum_{i=1}^{j} \nu_i$. Choose $r > 0$ such that all disks $D(z_i, r)$ are disjoint.

Using Theorem 5.12 and the fact that the number of distinct complex roots of $P(x', X_k)$ stays constant over S, we deduce that there exists a neighborhood V of a' in S such that for every $x' \in V$, each disk $D(z_i, r)$ contains one root of multiplicity μ_i of $P(x', X_k)$. Since the degree of $\gcd(P(x', X_k), Q(x', X_k))$ is equal to $\sum_{i=1}^{j} \nu_i$, this gcd must have exactly one root ζ_i, of multiplicity ν_i, in each disk $D(z_i, r)$ such that $\nu_i > 0$. If z_i is real, ζ_i is real (otherwise, its conjugate $\overline{\zeta_i}$ would be another root of $P(x', X_k)$ in $D(z_i, r)$). If z_i is not real, ζ_i is not real, since $D(z_i, r)$ is disjoint from its image by conjugation. Hence, if $x' \in V$, $P(x', X_k)$ has the same number of distinct real roots as $P(a', X_k)$. Since S is semi-algebraically connected, the number of distinct real roots of $P(x', X_k)$ is constant for $x' \in S$ according to Proposition 3.9. Let ℓ be this number. For $1 \le i \le \ell$, denote by $\xi_i: S \to \mathrm{R}$ the function which sends $x' \in S$ to the i-th real root (in increasing order) of $P(x', X_k)$. The above argument, with arbitrarily small r, also shows that the functions ξ_i are continuous.

It follows from the fact that S is semi-algebraically connected that each $\xi_i(x')$ has constant multiplicity as a root of $P(x', X_k)$ and as a root of $\gcd(P(x', X_k), Q(x', X_k))$ (cf Proposition 3.9). Moreover, if the multiplicity of $\xi_i(x')$ as a root of $\gcd(P(x', X_k), Q(x', X_k))$ is 0, the sign of Q is fixed on $\xi_i(x')$. If S is described by the formula $\Theta(X')$, the graph of ξ_i is described by the formula

$$\begin{aligned}&\Theta(X') \wedge ((\exists Y_1)\ldots(\exists Y_\ell)\,(Y_1 < \cdots < Y_\ell \wedge P(X', Y_1) = 0, \ldots, P(X, Y_\ell) = 0))\\ &\wedge ((\forall\, Y)\, P(X', Y) = 0 \Rightarrow (Y = Y_1 \vee \cdots \vee Y = Y_\ell)) \wedge X_k = Y_i\,,\end{aligned}$$

which shows that ξ_i is semi-algebraic. □

Algorithm 11.19. **[Restricted Elimination]**

- **Structure:** an ordered integral domain D contained in a real closed field R.
- **Input:** a variable X_k, a polynomial P, and a finite set $\mathcal{P} \subset \mathrm{D}[X_1, \ldots, X_k]$.
- **Output:** a finite set $\mathrm{RElim}_{X_k}(P, \mathcal{P}) \subset \mathrm{D}[X_1, \ldots, X_{k-1}]$. The set $\mathrm{RElim}_{X_k}(P, \mathcal{P})$ is such that the degree of P, the number of roots of P in R, the number of common roots of P and $Q \in \mathcal{P}$ in R, and the sign of $Q \in \mathcal{P}$ at the roots of P in R is fixed on each semi-algebraically connected component of the realization of a sign condition on $\mathrm{RElim}_{X_k}(P, \mathcal{P})$.
- **Complexity:** $s\, d^{O(k)}$, where s is a bound on the number of elements of $\mathcal{P}$ and d is a bound on the degrees of P and $\mathcal{P}$.

- **Procedure:**
  - Place in $\mathrm{RElim}_{X_k}(P, \mathcal{P})$ the following polynomials when they are not in D, using Algorithm 8.21 (Signed Subresultant) and Remark 8.50:
    - $\mathrm{sRes}_j(R, \partial R/\partial X_k$, $R \in \mathrm{Tru}(P)$, $j = 0, \ldots, \deg(R) - 2$ (see Definition 1.16).
    - $\mathrm{sRes}_j(\partial R/\partial X_k Q, R)$ for $Q \in \mathcal{P}$, $R \in \mathrm{Tru}(P)$, $j = 0, \ldots, \deg_{X_k}(R) - 1$.
    - $\mathrm{lcof}(R)$ for $R \in \mathrm{Tru}(P)$.

Proof of correctness: The correctness of the Restricted Elimination Algorithm follows from Proposition 11.25. □

Complexity analysis of Algorithm 11.1: Consider

$$\mathrm{D}[X_1, \ldots, X_k] = \mathrm{D}[X_1, \ldots, X_{k-1}][X_k].$$

There are at most $d + 1$ polynomials in $\mathrm{Tru}(P)$ and s polynomials in $\mathcal{P}$ so the number of signed subresultant sequences to compute is $O((s + d)d)$. Each computation of a signed subresultant sequence costs $O(d^2)$ arithmetic operations in the integral domain $\mathrm{D}[X_1, \ldots, X_{k-1}]$ by the complexity analysis of Algorithm 8.21 (Signed Subresultant). So the complexity is $O((s + d)d^3)$ in the integral domain $\mathrm{D}[X_1, \ldots, X_{k-1}]$. There are $O((s + d)d^2)$ polynomials output, of degree bounded by $2(d^2)$ in $\mathrm{D}[X_1, \ldots, X_{k-1}]$, by Proposition 8.45.

Since each multiplication and exact division of polynomials of degree $2(d^2)$ in $k - 1$ variables costs $O(d)^{4k}$ (see Algorithms 8.5 and 8.6), the complexity in D is $s\, d^{O(k)}$.

When $\mathrm{D} = \mathbb{Z}$, and the bitsizes of the coefficients of $\mathcal{P}$ are bounded by τ, the bitsizes of the intermediate computations and output are bounded by $\tau d^{O(k)}$, using Proposition 8.46. □

In some phases of our algorithms in the next chapters, we construct points whose coordinates belong to the field of algebraic Puiseux series $\mathrm{R}\langle\varepsilon\rangle$. We are going to see that it possible to replace these infinitesimal quantities by sufficiently small elements from the field R, using the preceding restricted elimination.

The next proposition makes it possible to replace infinitesimal quantities with sufficiently small elements from the field R, using RElim_T. For this, we need a bound on the smallest root of a polynomial in terms of its coefficients. Such a bound is given by Proposition 10.3.

We again use the notation introduced in Chapter 10 (Notation 10.1). Given a set of univariate polynomials $\mathcal{A}$, we define $c'(\mathcal{A}) = \min_{Q \in \mathcal{A}} c'(Q)$.

Proposition 11.26. *Let $f(\varepsilon, T) \in \mathrm{D}[\varepsilon, T]$ be a bivariate polynomial, $\mathcal{L}$ a finite subset of $\mathrm{D}[\varepsilon, T]$, and σ a sign condition on $\mathcal{L}$ such that f has a root $\bar{t} \in R\langle\varepsilon\rangle$ for which*

$$\bigwedge_{g \in \mathcal{L}} \mathrm{sign}(g(\varepsilon, \bar{t})) = \sigma(g).$$

Then, for any v in R, $0<v<c'(\mathrm{RElim}_T(f,\mathcal{L}))$, *there exists a root t of $f(v,T)$ having the same Thom encoding as $\bar{t}$ and such that*

$$\bigwedge_{g\in\mathcal{L}} \mathrm{sign}(g(v,t)=\sigma(g).$$

Proof: If $v < c'(\mathrm{RElim}_T(f,\,\mathcal{L}))$, then v is smaller than the absolute value of all roots of every Q in $c'(\mathrm{RElim}_T(f,\,\mathcal{L}))$ by Proposition 10.3. Hence, by the properties of the output of $\mathrm{RElim}_T(f,\,\mathcal{L})$, the number of roots of the polynomial $f(v,\,T)$ as well as the number of its common roots with the polynomials $g(v,\,T)$, $g\in\mathcal{L}$, and the Thom encodings of its roots remain invariant for all v satisfying $0<v<c'(\mathrm{RElim}_T(f,\mathcal{L}))$.

Since $\varepsilon<c'(\mathrm{RElim}_T(f,\mathcal{L}))$, it is clear that for all $g\in\mathcal{L}$,

$$\mathrm{sign}(g(v,t))=(\mathrm{sign}(g(\varepsilon,\bar{t}\,)).$$ □

Algorithm 11.20. **[Removal of Infinitesimals]**

- **Structure:** an ordered integral domain D contained in a real closed field R.
- **Input:** a polynomial $f(\varepsilon,T)\in \mathrm{D}[\varepsilon,T]$ and a finite set $\mathcal{L}\subset \mathrm{D}[\varepsilon,T]$.
- **Output:** a pair $(a,\,b)$ of elements of D such that, for all $v\in$ R satisfying $v\leq a/b$, the following remains invariant:
  - the number of roots of $f(v,T)$, and their Thom encodings,
  - the signs of the polynomials $g\in\mathcal{L}$ at these roots.
- **Complexity:** $m\,d^{O(1)}$, where m is a bound on the number of elements of $\mathcal{L}$ and d is a bound on the degrees of f and $\mathcal{L}$.
- **Procedure:** Compute $\mathrm{RElim}_T(f,\,\mathcal{L})\subset \mathrm{D}[\varepsilon]$ and $c'(\mathrm{RElim}_T(f,\mathcal{L}))$. Take a and b such that $a/b=c'(\mathrm{RElim}_T(f,\mathcal{L}))$.

Complexity analysis: According to the complexity analysis of Algorithm 11.19, the complexity is $m\,d^{O(1)}$ in D, since $k=2$.

Note that if $\mathrm{D}=\mathbb{Z}$ and the bitsizes of the coefficients of the polynomials f,g are bounded by τ then $c'(\mathrm{RElim}_T(f,\,\mathcal{L}))$ is bounded from below by rational numbers with numerators and denominators of bitsizes $\tau\ d^{O(1)}$, using the complexity analysis of Algorithm 11.19. In this case, we replace the infinitely small element with a rational number smaller than $c'(\mathrm{RElim}_T(f,\mathcal{L}))$. □

Remark 11.27. If the coefficients of f and $\mathcal{L}$ belong to $\mathrm{D}[w]$ where w is the root of a polynomial h of degree at most d with Thom encoding σ, Algorithm 11.20 (Removal of Infinitesimals) can be easily modified to output a pair $(a,\,b)$ of elements of D such that, for all $v\in$ R satisfying $v\leq a/b$, the following remains invariant:

- the number of roots of $f(v,T)$, and their Thom encodings,
- the signs of the polynomials $g\in\mathcal{L}$ at these roots.

We just replace $(a(w), b(w))$ in D$[w]$ computed by Algorithm 11.20 (Removal of Infinitesimals) by (α, β) where $\alpha = c'(A)$ and $\beta = c(B)$ where $A = \mathrm{Res}_Y(h(Y), T - a(Y))$ (resp. $B = \mathrm{Res}_Y(h(Y), T - a(Y))$. The complexity is clearly $m\, d^{O(1)}$. □

11.8 Bibliographical Notes

The cylindrical decomposition algorithm, due to Collins [45], is the first algorithm for quantifier elimination with a reasonable worst-case time bound. The complexity of the algorithm is polynomial in the degree and number of polynomials. However the complexity is doubly exponential in the number of variables [120]. The former proofs of quantifier elimination [156, 148, 43, 92] were effective, but the complexity of the associated algorithm is not elementary recursive, i.e. is not bounded by a tower of exponents of finite height [120]. The main reason for the improvement in complexity given by the cylindrical decomposition is the use of subresultant coefficients, since, using subresultants, the number of branches in the computation is better controlled.

12

Polynomial System Solving

This chapter is mainly devoted to algorithms for solving zero-dimensional polynomial systems and give some applications. In Section 12.1, we explain a few results on Gröbner bases. This enables us to decide in Section 12.2 whether a polynomial system is zero-dimensional. We use these results to design various algorithms for zero-dimensional systems, for instance computing the multiplication table for the quotient ring and using the multiplication table to compute information about the solutions of zero-dimensional systems. A special case is treated in details in Section 12.3. In Section 12.4, we define the univariate representations and use trace computations to express the solutions of a zero-dimensional system as rational functions of the roots of a univariate polynomial. In Section 12.5, we explain how to compute the limits of bounded algebraic Puiseux series which are zeros of polynomial systems. In Section 12.6, we introduce the notion of pseudo-critical points and design an algorithm for finding at least one point in every semi-algebraically connected component of a bounded algebraic set, using a variant of the critical point method. In Section 12.8, we describe an algorithm computing the Euler-Poincaré characteristic of an algebraic set.

Throughout this chapter, we assume that K is an ordered field contained in a real closed field R and that $\mathrm{C} = \mathrm{R}[i]$.

12.1 A Few Results on Gröbner Bases

Throughout Section 12.1 and Section 12.2, we identify a monomial $X^\alpha \in \mathcal{M}_k$ and the corresponding $\alpha \in \mathbb{N}^k$, and we fix a monomial ordering $<$ on $\mathcal{M}_k$ (see Definition 4.62). We consider the problem of computing the Gröbner bases of an ideal (see Definition 4.66), using the notation of Section 4.4.1.

It is useful to be able to decide whether a set of polynomials $\mathcal{G} \subset \mathrm{K}[X_1, \ldots, X_k]$ is a Gröbner basis. This is done using the notion of an S-polynomial. Given two polynomials P_1 and P_2, the **S-Polynomial** of P_1 and P_2 is defined as

$$\mathrm{S}(P_1, P_2) = \frac{\mathrm{lt}(P_2)}{g} P_1 - \frac{\mathrm{lt}(P_1)}{g} P_2,$$

where $g = \gcd(\mathrm{lmon}(P_1), \mathrm{lmon}(P_2))$. Note that

$$\begin{aligned}
\mathrm{lcm}(\mathrm{lmon}(P_1), \mathrm{lmon}(P_2)) &= \frac{\mathrm{lmon}(P_2)}{g} \mathrm{lmon}(P_1) \\
&= \frac{\mathrm{lmon}(P_1)}{g} \mathrm{lmon}(P_2), \\
\mathrm{lmon}(\mathrm{S}(P_1, P_2)) &< \mathrm{lcm}(\mathrm{lmon}(P_1), \mathrm{lmon}(P_2)).
\end{aligned}$$

Proposition 12.1. [Termination criterion] *Let $\mathcal{G} \subset \mathrm{K}[X_1, \ldots, X_k]$ be a finite set such that the leading monomial of any element of $\mathcal{G}$ is not a multiple of the leading monomial of another element in $\mathcal{G}$. Then $\mathcal{G}$ is a Gröbner basis if and only if the S-polynomial of any pair of polynomials in $\mathcal{G}$ is reducible to 0 modulo $\mathcal{G}$.*

Proof: If $\mathcal{G}$ is a Gröbner basis, for any $G, H \in \mathcal{G}$, we have $\mathrm{S}(G, H) \in I(\mathcal{G}, \mathrm{K})$, and thus $\mathrm{S}(G, H)$ is reducible to 0 modulo $\mathcal{G}$ by Proposition 4.67.

Conversely, suppose that the S-polynomial of any pair of polynomials in $\mathcal{G}$ is reducible to 0 modulo $\mathcal{G}$. Using Remark 4.65, this implies that for every pair $G, H \in \mathcal{G}$,

$$\exists\, G_1 \in \mathcal{G} \cdots \exists\, G_s \in \mathcal{G}\ \mathrm{S}(G, H) = \sum_{i=1}^{s} A_i G_i, \tag{12.1}$$

with $\mathrm{lmon}(A_i G_i) \le \mathrm{lmon}(\mathrm{S}(G, H))$ for all $i = 1, \ldots, s$.

Consider $P \in I(\mathcal{G}, \mathrm{K})$. We want to prove that $\mathrm{lmon}(P)$ is a multiple of one of the leading monomials of the elements of $\mathcal{G}$. We have $P = B_1 G_1 + \cdots + B_s G_s$ with $\mathrm{lmon}(P) \le_{\mathrm{grlex}} \sup\{\mathrm{lmon}(B_i G_i); 1 \le i \le s\} = X^\mu$, and we suppose without loss of generality that $\mathrm{lmon}(B_1 G_1) = X^\mu$ and $\mathrm{lcof}(G_i) = 1$, for all i.

There are two possibilities:

- $\mathrm{lmon}(P) = X^\mu$. Then the monomial X^μ of P is a multiple of $\mathrm{lmon}(G_1)$, and there is nothing to prove.
- $\mathrm{lmon}(P) <_{\mathrm{grlex}} X^\mu$. Then the number n_μ of i such that $\mathrm{lmon}(B_i G_i) = X^\mu$ is at least 2, and we can suppose without loss of generality that $\mathrm{lmon}(B_2 G_2) = X^\mu$.

We have

$$\begin{aligned}
B_1 G_1 &= b_\beta X^\beta G_1 + F_1 G_1, \\
B_2 G_2 &= c_\gamma X^\gamma G_2 + F_2 G_2,
\end{aligned}$$

with $\mathrm{lmon}(F_1 G_1) < X^\mu$, $\mathrm{lmon}(F_2 G_2) < X^\mu$.

We necessarily have that X^μ is a multiple of

$$X^\sigma = \mathrm{lcm}(\mathrm{lmon}(G_1), \mathrm{lmon}(G_2)).$$

We rewrite $B_1 G_1 + B_2 G_2$ as follows:

$$\begin{aligned} B_1 G_1 + B_2 G_2 &= (b_\beta + c_\gamma) X^\beta G_1 + c_\gamma (X^\gamma G_2 - X^\beta G_1) + F_1 G_1 + F_2 G_2 \\ &= (b_\beta + c_\gamma) X^\beta G_1 - c_\gamma X^{\mu-\sigma} \mathrm{S}(G_1, G_2) + F_1 G_1 + F_2 G_2. \end{aligned}$$

By hypothesis $\mathrm{S}(G_1, G_2) = \sum_{\ell=1}^{s} A_\ell G_\ell$ with

$$\mathrm{lmon}(A_\ell G_\ell) \le \mathrm{lmon}(\mathrm{S}(G_1, G_2)) <_{\mathrm{grlex}} X^\sigma,$$

and thus

$$\mathrm{lmon}(X^{\mu-\sigma} A_\ell G_\ell) < X^\mu.$$

It follows that we have written

$$P = \bar{B}_1 G_1 + \cdots + \bar{B}_s G_s,$$

and the number of terms $\bar{B}_i G_i$ with leading monomial μ has decreased, or

$$\sup \{\mathrm{lmon}(\bar{B}_i G_i); 1 \le i \le s\} < X^\mu.$$

This proves the result by induction on the lexicographical order of the pair (μ, n_μ) since either n_μ or ν has decreased. □

The basic idea for computing a Gröbner basis is thus quite simple: add to the original set the reduction of all possible S-polynomials as long as these are not all equal to 0.

Algorithm 12.1. **[Buchberger]**

- **Structure:** a field K.
- **Input:** a finite number of polynomials $\mathcal{P}$ of $\mathrm{K}[X_1, \ldots, X_k]$.
- **Output:** a Gröbner basis $\mathcal{G}$ of $I(\mathcal{P}, \mathrm{K})$.
- **Procedure:**
  - Reduce $\mathcal{P}$, which is done as follows:
  - While there is a polynomial P in $\mathcal{P}$ with a monomial X^α which is a multiple of the leading monomial of an element F of $\mathcal{P} \setminus \{P\}$,
    - If $\mathrm{Red}(P, X^\alpha, F) \neq 0$, update $\mathcal{P}$ by replacing P by $\mathrm{Red}(P, X^\alpha, F)$ in $\mathcal{P}$.
    - Otherwise, update $\mathcal{P}$ by removing P from $\mathcal{P}$.
  - Initialize $\mathcal{F} := \mathcal{P}$, $\mathcal{C} := \{(P, Q) \mid P \in \mathcal{P}, Q \in \mathcal{P}, P \neq Q\}$.
  - While $\mathcal{C} \neq \emptyset$
    - Remove a pair (P, Q) from $\mathcal{C}$.
    - Reduce $\mathrm{S}(P, Q)$ modulo $\mathcal{F}$, which is done as follows:
    - Initialize $R := \mathrm{S}(P, Q)$

- While there is a monomial of R that is a multiple of a leading monomial of an element of $\mathcal{F}$,
  - Pick the greatest monomial X^α of R which is a multiple of a leading monomial of an element of $\mathcal{F}$, and take any $F \in \mathcal{F}$ such that X^α is a multiple of $\text{lmon}(F)$. Replace R by $\text{Red}(R, X^\alpha, F)$.
- If $R \neq 0$, update $\mathcal{F} := \mathcal{F} \cup \{R\}$, and reduce $\mathcal{F}$:
- While there is a polynomial P in $\mathcal{F}$ with a monomial X^α which is a multiple of the leading monomial of an element F of $\mathcal{F} \setminus \{P\}$,
  - If $\text{Red}(P, X^\alpha, F) \neq 0$, update $\mathcal{F}$ by replacing P by $\text{Red}(P, X^\alpha, F)$ in $\mathcal{F}$.
  - Otherwise, update $\mathcal{F}$ by removing P from $\mathcal{F}$.
- Update $\mathcal{C} := \mathcal{C} \cup \{(P, Q) \mid P \in \mathcal{F}, Q \in \mathcal{F}, P \neq Q\}$.

- Output $\mathcal{G} = \mathcal{F}$.

Proof of correctness: It is first necessary to prove that the algorithm terminates. Denote by $\mathcal{F}_n$ the value of $\mathcal{F}$, obtained after having reduced n pairs of polynomials. Note that the ideal generated by $\text{lmon}(\mathcal{F}_{n-1})$ is contained in the ideal generated by $\text{lmon}(\mathcal{F}_n)$. Consider the ascending chain of ideals generated by the monomials $\text{lmon}(\mathcal{F}_n)$. This ascending chain is stationary by Corollary 4.70 and stops with the ideal generated by $\text{lmon}(\mathcal{F}_N)$. This means that for every pair (P, Q) of elements of $\mathcal{F}_N$, $\text{S}(P, Q)$ is reducible to 0 modulo $\mathcal{F}_N$. This ensures that the algorithm terminates. Finally, the leading monomial of any element of $\mathcal{F}_N$ is not a multiple of the leading monomial of another element in $\mathcal{F}_N$, because of the reductions performed at the end of the algorithm. So we conclude that $\mathcal{G} = \mathcal{F}_N$ is a Gröbner basis of $I(\mathcal{P}, \text{K})$ by Proposition 12.1. □

Note that the argument for the termination of Buchberger's algorithm does not provide a bound on the number or degrees of the polynomials output or the number of steps of the algorithm. Thus the complexity analysis of Buchberger's algorithm is a complicated problem which we do not consider here. The reader is referred to [104, 114] for complexity results for the problem of computing Gröbner basis of general polynomial ideals.

Remark 12.2. Note that Buchberger's thesis appears in [32] and that a lot of work has been done since then to find out how to compute Gröbner bases efficiently. It is impossible to give in a few words even a vague idea of all the results obtained in this direction (the bibliography [166] contains about 1000 references). In particular, the monomial ordering used plays a key role in the efficiency of the computation, and in many circumstances the reverse lexicographical ordering (Definition 4.63) is a good choice. Modern methods for computing Gröbner bases are often quite different from the original Buchberger's algorithm (see for example [58]). Very efficient Gröbner bases computation can be found at [59]. □

It turns out that certain special polynomial systems are automatically Gröbner bases. This will be very useful for the algorithms to be described in the next chapters.

Proposition 12.3. *A polynomial system* $\mathcal{G} = \{X_1^{d_1} + Q_1, \ldots, X_\ell^{d_\ell} + Q_\ell\}$ *with* $\mathrm{lmon}(Q_i) < X_i^{d_i}$ *is a Gröbner basis.*

Proof: Let $P_i = X_i^{d_i} + Q_i$. We need only prove that $\mathrm{S}(P_i, P_j)$ is reducible to 0 modulo $\mathcal{G}$. Clearly $-Q_i$ is a reduction of $X_i^{d_i}$ modulo $\mathcal{G}$. Note that if $i \neq j$, the polynomials $X_j^{d_j} Q_i$ and $X_i^{d_i} Q_j$ have no monomials in common. Hence, for $i \neq j$, $\mathrm{S}(P_i, P_j) = X_j^{d_j} Q_i - X_i^{d_i} Q_j$ is reducible to $X_j^{d_j} Q_i + Q_i Q_j = Q_i P_j$, which is reducible to 0 modulo $\mathcal{G}$. □

Let $\mathcal{G}$ be a Gröbner basis of an ideal I. The set of monomials which are multiple of the leading monomials of the polynomials in $\mathcal{G}$ coincides with the set of leading monomials of elements of I. The **corners of the staircase** of $\mathcal{G}$, $\mathrm{Cor}(\mathcal{G})$ are the leading monomials of the polynomials in $\mathcal{G}$. They are the minimal elements of the leading monomials of elements of I for the partial order of divisibility among monomials. The **orthant** generated by a corner X^α is $\alpha + \mathbb{N}^k$ and consists of multiples of X^α. The set of leading monomials of elements of I is the union of the orthants generated by the corners. The set of **monomials under the staircase** for $\mathcal{G}$, $\mathrm{Mon}(\mathcal{G})$, is the set of monomials that do not belong to the set of leading monomials of elements of I. The **border of the staircase** of $\mathcal{G}$ is the set $\mathrm{Bor}(\mathcal{G})$ of monomials which are leading monomials of elements of I and which are obtained by multiplying a monomial of $\mathrm{Mon}(\mathcal{G})$ by a variable.

Example 12.4. We illustrate these notions using an example in the plane. Let $k = 2$ and the monomial ordering be the reverse lexicographical ordering. Consider $\mathcal{G} := \{X_1^2 + 2X_2^2, X_2^4, X_1 X_2^2\}$. It is easy to check that $\mathcal{G}$ is a Gröbner basis. The set of leading monomials of elements of $\mathrm{Ideal}(\mathcal{G}, \mathrm{K})$ is the set of multiples of X_2^4, $X_1 X_2^2$ and X_1^2. It is the union of the three orthants generated by the corners X_2^4, $X_1 X_2^2$, X_1^2. The set of monomials under the staircase is the finite set of monomials

$$\mathrm{Mon}(\mathcal{G}) = \{1, X_2, X_2^2, X_2^3, X_1, X_1 X_2\}.$$

The border of $\mathcal{G}$ is

$$\mathrm{Bor}(\mathcal{G}) = \{X_1^2, X_1^2 X_2, X_1 X_2^2, X_1 X_2^3, X_2^4\}.$$

Here is the corresponding picture (the big black points are the corners, the other black points are the other elements of the border, the white points are under the staircase).

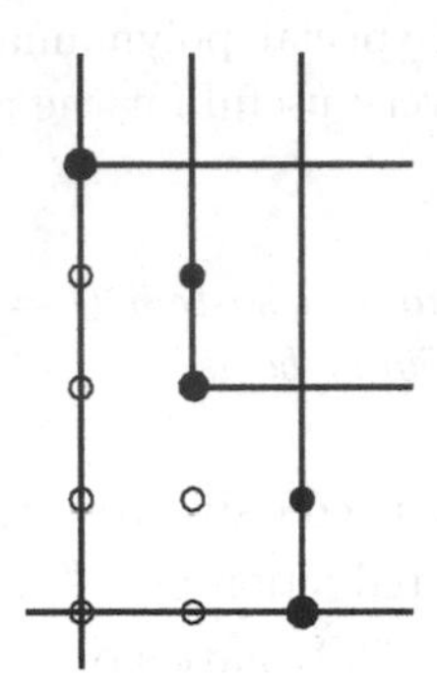

Fig. 12.1. Staircase

□

Proposition 12.5. *Let* $A = \mathrm{K}[X_1, ..., X_k]/I$ *for an ideal* I *and let* $\mathcal{G}$ *be a Gröbner basis of* I. *The monomials in* $\mathrm{Mon}(\mathcal{G})$ *constitute a basis of the* K-*vector space* A.

Proof: It is clear that any element of $\mathrm{K}[X_1, ..., X_k]$ is reducible modulo $\mathcal{G}$ to a linear combination of monomials in $\mathrm{Mon}(\mathcal{G})$. Conversely, a non-zero linear combination of monomials in $\mathrm{Mon}(\mathcal{G})$ cannot be reduced to 0 by $\mathcal{G}$, thus does not belong to I, and hence is not zero in A. □

Let $\mathcal{G}$ be a Gröbner basis. The **normal form** of $P \in \mathrm{K}[X_1, ..., X_k]$ modulo $\mathcal{G}$, denoted $\mathrm{NF}(P)$, is a linear combination Q of monomials in $\mathrm{Mon}(\mathcal{G})$ such that $P = Q \mod I(\mathcal{G}, \mathrm{K})$. Such a linear combination is unique by Proposition 12.5. Note that NF is a linear mapping from the K-vector space $\mathrm{K}[X_1, ..., X_k]$ to the K-vector space $\mathrm{K}[X_1, ..., X_k]/I(\mathcal{G}, \mathrm{K})$.

Algorithm 12.2. **[Normal Form]**

- **Structure:** a field K.
- **Input:** a Gröbner basis $\mathcal{G}$ and a polynomial $P \in \mathrm{K}[X_1, ..., X_k]$.
- **Output:** the normal form $\mathrm{NF}(P)$ of P modulo $\mathcal{G}$.
- **Procedure:**
  - Initialize $Q := P$.
  - While there are monomials of Q that are not in $\mathrm{Mon}(\mathcal{G})$,
    - Pick the greatest monomial X^α of Q which is not in $\mathrm{Mon}(\mathcal{G})$. Take any $G \in \mathcal{G}$ such that X^α is a multiple of $\mathrm{lmon}(G)$. Update Q, replacing it by $\mathrm{Red}(Q, X^\alpha, G)$.
  - Output $\mathrm{NF}(P) := Q$.

Example 12.6. Returning to example 12.4 the normal form of a polynomial in X_1 and X_2, is a linear combination of elements in $\mathrm{Mon}(\mathcal{G})$. For example, the normal form of $X_1^3 + X_1 + X_2$ is obtained by computing

$$\begin{aligned} \mathrm{Red}(X_1^3 + X_1 + X_2, X_1^3, X_1^2 + 2\,X_2^2) &= -2\,X_1 X_2^2 + X_1 + X_2, \\ \mathrm{Red}(-2\,X_1 X_2^2 + X_1 + X_2, X_1 X_2^2, X_1 X_2^2) &= X_1 + X_2. \end{aligned}$$

□

12.2 Multiplication Tables

An important property of a Gröbner basis is that it can be used to characterize zero-dimensional polynomial systems.

Proposition 12.7. *The finite set $\mathcal{P} \subset \mathrm{K}[X_1, \ldots, X_k]$ is a zero-dimensional polynomial system if and only if any Gröbner basis of* $\mathrm{Ideal}(\mathcal{P}, \mathrm{K})$ *contains a polynomial with leading monomial $X_i^{d_i}$ for each $i, 1 \le i \le k$.*

Proof: It is clear that the number of monomials under the staircase is finite if and only any Gröbner basis contains, for each i, $1 \le i \le k$, a polynomial with leading monomial $X_i^{d_i}$. We conclude by applying Proposition 12.5 and Theorem 4.85. □

As a consequence:

Corollary 12.8. *If $\mathcal{P} \subset \mathrm{K}[X_1, \ldots, X_k]$ is a zero-dimensional polynomial system and $\mathcal{G}$ is its Gröbner basis,* $\#(\mathrm{Zer}(\mathcal{P}, \mathrm{C}^k)) \le \#(\mathrm{Mon}(\mathcal{G}))$.

Proof: Use Proposition 12.5 and Theorem 4.85. □

Corollary 12.9. *A polynomial system $\mathcal{P} = \{X_1^{d_1} + Q_1, \ldots, X_k^{d_k} + Q_k\}$ with* $\mathrm{lmon}(Q_i) <_{\mathrm{grlex}} X_i^{d_i}$ *has a finite number of solutions.*

Proof: Apply Proposition 12.3 and Proposition 12.7. □

Suppose that $\mathcal{P} \subset \mathrm{K}[X_1, \ldots, X_k]$ is a zero-dimensional polynomial system and $\mathcal{B}$ is a basis of $\mathrm{A} = \mathrm{K}[X_1, \ldots, X_k]/\mathrm{Ideal}(\mathcal{P}, \mathrm{K})$. The **multiplication table** of A in $\mathcal{B}$ is, for every pair of elements a and b in $\mathcal{B}$, the expression of their product in A as a linear combination of elements in $\mathcal{B}$. Note that it happens often that $a\,b = a'\,b'$, with a, b, a', b' in $\mathcal{B}$. So in order to avoid repetitions we define

$$\mathrm{Tab}(\mathcal{B}) = \{a\,b \mid a, b \text{ in } \mathcal{B}\}.$$

The **size of the multiplication table**, denoted by T, is the number of elements of $\mathrm{Tab}(\mathcal{B})$. The number T can be significantly smaller than N^2 where N is the number of monomials in $\mathrm{Tab}(\mathcal{B})$. For example when $k=1$, there are at most $2N$ monomials in the multiplication table, taking $\mathcal{B}=\{1, X, \ldots, X^{N-1}\}$. The coefficients $\lambda_{c,d}$ such that, in A,

$$c = \sum_{d \in \mathcal{B}} \lambda_{c,d}\, d,$$

with $c \in \mathrm{Tab}(\mathcal{B})$, are the **entries of the multiplication table** $\mathrm{Mat}(\mathcal{B})$.

We explain now how to compute the multiplication table in the basis $\mathrm{Mon}(\mathcal{G})$.

Algorithm 12.3. **[Multiplication Table]**

- **Structure:** a field K.
- **Input:** a zero-dimensional polynomial system $\mathcal{P} \subset \mathrm{K}[X_1, \ldots, X_k]$, together with a Gröbner basis $\mathcal{G}$ of $I = \mathrm{Ideal}(\mathcal{P}, \mathrm{K})$.
- **Output:** the multiplication table $\mathrm{Mat}(\mathrm{Mon}(\mathcal{G}))$ of $\mathrm{A} = \mathrm{K}[X_1, \ldots, X_k]/I$ in the basis $\mathrm{Mon}(\mathcal{G})$. More precisely, for every $c \in \mathrm{Tab}(\mathrm{Mon}(\mathcal{G}))$, the coefficients $\lambda_{c,a}$, $a \in \mathrm{Mon}(\mathcal{G})$, such that $c = \sum_{a \in \mathrm{Mon}(\mathcal{G})} \lambda_{c,a}\, a$.
- **Complexity:** $O(k\, N^3 + T\, N^2)$, where N is the number of elements of $\mathrm{Mon}(\mathcal{G})$ and T is the number of elements of $\mathrm{Tab}(\mathrm{Mon}(\mathcal{G}))$.
- **Procedure:**
  - Step 1: For every $X^\alpha \in \mathrm{Bor}(\mathcal{G})$ in increasing order according to $<$ compute $\mathrm{NF}(X^\alpha) = \sum_{X^\delta \in \mathrm{Mon}(\mathcal{G})} \mu_{\alpha,\delta} X^\delta$ as follows:
  - If $X^\alpha \in \mathrm{Cor}(\mathcal{G})$, and $G \in \mathcal{G}$ is such that $\mathrm{lmon}(G) = X^\alpha$,

    $$\mathrm{NF}(X^\alpha) := G - X^\alpha.$$

  - If $X^\alpha \notin \mathrm{Cor}(\mathcal{G})$, and $X^\alpha = X_j\, X^\beta$ for some $j = 1, \ldots, k$, with $X^\beta \in \mathrm{Bor}(\mathcal{G})$, define

    $$\mathrm{NF}(X^\alpha) \;:=\; \sum_{(X^\gamma, X^\delta) \in \mathrm{Mon}(\mathcal{G})^2} \mu_{\beta,\gamma}\, \mu_{\gamma',\delta} X^\delta,$$

    with $X_j X^\gamma = X^{\gamma'}$, and

    $$\mathrm{NF}(X^\beta) \;=\; \sum_{X^\gamma \in \mathrm{Mon}(\mathcal{G})} \mu_{\beta,\gamma}\; X^\gamma,$$
    $$\mathrm{NF}(X^{\gamma'}) \;=\; \sum_{X^\delta \in \mathrm{Mon}(\mathcal{G})} \mu_{\gamma',\delta}\; X^\delta.$$

  - Step 2: Construct the matrices $M_1, \ldots, M_k$ corresponding to multiplication by $X_1, \ldots, X_k$, expressed in the basis $\mathrm{Mon}(\mathcal{G})$, using the normal forms of elements of $\mathrm{Bor}(\mathcal{G})$ already computed.
  - Step 3: For every $X^\alpha \in \mathrm{Tab}(\mathcal{G}) \setminus (\mathrm{Mon}(\mathcal{G}) \cup \mathrm{Bor}(\mathcal{G}))$ in increasing order according to $<$ compute $\mathrm{NF}(X^\alpha)$ as follows: since $X^\alpha = X_j\, X^\beta$, for some $j = 1, \ldots, k$, compute the vector $\mathrm{NF}(X^\alpha) = M_j \cdot \mathrm{NF}(X^\beta)$.

Proof of correctness: Note that

$$\begin{aligned}\mathrm{NF}(X^\alpha) &= \mathrm{NF}(X_j \mathrm{NF}(X^\beta)) \\ &= \sum_{X^\gamma \in \mathrm{Mon}(\mathcal{G})} \mu_{\beta,\gamma} X_j X^\gamma \\ &= \sum_{(X^\gamma, X^\delta) \in \mathrm{Mon}(\mathcal{G})^2} \mu_{\beta,\gamma} \mu_{\gamma',\delta} X^\delta,\end{aligned}$$

The only thing which remains to prove is that

$$\begin{aligned}\mathrm{NF}(X^\beta) &= \sum_{X^\gamma \in \mathrm{Mon}(\mathcal{G})} \mu_{\beta,\gamma} X^\gamma, \\ \mathrm{NF}(X^{\gamma'}) &= \sum_{X^\delta \in \mathrm{Mon}(\mathcal{G})} \mu_{\gamma',\delta} X^\delta,\end{aligned}$$

have been computed before, which follows from $X^\beta < X^\alpha$ and $X^{\gamma'} < X^\alpha$. □

Complexity analysis:

There are at most $k N$ elements in $\mathrm{Bor}(\mathcal{G})$. For each element of $\mathrm{Bor}(\mathcal{G})$, the computation of its normal form takes $O(N^2)$ operations. Thus, the complexity of Step 1 is $O(k N^3)$.

In Step 3 there is a matrix vector multiplication to perform for each element on $\mathrm{Tab}(\mathcal{G})$, so the total cost of Step 3 is $O(T N^2)$. □

Once a multiplication table is known, it is easy to perform arithmetic operations in the quotient ring $\mathrm{A} = \mathrm{K}[X_1, ..., X_k]/I$, and to estimate the complexity of these computations. All the arithmetic operations to perform take place in the ring generated by the entries of the multiplication table and the coefficients of the elements we want to add or multiply.

Algorithm 12.4. **[Zero-dimensional Arithmetic Operations]**

- **Structure:** a ring D contained in a field K.
- **Input:**
  - a zero-dimensional polynomial system $\mathcal{P} \subset \mathrm{D}[X_1, ..., X_k]$,
  - a basis $\mathcal{B}$ of $\mathrm{A} = \mathrm{K}[X_1, ..., X_k]/\mathrm{Ideal}(\mathcal{P}, \mathrm{K})$, such that the multiplication table $\mathrm{Mat}(\mathcal{B})$ has entries in D,
  - two elements $f, g \in \mathrm{A}$, given as a linear combination of elements of $\mathcal{B}$ with coefficients in D.
- **Output:** $f + g$ and $f g$ in A, specified by linear combinations with coefficients in D of elements of $\mathcal{B}$.
- **Complexity:** $O(N)$ for the addition, $O(T\ N^2)$ for the multiplication, where N is the number of elements of $\mathcal{B}$ and T is the number of elements of $\mathrm{Tab}(\mathcal{B})$.
- **Procedure:**
  - Let $f = \sum_{a \in \mathcal{B}} f_a\, a$, $g = \sum_{a \in \mathcal{B}} g_a\, a$.
  - Define $f + g := \sum_{\alpha \in \mathcal{B}} (a_\alpha + b_\alpha)\, \alpha$.

– If $c = \sum_{d \in \mathcal{B}} \lambda_{c,d}\, d$, for $c \in \mathrm{Tab}(\mathcal{B})$, define

$$fg := \sum_{d \in \mathcal{B}} \sum_{c \in \mathrm{Tab}(\mathcal{B})} \sum_{\substack{a \in \mathcal{B}, b \in \mathcal{B} \\ ab = c}} (f_a\, g_b) \lambda_{c,d}\, d.$$

Complexity analysis: It is clear that the complexity of the addition is $O(N)$, while the complexity of the multiplication is $O(TN^2)$. □

Lots of information can be obtained about the solutions of a zero-dimensional system once a multiplication table is known. We can compute traces, number of distinct zeroes and perform sign determination.

We use the notation introduced in Chapter 4 (Notation 4.95). For $f \in \mathrm{A}$ we denote by $L_f\colon \mathrm{A} \to \mathrm{A}$ the linear map defined by $L_f(g) = fg$ for $g \in \mathrm{A}$. We can compute the trace of L_f.

Algorithm 12.5. **[Trace]**

- **Structure:** a ring D contained in a field K.
- **Input:**
  - a zero-dimensional polynomial system $\mathcal{P} \subset \mathrm{D}[X_1, \ldots, X_k]$,
  - a basis $\mathcal{B}$ of $\mathrm{A} = \mathrm{K}[X_1, \ldots, X_k]/\mathrm{Ideal}(\mathcal{P}, \mathrm{K})$, such that the multiplication table $\mathrm{Mat}(\mathcal{B})$ has entries in D,
  - an element $f \in \mathrm{A}$, given as a linear combination of elements of $\mathcal{B}$ with coefficients in D.
- **Output:** the trace of the linear map L_f.
- **Complexity:** $O(N^2)$, where N is the number of elements of $\mathcal{B}$.
- **Procedure:** Let $f = \sum_{a \in \mathcal{B}} f_a\, a$. Compute

$$\mathrm{Tr}(L_f) := \sum_{b \in \mathcal{B}} \sum_{a \in \mathcal{B}} f_a\, \lambda_{ab,b}.$$

Proof of correctness: Since $fb = \sum_{c \in \mathcal{B}} \sum_{a \in \mathcal{B}} f_a\, \lambda_{ab,c}\, c$, the entry of the matrix of L_f corresponding to the elements b, c of the basis is $\sum_{a \in \mathcal{B}} f_a \lambda_{ab,c}$. The trace of L_f is obtained by summing the diagonal terms. □

Complexity analysis: The number of arithmetic operations in D is clearly $O(N^2)$. □

We can also compute the number of distinct zeroes.

Algorithm 12.6. **[Number of Distinct Zeros]**

- **Structure:** an integral domain D contained in a field K.
- **Input:**
  - a zero-dimensional polynomial system $\mathcal{P} \subset \mathrm{D}[X_1, \ldots, X_k]$,
  - a basis $\mathcal{B}$ of $\mathrm{A} = \mathrm{K}[X_1, \ldots, X_k]/\mathrm{Ideal}(\mathcal{P}, \mathrm{K})$, such that the multiplication table $\mathrm{Mat}(\mathcal{B})$ has entries in D,

- **Output:** $n = \#\mathrm{Zer}(\mathcal{P}, \mathrm{C}^k)$.
- **Complexity:** $O(TN^2)$, where N is the number of elements of $\mathcal{B}$ and T is the number of elements of $\mathrm{Tab}(\mathcal{B})$.
- **Procedure:** Apply Algorithm 12.5 (Trace) to the maps L_c for every element c of $\mathrm{Tab}(\mathcal{B})$. Then compute the rank of the matrix with entries $[\mathrm{Tr}(L_{ab})]$ for every a, b in $\mathcal{B}$, using Remark 8.18.

Proof of correctness: The matrix with entries $[\mathrm{Tr}(L_{ab})]$ is the matrix of $\mathrm{Her}(\mathcal{P})$ in the basis $\mathcal{B}$. Hence its rank is equal to $\#\mathrm{Zer}(\mathcal{P}, \mathrm{C}^k)$ by part a) of Theorem 4.100 (Multivariate Hermite). □

Complexity analysis: Let N be the dimension of the K-vector space A. There are T trace computations to perform, and then a rank computation for a matrix of size N. The number of arithmetic operations in D is thus $O(T\,N^2)$ using the complexity analysis of Algorithm 12.5 (Trace), Algorithm 8.16 (Jordan-Bareiss's method) and Remark 8.18. □

We can also compute Tarski-queries.

Algorithm 12.7. **[Multivariate Tarski-query]**

- **Structure:** an integral domain D contained in an ordered field K.
- **Input:**
  - a zero-dimensional polynomial system $\mathcal{P} \subset \mathrm{D}[X_1, \ldots, X_k]$,
  - a basis $\mathcal{B}$ of $\mathrm{A} = \mathrm{K}[X_1, \ldots, X_k]/\mathrm{Ideal}(\mathcal{P}, \mathrm{K})$, such that the multiplication table $\mathrm{Mat}(\mathcal{B})$ has entries in D,
  - an element $Q \in \mathrm{A}$, given as a linear combination of elements of $\mathcal{B}$ with coefficients in D.

  given as a linear combination of elements of $\mathcal{B}$ with coefficients in D.
- **Output:** the Tarski-query $\mathrm{TaQ}(Q, Z)$ with $Z = \mathrm{Zer}(\mathcal{P}, \mathrm{R}^k)$.
- **Complexity:** $O(TN^2)$, where N is the number of elements of $\mathcal{B}$ and T is the number of elements of $\mathrm{Tab}(\mathcal{B})$.
- **Procedure:** Apply the Trace Algorithm to the maps L_{Qc} for every $c \in \mathrm{Tab}(\mathcal{B})$. Then compute the signature of the Hermite quadratic form associated to Q by Algorithm 8.18 (Signature through Descartes).

Proof of correctness: The correctness of the Multivariate Tarski-query Algorithm follows from Theorem 4.100 (Multivariate Hermite). □

Complexity analysis: There are T traces to compute, and a signature computation to perform. The number of arithmetic operations in D is $O(TN^2)$, given the complexity analyses of Algorithms 12.5 (Trace) and 8.18 (Signature through Descartes). □

Algorithm 10.11 (Sign Determination) can be performed with Algorithm 12.7 (Multivariate Tarski-query) as a blackbox.

Algorithm 12.8. **[Multivariate Sign Determination]**

- **Structure:** an integral domain D contained in a field K.
- **Input:**
  - a zero-dimensional polynomial system $\mathcal{P} \subset \mathrm{D}[X_1, \ldots, X_k]$,
  - a basis $\mathcal{B}$ of $\mathrm{A} = \mathrm{K}[X_1, \ldots, X_k]/\mathrm{Ideal}(\mathcal{P}, \mathrm{K})$, such that the multiplication table $\mathrm{Mat}(\mathcal{B})$ has entries in D,
  - a list $\mathcal{Q} \subset \mathrm{A}$, given as a linear combinations of elements of $\mathcal{B}$ with coefficients in D.

  Denote by $Z = \mathrm{Zer}(\mathcal{P}, \mathrm{R}^k)$.
- **Output:** the list of sign conditions realized by $\mathcal{Q}$ on Z.
- **Complexity:** $s\, T N^3$, where N is the number of elements of $\mathcal{B}$, T is the number of elements of $\mathrm{Tab}(\mathcal{B})$ and s the number of elements of $\mathcal{Q}$.
- **Procedure:** Perform Algorithm 10.11 (Sign Determination) for Z, using Algorithm 12.7 (Multivariate Tarski-query) as a Tarski-query black box, the products of polynomials in $\mathcal{Q}$ being computed in A using Algorithm 12.4 (Zero-dimensional Arithmetic Operations).

Complexity analysis: Let N be the dimension of the K-vector space A and s the cardinality of $\mathcal{Q}$. Note that $\#(Z) \leq N$ by Theorem 4.85. According to the complexity of Algorithm 10.11 (Sign Determination), the number of calls to the Tarski-query black box is bounded by $1 + 2\, s\, N$. For each call to Algorithm 12.7 (Multivariate Tarski-query), a multiplication has to be performed by Algorithm 12.4 (Zero-dimensional Arithmetic Operations). The complexity is thus $O(s\, T N^3)$, using the complexity of Algorithm 12.7 (Multivariate Tarski-query). □

12.3 Special Multiplication Table

We now study a very special case of zero-dimensional system, where we start from a Gröbner basis with a very specific structure, used in the later chapters. In this section the only monomial ordering we consider is the graded lexicographical ordering (see Definition 2.15).

Let D be an integral domain. A Gröbner basis $\mathcal{G}$ is **special** if it is of the form

$$\mathcal{G} = \{b_1 X_1^{d_1} + Q_1, \ldots, b_k X_k^{d_k} + Q_k\} \subset \mathrm{D}[X_1, \ldots, X_k]$$

with $\deg(Q_i) < d_i$, $\deg_{X_j}(Q_i) < d_j$, $j \neq i$, $d_1 \geq \ldots \geq d_k$, $b_j \neq 0$. According to Proposition 12.3, a special Gröbner basis $\mathcal{G}$ is a Gröbner basis of $\mathrm{Ideal}(\mathcal{G}, \mathrm{K})$ for the graded lexicographical ordering on monomials. The monomials under the staircase $\mathrm{Mon}(\mathcal{G})$ are the monomials $X^\alpha = X_1^{\alpha_1} \ldots X_k^{\alpha_k}$ with $\alpha_i < d_i$. Note that, for every i, Q_i is a linear combination of $\mathrm{Mon}(\mathcal{G})$. Thus, $\mathrm{NF}(b_i X_i^{d_i}) = -Q_i$. The border of the staircase $\mathrm{Bor}(\mathcal{G})$ is the set of monomials such that $\alpha_i = d_i$ for some $i \in \{1, \ldots, k\}$ and $\alpha_j < d_j$ for all $j \neq i$.

We adapt Algorithm 12.3 (Multiplication Table) to this special case, the main change being that we manage to obtain coefficients in D. In order to be able to make the reduction inside the ring D, it is useful to multiply monomials in advance by a convenient product of the leading coefficients of the polynomials in $\mathcal{G}$. This is the reason why we introduce the following notation.

Notation 12.10. [Multiplication table] For $\alpha \in \mathbb{N}^k$, let $|\alpha| = \alpha_1 + \cdots + \alpha_k$ denote the total degree of X^α. Let b be a common multiple of $b_1, \ldots, b_k$ in D, i.e. for every $j = 1, \ldots, k$, $b = b_j \bar{b}_j$, $\bar{b}_j \in \mathrm{D}$. Denote by $\overline{\mathrm{Mon}}(\mathcal{G})$ the set of $b^{|\alpha|} X^\alpha$ such that $X^\alpha \in \mathrm{Mon}(\mathcal{G})$ and by $\overline{\mathrm{Bor}}(\mathcal{G})$ the set of $b^{|\alpha|} X^\alpha$ such that $X^\alpha \subset \mathrm{Bor}(\mathcal{G})$.□

We adapt Algorithm 12.3 (Multiplication Table) to this special case, the main change being that we manage to obtain coefficients in D.

Algorithm 12.9. **[Special Multiplication Table]**

- **Structure:** a ring D contained in a field K.
- **Input:** a special Gröbner basis
$$\mathcal{G} = \{b_1 X_1^{d_1} + Q_1, \ldots, b_k X_k^{d_k} + Q_k\} \subset \mathrm{D}[X_1, \ldots, X_k],$$
b a common multiple of $b_1, \ldots, b_k$ in D, with, for every j from 1 to k, $b = b_j \bar{b}_j$.
- **Output:** the multiplication table of $\mathrm{A} = \mathrm{K}[X_1, \ldots, X_k]/\mathrm{Ideal}(\mathcal{P}, \mathrm{K})$ in the basis $\overline{\mathrm{Mon}}(\mathcal{G})$. The multiplication table consists of polynomials in $\mathrm{D}[X_1, \ldots, X_k]$.
- **Complexity:** $O(2^k (d_1 \ldots d_k)^3)$.
- **Procedure:**
  - Step 1: For every $b^{|\alpha|} X^\alpha \in \overline{\mathrm{Bor}}(\mathcal{G})$ in increasing order according to $<_{\mathrm{deglex}}$ compute $\mathrm{NF}(b^{|\alpha|} X^\alpha) = \sum_{X^\delta \in \mathrm{Mon}(\mathcal{G})} \lambda_{\alpha,\delta} b^{|\alpha|} X^\delta$ as follows:
  - If $X^\alpha = X_i^{d_i}$, $\mathrm{NF}(b^{|\alpha|} X^\alpha) := -b^{d_i - 1} \bar{b}_i Q_i$.
  - Else, if $X^\alpha = X_j X^\beta$ for some $j = 1, \ldots, k$, with $X^\beta \in \mathrm{Bor}(\mathcal{G})$, define
$$\mathrm{NF}(b^{|\alpha|} X^\alpha) = \sum_{(X^\gamma, X^\delta) \in \mathrm{Mon}(\mathcal{G})^2} \lambda_{\beta,\gamma} \lambda_{\gamma',\delta} b^{|\gamma|} X^\delta,$$
with $X_j X^\gamma = X^{\gamma'}$, and
$$\mathrm{NF}(b^{|\beta|} X^\beta) = \sum_{X^\gamma \in \mathrm{Mon}(\mathcal{G})} \mu_{\beta,\gamma} b^{|\gamma|} X^\gamma,$$
$$\mathrm{NF}(b^{|\gamma'|} X^{\gamma'}) = \sum_{X^\delta \in \mathrm{Mon}(\mathcal{G})} \mu_{\gamma',\delta} b^{|\delta|} X^\delta.$$
  - Step 2: Construct the matrices $M_1', \ldots, M_k'$ corresponding to multiplication by $b X_1, \ldots, \mathrm{X}b_k$, expressed in the basis $\overline{\mathrm{Mon}}(\mathcal{G})$, using the normal forms of elements of $\overline{\mathrm{Bor}}(\mathcal{G})$ already computed.
  - Step 3: For every $X^\alpha \in \mathrm{Tab}(\mathcal{G}) \setminus (\mathrm{Mon}(\mathcal{G}) \cup \mathrm{Bor}(\mathcal{G}))$ in increasing order according to $<_{\mathrm{deglex}}$ compute $\mathrm{NF}(b^{|\alpha|} X^\alpha)$ as follows: since $X^\alpha = X_j X^\beta$, for some $j = 1, \ldots, k$, compute the vector
$$\mathrm{NF}(b^{|\alpha|} X^\alpha) = M_j' \cdot \mathrm{NF}(b^{|\beta|} X^\beta).$$

Proof of correctness: The correctness of the algorithm follows from the correctness of Algorithm 12.3 (Multiplication table). The fact that the multiplication table consists of polynomials with coefficients in D is clear. □

Complexity analysis: Follows from the complexity analysis of Algorithm 12.3 (Multiplication Table), noting that the number of elements of $\mathrm{Mon}(\mathcal{G})$ is $d_1 \cdots d_k$, and that the number of elements in $\mathrm{Tab}(\mathcal{G})$ is bounded by $2^k d_1 \cdots d_k$. □

It will be necessary in several subsequent algorithms to perform the same computation with parameters. The following paragraphs are quite technical but it seems to be unfortunately unavoidable. Let $Y = Y_1, \ldots, Y_\ell$. We say that $\mathcal{G}(Y)$ is a **parametrized special Gröbner basis** if it is of the form

$$\mathcal{G}(Y) = \{b_1(Y) X_1^{d_1} + Q_1(Y, X), \ldots, b_k(Y) X_k^{d_k} + Q_k(Y, X)\}$$

with $\deg(Q_i) < d_i$, where deg is the total degree with respect the variables $X_1, \ldots, X_k$, $\deg_{X_j}(Q_i) < d_i$, $j \neq i$, $d_1 \geq \ldots \geq d_k$, $b_j \neq 0 \in \mathrm{D}[Y]$ for $0 \leq j \leq k$. Let $b(Y)$ a common multiple of $b_1(Y), \ldots, b_k(Y)$ in $\mathrm{D}[Y]$, with $b(Y) = b_j(Y) \bar{b_j}(Y)$. Then, for any $y \in \mathrm{C}^\ell$ such that $b(y) \neq 0$,

$$\mathcal{G}(y) = \{b_1(y) X_1^{d_1} + Q_1(y, X), \ldots, b_k(y) X_k^{d_k} + Q_k(y, W)\} \subset \mathrm{C}[X_1, \ldots, X_k].$$

is a special Gröbner basis. Define $\overline{\mathrm{Mon}}(\mathcal{G})$ as the set of elements $b(Y)^{|\alpha|} X^\alpha = b(Y)^{|\alpha|} X_1^{\alpha_1} \cdots X_k^{\alpha_k}$ with $\alpha_i < d_i$ and $\overline{\mathrm{Bor}}(\mathcal{G})$ as the set of elements $b(Y)^{|\alpha|} X^\alpha$ such that $\alpha_i = d_i$ for some $i \in \{1, \ldots, k\}$ and $\alpha_i \leq d_i$ for any $i \in \{1, \ldots, k\}$.

Algorithm 12.10. **[Parametrized Special Multiplication Table]**

- **Structure:** a ring D contained in a field K.
- **Input:** a parametrized special Gröbner basis

  $$\mathcal{G} = \{b_1(Y) X_1^{d_1} + Q_1(Y, X), \ldots, b_k(Y) X_k^{d_k} + Q_k(Y, X)\} \subset \mathrm{D}[Y][X_1, \ldots, X_k]$$

  with $Y = (Y_1, \ldots, Y_\ell)$, and $b(Y)$ a common multiple of $b_1(Y), \ldots, b_k(Y)$ in $\mathrm{D}[Y]$, with $b(Y) = b_j(Y) \bar{b_j}(Y)$.
- **Output:** a parametrized multiplication table in the basis $\overline{\mathrm{Mon}}(\mathcal{G})$: i.e. for any two monomials $b(Y)^{|\alpha|} X^\alpha$ and $b(Y)^{|\beta|} X^\beta$ in $\overline{\mathrm{Mon}}(\mathcal{G})$ a linear combination $\mathrm{NF}(b(Y)^{|\alpha|+|\beta|} X^\alpha X^\beta)(Y)$ of monomials in $\overline{\mathrm{Mon}}(\mathcal{G})$ with coefficients in $\mathrm{D}[Y]$ such that for every $y \in \mathrm{C}^\ell$ such that $b(y) \neq 0$, the polynomial $\mathrm{NF}(b(Y)^{|\alpha|+|\beta|} X^\alpha X^\beta)(y)$ is the normal form of $b(y)^{|\alpha|} X^\alpha b(y)^{|\beta|} X^\beta$ modulo $\mathcal{G}(y)$.
- **Complexity:** $(d_1 \cdots d_k \lambda)^{O(\ell)}$.
- **Procedure:**
  - Step 1: For every $b(Y)^{|\alpha|} X^\alpha \in \overline{\mathrm{Bor}}(\mathcal{G})$ in increasing order according to $<_{\mathrm{deglex}}$ compute $\mathrm{NF}(b(Y)^{|\alpha|} X^\alpha) = \sum_{X^\delta \in \mathrm{Mon}(\mathcal{G})} \lambda_{\alpha,\delta}(Y)\, b(Y)^{|\alpha|} X^\delta$ as follows:
  - If $X^\alpha = X_i^{d_i}$, $\mathrm{NF}(b(Y)^{|\alpha|} X^\alpha) := -b(Y)^{d_i - 1} \bar{b_i}(Y) Q_i$.

- Else, if $X^\alpha = X_j X^\beta$ for some $j = 1, \ldots, k$, with $X^\beta \in \mathrm{Bor}(\mathcal{G})$, define
$$\mathrm{NF}(b(Y)^{|\alpha|} X^\alpha) := \sum_{(X^\gamma, X^\delta) \in \mathrm{Mon}(\mathcal{G})^2} \lambda_{\beta,\gamma}(Y) \lambda_{\gamma',\delta}(Y) b(Y)^{|\gamma'|} X^\delta,$$
with $X_j X^\gamma = X^{\gamma'}$, and
$$\begin{aligned} \mathrm{NF}(b(Y)^{|\beta|} X^\beta) &= \sum_{X^\gamma \in \mathrm{Mon}(\mathcal{G})} \lambda_{\beta,\gamma}(Y)\, b(Y)^{|\gamma|} X^\gamma, \\ \mathrm{NF}(b(Y)^{|\gamma'|} X^{\gamma'}) &= \sum_{X^\delta \in \mathrm{Mon}(\mathcal{G})} \lambda_{\gamma',\delta}(Y)\, b(Y)^{|\delta|} X^\delta. \end{aligned}$$

- Step 2: Construct the matrices $M'_1, \ldots, M'_k$ corresponding to multiplication by $b X_1, \ldots, \mathrm{X}b_k$, expressed in the basis $\overline{\mathrm{Mon}}(\mathcal{G})$, using the normal forms of elements of $\overline{\mathrm{Bor}}(\mathcal{G})$ already computed.
- Step 3: For every $X^\alpha \in \mathrm{Tab}(\mathcal{G}) \setminus (\mathrm{Mon}(\mathcal{G}) \cup \mathrm{Bor}(\mathcal{G}))$ in increasing order according to $<_{\mathrm{deglex}}$ compute $\mathrm{NF}(\, b(Y)^{|\alpha|} X^\alpha)$ as follows: since $X^\alpha = X_j X^\beta$, for some $j = 1, \ldots, k$, compute the vector
$$\mathrm{NF}(\, b(Y)^{|\alpha|} X^\alpha) = M'_j \cdot \mathrm{NF}(\, b(Y)^{|\beta|} X^\beta).$$

Proof of correctness: The correctness of the algorithm follows from the correctness of Algorithm 12.9 (Special Multiplication Table). □

The complexity analysis of Algorithm 12.10 (Parametrized Special Multiplication Table) uses the following lemmas.

Define

$$\begin{aligned} \overline{\mathrm{Mon}}_{<d} &= \left\{ b(Y)^{|\alpha|} X^\alpha \,\middle|\, |\alpha| < d \right\}, \\ \overline{\mathrm{Mon}}_{\leqslant d} &= \left\{ b(Y)^{|\alpha|} X^\alpha \,\middle|\, |\alpha| \leqslant d \right\}, \\ \overline{\mathrm{Mon}}_{d} &= \left\{ b(Y)^{|\alpha|} X^\alpha \,\middle|\, |\alpha| = d \right\}, \\ \overline{\mathrm{Mon}}_{<d}(\mathcal{G}) &= \left\{ b(Y)^{|\alpha|} X^\alpha \,\middle|\, X^\alpha \in \overline{\mathrm{Mon}}(\mathcal{G}), |\alpha| < d \right\} \\ \overline{\mathrm{Mon}}_{\leqslant d}(\mathcal{G}) &= \left\{ b(Y)^{|\alpha|} X^\alpha \,\middle|\, X^\alpha \in \overline{\mathrm{Mon}}(\mathcal{G}), |\alpha| \leqslant d \right\} \\ \overline{\mathrm{Mon}}_{d}(\mathcal{G}) &= \left\{ b(Y)^{|\alpha|} X^\alpha \,\middle|\, X^\alpha \in \overline{\mathrm{Mon}}(\mathcal{G}), |\alpha| = d \right\} \end{aligned}$$

Lemma 12.11. *The normal form of every $b(Y)^{|\alpha|} X^\alpha \in \overline{\mathrm{Mon}}_d \setminus \overline{\mathrm{Mon}}_d(\mathcal{G})$ is a linear combination of elements of $\overline{\mathrm{Mon}}(\mathcal{G})_{<d}$ with coefficients in $\mathrm{D}[Y]$.*

The normal form of every $b(Y)^{|\alpha|} X^\alpha \in \overline{\mathrm{Mon}}_d$ is a linear combination of elements of $\overline{\mathrm{Mon}}(\mathcal{G})_{\leqslant d}$ with coefficients in $\mathrm{D}[Y]$.

Proof: We prove the result by induction on d. Suppose that the result is true for $d' < d$, and take $b(Y)^{|\alpha|} X^\alpha \in \overline{\mathrm{Mon}}_d \setminus \overline{\mathrm{Mon}}_d(\mathcal{G})$

- If X^α is one of the $X_i^{d_i}$, the result is true, since $\mathrm{NF}(b(Y)^{d_i} X_i^{d_i}) = -b(Y)^{d_i - 1} \bar{b}_i(Y)\, Q_i$.

– If X^α is not one of the $X_i^{d_i}$, then $X^\alpha = X_i X^\beta$ with $b(Y)^{|\beta|} X^\beta \in \overline{\mathrm{Mon}}_{d-1} \setminus \overline{\mathrm{Mon}}(\mathcal{G})$. According to the induction hypothesis, the normal form of $b(Y)^{|\beta|} X^\beta$ is a linear combination with coefficients in D of elements $b(Y)^{|\gamma|} X^\gamma$ of $\overline{\mathrm{Mon}}(\mathcal{G})_{<d-1}$. Finally, if $b(Y)^{|\gamma|} X^\gamma \in \overline{\mathrm{Mon}}(\mathcal{G})_{<d-1}$, then $b(Y)^{|\gamma|+1} X_i X^\gamma \in \overline{\mathrm{Mon}}_{<d}$, and we can again use the induction hypothesis.

The last claim follows since when $X^\alpha \in \mathrm{Mon}(\mathcal{G})$, $\mathrm{NF}(X^\alpha) = X^\alpha$. □

Lemma 12.12. *Let λ be a bound on the degree in Y of $b_1, \ldots, b_k$ and the coefficients of $Q_1, \ldots, Q_k$. The entries of the matrix M_i' corresponding to multiplication by $b(Y)X_i$ have degrees in Y at most $k\, d_k\, (d_1 + \cdots + d_{k-1} - k + 1)\, \lambda$.*

Proof: Note that the degree in Y of $b(Y)$ is bounded by $k\,\lambda$. For $i = 1, \ldots, k$, let $f_{d,i}$ be the mapping sending a polynomial P of degree $< d$, with coefficients in D$[Y]$ in the basis $\overline{\mathrm{Mon}}_{<d}$, to $\mathrm{NF}(b(Y)\, X_i\, P)$. Note that $\mathrm{NF}(b(Y)\, X_i\, P)$ is a linear combination of monomials of $\overline{\mathrm{Mon}}_{\leqslant d}(\mathcal{G})$ by Lemma 12.11. We are going to estimate, for

$$d_k \leqslant d \leqslant d_1 + \cdots + d_k - k,$$

the degrees in Y of the entries of the matrix $M_{d,i}'$ of $f_{d,i}$ expressed in the bases $\overline{\mathrm{Mon}}_{<d}$ and $\overline{\mathrm{Mon}}_{\leqslant d}(\mathcal{G})$. We are going to prove by induction on d that the degrees in Y of the entries of the matrices $M_{d,i}'$ are bounded by $k\, d_1\, (d - d_k + 1)\, \lambda$.

If $d = d_k$, and $b(Y)^{|\alpha|} X^\alpha \in \overline{\mathrm{Mon}}_{<d}$, the normal form of $b(Y)^{|\alpha|+1} X_i X^\alpha$ is either

$$\begin{cases} -\,b(Y)^{d_i-1}\, \bar{b}_i(Y)\, Q_i & \text{if } d_i = d_k,\ X^\alpha = X_i^{d_i-1}, \\ b(Y)^{|\alpha|+1} X_i X^\alpha & \text{otherwise.} \end{cases}$$

Thus, the degrees in Y of the entries of $N_{d_k,i}$ are bounded by $k\, d_k\, \lambda \le k\, d_1\, \lambda$.

Consider $d_k < d < d_1 + \cdots + d_k - k$ and suppose by induction hypothesis that the degrees in Y of the entries of the $M_{d,i}'$ are bounded by

$$k\, d_1 (d - d_k + 1) \lambda.$$

Let $b(Y)^{|\alpha|} X^\alpha \in \overline{\mathrm{Mon}}_d \setminus \overline{\mathrm{Mon}}(\mathcal{G})$. Then

$$b(Y)^{|\alpha|} X^\alpha = b(Y)^{d_j} X_j^{d_j} b(Y)^{|\beta|} X^\beta$$

for some $j = 1, \ldots, k$. Replacing $b(Y)^{d_j} X_j^{d_j}$ by $-\,b(Y)^{d_j-1}\, \bar{b}_j(Y)\, Q_j$ gives a polynomial R of total degree in X $\leqslant d$ and with degree in Y bounded by $k\, d_j \lambda$. The normal form of $b(Y)^{|\alpha|+1} X_i X^\alpha$ is the normal form of $b(Y)\, X_i R$, and is computed by multiplying the matrix $M_{d,i}'$ with the vector of coefficients of R in the basis $\overline{\mathrm{Mon}}_{\leqslant d}$. Thus, since $d_j \leqslant d_1$, the degrees in Y of the entries of $M_{d+1,i}'$ are bounded by $k\, d_1\, ((d+1) - d_k + 1)\, \lambda$, using the complexity analysis of Algorithm 8.13 (Multiplication of matrices).

Finally we have proved that the entries of the matrix M_i' corresponding to multiplication by $b(Y)X_i$ in $\overline{\mathrm{Mon}}(\mathcal{G})$, which is a submatrix of $M'_{d_1+\cdots+d_k-k,i}$ have degrees in Y at most $k\,d_1\,(d_1+\cdots+d_{k-1}-k+1)\,\lambda$. □

Lemma 12.13. *Let λ be a bound on the degrees in Y of b_1, ..., b_k and the coefficients of $Q_1, \ldots, Q_k$, τ a bound on the bitsizes of $b_1, \ldots, b_k$ and the coefficients of Q_1, ..., Q_k, and ν' a bound on the bitsize of $((d_1+\cdots+d_k)\,\lambda+1)^{\ell}$. The entries of the matrix M_i' corresponding to multiplication by $b(Y)\,X_i$ in the basis $\overline{\mathrm{Mon}}(\mathcal{G})$ have degrees in Y at most $k\,d_1\,(d_1+\cdots+d_k-k)\,\lambda$ and bitsizes at most*

$$(d_1+\cdots+d_{k-1}-k+1)\,(k\,d_1+1)\,(\tau+\nu').$$

Proof: The claim about the degrees is already proved in Lemma 12.12. The bitsize estimate is proved using the same technique. Note that, using the complexity analysis of Algorithm 8.4, the bitsizes of the coefficients of b is bounded by $k\tau + \ell\,(\log_2(k\,\lambda)+1)$. For $i=1$, ..., k, let $f_{d,i}$ be the mapping sending a polynomial P, of degree $< d$, with coefficients in $\mathrm{D}[Y]$ in the basis $b^{|\alpha|}\,X^{\alpha}$, to $\mathrm{NF}(b\;X_i\;P)$. Note that $\mathrm{NF}(b\;X_i\;P)$ is a linear combination of monomials of $\overline{\mathrm{Mon}}_{\leqslant d}(\mathcal{G})$ by Lemma 12.11. We are going to estimate for $d_k \le d \le d_1+\cdots+d_k-k$ the bitsizes of the coefficients the entries of the matrix $M'_{d,i}$ of $f_{d,i}$ expressed in the bases $\overline{\mathrm{Mon}}_{<d}$ and $\overline{\mathrm{Mon}}_{\leqslant d}(\mathcal{G})$. We are going to prove by induction on d that the bitsizes of the entries of the matrices $M'_{d,i}$ are bounded by $(d-k+1)\,(k\,d_1+1)\,(\tau+\nu')$.

If $d=d_k$, and $b(Y)^{|\alpha|}\,X^{\alpha}\in\overline{\mathrm{Mon}}_{<d}$, the normal form of $b(Y)^{|\alpha|+1}\,X_i\,X^{\alpha}$ is

$$\begin{cases} -\,b(Y)^{d_i-1}\,\bar{b}_i(Y)\,Q_i & \text{if } d_i=d_k,\ X^{\alpha}=X_i^{d_i-1},\\ b(Y)^{|\alpha|+1}\,X_i\,X^{\alpha} & \text{otherwise.}\end{cases}$$

Thus the bitsizes of the coefficients of the entries of $M'_{d_k,i}$ are bounded by $k\,d_k\,(\tau+\nu')\le(k\,d_1+1)\,(\tau+\nu')$.

Consider $d_k < d < d_1+\cdots+d_k-k$ and suppose by induction hypothesis that the bitsizes of the entries of the matrices $M'_{d,i}$ are bounded by

$$(d-k+1)\,(k\,d_1+1)\,(\tau+\nu').$$

Let $b(Y)^{|\alpha|}\,X^{\alpha}\in\overline{\mathrm{Mon}}_d\setminus\overline{\mathrm{Mon}}(\mathcal{G})$, then $b(Y)^{|\alpha|}\,X^{\alpha}=b^{d_j}\,X_j^{d_j}\,b^{|\beta|}\,X^{\beta}$ for some $j=1,\ldots,k$. Replacing $b(Y)^{d_j}\,X_j^{d_j}$ by $-\,b(Y)^{d_j-1}\,\bar{b}_j\,Q_j$ gives a polynomial R of total degree in X $\le d$ with coefficients of bitsizes bounded by $k\,d_k\,(\tau+\nu')$. The normal form of $b(Y)^{|\alpha|+1}\,X_i\,X^{\alpha}$ is the normal form of $b(Y)\,X_i\,R$, and is computed by multiplying the matrix $M'_{d,i}$ with the vector of coefficients of R expressed in the basis $\overline{\mathrm{Mon}}_{\leqslant d}$. Thus, since $d_j\leqslant d_1$, the bitsizes of the entries of $N_{d+1,i}$ are bounded by

$$((d+1)-d_k+1)\,(k\,d_1+1)\,(\tau+\nu'),$$

using the complexity analysis of Algorithm 8.14 (Multiplication of several matrices).

Finally we have proved that the entries of the matrix M_i' corresponding to multiplication by $b(Y)X_i$ in $\overline{\mathrm{Mon}}(\mathcal{G})$, which is a submatrix of $M'_{d_1+\cdots+d_k-k,i}$ have bitsizes at most $(d_1+\cdots+d_{k-1}-k+1)\,(k\,d_1+1)\,(\tau+\nu')$. □

Complexity analysis:

The number of arithmetic operations in $\mathrm{D}[Y]$ of the algorithm is in $O(2^k\,d_1\cdots d_k)^3)$ using the complexity analysis of Algorithm 12.9.

If the coefficients of the polynomials in $\mathcal{G}$ are polynomials in Y of degree bounded by λ we estimate the degrees in Y of the normal forms computed through the algorithm. Closely following the algorithm would give a degree in Y exponential in d since the degree looks like it is doubled each time the degree is increased by 1. So we have to proceed in a more careful way, taking into account the special structure of the Gröbner basis and using Lemma 12.12.

By Lemma 12.12, the entries of the matrix M_i' of multiplication by $b(Y)X_i$ in $\overline{\mathrm{Mon}}(\mathcal{G})$, have degree in Y at most $k\,d_1\,(d_1+\cdots+d_{k-1}-k+1)\,\lambda$.

Multiplying at most $2\;(d_1+\cdots+d_k-k)$ times matrices of size $d_1...d_k$ with entries of degree $k\,d_1\,(d_1+\cdots+d_{k-1}-k+1)\,\lambda$ in Y produces matrices with entries of degree in Y bounded by $2\,k\,d_1\,(d_1+\cdots+d_k-k)^2\,\lambda$, using the complexity analysis of Algorithm 8.13 (Multiplication of matrices).

Finally, the number of arithmetic operations in D is $(d_1\cdots d_{k_*}\lambda)^{O(\ell)}$ since the number of arithmetic operations in $\mathrm{D}[Y]$ is $(d_1\cdots d_k)^{O(1)}$, and the degrees in Y of the polynomials appearing in the intermediate computations are bounded by $2\,k\,d_1\,(d_1+\cdots+d_k-k)^2\,\lambda$.

When $\mathrm{D}=\mathbb{Z}$, and the bitsizes of the coefficients of $\mathcal{G}$ are bounded by τ, multiplying at most $2\;(d_1+\cdots+d_k-k)$ times matrices of size $d_1...d_k$ with entries of degree in $Y\,k\,d_1\,(d_1+\cdots+d_{k-1}-k+1)\,\lambda$ and coefficients of bitsizes $(d_1+\cdots+d_{k-1}-k+1)\,(k\,d_1\,(\tau+\nu)+\nu')$ produces matrices with entries of degree in Y bounded by $2\,k\,d_1(d_1+\cdots+d_k-k)^2\lambda$, and coefficients of bitsizes $2\,(d_1+\cdots+d_k-k+1)^2\,(k\,d_1+1)\,(\tau+4\,\nu')$, using the complexity analysis of Algorithm 8.14 (Multiplication of several matrices), since ν' is a bound on the bitsize of $d_1...d_k$ and $2\,\nu'$ is a bound on the bitsize of $(k\,d_1\,(d_1+\cdots+d_k-k)^2\,\lambda+1)^\ell$. □

12.4 Univariate Representation

In this section, we describe a method, based on trace computations, for solving a system of polynomial equations, in the following sense. We are going to describe the coordinates of the solutions of a zero-dimensional polynomial system as rational functions of the roots of a univariate polynomial. As before, let $\mathcal{P}\subset\mathrm{K}[X_1,\ldots,X_k]$ be a zero-dimensional polynomial system, i.e. a finite set of polynomials such that $\mathrm{Zer}(\mathcal{P},\mathrm{C}^k)$ is finite. According to Theorem 4.85, $\mathrm{A}=\mathrm{K}[X_1,\ldots,X_k]/\mathrm{Ideal}(\mathcal{P},\mathrm{K})$ is a finite dimensional vector space over K having dimension $N\geq n=\#\mathrm{Zer}(\mathcal{P},\mathrm{C}^k)$.

For any element $a \in \mathrm{A}$, let $\chi(a, T)$ be the characteristic polynomial of the linear transformation L_a from A to A defined by $L_a(g) = a\, g$. Then, according to Theorem 4.97 (Stickelberger), Equation (4.8),

$$\chi(a,T) = \prod_{x \in \mathrm{Zer}(\mathcal{P},\mathrm{C}^k)} (T - a(x))^{\mu(x)}, \tag{12.2}$$

where $\mu(x)$ is the multiplicity of x as a root of $\chi(a,T)$. By Remark 4.98, we have $\chi(a,T) \in \mathrm{K}[T]$.

If a is separating (see Definition 4.88), for every root t of $\chi(a,T)$, there is a single point $x \in \mathrm{Zer}(\mathcal{P},\mathrm{C}^k)$ such that $a(x) = t$. Hence, it is natural to express the coordinates of the elements x in $\mathrm{Zer}(\mathcal{P},\mathrm{C}^k)$ as values of rational functions at the roots of $\chi(a,T)$ when a in A is a separating element.

Remark 12.14. An example of this situation has already been seen in the algebraic proof of the fundamental theorem of algebra seen in Chapter 2 (proof of $a) \Rightarrow b)$ in Theorem 2.11, see Remark 2.17). Using the notation of the proof of Theorem 2.11, the value z such that $D(z) \neq 0$ was such that the images of all the $\gamma_{i,j} = x_i + x_j + z\, x_i\, x_j$ where distinct and both $x_i + x_j$ and $x_i\, x_j$ where expressed as rational function of $\gamma_{i,j}$. □

For any a and f in A, we define

$$\varphi(a,f,T) = \sum_{x \in \mathrm{Zer}(\mathcal{P},\mathrm{C}^k)} \mu(x)\, f(x) \prod_{\substack{t \in \mathrm{Zer}(\chi(a,T),\mathrm{C}) \\ t \neq a(x)}} (T - t). \tag{12.3}$$

Note that $\chi(a,T)$ and $\varphi(a,1,T)$ are coprime.
If a is separating,

$$\varphi(a,f,T) = \sum_{x \in \mathrm{Zer}(\mathcal{P},\mathrm{C}^k)} \mu(x)\, f(x) \prod_{\substack{y \in \mathrm{Zer}(\mathcal{P},\mathrm{C}^k) \\ y \neq x}} (T - a(y)),$$

and thus if $x \in \mathrm{Zer}(\mathcal{P},\mathrm{C}^k)$

$$\varphi(a,f,a(x)) = \mu(x)\, f(x) \prod_{\substack{y \in \mathrm{Zer}(\mathcal{P},\mathrm{C}^k) \\ y \neq x}} (a(x) - a(y)).$$

Hence, if a is separating,

$$\frac{\varphi(a,f,a(x))}{\varphi(a,1,a(x))} = f(x). \tag{12.4}$$

Choosing the polynomial X_i for f in Equation (12.4), we see that

$$\frac{\varphi(a,X_i,a(x))}{\varphi(a,1,a(x))} = x_i. \tag{12.5}$$

In other words, if a is separating and x in $\mathrm{Zer}(\mathcal{P},\mathrm{C}^k)$, then x_i is the value of the rational function $\varphi(a,X_i,T)/\varphi(a,1,T)$ at the root $a(x)$ of the polynomial $\chi(a, T)$.

Note that, if a is separating, the roots of $\mathcal{P}$ in C^k are all simple if and only if $\chi(a,T)$ is separable, in which case

$$\varphi(a,1,T)=\chi(a,T)'. \tag{12.6}$$

Note that for any a, not necessarily separating,

$$\frac{\varphi(a,f,a(x))}{\varphi(a,1,a(x))}=\frac{\sum_{\substack{y\in\mathrm{Zer}(\mathcal{P},\mathrm{C})\\ a(y)=a(x)}}\mu(y)\,f(y)}{\sum_{\substack{y\in\mathrm{Zer}(\mathcal{P},\mathrm{C})\\ a(y)=a(x)}}\mu(y)}. \tag{12.7}$$

For any a (not necessarily separating) and f in A, $\varphi(a,f,T)$ belongs to $\mathrm{C}[T]$ by definition. In fact, as we now show, it belongs to $\mathrm{K}[T]$.

Lemma 12.15. *If a and f are in A, $\varphi(a,f,T)\in\mathrm{K}[T]$.*

Proof: Since $\chi(a,T)\in\mathrm{K}[T]$ by Remark 4.98, let $\bar{\chi}(a,T)\in\mathrm{K}[T]$ be the monic separable part of $\chi(a,T)$. It is clear that $\bar{\chi}(a,T)\in\mathrm{K}[T]$. Then

$$\begin{aligned}
\frac{\varphi(a,f,T)}{\bar{\chi}(a,T)} &= \sum_{x\in\mathrm{Zer}(\mathcal{P},\mathrm{C}^k)}\frac{\mu(x)\,f(x)}{T-a(x)},\\
\bar{\chi}(a,T) &= \prod_{t\in\mathrm{Z}(\chi(a,T),\mathrm{C})}(T-t),\\
\frac{\varphi(a,f,T)}{\bar{\chi}(a,T)} &= \prod_{t\in\mathrm{Z}(\chi(a,T),\mathrm{C})}(T-t),\\
&= \sum_{i\geq 0}\sum_{x\in\mathrm{Zer}(\mathcal{P},\mathrm{C}^k)}\frac{\mu(x)\,f(x)\,a(x)^i}{T^{i+1}}\\
&= \sum_{i\geq 0}\frac{\mathrm{Tr}(L_{fa^i})}{T^{i+1}}.
\end{aligned}$$

Let

$$\bar{\chi}(a,T)=\sum_{j=0}^{n'}c_{n'-j}T^{n'-j},$$

with $c_{n'-j}\in\mathrm{K}$, $c_{n'}=1$, $n'\leq n$. Note that if a is separating, then $n'=n$. Multiplying both sides by $\bar{\chi}(a,T)$, which is in $\mathrm{K}[T]$ and using the fact that $\varphi(a,f,T)$ is a polynomial in $\mathrm{C}[T]$, we have

$$\varphi(a,f,T)=\sum_{i=0}^{n'-1}\sum_{j=0}^{n'-i-1}\mathrm{Tr}(L_{fa^i})c_{n'-j}T^{n'-i-1-j} \tag{12.8}$$

Consider a Gröbner basis $\mathcal{G}$ of Ideal($\mathcal{P}$, K) and express L_{fa^i} in the basis Mon($\mathcal{G}$). Then $\mathrm{Tr}(L_{fa^i})$, which is the trace of a matrix with entries in K, is in K. This proves $\varphi(a,f,T)\in\mathrm{K}[T]$. □

The previous discussion suggests the following definition and proposition.

A k-**univariate representation** u is a $k+2$-tuple of polynomials in $\mathrm{K}[T]$,

$$u=(f(T),g(T)), \text{with } g=(g_0(T),g_1(T),\ldots,g_k(T)),$$

such that f and g_0 are coprime. Note that $g_0(t)\neq 0$ if $t\in\mathrm{C}$ is a root of $f(T)$. The **points associated** to a univariate representation u are the points

$$x_u(t)=\left(\frac{g_1(t)}{g_0(t)},\ldots,\frac{g_k(t)}{g_0(t)}\right)\in\mathrm{C}^k \tag{12.9}$$

where $t\in\mathrm{C}$ is a root of $f(T)$.

Let $\mathcal{P}\subset\mathrm{K}[X_1,\ldots,X_k]$ be a finite set of polynomials such that $\mathrm{Zer}(\mathcal{P},\mathrm{C}^k)$ is finite. The $k+2$-tuple $u=(f(T),g(T))$, **represents** $\mathrm{Zer}(\mathcal{P},\mathrm{C}^k)$ if u is a univariate representation and

$$\mathrm{Zer}(\mathcal{P},\mathrm{C}^k)=\{x\in\mathrm{C}^k\mid\exists t\in\mathrm{Zer}(f,\mathrm{C})\; x=x_u(t)\}.$$

A **real k-univariate representation** is a pair u,σ where u is a k-univariate representation and $\sigma$is the Thom encoding of a root of f, $t_\sigma\in\mathrm{R}$. The **point associated** to the real univariate representation u,σ is the point

$$x_u(t_\sigma)=\left(\frac{g_1(t_\sigma)}{g_0(t_\sigma)},\ldots,\frac{g_k(t_\sigma)}{g_0(t_\sigma)}\right)\in\mathrm{R}^k. \tag{12.10}$$

Let

$$\varphi(a,T)=(\varphi(a,1,T),\varphi(a,X_1,T),\ldots,\varphi(a,X_k,T)) \tag{12.11}$$

Proposition 12.16. *Let $\mathcal{P}\subset\mathrm{K}[X_1,\ldots,X_k]$ a zero-dimensional system and $a\in A$.*

The $k-$univariate representation $(\chi(a,T),\varphi(a,T))$ represents $\mathrm{Zer}(\mathcal{P},\mathrm{C}^k)$ if and only if a is separating.

If a is separating, the following properties hold.

- *The degree of the separable part of $\chi(a,T)$ is equal to the number of elements in $\mathrm{Zer}(\mathcal{P},\mathrm{C}^k)$.*
- *The bijection $x\mapsto a(x)$ from $\mathrm{Zer}(\mathcal{P},\mathrm{C}^k)$to $\mathrm{Zer}(\chi(a,T),\mathrm{R})$ respects the multiplicities.*

Proof: If a is separating, $(\chi(a,T),\varphi(a,T))$ represents $\mathrm{Zer}(\mathcal{P},\mathrm{C}^k)$ by (12.5). Conversely if $(\chi(a,T),\varphi(a,T))$ represents $\mathrm{Zer}(\mathcal{P},\mathrm{C}^k)$, then $a(x)=a(y)$ imply $x=y$, hence a is separating.

Since the degree of the separable part of $\chi(a,T)$ is equal to the number of distinct roots of $\chi(a,T)$, it coincides with $\mathrm{Zer}(\mathcal{P},\mathrm{C}^k)$ when a is separating.

Finally, if a is separating, the multiplicity of $a(x)$ is equal to $\mu(x)$ by (12.2). □

The following proposition gives a useful criterion for a to be separating.

Proposition 12.17. *Let*

The following properties are equivalent:

a) The element $a \in A$ is separating.
b) The $k+2$-tuple $(\chi(a,T), \varphi(a,T))$ represents $\mathrm{Zer}(\mathcal{P}, \mathrm{C}^k)$.
c) For every $k = 1, \ldots, k$,

$$\varphi(a,1,a)\, X_i - \varphi(a, X_i, a) \in \sqrt{\mathrm{Ideal}(\mathcal{P}, \mathrm{K})}.$$

Proof: By Proposition 12.16, a) and b) are equivalent.

Let us prove that b) and c) are equivalent. Since $\chi(a,T)$ and $\varphi(a,1,T)$ are coprime, for every $x \in \mathrm{Zer}(\mathcal{P}, \mathrm{C}^k)$, and every $i = 1, \ldots, k$

$$x_i = \frac{\varphi(a, X_i, a(x))}{\varphi(a, 1, a(x))}$$

is equivalent to the property (P): $\varphi(a,1,a)\, X_i - \varphi(a, X_i, a)$ vanishes at every $x \in \mathrm{Zer}(\mathcal{P}, \mathrm{C}^k)$, for every $i = 1, \ldots, k$. Property (P) is equivalent to

$$\varphi(a,1,a)\, X_i - \varphi(a, X_i, a) \in \sqrt{\mathrm{Ideal}(\mathcal{P}, \mathrm{K})}$$

for every $i = 1, \ldots, k$ by Hilbert Nullstellensatz (Theorem 4.78). □

Since $a \in \mathrm{A} = \mathrm{K}[X_1, \ldots, X_k]/\mathrm{Ideal}(\mathcal{P}, \mathrm{K})$ we also obtain,

Corollary 12.18.

If $a \in A$ is separating,

$$a(\mathrm{Zer}(\mathcal{P}, \mathrm{R}^k) = \mathrm{Zer}(\chi(a,T), \mathrm{R}).$$

In particular, $\#\mathrm{Zer}(\mathcal{P}, \mathrm{R}^k) = \#\mathrm{Zer}(\chi(a,T), \mathrm{R})$.

Proof: Since $a \in \mathrm{A}$, if $x \in \mathrm{Zer}(\mathcal{P}, \mathrm{R}^k)$, $a(x) \in \mathrm{Zer}(\chi(a,T), \mathrm{R})$. Conversely, if $\chi(a,t) = 0$, then $t = a(x)$ for $x \in \mathrm{Zer}(\mathcal{P}, \mathrm{C}^k)$, $x = x_{u(\mathcal{P},a)}(t)$ and

$$x_i = \frac{\varphi(a, X_i, a(x))}{\varphi(a, 1, a(x))},$$

by (12.5). Since $\varphi(a, X_i, T)$ and $\varphi(a, X_i, T)$ belong to $\mathrm{K}[T]$ by Lemma 12.15, $t = a(x) \in \mathrm{R}$ implies $x \in \mathrm{R}^k$. □

Let D be a ring contained in K, $\mathcal{P} \subset \mathrm{D}[X_1, \ldots, X_k]$ a zero-dimensional system and $\mathcal{B}$ a basis of A such that the multiplication table of A in $\mathcal{B}$ has entries in D. Consider $a \in \mathrm{A}$, $b \in \mathrm{D}$, and suppose that $a, b, b\,X_1, \ldots, b\,X_k$ have coordinates in D in the basis $\mathcal{B}$. Note that $\chi(a,T) \in \mathrm{D}[T]$, and let $\bar{\chi}(a,T)$ be a separable part of $\chi(a,T)$ with coefficients in D and leading coefficient c. We denote

$$\varphi_b(a,T) = (\varphi(a,b,T), \varphi(a, b\,X_1, T), \ldots, \varphi(a, b\,X_k, T)), \tag{12.12}$$

(using (12.2) and (?)).

Since

$$\frac{c\,\varphi(a,b\,X_i,a(x))}{c\,\varphi(a,b,a(x))} = \frac{\varphi(a,X_i,a(x))}{\varphi(a,1,a(x))} = x_i. \tag{12.13}$$

Proposition 12.16, Proposition 12.17 and Corollary 12.18 hold when replacing $\varphi(a,T)$ by $c\,\varphi_b(\mathrm{a,T})$. Introducing c, b plays a role in guaranteeing that the computations take place inside D.

Proposition 12.19. *Let* D *be a ring contained in* K, $\mathcal{P} \subset \mathrm{D}[X_1, \ldots, X_k]$ *a zero-dimensional system. Consider* $a \in A$, $b \neq 0 \in \mathrm{D}$.

Let $\mathcal{B}$ *a basis of* A *such that the multiplication table of* A *in* $\mathcal{B}$ *has entries in* D*. Suppose that* $a, b, b\,X_1, \ldots, b\,X_k$ *have coordinates in* D *in the basis* $\mathcal{B}$*, then, denoting by* c *the leading coefficient of a separable part* $\bar{\chi}$ *of* χ *in* $\mathrm{D}[T]$*, the components of* $c\,\varphi_b(a,T) \in \mathrm{D}[T]^{k+1}$.

Proof: The polynomial $\bar{\chi}(a,T)$ has coefficients in D and the various $\mathrm{Tr}(L_{a^j})$ $\mathrm{Tr}(L_{bX_ia^j})$ belong to D. Thus by Equation (12.8), $c\,\varphi_b(a,T) \in \mathrm{D}[T]^{k+1}$. □

Let $a \in \mathrm{A} = \mathrm{K}[X_1,\ldots,X_k]/\mathrm{Ideal}(\mathcal{P},\mathrm{K})$. The polynomial $\chi(a,T)$ is related to the traces of the powers of a as follows: The i-th Newton sum N_i associated to the polynomial $\chi(a,T)$ is the sum of the i-th powers of the roots of $\chi(a, T)$ and is thus

$$N_i = \sum_{x \in \mathrm{Zer}(\mathcal{P},\mathrm{C}^k)} \mu(x)\,a(x)^i.$$

According to Proposition 4.54, $N_i = \mathrm{Tr}(L_{a^i})$. Let

$$\chi(a,T) = \sum_{i=0}^{N} b_{N-i}T^{N-i}.$$

According to Newton's formula (Equation (4.1)),

$$(N-i)\,b_{N-i} = \sum_{j=0}^{i} \mathrm{Tr}(L_{a^j})\,b_{N-i+j}, \tag{12.14}$$

so that $\chi(a,T)$ can be computed from $\mathrm{Tr}(L_{a^i})$, for $i = 0, \ldots, N$. Moreover, a is separating when the number of distinct roots of $\chi(a,T)$ is $n = \#\mathrm{Zer}(\mathcal{P},\mathrm{C}^k)$.

We then compute a separable part of $\chi(a,T)$

$$\bar{\chi}(a,T) = \sum_{j=0}^{n'} c_{n'-j}T^{n'-j},$$

with leading coefficient $c = c_{n'}$ and write $c\,\varphi(a,f,T)$ as

$$c\,\varphi(a,f,T) = \sum_{i=0}^{n'-1} \mathrm{Tr}(L_{fa^i})\mathrm{Hor}_{n'-i-1}(\bar{\chi}(a,T),T), \tag{12.15}$$

where $\mathrm{Hor}_i(P,T)$ is the i-th Horner polynomial associated to P (see Notation 8.6).

So, $(\chi(a,T), c\,\varphi_b(a,T))$, can easily be obtained from the following traces

$$\mathrm{Tr}(a^i), i=0,\ldots,N,\ \mathrm{Tr}(a^i\, b\, X_j), i=0,\ldots,n',\ j=1,\ldots,k,$$

where $n' = \#\mathrm{Zer}(\chi(a,T),\mathrm{C}) \leq n = \#\mathrm{Zer}(\mathcal{P},\mathrm{C}^k)$, using Equation (12.14) and Equation (12.15).

Algorithm 12.11. **[Candidate Univariate Representation]**

- **Structure:** a ring D with division in $\mathbb{Z}$ contained in a field K.
- **Input:** a zero-dimensional polynomial system $\mathcal{P} \subset \mathrm{D}[X_1,\ldots,X_k]$, a basis $\mathcal{B}$ of $\mathrm{A} = \mathrm{K}[X_1, \ldots, X_k]/\mathrm{Ideal}(\mathcal{P}, \mathrm{K})$ such that the multiplication table $\mathcal{M}$ of A in $\mathcal{B}$ has entries in D, an element $a \in \mathrm{A}$ and $b \neq 0 \in \mathrm{D}$ such that $a, b, b\,X_1, \ldots, b\,X_k$ have coordinates in D in the basis $\mathcal{B}$.
- **Output:** $c \in \mathrm{D}$, and $(\chi(a,T), c\,\varphi_b(a,T)) \in \mathrm{D}[T^{k+2}]$.
- **Complexity:** $O(N^3 + k\,N^2)$, where N is the number of elements of $\mathcal{B}$, in the special case when $\mathcal{B} = \mathrm{Mon}(\mathcal{G})$.
- **Procedure:**
  - Step 1: Compute the traces of L_{a^i}, $i = 1, \ldots, N$ using Algorithm 12.5 (Trace). Then compute the coefficients of $\chi(a, T)$ using Algorithm 8.11(Newton sums) and Equation (12.14).
  - Step 2: Compute the separable part of $\chi(a,T)$ using Algorithm 10.1 (Gcd and Gcd-free Part), c its leading coefficient.
  - Step 3: Compute $c\,\varphi_b(a,T)$ using Algorithm 12.5 (Trace) and Equation (12.15).
  - Return $(\chi(a,T), c\varphi_b(a,b))$.

Proof of correctness: Immediate. Note that we know in advance that $\chi(a, T) \in \mathrm{D}[T]$ by Corollary 12.15, so exact division by an integer is possible. □

Complexity analysis: Let N be the dimension of the K-vector space A.

Before computing the traces, it is necessary to compute the normal forms of $1, a, \ldots, a^n$, which is done by multiplying at most N times the matrix of multiplication by a (which is a linear combination of the matrices M_i of multiplication by X_i) which takes $O(N^3)$ arithmetic operations. According to the complexity analyses of Algorithm 12.5 (Trace), Algorithm 8.11 (Newton sums), and Algorithm 10.1 (Gcd and Gcd-free part), using Equation (12.15), $c\varphi_b(a,T)$ can clearly be computed in $O(k\,N^2)$ arithmetic operations. □

Algorithm 12.12. **[Univariate Representation]**

- **Structure:** a ring D with division in $\mathbb{Z}$ contained in a field K.
- **Input:** a zero-dimensional polynomial system $\mathcal{P} \subset \mathrm{D}[X_1,\ldots,X_k]$, a basis $\mathcal{B}$ of $\mathrm{A} = \mathrm{K}[X_1, \ldots, X_k]/\mathrm{Ideal}(\mathcal{P}, \mathrm{K})$, such that the multiplication table $\mathcal{M}$ of A in $\mathcal{B}$ has entries in D, and $b \neq 0 \in \mathrm{D}$ such that $b, b\,X_1, \ldots, b\,X_k$ have coordinates in D in the basis $\mathcal{B}$.
- **Output:** a univariate representation u representing $\mathrm{Zer}(\mathcal{P}, \mathrm{C}^k)$.
- **Complexity:** $O(k\,N^2(N^3 + k\,N^2))$, where N is the number of elements of $\mathcal{B}$, in the special case when $\mathcal{B} = \mathrm{Mon}(\mathcal{G})$.

- **Procedure:**
  - Compute $n = \#\mathrm{Zer}(\mathcal{P}, \mathrm{C}^k)$ using Algorithm 12.6 (Number of distinct zeros).
  - Initialize $i := 0$.
  - ($\star$) Take $a := X_1 + i\, X_2 + i^2\, X_3 + \cdots + i^{k-1}\, X_k$. Compute $\chi(b\, a, T)$ using Step 1 of Algorithm 12.11 (Candidate Univariate Representation). Compute

$$n(b a) = \deg(\chi(b a, T)) - \deg(\gcd(\chi(b a, T), \chi'(b a, T)))$$

    using Algorithm 8.21 (Signed Subresultant).
  - While $n(b\, a) \neq n$, $i := i + 1$, return to ($\star$).
  - Compute c and $c\,\varphi_b(b\, a, T))$ using Step 3 of Algorithm 12.11 (Candidate Univariate Representation).
  - Return $u = (\chi(b a, T), c\,\varphi_b(b a, T))$.

Proof of correctness: Let N be the dimension of the K-vector space A. We know by Lemma 4.89 that there exists a separating element

$$a := X_1 + i\, X_2 + i^2\, X_3 + \cdots + i^{k-1}\, X_k \text{ f}$$

for $i \leq (k-1)\binom{n}{2}$, and by Theorem 4.85 that $n \leq N$. Note that $b\, a$ is separating as well since $b \neq 0$. The number $n(b\, a)$ is the number of distinct roots of $\chi(b a, T)$, and $n(b a) = n$ if and only if a is separating. □

Complexity analysis: The number of different a to consider is

$$(k-1)\binom{N}{2} + 1,$$

and for each a the cost of computation is $O(N^3 + k\, N^2)$ according to the complexity analysis of Algorithm 12.11 (Candidate Univariate Representation). Thus the complexity is $O(k\, N^2(N^3 + k\, N^2))$. □

Remark 12.20. Algorithm 12.12 (Univariate Representation) can be improved in various ways [134]. In particular, rather than looking for separating elements which are linear combination of variables, it is preferable to check first whether variables are separating. Second, the computations of Algorithm 12.6 (Number of Distinct Zeros) as well as the computation of a separating element can be performed using modular arithmetic, which avoids any growth of coefficients. Of course, if the prime modulo which the computations is performed is unlucky, the number of distinct elements and the separating element are not computed correctly. So it is useful to have a test that a candidate separating element is indeed separating; this can be checked quickly using Proposition 12.17 and Theorem 4.99. Similar remarks apply to Algorithm 12.13. Efficient computations of univariate representations can be found in [135]. □

When we know in advance that the zeroes of the polynomial system are all simple, which will be the case in many algorithms in the next chapters, the computation above can be simplified.

Algorithm 12.13. **[Simple Univariate Representation]**

- **Structure:** a ring D with division in $\mathbb{Z}$ contained in a field K.
- **Input:** a zero-dimensional polynomial system $\mathcal{P} \subset \mathrm{D}[X_1, \ldots, X_k]$, a basis $\mathcal{B}$ of $\mathrm{A} = \mathrm{K}[X_1, \ldots, X_k]/\mathrm{Ideal}(\mathcal{P}, \mathrm{K})$ such that the multiplication table $\mathcal{M}$ of A in $\mathcal{B}$ has entries in D, and $b \neq 0 \in \mathrm{D}$ such that $b, b\,X_1, \ldots, b\,X_k$ have coordinates in D in the basis $\mathcal{B}$. Moreover all the zeros of $\mathrm{Zer}(\mathcal{P}, \mathrm{C}^k)$ are simple, so that $N = n$.
- **Output:** a univariate representation u representing $\mathrm{Zer}(\mathcal{P}, \mathrm{C}^k)$.
- **Complexity:** $O(k\,N^2(N^3 + k\,N^2))$, where N is the number of elements of $\mathcal{B}$, in the special case when $\mathcal{B} = \mathrm{Mon}(\mathcal{G})$.
- **Procedure:**
  - Initialize $i := 0$.
  - $(\star)$ Take $a := X_1 + i\,X_2 + i^2\,X_3 + \cdots + i^{k-1}\,X_k$. Compute $\chi(a, T)$ using Step 1 of Algorithm 12.11 (Candidate Univariate Representation).
  - Compute gcd $(\chi(b\,a, T), \chi'(b\,a, T))$ by Algorithm 8.21 (Signed Subresultant).
  - While $\deg(\gcd(\chi(b\,a, T), \chi'(b\,a, T))) \neq 0$, $i := i + 1$, return to $(\star)$.
  - Compute $\varphi_b(b\,a, b, T)$ by Step 3 of Algorithm 12.11 (Candidate Univariate Representation).
  - Return $u = (\chi(b\,a, T), \varphi_b(b\,a, b, T))$.

Proof of correctness: Let N be the dimension of the K-vector space A. Since all the zeros of $\mathrm{Zer}(\mathcal{P}, \mathrm{C}^k)$ are simple, $n = N$ by Theorem 4.85. We know by Lemma 4.89 that there exists a separating element $i \leq (k-1)\binom{N}{2}$. Since all the zeros of $\mathrm{Zer}(\mathcal{P}, \mathrm{C}^k)$ are simple, and $b \neq 0$, a is separating if and only if $\chi(b\,a, T)$ is separable. □

Complexity analysis: The number of different a to consider is

$$(k-1)\binom{N}{2} + 1,$$

and for each a the computation is $O(N^3 + k\,N^2)$ according to the complexity analysis of Algorithm 12.11 (Univariate Representation). Thus the complexity is $O(k\,N^2(N^3 + k\,N^2))$. □

Remark 12.21. It is clear what we can use the rational univariate representation to give an alternative method to Algorithm 12.7 (Multivariate Tarski-query) (and to Algorithm 12.8 (Multivariate Sign Determination)) in the multivariate case. Given a univariate representation (f, g) representing $\mathrm{Zer}(\mathcal{P}, \mathrm{C}^k)$ we simply replace the Tarski-query $\mathrm{TaQ}(Q, \mathrm{Zer}(\mathcal{P}, \mathrm{R}^k))$ by the Tarski-query $\mathrm{TaQ}(Q_u, \mathrm{Zer}(f, \mathrm{R}))$, with

$$Q_u = g_0^e\, Q\left(\frac{g_k}{g_0}, \ldots, \frac{g_k}{g_0}\right). \tag{12.16}$$ □

Remark 12.22. Note that in the two last sections, the computation of the traces of various L_f was crucial in most algorithms. These are easy to compute once a multiplication table is known. However, in big examples, the size of the multiplication table can be the limiting factor in the computations. Efficient ways for computing the traces without storing the whole multiplication table are explained in [134, 136]. □

12.5 Limits of the Solutions of a Polynomial System

In the next chapters, it will be helpful for complexity reasons to perturb polynomials, making infinitesimal deformations. The solutions to systems of perturbed equations belong to fields of algebraic Puiseux series. We will have to deal with the following problem: given a finite set of points with algebraic Puiseux series coordinates, compute the limits of the points as the infinitesimal quantities tend to zero.

Notation 12.23. [Limit] Let $\varepsilon = \varepsilon_1, \ldots, \varepsilon_m$ be variables. As usual, we denote by $\mathrm{K}[\varepsilon] = \mathrm{K}[\varepsilon_1, \ldots, \varepsilon_m]$ the ring of polynomials in $\varepsilon_1, \ldots, \varepsilon_m$, and by $\mathrm{K}(\varepsilon) = \mathrm{K}(\varepsilon_1, \ldots, \varepsilon_m)$, the field of rational functions in $\varepsilon_1, \ldots, \varepsilon_m$, which is the fraction field of $\mathrm{K}[\varepsilon]$. If K is a field of characteristic 0, and δ is *a* variable we denote as in Chapter 2 by $\mathrm{K}\langle\delta\rangle$ the field of algebraic Puiseux series in δ with coefficients in K. We denote by $\mathrm{K}\langle\varepsilon\rangle$ the field $\mathrm{K}\langle\varepsilon_1\rangle\ldots\langle\varepsilon_m\rangle$. If $\nu = (\nu_1, \ldots, \nu_m) \in \mathbb{Q}^m$, ε^ν denotes $\varepsilon_1^{\nu_1}\ldots\varepsilon_m^{\nu_m}$. It follows from Theorem 2.91 and Theorem 2.92 that $\mathrm{R}\langle\varepsilon\rangle$ is real closed and $\mathrm{C}\langle\varepsilon\rangle$ is algebraically closed. Note that in $\mathrm{R}\langle\varepsilon\rangle$, $\varepsilon^\nu < \varepsilon^\mu$ if and only if $(\nu_m, \ldots, \nu_1) >_{\mathrm{lex}} (\mu_m, \ldots, \mu_1)$ (see Definition 2.14). In particular $\varepsilon_m < \cdots < \varepsilon_1$ in $\mathrm{R}\langle\varepsilon\rangle$. The preceding order on elements of $\mathbb{Q}^m$ and their corresponding monomials is denoted by $<_\varepsilon$ to avoid confusions. We denote by $\mathrm{K}(\varepsilon)_b$ and $\mathrm{K}\langle\varepsilon\rangle_b$ the subrings of $\mathrm{K}(\varepsilon)$ and $\mathrm{K}\langle\varepsilon\rangle$ which are sums of ε^ν with $\varepsilon^\nu \leq_\varepsilon 1$.

The elements of $\mathrm{R}(\varepsilon)_b$ and $\mathrm{R}\langle\varepsilon\rangle_b$ (resp. $\mathrm{C}(\varepsilon)_b$ and $\mathrm{C}\langle\varepsilon\rangle_b$) are the elements of $\mathrm{R}(\varepsilon)$ and $\mathrm{R}\langle\varepsilon\rangle$ (resp. $\mathrm{C}(\varepsilon)$ and $\mathrm{C}\langle\varepsilon\rangle$) bounded over R i.e. whose absolute value (resp. norm) is bounded by a positive element of R.

An element $\tau \neq 0$ of $\mathrm{K}\langle\varepsilon\rangle$ can be written uniquely as $\varepsilon^{o(\tau)}\,(\mathrm{In}(\tau) + \tau')$ with $\varepsilon^{o(\tau)}$ the biggest monomial of τ for the order of $<_\varepsilon$, $\mathrm{In}(\tau) \neq 0 \in \mathrm{K}$ and $\tau' \in K\langle\varepsilon\rangle_b$ with biggest monomial $<_\varepsilon 1$. The m-tuple $o(\tau)$ is the **order** of τ and $\mathrm{In}(\tau)$ is its initial coefficient. We have

$$\begin{aligned} o(\tau\tau') &= o(\tau) + o(\tau'), \\ o(\tau) <_\varepsilon o(\tau') &\Rightarrow o(\tau + \tau') = o(\tau'), \\ o(\tau) = o(\tau') &\Rightarrow o(\tau) \leq_\varepsilon o(\tau + \tau'). \end{aligned}$$

We define $\lim_\varepsilon(\tau)$ from $\mathrm{K}\langle\varepsilon\rangle_b$ to K as follows:

$$\begin{cases} \lim_\varepsilon(\tau) = \mathrm{In}(\tau) & \text{if } \varepsilon^{o(\tau)} = 1, \\ \lim_\varepsilon(\tau) = 0 & \text{otherwise.} \end{cases}$$

□

Example 12.24. Let $m=2$ and consider $\mathrm{K}\langle\varepsilon_1,\varepsilon_2\rangle$. Note that $\varepsilon_1/\varepsilon_2 \notin \mathrm{K}\langle\varepsilon_1,\varepsilon_2\rangle_b$. and $\varepsilon_2/\varepsilon_1 \in \mathrm{K}\langle\varepsilon_1,\varepsilon_2\rangle_b$. Then $\lim_\varepsilon (\varepsilon_2/\varepsilon_1 + 2\varepsilon_1) = 0$, and $\lim_\varepsilon (\varepsilon_2/\varepsilon_1 + 2) = 2$, while $\lim_\varepsilon$ is not defined for $\varepsilon_1/\varepsilon_2 \notin \mathrm{K}\langle\varepsilon_1,\varepsilon_2\rangle_b$. □

We first discuss how to find the limits of the roots of a univariate monic polynomial $F(T) \in \mathrm{K}(\varepsilon)[T]$. Note that in our computations, we are going to compute polynomials in $\mathrm{K}(\varepsilon)[T]$, with roots in $\mathrm{C}\langle\varepsilon\rangle[T]$.

We denote $\mathrm{Zer}_b(F(T), \mathrm{C}\langle\varepsilon\rangle)$ the set $\{\tau \in \mathrm{C}\langle\varepsilon\rangle_b \mid F(\tau) = 0\}$.

Notation 12.25. [Order of a polynomial] Given $F(T) \in \mathrm{K}(\varepsilon)[T]$, we denote by $o(F)$ the maximal value of $o(c)$ with respect to the ordering $<_\varepsilon$ for c coefficient of F. In other words, $\varepsilon^{o(F)}$ is the minimal monomial with respect to the ordering $<_\varepsilon$ such that $\varepsilon^{-o(F)} F(T)$ belongs to $\mathrm{K}(\varepsilon)_b[T]$. □

Denote by $f(T) = \lim_\varepsilon (\varepsilon^{-o(F)} F(T))$ the univariate polynomial obtained by replacing the coefficients of $\varepsilon^{-o(F)} F(T)$ by their limit under $\lim_\varepsilon$.

Now we relate the roots of $F(T)$ in $\mathrm{C}\langle\varepsilon\rangle$ and the roots of $f(T)$ in C.

Lemma 12.26. *Let Z_b be the set of roots of $F(T)$ in $\mathrm{C}\langle\varepsilon\rangle_b$, and let Z_u be the roots of $F(T)$ in $\mathrm{C}\langle\varepsilon\rangle \setminus \mathrm{C}\langle\varepsilon\rangle_b$. We denote by $\mu(\tau)$ the multiplicity of a root τ of*

a) We have

$$\begin{aligned} o(F) &= \sum_{\tau \in Z_u}^{p} \mu(\tau)\, o(\tau), \\ \mathrm{Zer}(f(T), \mathrm{C}) &= \lim_\varepsilon (\mathrm{Zer}_b(F(T), \mathrm{C}\langle\varepsilon\rangle)). \end{aligned}$$

b) If t is a root of multiplicity μ of $f(T)$ in C,

$$\mu = \sum_{\substack{\tau \in Z_b \\ \lim_\varepsilon(\tau) = t}} \mu(\tau).$$

Proof: a) We have

$$F(T) = \prod_{\tau \in Z_b} (T - \tau) \prod_{\tau \in Z_u} (T - \tau) \in \mathrm{K}(\varepsilon)[T],$$

with $o(\tau) \leq_\varepsilon 0$, for $\tau \in Z_b$, $o(\tau) >_\varepsilon 0$, for $\tau \in Z_u$. Using the properties of the order listed in Notation 12.23, and denoting $\ell = \sum_{\tau \in Z_b} \mu(\tau)$, the order of the coefficient of T^ℓ in $F(T)$ is exactly $\sum_{\tau \in Z_u} \mu(\tau)\, o(\tau)$. Moreover, the order of any other coefficient of $F(T)$ is at most $\sum_{\tau \in Z_u} \mu(\tau)\, o(\tau)$ for $<_\varepsilon$. Thus $o(F) = \sum_{\tau \in Z_u}^{p} \mu(\tau)\, o(\tau)$ and

$$\varepsilon^{-o(P)} F(T) = \prod_{\tau \in Z_b} (T - \tau) \prod_{\tau \in Z_u} (\varepsilon^{-o(\tau)} T - \varepsilon^{-o(\tau)} \tau) \in \mathrm{K}(\varepsilon)_b[T].$$

Taking $\lim_\varepsilon$ on both sides, we get

$$f(T) = \prod_{\tau \in Z_u} (-\mathrm{In}(\tau)) \prod_{\tau \in Z_u} (T - \lim_\varepsilon(\tau)).$$

b) is an immediate consequence of the last equality. □

Corollary 12.27. *If $F(T)$ is separable, the number* $\deg_T(F(T)) - \deg_T(f(T))$ *is the number of unbounded roots of P.*

Now, let $\mathcal{P} \subset \mathrm{K}(\varepsilon)[X_1, \ldots, X_k]$ be a zero-dimensional polynomial system, so that $\mathrm{Zer}(\mathcal{P}, \mathrm{C}\langle\varepsilon\rangle^k)$ is non-empty and finite, and

$$\mathrm{A} = K(\varepsilon)[X_1, \ldots, X_k]/\mathrm{Ideal}(\mathcal{P}, \mathrm{K}(\varepsilon)).$$

Suppose, moreover, for the rest of this section that all the zeros of $\mathcal{P}$ are simple. This assumption leads to technical simplifications and will be satisfied whenever we apply the results of this section in the future. We define

$$\begin{aligned} \mathrm{Zer}_b(\mathcal{P}, \mathrm{R}\langle\varepsilon\rangle^k) &= \mathrm{Zer}(\mathcal{P}, \mathrm{R}\langle\varepsilon\rangle^k) \cap \mathrm{R}\langle\varepsilon\rangle^k_b, \\ \mathrm{Zer}_b(\mathcal{P}, \mathrm{C}\langle\varepsilon\rangle^k) &= \mathrm{Zer}(\mathcal{P}, \mathrm{C}\langle\varepsilon\rangle^k) \cap \mathrm{C}\langle\varepsilon\rangle^k_b. \end{aligned}$$

These are the points of $\mathrm{Zer}(\mathcal{P}, \mathrm{R}\langle\varepsilon\rangle^k)$ and $\mathrm{Zer}(\mathcal{P}, \mathrm{C}\langle\varepsilon\rangle^k)$ that are bounded over R. Note that $\lim_\varepsilon(\mathrm{Zer}_b(\mathcal{P}, \mathrm{R}\langle\varepsilon\rangle^k)) \subset \lim_\varepsilon(\mathrm{Zer}_b(\mathcal{P}, \mathrm{C}\langle\varepsilon\rangle^k)) \cap \mathrm{R}^k$. Observe that this inclusion might be strict, since there may be algebraic Puiseux series with complex coefficients and real $\lim_\varepsilon$. If $a \in \mathrm{K}[X_1, \ldots, X_k]$, a defines a mapping from $\mathrm{Zer}_b(\mathcal{P}, \mathrm{C}\langle\varepsilon\rangle^k)$ to $\mathrm{C}\langle\varepsilon\rangle_b$, also denoted by a, associating to x the element $a(x)$.

We are going to describe $\lim_\varepsilon(\mathrm{Zer}_b(\mathcal{P}, \mathrm{C}\langle\varepsilon\rangle^k))$ by using a univariate representation of $\mathrm{Zer}(\mathcal{P}, \mathrm{C}\langle\varepsilon\rangle^k)$ and taking its limit. In order to give such a description of $\lim_\varepsilon(\mathrm{Zer}_b(\mathcal{P}, \mathrm{C}\langle\varepsilon\rangle^k))$, it is useful to define the notion of well-separating element.

A **well-separating element** a is an element of $\mathrm{K}[X_1, \ldots, X_k]$ that is a separating element for $\mathcal{P}$, such that a sends unbounded elements of $\mathrm{Zer}(\mathcal{P}, \mathrm{C}\langle\varepsilon\rangle^k)$ to unbounded elements of $\mathrm{C}\langle\varepsilon\rangle$, and such that a sends two non-infinitesimally close elements of $\mathrm{Zer}_b(\mathcal{P}, \mathrm{C}\langle\varepsilon\rangle^k)$ on two non-infinitesimally close elements of $\mathrm{C}\langle\varepsilon\rangle_b$.

To illustrate how the notions of separating element and well-separating element can differ, consider the following examples:

Example 12.28.

a) Consider the polynomial system $X\,Y = 1$, $X = \varepsilon$. The only solution is $(\varepsilon, 1/\varepsilon)$ which is unbounded. The image of this solution by X is ε, which is bounded. Thus X is separating, but not well-separating.
b) Consider the polynomial system $X^2 + Y^2 - 1 = 0$, $\varepsilon Y = X$. The only solutions are $(\varepsilon/(1+\varepsilon^2)^{1/2}, 1/(1+\varepsilon^2)^{1/2}), (-\varepsilon/(1+\varepsilon^2)^{1/2}, -1/(1+\varepsilon^2)^{1/2})$ which are bounded and not infinitesimally close (see Figure 12.2).

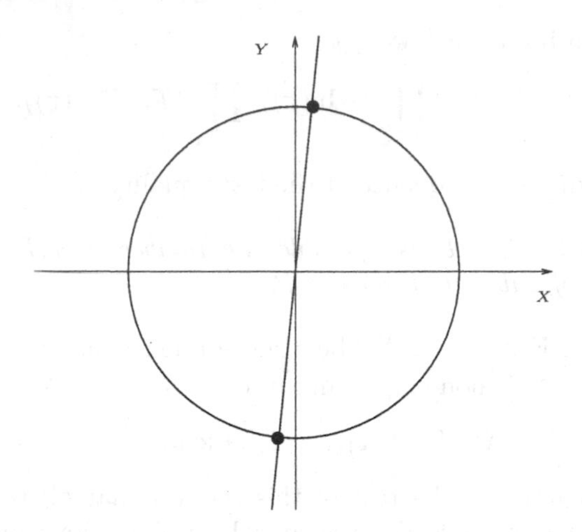

Fig. 12.2. Separating but not well-separating

The image of these solutions by X are $\varepsilon/(1+\varepsilon^2)^{1/2}, -\varepsilon/(1+\varepsilon^2)^{1/2}$, which are infinitesimally close. Thus X is separating, but not well-separating. □

Let $a \in \mathrm{K}[X_1, ..., X_k]$ be a separating element for $\mathcal{P}$. Since the polynomial system $\mathcal{P}$ is contained in $\mathrm{K}(\varepsilon)[X_1, ..., X_k]$, the polynomials of the univariate representation $(\chi(a,T), \varphi(a,T))$ are elements of $\mathrm{K}(\varepsilon)[T]$. Note that $\chi(a,T)$ is monic and separable, since we have supposed that all the zeros of $\mathcal{P}$ in $\mathrm{C}\langle\varepsilon\rangle^k$ are simple. Using Notation 12.25, note that $\varepsilon^{-o(\chi(a,T))}\varphi(a,1,T) \in \mathrm{K}(\varepsilon)_b[T]$ since it is the derivative of $\varepsilon^{-o(\chi(a,T))}\chi(a,T) \in \mathrm{K}(\varepsilon)_b[T]$, by Equation (12.6). However it may happen that some $\varepsilon^{-o(\chi(a,T))}\varphi(a, X_i, T)$ do not belong to $\mathrm{K}(\varepsilon)_b[T]$. In other words, denoting by $o(\varphi(a,T))$ the maximum value for $<_\varepsilon$ of $o(c)$ for c a coefficient of $\chi(a,T)$, $\varphi(a,X_i,T)$, $i=1, ..., k$, it may happen that $\varepsilon^{o(\varphi(a,T))} >_\varepsilon \varepsilon^{o(\chi(a,T))}$.

Example 12.29. In Example 12.28 a), with $a = X$,

$$\varphi(a,1,T) = 1, \varphi(a,X_1,T) = \varepsilon, \varphi(a,X_2,T) = 1/\varepsilon.$$

Thus $o(\chi(a,T)) = 0, o(\varphi(a,T)) = -1$, and $\varphi(a,X_2,T) \notin \mathrm{K}(\varepsilon)_b[T]$. □

Notation 12.30. When $o(\chi(a,T)) = o(\varphi(a,T))$ denote by $(\hat{\chi}(a,T), \hat{\varphi}(a, T))$ the $j+2$-tuple defined by

$$\begin{aligned} \hat{\chi}(a,T) &= \lim_\varepsilon (\varepsilon^{-o(\chi(a,T))}\chi(a,T)), \\ \hat{\varphi}(a, T) &= \lim_\varepsilon (\varepsilon^{-o(\chi(aT))}\varphi(a,T)), \end{aligned}$$

with $\hat{\varphi}(a, T) = (\hat{\varphi}(a,1,t), \hat{\varphi}(a,X_1,t), ..., \hat{\varphi}(a,X_k,t))$. □

Lemma 12.31. *Suppose that $a \in \mathrm{K}[X_1, \ldots, X_k]$ is well-separating for $\mathcal{P}$ and such that $o(\chi(a,T)) = o(\varphi(a,T))$. Then, for every root τ of $\chi(a,T)$ in $\mathrm{C}\langle\varepsilon\rangle_b$ such that $\lim_\varepsilon(\tau) = t$ is a root of multiplicity μ of $\hat{\chi}(a,T)$,*

$$\lim_\varepsilon \left(\frac{\varphi(a, X_i, \tau)}{\varphi(a, 1, \tau)} \right) = \frac{\hat{\varphi}^{(\mu-1)}(a, X_i, t)}{\hat{\varphi}^{(\mu-1)}(a, 1, t)}.$$

Proof: Let $\tau_1, \ldots, \tau_\ell$ be the roots of $\chi(a,T)$ in $\mathrm{C}\langle\varepsilon\rangle_b$ and let $\tau_{\ell+1}, \ldots, \tau_N$ be the roots of $\chi(a,T)$ in $\mathrm{C}\langle\varepsilon\rangle \setminus \mathrm{C}\langle\varepsilon\rangle_b$. Let $t_j = \lim_\varepsilon(\tau_j)$ for $j = 1, \ldots, \ell$. Suppose that $t = \lim_\varepsilon(\tau_1)$ is a root of multiplicity μ of $\hat{\chi}(a,T)$. By Lemma 12.26 b), there exist $\mu - 1$ roots of $\chi(a,T)$ in $\mathrm{C}\langle\varepsilon\rangle_b$, numbered $\tau_2, \ldots, \tau_\mu$ without loss of generality, such that

$$t = \lim_\varepsilon(\tau_1) = \cdots = \lim_\varepsilon(\tau_\mu),$$

and for every $\mu < j \le \ell$, $t_j = \lim_\varepsilon(\tau_j) \neq t$.

We have

$$\begin{aligned} \chi(a,T) &= \prod_{m \in \{1,\ldots\ell\}} (T - \tau_m) \prod_{m \in \{\ell+1,\ldots N\}} (T - \tau_m), \\ \varphi(a,1,T) &= \sum_{j=1}^{N} \prod_{m \in \{1,\ldots N\} \setminus \{j\}} (T - \tau_m), \end{aligned}$$

and $o(\chi(a,T)) = \sum_{i=\ell+1}^{N} o(\tau_j)$ by Lemma 12.26. Thus,

$$\begin{aligned} &\varepsilon^{-o(\chi(a,T))}\varphi(a,1,T) \\ &= \left(\sum_{j=1}^{\mu} \prod_{m \in \{1,\ldots\ell\} \setminus \{j\}} (T - \tau_m) \right) \prod_{m=\ell+1}^{N} (\varepsilon^{-o(\tau_m)}T - \varepsilon^{-o(\tau_m)}\tau_m) \\ &+ \left(\sum_{j=\mu+1}^{\ell} \prod_{m \in \{1,\ldots\ell\} \setminus \{j\}} (T - \tau_m) \right) \prod_{m=\ell+1}^{N} (\varepsilon^{-o(\tau_m)}T - \varepsilon^{-o(\tau_m)}\tau_m) \\ &+ \prod_{m=1}^{\ell} (T - \tau_m) \left(\sum_{j=\ell+1}^{N} \varepsilon^{-o(\tau_j)} \prod_{m \in J} (\varepsilon^{-o(\tau_m)}T - \varepsilon^{-o(\tau_m)}\tau_m) \right). \end{aligned}$$

with $J = \{\ell+1, \ldots N\} \setminus \{j\}$.

Since $-o(\tau_j) <_\varepsilon 0$, it follows, taking $\lim_\varepsilon$, that

$$\begin{aligned} &\hat{\varphi}(a,1,T) \\ &= c\mu(T-t)^{\mu-1} \prod_{m=\mu+1}^{\ell} (T - t_m) \\ &+ c \sum_{j=\mu+1}^{\ell} (T-t)^{\mu} \prod_{m \in \{\mu+1,\ldots N\} \setminus \{j\}} (T - t_m), \end{aligned}$$

with $c = \prod_{m=\ell+1}^{N} (-\mathrm{In}(\tau_j))$. Thus,

$$\hat{\varphi}^{(\mu-1)}(a,1,t) = c\,\mu! \prod_{m=\mu+1}^{\ell} (t - t_m).$$

Denoting by ξ_j the unique point of $\mathrm{Zer}(\mathcal{P}, \mathrm{C}\langle\varepsilon\rangle^k)$ such that $a(\xi_j) = \tau_j$, and by ξ_{ji} the i-th coordinate of ξ_j, we have similarly

$$\begin{aligned}&\varphi(a, X_i, T)\\ &= \sum_{j=1}^{N} \xi_{ji} \prod_{m\in\{1,\ldots N\}\setminus\{j\}} (T - \tau_m)\end{aligned}$$

and

$$\begin{aligned}&\varepsilon^{-o(\chi(a,T))}\,\varphi(a, X_i, T)\\ &= \left(\sum_{j=1}^{\ell} \xi_{ji} \prod_{m\in\{1,\ldots\ell\}\setminus\{j\}} (T - \tau_m)\right) \prod_{m=\ell+1}^{N} (\varepsilon^{-o(\tau_m)}\,T - \varepsilon^{-o(\tau_m)}\,\tau_m)\\ &\quad + \prod_{m=1}^{\ell} (T - \tau_m) C.\end{aligned}$$

with

$$C = \sum_{j=\ell+1}^{N} \varepsilon^{-o(\tau_j)}\,\xi_{ji} \prod_{m\in J} (\varepsilon^{-o(\tau_m)}\,T - \varepsilon^{-o(\tau_m)}\,\tau_m).$$

Since a is well-separating and $a(\xi_j)$ is bounded, it follows that for all $i = 1, \ldots, \ell$, $j = 1, \ldots, k$, $\xi_{ji} \in \mathrm{C}\langle\varepsilon\rangle_b$. So we have $A \in \mathrm{C}\langle\varepsilon\rangle_b[T]$, with

$$A = \left(\sum_{j=1}^{\ell} \xi_{ji} \prod_{m\in\{1,\ldots\ell\}\setminus\{j\}} (T - \tau_m)\right) \prod_{m=\ell+1}^{N} (\varepsilon^{-o(\tau_m)}\,T - \varepsilon^{-o(\tau_m)}\,\tau_m).$$

It is also clear that $B = \prod_{m=1}^{\ell} (T - \tau_m) \in \mathrm{C}\langle\varepsilon\rangle_b[T]$.
Since $\varepsilon^{-o(\chi(a,T))}\,\varphi(a, X_i, T) \in \mathrm{K}(\varepsilon)_b[T]$, $A \in \mathrm{C}\langle\varepsilon\rangle_b[T]$, $B \in \mathrm{C}\langle\varepsilon\rangle_b[T]$, and B is monic,

$$C = \sum_{j=\ell+1}^{N} \varepsilon^{-o(\tau_j)}\,\xi_{ji} \prod_{m\in J} (\varepsilon^{-o(\tau_m)}\,T - \varepsilon^{-o(\tau_m)}\,\tau_m) \in \mathrm{C}\langle\varepsilon\rangle_b[T].$$

So finally

$$\varepsilon^{-o(\chi(a,T))}\,\varphi(a, X_i, T) = A + \prod_{m=1}^{\mu} (T - \tau_m) \prod_{m=\mu+1}^{\ell} (T - \tau_m) C.$$

Since

$$t = \lim_{\varepsilon} (\tau_1) = \cdots = \lim_{\varepsilon} (\tau_\mu), t \neq t_j = \lim_{\varepsilon} (\tau_j), j > \mu,$$

and a is well-separating,

$$\lim_\varepsilon (\xi_1) = \cdots = \lim_\varepsilon (\xi_\mu).$$

Denoting by

$$\begin{aligned} x = (x_1, \ldots, x_k) &= \lim_\varepsilon (\xi_1) = \cdots = \lim_\varepsilon (\xi_\mu), \\ y_j = (y_{j,1}, \ldots, y_{j,k}) &= \lim_\varepsilon (\xi_j) \end{aligned}$$

and by D the polynomial obtained by replacing successively $\varepsilon_m, \varepsilon_{m-1} \ldots \varepsilon_1$ by 0 in $\prod_{m=\mu+1}^{\ell} (T - \tau_m)\, C$, we get

$$\begin{aligned} \hat{\varphi}(a, X_i, T) &= c\,\mu\, x_i (T-t)^{\mu-1} \prod_{m=\mu+1}^{\ell} (T - t_m) \\ &+ (T-t)^\mu \left(c \sum_{j=\mu+1}^{\ell} y_{j,i} \prod_{m \in \{\mu+1, \ldots, \ell\} \setminus \{j\}} (T - t_m) + D \right), \\ \hat{\varphi}^{(\mu-1)}(a, X_i, t) &= c\,\mu!\, x_i \prod_{j=\mu+1}^{\ell} (t - t_j) \end{aligned}$$

with $c = \prod_{m=\ell+1}^{N} (-\mathrm{In}(\tau_j))$. Finally,

$$\begin{aligned} \lim_\varepsilon \left(\frac{\varphi(a, X_i, \tau_j)}{\varphi(a, 1, \tau_j)} \right) &= \lim_\varepsilon (\xi_{j,i}) \\ &= x_i \\ &= \frac{\hat{\varphi}^{(\mu-1)}(a, X_i, t)}{\hat{\varphi}^{(\mu-1)}(a, 1, t)}, \end{aligned}$$

for every $j = 1, \ldots, \mu$, $i = 1, \ldots, k$. □

As is the case for separating elements (see Lemma 4.89), well-separating elements can be chosen in a set defined in advance.

Lemma 12.32. *If* $\#\mathrm{Zer}(\mathcal{P}, \mathrm{C}\langle\varepsilon\rangle^k) = N$, *then at least one element* a *of*

$$\mathcal{A} = \{X_1 + i\, X_2 + \cdots + i^{k-1} X_k \mid 0 \le i \le (k-1) N^2\}$$

is well-separating and such that $o(\chi(a, T)) = o(\varphi(a, T))$.

Proof: Define

- $\mathcal{W}_1$, of cardinality $\le N\,(N-1)/2$, to be the set of vectors $\xi - \eta$ with ξ and η distinct solutions of $\mathcal{P}$ in $\mathrm{C}\langle\varepsilon\rangle^k$,
- $\mathcal{W}_2$, of cardinality $\le N\,(N-1)/2$, to be the set of vectors $\lim_\varepsilon(\xi) - \lim_\varepsilon(\eta)$ with ξ and η distinct non-infinitesimally close bounded solutions of $\mathcal{P}$ in $\mathrm{C}\langle\varepsilon\rangle^k$,

- $\mathcal{W}_3$, of cardinality $\leq N$, to be the set of vectors $c = (c_1, \ldots, c_k)$ with c_i the coefficient of $\varepsilon^{\max_{i=1,\ldots,k}(o(\xi_i))}$ in ξ_i, for $\xi = (\xi_1, \ldots, \xi_k) \in \mathrm{Zer}(\mathcal{P}, \mathrm{C}\langle\varepsilon\rangle^k)$,
- $\mathcal{W} = \mathcal{W}_1 \cup \mathcal{W}_2 \cup \mathcal{W}_3$. Note that $\mathcal{W}$ is of cardinality $\leq N^2$

If j is such that, for every $c \in \mathcal{W}_3$, $c_1 + \cdots + j^{k-1} c_k \neq 0$ and

$$a = X_1 + \cdots + j^{k-1} X_k,$$

then for every $\xi \in \mathrm{Zer}(\mathcal{P}, \mathrm{C}\langle\varepsilon\rangle^k)$,

$$\mathrm{In}(a(\xi)) = c_1 + \cdots + j^{k-1} c_k$$

and $o(a(\xi)) = \max_{i=1,\ldots,k}(o(\xi_i))$.

If, for every $\xi = (\xi_1, \ldots, \xi_k) \in \mathrm{Zer}(\mathcal{P}, \mathrm{C}\langle\varepsilon\rangle^k)$, $o(a(\xi)) = \max_{i=1,\ldots,k}(o(\xi_i))$ then a maps unbounded elements of $\mathrm{Zer}(\mathcal{P}, \mathrm{C}\langle\varepsilon\rangle^k)$ to unbounded elements of $\mathrm{C}\langle\varepsilon\rangle$. Denote by $\xi_1, \ldots, \xi_\ell$ the elements of $\mathrm{Zer}_b(\mathcal{P}, \mathrm{C}\langle\varepsilon\rangle^k)$ and by $\xi_{\ell+1}, \ldots, \xi_N$ the elements of $\mathrm{Zer}(\mathcal{P}, \mathrm{C}\langle\varepsilon\rangle^k) \setminus \mathrm{Zer}_b(\mathcal{P}, \mathrm{C}\langle\varepsilon\rangle^k)$. Then $\tau_1 = a(\xi_1), \ldots, \tau_\ell = a(\xi_\ell)$ are the roots of $\chi(a, T)$ in $\mathrm{C}\langle\varepsilon\rangle_b$ and $\tau_{\ell+1} = a(\xi_{\ell+1}), \ldots, \tau_N = a(\xi_N)$ are the roots of $\chi(a, T)$ in $\mathrm{C}\langle\varepsilon\rangle \setminus \mathrm{C}\langle\varepsilon\rangle_b$. For $i = 1, \ldots k$,

$$\begin{aligned}
&\varepsilon^{-o(\chi(a,T))} \varphi(a, X_i, T) \\
&= \left(\sum_{j=1}^{\ell} \xi_{ji} \prod_{m \in \{1,\ldots\ell\} \setminus \{j\}} (T - \tau_m) \right) \prod_{m=\ell+1}^{N} (\varepsilon^{-o(\tau_m)} T - \varepsilon^{-o(\tau_m)} \tau_m) \\
&\quad + \prod_{m=1}^{\ell} (T - \tau_m) \left(\sum_{j=\ell+1}^{N} \varepsilon^{-o(\tau_j)} \xi_{ji} C \right)
\end{aligned}$$

with

$$C = \sum_{j=\ell+1}^{N} \varepsilon^{-o(\tau_j)} \xi_{ji} \prod_{m \in \{\ell+1,\ldots N\} \setminus \{j\}} (\varepsilon^{-o(\tau_m)} T - \varepsilon^{-o(\tau_m)} \tau_m$$

belongs to $K\langle\varepsilon\rangle_b[T]$, since $o(\tau_j) = o(a(\xi_j)) \geq_\varepsilon o(\xi_{j,i})$ and $o(\varepsilon^{-o(\tau_j)} \xi_{ji}) \leq_\varepsilon 0$.

So, if j is such that, for every $w \in \mathcal{W}$, $w_1 + \cdots + j^{k-1} w_k \neq 0$, then $a = X_1 + \cdots + j^{k-1} X_k$ is well-separating and such that

$$o(\chi(a, T)) = o(\phi(a, T)).$$

For a fixed $w \in \mathcal{W}$, there are at most $k - 1$ elements of $\mathcal{A}$ that satisfy $w_1 + \cdots + j^{k-1} w_k = 0$. This is because an element

$$X_1 + j X_2 + \cdots + j^{k-1} X_k$$

satisfying $w_1 + \cdots + j^{k-1} w_k = 0$ is such that $P_w(j) = 0$, with

$$P_w(T) = w_1 + T w_2 + \cdots + T^{k-1} w_k.$$

But $P_w(T)$, which is non-zero, has at most $k-1$ roots. So the result is clear by the pigeon-hole principle. □

According to the preceding results, the set $\lim_\varepsilon (\mathrm{Zer}_b(\mathcal{P}, \mathrm{C}\langle\varepsilon\rangle^k)) \cap \mathrm{R}^k$ can be obtained as follows:

- Determine a well-separating element $a = X_1 + \cdots + j^{k-1}X_k$ such that $o(\chi(a,T)) = o(\varphi(a,T))$ as follows:
  - List all $a \in \mathcal{A}$ that are separating and compute the corresponding $\hat{\chi}(a,T)$.
  - Among these list the a such that the degree of $\hat{\chi}(a, T)$ is minimal. This condition guarantees that a maps unbounded elements of $\mathrm{Zer}(\mathcal{P}, \mathrm{C}\langle\varepsilon\rangle^k)$ to unbounded roots of $\chi(a, T)$ since, by Corollary 12.27, $\deg(\chi(a,T)) - \deg(\hat{\chi}(a,T))$ is the number of unbounded roots of $\chi(a,T)$ and is maximal when all unbounded elements of $\mathrm{Zer}(\mathcal{P}, \mathrm{C}\langle\varepsilon\rangle^k)$ have unbounded images by a.
  - Among these list those such that $o(\chi(a,T)) = o(\varphi(a,T))$.
  - Among these find an a such that the number of distinct roots of $\hat{\chi}(a,T)$ is maximal, i.e. such that $\deg(\gcd(\hat{\chi}(a,T), \hat{\chi}'(a,T)))$ is minimal. This guarantees that no two non-infinitesimally close elements of $\mathrm{Zer}_b(\mathcal{P}, \mathrm{C}\langle\varepsilon\rangle^k)$ are sent by a to two infinitesimally close elements of $\mathrm{C}\langle\varepsilon\rangle_b$.
- Lemma 12.32 guarantees that there exists such an a in $\mathcal{A}$.
- For such an a and every root t of $\hat{\chi}(a,T)$ in R with multiplicity μ consider

$$x_i = \frac{\hat{\varphi}^{(\mu-1)}(a, X_i, t)}{\hat{\varphi}^{(\mu-1)}(a, 1, t)}.$$

  The root of $\hat{\chi}(a, T)$ in R can be described by its Thom encoding. All elements of $\lim_\varepsilon (\mathrm{Zer}_b(\mathcal{P}, \mathrm{C}\langle\varepsilon\rangle^k)) \cap \mathrm{R}^k$ are obtained this way since if $x \in \lim_\varepsilon (\mathrm{Zer}_b(\mathcal{P}, \mathrm{C}\langle\varepsilon\rangle^k)) \cap \mathrm{R}^k$, $a(x) \in \mathrm{R}$ is a root of $\hat{\chi}(a,T)$. Conversely if $t \in \mathrm{R}$ is a root of $\hat{\chi}(a,T)$ of multiplicity μ,

$$x_i = \frac{\hat{\varphi}^{(\mu-1)}(a, X_i, t)}{\hat{\varphi}^{(\mu-1)}(a, 1, t)} \in \mathrm{R},$$

  since $\hat{\varphi}(a, X_i, T) \in \mathrm{K}[T], t \in \mathrm{R}$.

We can now describe an algorithm for computing the limit of the bounded solutions of a polynomial system. Since we want to perform the computations in a ring rather than in a field, the following remark will be useful.

Remark 12.33. Let $\#\mathrm{Zer}(\mathcal{P}, \mathrm{C}\langle\varepsilon\rangle^k) = N$. Consider $b \neq 0 \in \mathrm{K}(\varepsilon)$. Using Notation 12.30, we have

$$(\chi(b\,a, b\,T), \varphi_b(b\,a, b\,T)) = b^N(\chi(a,T), \varphi(a,T)).$$

Thus, $o(\chi(a,T))=o(\varphi(a,T))$ if and only if $o(\chi(b\,a,b\,T))=o(\varphi_b(b\,a,b\,T))$. □

Algorithm 12.14. **[Limits of Bounded Points]**

- **Structure:** an ordered ring D with division in $\mathbb{Z}$ contained in an ordered field K.
- **Input:** $\varepsilon=(\varepsilon_1,\ldots,\varepsilon_m)$, a zero-dimensional polynomial system with only simple zeroes $\mathcal{P}\subset \mathrm{D}[\varepsilon][X_1,\ldots,X_k]$, a basis $\mathcal{B}$ of
$$\mathrm{A}=\mathrm{K}(\varepsilon)[X_1,\ldots,X_k]/\mathrm{Ideal}(\mathcal{P},\mathrm{K}(\varepsilon))$$
such that the multiplication table $\mathcal{M}$ of A in $\mathcal{B}$ has entries in $\mathrm{D}[\varepsilon]$, an element $b\neq 0\in \mathrm{D}[\varepsilon]$ such that $b, b\,X_1,\ldots,b\,X_k$ have coordinates in D in the basis $\mathcal{B}$.
- **Output:** a set $\mathcal{U}$ of real univariate representations, such that the set of points in R^k associated to these $k+2$ tuples are the elements of $\lim_\varepsilon(\mathrm{Zer}_b(\mathcal{P},\mathrm{C}\langle\varepsilon\rangle^k))\cap\mathrm{R}^k$.
- **Complexity:**
$$(\lambda\, d_1\cdots d_k)^{O(m)}$$
when $\mathcal{P}$ is a special Gröbner basis
$$\mathcal{P}=\{b_1X_1^{d_1}+Q_1,\ldots,b_kX_k^{d_k}+Q_k\}\subset \mathrm{D}[\varepsilon][X_1,\ldots,X_k],$$
$\deg_X(Q_i)<d_i$, $\deg_{X_j}(Q_i)<d_j$ $\deg_\varepsilon(Q_i)\le\lambda$, $\deg_\varepsilon(b_i)\le\lambda$, $b=b_1\cdots b_k$.
- **Procedure:**
  - For every $a=X_1+j\,X_2+\cdots+j^{k-1}X_k$, $j=0,\ldots,(k-1)\,N^2$ compute $(\chi(b\,a,T),\ \varphi_b(b\,a,\,b,\,T))$using Algorithm 12.11 (Candidate Univariate Representation).
  - Keep the values of a such that $o(\chi(b\,a,b\,T))=o(\varphi_b(b\,a,b\,T))$.
  - Compute
$$\begin{aligned}\hat{\chi}(b\,a,b\,T) &= \lim_\varepsilon(\varepsilon^{-o(\chi(ba,bT))}\chi(b\,a,b\,T))\\ \hat{\varphi}(b\,a,b\,T) &= \lim_\varepsilon(\varepsilon^{-o(\chi(ba,bT))}c\,\varphi_b(b\,a,b\,T)).\end{aligned}$$
  - Choose an a among those for which $\deg(\hat{\chi}(b\,a,b\,T))$ is minimal and for which $\deg(\gcd(\hat{\chi}(b\,a,\,b\,T),\ \hat{\chi}'(b\,a,\,b\,T)))$ is minimal (Notation 12.30), computing $\gcd(\hat{\chi}(b\,a,b\,T),\hat{\chi}'(b\,a,b\,T))$ using Remark 10.18.
  - Return $(\hat{\chi}(b\,a,b\,T),\hat{\varphi}(b\,a,b\,T))$.
  - Compute the list of Thom encodings of the roots of $\hat{\chi}(b\,a,\,b\,T)$ in R using Algorithm 10.14 (Thom Encoding) and Remark 10.76. Read from the Thom encoding σ the multiplicity μ of the associated root t_σ. For every such Thom encoding σ, place $(\hat{\chi}(b\,a,b\,T),\hat{\varphi}^{(\mu-1)}(b\,a,b\,T),\sigma)$ in $\mathcal{U}$.

Proof of correctness: The correctness follows from the discussion preceding the algorithm, using Remark 12.33. □

Complexity analysis: We estimate the complexity only in the case where $\mathcal{P}$ is a special Gröbner basis contained in $\mathrm{D}[\varepsilon][X_1,\ldots,X_k]$, since this is the only way we are going to use it later. Let

$$\mathcal{P}=\{b_1 X_1^{d_1}+Q_1,\ldots,b_k X_k^{d_k}+Q_k\}\subset \mathrm{D}[\varepsilon][X_1,\ldots,X_k],$$

with $\deg_X(Q_i)<d_i$, $\deg_{X_j}(Q_i)<d_j$ $\deg_\varepsilon(Q_i)\leq\lambda$, $\deg_\varepsilon(b_i)\leq\lambda$, $b=b_1\cdots b_k$.

Then the number of arithmetic operations in $\mathrm{D}[\varepsilon]$ is $(d_1\cdots d_k)^{O(1)}$ according to the complexity analysis of Algorithm 12.11 (Univariate Representation), Remark 10.18, and Remark 10.76. The degrees in ε of the polynomials occurring in the multiplication table are

$$\lambda\, k\, d_1\, (d_1+\cdots+d_{k-1}-k+1)$$

according to the complexity analysis of Algorithm 12.10 (Parametrized Special Multiplication Table). Finally, using the complexity of Algorithm 8.4 (Addition of multivariate polynomials) and Algorithm 8.5 (Multiplication of multivariate polynomials), the complexity in D is

$$(\lambda\, d_1\cdots d_k)^{O(m)}.$$

The degree in T and number of real univariate representations output is $d_1\cdots d_k$.

When $\mathrm{D}=\mathbb{Z}$, and the bitsizes of the coefficients of $\mathcal{P}$ are bounded by τ, the bitsizes of the coefficients of the polynomials occurring in the computation of the multiplication table and its output are

$$(d_1+\cdots+d_{k-1}-k+1)\,(k\, d_1+1)\,(\tau+4\,\nu'),$$

where ν' is the bitsize of $(\lambda\,(d_1+\cdots+d_k)+1)^m$, according to the complexity analysis of Algorithm 12.10 (Parametrized Special Multiplication Table). □

In later chapters of the book, we need a parametrized version of this algorithm in the case of a parametrized special system.
A **parametrized univariate representation** with parameters Y is a $k+2$-tuple

$$\begin{aligned} u(Y) &= (f(Y,T), g(Y,T)\in \mathrm{D}[Y][T]^{k+2},\\ g(Y,T) &= (g_0(Y,T), g_1(Y,T)\ldots, g_k(Y,T)). \end{aligned}$$

We need a notation. Let $\varepsilon=(\varepsilon_1,\ldots,\varepsilon_m)$. If $f\in\mathrm{A}[\varepsilon]$ and $\alpha=(\alpha_1,\ldots,\alpha_m)\in\mathbb{N}^m$, we denote by $f_\alpha\in\mathrm{A}$ the coefficient of ε^α in f and by $g_\alpha=(g_{0\alpha},g_{1\alpha},\ldots,g_{k\alpha})$

Algorithm 12.15. **[Parametrized Limit of Bounded Points]**

- **Structure:** an ordered ring D with division in $\mathbb{Z}$ contained in an ordered field K.
- **Input:** $\varepsilon=(\varepsilon_1,\ldots,\varepsilon_m)$, a parametrized special Gröbner basis

$$\mathcal{P}=\{b_1 X_1^{d_1}+Q_1(Y,X),\ldots,b_k X_k^{d_k}+Q_k(Y,X)\}\subset \mathrm{D}[Y,\varepsilon][X_1,\ldots,X_k]$$

with $Y = (Y_1, \ldots, Y_\ell)$, and the corresponding parametrized multiplication table $\mathcal{M}$ with entries in $\mathrm{D}[Y, \varepsilon]$, with $b_i \in \mathrm{D}[\varepsilon]$.

- **Output:** a set $\mathcal{U}$ of parametrized univariate representations of the form, $u(Y) = (f, g)$, where $(f, g) \in \mathrm{D}[Y][T]^{k+2}$. The set $\mathcal{U}$ has the property that for any point $y \in \mathrm{R}^\ell$, denoting by $\mathcal{U}(y)$ the subset of $\mathcal{U}$ such that $f(y, T)$ and $g_0(y, T)$ are coprime, the points associated to the univariate representations $u(y)$ in $\mathcal{U}(y)$ contain $\lim_\varepsilon (\mathrm{Zer}_b(\mathcal{G}_y, \mathrm{C}\langle\varepsilon\rangle^k)) \cap \mathrm{R}^k$.
- **Complexity:** $((\lambda + t)\, d_1 \cdots d_k)^{O(m+\ell)}$.
- **Procedure:**
  - $b := b_1 \ldots b_k$.
  - For every
  $$a = X_1 + j X_2 + \cdots + j^{k-1} X_k,\ j = 0, \ldots, (k-1) N^2,$$
  compute the parametrized univariate representation
  $$(\chi(b\,a, b\,T), \varphi_b(b\,a, b\,T))$$
  by performing the computations of Algorithm 12.11 (Candidate Univariate Representation) in $\mathrm{D}[Y, \varepsilon]$.
  - For every $\alpha \in \mathbb{Z}^m$ such that ε^α appears in χ, for every $\mu \leq \deg_T(\chi_\alpha)$, include $(\chi_\alpha(b\,a, b\,T), \varphi_{b,\alpha}^{(\mu-1)}(b\,a, b\,T))$ in the set $\mathcal{U}$.
  - Output $\mathcal{U}$.

Proof of correctness: The correctness follows from the discussion preceding Algorithm 12.14. In this parametric situation, the choice of the well-separating element and the order of the univariate representation depends on the parameters, as well as the multiplicities of the roots. This is the reason why we place all the possibilities in $\mathcal{U}$. □

Complexity analysis: Let

$$\mathcal{P} = \{b_1 X_1^{d_1} + Q_1, \ldots, b_k X_k^{d_k} + Q_k\} \subset \mathrm{D}[Y][\varepsilon][X_1, \ldots, X_k],$$

with $\deg_X(Q_i) < d_i$, $\deg_{X_j}(Q_i) < d_j$, $\deg_\varepsilon(Q_i) \leq \lambda$, $\deg_Y(Q_i) \leq t$, and $\deg_\varepsilon(b_i) \leq \lambda$, $b = b_1 \cdots b_k$. The number of arithmetic operations in $\mathrm{D}[Y][\varepsilon]$ is $(d_1 \cdots d_k)^{O(1)}$ according to the complexity analysis of Algorithm 12.11 (Univariate Representation), Remark 10.18 and Remark 10.76. The degrees in ε and Y of the polynomials occurring in the multiplication table are respectively $O(\lambda\, k\, d_1\, (d_1 + \cdots + d_k))$ and $O(t\, k\, d_1\, (d_1 + \cdots + d_k))$ according to the complexity analysis of 12.10 (Parametrized Special Multiplication Table). Finally, using the complexity of Algorithm 8.4 (Addition of multivariate polynomials) and Algorithm 8.5 (Multiplication of multivariate polynomials), the complexity in D is

$$((\lambda + t)\, d_1 \cdots d_k)^{O(m+\ell)}.$$

The degrees in T of the of real univariate representations output is $d_1 \cdots d_k$, and their degrees in Y is $(d_1 \cdots d_k)^{O(1)}$. The number of real univariate representations output is $(d_1 \cdots d_k)^{O(1)}$.

When $\mathrm{D} = \mathbb{Z}$, and the bitsizes of the coefficients of $\mathcal{P}$ are bounded by τ, the bitsizes of the coefficients of the polynomials occurring computing the multiplication table and its output are $(d_1 + \cdots + d_{k-1} - k + 1)(k d_1 + 1)(\tau + 4\nu')$, where ν' is the bitsize of $((\lambda + t)(d_1 + \cdots + d_k) + 1)^{m+\ell}$, according to the complexity analysis of 12.10 (Parametrized Special Multiplication Table). □

12.6 Finding Points in Connected Components of Algebraic Sets

We are going to describe a method for finding at least one point in every semi-algebraically connected component of an algebraic set. We know by Proposition 7.9 that when we consider a bounded nonsingular algebraic hypersurface, it is possible to change coordinates so that its projection to the X_1-axis has a finite number of non-degenerate critical points. These points provide at least one point in every semi-algebraically connected component of the bounded nonsingular algebraic hypersurface by Proposition 7.4. Unfortunately this result is not very useful in algorithms since it provides no method for performing this linear change of variables. Moreover when we deal with the case of a general algebraic set, which may be unbounded or singular, this method no longer works.

We first explain how to associate to a possibly unbounded algebraic set $Z \subset \mathrm{R}^k$ a bounded algebraic set $Z' \subset \mathrm{R}\langle\varepsilon\rangle^{k+1}$, whose semi-algebraically connected components are closely related to those of Z.

Let $Z = \mathrm{Zer}(Q, \mathrm{R}^k)$ and consider

$$Z' = \mathrm{Zer}(Q^2 + (\varepsilon^2(X_1^2 + \cdots + X_{k+1}^2) - 1)^2, \mathrm{R}\langle\varepsilon\rangle^{k+1}).$$

The set Z' is the intersection of the sphere S_ε^k of center 0 and radius $1/\varepsilon$ with a cylinder based on the extension of Z to $\mathrm{R}\langle\varepsilon\rangle$. The intersection of Z' with the hyperplane $X_{k+1} = 0$ is the intersection of Z with the sphere S_ε^{k-1} of center 0 and radius $1/\varepsilon$. Denote by π the projection from $\mathrm{R}\langle\varepsilon\rangle^{k+1}$ to $\mathrm{R}\langle\varepsilon\rangle^k$.

Proposition 12.34. *Let N be a finite set of points meeting every semi-algebraically connected component of Z'. Then $\pi(N)$ meets every semi-algebraically connected component of the extension* $\mathrm{Ext}(Z, \mathrm{R}\langle\varepsilon\rangle)$ *of Z to* $\mathrm{R}\langle\varepsilon\rangle$.

Proof: Let D a semi-algebraically connected components of Z. If D is bounded, $\mathrm{Ext}(D, \mathrm{R}\langle\varepsilon\rangle)$ does not intersect S_ε^{k-1}, and $\pi^{-1}(\mathrm{Ext}(D, \mathrm{R}\langle\varepsilon\rangle))$ is semi-algebraically homeomorphic to two copies of $\mathrm{Ext}(D, \mathrm{R}\langle\varepsilon\rangle)$, one in each of the half-spaces defined, respectively, by $X_{k+1} > 0$ and by $X_{k+1} < 0$. Thus, since N intersects every semi-algebraically connected component of Z', N intersects $\pi^{-1}(\mathrm{Ext}(D, \mathrm{R}\langle\varepsilon\rangle))$ and $\pi(N)$ intersects $\mathrm{Ext}(D, \mathrm{R}\langle\varepsilon\rangle)$.

If D is unbounded, the set A of elements $r \in \mathrm{R}$ such that D intersects the sphere $S^{k-1}(0, r)$ of center 0 and radius r is semi-algebraic and unbounded and contains an open interval $(a, +\infty)$. Thus $1/\varepsilon \in \mathrm{Ext}(A, \mathrm{R}\langle\varepsilon\rangle)$, and $\mathrm{Ext}(D, \mathrm{R}\langle\varepsilon\rangle)$ intersects S_ε^{k-1}. Take $z \in \mathrm{Ext}(D, \mathrm{R}\langle\varepsilon\rangle) \cap S_\varepsilon^{k-1}$, and denote by D' the semi-algebraically connected component of Z' containing $z' = (z, 0) \in Z'$. Take $x \in D' \cap N$ and consider a semi-algebraic path γ connecting z' to x inside D'. Then, $\pi(\gamma)$ is a semi-algebraic path connecting z to $\pi(x)$ inside $\mathrm{Ext}(Z, R\langle\epsilon\rangle)$, thus $\pi(x)$ and z belong to the same semi-algebraically connected component of $\mathrm{Ext}(Z, \mathrm{R}\langle\varepsilon\rangle)$. Since $z \in \mathrm{Ext}(D, \mathrm{R}\langle\varepsilon\rangle)$, then $\pi(x) \in \mathrm{Ext}(D, \mathrm{R}\langle\varepsilon\rangle)$, and $\pi(N)$ intersects $\mathrm{Ext}(D, \mathrm{R}\langle\varepsilon\rangle)$. □

Let us illustrate this result. If $Q = X_2^2 - X_1(X_1 - 1)(X_1 + 1)$, then $Z = \mathrm{Zer}(Q, \mathrm{R}^2)$ is a cubic curve with one bounded semi-algebraically connected component and one unbounded semi-algebraically connected component (see Figure 12.3).

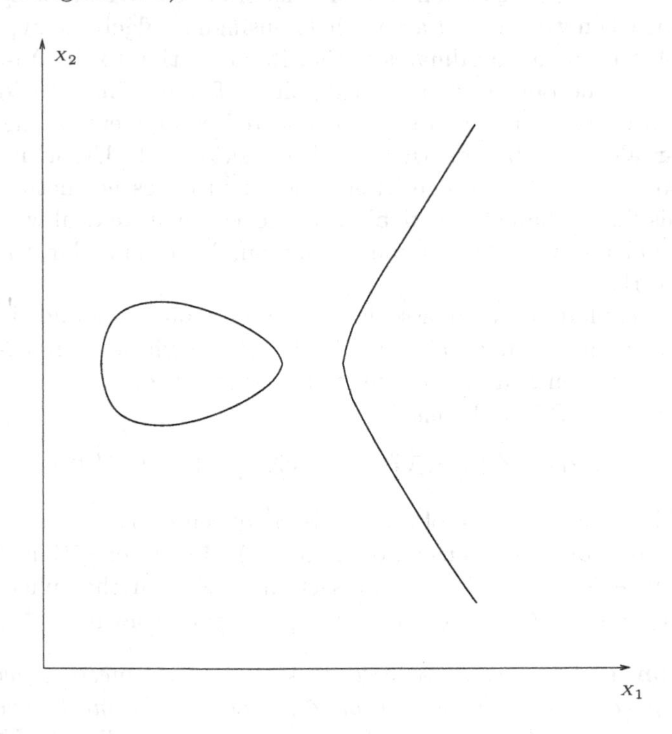

Fig. 12.3. Cubic curve in the plane

The corresponding $Z' \subset \mathrm{R}\langle\varepsilon\rangle^3$ (see Figure 12.4) has two semi-algebraically connected components above the bounded semi-algebraically connected component of the cubic curve, and one semi-algebraically connected component above the unbounded semi-algebraically connected component of the cubic curve.

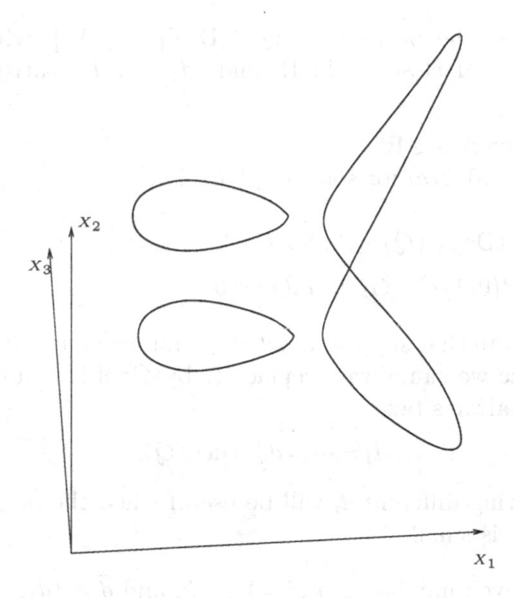

Fig. 12.4. Cubic curve lifted to a big sphere

So, if we have a method for finding a point in every semi-algebraically connected component of a bounded algebraic set, we obtain immediately, using Proposition 12.34, a method for finding a point in every connected component of an algebraic set. Note that these points have coordinates in the extension $\mathrm{R}\langle\varepsilon\rangle$ rather than in the real closed field R we started with. However, the extension from R to $\mathrm{R}\langle\varepsilon\rangle$ preserves semi-algebraically connected components (Proposition 5.24).

We are going to define X_1-pseudo-critical points of $\mathrm{Zer}(Q, \mathrm{R}^k)$ when $\mathrm{Zer}(Q, \mathrm{R}^k)$ is a bounded algebraic set. These pseudo-critical points are a finite set of points meeting every semi-algebraically connected component of $\mathrm{Zer}(Q, \mathrm{R}^k)$. They are the limits of the critical points of the projection to the X_1 coordinate of a bounded nonsingular algebraic hypersurface defined by a particular infinitesimal perturbation of the polynomial Q. Moreover, the equations defining the critical points of the projection on the X_1 coordinate on the perturbed algebraic set have the special algebraic structure considered in Proposition 12.7.

Given a polynomial $Q \in \mathrm{R}[X_1, ..., X_k]$ we define $\mathrm{tDeg}_{X_i}(Q)$, the **total degree of** Q **in** X_i, as the maximal total degree of the monomials in Q containing the variable X_i.

Notation 12.35. [Deformation] Let $\bar{d} = (\bar{d}_1, ..., \bar{d}_k)$, and c

$$\begin{aligned} G_k(\bar{d}, c) &= c^{\bar{d}_1}(X_1^{\bar{d}_1} + \cdots + X_k^{\bar{d}_k} + X_2^2 + \cdots + X_k^2) - (2k-1), \\ \mathrm{Def}(Q, \zeta) &= \zeta G_k(\bar{d}, c) + (1-\zeta) Q. \end{aligned} \tag{12.17}$$

□

In the next pages, the polynomial $Q \in \mathrm{D}[X_1, \ldots, X_k]$, where D is a ring contained in the real closed field R, and $(d_1, \ldots, d_k)$ satisfy the following conditions:

- $Q(x) \geq 0$ for every $x \in \mathrm{R}^k$,
- $\mathrm{Zer}(Q, \mathrm{R}^k) \subset B(0, 1/c)$ for some $c \leq 1, c \in \mathrm{D}$,
- $d_1 \geq d_2 \cdots \geq d_k$,
- $\deg(Q) \leq d_1$, $\mathrm{tDeg}_{X_i}(Q) \leq d_i$, for $i = 2, \ldots, k$.

Note that $\forall\, x \in B(0, 1/c) \quad G_k(\bar{d}\,, c)(x) < 0$.

Remark 12.36. Note that supposing $Q(x) \geq 0$ for every $x \in \mathrm{R}^k$ is not a big loss of generality since we can always replace Q by Q^2 if it is not the case. Note also that we can always take

$$d_1 = \cdots = d_k = \deg(Q).$$

However considering different d_i will be useful when the degree with respect to some variables is small. □

Let $\bar{d}_i$ be an even number $> d_i, i = 1, \ldots, k$, and $\bar{d} = (\bar{d}_1, \ldots, \bar{d}_k)$.

Let ζ be a variable and $\mathrm{R}\langle\zeta\rangle$ be as usual the field of algebraic Puiseux series in ζ with coefficients in R.

Proposition 12.37.

$$\lim_{\zeta} (\mathrm{Zer}(\mathrm{Def}(Q, \zeta), \mathrm{R}\langle\zeta\rangle^k)) = \mathrm{Zer}(Q, \mathrm{R}^k).$$

Moreover $\mathrm{Zer}(\mathrm{Def}(Q, \zeta), \mathrm{R}\langle\zeta\rangle^k) \subset B(0, 1/c)$.

Proof: Since $\lim_{\zeta}$ is a ring homomorphism from $\mathrm{R}\langle\zeta\rangle_b$ to R, it is clear that $\lim_{\zeta}(\mathrm{Zer}(\mathrm{Def}(Q, \zeta), \mathrm{R}\langle\zeta\rangle^k)) \subset \mathrm{Zer}(Q, \mathrm{R}^k)$. We show that

$$\mathrm{Zer}(Q, \mathrm{R}^k) \subset \lim_{\zeta}(\mathrm{Zer}(\mathrm{Def}(Q, \zeta), \mathrm{R}\langle\zeta\rangle^k)).$$

Let $x \in \mathrm{Zer}(Q, \mathrm{R}^k)$. Since $\mathrm{Zer}(Q, \mathrm{R}^k)$ is bounded, for every $r > 0$ in R there is a $y \in B(x, r)$ such that $Q(y) > 0$. Thus, using Theorem 3.19 (Curve selection lemma), there exists a semi-algebraic path γ from $[0, 1]$ to R^k starting from x such that $Q(\gamma(t)) \neq 0$ for $t \in (0, 1]$. By Theorem 3.20, the set $\gamma([0, 1])$ is a bounded subset of R^k. Denote by $\bar{\gamma}$ the extension of γ to $\mathrm{R}\langle\zeta\rangle$, and note that $\bar{\gamma}([0, 1])$ is a bounded subset of $\mathrm{R}\langle\zeta\rangle^k$, using Proposition 2.87. Since $\mathrm{Def}(Q, \zeta)(\bar{\gamma}(0)) < 0$ and $\mathrm{Def}(Q, \zeta)(\bar{\gamma}(t)) > 0$, for every $t \in \mathrm{R}$, with $0 < t < 1$, there exists $\tau \in \mathrm{R}\langle\zeta\rangle$, $\lim_{\zeta}(\tau) = 0$, such that $\mathrm{Def}(Q, \zeta)(\bar{\gamma}(\tau)) = 0$ by Proposition 3.4. Since $\lim_{\zeta}$ is a ring homomorphism

$$Q(\lim_{\zeta} (\bar{\gamma}(\tau))) = \lim_{\zeta} (Q(\bar{\gamma}(\tau))) = \lim_{\zeta} (\mathrm{Def}(Q, \zeta)(\bar{\gamma}(\tau))) = 0.$$

Since $\lim_{\zeta}(\tau) = 0$, $\gamma(0) = x$, and γ is continuous, we have

$$\lim_{\zeta}(\bar{\gamma}(\tau)) = \gamma(0) = x,$$

using Lemma 3.21. Thus we have found

$$\bar{\gamma}(\tau) \in \mathrm{Zer}(\mathrm{Def}(Q, c), \mathrm{R}\langle\zeta\rangle^k)$$

such that $\lim_\zeta(\bar{\gamma}(\tau)) = x$.

Since

$$\lim_\zeta (\mathrm{Zer}(\mathrm{Def}(Q, \zeta), \mathrm{R}\langle\zeta\rangle^k)) = \mathrm{Zer}(Q, \mathrm{R}^k)$$

and every point $x = (x_1, \ldots, x_k) \in \mathrm{Zer}(Q, \mathrm{R}^k)$ satisfies $x_1^2 + \cdots + x_k^2 < 1/c$, is follows clearly that every point $y = (y_1, \ldots, y_k) \in \mathrm{Zer}(\mathrm{Def}(Q, \zeta), \mathrm{R}\langle\zeta\rangle^k)$ satisfies $y_1^2 + \cdots + y_k^2 < 1/c$. □

Proposition 12.38. *The algebraic set* $\mathrm{Zer}(\mathrm{Def}(Q, \zeta), \mathrm{R}\langle\zeta\rangle^k)$ *is a non-singular algebraic hypersurface bounded over* R.

Proof: The fact that $\mathrm{Zer}(\mathrm{Def}(Q, \zeta), \mathrm{R}\langle\zeta\rangle^k)$ is bounded follows from Proposition 12.37.

To prove that $\mathrm{Zer}(\mathrm{Def}(Q, \zeta), \mathrm{R}\langle\zeta\rangle^k)$ is a non-singular hypersurface, consider the function

$$\Phi(x) = \frac{Q(x)}{Q(x) - G_k(\bar{d}, c)(x)}$$

from $\mathrm{R}^k \setminus \mathrm{Zer}(Q - G_k(\bar{d}, c), \mathrm{R}^k)$ to R. By Sard's Theorem (Theorem 5.56) the set of critical values of Φ is finite. So there is an $a \in \mathrm{R}$, $a > 0$, such that for every $b \in (0, a)$ the function Φ has no critical value.

Since $\mathrm{Zer}(\mathrm{Def}(Q, b), \mathrm{R}^k) \cap \mathrm{Zer}(Q - G_k(\bar{d}, c), \mathrm{R}^k) = \emptyset$,

$$\mathrm{Zer}(\mathrm{Def}(Q, b), \mathrm{R}^k) = \{x \in \mathrm{R}^k \mid \Phi(x) = b\}.$$

The set $\mathrm{Zer}(\mathrm{Def}(Q, b), \mathrm{R}^k)$ is a non-singular algebraic hypersurface, since $\mathrm{Grad}(\mathrm{Def}(Q, b))(x) = 0$ on $\mathrm{Zer}(\mathrm{Def}(Q, b), \mathrm{C}^k)$ implies that $\mathrm{Grad}(\Phi)(x) = 0$. So the formula $\Psi(a)$ defined by

$$\forall b \ \forall x \ \ (0 < b < a \wedge \mathrm{Def}(Q, b)(x) = 0) \Rightarrow \mathrm{Grad}(\mathrm{Def}(Q, b))(x) \neq 0$$

is true in R. Using Theorem 2.80 (Tarski-Seidenberg principle), $\Psi(a)$ is true in $\mathrm{R}\langle\zeta\rangle$ which contains R. Hence, since $0 < \zeta < a$, $\mathrm{Zer}(\mathrm{Def}(Q, \zeta), \mathrm{R}\langle\zeta\rangle^k)$ is a non-singular algebraic hypersurface. □

Notation 12.39. Let $\bar{d} = (\bar{d}_1, \ldots, \bar{d}_k)$, and using Notation 12.35, consider

$$\begin{aligned}
\mathrm{Cr}(Q, \zeta) &= \left\{\mathrm{Def}(Q, \zeta), \frac{\partial \mathrm{Def}(Q, \zeta)}{\partial X_2}, \ldots, \frac{\partial \mathrm{Def}(Q, \zeta)}{\partial X_k}\right\}, \\
\mathrm{Def}_+(Q, \zeta) &= \mathrm{Def}(Q, \zeta) + X_{k+1}^2, \\
\mathrm{Cr}_+(Q, \zeta) &= \left\{\mathrm{Def}(Q, \zeta), \frac{\partial \mathrm{Def}(Q, \zeta)}{\partial X_2}, \ldots, \frac{\partial \mathrm{Def}(Q, \zeta)}{\partial X_k}, 2X_{k+1}\right\}.
\end{aligned}$$

□

Note that

$$\mathrm{Zer}(\mathrm{Cr}(Q, \zeta), \mathrm{R}\langle\zeta\rangle^k)$$

is the set of X_1-critical points on

$$\mathrm{Zer}(\mathrm{Def}(Q, \zeta), \mathrm{R}\langle\zeta\rangle^k)$$

i.e. the critical points on $\mathrm{Zer}(\mathrm{Def}(Q, \zeta), \mathrm{R}\langle\zeta\rangle^k)$ of the projection map to the X_1 coordinate.

The following lemma is easy to prove using the arguments in the proofs of Propositions 12.38, and 12.44.

Lemma 12.40. *The algebraic set* $\mathrm{Zer}(\mathrm{Def}_+(Q, \zeta), \mathrm{R}\langle\zeta\rangle^{k+1})$ *is a non-singular algebraic hypersurface which is bounded over* R. *Moreover,*

$$\lim_{\zeta} (\mathrm{Zer}(\mathrm{Def}_+(Q, \zeta), \mathrm{R}\langle\zeta\rangle^{k+1})) = \mathrm{Zer}(Q, \mathrm{R}^k) \times \{0\},$$

and π *(the projection of* $(x_1, \ldots, x_{k+1}) \in \mathrm{R}\langle\zeta\rangle^{k+1}$ *to* $x_1 \in \mathrm{R}\langle\zeta\rangle$*) has a finite number of critical points on* $\mathrm{Zer}(\mathrm{Def}_+(Q, \zeta), \mathrm{R}\langle\zeta\rangle^{k+1})$.

Note that an X_1-critical point on $\mathrm{Zer}(\mathrm{Def}_+(Q, \zeta,), \mathrm{R}\langle\zeta\rangle^{k+1})$ must have its last coordinate 0 and thus its first k coordinates define an X_1-critical point on $\mathrm{Zer}(\mathrm{Def}(Q, \zeta), \mathrm{R}\langle\zeta\rangle^k)$.

Definition 12.41. An X_1**-pseudo-critical point** on $\mathrm{Zer}(Q, \mathrm{R}^k)$ is the $\lim_\zeta$ of an X_1-critical point on $\mathrm{Zer}(\mathrm{Def}(Q, \zeta), \mathrm{R}\langle\zeta\rangle^k)$.

An X_1**-pseudo-critical value** on $\mathrm{Zer}(Q, \mathrm{R}^k)$ is the projection to the X_1-axis of an X_1-pseudo-critical point on $\mathrm{Zer}(Q, \mathrm{R}^k)$. □

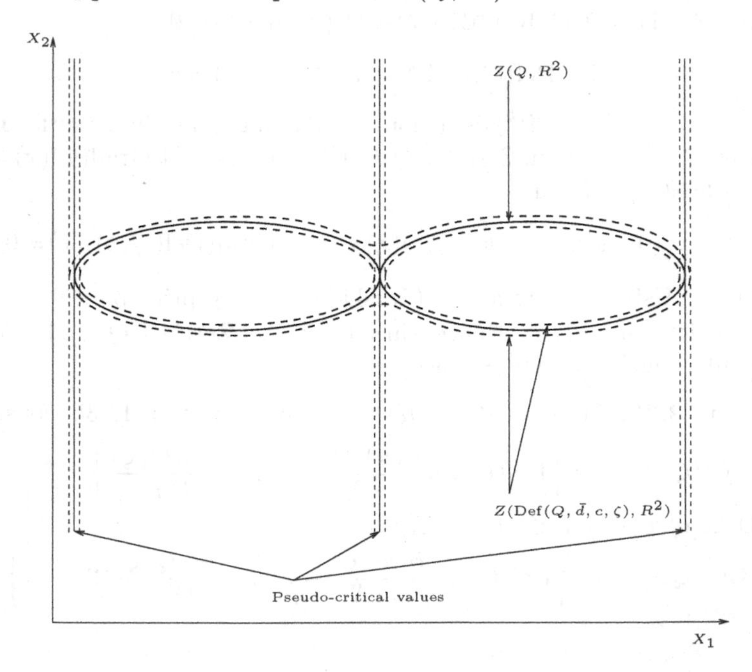

Fig. 12.5. Pseudo-critical values of an algebraic set in R^2

According to Definition 12.41, an X_1-pseudo-critical point of $\mathrm{Zer}(Q, \mathrm{R}^k)$ is the $\lim_\zeta$ of an X_1-critical point on $\mathrm{Zer}(\mathrm{Def}(Q, \zeta), \mathrm{R}\langle\zeta\rangle^k)$, so that an X_1-pseudo-critical point on $\mathrm{Zer}(Q, \mathrm{R}^k)$ is also the $\lim_\zeta$ of an X_1-critical point on $\mathrm{Zer}(\mathrm{Def}_+(Q,\zeta), \mathrm{R}\langle\zeta\rangle^{k+1})$.

Proposition 12.42. *The set of X_1 -pseudo-critical points on $\mathrm{Zer}(Q, \mathrm{R}^k)$ meets every semi-algebraically connected component of $\mathrm{Zer}(Q, \mathrm{R}^k)$.*

The proof of Proposition 12.42 will use the following result.

Proposition 12.43. *If $S' \subset \mathrm{R}\langle\zeta\rangle^k$ is a semi-algebraic set, then $\lim_\zeta (S')$ is a closed semi-algebraic set. Moreover, if $S' \subset \mathrm{R}\langle\zeta\rangle^k$ is a semi-algebraic set bounded over* R *and semi-algebraically connected, then $\lim_\zeta (S')$ is semi-algebraically connected.*

Proof: Using Proposition 2.82, we can suppose that $S' \subset \mathrm{R}\langle\zeta\rangle^k$ is described by a quantifier free formula $\Phi(X, \zeta)$ with coefficients in $\mathrm{R}[\zeta]$. Introduce a new variable X_{k+1} and denote by $\Phi(X, X_{k+1})$ the result of substituting X_{k+1} for ζ in $\Phi(X, \zeta)$. Embed R^k in R^{k+1} by sending X to $(X, 0)$.

We prove that $\lim_\zeta (S') = \overline{T} \cap \mathrm{Zer}(X_{k+1}, \mathrm{R}^{k+1})$, where

$$T = \{(x, x_{k+1}) \in R^{k+1} | \Phi(x, x_{k+1}) \wedge\ x_{k+1} > 0\}$$

and $\overline{T}$ is the closure of T. If $x \in \lim_\zeta (S')$, then there exists $z \in S'$ such that $\lim_\zeta (z) = x$. Since (z, ζ) belongs to the extension of $B(x, r) \cap T$ to $\mathrm{R}\langle\zeta\rangle$, it follows that $B(x, r) \cap T$ is non-empty for every $r \in \mathrm{R}$, $r > 0$, and hence that $x \in \overline{T}$. Conversely, let x be in $\overline{T} \cap \mathrm{Zer}(X_{k+1}, \mathrm{R}^{k+1})$. For every $r \in \mathrm{R}$, with $r > 0$, $B(x, r) \cap T \cap \mathrm{Zer}(X_{k+1}, \mathrm{R}^{k+1})$ is non-empty, and hence, according to Theorem 2.80, $B(x, \zeta) \cap \mathrm{Ext}(T, \mathrm{R}\langle\zeta\rangle) \cap \mathrm{Zer}(X_{k+1}, \mathrm{R}\langle\zeta\rangle^{k+1})$ is non-empty and contains an element z. It is clear that $\lim_\zeta (z) = x$.

If S' is bounded over R by M and semi-algebraically connected, then, by Theorem 5.46 (Semi-algebraic triviality), there exists a positive t in R such that $T_{(0,2t)}$ is semi-algebraically homeomorphic to $T_t \times (0, 2t)$. Thus $\mathrm{Ext}(T_t, \mathrm{R}\langle\zeta\rangle)$ is semi-algebraically homeomorphic to $T_\zeta = S$, which is semi-algebraically connected. Thus T_t and $T_{(0,t)}$ are semi-algebraically connected. It follows that

$$S = \overline{T} \cap \mathrm{Zer}(X_{k+1}, \mathrm{R}^{k+1}) = \overline{T} \cap (B(0, M) \times [0, t]) \cap \mathrm{Zer}(X_{k+1}, \mathrm{R}^{k+1})$$

is semi-algebraically connected. □

Proof of Proposition 12.42: The proposition follows from

$$\lim_\zeta (\mathrm{Zer}(\mathrm{Def}(Q, \zeta), \mathrm{R}\langle\zeta\rangle^k)) = \mathrm{Zer}(Q, \mathrm{R}^k),$$

since $\mathrm{Zer}(\mathrm{Cr}(Q, \zeta), \mathrm{R}\langle\zeta\rangle^k)$ meets every connected component of

$$\mathrm{Zer}(\mathrm{Def}(Q, \zeta), \mathrm{R}\langle\zeta\rangle^k)$$

by Proposition 7.4 and the image of a bounded semi-algebraically connected semi-algebraic set under $\lim_\zeta$ is again semi-algebraically connected by Proposition 12.43. □

Moreover, the polynomial system $\mathrm{Cr}(Q, \zeta)$ has good algebraic properties.

Proposition 12.44.

a) The polynomial system $\mathrm{Cr}(Q, \zeta)$ *is a Gröbner basis for the graded lexicographical ordering with* $X_1 >_{\mathrm{grlex}} \cdots >_{\mathrm{grlex}} X_k$.
b) The set $\mathrm{Zer}(\mathrm{Cr}(Q, \zeta), \mathrm{C}\langle\zeta\rangle^k)$ *is finite.*
c) The zeros of the polynomial system $\mathrm{Cr}(Q, \zeta)$ *are simple.*

For the proof of the proposition, we need the following lemma.

Lemma 12.45. *The polynomial system*

$$\mathrm{Cr}(Q, 1) = \left\{ G_k(\bar{d}, c), \frac{\partial G_k(\bar{d}, c)}{\partial X_2}, \ldots, \frac{\partial G_k(\bar{d}, c)}{\partial X_k} \right\}$$

has a finite number of zeros in C^k *all of which are simple.*

Proof: Since

$$\frac{\partial G_k(\bar{d}, c)}{\partial X_i} = c^{\bar{d}_1} (\bar{d}_i X_i^{\bar{d}_i - 1} + 2 X_i),$$

for $i > 1$, and the zeros of $\bar{d}_i X_i^{\bar{d}_i - 1} + 2 X_i$ in C are simple, the zeros of

$$\left\{ \frac{\partial G_k(\bar{d}, c)}{\partial X_2}, \ldots, \frac{\partial G_k(\bar{d}, c)}{\partial X_k} \right\}$$

in C^{k-1} are simple and finite in number. A zero of $\mathrm{Cr}(Q, 1)$ in C^k corresponds to a zero $(x_2, \ldots, x_k)$ of

$$\frac{\partial G_k(\bar{d}, c)}{\partial X_2}, \ldots, \frac{\partial G_k(\bar{d}, c)}{\partial X_k}$$

in C^{k-1} and a zero of $G_k(\bar{d}, c)(X_1, x_2, \ldots, x_k)$ in C. Since x_i, $i = 2, \ldots, k$, has norm less than 1 and $c \leq 1$,

$$G_k(\bar{d}, c)(X_1, x_2, \ldots, x_k) = c^{\bar{d}_1} X_1^{\bar{d}_1} + a,$$

with a non-zero, has a finite number of zeros, and all its zeros are simple. This proves the claim. □

Proof of Proposition 12.44: The polynomial system $\mathrm{Cr}(Q, \zeta)$ is a Gröbner basis for the graded lexicographical ordering according to Proposition 12.3. The set $\mathrm{Zer}(\mathrm{Cr}(Q, \zeta), \mathrm{C}\langle\zeta\rangle^k)$ is finite according to Corollary 12.9. Consider, for every $b \neq 0 \in \mathrm{C}$,

$$\mathrm{Def}(Q, b) = b\, G_k(\bar{d}, c) + (1 - b)\, Q.$$

The polynomial system

$$\mathrm{Cr}(Q,b)=\left\{\mathrm{Def}(Q,b),\frac{\partial\mathrm{Def}(Q,b)}{\partial X_2},\ldots,\frac{\partial\mathrm{Def}(Q,b)}{\partial X_k}\right\}$$

is a Gröbner basis for the graded lexicographical ordering according to Proposition 12.3. The set $\mathrm{Zer}(\mathrm{Cr}(Q,b),\mathrm{C}^k)$ is finite according to Corollary 12.9. We denote by A_b the finite dimensional vector space

$$\mathrm{A}_b=\mathrm{R}[X_1,\ldots,X_k]/\mathrm{Ideal}(\mathrm{Cr}(Q,b),\mathrm{R}).$$

Let a be a separating element of $\mathrm{Zer}(\mathrm{Cr}(Q,1),\mathrm{C}^k)$. According to Lemma 12.45, the zeros of $\mathrm{Cr}(Q,1)$ are simple, thus the characteristic polynomial $\chi_1(T)$ of the linear map L_a from A_1 to A_1 has only simple roots by Proposition 12.16.

Denoting the characteristic polynomial of the linear map L_a from A_b to A_b by $\chi_b(T)$,

$$B=\{b\in\mathrm{C}\mid b=0 \text{ or } b\neq 0 \text{ and } \mathrm{Disc}_T(\chi_b(T))=0\}$$

is an algebraic subset of C which does not contain 1 and is thus finite (see Exercise 1.1). It is clear that $\zeta\notin\mathrm{Ext}(B,\mathrm{C}\langle\zeta\rangle)$ (see Exercise 1.12). So, $\mathrm{Disc}_T(\chi_\zeta(T))\neq 0$, and by Proposition 4.18, the characteristic polynomial of the linear map L_a from

$$\mathrm{A}_\zeta=\mathrm{R}\langle\zeta\rangle[X_1,\ldots,X_k]/\mathrm{Ideal}(\mathrm{Cr}(Q,\zeta),\mathrm{R}\langle\zeta\rangle)$$

to A_ζ has only simple zeros. Hence, by Theorem 4.97 (Stickelberger), the zeros of $\mathrm{Cr}(Q,\zeta)$ are simple. □

Notation 12.46. We need to modify slightly the polynomial system

$$\mathrm{Cr}(Q,\zeta)=\left\{\mathrm{Def}(Q,\zeta),\frac{\partial\mathrm{Def}(Q,\zeta)}{\partial X_2},\ldots,\frac{\partial\mathrm{Def}(Q,\zeta)}{\partial X_k}\right\}$$

defined in Notation 12.35 in order to obtain a special Gröbner basis.

Note that defining Q_i, $1<i\le k$, by

$$\frac{\partial\mathrm{Def}(Q,\zeta)}{\partial X_i}=\bar{d}_i\zeta c^{\bar{d}_1}X_i^{\bar{d}_i-1}+Q_i,$$

we have $\deg(Q_i)<\bar{d}_i-1$, $\deg_{X_j}(Q_i)<\bar{d}_j-1, j\neq i, 1\le j\le k$, so that $\mathrm{Cr}(Q,\zeta)$ is nearly a special Gröbner basis. The only properties that are not satisfied are that, defining R by $\mathrm{Def}(Q,\zeta)=\zeta c^{\bar{d}_1}X_1^{\bar{d}_1}+R$, we do not have $\deg(R)<\bar{d}_1$, and $\deg_{X_j}(R)<\bar{d}_j-1, 2\le j\le k$. With $d=\bar{d}_2\ldots\bar{d}_k$, we only have to reduce

$$d^2\zeta^{2k-3}c^{(2k-3)\bar{d}_1}\mathrm{Def}(Q,c)$$

twice modulo each polynomial

$$\frac{\partial\mathrm{Def}(Q,\zeta)}{\partial X_2},\ldots,\frac{\partial\mathrm{Def}(Q,\zeta)}{\partial X_k}$$

to obtain a polynomial

$$\overline{\mathrm{Def}}(Q,\zeta)=b\,X_1^{\bar{d}_1}+R_1\in\mathrm{D}[X_1,\ldots,X_k]$$

with $\deg(R_1) < \bar{d}_i - 1$, $\deg_{X_j}(R_1) < \bar{d}_j - 1, j \neq 1$.

Let

$$\overline{\mathrm{Cr}}(Q,\zeta) = \left\{ \overline{\mathrm{Def}}(Q,\zeta), \frac{\partial \mathrm{Def}(Q,\zeta)}{\partial X_2}, \ldots, \frac{\partial \mathrm{Def}(Q,\zeta)}{\partial X_k} \right\}.$$

□

It is clear that $\overline{\mathrm{Cr}}(Q,\zeta)$ is a special Gröbner basis.

Note that $\mathrm{Cr}(Q,\zeta)$ and $\overline{\mathrm{Cr}}(Q,\zeta)$ have the same set of zeros.

We are now ready to describe an algorithm giving a point in every connected component of a bounded algebraic set. We simply compute pseudo-critical values and their limits.

Algorithm 12.16. **[Bounded Algebraic Sampling]**

- **Structure**: an ordered integral domain D contained in a real closed field R.
- **Input**: a polynomial $Q \in \mathrm{D}[X_1, \ldots, X_k]$ such that $Q(x) \geq 0$ for every $x \in \mathrm{R}^k$ and such that $\mathrm{Zer}(Q, \mathrm{R}^k)$ is contained in $B(0, 1/c)$.
- **Output**: a set $\mathcal{U}$ of real univariate representations of the form

$$(f(T), g(T), \sigma), \text{with } (f, g) \in \mathrm{D}[T]^{k+2}$$

  The set of points associated to these univariate representations meets every semi-algebraically connected component of $\mathrm{Zer}(Q, \mathrm{R}^k)$ and contains the set of X_1-pseudo-critical points on $\mathrm{Zer}(Q, \mathrm{R}^k)$.
- **Complexity:** $d^{O(k)}$, where d is a bound on the degree of Q.
- **Procedure**:
  - Choose $(d_1, \ldots, d_k)$ such that $d_1 \geq \ldots \geq d_k$, $\deg(Q) \leq d_1$, $\mathrm{tDeg}_{X_i}(Q) \leq d_i$, for $i = 2, \ldots, k$. Take as $\bar{d}_i$ the smallest even number $> d_i, i = 1, \ldots, k$, and $\bar{d} = (\bar{d}_1, \ldots, \bar{d}_k)$.
  - Compute $\overline{\mathrm{Cr}}(Q,\zeta)$ (Notation 12.46).
  - Compute the multiplication table $\mathcal{M}$ of $\overline{\mathrm{Cr}}(Q,\zeta)$ by Algorithm 12.9 (Special Multiplication Table).
  - Apply the $\lim_\zeta$ map using Algorithm 12.14 (Limit of Real Bounded Points) with input $\mathcal{M}$, and obtain a set $\mathcal{U}$ of real univariate representations v with

$$v = (f(T), g(T), \sigma), (f(T), g(T)) \in \mathrm{D}[T]^{k+2}.\}$$

Proof of correctness: This follows from Proposition 12.42 and the correctness of Algorithm 12.9 (Special Multiplication Table) and Algorithm 12.14 (Limit of Real Bounded Points). □

Complexity analysis: Using the complexity analysis of Algorithm 12.9 (Special Multiplication Table) and Algorithm 12.14 (Limit of Real Bounded Points), the complexity is $(d_1 \ldots d_k)^{O(1)}$ in the ring D. The polynomials output are of degree $O(d_1) \ldots O(d_k)$ in T.

When $D=\mathbb{Z}$, and the bitsizes of the coefficients of Q are bounded by τ, the bitsizes of the coefficients of the polynomials occurring in the computations of the multiplication table and its output are

$$O(d_1+\cdots+d_{k-1})k\,d_1(\tau+\nu'),$$

where ν' is the bitsize of $O(d_1+\cdots+d_k)$, according to the complexity analysis of Algorithm 12.10 (Parametrized Special Multiplication Table).

Finally the complexity is $d^{O(k)}$, the degree of the univariate representations output are $O(d)^k$ and the bitsizes of the output are bounded by $\tau d^{O(k)}$. □

Algorithm 12.17. **[Algebraic Sampling]**

- **Structure:** an ordered integral domain D contained in a real closed field R.
- **Input**: a polynomial $Q\in D[X_1,\ldots,X_k]$.
- **Output**: a set $\mathcal{U}$ of real univariate representations of the form

  $$(f,g,\sigma),\text{with}\,(f,g)\in D[\varepsilon][T]^{k+2}.$$

  The set of points associated to these univariate representations meets every semi-algebraically connected component of $\mathrm{Zer}(Q,R\langle\varepsilon\rangle^k)$.
- **Complexity:** $d^{O(k)}$, where d is a bound on the degree of Q.
- **Procedure**:
  - Define

    $$R:=Q^2+(\varepsilon(X_1^2+\cdots+X_{k+1}^2)-1)^2.$$

  - Apply Algorithm 12.16 (Bounded Algebraic Sampling) to R, and obtain a set $\mathcal{V}$ of real univariate representations v with

    $$v=(f(T),h(T),\sigma),\text{with}\,(f,h)\in D[\varepsilon][T]^{k+2}.$$

    Define $\pi(v)$ by u, with

    $$u=(f(T),h_0(T),\ldots,h_k(T),\sigma).$$

    and $\mathcal{U}=\pi(\mathcal{V})$.

Proof of correctness: This follows from Proposition 12.34 and the correctness of Algorithm 12.16 (Bounded Algebraic Sampling). □

Complexity analysis: Using the complexity analysis of Algorithm 12.16 (Bounded Algebraic Sampling), and since the degree of R with respect to X_{k+1} is 4, the complexity is $(d_1\ldots d_k)^{O(1)}$ in the ring $D[\varepsilon]$. The polynomials output are of degree $O(d_1)\ldots O(d_k)$ in T. Moreover the degrees with respect to ε occurring in the computations of the multiplication table are bounded by

$$O(d_1+\cdots+d_{k-1})k\,d_k,$$

according to the multiplicity analysis of Algorithm 12.10 (Parametrized Special Multiplication Table).

When $D=\mathbb{Z}$, and the bitsizes of the coefficients of Q are bounded by τ, the bitsizes of the coefficients of the polynomials occurring in the computations of the multiplication table and its output are

$$O(d_1+\cdots+d_{k-1})k\,d_1(\tau+\nu'),$$

where ν' is the bitsize of $O(d_1+\cdots+d_k)$, according to the complexity analysis of Algorithm 12.10 (Parametrized Special Multiplication Table).

Finally the complexity is $d^{O(k)}$, the degree of the univariate representations output in T and ε are $O(d)^k$ and the bitsizes of the output are bounded by $d^{O(k)}$. □

The following parametrized version of Algorithm 12.16 (Bounded Algebraic Sampling) will be useful in later chapters.

Algorithm 12.18. **[Parametrized Bounded Algebraic Sampling]**

- **Structure**: an ordered integral domain D.
- **Input**: a polynomial $Q\in D[Y,X_1,\ldots,X_k]$, such that $Q(y,x)\geq 0$ for every $x\in R^k$, $y\in R^\ell$, and for every $y\in R^\ell$ $\mathrm{Zer}(Q(y),R^k)$ is contained in $B(0,1/c)$.
- **Output**: a set $\mathcal{U}$ of parametrized univariate representations of the form

$$(f,g)\in D[Y,T]^{k+2}.$$

  For every $y\in R^\ell$, the set of points associated to these univariate representations meets every semi-algebraically connected component of $\mathrm{Zer}(Q(y),R^k)$ and contains the set of X_1-pseudo-critical points on $\mathrm{Zer}(Q(y),R^k)$.
- **Complexity:** $(\lambda d^k)^{O(\ell)}$, where d is a bound on the degree of Q with respect to X and λ is a bound on the degree of Q with respect to Y.
- **Procedure**:
  - Choose $(d_1,\ldots,d_k)$ such that $d_1\geq\ldots\geq d_k$, $\deg(Q)\leq d_1$, $\mathrm{tDeg}_{X_i}(Q)\leq d_i$, for $i=2,\ldots,k$. Take as $\bar{d}_i$ the smallest even number $>d$, $i=1,\ldots,k$, and $\bar{d}=(\bar{d}_1,\ldots,\bar{d}_k)$.
  - Consider $\overline{\mathrm{Cr}(Q,\zeta)}$, using Notation 12.46.
  - Compute the parametrized multiplication table $\mathcal{M}$ of $\overline{\mathrm{Cr}}(Q,\zeta)$ by Algorithm 12.10 (Parametrized Special Multiplication Table).
  - Apply Algorithm 12.15 (Parametrized Limit of Bounded Points) with input $\mathcal{M}$ and ζ and obtain a set $\mathcal{U}$ of parametrized univariate representations (v,σ) with

$$u=f(T),g(T)\in D[Y,T]^{k+2}.$$

Proof of correctness: Follows from Proposition 12.42, and the correctness of Algorithm 12.10 (Parametrized Special Multiplication Table) and Algorithm 12.15 (Parametrized Limit of Bounded Points). □

Complexity analysis: Using the complexity analysis of Algorithm 12.10 (Parametrized Special Multiplication Table) and Algorithm 12.15 (Parametrized Limit of Bounded Points), the complexity is $(d_1..., d_k)^{O(1)}$ in the ring $\mathrm{D}[Y]$. The polynomials output are of degree $O(d_1)...O(d_k)$ in T and, if λ is a bound on the total degree in $Y=(Y_1, ..., Y_\ell)$ of Q, of degrees $\lambda(d_1...d_k)^{O(1)}$ in Y. Finally, the complexity is $(\lambda d_1...d_k)^{O(\ell)}$ in the ring D. The number of elements of $\mathcal{U}$ is $O(d_1)...O(d_k)$.

When $\mathrm{D}=\mathbb{Z}$, and the bitsizes of the coefficients of Q are bounded by τ, the bitsizes of the coefficients of the polynomials occurring in the computations of the multiplication table and its output are

$$O(d_1+\cdots+d_{k-1})k\,d_1(\tau+\nu'),$$

where ν' is the bitsize of $O(\lambda(d_1+\cdots+d_k))^{\ell+1}$, according to the complexity analysis of 12.10 (Parametrized Special Multiplication Table). Finally the complexity is $(\lambda d^k)^{O(\ell)}$, the degree of the univariate representations output are $O(d)^k$ and the bitsizes of the output are bounded by $O(k^2d^2(\tau+\ell\log_2(k\,d)))$. □

12.7 Triangular Sign Determination

We now give algorithms for sign determination and Thom's encodings of triangular systems (Definition 11.3). These algorithms have a slightly better complexity than the similar recursive ones in Chapter 11 and will also be easier to generalize to a parametrized setting in later chapters.

We need a notation: let $u=(f, g)\in \mathrm{K}[T]^{k+2}$, $g=(g_0, ..., g_k)$ be a k-univariate representation and $Q\in \mathrm{K}[X_1, ..., X_k]$. Set

$$Q_u=g_0^e\,Q\left(\frac{g_k}{g_0}, ..., \frac{g_k}{g_0}\right), \tag{12.18}$$

where e is the least even number not less than the degree of Q.

Algorithm 12.19. **[Triangular Sign Determination]**

- **Structure:** an ordered integral domain D contained in a real closed field K.
- **Input:** a triangular system $\mathcal{T}$, and a list $\mathcal{Q}$ of elements of $\mathrm{D}[X_1, ..., X_k]$. Denote by $Z=\mathrm{Zer}(\mathcal{T}, \mathrm{R}^k)$.
- **Output:** the set $\mathrm{SIGN}(\mathcal{Q}, \mathcal{T})$ of sign conditions realized by $\mathcal{Q}$ on Z.
- **Complexity:** $s\,d'^{O(k)}\,d^{O(1)}$, where s is a bound on the number of elements of $\mathcal{Q}$, d' is a bound on the degrees on the elements of $\mathcal{T}$, and d is a bound on the degrees of $Q\in\mathcal{Q}$.

- **Procedure:** Apply Algorithm 12.16 (Bounded Algebraic Sampling) to $P = \sum_{A \in \mathcal{T}} A^2$. Let $\mathcal{U}$ be the set of real univariate representations output. Keep those $(u, \sigma) \in \mathcal{U}$, with $u = (f, g_0, g_1, \ldots, g_k)$, such that P_u is zero at the root t_σ of f with Thom encoding σ, using Algorithm 10.15 (Sign at the Root in a real closed field). For every such real univariate representation (u, σ) and for every $Q \in \mathcal{Q}$, compute the sign of Q_u at t_σ, using Algorithm 10.15 (Sign at the Roots in a real closed field).

Complexity analysis: Let d' be a bound on the degrees of P_i, d a bound on the degrees of $Q \in \mathcal{Q}$, and let s be a bound on the cardinality of $\mathcal{Q}$. The number of arithmetic operations in D is $s\, d'^{O(k)}\, d^{O(1)}$, using the complexity analysis of Algorithm 12.16 (Bounded Algebraic Sampling) and Algorithm 10.15 (Sign at the Roots in a real closed field).

When $\mathrm{D} = \mathbb{Z}$, and the bitsizes of the coefficients of $\mathcal{T}$ and $\mathcal{Q}$ are bounded by τ, the bitsizes of the intermediate computations and the output are bounded by $\tau\, d'^{O(k)}\, d^{O(1)}$, using the complexity analysis of Algorithm 12.16 (Bounded Algebraic Sampling) and Algorithm 10.15 (Sign at the Roots in a real closed field). □

Algorithm 12.20. **[Triangular Thom Encoding]**

- **Structure:** an ordered integral domain D contained in a real closed field R.
- **Input:** a zero-dimensional system of equations $\mathcal{T}$.
- **Output:** the list $\mathrm{Thom}(\mathcal{T})$ of Thom encodings of the roots of $\mathcal{T}$ (Definition).
- **Complexity:** $d^{O(k)}$, where d is a bound on the degrees of $P \in \mathcal{T}$.
- **Procedure:** Apply Algorithm 12.19 (Triangular Sign Determination) to $\mathcal{T}$ and $\mathrm{Der}(\mathcal{T})$.

Complexity analysis: Since there are $k\,d$ polynomials of degrees bounded by d, in $\mathrm{Der}(\mathcal{T})$, the number of arithmetic operations in D is $d^{O(k)}$, using the complexity analysis of Algorithm 12.19 (Triangular Sign Determination).

When $\mathrm{D} = \mathbb{Z}$, and the bitsizes of the coefficients of $\mathcal{T}$ are bounded by τ, the bitsizes of the intermediate computations and the output are bounded by $\tau d^{O(k)}$, using the complexity analysis of Algorithm 12.19 (Triangular Sign Determination). □

Algorithm 12.21. **[Triangular Comparison of Roots]**

- **Structure:** an ordered integral domain D, contained in a real closed field R.
- **Input:** a Thom encoding $\mathcal{T}$, σ specifying $z \in \mathrm{R}^{k-1}$, and two non-zero polynomials P and Q in $\mathrm{D}[X_1, \ldots, X_k]$.
- **Output:** the ordered list of the Thom encodings of the roots of P and Q above σ (Definition 11.6).

- **Complexity:** $d^{O(k)}$, where d is a bound on the degrees of P, Q and the polynomials in $\mathcal{T}$.
- **Procedure:** Apply Algorithm 12.19 (Triangular Sign Determination) to $\mathcal{T}, P$, $\mathrm{Der}(\mathcal{T}) \cup \mathrm{Der}(P) \cup \mathrm{Der}(Q)$, then to $\mathcal{T}, Q, \mathrm{Der}(\mathcal{T}) \cup \mathrm{Der}(Q) \cup \mathrm{Der}(P)$. Compare the roots using Proposition 2.28.

Complexity analysis: Since there are $(k+1)d$ polynomials of degrees bounded by d in $\mathrm{Der}(\mathcal{T}) \cup \mathrm{Der}(P) \cup \mathrm{Der}(Q)$, the number of arithmetic operations in D is $d^{O(k)}$, using the complexity analysis of Algorithm 12.19 (Triangular Sign Determination).

When $\mathrm{D} = \mathbb{Z}$, and the bitsizes of the coefficients of $\mathcal{T}$, P and Q are bounded by τ, the bitsizes of the intermediate computations and the output are bounded by $\tau\, d^{O(k)}$, using the complexity analysis of Algorithm 12.19 (Triangular Sign Determination). □

We can also construct points between two consecutive roots.

Algorithm 12.22. **[Triangular Intermediate Points]**

- **Structure:** an ordered integral domain D contained in a real closed field R.
- **Input:** a Thom encoding $\mathcal{T}$, σ specifying $z \in \mathrm{R}^{k-1}$, and two non-zero polynomials P and Q in $\mathrm{D}[X_1, \ldots, X_k]$.
- **Output:** Thom encodings specifying values y intersecting intervals between two consecutive roots of $P(z, X_k)$ and $Q(z, X_k)$.
- **Complexity:** $d^{O(k)}$, where d is a bound on the degrees of P, Q and the polynomials in $\mathcal{T}$.
- **Procedure:** Compute the Thom encodings of the roots of
$$\partial(PQ)/\partial X_k(z, X_k)$$
above $\mathcal{T}$, σ using Algorithm 12.20 (Triangular Thom Encoding) and compare them to the roots of P and Q above σ using Algorithm 12.21 (Triangular Comparison of Roots). Keep one intermediate point between two consecutive roots of PQ.

Complexity analysis: Then the degree of $\partial(PQ)/\partial X_k(z, X_k)$ is $O(d)$. Using the complexity of Algorithms 12.20 (Triangular Thom Encoding) and Algorithm 12.21 (Triangular Comparison of Roots), the complexity is $d^{O(k)}$.

When $\mathrm{D} = \mathbb{Z}$, and the bitsizes of the coefficients of $\mathcal{T}$ P and Q are bounded by τ, the bitsizes of the intermediate computations and the output are bounded by $\tau\, d^{O(k)}$, using the complexity analysis of Algorithms 12.20 (Triangular Thom Encoding) and Algorithm 12.21 (Triangular Comparison of Roots). □

Finally we can compute sample points on a line.

Algorithm 12.23. **[Triangular Sample Points]**

- **Structure:** an ordered integral domain D contained in a real closed field R.
- **Input:** a Thom encoding $\mathcal{T}$, σ specifying $z \in \mathrm{R}^{k-1}$, and a family of polynomials $\mathcal{P} \subset \mathrm{D}[X_1, \ldots, X_k]$.
- **Output:** an ordered list L of Thom encodings specifying the roots in R of the non-zero polynomials $P(z, X_k)$, $P \in \mathcal{P}$, an element between two such consecutive roots, an element of R smaller than all these roots, and an element of R greater than all these roots. Moreover (τ_1) appears before (τ_2) in L if and only if $x_k(\tau_1) \leq x_k(\tau_2)$. The sign of $Q(z, x_k(\tau))$ is also output for every $Q \in \mathcal{P}, \tau \in L$.
- **Complexity:** $s^2 d^{O(k)}$, where s is a bound on the number of elements of $\mathcal{P}$ and d is a bound on the degrees of the polynomials in $\mathcal{T}$ and $\mathcal{Q}$.
- **Procedure:** Characterize the roots of the polynomials in R using Algorithm 12.20 (Triangular Thom Encoding). Compare these roots using Algorithm 12.21 (Triangular Comparison of Roots) for every pair of polynomials in $\mathcal{P}$. Characterize a point in each interval between the roots by Algorithm 12.22 (Triangular Intermediate Points). Use Proposition 10.5 to find an element of R smaller and bigger than any root of any polynomial in $\mathcal{P}$ above z.

Complexity analysis: Using the complexity analyses of Algorithm 12.20 (Triangular Thom Encoding), Algorithm 12.21 (Triangular Comparison of Roots) and Algorithm 12.22 (Triangular Intermediate Points), the complexity is $s^2 d^{O(k)}$.

When $\mathrm{D} = \mathbb{Z}$, and the bitsizes of the coefficients of $\mathcal{T}$ and $\mathcal{P}$ are bounded by τ, the bitsizes of the intermediate computations and the output are bounded by $\tau d^{O(k)}$, using the complexity analysis of Algorithm 12.20 (Triangular Thom Encoding), Algorithm 12.21 (Triangular Comparison of Roots) and Algorithm 12.22 (Triangular Intermediate Points). □

12.8 Computing the Euler-Poincaré Characteristic of an Algebraic Set

In this section we first describe an algorithm for computing the Euler-Poincaré characteristic of an algebraic set. The complexity of this algorithm is asymptotically the same as that of Algorithm 12.16 (Bounded Algebraic Sampling) for computing sample points in every connected component of a bounded algebraic set described in the Section 12.6.

We first describe an algorithm for computing the Euler-Poincaré characteristic of a bounded algebraic set and then use this algorithm for computing the Euler-Poincaré characteristic of a general algebraic set.

From now on we consider a polynomial $Q \in \mathrm{D}[X_1, \ldots, X_k]$, where D is a ring contained in the real closed field R, satisfying $\mathrm{Zer}(Q, \mathrm{R}^k) \subset B(0, 1/c)$ for some $0 < c \leq 1, c \in \mathrm{D}$. Let $\bar{d} = (\bar{d}_1, \ldots, \bar{d}_k)$ with $\bar{d}_i$ even and $\bar{d}_i > d_i$, where d_i is the total degree of Q^2 in X_i.

Notation 12.47. We denote

$$\begin{aligned}
G_k(\bar{d}, c) &= c^{\bar{d}_1}(X_1^{\bar{d}_1} + \cdots + X_k^{\bar{d}_k} + X_2^2 + \cdots + X_k^2) - (2k-1),\\
\mathrm{Def}(Q^2, \zeta) &= \zeta G_k(\bar{d}, c) + (1-\zeta) Q^2,\\
\mathrm{Def}_+(Q^2, \zeta) &= \mathrm{Def}(Q^2, \zeta) + X_{k+1}^2,\\
\mathrm{Cr}(Q^2, \zeta) &= \left\{\mathrm{Def}(Q^2, \zeta), \frac{\partial \mathrm{Def}(Q^2, \zeta)}{\partial X_2}, \ldots, \frac{\partial \mathrm{Def}(Q^2, \zeta)}{\partial X_k}\right\},\\
\mathrm{Cr}_+(Q^2, \zeta) &= \left\{\mathrm{Def}_+(Q^2, \zeta), \frac{\partial \mathrm{Def}_+(Q^2, \zeta)}{\partial X_2}, \ldots, \frac{\partial \mathrm{Def}_+(Q^2, \zeta)}{\partial X_k}, 2X_{k+1}\right\},\\
\overline{\mathrm{Cr}}(Q^2, \zeta) &= \left\{\overline{\mathrm{Def}}(Q^2, \zeta), \frac{\partial \mathrm{Def}(Q^2, \zeta)}{\partial X_2}, \ldots, \frac{\partial \mathrm{Def}(Q^2, \zeta)}{\partial X_k}\right\},\\
\overline{\mathrm{Cr}_+}(Q^2, \zeta) &= \left\{\overline{\mathrm{Def}_+}(Q^2, \zeta), \frac{\partial \mathrm{Def}_+(Q^2, \zeta)}{\partial X_2}, \ldots, \frac{\partial \mathrm{Def}_+(Q^2, \zeta)}{\partial X_k}\right\}
\end{aligned}$$

where, $\overline{\mathrm{Def}_+}(Q^2, \zeta)$ is obtained from $\mathrm{Def}_+(Q^2, \zeta)$ as in Notation 12.46. □

It is clear that $\overline{\mathrm{Cr}}(Q^2, \zeta)$ as well as $\overline{\mathrm{Cr}_+}(Q^2, \zeta)$ are both special Gröbner bases. Note that $\mathrm{Cr}(Q^2, \zeta)$ and $\overline{\mathrm{Cr}}(Q^2, \zeta)$ (resp. $\mathrm{Cr}_+(Q^2, \zeta)$ and $\overline{\mathrm{Cr}_+}(Q^2, \zeta)$) have the same set of zeros.

Algorithm 12.24. **[Euler-Poincaré Characteristic of a Bounded Algebraic Set]**

- **Structure**: an ordered domain D contained in a real closed field R.
- **Input**: a polynomial $Q \in \mathrm{D}[X_1, \ldots, X_k]$ for which $\mathrm{Zer}(Q, \mathrm{R}^k) \subset B(0, 1/c)$.
- **Output**: the Euler-Poincaré characteristic $\chi(\mathrm{Zer}(Q, \mathrm{R}^k))$.
- **Complexity:** $d^{O(k)}$, where d is a bound on the degree of Q.
- **Procedure**:
  - Choose $(d_1, \ldots, d_k)$ such that $d_1 \geq \ldots \geq d_k$, $\deg(Q^2) \leq d_1$, and $\mathrm{tDeg}_{X_i}(Q^2) \leq d_i$, for $i = 2, \ldots, k$. Take $\bar{d}_i$ the smallest even number $> d_i, i = 1, \ldots, k$, and $\bar{d} = (\bar{d}_1, \ldots, \bar{d}_k)$.
  - Consider $\overline{\mathrm{Cr}}(Q^2, \zeta)$ and $\overline{\mathrm{Cr}_+}(Q^2, \zeta)$, using Notation 12.47.
  - Compute the multiplication tables $\mathcal{M}$ and $\mathcal{M}_+$ of $\overline{\mathrm{Cr}}(Q^2, \zeta)$ and $\overline{\mathrm{Cr}_+}(Q^2, \zeta)$ using Algorithm 12.10 (Parametrized Special Multiplication Table), with parameter ζ.
  - Compute the characteristic polynomial of the matrices

$$H_1 = \left[\frac{\partial^2 \mathrm{Def}(Q^2, \zeta)}{\partial X_i \partial X_j}\right]_{2 \leq i,j \leq k}$$

and,

$$H_2 = \left[\frac{\partial^2 \mathrm{Def}_+(Q^2,\zeta)}{\partial X_i \partial X_j}\right]_{2\le i,j\le k+1}$$

using Algorithm 8.17 (Characteristic Polynomial).

- Compute the signature $\mathrm{Sign}(H_1)$ (resp. $\mathrm{Sign}(H_2)$), of the matrix H_1 (resp. H_2) at the real roots of $\mathrm{Cr}(Q^2)$ (resp. $\mathrm{Cr}_+(Q^2)$) using Algorithm 12.8 (Multivariate Sign Determination) with input the list of coefficients of the characteristic polynomial of H_1 (resp. H_2) and the multiplication table $\mathcal{M}$ (resp. $\mathcal{M}_+$) for the zero-dimensional system $\overline{\mathrm{Cr}}(Q^2)$ (resp. $\overline{\mathrm{Cr}_+}(Q^2)$), to determine the signs of the coefficients of characteristic polynomials of H_1 (resp. H_2) at the real roots of the corresponding system.
- For i from 0 to $k-1$ let,

$$\ell_i := \#\{x \in \mathrm{Zer}(\mathrm{Cr}(Q^2), \mathrm{C}\langle\zeta\rangle^k) \mid k-1+\mathrm{Sign}(H_1(x))/2 = i\}.$$

- For i from 0 to k, let

$$m_i := \#\{x \in \mathrm{Zer}(\mathrm{Cr}_+(Q^2), \mathrm{C}\langle\zeta\rangle^{k+1}) \mid k+\mathrm{Sign}(H_2(x))/2 = i\}.$$

- Output

$$\chi(\mathrm{Zer}(Q,\mathrm{R}^k)) = \frac{1}{2}\left(\sum_{i=0}^{k-1} (-1)^{k-1-i}\ell_i + \sum_{i=0}^{k} (-1)^{k-i} m_i\right).$$

We need the following lemma.

Lemma 12.48. *The Hessian matrices H_1 (resp. H_2) are non-singular at the points of* $\mathrm{Zer}(\mathrm{Cr}(Q^2,\zeta), \mathrm{C}\langle\zeta\rangle^k)$ *(resp.* $\mathrm{Zer}(\mathrm{Cr}_+(Q^2,\zeta), \mathrm{C}\langle\zeta\rangle^{k+1})$*).*

Proof: Let

$$\begin{aligned}\mathrm{Def}(Q^2,\lambda,\mu) &= \lambda Q^2 + \mu G(\bar{d}\,, c),\\ \mathrm{Def}^h(Q^2,\lambda,\mu) &= \lambda (Q^2)^h + \mu G(\bar{d}\,, c)^h\end{aligned}$$

being its homogenization in degree $\bar{d}_1$.

Moreover, let

$$H_1(\lambda,\mu) = \left[\frac{\partial^2 \mathrm{Def}(Q^2,\lambda,\mu)}{\partial X_i \partial X_j}\right]_{2\le i,j\le k}$$

be the corresponding Hessian matrix and $\mathrm{Cr}^h(Q^2,\lambda,\mu,)$ the corresponding system of equations for the polynomial $\mathrm{Def}^h(Q^2,\lambda,\mu,)$.

Now, $H_1(0,1)$ is the diagonal matrix with entries

$$\bar{d}_i(\bar{d}_i-1)X_i^{\bar{d}_i-2}+2, 2\le i,j\le k.$$

Also, if $(x_1,\ldots,x_k)\in \mathrm{Zer}(\mathrm{Cr}(Q^2,0,1),\mathrm{C}^k)$, then each x_i, for $i=2,\ldots,k$ has norm less than 1 (see proof of Lemma 12.45). Also, $\mathrm{Zer}(\mathrm{Cr}^h(Q^2,0,1),\mathbb{P}_k(\mathrm{C}))$ has no points at infinity. Hence, it is clear that $\det^h(H_1(0,1))\neq 0$ at any point of $\mathrm{Zer}(\mathrm{Cr}^h(Q^2,0,1),\mathbb{P}_k(\mathrm{C}))$.

Thus, the set D of $(\lambda{:}\ \mu)\in\mathbb{P}_1(\mathrm{C})$ such that $\det^h(H_1(\lambda,\mu))=0$ at a point of $\mathrm{Zer}(\mathrm{Cr}^h(Q^2,\lambda,\mu),\mathbb{P}_k(\mathrm{C}))$ does not contain $(0{:}\,1)$.

Moreover, D is the projection on $\mathbb{P}_1(\mathrm{C})$ of an algebraic subset of $\mathbb{P}_k(\mathrm{C})\times\mathbb{P}_1(\mathrm{C})$ and is thus algebraic by Theorem 4.102. Since, D does not contain the point $(0{:}\ 1)$ it is a finite subset of $\mathbb{P}_1(\mathrm{C})$ by Lemma 4.101. Hence, the set of $t\in\mathrm{C}$ such that $\det^h(H_1(1-t,\ t))=0$ at a point of $\mathrm{Zer}(\mathrm{Cr}(Q^2,1-t,t),\mathrm{C})$ is finite and its extension to $\mathrm{C}\langle\zeta\rangle$ is a finite number of elements of C which does not contain ζ.

This claim now follows for H_1 and the proof is identical for H_2. □

Proof of correctness of Algorithm 12.24: It follows from Proposition 12.38 and Lemma 12.40 that the algebraic sets $\mathrm{Zer}(\mathrm{Def}(Q^2,{}^-\zeta),\mathrm{R}\langle\zeta\rangle^k)$ and $\mathrm{Zer}(\mathrm{Def}_+(Q^2,\zeta),\mathrm{R}\langle\zeta\rangle^{k+1})$ are non-singular algebraic hypersurfaces bounded over R.

Moreover, by Proposition 12.44, the zeros of the polynomial system $\mathrm{Cr}(Q^2,\zeta)$ are simple. The same holds for $\mathrm{Cr}_+(Q^2,\zeta)$.

It follows from Lemma 12.48 that the projection map onto the X_1 coordinate has non-degenerate critical points on the hypersurfaces $\mathrm{Zer}(\mathrm{Def}(Q^2,\zeta),\mathrm{R}\langle\zeta\rangle^k)$ and $\mathrm{Zer}(\mathrm{Def}_+(Q^2,\zeta),\mathrm{R}\langle\zeta\rangle^{k+1})$.

Now, the algebraic set $\mathrm{Zer}(\mathrm{Def}_+(Q^2,\ \zeta),\ \mathrm{R}\langle\zeta\rangle^{k+1})$ is semi-algebraically homeomorphic to two copies of the set, S defined by $\mathrm{Def}(Q^2,\ \zeta)\le 0$, glued along $\mathrm{Zer}(\mathrm{Def}(Q^2,\zeta),\mathrm{R}\langle\zeta\rangle^k)$.

It follows from Equation 6.36 that,

$$\chi(\mathrm{Zer}(\mathrm{Def}_+(Q^2,\zeta),\mathrm{R}\langle\zeta\rangle^{k+1}))=2\chi(S)-\chi(\mathrm{Zer}(\mathrm{Def}(Q^2,\zeta),\mathrm{R}\langle\zeta\rangle^k)).$$

We now claim that the closed and bounded set S has the same homotopy type (and hence the same Euler-Poincaré characteristic) as $\mathrm{Zer}(Q,\mathrm{R}\langle\zeta\rangle^k)$.

We replace ζ in the definition of the set S by a new variable T, and consider the set $K\subset\mathrm{R}^{k+1}$ defined by $\{(x,t)\in\mathrm{R}^{k+1}|\mathrm{Def}(Q^2,T)\le 0\}$ and let for $b>0$, $K_b=\{x|(x,b)\in K\}$. Note that $\mathrm{Ext}(K,\mathrm{R}\langle\zeta\rangle)_\zeta=S$. Let π_X (resp. aπ_T) denote the projection map onto the X (resp. T) coordinates.

Clearly, $\mathrm{Zer}(Q,\ \mathrm{R}^k)\subset K_b$ for all $b>0$. By Theorem 5.46 (Semi-algebraic triviality), for all small enough $b>0$, there exists a semi-algebraic homeomorphism, $\phi{:}\ K_b\times(0,b]\to K\cap\pi_T^{-1}((0,b])$, such that $\pi_T(\phi(x,s))=s$ and $\phi(\mathrm{Zer}(Q,\mathrm{R}^k),s)=\mathrm{Zer}(Q,\mathrm{R}^k)$ for all $s\in(0,b]$.

Let $G{:}\ K_b\times[0,\ b]\to K_b$ be the map defined by $G(x,\ s)=\pi_X(\phi(x,\ s))$ for $s>0$ and $G(x,0)=\lim_{s\to 0+}\pi_X(\phi(x,s))$. Let $g{:}\,K_b\to$ be the map $G(x,0)$ and $i{:}\,\mathrm{Zer}(Q,\mathrm{R}^k)\to K_b$ the inclusion map. Using the homotopy G, we see that $i\circ g\sim\mathrm{Id}_{K_b}$, and $g\circ i\sim\mathrm{Id}_{\mathrm{Zer}(Q,\mathrm{R}^k)}$, which shows that $\mathrm{Zer}(Q,\ \mathrm{R}^k)$ is homotopy equivalent to K_b for all small enough $b>0$. Now, specialize b to ζ.

Finally, the correctness of the computations of

$$\chi(\mathrm{Zer}(\mathrm{Def}_+(Q^2,\zeta),\mathrm{R}\langle\zeta\rangle^{k+1})),\ \chi(\mathrm{Zer}(\mathrm{Def}(Q^2,\zeta),\mathrm{R}\langle\zeta\rangle^{k}))$$

is a consequence of Lemma 7.25, using Tarski-Seidenberg principle (Theorem 2.80). □

Complexity analysis of Algorithm 12.24: The complexity of the algorithm is $d^{O(k)}$, according to the complexity analysis of Algorithm 12.10 (Special Multiplication Table), Algorithm 8.17 (Characteristic polynomial), Algorithm 12.8 (Multivariate Sign Determination).

When $\mathrm{D}=\mathbb{Z}$ and the bitsizes of the coefficients of Q are bounded by τ, the bitsizes of the intermediate computations and the output are bounded by $\tau d^{O(k)}$. □

Algorithm 12.25. **[Euler-Poincaré Characteristic of an Algebraic Set]**

- **Structure**: an ordered domain D contained in a real closed field R.
- **Input**: a polynomial $Q \in \mathrm{D}[X_1,\ldots,X_k]$.
- **Output**:the Euler-Poincaré characteristic $\chi(\mathrm{Zer}(Q,\mathrm{R}^k))$.
- **Complexity:** $d^{O(k)}$, where d is a bound on the degree of Q.
- **Procedure:**
  - Define

$$\begin{aligned} Q_1 &= Q^2+(\varepsilon^2(X_1^2+\cdots+X_k^2)-1)^2,\\ Q_2 &= Q^2+(\varepsilon^2(X_1^2+\cdots+X_{k+1}^2)-1)^2, \end{aligned}$$

  - Using Algorithm 12.24 (Euler-Poincaré Characteristic of a Bounded Algebraic Set) compute $\chi(\mathrm{Zer}(Q_1,\mathrm{R}\langle\varepsilon\rangle^k))$ and $\chi(\mathrm{Zer}(Q_2,\mathrm{R}\langle\varepsilon\rangle^{k+1}))$.
  - Output,

$$\chi(\mathrm{Zer}(Q,\mathrm{R}^k))=(\chi(\mathrm{Zer}(Q_2,\mathrm{R}\langle\varepsilon\rangle^{k+1}))-\chi(\mathrm{Zer}(Q_1,\mathrm{R}\langle\varepsilon\rangle^{k})))/2.$$

Proof of correctness: It is clear that the algebraic sets $\mathrm{Zer}(Q_1,\ \mathrm{R}\langle\varepsilon\rangle^k)$ and $Z(Q_2,\ \mathrm{R}\langle\varepsilon\rangle^{k+1}))$ are bounded over $\mathrm{R}\langle\varepsilon\rangle$ and hence we can apply Algorithm 12.24 to compute their Euler-Poincaré characteristics.

Moreover, $\mathrm{Zer}(Q_2,\ \mathrm{R}\langle\varepsilon\rangle^{k+1})$ is semi-algebraically homeomorphic to two copies of $\mathrm{Zer}(Q,\ \mathrm{R}\langle\varepsilon\rangle^k)\ \cap\ B(0,\ 1/\varepsilon)$ glued along the algebraic set $\mathrm{Zer}(Q_1,\mathrm{R}\langle\varepsilon\rangle^k)$. Hence, using Equation 6.36 we obtain that,

$$\chi(\mathrm{Zer}(Q,\mathrm{R}\langle\varepsilon\rangle^k)\cap B(0,1/\varepsilon))=(\chi(\mathrm{Zer}(Q_2,\mathrm{R}\langle\varepsilon\rangle^{k+1}))+\chi(\mathrm{Zer}(Q_1,\mathrm{R}\langle\varepsilon\rangle^{k})))/2.$$

The correctness of the algorithm now follows from the fact that,

$$\chi(\mathrm{Zer}(Q,\mathrm{R}^k))=\mathrm{Zer}(Q,\mathrm{R}\langle\varepsilon\rangle^k)=\chi(\mathrm{Zer}(Q,\mathrm{R}\langle\varepsilon\rangle^k)\cap B(0,1/\varepsilon)),$$

since $\mathrm{Zer}(Q,\ \mathrm{R}\langle\varepsilon\rangle^k)$ and $\mathrm{Zer}(Q,\ \mathrm{R}\langle\varepsilon\rangle^k)\ \cap\ B(0,\ 1/\varepsilon)$ are semi-algebraically homeomorphic. □

Complexity Analysis: The complexity of the algorithm is $d^{O(k)}$ using the complexity analysis of Algorithm 12.24, and the bound $d^{O(k)}$ on the degree in ε, ζ obtained in the complexity analysis of Algorithm 12.10 (Parametrized Special Multiplication Table).

When $\mathrm{D} = \mathbb{Z}$ and the bitsizes of the coefficients of Q are bounded by τ, the bitsizes of the intermediate computations and the output are bounded by $d^{O(k)}$. □

12.9 Bibliographical Notes

Gröbner basis have been introduced and studied by B. Buchberger [32, 33]. They are a major tool in computational algebraic geometry and polynomial system solving.

The idea of representing solutions of polynomial systems by Rational Univariate Representation seems to appear first in the work of Kronecker [100]. The use of these representations in computer algebra starts with [4, 134] and is now commonly used. The first single exponential algorithms in time $d^{O(k)}$ for finding a point in every connected component of an algebraic set can be found in [37, 133]. The notion of pseudo-critical point is introduced in [75]. The notion of well-separating element and the limit process comes from [137]. The algorithm for computing the Euler-Poincaré characteristic for algebraic sets appears in [11].

13

Existential Theory of the Reals

The **decision problem for the existential theory of the reals** is to decide the truth or falsity of a sentence $(\exists X_1) \ldots (\exists X_k)\, F(X_1, \ldots, X_k)$, where $F(X_1, \ldots, X_k)$ is a quantifier free formula in the language of ordered fields with coefficients in a real closed field R. This problem is equivalent to deciding whether or not a given semi-algebraic set is empty. It is a special case of the general decision problem seen in Chapter 11.

When done by the Cylindrical Decomposition Algorithm of Chapter 11, deciding existential properties of the reals has complexity doubly exponential in k, the number of variables. But the existential theory of the reals has a special logical structure, since the sentence to decide has a single block of existential quantifiers. We take advantage of this special structure to find an algorithm which is singly exponential in k.

Our method for solving the existential theory of the reals is to compute the set of realizable sign conditions of the set of polynomials $\mathcal{P}$ appearing in the quantifier free formula F. We have already seen in Proposition 7.35 that the set of realizable sign condition of $\mathcal{P}$ is polynomial in the degree d and the number s of polynomials and singly exponential in the number of variables k. The proof of Proposition 7.35 used Mayer-Vietoris sequence and Theorem 7.23. Our technique here will be quite different, though the main ideas are inspired by the critical point method already used in Chapter 7 and Chapter 12.

In Section 13.1, we describe an algorithm for computing the set of realizable sign conditions, as well as sample points in their realizations, whose complexity is polynomial in s and d and singly exponential in k. This algorithm uses pseudo-critical points introduced in Chapter 12 and additional techniques for achieving general position by infinitesimal perturbations.

In Section 13.1, we describe some applications of the preceding results related to bounding the size of a ball meeting every semi-algebraically connected component of the realization of every realizable sign condition, as well as to certain real and complex decision problems.

In Section 13.3, we describe an algorithm for computing sample points in realizations of realizable sign conditions on an algebraic set taking advantage of the (possibly low) dimension of the algebraic set.

Finally, in Section 13.4 we describe a method for computing the Euler-Poincaré characteristic of all possible sign conditions defined by a family of polynomials.

13.1 Finding Realizable Sign Conditions

In this section, let $\mathcal{P} = \{\mathcal{P}_1, \ldots, \mathcal{P}_s\} \subset \mathrm{R}[X_1, \ldots, X_k]$. Recall that we denote by $\mathrm{SIGN}(\mathcal{P}) \subset \{0, 1, -1\}^{\mathcal{P}}$ the set of all realizable sign conditions for $\mathcal{P}$ (see Notation 7.29). We are now going to present an algorithm which computes $\mathrm{SIGN}(\mathcal{P})$.

We first prove that we can reduce the problem of computing a set of sample points meeting the realizations of every realizable sign conditions of a family of polynomials to the problem already considered in Chapter 12, namely finding points in every semi-algebraically connected component of certain algebraic sets.

Proposition 13.1. *Let $D \subset \mathrm{R}^k$ be a non-empty semi-algebraically connected component of a basic closed semi-algebraic set defined by*

$$P_1 = \cdots = P_\ell = 0, P_{\ell+1} \geq 0, \cdots, P_s \geq 0.$$

There exists an algebraic set W defined by equations

$$P_1 = \cdots = P_\ell = P_{i_1} = \cdots P_{i_m} = 0,$$

(with $\{i_1, \ldots, i_m\} \subset \{\ell + 1, \ldots, s\}$) such that a semi-algebraically connected component D' of W is contained in D.

Proof: Consider a maximal set of polynomials

$$\{P_1, \ldots, P_\ell, P_{i_1}, \ldots, P_{i_m}\},$$

where

$$m = 0 \text{ or } \ell < i_1 < \cdots < i_m \leq s,$$

with the property that there exists a point $p \in D$ where

$$P_1 = \cdots = P_\ell = P_{i_1} = \cdots = P_{i_m} = 0.$$

Consider the semi-algebraically connected component D' of the algebraic set defined by

$$P_1 = \cdots = P_\ell = P_{i_1} = \cdots = P_{i_m} = 0,$$

which contains p. We claim that $D' \subset D$. Suppose that there exists a point $q \in D'$ such that $q \notin D$. Then by Proposition 5.23, there exists a semi-algebraic path γ: $[0, 1] \to D'$ joining p to q in D'. Denote by q' the first point of the path γ on the boundary of D. More precisely, note that

$$A = \{t \in [0, 1] \mid \gamma([0, t]) \subset D\}$$

is a closed semi-algebraic subset of $[0,1]$ which does not contain 1. Thus A is the union of a finite number of closed intervals

$$A = [0, b_1] \cup \ldots \cup [a_\ell, b_\ell].$$

Take $q' = \gamma(b_1)$. At least one of the polynomials, say P_j, $j \notin \{1, \ldots, \ell, i_1, \ldots, i_m\}$ must be 0 at q'. This violates the maximality of the set

$$\{P_1, \ldots, P_\ell, P_{i_1}, \ldots, P_{i_m}\}.$$

It is clear that if D is bounded, D' is bounded. □

Proposition 13.2. *Let $D \subset \mathrm{R}^k$ be a non-empty semi-algebraically connected component of a semi-algebraic set defined by*

$$P_1 = \cdots = P_\ell = 0, P_{\ell+1} > 0, \cdots, P_s > 0.$$

There exists an algebraic set $W \subset \mathrm{R}\langle\varepsilon\rangle^k$ defined by equations

$$P_1 = \cdots = P_\ell = 0, P_{i_1} = \cdots P_{i_m} = \varepsilon$$

(with $\{i_1, \ldots, i_m\} \subset \{\ell+1, \ldots, s\}$) such that there exists a semi-algebraically connected component D' of W which is contained in $\mathrm{Ext}(D, \mathrm{R}\langle\varepsilon\rangle)$.

Proof: Consider two points x and y in D. By Proposition 5.23, there is a semi-algebraic path γ from x to y inside D. Since γ is closed and bounded, the semi-algebraic and continuous function $\min_{\ell+1 \le i \le s}(P_i)$ has a strictly positive minimum on γ. The extension of the path γ to $\mathrm{R}\langle\varepsilon\rangle$ is thus entirely contained inside the subset S of $\mathrm{R}\langle\varepsilon\rangle^k$ defined by

$$P_1 = \cdots = P_\ell = 0, P_{\ell+1} - \varepsilon \ge 0, \cdots, P_s - \varepsilon \ge 0.$$

Thus, there is one non-empty semi-algebraically connected component $\bar{D}$ of S containing D. Applying Proposition 13.1 to $\bar{D}$ and S, we get a semi-algebraically connected component D' of some

$$P_1 = \cdots = P_\ell = 0, P_{i_1} = \cdots P_{i_m} = \varepsilon,$$

contained in $\bar{D}$. Then $D' \subset \mathrm{Ext}(D, \mathrm{R}\langle\varepsilon\rangle)$. □

Remark 13.3. Proposition 13.2, Algorithm 12.16 (Bounded Algebraic Sampling), and Algorithm 11.20 (Removal of Infinitesimals) provide an algorithm outputting a set of points meeting every semi-algebraically connected component of the realization of a realizable sign condition of a family $\mathcal{P}$ of s polynomials on a bounded algebraic set $\mathrm{Zer}(Q, \mathrm{R}^k)$ with complexity $2^s d^{O(k)}$ (where d is a bound on the degree of Q and the $P \in \mathcal{P}$), considering all possible subsets of $\mathcal{P}$. Note that this algorithm does not involve polynomials of degree doubly exponential in k, in contrast to Algorithm 11.2 (Cylindrical Decomposition). □

Exercise 13.1.

a) Describe precisely the algorithm outlined in the preceding remark and prove its complexity.

b) Describe an algorithm with the same complexity without the hypothesis that $\mathrm{Zer}(Q, \mathrm{R}^k)$ is bounded.

When s is bigger than the dimension k of the ambient space, the algorithm proposed in the preceding remark does not give a satisfactory complexity bound, since the complexity is exponential in s. Reduction to general position, using infinitesimal deformations, will be the key for a better complexity result.

Let us define precisely the notion of general position that we consider. Let $\mathcal{P}^\star = \{\mathcal{P}_1^\star, \ldots, \mathcal{P}_s^\star\}$, where for every $i = 1, \ldots, s$, $\mathcal{P}_i^\star \subset \mathrm{R}[X_1, \ldots, X_k]$ is finite, and such that two distinct elements of $\mathcal{P}_i^\star$ have no common zeros in R^k. The family $\mathcal{P}^\star$ is in **ℓ-general position** with respect to $Q \in \mathrm{R}[X_1, \ldots, X_k]$ in R^k if no $\ell + 1$ polynomials belonging to different $\mathcal{P}_i^\star$ have a zero in common with Q in R^k.

The family $\mathcal{P}^\star$ is in **strong ℓ-general position** with respect to $Q \in \mathrm{R}[X_1, \ldots, X_k]$ in R^k if moreover any ℓ polynomials belonging to different $\mathcal{P}_i^\star$ have at most a finite number of zeros in common with Q in R^k.

When $Q = 0$, we simply say that $\mathcal{P}^\star \subset \mathrm{R}[X_1, \ldots, X_k]$ is in ℓ-general position (resp. strong ℓ-general position) in R^k.

We also need the notion of a family of homogeneous polynomials in general position in $\mathbb{P}_k(\mathrm{C})$. The reason for considering common zeros in $\mathbb{P}_k(\mathrm{C})$ is that we are going to use in our proofs the fact that, in the context of complex projective geometry, the projection of an algebraic set is algebraic. This was proved in Theorem 4.102.

Let $\mathcal{P}^\star = \{\mathcal{P}_1^\star, \ldots, \mathcal{P}_s^\star\}$ where for every $i = 1, \ldots, s$, $\mathcal{P}_i^\star \in \mathrm{R}[X_0, X_1, \ldots, X_k]$ is homogeneous. The family $\mathcal{P}^\star$ is in **ℓ-general position** with respect to a homogeneous polynomial $Q^h \in \mathrm{R}[X_0, X_1, \ldots, X_k]$ in $\mathbb{P}_k(\mathrm{C})$ if no more than ℓ polynomials in $\mathcal{P}_i^\star$ have a zero in common with Q^h in $\mathbb{P}_k(\mathrm{C})$.

We first give an example of a finite family of polynomials in general position and then explain how to perturb a finite set of polynomials to get a family in strong general position.

Notation 13.4. Define

$$H_k(d, i) = 1 + \sum_{1 \le j \le k} i^j X_j^d,$$
$$H_k^h(d, i) = X_0^d + \sum_{1 \le j \le k} i^j X_j^d.$$

Note that when d is even, $H_k(d, i)(x) > 0$ for every $x \in \mathrm{R}^k$. □

Lemma 13.5. *For any positive integer d, the polynomials $H_k^h(d, i)$, $0 \le i \le s$, are in k-general position in $\mathbb{P}_k(\mathrm{C})$.*

Proof: Take $P(T, X_0, \ldots, X_k) = X_0^d + \sum_{1 \le j \le k} T^j X_j^d$. If $k + 1$ of the $H_k^h(d, i)$ had a common zero $\bar{x}$ in $\mathbb{P}_k(\mathrm{C})$, substituting homogeneous coordinates of this common zero in P would give a non-zero univariate polynomial in T of degree at most k with $k + 1$ distinct roots, which is impossible. □

Consider three variables $\varepsilon, \delta, \gamma$ and $\mathrm{R}\langle\varepsilon, \delta, \gamma\rangle$. Note that $\varepsilon, \delta, \gamma$ are three infinitesimal quantities in $\mathrm{R}\langle\varepsilon, \delta, \gamma\rangle$ with $\varepsilon > \delta > \gamma > 0$. The reason for using these three infinitesimal quantities is the following. The variable ε is used to get bounded sets, the variables δ, γ are used to reach general position, and describe sets which are closely related to realizations of sign conditions on the original family.

Let $\mathcal{P} = \{P_1, \ldots, P_s\} \subset \mathrm{R}[X_1, \ldots, X_k]$ be polynomials of degree bounded by d. With $d' > d$, let $\mathcal{P}^\star$ be the family $\{P_1^\star, \ldots, P_s^\star\}$ with

$$\begin{aligned} P_i^\star &= \{(1-\delta)\, P_i + \delta\, H_k(d', i), (1-\delta)\, P_i - \delta\, H_k(d', i), \\ &\quad (1-\delta)\, P_i + \delta\gamma\, H_k(d', i), (1-\delta)\, P_i - \delta\gamma\, H_k(d', i)\}. \end{aligned}$$

We prove

Proposition 13.6. *The family $\mathcal{P}^\star$ is in strong k-general position in $\mathrm{R}\langle\epsilon, \delta, \gamma\rangle^k$.*

Proof: For $P_i \in \mathcal{P}$ we write

$$P_i^h = X_0^{d'}\, P_i\left(\frac{X_1}{X_0}, \ldots, \frac{X_k}{X_0}\right).$$

Consider

$$\begin{aligned} \bar{P}_i^\star(\lambda, \mu) &= \{\lambda\, P_i^h + \mu\, H_k^h(d', i), \lambda\, P_i^h - \mu\, H_k^h(d', i), \\ &\quad \lambda\, P_i^h + \mu\gamma\, H_k^h(d', i), \lambda\, P_i^h - \mu\gamma\, H_k^h(d', i)\}. \end{aligned}$$

Let $I = \{i_1, \ldots, i_{k+1}\}$, and $Q_{i_j} \in \bar{P}_{i_j}^\star$, $j = 1, \ldots, k+1$. The set D_I of $(\lambda : \mu) \in \mathbb{P}_1(\mathrm{C}\langle\gamma\rangle)$ such that

$$Q_{i_1}^h(\lambda, \mu), \ldots, Q_{i_{k+1}}^h(\lambda, \mu)$$

have a common zero is the projection on $\mathbb{P}_1(\mathrm{C}\langle\gamma\rangle)$ of an algebraic subset of $\mathbb{P}_k(\mathrm{C}\langle\gamma\rangle) \times \mathbb{P}_1(\mathrm{C}\langle\gamma\rangle)$ and is thus algebraic by Theorem 4.102. Since $d' > d$, Lemma 13.5 and Proposition 1.27 imply that $(0 : 1) \notin D_I$. So D_I is a finite subset of $\mathbb{P}_1(\mathrm{C}\langle\gamma\rangle)$ by Lemma 4.101.

Thus the set of $t \in \mathrm{C}\langle\gamma\rangle$ such that $k+1$ polynomials each in $P_i^\star(1-t, t)$, have a common zero in $\mathrm{C}\langle\gamma\rangle^k$ is finite and its extension to $\mathrm{C}\langle\varepsilon, \delta, \gamma\rangle$ is a finite number of elements of $\mathrm{C}\langle\gamma\rangle$ which does not contain δ.

It remains to prove that k polynomials $Q_{i_j} \in \bar{P}_{i_j}^\star$, $j = 1, \ldots, k$ have a finite number of common zeroes in $\mathrm{R}\langle\epsilon, \delta, \gamma\rangle^k$, which is an immediate consequence of Proposition 12.3,since $d' > d$. □

There is a close relationship between the sign conditions on $\mathcal{P}$ and certain weak sign conditions on the polynomials in $\mathcal{P}^*$ described by the following proposition. The role of the two infinitesimal quantities δ and γ is the following: δ is used to replace strict inequalities by weak inequalities and γ to replace equations by weak inequalities.

Proposition 13.7. *Let* $\mathcal{P}=\{P_1,\ldots,P_s\}\subset \mathrm{R}[X_1,\ldots,X_k]$ *be such that* $\deg P_i \le d$ *for all* i, *and suppose* $d' > d$, d' *even. Let* $D \subset \mathrm{R}^k$ *be a semi-algebraically connected component of the realization of the sign condition*

$$\begin{aligned} P_i &= 0, i\in I\subset\{1,\ldots,s\},\\ P_i &> 0, i\in\{1,\ldots,s\}\setminus I. \end{aligned}$$

Then there exists a semi-algebraically connected component D' *of the subset* $\bar{D} \subset \mathrm{R}\langle\varepsilon,\delta,\gamma\rangle^k$ *defined by the weak sign condition*

$$-\gamma\,\delta\, H_k(d',i) \le (1-\delta)\,P_i \le \gamma\,\delta\, H_k(d',i),\ i\in I,$$

$$(1-\delta)\,P_i \ge \delta\, H_k(d',i),\ i\in\{1,\ldots,s\}\setminus I$$

$$\varepsilon^2\,(X_1^2+\cdots+X_k^2)\le 1$$

such that $\lim_\gamma(D')$ *is contained in the extension of* D *to* $\mathrm{R}\langle\varepsilon,\delta\rangle$.

Proof: If $x\in D\subset \mathrm{R}^k$, then $x\in\bar{D}$. Let D' be the semi-algebraically connected component of $\bar{D}$ which contains x. Since $\lim_\gamma$ is a ring homomorphism and d' is even, it is clear that $\lim_\gamma(D')$ is contained in the realization of the conjunction of $P_i=0$, for $i\in I$, and $P_i>0$, for $i\in\{1,\ldots,s\}\setminus I$ in $\mathrm{R}\langle\varepsilon,\delta\rangle^k$ and that it also contains $x\in D$. Since $\bar{D}$ is bounded, by Proposition 12.43, $\lim_\gamma(D')$ is also semi-algebraically connected. The statement of the proposition follows. □

Corollary 13.8. *Let* $\mathcal{P}=\{P_1,\ldots,P_s\}\subset \mathrm{R}[X_1,\ldots,X_k]$ *be a finite subset of polynomials of degree less than* d *and suppose* $d'>d$, d' *even. Let* D *be a semi-algebraically connected component of the realization of the sign condition*

$$\begin{aligned} P_i &= 0, i\in I\subset\{1,\ldots,s\},\\ P_i &> 0, i\in\{1,\ldots,s\}\setminus I. \end{aligned}$$

Then there exists a semi-algebraically connected component E' *of the realization* $E\subset \mathrm{R}\langle\varepsilon,\delta,\gamma\rangle^{k+1}$ *of*

$$-\gamma\,\delta\, H_k(d',i) \le (1-\delta)\,P_i \le \gamma\,\delta\, H_k(d',i),\ 1i\in\{1,\ldots,s\}\setminus I\in I,$$

$$(1-\delta)\,P_i \ge \delta\, H_k(d',i),$$

$$\varepsilon^2\,(X_1^2+\cdots+X_k^2+X_{k+1}^2)=1$$

such that $\Pi(\lim_\gamma(E'))$ *is contained in the extension of* D *to* $\mathrm{R}\langle\varepsilon,\delta\rangle$, *where* Π *is the projection of* R^{k+1} *to* R^k *forgetting the last coordinate.*

As a consequence of Corollary 13.8, in order to compute all realizable sign conditions on $\mathcal{P}$ it will be enough, using Proposition 13.1 and Proposition 13.6, to consider equations of the form

$$Q=Q_{i_1}^2+\cdots+Q_{i_j}^2+(\varepsilon^2\,(X_1^2+\cdots+X_k^2+X_{k+1}^2)-1)^2=0,$$

where $j \le k, Q_{i_1} \in P_{i_1}^\star, \ldots, 1 \le i_1 < \cdots < i_j \le s, Q_{i_j} \in P_{i_j}^\star$, to find a point in each of the semi-algebraically connected components of their zero sets and to take their limit under $\lim_\gamma$.

A finite set $\mathcal{S} \subset \mathrm{R}^k$ is a **set of sample points for** $\mathcal{P}$ in R^k if $\mathcal{S}$ meets the realizations of all $\sigma \in \mathrm{SIGN}(\mathcal{P})$ (Notation 7.29). Note that the sample points output by Algorithm 11.2 (Cylindrical Decomposition) are a set of sample points for $\mathcal{P}$ in R^k, since the cells of a cylindrical decomposition of R^k adapted to $\mathcal{P}$ are $\mathcal{P}$ invariant and partition R^k. We are going to produce a set of sample points much smaller than the one output by Algorithm 11.2 (Cylindrical Decomposition), which was doubly exponential in the number of variables.

We present two versions of the Sampling algorithm. In the first one, the coordinates of the sample points belong to an extension of R while in the second one the coordinates of the sample points belong to R. The reason for presenting these two versions is technical: in Chapter 14, when we perform the same computation in a parametrized situation, the first version of Sampling will be easier to generalize, while in Chapter 15 it will be convenient to have sample points in R^k. The two algorithms differ only in their last step.

We use the notation (12.18): let $u = (f, g) \in \mathrm{K}[T]^{k+2}$, $g = (g_0, \ldots, g_k)$ be a k-univariate representation and $P \in \mathrm{K}[X_1, \ldots, X_k]$.

$$P_u = g_0^e P\left(\frac{g_k}{g_0}, \ldots, \frac{g_k}{g_0}\right), \tag{13.1}$$

where e is the least even number not less than the degree of P.

Algorithm 13.1. **[Computing Realizable Sign Conditions]**

- **Structure:** an ordered integral domain D contained in a real closed field R.
- **Input:** a set of s polynomials,

  $$\mathcal{P} = \{P_1, \ldots, P_s\} \subset \mathrm{D}[X_1, \ldots, X_k],$$

  each of degree at most d.
- **Output:** a set of real univariate representations in $\mathrm{D}[\varepsilon, \delta, T]^{k+2}$ such that the associated points form a set of sample points for $\mathcal{P}$ in $\mathrm{R}\langle\varepsilon, \delta\rangle^k$, meeting every semi-algebraically connected component of $\mathrm{Reali}(\sigma)$ for every $\sigma \in \mathrm{SIGN}(\mathcal{P})$ and the signs of the elements of $\mathcal{P}$ at these points.
- **Complexity:** $s^{k+1} d^{O(k)}$.
- **Procedure:**
  - Initialize $\mathcal{U}$ to the empty set.
  - Take as d' the smallest even natural number $> d$.
  - Define

$$\begin{aligned} P_i^\star &= \{(1-\delta) P_i + \delta H_k(d', i), (1-\delta) P_i - \delta H_k(d', i), \\ &\quad (1-\delta) P_i + \delta \gamma H_k(d', i), (1-\delta) P_i - \delta \gamma H_k(d', i)\} \\ \mathcal{P}^\star &= \{P_1^\star, \ldots, P_s^\star\} \text{ for } 0 \le i \le s, \text{ using Notation 13.4.} \end{aligned}$$

– For every subset of $j \le k$ polynomials $Q_{i_1} \in P_{i_1}^\star, \ldots, Q_{i_j} \in P_{i_j}^\star$,
– Let

$$Q = Q_{i_1}^2 + \cdots + Q_{i_j}^2 + (\varepsilon^2 (X_1^2 + \cdots + X_k^2 + X_{k+1}^2) - 1)^2.$$

  – For $i = 1, \ldots, k$, let $\bar{d}_i$ be the smallest even natural number greater than $\deg_{X_i}(Q)$, $i = 1, \ldots, k$, and let $\bar{d}_{k+1} = 6$, $\bar{d} = (\bar{d}_1, \ldots, \bar{d}_k, \bar{d}_{k+1})$ and $c = \varepsilon$.
  – Compute the multiplication table $\mathcal{M}$ of $\overline{\mathrm{Cr}}(Q, \zeta)$ (Notation 12.46) using Algorithm 12.9 (Special Multiplication Table).
  – Apply the $\lim_{\gamma,\zeta}$ map using Algorithm 12.14 (Limit of Real Bounded Points) with input $\mathcal{M}$, and obtain a set of real univariate representations (v, σ) with

$$v = (f(T), g_0(T), \ldots, g_k(T), g_{k+1}(T)) \in \mathrm{D}[\varepsilon, \delta][T]^{k+3}.$$

  – Ignore $g_{k+1}(T)$ and consider only the real univariate representations (u, σ)

$$u = (f(T), g_0(T), \ldots, g_k(T)) \in \mathrm{D}[\varepsilon, \delta][T]^{k+2}.$$

  Add u to $\mathcal{U}$.
– Compute the signs of $P \in \mathcal{P}$ at the points associated to the real univariate representations in $\mathcal{U}$, using Algorithm 10.13 (Univariate Sign Determination) with input f and its derivatives and the P_u, $P \in \mathcal{P}$.

Proof of correctness: The correctness follows from Proposition 13.1, Proposition 12.42, Proposition 13.6, Corollary 13.8 and the correctness of Algorithm 12.9 (Special Multiplication Table), Algorithm 12.14 (Limit of Real Bounded Points) and Algorithm 10.13 (Univariate Sign Determination). □

Complexity analysis: The total number of $j \le k$-tuples examined is $\sum_{j \le k} 4^j \binom{s}{j}$. Hence, the number of calls to Algorithm 12.9 (Special Multiplication Table) and Algorithm 12.13 (Simple Univariate Representation) is also bounded by $\sum_{j \le k} 4^j \binom{s}{j}$.} Each such call costs $d^{O(k)}$ arithmetic operations in $\mathrm{D}[\varepsilon, \delta, \gamma, \zeta]$, using the complexity analysis of Algorithm 12.9 (Special Multiplication Table). Since there is a fixed number of infinitesimal quantities appearing with degree one in the input equations, the number of arithmetic operations in D is also $d^{O(k)}$, using the complexity analysis of Algorithm 12.10 (Parametrized Special Multiplication Table). Thus the total number of real univariate representations produced is bounded by $\sum_{j \le k} 4^j \binom{s}{j} O(d)^k$, while the number of arithmetic operations performed for outputting sample points in $\mathrm{R}\langle\varepsilon, \delta\rangle^k$, is bounded by $\sum_{j \le k} 4^j \binom{s}{j} d^{O(k)} = s^k d^{O(k)}$. The sign determination takes $s \sum_{j \le k} 4^j \binom{s}{j} d^{O(k)} = s^{k+1} d^{O(k)}$ arithmetic operations, using the complexity analysis of Algorithm 10.11 (Sign Determination).

If $D = \mathbb{Z}$ and the bitsizes of the coefficients of the input polynomials are bounded by τ, the size of the integer coefficients of the univariate representations are bounded by $\tau\, d^{O(k)}$, using the binary complexity analysis of Algorithm 12.10 (Parametrized Special Multiplication Table). □

Remark 13.9. **[Hardness of the Existential Theory of the Reals]**

In computational complexity theory [127], the class of problems which can be solved in polynomial time is denoted by P. A problem is said to belong to the class NP if it can be solved by a non-deterministic Turing machine in polynomial time. Clearly, $P \subset NP$ but it is unknown whether $P \neq NP$. A problem is NP-complete if it belongs to the class NP, and every other problem in NP can be reduced in polynomial time to an instance of this problem. A problem having only the latter property is said to be NP-hard. Since it is strongly believed that $P \neq NP$, it is very unlikely that an NP-hard problem will have a polynomial time algorithm.

It is a classical result in computational complexity that the Boolean satisfiability problem is NP-complete (see [127], Theorem 8.2 p. 171). The Boolean satisfiability problem is the following: given a Boolean formula, $\phi(X_1, \ldots, X_n)$, written as a conjunction of disjunctions to decide whether it is satisfiable.

Since the Boolean satisfiability problem is NP-complete, it is very easy to see that the problem of existential theory of the reals is an NP-hard problem. Given an instance of a Boolean satisfiability problem, we can reduce it to an instance of the problem of the existential theory of the reals, by replacing each Boolean variable X_i by a real variable Y_i and adding the equation $Y_i^2 - Y_i = 0$ and replacing each Boolean disjunction, $X_{i_1} \vee X_{i_2} \vee \cdots \vee X_{i_m}$ by the real inequality,

$$Y_{i_1} + Y_{i_2} + \cdots + Y_{i_m} \geqslant 1.$$

It is clear that the original Boolean formula is satisfiable if and only if the semi-algebraic set defined by the corresponding real inequalities defined above is non-empty. This shows it is quite unlikely (unless $P = NP$) that there exists any algorithm with binary complexity polynomial in the input size for the existential theory of the reals. □

Remark 13.10. **[Polynomial Space]**

Since Algorithm 13.1 (Computing Realizable Sign Conditions) is based essentially on computations of determinants of size $O(d)^k$, Remark 8.14 implies that it is possible to find the list of realizable sign conditions with complexity $(s\, d)^{O(k)}$ using only $(k \log_2(d))^{O(1)}$ amount of space at any time during the computation. In other words, the existential theory of the reals is in PSPACE [127]. The same remark applies to all the algorithms in Chapter 13, Chapter 15 and Chapter 16, as well as to all the algorithms in Chapter 14 when the number of block of quantifiers is bounded. □

Algorithm 13.2. **[Sampling]**

- **Structure:** an ordered integral domain D contained in a real closed field R.
- **Input:** a set of s polynomials,

$$\mathcal{P} = \{P_1, \ldots, P_s\} \subset \mathrm{D}[X_1, \ldots, X_k],$$

each of degree at most d.
- **Output:** a set $\mathcal{U}$ of real univariate representations in $\mathrm{D}[T]^{k+2}$ such that the associated points form a set of sample points for $\mathcal{P}$ in R^k, meeting every semi-algebraically connected component of $\mathrm{Reali}(\sigma)$ for every $\sigma \in \mathrm{SIGN}(\mathcal{P})$, and the signs of the elements of $\mathcal{P}$ at these points.
- **Complexity:** $s^{k+1} d^{O(k)}$.
- **Procedure:**
  - Perform Algorithm 13.1 (Computing Realizable Sign Conditions) with input $\mathcal{P}$ and output $\mathcal{U}$.
  - For every $u \in \mathcal{U}$, $u = (f, g)$, replace δ and ε by appropriately small elements from the field of quotients of D using Algorithm 11.20 (Removal of Infinitesimals) with input f, its derivatives and the P_u, $P \in \mathcal{P}$. Then clear denominators to obtain univariate representation with entries in $\mathrm{D}[T]$.

Proof of correctness: The correctness follows from the correctness of Algorithm 13.1 (Computing Realizable Sign Conditions) and Algorithm 11.20 (Removal of Infinitesimals) □

Complexity analysis: According to the complexity of Algorithm 13.1 (Computing Realizable Sign Conditions), the total number of real univariate representations produced is bounded by $\sum_{j \le k} 4^j \binom{s}{j} O(d)^k$, while the number of arithmetic operations performed for outputting sample points in $\mathrm{R}\langle\varepsilon, \delta\rangle^k$, is bounded by

$$\sum_{j \le k} 4^j \binom{s}{j} d^{O(k)} = s^k d^{O(k)}.$$

The sign determination takes $s \sum_{j \le k} 4^j \binom{s}{j} d^{O(k)} = s^{k+1} d^{O(k)}$ arithmetic operations.

Using Algorithm 11.20 (Removal of Infinitesimals) requires a further overhead of $s\ d^{O(k)}$ arithmetic operations for every univariate representation output. Thus the number of arithmetic operations is bounded by

$$s \sum_{j \le k} 4^j \binom{s}{j} d^{O(k)} = s^{k+1} d^{O(k)}.$$

However, the number of points actually constructed is only

$$\sum_{j \le k} 4^j \binom{s}{j} O(d)^k.$$

If $\mathrm{D}=\mathbb{Z}$ and the bitsizes of the coefficients of the input polynomials are bounded by τ, from the complexity of Algorithm 13.1 (Computing Realizable Sign Conditions), the size of the integer coefficients of the univariate representations in $\mathcal{U}$ are bounded by $\tau d^{O(k)}$. In Algorithm 11.20 (Removal of Infinitesimals), we substitute a rational number, with numerator and denominator of bitsize $\tau d^{O(k)}$, in place of the variables ε and δ and thus get points defined over $\mathbb{Z}$ by polynomials with coefficients of bitsize $\tau d^{O(k)}$. □

Finally, we have proved the following theorem:

Theorem 13.11. *Let $\mathcal{P}$ be a set of s polynomials each of degree at most d in k variables with coefficients in a real closed field* R. *Let* D *be the ring generated by the coefficients of $\mathcal{P}$. There is an algorithm that computes a set of $\sum_{j\le k} 4^j \binom{s}{j} O(d)^k$ points meeting every semi-algebraically connected component of the realization of every realizable sign condition on $\mathcal{P}$ in $\mathrm{R}\langle\varepsilon,\delta\rangle^k$. The algorithm has complexity $\sum_{j\le k} 4^j \binom{s}{j} d^{O(k)} = s^k O(d)^k$ in* D. *There is also an algorithm computing the signs of all the polynomials in $\mathcal{P}$ at each of these points with complexity $s\sum_{j\le k} 4^j \binom{s}{j} d^{O(k)} = s^{k+1} d^{O(k)}$ in* D. *The degrees of the univariate representations output are bounded by $O(d)^k$. If the polynomials in $\mathcal{P}$ have coefficients in $\mathbb{Z}$ of bitsizes at most τ, the bitsizes of the coefficients of these univariate representations are bounded by $\tau d^{O(k)}$.*

Note that if we want the points to have coordinates in R^k, the complexity of finding sample points is also $s\sum_{j\le k} 4^j \binom{s}{j} \$ d^{O(k)} = s^{k+1} d^{O(k)}$ in D, using Algorithm 13.2 (Sampling).

As a corollary,

Theorem 13.12. *Let* R *be a real closed field. Given a finite set, $\mathcal{P}\subset \mathrm{R}[X_1,\ldots,X_k]$ of s polynomials each of degree at most d, then there exists an algorithm computing the set of realizable sign conditions* $\mathrm{SIGN}(\mathcal{P})$ *with complexity $s\sum_{j\le k} 4^j \binom{s}{j} d^{O(k)} = s^{k+1} d^{O(k)}$ in* D, *where* D *is the ring generated by the coefficients of the polynomials in $\mathcal{P}$.*

Recall that a $\mathcal{P}$- atom is one of $P=0, P\neq 0, P>0, P<0$, where P is a polynomial in $\mathcal{P}$ and a $\mathcal{P}$-formula is a formula written with $\mathcal{P}$-atoms. Since the truth or falsity of a sentence

$$(\exists X_1)\ldots(\exists X_k)\ F(X_1,\ldots,X_k),$$

where $F(X_1,\ldots,X_k)$ is a quantifier free $\mathcal{P}$-formula, can be decided by reading the list of realizable sign conditions on $\mathcal{P}$, the following theorem is an immediate corollary of Theorem 13.11.

Theorem 13.13. [Existential Theory of the Reals] *Let* R *be a real closed field. Let* $\mathcal{P} \subset \mathrm{R}[X_1, \ldots, X_k]$ *be a finite set of* s *polynomials each of degree at most* d*, and let*

$$(\exists X_1)\ldots(\exists X_k)\; F(X_1, \ldots, X_k),$$

be a sentence, where $F(X_1, \ldots, X_k)$ *is a quantifier free* $\mathcal{P}$*-formula. There exists an algorithm to decide the truth of the sentence with complexity* $s \sum_{j \le k} 4^j \binom{s}{j} d^{O(k)} = s^{k+1} d^{O(k)}$ *in* D *where* D *is the ring generated by the coefficients of the polynomials in* $\mathcal{P}$.

Remark 13.14. Note that Theorem 13.11 implies that the total number of semi-algebraically connected components of realizable sign conditions defined by $\mathcal{P}$ is bounded by $\sum_{j \le k} 4^j \binom{s}{j} O(d)^k$. This bound is slightly less precise than the bound $\sum_{j \le k} \binom{s}{j} 4^j d\,(2\,d-1)^{k-1}$ given in Proposition 7.35, but does not require to use homology. □

13.2 A Few Applications

As a first application of the preceding results, we prove a bound on the radius of a ball meeting every semi-algebraically connected component of the realizations of the realizable sign conditions on a family of polynomials.

Theorem 13.15. *Given a set* $\mathcal{P}$ *of* s *polynomials of degree at most* d *in* k *variables with coefficients in* $\mathbb{Z}$ *of bitsizes at most* τ*, there exists a ball of radius* $2^{\tau O(d)^k}$ *intersecting every semi-algebraically connected component of the realization of every realizable sign condition on* $\mathcal{P}$.

Proof: The theorem follows from Theorem 13.11 together with Lemma 10.3 (Cauchy). □

We also have the following result.

Theorem 13.16. *Given a set* $\mathcal{P}$ *of* s *polynomials of degree at most* d *in* k *variables with coefficients in* $\mathbb{Z}$ *of bitsizes at most* τ *such that* $S = \{x \in \mathrm{R}^k \mid P(x) > 0, P \in \mathcal{P}\}$ *is a non-empty set, then in each semi-algebraically connected component of* S *there exists a point whose coordinates are rational numbers* a_i/b_i *where* a_i *and* b_i *have bitsizes* $\tau d^{O(k)}$.

Proof: This is a consequence of Theorem 13.11. We consider a point x belonging to S associated to a univariate representation

$$u = (f(T), g_0(T), \ldots, g_k(T))$$

output by the algorithm, Algorithm 13.2 (Sampling), so that $x_i = g_i(t)/g_0(t)$, with t a root of f in R known by its Thom encoding. Using Notation 13.8, each $P_u(T)$ is of degree $O(d)^k$, and the bitsizes of its coefficients are bounded by $\tau d^{O(k)}$. Moreover, for every $P \in \mathcal{P}$, $P_u(t) > 0$. Since the minimal distance between two roots of a univariate polynomial of degree $O(d)^k$ with coefficients in $\mathbb{Z}$ of bitsize $\tau d^{O(k)}$ is at least $2^{\tau d^{O(k)})}$ by Proposition 10.21, we get, considering polynomials $P_u(T)$ for $P \in \mathcal{P}$, that there exists a rational number c/d with c and d of bitsizes $\tau d^{O(k)}$ such that $P_u(c/d)$ is positive. Thus defining the k-tuple $a\, b$ by $a_i/b_i = (g_i/g_0)(c/d)$, we get $P(a/b) > 0$ for all $P \in \mathcal{P}$ with bitsizes as claimed. □

We also apply our techniques to the algorithmic problem of checking whether an algebraic set has real dimension zero. We prove the following theorem.

Note that the only assumption we require in the second part of the theorem is that the real dimension of the algebraic set is 0 (the dimension of the complex part could be bigger).

Theorem 13.17. *Let $Q \in \mathrm{R}[X_1, \ldots X_k]$ have degree at most d, and let* D *be the ring generated by the coefficients of Q. There is an algorithm which checks if the real dimension of the algebraic set* $\mathrm{Zer}(Q, \mathrm{R}^k)$ *is 0 with complexity $d^{O(k)}$ in* D.

If the real dimension of $\mathrm{Zer}(Q, \mathrm{R}^k)$ *is 0, the algorithm outputs a univariate representation of its points with complexity $d^{O(k)}$ in* D. *Moreover, if* $\mathrm{D} = \mathbb{Z}$ *and the bitsizes of the coefficients are bounded by τ, these points are contained in a ball of radius a/b with a and b in $\mathbb{Z}$ of bitsizes $\tau d^{O(k)}$.*

Let $\mathcal{P} \subset \mathrm{R}[X_1, \ldots X_k]$ be s polynomials of degrees at most d, and let D *the ring generated by the coefficients of the polynomials in $\mathcal{P}$. Then the signs of all the polynomials in $\mathcal{P}$ at the points of* $\mathrm{Zer}(Q, \mathrm{R}^k)$ *can be computed with complexity $s\, d^{O(k)}$ in* D.

Proof: In order to check whether the algebraic set $\mathrm{Zer}(Q, \mathrm{R}^k)$ is zero- dimensional, we apply Algorithm 13.1 (Sampling) to Q. Let $\mathcal{U}$ be the set of real univariate representations output. Denote by K the finite set of points output, which intersects every semi- algebraically connected component of the algebraic set $\mathrm{Zer}(Q, \mathrm{R}^k)$. Now, $\mathrm{Zer}(Q, \mathrm{R}^k)$ is zero-dimensional, if and only if every point in K has a sufficiently small sphere centered around it, which does not intersect $\mathrm{Zer}(Q, \mathrm{R}^k)$. For every $(f(T), g_0(T), \ldots, g_k(T), \sigma) \in \mathcal{U}$, with associated point x, we introduce a new polynomial,

$$\begin{aligned} P(X_1, \ldots, X_k, T) &= Q^2(X_1, \ldots, X_k) + f^2(T) \\ &\quad + ((g_0(T) X_1 - g_1(T))^2 \\ &\quad + \cdots \\ &\quad + (g_0(T) X_k - g_k(T))^2 - g_0^2(T)\, \beta)^2 \end{aligned}$$

where β is a new variable. We apply Algorithm 12.16 (Bounded Algebraic Sampling) to this $(k+1)$-variate polynomial and check whether the corresponding zero set, intersected with the realization of the sign conditions σ on $f(T)$ and its derivatives, is empty or not in $\mathrm{R}\langle\beta\rangle^k$.

If $\mathrm{D} = \mathbb{Z}$, the bounds claimed follow from the fact that the polynomials in the univariate representations computed have integer coefficients of bitsizes $\tau d^{O(k)}$.

For the third part we apply Algorithm 10.13 (Univariate Sign Determination) to the output of Algorithm 13.1 (Sampling). There are $d^{O(k)}$ calls to Algorithm 10.13 (Univariate Sign Determination). The number of arithmetic operations is $s\, d^{O(k)}$ in D. □

The following corollary follows immediately from the proof of Theorem 13.17.

Corollary 13.18. *Let $\mathcal{Q}$ be a finite set of m polynomials in $\mathrm{R}[X_1, ...X_k]$ of degree at most d. Then the coordinates of the isolated points of $\mathrm{Zer}(\mathcal{Q}, \mathrm{R}^k)$ are zeros of polynomials of degrees $O(d)^k$. Moreover if $\mathrm{D} = \mathbb{Z}$, and the bitsizes of the coefficients of the polynomials is bounded by τ, then these points are contained in a ball of radius a/b with a and b in $\mathbb{Z}$ of bitsizes $(\tau + \log(m))\, d^{O(k)}$ in R^k.*

Our techniques can also be applied to decision problems in complex geometry.

Proposition 13.19. *Given a set $\mathcal{P}$ of m polynomials of degree d in k variables with coefficients in C, we can decide with complexity $m\, d^{O(k)}$ in D (where D is the ring generated by the real and imaginary parts of the coefficients of the polynomials in $\mathcal{P}$) whether $\mathrm{Zer}(\mathcal{P}, \mathrm{C}^k)$ is empty. Moreover if $\mathrm{D} = \mathbb{Z}$, and the bitsizes of the coefficients of the polynomials is bounded by τ, then bitsizes of the integers appearing in the intermediate computations and the output are bounded by $(\tau + \log(m))\, d^{O(k)}$.*

Proof: Define Q as the sums of squares of the real and imaginary parts of the polynomials in $\mathcal{P}$ and apply the Algorithm 13.1 (Computing realizable sign conditions). □

Proposition 13.20. *Given a set $\mathcal{P}$ of m polynomials of degree d in k variables with coefficients in C, we can decide with complexity $m\, d^{O(k)}$ in D (where D is the ring generated by the real and imaginary parts of the coefficients of polynomials in $\mathcal{P}$) whether the set of zeros of $\mathcal{P}$ is zero-dimensional in C^k. Moreover if $\mathrm{D} = \mathbb{Z}$, and the bitsizes of the coefficients of the polynomials are bounded by τ, then these points are contained in a ball of radius a/b with a and b in $\mathbb{Z}$ of bitsize $(\tau + \log(m))\, d^{O(k)}$ in R^k.*

Proof: Define Q as the sums of squares of the real and imaginary parts of the polynomials in $\mathcal{P}$ and apply Theorem 13.17 □

Proposition 13.21. *Given an algebraic set* $\mathrm{Zer}(\mathcal{P}, \mathrm{C}^k)$ *where* $\mathcal{P}$ *is a set of* m *polynomials of degree* d *in* k *variables with coefficients in* C, *the real and imaginary parts of the coordinates of the isolated points of* $\mathrm{Zer}(\mathcal{P}, \mathrm{C}^k)$ *are zeros of polynomials with coefficients in* D *(where* D *is the ring generated by the real and imaginary parts of the coefficients of polynomials in* $\mathcal{P}$*) of degrees bounded by* $O(d)^k$. *Moreover if* $\mathrm{D} = \mathbb{Z}$, *and the bitsizes of the coefficients of the polynomials are bounded by* τ, *then these points are contained in a ball of radius* a/b *with* a *and* b *in* $\mathbb{Z}$ *of bitsize* $(\tau + \log(m))\, d^{O(k)}$ *in* $\mathrm{R}^{2k} = \mathrm{C}^k$.

Proof: As in Proposition 13.18. □

13.3 Sample Points on an Algebraic Set

In this section we consider the problem of computing a set of sample points meeting the realizations of every realizable sign conditions of a family of polynomials restricted to an algebraic set. The goal is to have an algorithm whose complexity depends on the (possibly low) dimension of the algebraic set, which would be better than the complexity of Algorithms 13.1 and 13.2.

We prove the following theorem.

Theorem 13.22. *Let* $\mathrm{Zer}(Q, \mathrm{R}^k)$ *be an algebraic set of real dimension* k', *where* Q *is a polynomial in* $\mathrm{R}[X_1, \ldots, X_k]$ *of degree at most* d, *and let* $\mathcal{P} \subset \mathrm{R}[X_1, \ldots, X_k]$ *be a finite set of* s *polynomials with each* $P \in \mathcal{P}$ *also of degree at most* d. *Let* D *be the ring generated by the coefficients of* Q *and the polynomials in* $\mathcal{P}$. *There is an algorithm which computes a set of points meeting every semi-algebraically connected component of every realizable sign condition on* $\mathcal{P}$ *in* $\mathrm{Zer}(Q, \mathrm{R}\langle\varepsilon, \delta\rangle^k)$. *The algorithm has complexity*

$$(k'(k-k')+1) \sum_{j \leq k'} 4^j \binom{s}{j} d^{O(k)} = s^{k'} d^{O(k)}$$

in D. There is also an algorithm providing the signs of the elements of $\mathcal{P}$ *at these points with complexity*

$$(k'(k-k')+1) \sum_{j \leq k'} 4^j \binom{s}{j} s\, d^{O(k)} = s^{k'+1} d^{O(k)}$$

in D.*The degrees in* T *of the univariate representations output is* $O(d)^k$. *If* D=Z, *and the bitsizes of the coefficients of the polynomials are bounded by* τ *then the bitsizes of the integers appearing in the intermediate computations and these univariate representations are bounded by* $\tau d^{O(k)}$.

Remark 13.23. This result is rather satisfactory since it fits with the bound on the number of realizable sign conditions proved in Proposition 7.35. □

We consider now a bounded algebraic set $\mathrm{Zer}(Q, \mathrm{R}^k) \subset B(0, 1/c)$, $c \leq 1$, with $Q \in \mathrm{R}[X_1, \ldots, X_k]$, $Q(x) \geq 0$ for every $x \in \mathrm{R}^k$, of degree bounded by d and of real dimension k'.

Let x be a smooth point of dimension k' of $\mathrm{Zer}(Q, \mathrm{R}^k)$. We denote by T_x the tangent plane to $\mathrm{Zer}(Q, \mathrm{R}^k)$ at x and suppose that T_x is **transversal** to the $k-k'$-plane L defined by $X_{k-k'+1} = \cdots = X_k = 0$, i.e. the intersection of T_x and L is $\{0\}$.

We will construct an algebraic set $\mathrm{Zer}(\mathcal{Q}, \mathrm{R}\langle\eta\rangle^k)$ covering $\mathrm{Zer}(Q, \mathrm{R}^k)$ in the neighborhood of x. The algebraic set $\mathrm{Zer}(\mathcal{Q}, \mathrm{R}\langle\eta\rangle^k)$ has two important properties. Firstly, $\mathrm{Zer}(\mathcal{Q}, \mathrm{R}\langle\eta\rangle^k)$ is defined by $k - k'$ equations and has dimension k'. Secondly, any point z from a neighborhood of x in $\mathrm{Zer}(Q, \mathrm{R}^k)$ is infinitesimally close to a point from $\mathrm{Zer}(\mathcal{Q}, \mathrm{R}\langle\eta\rangle^k)$.

Let η be a variable. Define, for $d' > d$ and even, and $\bar{d} = (d', \ldots, d')$,

$$\begin{aligned} G(\bar{d}, c) &= c^{d'} X_1^{d'} + \cdots + X_{k-k'}^{d'} + X_2^2 + \cdots + X_{k-k'}^2) - (2(k-k') - 1) \\ \mathrm{Def}(Q, \eta) &= \eta\, G(\bar{d}, c) + (1-\eta)\, Q \\ \mathrm{App}(Q, \eta) &= \left\{ \mathrm{Def}(Q, \eta), \frac{\partial \mathrm{Def}(Q, \eta)}{\partial X_2}, \ldots, \frac{\partial \mathrm{Def}(Q, \eta)}{\partial X_{k-k'}} \right\}. \end{aligned}$$

Lemma 13.24. $\dim(\mathrm{Zer}(\mathrm{App}(Q, \eta), \mathrm{R}\langle\eta\rangle^k)) \le k'$.

Proof: For every choice of $z = (z_{k-k'+1}, \ldots, z_k)$ in $\mathrm{R}\langle\eta\rangle^{k'}$, the affine $(k-k')$-plane L' defined by $X_{k-k'+1} = z_{k-k'+1}, \ldots, X_k = z_k$ intersects the algebraic set $\mathrm{Zer}(\mathrm{App}(Q, \eta), \mathrm{C}^k\langle\eta\rangle)$ in at most a finite number of points. Indeed, consider the graded lexicographical ordering on the monomials for which $X_1 < \cdots < X_{k-k'}$. Denoting $\bar{X} = (X_1, \ldots, X_{k-k'})$, by Proposition 12.3,

$$\mathcal{G} = \mathrm{App}(Q(\bar{X}, z), \eta)$$

is a Gröbner basis of the ideal $\mathrm{Ideal}(\mathcal{G}, \mathrm{R}\langle\eta\rangle)$ for the graded lexicographical ordering, since the leading monomials of elements of $\mathcal{G}$ are pure powers of different X_i. Moreover, the quotient $\mathrm{R}\langle\eta\rangle[\bar{X}]/\mathrm{Ideal}(\mathcal{G}, \mathrm{R}\langle\eta\rangle)$ is a finite dimensional vector space and thus $\mathcal{G}$ has a finite number of solutions in $\mathrm{C}\langle\eta\rangle$ according to Proposition 12.7. The conclusion follows clearly by Corollary 5.28. □

Proposition 13.25. *There exists* $y \in \mathrm{Zer}(\mathrm{App}(Q, \eta), \mathrm{R}\langle\eta\rangle^k)$ *such that* $\lim_\eta (y) = x$.

Proof: Since the tangent plane T_x to $\mathrm{Zer}(Q, \mathrm{R}^k)$ at x is transversal to L, the point x is an isolated point of the algebraic set

$$\mathrm{Zer}(Q(\bar{X}, x_{k-k'+1}, \ldots, x_k), \mathrm{R}\langle\eta\rangle^{k-k'}).$$

We can apply Proposition 12.42 to $Q(\bar{X}, x_{k-k'+1}, \ldots, x_k)$. □

Using the preceding construction, we are able to approximate any smooth point such that T_x is transversal to the $k-k'$-plane defined by

$$X_{k-k'+1} = \cdots = X_k = 0.$$

In order to approximate every point in $\mathrm{Zer}(Q,\mathrm{R}^k)$, we are going to construct a family $\mathcal{L}_{k,k-k'}$ of $k-k'$-planes with the following property: any linear subspace T of R^k of dimension k' is transversal to at least one element of the family $\mathcal{L}_{k,k-k'}$, i.e. there is an element L of $\mathcal{L}_{k,k-k'}$ such that $T\cap L=\{0\}$. The construction of $\mathcal{L}_{k,k-k'}$ is based on properties of Vandermonde matrices.

Notation 13.26. Denoting by $v_k(x)$ the Vandermonde vector $(1,x,\ldots,x^{k-1})$, and by V_ℓ the vector subspace of R^k generated by $v_k(\ell),v_k(\ell+1),\ldots,v_k(\ell+k-k'-1)$, it is clear that V_ℓ is of dimension $k-k'$ since the matrix of coordinates of vectors

$$v_{k-k'}(\ell),v_{k-k'}(\ell+1),\ldots,v_{k-k'}(\ell+k-k'-1)$$

is an invertible Vandermonde matrix of dimension $k-k'$.

We now describe equations for V_ℓ. Let, for $k-k'+1\le j\le k$,

$$\begin{aligned}\bar{X}_j &= (X_1,\ldots,X_{k-k'},X_j),\\ v_{k-k',j}(\ell) &= (1,\ldots,\ell^{k-k'-1},\ell^{j-1})\\ f_{\ell,j} &= \det(v_{k-k',j}(\ell),\ldots,v_{k-k',j}(\ell+k-k'-1),\bar{X}_j),\\ L_{k',\ell}(X_1,\ldots,X_k) &= (X_1,\ldots,X_{k-k'},f_{\ell,k-k'+1},\ldots,f_{\ell,k}).\end{aligned}$$

Note that the zero set of the linear forms $f_{\ell,j},k-k'+1\le j\le k$ is the vector space V_ℓ and that $L_{k',\ell}$ is a linear bijection such that $L_{k',\ell}(V_\ell)$ consists of vectors of R^k having their last k' coordinates equal to 0. We denote also by $M_{k',\ell}=(d_{k-k',\ell})^{k'}L_{k',\ell}^{-1}$, with

$$d_{k-k',\ell}=\det(v_{k-k'}(\ell),\ldots,v_{k-k'}(\ell+k-k'-1)).$$

Note that $M_{k',\ell}$ plays the same role as the inverse of $L_{k',\ell}$ but is with integer coordinates, since, for $k-k'+1\le j\le k$, $d_{k-k',\ell}$ is the coefficient of X_j in $f_{\ell,j}$.

Define $\mathcal{L}_{k,k-k'}=\{V_\ell\,|\,0\le\ell\le k'(k-k')\}$. □

Proposition 13.27. *Any linear subspace T of R^k of dimension k' is transversal to at least one element of the family $\mathcal{L}_{k,k-k'}$.*

Corollary 13.28. *Any linear subspace T of R^k of dimension $j\ge k'$ is such that there exists $0\le\ell\le k'(k-k')$ such that V_ℓ and T span R^k.*

In order to prove the proposition, we need the following lemma. Given a polynomial $f(X)\in\mathrm{R}[X]$, we denote by $f^{\{n\}}(X)$ the n-th iterate of f, defined by

$$f^{\{0\}}(X)=X,\ f^{\{n+1\}}(X)=f(f^{\{n\}}(X)).$$

We denote by $V^r(X)$ the vector subspace of $\mathrm{R}(X)^k$ generated by

$$v_k(X),v_k(f(X)),\ldots,v_k(f^{\{r-1\}}(X)).$$

By convention, $V^0(X) = \{0\}$.

Lemma 13.29. *Let T be a linear subspace of $\mathrm{R}(X)^k$ of dimension $\leq k'$. Let $f \in \mathrm{R}[X]$ be such that $f^{\{i\}}(X) \neq f^{\{j\}}(X)$, if $i \neq j$. Then the vector space $V^{k-k'}(X)$ is transversal to T in $\mathrm{R}(X)^k$.*

Proof: The proof is by induction on $k - k'$. If $k - k' = 0$, the claim is clear since $V^0(X) = \{0\}$. Assume now by contradiction that $k - k' \geq 1$ and $V^{k-k'}(X)$ is not transversal to T. By induction hypothesis, $V^{k-k'-1}(X)$ is transversal to T. Hence $v(f^{\{k-k'-1\}}(X))$ belongs to the vector space generated by T and $V^{k-k'-1}(X)$.

It follows by induction on j that for every $j \geq k - k'$, $v(f^{\{j-1\}}(X))$ belongs to the vector space generated by T and $V^{k-k'-1}(X)$. Consider the Vandermonde matrix $V(X, \ldots, f^{\{k-1\}}(X))$. Since

$$\det(V(X, \ldots, f^{\{k-1\}}(X))) = \prod_{k-1 \geq i > j \geq 0} (f^{\{i\}}(X) - f^{\{j\}}(X)) \neq 0,$$

and the dimension of the vector space generated by T and $V^{k-k'-1}(X)$ is $< k$, we obtained a contradiction. □

Proof of Proposition 13.27: We apply Lemma 13.29 to $f(X) = X + 1$. Denoting by $e_1, \ldots, e_{k'}$ a basis of T, and applying Lemma 13.29

$$D = \det(e_1, \ldots, e_{k'}, v_k(X), v_k(X+1), \ldots, v_k(X + k - k' - 1))$$

is not identically 0. Since

$$\begin{aligned} D' &= \det(e_1, \ldots, e_{k'}, v_k(X_1), v_k(X_2), \ldots, v_k(X_{k-k'})) \\ &= \left(\prod_{1 \leq i < j \leq k-k'} (X_i - X_j) \right) S(X_1, \ldots, X_{k-k'}) \end{aligned}$$

with $S(X_1, \ldots, X_{k-k'}) \in \mathrm{R}[X_1, \ldots, X_{k-k'}]$, and the degree of D' is bounded by $\sum_{k'}^{k-1} i = (1/2)(k - k')\ (k + k' - 1)$ the degree of S is bounded by $(1/2)\,(k - k')\,(k + k' - 1) - \binom{k-k'}{2} = k'(k - k')$. Since $(X + i) - (X + j)$ is a constant, it is clear that the degree of $D = S(X, \ldots, X + k - k' - 1)$ is also bounded by $k'(k - k')$. Hence, there exists $\ell \in \{0, \ldots, k'\ (k - k')\}$ which is not a root of D. The corresponding V_ℓ is transversal to T. □

Notation 13.30. Let η be a variable, d' an even natural number, $d' > d$, and $\bar{d} = (d', \ldots, d')$ and $0 \leq \ell \leq k'(k - k')$. Using Notation 13.26, let

$$\begin{aligned} Q_\ell(X_1, \ldots, X_k) &= Q(M_{k',\ell}(X_1, \ldots, X_k)), \\ \mathrm{Def}(Q_\ell, \eta) &= \eta\, G(\bar{d}\,, c) + (1 - \eta)\, Q_\ell, \\ \mathrm{App}(Q_\ell, \eta) &= \left\{ \mathrm{Def}(Q_\ell, \eta), \frac{\partial \mathrm{Def}(Q_\ell, \eta)}{\partial X_2}, \ldots, \frac{\partial \mathrm{Def}(Q_\ell, \eta)}{\partial X_{k-k'}} \right\}. \end{aligned}$$

Define $Z_\ell = M_{k',\ell}(\mathrm{Zer}(\mathrm{App}(Q_\ell,\eta),\mathrm{R}\langle\eta\rangle^k))$, $0\le\ell\le k'(k-k')$, the **approximating varieties** of $\mathrm{Zer}(Q,\mathrm{R}^k)$. □

This terminology is justified by the following result.

Proposition 13.31.

$$\lim_\eta\left(\bigcup_{\ell=0}^{k'(k-k')} Z_\ell\right)=\mathrm{Zer}(Q,\mathrm{R}^k).$$

Proof: It is clear that

$$\lim_\eta\,(\mathrm{Zer}(\mathrm{App}(Q_\ell,\eta),\mathrm{R}\langle\eta\rangle^k))\subset\mathrm{Zer}(Q_\ell,\mathrm{R}^k).$$

Denote by S_ℓ the set of all smooth points of $\mathrm{Zer}(Q,\mathrm{R}^k)$ having a tangent k''-plane ($k''\le k'$) transversal to V_ℓ. Using Notation 13.26, Proposition 13.25 and Proposition 13.27, the union of the S_ℓ for $0\le\ell\le k'(k-k')$ is a semi-algebraic subset of $\mathrm{Zer}(Q,\ \mathrm{R}^k)$ whose closure is $\mathrm{Zer}(Q,\ \mathrm{R}^k)$, using Proposition 5.54. The image under $\lim_\eta$ of $\bigcup_{\ell=0}^{k'(k-k')} Z_\ell$ contains $\bigcup_{\ell=0}^{k'(k-k')} S_\ell$. Moreover, by Proposition 12.43, the image under $\lim_\eta$ of a semi-algebraic set is closed. Hence,

$$\lim_\eta\left(\bigcup_{\ell=0}^{k'(k-k')} Z_\ell\right)\supset\mathrm{Zer}(Q,\mathrm{R}^k).$$ □

Notation 13.32. Let $\mathcal{P}=\{P_1,\ldots,P_s\}\subset\mathrm{R}[X_1,\ldots,X_k]$ be polynomials of degree bounded by d, d' an even natural number, $d'>d$, $\bar d=(d',\ldots,d')$. In order to perturb the polynomials in $\mathcal{P}=\{P_1,\ldots,P_s\}$ to get a family in k'-general position with $\mathrm{App}(Q,\eta)$, we use polynomials $H(d'',i)=1+\sum_{1\le j\le k-k'} i^j X_{k-k'+j}^{d''}$, for $1\le i\le s$, where d'' is an even natural number, $d''>d'$. We consider two variables δ,γ. Let

$$\begin{aligned}P_i^\star &= \{(1-\delta)\,P_i+\delta\,H(d'',i),(1-\delta)\,P_i-\delta\,H(d'',i),\\ &\quad (1-\delta)\,P_i+\gamma\delta\,H(d'',i),(1-\delta)\,P_i-\gamma\delta\,H(d'',i).\}\\ \mathcal{P}^\star &= \{P_1^\star,\ldots,P_s^\star\}.\end{aligned}$$

□

Proposition 13.33. *The family $\mathcal{P}^\star$ is in strong k'-general position with respect to $\mathrm{App}(Q,\eta)$ in $\mathrm{R}\langle\delta,\gamma,\eta\rangle^k$.*

The proof of the proposition uses the following lemma.

Lemma 13.34. *The polynomials $H(d'',i)$, $0\le i\le s$, are in k'-general position with respect to $\mathrm{App}(Q,\eta)$ in $\mathrm{R}\langle\eta\rangle^k$.*

Proof: Let

$$\begin{aligned}\mathrm{Def}(Q,\lambda,\mu) &= \lambda Q+\mu G(\bar{d}\,,c)\\ \mathrm{App}(Q,\mu) &= \left\{\mathrm{Def}(Q,\lambda,\mu),\frac{\partial \mathrm{Def}(Q,\lambda,\mu)}{\partial X_{k'+2}},\ldots,\frac{\partial \mathrm{Def}(Q,{}^{-}\lambda,\mu)}{\partial X_k}\right\}\\ \mathrm{App}^h(Q,\lambda,\mu) &= \left\{\mathrm{Def}^h(Q,\lambda,\mu),\frac{\partial \mathrm{Def}^h(Q,\lambda,\mu)}{\partial X_{k'+2}},\ldots,\frac{\partial \mathrm{Def}^h(Q,\lambda,\mu)}{\partial X_k}\right\}\end{aligned}$$

where

$$\mathrm{Def}^h(Q,\lambda,\mu)=X_0^{d'}\,\mathrm{Def}(Q,\lambda,\mu)\left(\frac{X_1}{X_0},\ldots,\frac{X_k}{X_0}\right).$$

The system $\mathrm{App}^h(Q,\,0,\,1)$ has only the solution $(1{:}\ 0{:}\ \ldots{:}\ 0)$ in $\mathbb{P}_{k-k'}(\mathrm{C})$. The polynomials $H^h(d'',\,i)$, $0\le i\le s$, are in k'-general position in $\mathbb{P}_{k'}(\mathrm{C})$ by Lemma 13.5. Thus, with $J=\{j_1,\,\ldots,\,j_{k'+1}\}\subset\{1,\,\ldots,\,s\}$, the set D_J of $(\lambda{:}\,\mu)\in\mathbb{P}_1(\mathrm{C})$ such that $H^h(d'',j_1),\ldots,H^h(d'',j_{k'+1})$ have a common zero on $\mathrm{App}^h(Q,\lambda,\mu)$ does not contain $(0{:}\,1)$.

Moreover, the set D_J is the projection to $\mathbb{P}_1(\mathrm{C})$ of an algebraic subset of $\mathbb{P}_k(\mathrm{C})\,\times\,\mathbb{P}_1(\mathrm{C})$ and is thus algebraic by Theorem 4.102. Since $d''>d'$ and $(0{:}\,1)\notin D_J$, D_J is a finite subset of $\mathbb{P}_1(\mathrm{C})$. Thus the set of $t\in\mathrm{C}$ such that $k'+1$ polynomials among $H_T(d'',\,j)$, $j\le s$, have a common zero on $\mathrm{Zer}(\mathrm{App}(Q,\,t,\,1-t),\,\mathrm{C}^k)$ is finite, and its extension to $\mathrm{C}\langle\eta\rangle$ is a finite set of elements which does not contain η. □

Proof of Proposition 13.33: Consider

$$\begin{aligned}\bar{P}_i^{\star h} &= \{\lambda\,P_i^h+\mu\,H^h(d'',i),\lambda\,P_i^h-\mu\,H^h(d'',i),\\ &\quad\ \lambda\,P_i^h+\mu\,\gamma\,H^h(d'',i),\lambda\,P_i^h-\mu\,\gamma\,H^h(d'',i)\},\end{aligned}$$

$0\le i\le s$. Let $J=\{j_1,\ldots,j_{k'+1}\}\subset\{1,\ldots,s\}$ and $A_{j_i}\in\bar{P}_i^{\star h}$. The set D_J of $(\lambda{:}\,\mu)$ such that $A_{j_1}(\lambda,\,\mu),\ldots,A_{j_{k'+1}}(\lambda,\,\mu)$ have a common zero with

$$\mathrm{Zer}(\mathrm{App}^h(Q,\lambda,\mu),\mathbb{P}_k(\mathrm{C}))$$

in $\mathbb{P}_k(\mathrm{C}\langle\gamma,\ \eta\rangle)$ is the projection to $\mathbb{P}_1(\mathrm{C}\langle\gamma,\ \eta\rangle)$ of an algebraic subset of $\mathbb{P}_k(\mathrm{C}\langle\gamma,\ \eta\rangle)\,\times\,\mathbb{P}_1(\mathrm{C}\langle\gamma,\ \eta\rangle)$ and is thus algebraic by Theorem 4.102. Since $d''>d'$, $(0{:}\ 1)\notin D_J$ by Lemma 13.34, and D_J is a finite subset of $\mathbb{P}_1(\mathrm{C}\langle\gamma,\eta\rangle)$. Thus the set of $t\in\mathrm{C}\langle\gamma,\ \eta\rangle$ such that $k'+1$ polynomials among $(1-t)\,P_i+t\,H(d'',j)$, $j\le s$, have a common zero on

$$\mathrm{Zer}(\mathrm{App}(Q,\eta),\mathrm{R}\langle\gamma,\eta\rangle^k)$$

is finite, and its extension to $\mathrm{C}\langle\delta,\gamma,\eta\rangle$ is a finite set of elements of $\mathrm{C}\langle\delta,\gamma,\eta\rangle$ which does not contain δ. It remains to prove that k' polynomials

$$A_{j_1}(\lambda,\mu),\ldots,A_{j_{k'}}(\lambda,\mu)$$

have a finite number of common zeroes in $\mathrm{R}\langle\delta,\eta,\gamma\rangle^k$, which is an immediate consequence of Proposition 12.3,since $d'>d$. □

We consider now a polynomial $Q \in \mathrm{R}[X_1, \ldots, X_k]$, with $\mathrm{Zer}(Q, \mathrm{R}^k)$ not necessarily bounded.

The following proposition holds.

Proposition 13.35. *Let $Q \in \mathrm{R}[X_1, \ldots, X_k]$ and $\mathcal{P} = \{P_1, \ldots, P_s\}$ be a finite subset of $\mathrm{R}[X_1, \ldots, X_k]$. Let d be a bound on the degrees of Q and the elements of $\mathcal{P}$, d' an even number $> 2d$, and d''an even number $> d'$. Let D be a connected component of the realization of the sign condition*

$$\begin{aligned} Q &= 0 \\ P_i &= 0, i \in I \subset \{1, \ldots, s\} \\ P_i &> 0, i \in \{1, \ldots, s\} \setminus I. \end{aligned}$$

Let $\bar{Q} = Q^2 + (\varepsilon^2 (X_1^2 + \cdots + X_k^2 + X_{k+1}^2) - 1)^2$. If the set $E \subset \mathrm{R}\langle \varepsilon, \delta, \gamma, \eta \rangle^k$ described by

$$\bigwedge_{R \in \mathrm{App}(\bar{Q}, \eta, \bar{d}, c)} R = 0$$

$$-\gamma \delta H_k(\bar{d}, i) \leq (1 - \delta) P_i \leq \gamma \delta H_k(\bar{d}, i),\ i \in I,$$

$$(1 - \delta) P_i \geq \delta H_k(\bar{d}, i),\ i \in \{1, \ldots, s\} \setminus I$$

$$\varepsilon^2 (X_1^2 + \cdots + X_k^2) \leq 1$$

is non-empty, there exists a connected component E' of E such that $\pi(\lim_{\gamma, \eta} (E'))$ is contained in the extension of D to $\mathrm{R}\langle \varepsilon, \delta \rangle$, where π is the projection of R^{k+1} to R^k forgetting the last coordinate.

Proof: The proof is similar to the proof of Proposition 13.7, using Proposition 13.31. □

Notation 13.36. The set $\mathrm{SIGN}(\mathcal{P}, \mathcal{Q}) \subset \{0, 1, -1\}^{\mathcal{P}}$ is the set of all realizable sign conditions for $\mathcal{P}$ on $\mathrm{Zer}(\mathcal{Q}, \mathrm{R}^k)$. If $\sigma \in \mathrm{SIGN}(\mathcal{P}, \mathcal{Q})$ we denote

$$\mathrm{Reali}(\sigma, \mathcal{Q}) = \{x \in \mathrm{R}^k \mid \bigwedge_{Q \in \mathcal{Q}} Q(x) = 0 \wedge \bigwedge_{P \in \mathcal{P}} \mathrm{sign}(P(x)) = \sigma\}.$$

For $0 \leq \ell \leq k'(k - k')$, and $P \in \mathrm{R}[X_1, \ldots, X_k]$ we denote by

$$P_\ell(X_1, \ldots, X_k) = P(M_{k',\ell}(X_1, \ldots, X_k))$$

If $\mathcal{P} \subset \mathrm{R}[X_1, \ldots, X_k]$, $\mathcal{P}_\ell = \{P_\ell \mid P \in \mathcal{P}\}$.

Given a real univariate representation

$$v = (f(T), g_0(T), g_1(T), \ldots, g_k(T)), \sigma),$$

with associated point z, we denote by

$$M_{k',\ell}(v) = (f(T), g_0(T), h_1(T) \ldots, h_k(T)),$$

with $h_1(T),\ldots,h_k(T)=M_{k'\ell}(g_1(T),\ldots,g_k(T))$, the real univariate representation with associated point $M_{k',\ell}(z)$. □

Algorithm 13.3. **[Sampling on an Algebraic Set]**

- **Structure:** an ordered integral domain D contained in a real closed field R.
- **Input:**
  - a polynomial $Q\in \mathrm{D}[X_1,\ldots,X_k]$ of degree at most d, with $\mathrm{Zer}(Q,\mathrm{R}^k)$ of real dimension k',
  - a set of s polynomials , $\mathcal{P}=\{P_1,\ldots,P_s\}\subset \mathrm{D}[X_1,\ldots,X_k]$, each of degree at most d.
- **Output:** a set $\mathcal{U}$ of real univariate representations in $\mathrm{D}[\varepsilon,\delta][T]^{k+2}$ such that for every $\sigma\in\mathrm{SIGN}(\mathcal{P},Q)$, the associated points meet every semi-algebraically connected component of the extension of $\mathrm{Reali}(\sigma,Q)$ to $\mathrm{R}\langle\varepsilon,\delta\rangle^k$.
- **Complexity:** $s^{k'+1}d^{O(k)}$.
- **Procedure:**
  - Take $d'=2(d+1)$,$\bar{d}=(d',\ldots,d')$, $d''=2(d+2)$.
  - For every $0\le\ell\le k'(k-k')$, define
$$\bar{Q}_\ell=Q_\ell^2+(\varepsilon^2(X_1^2+\cdots+X_k^2+X_{k+1}^2)-1)^2,$$
and define $\mathrm{App}(\bar{Q}_\ell,\eta,\bar{d},\varepsilon)$ and $\mathcal{P}_\ell^\star$, using Notation 13.30 and Notation 13.32.
  - For every $j\le k'$-tuple of polynomials $A_{t_1}\in P_{t_1}^\star,\ldots,A_{t_j}\in P_{t_j}^\star$ let
$$R=\sum_{P\in\mathrm{App}(\bar{Q}_\ell,\eta,\bar{d},\varepsilon)} P^2+A_{i_1}^2+\cdots+A_{i_j}^2.$$
  - Take for $i=1,\ldots,k$, $\bar{d}_i$ equal to the smallest even natural number $>\deg_{X_i}(R)$, $\bar{d}_{k+1}=8$, $\bar{d}=(\bar{d}_1,\ldots,\bar{d}_k,\bar{d}_{k+1})$, $c=\varepsilon$.
  - Compute the multiplication table $\mathcal{M}$ of $\overline{\mathrm{Cr}}(R,\zeta)$ (Notation 12.46) using Algorithm 12.9 (Special Multiplication Table). Apply the $\lim_{\gamma,\eta,\zeta}$ map using Algorithm 12.14 (Limit of Real Bounded Points) with input $\mathcal{M}$, and obtain a set $\mathcal{U}_\ell$ of real univariate representations v with
$$\begin{aligned} v &= ((f(T),g_0(T),\ldots,g_k(T)),\sigma)\\ &\quad (f(T),g_0(T),\ldots,g_k(T))\in\mathrm{D}[\varepsilon,\delta][T]^{k+2}.\end{aligned}$$
  - Define $\mathcal{U}=\bigcup_{\ell=0}^{k'(k-k')} M_{k',\ell}(\mathcal{U}_\ell)$. Compute the signs of $P\in\mathcal{P}$ at the points associated to the real univariate representations v in $\mathcal{U}$,
$$v=(f(T),g_0(T),\ldots,g_k(T)),\sigma)$$
using Algorithm 10.13 (Univariate Sign Determination) with input f and its derivatives and $\mathcal{P}$.

Proof of correctness: Follows from Proposition 13.1, Proposition 13.31, Proposition 13.33, and Proposition 13.35. □

Complexity analysis: It is clear that $\sum_{j \le k'} 4^j \binom{s}{j}$ tuples of polynomials are considered for each $0 \le \ell \le k'(k-k')$. The cost for each such tuple is $d^{O(k)}$ using the complexity analysis of Algorithm 12.18 (Parametrized Bounded Algebraic Sampling), since we are using a fixed number of infinitesimal quantities. Hence, the complexity for finding sample points in $\mathrm{R}\langle\varepsilon,\delta\rangle$ is bounded by $(k'(k-k')+1)\sum_{j\le k'} 4^j \binom{s}{j} d^{O(k)} = s^{k'} d^{O(k)}$.Note that the degrees of the polynomials output are bounded by $O(d)^k$ and that when $\mathrm{D} = \mathbb{Z}$, and the bitsizes of the coefficients of Q and $P \in \mathcal{P}$ are bounded by τ, the bitsizes of the coefficients of the polynomials occurring in the multiplication table are $\tau d^{O(k)}$. Moreover the number of real univariate representations output is $s^{k'} O(d)^k$.

The cost of computing the signs is $s\, d^{O(k)}$ per point associated to a real univariate representation. Hence, the complexity of the sign determination at the end of the algorithm is bounded by

$$(k'(k-k')+1)\sum_{j \le k'} 4^j \binom{s}{j} s\, d^{O(k)} = s^{k'+1} d^{O(k)}.$$

Note that if we want the points to have coordinates in R^k, the complexity of finding sample points is still $s^{k'+1} d^{O(k)}$ in D, using Algorithm 11.20 (Removal of Infinitesimals).

If $\mathrm{D} = \mathbb{Z}$, and the bitsizes of the coefficients of the polynomials are bounded by τ, then the bitsizes of the integers appearing in the intermediate computations and the output are bounded by $\tau d^{O(k)}$. □

Proof of Theorem 13.22: The claim is an immediate consequence of the complexity analysis of Algorithm 13.3 (Sampling on an Algebraic Set). □

Remark 13.37. The complexity of Algorithm 13.3 is rather satisfactory since it fits with the bound on the number of realizable sign conditions proved in Proposition 7.35. □

The following result is an immediate corollary of Theorem 13.22.

Theorem 13.38. *Let* $\mathrm{Zer}(Q, \mathrm{R}^k)$ *be an algebraic set of real dimension* k', *where* Q *is a polynomial in* $\mathrm{R}[X_1, \ldots, X_k]$ *of degree at most* d, *and let* $\mathcal{P} \subset \mathrm{R}[X_1, \ldots, X_k]$ *be a finite set of* s *polynomials with each* $P \in \mathcal{P}$ *also of degree at most* d. *Let* D *be the ring generated by the coefficients of* Q *and the polynomials in* $\mathcal{P}$. *There is an algorithm that takes as input* Q, k', *and* $\mathcal{P}$ *and computes* $\mathrm{SIGN}(\mathcal{P}, Q)$ *with complexity*

$$(k'(k-k')+1)\sum_{j \le k'} 4^j \binom{s}{j} s\, d^{O(k)} = s^{k'+1} d^{O(k)}$$

in D. If D$=$Z*, and the bitsizes of the coefficients of the polynomials are bounded by* τ*, then the bitsizes of the integers appearing in the intermediate computations and the output are bounded by* $\tau d^{O(k)}$.

Remark 13.39. Note that the dimension of the algebraic set is part of the input. A method for computing the dimension of an algebraic set is given at the end of Chapter 14. □

13.4 Computing the Euler-Poincaré Characteristic of Sign Conditions

Our aim is to give a method for determining the Euler-Poincaré characteristic of the realization of sign conditions realized by a finite set $\mathcal{P} \subset \mathrm{R}[X_1, \ldots, X_k]$ on an algebraic set $Z = \mathrm{Zer}(Q, \mathrm{R}^k)$, with $Q \in \mathrm{R}[X_1, \ldots, X_k]$.

This is done by a method very similar to Algorithm 10.11 (Sign Determination): we compute Euler-Poincaré characteristics of realizations of sign conditions rather than cardinalities of sign conditions on a finite set, using the notion of Euler-Poincaré-query rather than that of Tarski-query.

We recall the following definitions already introduced in Section 6.3.

Given S a locally closed semi-algebraic set contained in Z, we denote by $\chi(S)$ the Euler-Poincaré characteristic of S.

Given $P \in \mathrm{R}[X_1, \ldots, X_k]$, we denote

$$\begin{aligned} \mathrm{Reali}(P=0, S) &= \{x \in S \mid P(x)=0\}, \\ \mathrm{Reali}(P>0, S) &= \{x \in S \mid P(x)>0\}, \\ \mathrm{Reali}(P>0, S) &= \{x \in S \mid P(x)<0\}, \end{aligned}$$

and $\chi(P=0, S), \chi(P>0, S), \chi(P<0, S)$ the Euler-Poincaré characteristics of the corresponding sets The Euler-Poincaré-query of P for S is

$$\mathrm{EuQ}(P, S) = \chi(P>0, S) - \chi(P<0, S).$$

Let $\mathcal{P} = P_1, \ldots, P_s$ be a finite list of polynomials in $\mathrm{R}[X_1, \ldots, X_k]$.

Let σ be a sign condition on $\mathcal{P}$. The **realization of the sign condition** σ **at** S is

$$\mathrm{Reali}(\sigma, S) = \{x \in S \mid \bigwedge_{P \in \mathcal{P}} \mathrm{sign}(P(x)) = \sigma(P)\},$$

and its Euler-Poincaré characteristic is denoted $\chi(\sigma, S)$.

Notation 13.40. Let $Q \in \mathrm{R}[X_1, \ldots, X_k]$, $Z = \mathrm{Zer}(Q, \mathrm{R}^k)$. We denote as usual by $\mathrm{SIGN}(\mathcal{P}, Z)$ the list of $\sigma \in \{0, 1, -1\}^{\mathcal{P}}$ such that $\mathrm{Reali}(\sigma, Z)$ is non-empty. We denote by $\chi(\mathcal{P}, Z)$ the list of Euler-Poincaré characteristics $\chi(\sigma, Z) = \chi(\mathrm{Reali}(\sigma, Z))$ for $\sigma \in \mathrm{SIGN}(\mathcal{P}, Z)$. We are going to compute $\chi(\mathcal{P}, Z)$, using Euler-Poincaré-queries of products of elements of $\mathcal{P}$. □

We use Notation 10.67, and order lexicographically $\{0, 1, -1\}^{\mathcal{P}}$ and $\{0,1,2\}^{\mathcal{P}}$. Given $A = \alpha_1, \ldots, \alpha_m$ a list of elements of $\{0, 1, 2\}^{\mathcal{P}}$, with $\alpha_1 <_{\text{lex}} \ldots <_{\text{lex}} \alpha_m$, we write $\mathcal{P}^A$ for $\mathcal{P}^{\alpha_1}, \ldots, \mathcal{P}^{\alpha_m}$, and $\text{EuQ}(\mathcal{P}^A, S)$ for $\text{EuQ}(\mathcal{P}^{\alpha_1}, S), \ldots, \text{EuQ}(\mathcal{P}^{\alpha_m}, S)$.

We denote by $\text{Mat}(A, \Sigma)$ the matrix of signs of $\mathcal{P}^A$ on Σ (see Definition 10.3).

Proposition 13.41. *If* $\cup_{\sigma\in\Sigma} \text{Reali}(\sigma, S) = S$, *then*

$$\text{Mat}(A, \Sigma) \cdot \chi(\Sigma, S) = \text{EuQ}(\mathcal{P}^A, S).$$

Proof: The proof is by induction on the number s of polynomials in $\mathcal{P}$. The statement when $s=1$ follows from Proposition 6.60, since the Euler-Poincaré characteristic of an empty sign condition is zero. Suppose the statement holds for $\mathcal{P}' = P_1, \ldots, P_{s-1}$ and consider $\mathcal{P} = P_1, \ldots, P_s$. Define

$$\begin{aligned}
\Sigma_0 &= \{\sigma \in \Sigma \mid \sigma(P_s) = 0\}\\
\Sigma_1 &= \{\sigma \in \Sigma \mid \sigma(P_s) = 1\}\\
\Sigma_{-1} &= \{\sigma \in \Sigma \mid \sigma(P_s) = -1\},\\
T &= \bigcup_{\sigma\in\Sigma_0} \text{Reali}(\sigma, S)\\
U &= \bigcup_{\sigma\in\Sigma_1} \text{Reali}(\sigma, S)\\
V &= \bigcup_{\sigma\in\Sigma_{-1}} \text{Reali}(\sigma, S).
\end{aligned}$$

Note that T, U, and V are all locally closed whenever S is locally closed. Let $\alpha \in \{0, 1, 2\}^{\mathcal{P}}$ and $\alpha' \in \{0, 1, 2\}^{\mathcal{P}'}$ defined by $\alpha'(P_j) = \alpha(P_j)$, for $1 \le j \le s-1$. Using the additive property of Euler-Poincaré characteristic (Proposition 6.57),

$$\begin{aligned}
\chi(\mathcal{P}^\alpha = 0, S) &= \chi(\mathcal{P}^\alpha = 0, T) + \chi(\mathcal{P}^\alpha = 0, U) + \chi(\mathcal{P}^\alpha = 0, V),\\
\chi(\mathcal{P}^\alpha > 0, S) &= \chi(\mathcal{P}^\alpha > 0, T) + \chi(\mathcal{P}^\alpha > 0, U) + \chi(\mathcal{P}^\alpha > 0, V),\\
\chi(\mathcal{P}^\alpha < 0, S) &= \chi(\mathcal{P}^\alpha < 0, T) + \chi(\mathcal{P}^\alpha < 0, U) + \chi(\mathcal{P}^\alpha < 0, V).
\end{aligned}$$

- If $\alpha(P_s) = 0$,

$$\text{EuQ}(\mathcal{P}^\alpha, S) = \text{EuQ}(\mathcal{P}'^{\alpha'}, T) + \text{EuQ}(\mathcal{P}'^{\alpha'}, U) + \text{EuQ}(\mathcal{P}'^{\alpha'}, V).$$

- If $\alpha(P_s) = 1$,

$$\text{EuQ}(\mathcal{P}^\alpha, S) = \text{EuQ}(\mathcal{P}'^{\alpha'}, U) - \text{EuQ}(\mathcal{P}'^{\alpha'}, V).$$

- If $\alpha(P_s) = 2$,

$$\text{EuQ}(\mathcal{P}^\alpha, S) = \text{EuQ}(\mathcal{P}'^{\alpha'}, U) + \text{EuQ}(\mathcal{P}'^{\alpha'}, V).$$

The claim follows from the induction hypothesis applied to T, U and V, the definition of $\text{Mat}(A, \Sigma)$ (Definition 2.66) and the additive property of Euler-Poincaré characteristic (Proposition 6.57), which implies, for every $\sigma \in \Sigma$,

$$\chi(\sigma, S) = \chi(\sigma, T) + \chi(\sigma, U) + \chi(\sigma, V).$$

□

Let $Q \in \mathrm{R}[X_1, \ldots, X_k]$, $Z = \mathrm{Zer}(Q, \mathrm{R}^k)$. We consider a list $A(Z)$ of elements in $\{0, 1, 2\}^{\mathcal{P}}$ **adapted to sign determination** for $\mathcal{P}$ on Z, i.e. such that the matrix of signs of $\mathcal{P}^A$ over $\mathrm{SIGN}(\mathcal{P}, Z)$ is invertible. If $\mathcal{P} = P_1, \ldots, P_s$, let $\mathcal{P}_i = P_i, \ldots, P_s$, for $0 \leq i \leq s$. A method for determining a list $A(\mathcal{P}, Z)$ of elements in $\{0,1,2\}^{\mathcal{P}}$ adapted to sign determination for $\mathcal{P}$ on Z from $\mathrm{SIGN}(\mathcal{P}, Z)$ has been given in Algorithm 10.12 (Family adapted to Sign Determination).

We are ready for describing the algorithm computing the Euler-Poincaré characteristic. We start with an algorithm for the Euler-Poincaré-query.

Algorithm 13.4. **[Euler-Poincaré-query]**

- **Structure:** an ordered domain D contained in a real closed field R.
- **Input:** a polynomial $Q \in \mathrm{D}[X_1, \ldots, X_k]$, with $Z = \mathrm{Zer}(Q, \mathrm{R}^k)$, a polynomial $P \in \mathrm{D}[X_1, \ldots, X_k]$.
- **Output:** the Euler-Poincaré-query
$$\mathrm{EuQ}(P, Z) = \chi(P > 0, Z) - \chi(P < 0, Z).$$
- **Complexity:** $d^{O(k)}$, where d is a bound on the degree of Q and the degree of P.
- **Procedure:**
  - Introduce a new variable X_{k+1}, and let
$$\begin{aligned} Q_+ &= Q^2 + (P - X_{k+1}^2)^2, \\ Q_- &= Q^2 + (P + X_{k+1}^2)^2. \end{aligned}$$
  Using Algorithm 12.25 compute $\chi(\mathrm{Zer}(Q_+, \mathrm{R}^{k+1}))$ and $\chi(\mathrm{Zer}(Q_-, \mathrm{R}^{k+1}))$. Output
$$(\chi(\mathrm{Zer}(Q_+, \mathrm{R}^{k+1})) - \chi(\mathrm{Zer}(Q_-, \mathrm{R}^{k+1})))/2.$$

Proof of correctness: The algebraic set $\mathrm{Zer}(Q_+, \mathrm{R}^{k+1})$ is semi-algebraically homeomorphic to the disjoint union of two copies of the semi-algebraic set defined by $(P > 0) \wedge (Q = 0)$, and the algebraic set defined by $(P = 0) \wedge (Q = 0)$. Hence, using Proposition 6.57, we have that
$$2\,\chi(P > 0, Z) = \chi(\mathrm{Zer}(Q_+, \mathrm{R}^{k+1})) - \chi(\mathrm{Zer}((Q, P), \mathrm{R}^k)).$$
Similarly, we have that
$$2\,\chi(P < 0, Z) = \chi(\mathrm{Zer}(Q_-, \mathrm{R}^{k+1})) - \chi(\mathrm{Zer}((Q, P), \mathrm{R}^k)).$$
□

Complexity Analysis: The complexity of the algorithm is $d^{O(k)}$ using the complexity analysis of Algorithm 12.25.

When $\mathrm{D} = \mathbb{Z}$ and the bitsizes of the coefficients of P are bounded by τ, the bitsizes of the intermediate computations and the output are bounded by $O(k^2 d^2(\tau + \log_2(k d))$. □

We are now ready to describe an algorithm for computing the Euler-Poincaré characteristic of the realizations of sign conditions.

Algorithm 13.5. **[Euler-Poincaré Characteristic of Sign Conditions]**

- **Structure:** an ordered domain D contained in a real close field R.
- **Input:** an algebraic set $Z = \mathrm{Zer}(Q, \mathrm{R}^k) \subset \mathrm{R}^k$ and a finite list $\mathcal{P}$ of polynomials in $\mathrm{D}[X_1, \ldots, X_k]$.
- **Output:** the list $\chi(\mathcal{P}, Z)$.
- **Complexity:** $s^{k'+1} O(d)^k + s^{k'}((k' \log_2(s) + k \log_2(d))\, d)^{O(k)}$, where k' is the dimension of Z, s is a bound on the number of elements of $\mathcal{P}$ and d is a bound on the degree of Q and the elements of $\mathcal{P}$.
- **Procedure:**
  - Let $\mathcal{P} = P_1, \ldots, P_s$, $\mathcal{P}_i = P_1, \ldots, P_i$. Compute $\mathrm{SIGN}(\mathcal{P}, Z)$ using Algorithm 13.3 (Sampling on an Algebraic Set).
  - Determine a list $A(\mathcal{P}, Z)$ adapted to sign determination for $\mathcal{P}$ on Z using Algorithm 10.12 (Family adapted to Sign Determination).
  - Define $A = A(\mathcal{P}, Z)$, $M = M(\mathcal{P}^A, \mathrm{SIGN}(\mathcal{P}, Z))$.
  - Compute $\mathrm{EuQ}(\mathcal{P}^A, Z)$ using repeatedly Algorithm 13.4 (Euler-Poincaré-query).
  - Using
  $$M \cdot \chi(\mathcal{P}, Z) = \mathrm{EuQ}(\mathcal{P}^A, Z),$$
  and the fact that M is invertible, compute $\chi(\mathcal{P}, Z)$.

Proof of correctness: Immediate from Proposition 13.41. □

Complexity analysis: By Proposition 7.35,

$$\#(\mathrm{SIGN}(\mathcal{P}, Z)) \leq \sum_{0 \leq j \leq k'} \binom{s}{j} 4^j d(2d-1)^{k-1} = s^{k'} O(d)^k.$$

The number of calls to to Algorithm 13.4 (Euler-Poincaré-query) is equal to $\#(\mathrm{SIGN}(\mathcal{P}, Z))$. The calls to Algorithm 13.4 (Euler-Poincaré-query) are done for polynomials which are products of at most

$$\log_2(\#(\mathrm{SIGN}(\mathcal{P}, Z))) = k' \log_2(s) + k\,(\log_2(d) + O(1).$$

products of polynomials of the form P or P^2, $P \in \mathcal{P}$ by Proposition 10.71, hence of degree $(k' \log_2(s) + k\ (\log_2(d) + O(1))\ d$. Using the complexity analysis of Algorithm 13.3 (Sampling on an Algebraic Set) and the complexity analysis of Algorithm 13.4 (Euler-Poincaré-query), the number of arithmetic operations is

$$s^{k'+1} O(d)^k + s^{k'}\,((k' \log_2(s) + k \log_2(d))\, d)^{O(k)}.$$

The algorithm also involves the inversion matrices of size $s^{k'} O(d)^k$ with integer coefficients.

If $\mathrm{D} = \mathbb{Z}$, and the bitsizes of the coefficients of the polynomials are bounded by τ, then the bitsizes of the integers appearing in the intermediate computations and the output are bounded by $\tau d^{O(k)}$. □

13.5 Bibliographical Notes

Grigor'ev and Vorobjov [77] gave the first algorithm to solve the decision problem for the existential theory of the reals whose time complexity is singly exponential in the number of variables. Canny [37], Heintz, Roy, and Solerno [85], and Renegar [133] improved their result in several directions. Renegar's [133] algorithms solved the existential theory of the reals in time $(s\ d)^{O(k)}$ (where d is the degree, k the number of variables, and s the number of polynomials). The first single exponential complexity computation for the Euler-Poincaré characteristic appears in [11].

The results presented in the three first sections are based on [13, 15]. The construction of the family $\mathcal{L}_{k,k-k'}$ described in Section 13.3, is on the work of Chistov, Fournier, Gurvits, and Koiran [42]. In terms of algebraic complexity (the degree of the equations), they are similar to [133]. They are more precise in terms of combinatorial complexity (the dependence on the number of equations), particularly for the computation of the realizable sign conditions on a lower dimensional algebraic set.

14

Quantifier Elimination

The principal problem we consider in this chapter is the quantifier elimination problem. This problem was already studied in Chapter 11, where we obtained doubly exponential complexity in the number of variables. On the other hand, we have seen in Chapter 13 an algorithm for the existential theory of the reals (which is to decide the truth or the falsity of a sentence with a single block of existential quantifiers) with complexity singly exponential in the number of variables (see Theorem 13.13). In this chapter, we pay special attention to the structure of the blocks of variables in a formula in order to take into account this block structure in the complexity estimates and improve the results obtained in Chapter 11.

If $Z=(Z_1,\ldots,Z_\ell)$, Φ is a formula, and $\mathrm{Qu}\in\{\forall,\exists\}$, we denote the formula $(\mathrm{Qu}\ Z_1)\ldots(\mathrm{Qu}\ Z_\ell)\ \Phi$ by the abbreviation $(\mathrm{Qu}\ Z)\ \Phi$.

Let $\mathcal{P}\subset \mathrm{R}[X_1,\ldots,X_k,Y_1,\ldots,Y_\ell]$ be finite, and let Π denote a partition of the list of variables $X=(X_1,\ldots,X_k)$ into blocks, $X_{[1]},\ldots,X_{[\omega]}$, where the block $X_{[i]}$ is of size $k_i, 1\le i\le\omega$, $\sum_{1\le i\le\omega} k_i=k$.

A $(\mathcal{P},\Pi)$-formula $\Phi(Y)$ is a formula of the form

$$\Phi(Y)=(\mathrm{Qu}_1 X_{[1]})\ldots(\mathrm{Qu}_\omega X_{[\omega]})F(X,Y),$$

where $\mathrm{Qu}_i\in\{\forall,\exists\}$, $Y=(Y_1,\ldots,Y_\ell)$, and $F(X,Y)$ is a quantifier free $\mathcal{P}$-formula.

In Section 14.1, we describe an algorithm for solving the general decision problem, that is a procedure to decide the truth or falsity of a $(\mathcal{P},\Pi)$-sentence. The key notion here is the tree of realizable sign conditions of a family of polynomials with respect to a block structure Π on the set of variables. This is a generalization of the set of realizable sign conditions, seen in Chapter 7, which corresponds to one single block of variables. It is also a generalization of the tree of cylindrical realizable sign conditions, seen in Chapter 11, which correspond to k blocks of one variable each. The basic idea of this algorithm is to perform parametrically the algorithm in Chapter 13, using the critical point method.

Section 14.2 is devoted to the more general problem of quantifier elimination for a $(\mathcal{P},\Pi)$-formula.

Section 14.3 is devoted to a variant of Quantifier Elimination exploiting better the logical structure of the formula.

Finally, the block elimination technique is used to perform global optimization and compute the dimension of a semi-algebraic set in Section 14.4 and Section 14.5.

14.1 Algorithm for the General Decision Problem

We first study the general decision problem, which is to decide the truth or falsity of a $(\mathcal{P},\Pi)$-sentence (which is a $(\mathcal{P},\Pi)$-formula without free variables). In order to decide the truth or falsity of a sentence, we construct a certain tree of sign conditions adapted to the block structure Π of the sentence, which we define below.

The following definition generalizes the definition of the tree of cylindrical realizable sign conditions (Notation 11.7). The importance of this notion is that the truth or falsity of any $(\mathcal{P}, \Pi)$-sentence can be decided from $\mathrm{SIGN}_\Pi(\mathcal{P})$.

Notation 14.1. [Block realizable sign conditions] Let $\mathcal{P}$ be a set of s polynomials in k variables $X_1,\ldots,X_k$, and let Π denote a partition of the list of variables $X_1,\ldots,X_k$ into blocks, $X_{[1]},\ldots,X_{[\omega]}$, where the block $X_{[i]}$ is of size k_i, for $1\le i\le\omega$, $\sum_{1\le i\le\omega} k_i=k$. Let $\mathrm{R}^{[i]}=\mathrm{R}^{k_1+\cdots+k_i}$, and let $\pi_{[i]}$ be the projection from $\mathrm{R}^{[i+1]}$ to $\mathrm{R}^{[i]}$ forgetting the last k_{i+1}-coordinates. Note that $\mathrm{R}^{[\omega]}=\mathrm{R}^k$. By convention, $\mathrm{R}^{[0]}=\{0\}$.

We are going to define inductively the tree of realizable sign conditions of $\mathcal{P}$ with respect to Π.

For $z\in\mathrm{R}^{[\omega]}$, let $\mathrm{SIGN}_{\Pi,\omega}(\mathcal{P})(z)=\mathrm{sign}(\mathcal{P})(z)$, where $\mathrm{sign}(\mathcal{P})(z)$ is the sign condition on $\mathcal{P}$ mapping $P\in\mathcal{P}$ to $\mathrm{sign}(P)(z)\in\{0,1,-1\}$ (Notation 11.7).

For all i, $0\le i<\omega$, and $y\in\mathrm{R}^{[i]}$, we inductively define,

$$\mathrm{SIGN}_{\Pi,i}(\mathcal{P})(y)=\{\mathrm{SIGN}_{\Pi,i+1}(\mathcal{P})(z)|z\in\mathrm{R}^{[i+1]},\pi_{[i]}(z)=y\}.$$

Finally, we define

$$\mathrm{SIGN}_\Pi(\mathcal{P})=\mathrm{SIGN}_{\Pi,0}(\mathcal{P})(0).$$

Note that $\mathrm{SIGN}_\Pi(\mathcal{P})$ is naturally equipped with a tree structure. We call $\mathrm{SIGN}_\Pi(\mathcal{P})$ the **tree of realizable sign conditions of $\mathcal{P}$ with respect to Π.** □

When there is only one block of variables, we recover $\mathrm{SIGN}(\mathcal{P})$ (Notation 7.29). When $\Pi=\{X_1\},\ldots,\{X_k\}$, we recover $\mathrm{CSIGN}(\mathcal{P})$ (Notation 11.7).

We will see that the truth or falsity of a $(\mathcal{P},\Pi)$-sentence can be decided from the set $\mathrm{SIGN}_\Pi(\mathcal{P})$. We first consider an example.

Example 14.2. Let $P = X_1^2 + X_2^2 + X_3^2 - 1$, $\mathcal{P} = \{P\}$. Let Π consist of two blocks of variables, defined by $X_{[1]} = X_1$ and $X_{[2]} = \{X_2, X_3\}$. Note that $\pi_{[1]}$ projects $\mathrm{R}^{[2]} = \mathrm{R}^3$ to $\mathrm{R}^{[1]} = \mathrm{R}$ by forgetting the last two coordinates. We now determine $\mathrm{SIGN}_\Pi(\mathcal{P})$.

For $x \in \mathrm{R} = \mathrm{R}^{[1]}$,

$$\mathrm{SIGN}_{\Pi,1}(\mathcal{P})(x) = \{\mathrm{sign}(P(z)) \mid z \in \mathrm{R}^{[2]}, \pi_{[1]}(z) = x\}.$$

Thus

$$\mathrm{SIGN}_{\Pi,1}(\mathcal{P})(x) = \begin{cases} \{0,1,-1\} & \text{if } x \in (-1,1) \\ \{0,1\} & \text{if } x \in \{-1,1\} \\ \{1\} & \text{otherwise.} \end{cases}$$

Finally,

$$\mathrm{SIGN}_\Pi(\mathcal{P}) = \{\mathrm{SIGN}_{\Pi,1}(\mathcal{P})(x) \mid x \in \mathrm{R}\}.$$

Thus

$$\mathrm{SIGN}_\Pi(\mathcal{P}) = \{\{1\}, \{0,1\}, \{0,1,-1\}\}.$$

This means that there are three cases:

- there are values of x_1 for which the only sign taken by $P(x_1, x_2, x_3)$ when (x_2, x_3) varies in R^2 is 1,
- there are values of x_1 for which the only sign taken by $P(x_1, x_2, x_3)$ when (x_2, x_3) varies in R^2 are 0 and 1,
- there are values of x_1 for which the signs taken by $P(x_1, x_2, x_3)$ when (x_2, x_3) varies in R^2 are 0, 1 and -1,
- and these exhaust all choices of $x_1 \in \mathrm{R}$.

So, the sentence $(\forall X_1)\ (\exists (X_2, X_3))\ X_1^2 + X_2^2 + X_3^2 - 1 > 0$ is certainly true.

Since there are values of x_1 for which the only sign taken by $P(x_1, x_2, x_3)$ for every $(x_2, x_3) \in \mathrm{R}^2$ is 1 it is equally clear that the sentence $(\exists X_1)\ (\forall (X_2, X_3))\ X_1^2 + X_2^2 + X_3^2 - 1 > 0$ is true.

On the other hand, the sentence $(\forall X_1)\ (\exists (X_2, X_3))\ X_1^2 + X_2^2 + X_3^2 - 1 = 0$ is false: there are values of x_1 for which the only sign taken by $P(x_1, x_2, x_3)$ is 1.

This differs from what was done in Example 11.10 in that here we do not decompose the (X_2, X_3) space: this is because the variables $\{X_2, X_3\}$ belong to the same block of quantifiers. So the information provided by $\mathrm{SIGN}_\Pi(\mathcal{P})$ is weaker than the information provided by $\mathrm{CSIGN}(\mathcal{P})$ (Notation 11.7). Note that $\mathrm{SIGN}_\Pi(\mathcal{P})$ does not provide the information necessary to decide the truth or falsity of the sentence

$$\Phi = (\exists X_1)\ (\forall X_2)\ (\exists X_3)\ X_1^2 + X_2^2 + X_3^2 - 1 = 0$$

since we do not have information for the corresponding block structure, while we have able to decide that Φ is false using

$\mathrm{CSIGN}(\mathcal{P}) = \{\{\{\{1\}, \{0,1\}, \{0,1,-1\}\}, \{\{1\}, \{0,1\}\}, \{\{1\}\}\}$ in Example 11.16.

If we take $\mathcal{Q}=\{X_1-X_3^2\}$, it is easy to check that

$$\mathrm{SIGN}_\Pi(\mathcal{Q})=\{\{1\},\{0,1\},\{0,1,-1\}\}=\mathrm{SIGN}_\Pi(\mathcal{P}).$$

On the other hand we can determine

$$\mathrm{CSIGN}(\mathcal{Q})=\{\{\{1\}\},\{\{0,1\}\},\{\{\{0,1,-1\}\}\}$$

and notice that

$$\mathrm{CSIGN}(\mathcal{Q})\neq\mathrm{CSIGN}(\mathcal{P}).$$

Using $\mathrm{CSIGN}(\mathcal{Q})$, we can check that the sentence

$$\Phi'=(\exists X_1)\ (\forall X_2)\ (\exists X_3)\ X_1-X_3^2=0$$

while the corresponding Φ, discussed above, is false. □

We use again Notation 11.12.

Proposition 14.3. *The* $(\mathcal{P},\Pi)$*-sentence*

$$(\mathrm{Qu}_1 X_{[1]})\ (\mathrm{Qu}_2 X_{[2]})\ \ldots(\mathrm{Qu}_\omega X_{[\omega]})\ F(X),$$

is true if and only if

$$\mathrm{Qu}_1\sigma_1\in\mathrm{SIGN}_\Pi(\mathcal{P})\quad \mathrm{Qu}_2\sigma_2\in\sigma_1\ldots\mathrm{Qu}_\omega\sigma_\omega\in\sigma_{\omega-1}\quad F^\star(\sigma_\omega).$$

Proof: The proof is by induction on the number ω of blocks of quantifiers, starting from the one outside.

Since $(\forall\ X)\ \Phi$ is equivalent to $\neg\ (\exists X)\ \neg\Phi$, we can suppose without loss of generality that Qu_1 is $\exists$.

The claim is certainly true when there is one block of existential quantifiers, by definition of $\mathrm{sign}(\mathcal{P})$.

Suppose that

$$(\exists X_{[1]})\ (\mathrm{Qu}_2 X_{[2]})\ldots\ (\mathrm{Qu}_\omega X_{[\omega]})F(X)$$

is true, and choose $a_{[1]}\in\mathrm{R}^{k_1}$ such that

$$(\mathrm{Qu}_2 X_{[2]})\ldots\ (\mathrm{Qu}_\omega X_{[\omega]})F(a_{[1]},X_{[2]},\ldots,X_{[\omega]})$$

is true. Note that if $\mathcal{P}_{a_{[1]}}$ is the set of polynomials obtained by substituting $a_{[1]}\in\mathrm{R}^{k_1}$ for $X_{[1]}$ in $\mathcal{P}$ and $\Pi'=X_{[2]},\ldots,X_{[\omega]}$,

$$\mathrm{SIGN}_{\Pi,1}(\mathcal{P})(a_{[1]})=\mathrm{SIGN}_{\Pi'}(\mathcal{P}_{a_{[1]}}).$$

By induction hypothesis,

$$\mathrm{Qu}_2\sigma_2\in\mathrm{SIGN}_{\Pi'}(\mathcal{P}_{a_{[1]}})\ldots\mathrm{Qu}_\omega\sigma_\omega\in\sigma_{\omega-1}\quad F^\star(\sigma_\omega)$$

is true. So taking $\sigma_1 = \mathrm{SIGN}_{\Pi,1}(\mathcal{P})(a_{[1]}) = \mathrm{SIGN}_{\Pi'}(\mathcal{P}_{a_{[1]}}) \in \mathrm{SIGN}_\Pi(\mathcal{P})$,

$$\exists \sigma_1 \in \mathrm{SIGN}_\Pi(\mathcal{P}) \quad \mathrm{Qu}_2\sigma_2 \in \sigma_1 ... \mathrm{Qu}_\omega\sigma_\omega \in \sigma_{\omega-1} \quad F^\star(\sigma_\omega)$$

is true.

Conversely, suppose

$$\exists \sigma_1 \in \mathrm{SIGN}_\Pi(\mathcal{P}) \quad \mathrm{Qu}_2\sigma_2 \in \sigma_1 ... \mathrm{Qu}_\omega\sigma_\omega \in \sigma_{\omega-1} \quad F^\star(\sigma_\omega)$$

is true and choose $\sigma_1 \in \mathrm{SIGN}_\Pi(\mathcal{P})$ such that

$$\mathrm{Qu}_2\sigma_2 \in \sigma_1 ... \mathrm{Qu}_\omega\sigma_\omega \in \sigma_{\omega-1} \quad F^\star(\sigma_\omega)$$

is true. By definition of $\mathrm{SIGN}_\Pi(\mathcal{P})$, $\sigma_1 = \mathrm{SIGN}_{\Pi'}(\mathcal{P})(a_{[1]})$ for some $a_{[1]} \in \mathrm{R}^{k_1}$, and hence

$$\mathrm{Qu}_2\sigma_2 \in \mathrm{SIGN}_{\Pi'}(\mathcal{P}_{a_{[1]}}) ... \mathrm{Qu}_\omega\sigma_\omega \in \sigma_{\omega-1} \quad F^\star(\sigma_\omega)$$

is true. By induction hypothesis,

$$(\mathrm{Qu}_2 X_{[2]}) ... (\mathrm{Qu}_\omega X_\omega)\ F(a_{[1]}, X_{[2]}, ..., X_{[\omega]})$$

is true. Thus

$$(\exists X_{[1]})\ (\mathrm{Qu}_2 X_{[2]}) ... (\mathrm{Qu}_\omega X_{[\omega]}) F(X)$$

is true. □

In the cylindrical situation studied in Chapter 11, $\mathrm{CSIGN}(\mathcal{P})$ was obtained from a cylindrical set of sample points of a cylindrical decomposition adapted to $\mathcal{P}$. We generalize this approach to a general block structure.

A **Π-set** $\mathcal{A} = \mathcal{A}_1, ..., \mathcal{A}_\omega$ is such that $\mathcal{A}_i$ is a finite set contained in $\mathrm{R}^{[i]}$ and $\pi_{[i]}(\mathcal{A}_{i+1}) = \mathcal{A}_i$.

We define inductively the **tree of realizable sign conditions of $\mathcal{P}$ for $\mathcal{A}$ with respect to** Π, $\mathrm{SIGN}_\Pi(\mathcal{P}, \mathcal{A})$, as follows:

- For $z \in \mathcal{A}_\omega$, let $\mathrm{SIGN}_{\Pi,\omega}(\mathcal{P})(z) = \mathrm{sign}(\mathcal{P})(z)$, where $\mathrm{sign}(\mathcal{P})(z)$ is the sign condition on $\mathcal{P}$ mapping $P \in \mathcal{P}$ to $\mathrm{sign}(P)(z) \in \{0, 1, -1\}$ (Notation 11.7).
- For all i, $1 \le i < \omega$, and all $y \in \mathcal{A}_i$, we inductively define,

$$\mathrm{SIGN}_{\Pi,i}(\mathcal{P}, \mathcal{A})(y) = \{\mathrm{SIGN}_{\Pi,i+1}(\mathcal{P}, \mathcal{A})(z) | z \in \mathcal{A}_{i+1}, \pi_{[i]}(z) = y\}.$$

Finally, we define

$$\mathrm{SIGN}_\Pi(\mathcal{P}, \mathcal{A}) = \mathrm{SIGN}_{\Pi,0}(\mathcal{P}, \mathcal{A})(0).$$

Note that $\mathrm{SIGN}_\Pi(\mathcal{P}) = \mathrm{SIGN}_\Pi(\mathcal{P}, \mathrm{R}^k)$. Note also that $\mathrm{SIGN}_\Pi(\mathcal{P}, \mathcal{A})$ is a subtree of $\mathrm{SIGN}_\Pi(\mathcal{P})$.

A **Π-set of sample points** for $\mathcal{P}$ is a Π-set $\mathcal{A} = \mathcal{A}_1, ..., \mathcal{A}_\omega$ such that

$$\mathrm{SIGN}_\Pi(\mathcal{P}, \mathcal{A}) = \mathrm{SIGN}_\Pi(\mathcal{P}).$$

A **Π-partition adapted to** $\mathcal{P}$ is given by $\mathcal{S} = \mathcal{S}_1, \ldots, \mathcal{S}_\omega$, where $\mathcal{S}_i$ is a partition of $\mathrm{R}^{[i]}$ into a finite number of semi-algebraically connected semi-algebraic sets such that for every $S \in \mathcal{S}_{i+1}$, $\pi_{[i]}(S) \in \mathcal{S}_i$, and such that every $S \in \mathcal{S}_\omega$ is $\mathcal{P}$- invariant. A **Π-set of sample points** for a Π-partition $\mathcal{S}$ is a Π-set $\mathcal{A} = \mathcal{A}_1, \ldots, \mathcal{A}_\omega$ such that

- for every i, $1 \le i \le \omega$, $\mathcal{A}_i$ intersects every $S \in \mathcal{S}_i$,
- for every i, $1 \le i \le \omega - 1$, $\pi_{[i]}(\mathcal{A}_{i+1}) = \mathcal{A}_i$.

Note that the partition of R^k by the semi-algebraically connected components of realizable sign conditions of $\mathcal{P}$ is a Π-partition with the block structure $\Pi = \{X_1 \ldots, X_k\}$ (i.e. with a single block), and a set of sample points for $\mathcal{P}$ is a Π-set of sample points for this block structure. Note also that a cylindrical decomposition $\mathcal{S}$ adapted to $\mathcal{P}$ is a Π-partition for the block structure $X_1, \ldots, X_k$ (k-blocks of one variable) and a cylindrical set of sample points for $\mathcal{S}$ is a Π-set of sample points for $\mathcal{S}$ for this block structure.

We are going to prove a result generalizing Proposition 11.9 to the case of a general block structure.

Proposition 14.4. *Let $\mathcal{S} = \mathcal{S}_1, \ldots, \mathcal{S}_\omega$ be a Π-partition of R^k adapted to $\mathcal{P}$ and $\mathcal{A} = \mathcal{A}_1, \ldots, \mathcal{A}_\omega$ be a Π-set of sample points for $\mathcal{S}$. Then $\mathcal{A}$ is a Π-set of sample points for $\mathcal{P}$.*

The proof is similar to the proof of Proposition 11.9 and uses the following generalization of Proposition 11.11.

Proposition 14.5. *Let $\mathcal{S} = \mathcal{S}_1, \ldots, \mathcal{S}_\omega$ be a Π-partition of R^k adapted to $\mathcal{P}$. For every $1 \le i \le \omega$ and every $S \in \mathcal{S}_i$, $\mathrm{SIGN}_{\Pi,i}(y)$ is constant as y varies in S.*

Proof: The proof is by induction on $\omega - i$.

If $i = \omega$, the claim is true since the sign of every $P \in \mathcal{P}$ is fixed on $S \in \mathcal{S}_\omega$.

Suppose that the claim is true for $i+1$ and consider $S \in \mathcal{S}_i$. Let $T_1, \ldots, T_\ell$ be the elements of $\mathcal{S}_{i+1}$ such that $\pi_{[i]}(T_j) = S$. By induction hypothesis, $\mathrm{SIGN}_{\Pi,i+1}(\mathcal{P})(z)$ is constant as z varies in T_j. Since $\mathcal{S}$ is a Π-partition, $\bigcup_{j=1}^{\ell} T_j = S \times \mathrm{R}^{k_{i+1}}$. Thus

$$\mathrm{SIGN}_{\Pi,i}(\mathcal{P})(y) = \{\mathrm{SIGN}_{\Pi,i+1}(\mathcal{P})(z) | z \in \mathrm{R}^{[i+1]}, \pi_{[i]}(z) = y\}$$

is constant as y varies in S. □

Proof of Proposition 14.4: Let $\mathcal{A}_0 = \{0\}$. We are going to prove that for every $y \in \mathcal{A}_i$,

$$\mathrm{SIGN}_{\Pi,i}(\mathcal{P})(y) = \mathrm{SIGN}_{\Pi,i}(\mathcal{P}, \mathcal{A})(y).$$

The proof is by induction on $\omega - i$.

If $i = \omega$, the claim is true since $\mathcal{A}_\omega$ meets every element of $\mathcal{S}_\omega$.

Suppose that the claim is true for $i+1$ and consider $y \in \mathcal{A}_i$. Let S be the element of $\mathcal{S}_i$ containing y, and let $T_1, \ldots, T_\ell$ be the elements of $\mathcal{S}_{i+1}$ such that $\pi_{[i]}(T_j) = S$. Denote by z_j a point of $T_j \cap \mathcal{A}_{i+1}$ such that $\pi_{[i]}(z_j) = y$. By induction hypothesis,

$$\mathrm{SIGN}_{\Pi,i+1}(\mathcal{P})(z_j) = \mathrm{SIGN}_{\Pi,i+1}(\mathcal{P}, \mathcal{A})(z_j).$$

Since $T_1 \cup \ldots \cup T_\ell = S \times \mathrm{R}^{k_{i+1}}$ and $\mathrm{SIGN}_{\Pi,i+1}(\mathcal{P})(z)$ does not change as z varies over T_j,

$$\begin{aligned} \mathrm{SIGN}_{\Pi,i}(\mathcal{P})(y) &= \{\mathrm{SIGN}_{\Pi,i+1}(\mathcal{P})(z) | z \in \mathrm{R}^{[i+1]}, \pi_{[i]}(z) = y\} \\ &= \{\mathrm{SIGN}_{\Pi,i+1}(\mathcal{P}, \mathcal{A})(z) | z \in \mathcal{A}_{i+1}, \pi_{[i]}(z) = y\} \\ &= \mathrm{SIGN}_{\pi,i}(\mathcal{P}, \mathcal{A})(y). \end{aligned}$$

□

We now construct a Π-partition of R^k adapted to $\mathcal{P}$, generalizing Theorem 5.6. Note that a cylindrical decomposition adapted to $\mathcal{P}$ gives a Π-partition of R^k adapted to $\mathcal{P}$, so the issue here is not an existence theorem similar to Theorem 5.6 but rather a complexity result taking into account the block structure Π. The construction of a cylindrical decomposition adapted to $\mathcal{P}$ in Chapter 5 and Chapter 11 was based on a recursive call to an Elimination procedure eliminating one variable (see Algorithm 11.1 (Subresultant Elimination)). In the general block structure context, we define a Block Elimination procedure which replaces a block of variables by one single variable and computes parametrized univariate representations, giving in a parametric way sample points for every non-empty sign condition. Finally we eliminate this variable.

Algorithm 14.1. **[Block Elimination]**

- **Structure:** an ordered domain D contained in a real closed field R.
- **Input:** a block of variables $X = (X_1, \ldots, X_k)$ and a set of polynomials

$$\mathcal{P}(Y) \subset \mathrm{D}[Y_1, \ldots, Y_\ell, X_1, \ldots, X_k].$$

- **Output:**
  - a set $\mathrm{BElim}_X(\mathcal{P}) \subset \mathrm{D}[Y]$ such that $\mathrm{SIGN}(\mathcal{P}(y, X_1, \ldots, X_k))$ (Notation 7.29) is fixed as y varies over a semi-algebraically connected component of a realizable sign condition of $\mathrm{BElim}_X(\mathcal{P})$,
  - a set $\mathrm{UR}_X(\mathcal{P})$ of parametrized univariate representations of the form

$$u(Y, \varepsilon, \delta) = (f, g_0, \ldots, g_k),$$

  where $f, g_i \in \mathrm{D}[Y, \varepsilon, \delta][T]$. For any point $y \in \mathrm{R}^\ell$, denoting by $\mathrm{UR}_X(\mathcal{P})(y)$ the subset of $\mathrm{UR}_X(\mathcal{P})$ such that $f(y, T)$ and $g_0(y, T)$ are coprime, the points associated to the univariate representations in $\mathrm{UR}_X(\mathcal{P})(y)$ intersect every semi-algebraically connected component of every realizable sign condition of the set $\mathcal{P}(y)$ in $\mathrm{R}\langle\varepsilon, \delta\rangle^k$.

- **Complexity:** $s^{k+1} d^{O(\ell k)}$, where s is a bound on the number of elements of $\mathcal{P}$ and d is a bound on their degree.
- **Procedure:**
  - Initialize $\mathrm{UR}_X(\mathcal{P})$ to the empty set.
  - Take as d' the smallest even natural number $>d$.
  - Define

$$\begin{aligned} P_i^\star &= \{(1-\delta)\, P_i + \delta\, H_k(d', i), (1-\delta)\, P_i - \delta\, H_k(d', i), \\ &\quad (1-\delta)\, P_i + \delta\, \gamma\, H_k(d', i), (1-\delta)\, P_i - \delta\, \gamma\, H_k(d', i)\} \\ \mathcal{P}^\star &= \{P_1^\star, \ldots, P_s^\star\} \end{aligned}$$

  for $0 \le i \le s$ using Notation 13.4.
  - For every subset $\mathcal{Q}$ of $j \le k$ polynomials $Q_{i_1} \in P_{i_1}^\star, \ldots, Q_{i_j} \in P_{i_j}^\star$,
    - let $Q = Q_{i_1}^2 + \cdots + Q_{i_j}^2 + (\varepsilon^2 (X_1^2 + \cdots + X_k^2 + X_{k+1}^2) - 1)^2$.
    - Take for $i = 1, \ldots, k$, $\bar{d}_i$ the smallest even natural number $> \deg(Q)$, $\bar{d}_{k+1} = 6$, $\bar{d} = (\bar{d}_1, \ldots, \bar{d}_k, \bar{d}_{k+1})$, and $c = \varepsilon$.
    - Perform Algorithm 12.10 (Parametrized Multiplication Table) with input $\overline{\mathrm{Cr}}(Q, \zeta)$ (using Notation 12.46). Output $\mathcal{M}$.
    - Perform Algorithm 12.15 (Parametrized Limit of Bounded Points) with input $\gamma, \zeta, \overline{\mathrm{Cr}}(Q, \zeta)$, and $\mathcal{M}$. Add the parametrized univariate representations (belonging to $\mathrm{D}[Y, \varepsilon, \delta][T]^{k+2}$) output to $\mathrm{UR}_X(\mathcal{P})$.
  - For every $v = (f, g_0, \ldots, g_k) \in \mathrm{UR}_X(\mathcal{P})$, consider the family of univariate polynomials $\mathcal{F}_v$ consisting of f, its derivatives with respect to T, and P_v (see Notation 13.8), for every $P \in \mathcal{P}$. Compute $\mathrm{RElim}_T(f, \mathcal{F}_v)$ using Algorithm 11.19 (Restricted Elimination). Denote by $\mathcal{B}_v$ the family of polynomials in Y that are the coefficients of the polynomials in

$$\mathrm{RElim}_T(\mathcal{F}_v) \subset \mathrm{D}[Y, \varepsilon, \delta].$$

  - Define $\mathrm{BElim}_X(\mathcal{P})$ to be the union of the sets $\mathcal{B}_v \subset \mathrm{D}[Y]$ for every $v \in \mathrm{UR}_X(\mathcal{P})$.
  - Output $\mathrm{BElim}_X(\mathcal{P})$ and $\mathrm{UR}_X(\mathcal{P})$.

The proof of correctness of Algorithm 14.1 (Block Elimination) uses the following results, which describe how to get rid of infinitesimal quantities.

Notation 14.6. Let $\varepsilon_1, \varepsilon_2, \cdots, \varepsilon_m$ be variables and consider the real closed field $\mathrm{R}\langle\varepsilon_1, \varepsilon_2, \cdots, \varepsilon_m\rangle$. Let $S \subset \mathrm{R}\langle\epsilon_1, \ldots, \epsilon_m\rangle^k$ be a semi-algebraic set defined by a quantifier-free $\mathcal{P}$-formula Φ with $\mathcal{P} \subset \mathrm{D}[\epsilon_1, \ldots, \epsilon_m, X_1, \ldots, X_k]$Let $P \in \mathcal{P}$. We write P as a polynomial in $\epsilon = (\epsilon_1, \ldots, \epsilon_m)$ and order the monomials with the order induced by the order $<_\varepsilon$ on $\mathrm{R}\langle\varepsilon_1, \varepsilon_2, \cdots, \varepsilon_m\rangle$ with $\epsilon_1 >_\varepsilon \ldots >_\varepsilon \epsilon_m$. Let

$$P = P_0 \epsilon^{\alpha_0} + P_1 \epsilon^{\alpha_1} + \cdots + P_m \epsilon^{\alpha_m},$$

where, $P_i \in \mathrm{D}[X_1, \ldots, X_k]$, $\alpha_i \in \mathbb{N}^\ell$, and $\epsilon^{\alpha_i} >_\varepsilon \epsilon^{\alpha_{i+1}}$, for $0 \le i \le m$.

Define

$$\begin{aligned}
\mathrm{Remo}_\varepsilon(P=0) &= \bigwedge_{i=0}^{m} (P_i=0),\\
\mathrm{Remo}_\varepsilon(P>0) &= (P_0>0) \vee (P_0=0 \wedge P_1>0) \vee \cdots \vee \Big(\bigwedge_{i=0}^{m-1} P_i=0 \wedge P_m>0\Big),\\
\mathrm{Remo}_\varepsilon(P<0) &= (P_0<0) \vee (P_0=0 \wedge P_1<0) \vee \cdots \vee \Big(\bigwedge_{i=0}^{m-1} P_i=0 \wedge P_m<0\Big).
\end{aligned}$$

Let $\mathrm{Remo}_\varepsilon(\Phi)$ be the formula obtained from Φ by replacing every atom, $P=0$, $P>0$, or $P<0$ in Φ by the corresponding formula

$$\mathrm{Remo}_\varepsilon(P=0), \mathrm{Remo}_\varepsilon(P>0), \mathrm{Remo}_\varepsilon(P<0). \qquad \square$$

Proposition 14.7. *Let $S \subset \mathrm{R}\langle \epsilon_1, \ldots, \epsilon_m\rangle^k$ be a semi-algebraic set defined by a quantifier-free $\mathcal{P}$-formula Φ with $\mathcal{P} \subset \mathrm{D}[\epsilon_1, \ldots, \epsilon_m, X_1, \ldots, X_k]$. Let $S' \subset \mathrm{R}^k$ be the semi-algebraic set defined by $\mathrm{Remo}_\varepsilon(\Phi)$. Then, $S' = S \cap \mathrm{R}^k$.*

Proof: Let $x \in \mathrm{R}^k$ satisfy $\mathrm{Remo}_\varepsilon(\Phi)$. It is clear by construction that x also satisfies Φ. Conversely, if $x \in S \cap \mathrm{R}^k$, then for any polynomial

$$P \in \mathrm{D}[\epsilon_1, \ldots, \epsilon_m, X_1, \ldots, X_k],$$

the sign of $P(x)$ is determined by the sign of the coefficient of the biggest monomial in the lexicographical ordering, when $P(x)$ is expressed as a polynomial in $\epsilon_1, \ldots, \epsilon_m$. This immediately implies that x satisfies the formula $\mathrm{Remo}_\varepsilon(\Phi)$. $\square$

Proof of correctness of Algorithm 14.1: The result follows from the correctness of Algorithm 13.1 (Computing Realizable Sign Conditions) and Algorithm 11.19 (Restricted Elimination). Consider a semi-algebraically connected component S of a realizable sign condition on $\mathrm{BElim}_X(\mathcal{P})$. Then, the following remain invariant as y varies over S: the set $\mathrm{UR}_X(\mathcal{P})(y)$, for every

$$u(Y, \varepsilon, \delta) = (f(Y, \varepsilon, \delta, T), g_0(Y, \varepsilon, \delta, T), \ldots, g_k(Y, \varepsilon, \delta, T)) \in \mathrm{UR}_X(\mathcal{P})(y),$$

the number of roots of $f(y, \varepsilon, \delta, T)$ in $\mathrm{R}\langle \varepsilon, \delta\rangle$ and their Thom encodings, as well as the number of roots in $\mathrm{R}\langle \varepsilon, \delta\rangle$ that are common to $f(y, \varepsilon, \delta, T)$ and $P_u(y, \varepsilon, \delta, T)$ for all $P \in \mathcal{P}$. These are consequences of the properties of RElim_T (see Algorithm 11.19 (Restricted Elimination)). It is finally clear that $\mathrm{SIGN}(\mathcal{P}(y, X))$ is constant as y varies in a semi-algebraically connected component S of a realizable sign condition on $\mathrm{BElim}_X(\mathcal{P})$, using Proposition 14.7. $\square$

Complexity analysis of Algorithm 14.1: The number of arithmetic operations in $\mathrm{D}[Y, \varepsilon, \delta, \gamma, \zeta]$ for computing

$$\mathrm{UR}_X(\mathcal{P}) \subset \mathrm{D}[Y, \varepsilon, \delta][T]$$

is $\sum_{j\leq k} 4^j \binom{s}{j} d^{O(k)}$, using the complexity analysis of Algorithm 13.1 (Computing Realizable Sign Conditions). The degrees of the polynomials in T generated in this process are bounded by $O(d)^k$ (independent of ℓ), and the degree in the variables Y as well as in the variables ε and δ is $d^{O(k)}$, using the complexity analysis of Algorithm 12.10 (Parametrized Multiplication Table) and Algorithm 12.15 (Parametrized Limit of Bounded Points).

The complexity in D for computing $\mathrm{UR}_X(\mathcal{P})$ is $\sum_{j\leq k} 4^j \binom{s}{j} d^{O(\ell k)}$, using the complexity analysis of Algorithm 8.4 (Addition of multivariate polynomials) and Algorithm 8.5 (Multiplication of multivariate polynomials).

Using the complexity of Algorithm 11.19 (Restricted Elimination), the size of the set $\mathrm{BElim}_X(\mathcal{P})$ is $s\sum_{j\leq k} 4^j \binom{s}{j} d^{O(k)} = s^{k+1} d^{O(k)}$, and the degrees of the elements of $\mathrm{BElim}_X(\mathcal{P})$ is $d^{O(k)}$.

The complexity in D is finally $s\sum_{j\leq k} 4^j \binom{s}{j} d^{O(\ell k)} = s^{k+1} d^{O(\ell k)}$.

If $\mathrm{D} = \mathbb{Z}$, and the bitsizes of the coefficients of the polynomials are bounded by τ, then the bitsizes of the integers appearing in the intermediate computations and the output are bounded by $\tau d^{O(k)}$. □

We construct a Π-partition adapted to $\mathcal{P}$ using recursive calls to Algorithm 14.1 (Block Elimination).

Notation 14.8. Defining $\mathrm{B}_{\Pi,\omega}(\mathcal{P}) = \mathcal{P}$, we denote, for $1 \leq i \leq \omega - 1$,

$$\mathrm{B}_{\Pi,i}(\mathcal{P}) = \mathrm{BElim}_{X_{[i+1]}}(\mathrm{B}_{\Pi,i+1}(\mathcal{P})),$$

so that $\mathrm{B}_{\Pi,i}(\mathcal{P}) \subset \mathrm{R}[X_{[1]}, \ldots, X_{[i]}]$. □

For every i, $1 \leq i \leq \omega$, let $\mathcal{S}_i$ be the set of semi-algebraically connected components of non-empty realizations of sign conditions on $\bigcup_{j=1}^{i} \mathrm{B}_{\Pi,i}(\mathcal{P})$.

The following proposition follows clearly from the correctness of Algorithm 14.1 (Block Elimination).

Proposition 14.9. *The list $\mathcal{S} = \mathcal{S}_1, \ldots, \mathcal{S}_\omega$ is a Π-partition adapted to $\mathcal{P}$.*

In order to describe a Π-set of sample points for $\mathcal{S}$, we are going to use the parametrized univariate representations computed in Algorithm 14.1 (Block Elimination).

Notation 14.10. *Note that for every $i = \omega - 1, \ldots, 0$,*

$$\mathrm{UR}_{\Pi,i}(\mathcal{P}) = \mathrm{UR}_{X_{[i+1]}}\mathrm{B}_{\Pi,i+1}(\mathcal{P})).$$

The elements of $\mathrm{UR}_{\Pi,i}(\mathcal{P})$ are parametrized univariate representations in the variable T_{i+1}, contained in $\mathrm{D}[X_{[1]}, \ldots, X_{[i]}, \varepsilon_{i+1}, \delta_{i+1}][T_{i+1}]^{k_{i+1}+2}$. Let

$$u = (u_0, \ldots, u_{\omega-1}) \in \mathcal{U} = \prod_{i=0}^{\omega-1} \mathrm{UR}_{\Pi,i}(\mathcal{P}),$$

with

$$u_{i-1} = (f^{[i]}, g_0^{[i]}, g_1^{[i]}, \ldots, g_{k_i}^{[i]}).$$

For a polynomial $P(X_{[1]}, \ldots, X_{[i]})$, let $P_{u,i..j}(X_{[1]}, \ldots, X_{[j-1]}, T_j, \ldots, T_i)$ denote the polynomial obtained by successively replacing the blocks of variables $X_{[\ell]}$, with the rational fractions associated with the tuple $u_{\ell-1}$ (using Notation ?), for ℓ from i to j. Denoting $P_{u,i}(T_1, \ldots, T_i) = P_{u,i..1}(T_1, \ldots, T_i)$, define

$$\begin{aligned} \mathcal{T}_{u,i} &= (f^{[1]}(T_1), f^{[2]}_{u,1}(T_1, T_2), \ldots, f^{[i]}_{u,i-1}(T_1, T_2, \ldots, T_i)), \\ \mathcal{T}_u &= (f^{[1]}(T_1), f^{[2]}_{u,1}(T_1, T_2), \ldots, f^{[\omega]}_{u,\omega-1}(T_1, T_2, \ldots, T_\omega)), \\ \bar{u}_{i-1} &= (f^{[i]}_{u,i-1}, g^{[i]}_{0\ u,i-1}, g^{[i]}_{1\ u,i-1}, \ldots, g^{[i]}_{k_i u,i-1}) \end{aligned}$$

Note that $\bar{u}_{i-1}$ are univariate representations contained in

$$\mathrm{D}[T_1, \ldots, T_{i-1}, \varepsilon_1, \delta_1, \ldots, \varepsilon_i, \ \delta_i][T_i]^{k_i+2}.$$

For $u \in \mathcal{U}$ and $t_\sigma \in \mathrm{Zer}(\mathcal{T}_u, \mathrm{R}\langle\varepsilon_1, \delta_1, \ldots, \varepsilon_\omega, \delta_\omega\rangle)$, with Thom encoding σ let $x_{u,\sigma,i} \in \mathrm{R}\langle\varepsilon_1, \delta_1, \ldots, \varepsilon_\omega, \delta_\omega\rangle^{[i]}$ be the point obtained by substituting t_σ in the rational functions associated to $\bar{u}_{j-1}$, $j \leq i$. Let $\mathcal{A}_i$ be the set of points $x_{u,\sigma,i} \in \mathrm{R}\langle\varepsilon_1, \delta_1, \ldots, \varepsilon_\omega, \delta_\omega\rangle^{[i]}$ obtained by considering all $u \in \mathcal{U}$ and $t_\sigma \in \mathrm{Zer}(\mathcal{T}_u, \mathrm{R}\langle\varepsilon_1, \delta_1, \ldots, \varepsilon_\omega, \delta_\omega\rangle)$. Then $\mathcal{A} = \mathcal{A}_1, \ldots, \mathcal{A}_\omega$ is a Π-set, specified by $\mathcal{V}$ where the elements of $\mathcal{V}$ are pairs of an element $u \in \mathcal{U}$ and a Thom encoding σ of an element of $\mathrm{Zer}(\mathcal{T}_u, \mathrm{R}\langle\varepsilon_1, \delta_1, \ldots, \varepsilon_\omega, \delta_\omega\rangle)$. □

The correctness of Algorithm 14.1 (Block Elimination) implies the following proposition.

Proposition 14.11. *The set $\mathcal{A}$ is a Π-set of sample points for $\mathcal{P}$.*

Thus, in order to construct the set $\mathrm{SIGN}_\Pi(\mathcal{P})$, it suffices to compute the signs of $\mathcal{P}_{u,\omega}$ at the zeros of $\mathcal{T}_u$, $u \in \mathcal{U}$.

The algorithm is as follows, using the notation of Algorithm 14.1 (Block Elimination):

Algorithm 14.2. **[Block Structured Signs]**

- **Structure:** an ordered domain D contained in a real closed field R.
- **Input:** a set $\mathcal{P} \subset \mathrm{R}[X_1, \ldots, X_k]$, and a partition, Π, of the variables $X_1, \ldots,$ X_k into blocks, $X_{[1]}, \ldots, X_{[\omega]}$.
- **Output:** the tree $\mathrm{SIGN}_\Pi(\mathcal{P})$ of realizable sign conditions of $\mathcal{P}$ with respect to Π.
- **Complexity:** $s^{(k_\omega+1)\cdots(k_1+1)}\ d^{O(k_\omega)\cdots O(k_1)}$, where s is bound on the number of elements of $\mathcal{P}$, d is a bound on their degree, and $k_{[i)}$ is the number of elements of $X_{[i]}$.
- **Procedure:**
  - Initialize $\mathrm{B}_{\Pi,\omega}(\mathcal{P}) := \mathcal{P}$.
  - Block Elimination Phase: Compute

$$\mathrm{B}_{\Pi,i}(\mathcal{P}) = \mathrm{BElim}_{X_{[i+1]}}(\mathrm{B}_{\Pi,i+1}(\mathcal{P})),$$

for $1 \le i \le \omega - 1$, applying repeatedly $\mathrm{BElim}_{X_{[i+1]}}$, using Algorithm 14.1 (Block Elimination). Define $\mathrm{B}_{\Pi,0}(\mathcal{P}) = \{1\}$. Compute $\mathrm{UR}_{\Pi,i}(\mathcal{P})$, for every $i = \omega - 1, ..., 0$, using Algorithm 14.1 (Block Elimination). The elements of $\mathrm{UR}_{\Pi,i}(\mathcal{P})$ are parametrized univariate representations in the variable T_{i+1}, contained in

$$\mathrm{D}[X_{[1]}, ..., X_{[i]}, \varepsilon_{i+1}, \delta_{i+1}][T_{i+1}]^{k_{i+1}+2}.$$

- Substitution Phase: Compute the set of pairs $\{(\mathcal{T}_u, \mathcal{P}_{u,\omega}) \mid u \in \mathcal{U}\}$, using their definition in Notation 14.10.
- Sign Determination Phase: Compute the signs of the set of the polynomials in $\mathcal{P}_{u,\omega}$ on $\mathrm{Zer}(\mathcal{T}_u, \mathrm{R}\langle\varepsilon_1, \delta_1, ..., \varepsilon_\omega, \delta_\omega\rangle^\omega)$ using Algorithm 12.19 (Zero-dimensional Sign Determination).
- Construct the set $\mathrm{SIGN}_\Pi(\mathcal{P})$ from these signs.

Proof of correctness: The correctness of the algorithm follows from Proposition 14.11. □

Complexity analysis: Using the complexity of Algorithm 14.1 (Block Elimination), the degrees and number of the parametrized univariate representations in $\mathrm{UR}_{\Pi,\omega-1}(\mathcal{P})$ produced after eliminating the first block of variables $X_{[\omega]}$ are bounded respectively by $O(d)^{k_\omega}$ and $s^{k_\omega} O(d)^{k_\omega}$. The number of arithmetic operations in this step is bounded by $s^{k_\omega} d^{O((k-k_\omega)k_\omega)}$, and the size of the set $\mathrm{B}_{\Pi,\omega-1}(\mathcal{P})$ is $s^{k_\omega+1} d^{O(k_\omega)}$. Since the cardinality of $\mathrm{SIGN}_{\Pi,\omega-1}(\mathcal{P})(z)$ is, for every $z \in \mathrm{R}^{[\omega-1]}$, bounded by the number of points associated to the univariate representations obtained by substituting z to the parameters in the elements of $\mathrm{UR}_{\Pi,\omega-1}(\mathcal{P})$, $\#(\mathrm{SIGN}_{\Pi,\omega-1}(\mathcal{P})(z))$ is $s^{k_\omega} O(d)^{k_\omega}$.

An easy inductive argument shows that the number of univariate representations in $\mathrm{UR}_{\Pi,i}(\mathcal{P})$ produced after eliminating the $(i+1)$-th block of variables is bounded by

$$s^{(k_\omega+1)\cdots(k_{i+2}+1)k_{i+1}} d^{O(k_\omega)\cdots O(k_{i+1})}.$$

By a similar argument, one can show that the degrees of the parametrized univariate representations in $\mathrm{UR}_{\Pi,i}(\mathcal{P})$ are bounded by $d^{O(k_\omega)\cdots O(k_{i+1})}$. The complexity in D is bounded by

$$s^{(k_\omega+1)\cdots(k_{i+1}+1)} d^{(k_1+\cdots+k_i+2(\omega-i))O(k_\omega)\cdots O(k_{i+1})},$$

since the arithmetic is done in a polynomial ring with $k_1 + \cdots + k_i + 2(\omega - i)$ variables.

A similar inductive argument shows that the the size of the set $\mathrm{B}_{\Pi,i}(\mathcal{P})$ is bounded by $s^{(k_\omega+1)\cdots(k_{i+1}+1)} d^{O(k_\omega)\cdots O(k_{i+1})}$, and their degrees are bounded by $d^{O(k_\omega)\cdots O(k_i)}$.

The above analysis shows that the size of the set of pairs $(\mathcal{T}_u, \mathcal{P}_u)$, constructed at the end of the Substitution Phase is

$$s^{(k_\omega+1)\cdots(k_1+1)} d^{O(k_\omega)\cdots O(k_1)},$$

and the degrees are bounded by $d^{O(k_\omega)\cdots O(k_1)}$. It should also be clear that the number of arithmetic operations in D for the Substitution Phase is equally bounded by

$$s^{(k_\omega+1)\cdots(k_1+1)}d^{O(k_\omega)\cdots O(k_1)}.$$

Since the number of triangular systems $\mathcal{T}$ is

$$s^{(k_\omega+1)\cdots(k_1+1)}d^{O(k_\omega)\cdots O(k_1)}.$$

and each call to Algorithm 12.19 (Triangular Sign Determination) takes time

$$d^{\omega O(k_\omega)\cdots O(k_1)} = d^{O(k_\omega)\cdots O(k_1)},$$

the time taken for the Sign Determination Phase, is

$$s^{(k_\omega+1)\cdots(k_1+1)}d^{O(k_\omega)\cdots O(k_1)}.$$

The time required to construct $\mathrm{SIGN}_\Pi(\mathcal{P})$ is again bounded by

$$s^{(k_\omega+1)\cdots(k_1+1)}d^{O(k_\omega)\cdots O(k_1)}.$$

Thus the total time bound for the elimination and sign determination phase is

$$s^{(k_\omega+1)\cdots(k_1+1)}d^{O(k_\omega)\cdots O(k_1)}.$$

If $\mathrm{D} = \mathbb{Z}$, and the bitsizes of the coefficients of the polynomials are bounded by τ, then the bitsizes of the integers appearing in the intermediate computations and the output are bounded by $\tau d^{O(k_\omega)\cdots O(k_1)}$. □

Remark 14.12. In fact, Algorithm 14.2 (Block Structured Signs) does not only computes $\mathrm{SIGN}_\Pi(\mathcal{P})$*, it also produces the set* $\mathcal{V}$ *specifying a* Π*-set of sampling points for* $\mathcal{P}$ *described at the end of Notation 14.10.* □

We have proved the following result:

Theorem 14.13. *Let* $\mathcal{P}$ *be a set of at most* s *polynomials each of degree at most* d *in* k *variables with coefficients in a real closed field* R*, and let* Π *denote a partition of the list of variables* $(X_1,\ldots,X_k)$ *into blocks* $X_{[1]},\ldots,X_{[\omega]}$*, where the block* $X_{[i]}$ *has size* $k_i, 1 \le i \le \omega$*. Then the size of the set* $\mathrm{SIGN}_\Pi(\mathcal{P})$ *is bounded by*

$$s^{(k_\omega+1)\cdots(k_1+1)}d^{O(k_\omega)\cdots O(k_1)}.$$

Moreover, there exists an algorithm which computes this set with complexity

$$s^{(k_\omega+1)\cdots(k_1+1)}d^{O(k_\omega)\cdots O(k_1)}$$

in D*, where* D *is the ring generated by the coefficients of* $\mathcal{P}$*.*

If $\mathrm{D} = \mathbb{Z}$*, and the bitsizes of the coefficients of the polynomials are bounded by* τ*, then the bitsizes of the integers appearing in the intermediate computations and the output are bounded by* $\tau d^{O(k_\omega)\cdots O(k_1)}$*.*

Using the set $\mathrm{SIGN}_\Pi(\mathcal{P})$, it is now easy to solve the general decision problem, which is to design a procedure to decide the truth or falsity of a $(\mathcal{P},\Pi)$-sentence.

Algorithm 14.3. **[General Decision]**

- **Structure:** an ordered domain D contained in a real closed field R
- **Input:** a $(\mathcal{P},\Pi)$-sentence Φ, where $\mathcal{P} \subset \mathrm{D}[X_1,\ldots,X_k]$, and Π is a partition of the variables $X_1,\ldots,X_k$ into blocks $X_{[1]},\ldots,X_{[\omega]}$.
- **Output:** 1 if Φ is true and 0 otherwise.
- **Complexity:** $s^{(k_\omega+1)\cdots(k_1+1)}\ d^{O(k_\omega)\cdots O(k_1)}$, where s is a bound on the number of elements of $\mathcal{P}$, d is a bound on their degree, and k_i is the number of elements of $X_{[i]}$.
- **Procedure:**
  - Compute $\mathrm{SIGN}_\Pi(\mathcal{P})$.
  - Trying all possibilities, decide whether

    $$\mathrm{Qu}_1\sigma_1 \in \mathrm{SIGN}_\Pi(\mathcal{P}) \quad \mathrm{Qu}_2\sigma_2 \in \sigma_1 \ldots \mathrm{Qu}_\omega\sigma_\omega \in \sigma_{\omega-1} \quad F^\star(\sigma_\omega) = \mathrm{True},$$

    which is clearly a finite verification.

Proof of correctness: Follows from the properties of $\mathrm{SIGN}_\Pi(\mathcal{P})$. □

Complexity analysis: Given the complexity of Algorithm 14.2 (Block Structured Signs), the complexity for the general decision algorithm is

$$s^{(k_\omega+1)\cdots(k_1+1)} d^{O(k_\omega)\cdots O(k_1)}$$

in D. Note that the evaluation of the boolean formulas are not counted in this model of complexity since we count only arithmetic operations in D.

If $\mathrm{D} = \mathbb{Z}$, and the bitsizes of the coefficients of the polynomials are bounded by τ, then the bitsizes of the integers appearing in the intermediate computations and the output are bounded by $\tau d^{O(k_\omega)\cdots O(k_1)}$. □

Note that the first step of the computation depend only on $(\mathcal{P}, \Pi)$ and not on Φ. As noted before $\mathrm{SIGN}_\Pi(\mathcal{P})$ allows to decide the truth or falsity of every $(\mathcal{P},\Pi)$-sentence.

We have proved the following result.

Theorem 14.14. [General Decision] *Let $\mathcal{P}$ be a set of at most s polynomials each of degree at most d in k variables with coefficients in a real closed field* R, *and let Π denote a partition of the list of variables $(X_1,\ldots,X_k)$ into blocks $X_{[1]},\ldots,X_{[\omega]}$, where the block $X_{[i]}$ has size $k_i, 1 \le i \le \omega$. Given a $(\mathcal{P},\Pi)$-sentence Φ, there exists an algorithm to decide the truth of Φ with complexity*

$$s^{(k_\omega+1)\cdots(k_1+1)} d^{O(k_\omega)\cdots O(k_1)}$$

in D, *where* D *is the ring generated by the coefficients of $\mathcal{P}$. If $\mathrm{D} = \mathbb{Z}$, and the bitsizes of the coefficients of the polynomials are bounded by τ, then the bitsizes of the integers appearing in the intermediate computations and the output are bounded by $\tau d^{O(k_\omega)\cdots O(k_1)}$.*

14.2 Quantifier Elimination

In our Quantifier Elimination Algorithm, we use a parametrized version of Algorithm 12.8 (Multivariate Sign Determination) to solve the following problem.

Notation 14.15. *Let* D *be a ring contained in a real closed field* R. *A* ***parametrized triangular system*** *with parameters* $Y = (Y_1, \ldots, Y_\ell)$ *and variables* $T_1, \ldots, T_\omega$ *is a list* $\mathcal{T} = \mathcal{T}_1, \mathcal{T}_2, \ldots, \mathcal{T}_\omega$ *where*

$$\begin{aligned} \mathcal{T}_1(Y) &\in \mathrm{D}[Y, T_1] \\ \mathcal{T}_2(Y) &\in \mathrm{D}[Y, T_1, T_2] \\ &\vdots \\ \mathcal{T}_\omega(Y) &\in \mathrm{D}[Y, T_1, \ldots, T_\omega]. \end{aligned}$$

Given a parametrized triangular system $\mathcal{T} = \mathcal{T}_1, \mathcal{T}_2, \ldots, \mathcal{T}_\omega$ with parameters $Y = (Y_1, \ldots, Y_\ell)$, a set of polynomials $\mathcal{P} \subset \mathrm{D}[Y, T_1, \ldots, T_i]$ and a point $y \in \mathrm{R}^\ell$ such that $\mathcal{T}(y)$ is zero-dimensional, wc denote by $\mathrm{SIGN}(\mathcal{P}(y), \mathcal{T}(y))$ the list of sign conditions satisfied by $\mathcal{P}(y)$ at the zeros of $\mathcal{T}(y)$. We want to compute a quantifier free formula such that $\Phi(z)$ holds if and only if

$$\mathrm{SIGN}(\mathcal{P}(z), \mathcal{T}(z)) = \mathrm{SIGN}(\mathcal{P}(y), \mathcal{T}(y)). \qquad \square$$

Algorithm 14.4. **[Inverse Sign Determination]**

- **Structure:** an ordered domain D contained in a real closed field R.
- **Input:**
  - a parametrized triangular system of polynomials, $\mathcal{T}$ with parameters $Y = (Y_1, \ldots, Y_\ell)$,
  - a point $y \in \mathrm{R}^\ell$, specified by a Thom encoding, such that $\mathcal{T}(y)$ is zero-dimensional,
  - a subset $\mathcal{P} \subset \mathrm{D}[Y, T_1, \ldots, T_\omega]$.
- **Output:**
  - a family $\mathcal{A}(y) \subset \mathrm{D}[Y]$,
  - a quantifier free $\mathcal{A}(y)$-formula $\Phi(y)(Y)$ such that for any $z \in \mathrm{R}^\ell$, the formula $\Phi(y)(z)$ is true if and only if $\mathcal{T}(z)$ is zero-dimensional and

$$\mathrm{SIGN}(\mathcal{P}(y), \mathcal{T}(y)) = \mathrm{SIGN}(\mathcal{P}(z), \mathcal{T}(z)).$$

- **Complexity:** $s^{\ell+1}(d'^\omega d)^{O(\ell)}$, where s is a bound on the number of elements of $\mathcal{P}$ and d is a bound on the degrees of the polynomials in $\mathcal{T}$ and $\mathcal{P}$.
- **Procedure:**
  - Use Algorithm 12.19 (Triangular Sign Determination) to compute $\mathrm{SIGN}(\mathcal{Q}(y), \mathcal{T}(y))$. Form the list

$$B(\mathrm{SIGN}(\mathcal{Q}(y), \mathcal{T}(y))) \subset \{0, 1, 2\}^{\mathcal{Q}},$$

  using Remark 10.69 and its notation.

– Using Algorithm 12.18 (Parametrized Bounded Algebraic Sampling) with input $\mathcal{T}_1^1 + \cdots + \mathcal{T}_k^2$, output a finite set $\mathcal{U}$ of parametrized univariate representations.
– For every $\alpha \in B(\mathrm{SIGN}(\mathcal{Q}(y) \cup \mathrm{Der}(\mathcal{T}(y)), \mathcal{T}(y)))$ and every $u = (f, g_0, \ldots, g_k) \in \mathcal{U}$, compute the signed subresultant coefficients of f and $\mathcal{Q}_u^\alpha$, using Algorithm 8.21 (Signed subresultant) and place them in a set $\mathcal{A}(y) \subset \mathrm{D}[Y]$.
– Using Algorithm 13.1 (Computing realizable sign conditions), output the set $\mathrm{SIGN}(\mathcal{A}(y))$ of realizable sign conditions on $\mathcal{A}(y)$ and the subset $\Sigma(y)$ of $\mathrm{SIGN}(\mathcal{A}(y))$ of ρ such that for every z in the realization of ρ, the Tarski-queries of $f(z, T)$ and $\mathcal{Q}_u^\alpha(z, T)$ give rise to a list of non-empty sign conditions $\mathrm{SIGN}(\mathcal{P}(z), \mathcal{T}(z))$ that coincides with $\mathrm{SIGN}(\mathcal{P}(y), \mathcal{T}(y))$.
– Output $\mathcal{A}(y)$ and

$$\Phi(y)(Y) = \bigvee_{\sigma \in \Sigma(y)} \bigwedge_{Q \in \mathcal{A}(y)} \mathrm{sign}(Q(Y)) = \sigma(Q).$$

Proof of correctness: It follows from the correctness of Algorithm 12.19 (Triangular Sign Determination), Remark 10.69, Algorithm 12.18 (Parametrized Bounded Algebraic Sampling), Algorithm 8.21 (Signed subresultant) and Algorithm 13.1 (Computing realizable sign conditions). □

Complexity analysis: Suppose that the degree of f_i is bounded by d' and the degrees of all the polynomials in $\mathcal{P}$ are bounded by d, and that the number of polynomials in $\mathcal{P}$ is s. Using the complexity of Algorithm 12.19 (Triangular Sign Determination), the number of arithmetic operations in D in Step 1 is bounded by $s\, d'^{O(\omega)}$. The number of elements of $B(\mathrm{SIGN}(\mathcal{Q}(y), \mathcal{T}(y)))$ is bounded by $s\, O(d')^\omega d$, using Remark 10.69. The number of arithmetic operations in $\mathrm{D}[Y]$ is bounded by $s\, d'^{O(\omega)} d^{O(1)}$. The degree in Y in the intermediate computations is bounded by $d'^{O(\omega)} d^{O(1)}$, using the complexity of Algorithm 12.19 (Triangular Sign Determination). Using the complexity analyses of Algorithms 8.4 (Addition of multivariate polynomials), 8.5 (Multiplication of multivariate polynomials), and 8.6 (Exact division of multivariate polynomials), the number of arithmetic operations in D is bounded by $s(d'^\omega d)^{O(\ell)}$. The number of elements in $\mathcal{A}(y)$ is $s\, d'^{O(\omega)} d^{O(1)}$. Using the complexity of Algorithm 13.1 (Computing realizable sign conditions), the final complexity is $s^{\ell+1}(d'^\omega d)^{O(\ell)}$.

If $\mathrm{D} = \mathbb{Z}$, and the bitsizes of the coefficients of the polynomials are bounded by τ, then the bitsizes of the integers appearing in the intermediate computations and the output are bounded by $\tau(d'^\omega d)^{O(\ell)}$. □

We now describe our algorithm for the quantifier elimination problem. We make use of Algorithm 14.2 (Block Structured Signs) and Algorithm 14.4 (Inverse Sign Determination).

Let $\mathcal{P} \subset \mathrm{R}[X_1, \ldots, X_k, Y_1, \ldots, Y_\ell]$ be finite and let Π denote a partition of the list of variables $X = (X_1, \ldots, X_k)$ into blocks, $X_{[1]}, \ldots, X_{[\omega]}$, where the block $X_{[i]}$ is of size k_i, $1 \le i \le \omega$, $\sum_{1 \le i \le \omega} k_i = k$. We proceed in the same manner as the algorithm for the general decision problem, starting with the set $\mathcal{P}$ of polynomials and eliminating the blocks of variables to obtain a set of polynomials $\mathrm{B}_\Pi(\mathcal{P})$ in the variables Y. For a fixed $y \in \mathrm{R}^\ell$, the truth or falsity of the formula $\Phi(y)$ can be decided from the set $\mathrm{SIGN}_\Pi(\mathcal{P})(y)$. We next apply Algorithm 13.1 (Sampling) to the set of polynomials $\mathrm{B}_\Pi(\mathcal{P}) \subset D[Y]$, to obtain points in every semi-algebraically connected component of a realizable sign condition of $\mathrm{B}_\Pi(\mathcal{P})$. For each sample point y so obtained, we determine whether or not y satisfies the given formula using the set $\mathrm{SIGN}_\Pi(\mathcal{P})(y)$. If it does, then we use the Inverse Sign Determination Algorithm with the various $\mathcal{T}_u, \mathcal{P}_{u,\omega}, y$ as inputs to construct a formula $\Psi_y(Y)$. The only problem left is that this formula contains the infinitesimal quantities introduced by the general decision procedure. However we can replace each equality, or inequality in $\Psi_y(Y)$, by an equivalent larger formula without the infinitesimal quantities by using the ordering amongst the infinitesimal quantities. We output the disjunction of the formulas $\Psi_y(Y)$ constructed above.

We now give a more formal description of the algorithm and prove the bounds on the time complexity and the size of the output formula.

Algorithm 14.5. **[Quantifier Elimination]**

- **Structure:** an ordered domain D contained in a real closed field R.
- **Input:** a finite subset $\mathcal{P} \subset \mathrm{D}[X_1, \ldots, X_k, Y_1, \ldots, Y_\ell]$ of s polynomials of degree at most d, a partition Π of the list of variables $X = (X_1, \ldots, X_k)$ into blocks, $X_{[1]}, \ldots, X_{[\omega]}$, where the block $X_{[i]}$ is of size $k_i, 1 \le i \le \omega$, with $\sum_{1 \le i \le \omega} k_i = k$ and a $(\mathcal{P}, \Pi)$-formula $\Phi(Y)$.
- **Output:** a quantifier free formula $\Psi(Y)$ equivalent to $\Phi(Y)$.
- **Complexity:** $s^{(k_\omega+1)\cdots(k_1+1)(\ell+1)} d^{O(k_\omega)\cdots O(k_1) O(\ell)}$.
- **Procedure:**
  - Block Elimination Phase: Perform the Block Elimination Phase of Algorithm 14.2 (Block Structured Signs) on the set of polynomials $\mathcal{P}$, with $\omega + 1$ blocks of variables $(Y, X_{[1]}, \ldots, X_{[\omega]}$ to obtain the set $\mathcal{U}$ consisting of triangular systems $\mathcal{T}_u$ and the set of polynomials $\mathcal{P}_{u,\omega+1}$.
  - Formula Building Phase: For every $u = (u_1, \ldots, u_{\omega+1}) \in \mathcal{U}$ and every point y associated to u_1, compute $\mathrm{SIGN}(\mathcal{T}_u(y), \mathcal{P}_{u,\omega}(y))$, using Algorithm 12.19 (Triangular Sign Determination). Output the set $\mathrm{SIGN}_\Pi(\mathcal{P})(y)$ from the set $\{\mathrm{SIGN}(\mathcal{T}_u(y), \mathcal{P}_{u,\omega}(y)) \mid u \in \mathcal{U}\}$, and hence decide whether the formula $\Phi(y)$ is true.
  - If $\Phi(y)$ is true, apply Algorithm 14.4 (Inverse Sign Determination) with

$$\mathcal{T}_u, \mathcal{P}_{u,\omega}, y$$

as inputs to get the formulas $\Psi_{u,y}(Y)$. Let $\Psi_y(Y) = \bigwedge_u \Psi_{u,y}(Y)$, and let $\overline{\Psi(Y)} = \bigvee_y \Psi_y(Y)$, where the disjunction is over all the y for which $\Phi(y)$ is true in the previous step.
– Output $\Psi(Y) := \mathrm{Remo}_{\varepsilon_1,\delta_1,\ldots,\varepsilon_{\omega+1},\delta_{\omega+1}}(\overline{\Psi(Y)})$ (Notation 14.6).

Proof of correctness: The correctness of the algorithm follows from the correctness of Algorithm 14.3 ([General Decision), Algorithm 14.4 (Inverse Sign Determination), and Proposition 14.7. □

Complexity analysis: The elimination phase takes at most

$$s^{(k_\omega+1)\cdots(k_1+1)(\ell+1)}d^{O(k_\omega)\cdots O(k_1)O(\ell)}$$

arithmetic operations, and the number of sign conditions produced is also bounded by

$$s^{(k_\omega+1)\cdots(k_1+1)(\ell+1)}d^{O(k_\omega)\cdots O(k_1)O(\ell)}.$$

The degrees in the variables $T_1, \ldots, T_\omega, T_{\omega+1}, \varepsilon_1, \delta_1, \ldots, \varepsilon_{\omega+1}, \delta_{\omega+1}$ in the polynomials produced, are all bounded by $d^{O(k_\omega)\cdots O(k_1)O(\ell)}$.

Invoking the bound on the Algorithm 14.4 (Inverse Sign Determination), and the bound on the number of tuples produced in the elimination phase, which is $s^{(k_\omega+1)\cdots(k_1+1)\ell}d^{O(k_\omega)\cdots O(k_1)O(\ell)}$ we see that the formula building phase takes no more than

$$s^{(k_\omega+1)\cdots(k_1+1)\ell+\ell}d^{O(k_\omega)\cdots O(k_1)O(\ell)}$$

operations. Since the degrees of the variables $\varepsilon_{\omega+1}, \delta_{\omega+1}, \ldots, \varepsilon_1, \delta_1,$ are all bounded by $d^{O(k_\omega)\cdots O(k_1)O(\ell)}$, each atom is expanded to a formula of size at most $d^{(O(k_\omega)\cdots O(k_1)O(\ell)}$.

The bound on the size of the formula is an easy consequence of the bound on the number of tuples produced in the elimination phase, and the bound on the formula size produced by Algorithm 14.4 (Inverse Sign Determination).

If $\mathrm{D} = \mathbb{Z}$, and the bitsizes of the coefficients of the polynomials are bounded by τ, then the bitsizes of the integers appearing in the intermediate computations and the output are bounded by $\tau d^{O(k_\omega)\cdots O(k_1)O(\ell)}$. □

This proves the following result.

Theorem 14.16. [Quantifier Elimination] *Let $\mathcal{P}$ be a set of at most s polynomials each of degree at most d in $k + \ell$ variables with coefficients in a real closed field* R, *and let Π denote a partition of the list of variables $(X_1, \ldots, X_k)$ into blocks, $X_{[1]}, \ldots, X_{[\omega]}$, where the block $X_{[i]}$ has size k_i,* for $1 \le i \le \omega$. *Given $\Phi(Y)$, a $(\mathcal{P}, \Pi)$-formula, there exists an equivalent quantifier free formula,*

$$\Psi(Y) = \bigvee_{i=1}^{I} \bigwedge_{j=1}^{J_i} \left(\bigvee_{n=1}^{Ni,j} \mathrm{sign}(P_{ijn}(Y)) = \sigma_{ijn}\right),$$

where $P_{ijn}(Y)$ *are polynomials in the variables* Y, $\sigma_{ijn} \in \{0, 1, -1\}$,

$$\begin{aligned} I &\leq s^{(k_\omega+1)\cdots(k_1+1)(\ell+1)} d^{O(k_\omega)\cdots O(k_1)O(\ell)}, \\ J_i &\leq s^{(k_\omega+1)\cdots(k_1+1)} d^{O(k_\omega)\cdots O(k_1)}, \\ N_{ij} &\leq d^{O(k_\omega)\cdots O(k_1)}, \end{aligned}$$

and the degrees of the polynomials $P_{ijk}(y)$ *are bounded by* $d^{O(k_\omega)\cdots O(k_1)}$. *Moreover, there is an algorithm to compute* $\Psi(Y)$ *with complexity*

$$s^{(k_\omega+1)\cdots(k_1+1)(\ell+1)} d^{O(k_\omega)\cdots O(k_1)O(\ell)}$$

in D, *denoting by* D *the ring generated by the coefficients of* $\mathcal{P}$.

If $\mathrm{D} = \mathbb{Z}$, *and the bitsizes of the coefficients of the polynomials are bounded by* τ, *then the bitsizes of the integers appearing in the intermediate computations and the output are bounded by* $\tau d^{O(k_\omega)\cdots O(k_1)O(\ell)}$.

Remark 14.17. Note that, for most natural geometric properties that can be expressed by a formula in the language of ordered fields, the number of alternations of quantifiers in the formula is small (say at most five or six) while the number of variables can be arbitrarily big. A typical illustrative example is the formula describing the closure of a semi-algebraic set. In such situations, using Theorem 14.16, the complexity of quantifier elimination is singly exponential in the number of variables. □

Exercise 14.1. Design an algorithm computing the minimum value (maybe $-\infty$) of a polynomial of degree d defined on R^k with complexity $d^{O(k)}$. Make precise how this minimum value is described.

14.3 Local Quantifier Elimination

In this section we discuss a variant of Algorithm 14.5 (Quantifier Elimination) whose complexity is slightly better. A special feature of this algorithm is that the quantifier-free formula output will not necessarily be a disjunction of sign conditions, but will have a more complicated nested structure reflecting the logical structure of the input formula.

For this purpose, we need a parametrized version of Algorithm 12.20 (Triangular Thom Encoding). This algorithm will be based on Algorithm 14.6 (Parametrized Sign Determination).

Algorithm 14.6. **[Parametrized Sign Determination]**

- **Structure:** an ordered domain D contained in a real closed field R.
- **Input**: a parametrized triangular system $\mathcal{T}$ with parameters $(Y_1, \ldots, Y_\ell)$, and variables $(X_1, \ldots, X_k)$ and a finite set $\mathcal{Q} \subset \mathrm{D}[Y_1, \ldots, Y_\ell, X_1, \ldots, X_k]$.
- **Output**:
  - a finite set $\mathcal{A} \subset \mathrm{D}[Y]$, with $Y = (Y_1, \ldots, Y_k)$.

  - for every $\rho \in \text{SIGN}(\mathcal{A})$, a list $\text{SIGN}(\mathcal{Q},\mathcal{T})(\rho)$ of sign conditions on $\mathcal{Q}$ such that, for every y in the realization $\text{Reali}(\rho)$ of ρ, $\text{SIGN}(\mathcal{Q},\mathcal{T})(\rho)$ is the list of sign conditions realized by $\mathcal{Q}(y)$ on the zero set $Z(y)$ of $\mathcal{T}(y)$.
- **Complexity**:] $s^{\ell(\ell+1)+1} d'^{O(k\ell)} d^{O(\ell)}$, where s is a bound on the number of polynomials in $\mathcal{Q}$ and d is a bound on the degrees of the polynomials in $\mathcal{T}$ and $\mathcal{Q}$.
- **Procedure**:
  - Step 1: Perform Algorithm 12.18 (Parametrized Bounded Algebraic Sampling) with input $\mathcal{T}_1^2 + \ldots, + \mathcal{T}_k^2$, for $\mathcal{T}_i \in \mathcal{T}$ and output $\mathcal{U}$.
  - Step 2: Consider for every $u = (f, g_0, \ldots, g_k) \in \mathcal{U}$ and every $Q \in \mathcal{Q}$ the finite set $\mathcal{F}_{u,Q}$ containing Q_u (Notation 13.8) and all the derivatives of f with respect to T, and compute

$$\mathcal{D}_{u,Q} = \text{RElim}_T(f, \mathcal{F}_{u,Q}) \subset \text{D}[Y],$$

    using Algorithm 11.19 (Restricted Elimination).
  - Step 3: Define $\mathcal{D} = \bigcup_{u \in \mathcal{U}, Q \in \mathcal{Q}} \mathcal{D}_{u,Q}$. Perform Algorithm 13.1 (Sampling) with input $\mathcal{D}$. Denote by $\mathcal{S}$ the set of sample points output.
  - Step 4: For every sample point y, perform Algorithm 14.4 (Inverse Sign Determination) and output the set $\mathcal{A}(y) \subset \text{D}[Y]$, as well as $\text{SIGN}(\mathcal{Q}(y), \mathcal{T}(y))$ and $\Phi(y)(Y)$.
  - Step 5: Define $\mathcal{A} = \mathcal{D} \cup \bigcup_{y \in \mathcal{S}} \mathcal{A}(y)$. Compute the set of realizable sign conditions on $\mathcal{A}$ using Algorithm 13.1 (Sampling).
  - Step 6: For every $\rho \in \text{SIGN}(\mathcal{A})$ denote by y the sample point of $\text{Reali}(\rho)$. Define $\text{SIGN}(\mathcal{Q}, \mathcal{T})(\rho)$ as $\text{SIGN}(\mathcal{Q}(y), \mathcal{T}(y))$, computed by Algorithm 12.19 (Triangular Sign Determination).

Proof of correctness: Follows from the correctness of Algorithm 12.18 (Parametrized Bounded Algebraic Sampling), Algorithm 11.19 (Restricted Elimination), Algorithm 13.1 (Sampling), Algorithm 14.4 (Inverse Sign Determination), Algorithm 13.1 (Sampling) and Algorithm 12.19 (Triangular Sign Determination). □

Complexity analysis: We estimate the complexity in terms of the number of parameters ℓ, the number of variables k, the number s of polynomials in $\mathcal{P}$, a bound d' on the degrees of the polynomials in $\mathcal{T}$ and a bound d on the degrees of the polynomials in $\mathcal{P}$.

- Step 1: Using the complexity analysis of Algorithm 12.18 (Parametrized Bounded Algebraic Sampling), the complexity of this step is $d'^{O(k)}$ in the ring $\text{D}[Y]$. The polynomials output are of degree $O(d')^k$ in T and of degrees $d'^{O(k)}$ in Y. Finally, the complexity is $d'^{O(k\ell)}$ in the ring D. The number of elements of $\mathcal{U}$ is $O(d')^k$.
- Step 2: The complexity of this step is $s\, d'^{O(k\ell)} d^{O(\ell)}$, using the complexity analysis of Algorithm 11.19 (Restricted Elimination). The number of polynomials output is $s\, d'^{O(k)} d^{O(1)}$.

– Step 3: The complexity of this step is $s^\ell d'^{O(k\ell)} d^{O(1)}$, using the complexity analysis of Algorithm 13.1 (Sampling). There are $s^\ell d'^{O(k\ell)} d^{O(\ell)}$ points output.
– Step 4: For each sample point, the complexity is $s^{\ell+1} d'^{O(k\ell)} d^{O(\ell)}$ using the complexity analysis of Algorithm 14.4 (Inverse Sign Determination). So the complexity of this step is $s^{2\ell+1} d'^{O(k\ell)} d^{O(\ell)}$. The number of elements of $\mathcal{A}(y)$ is bounded by $s\, d'^{O(k)} d^{O(1)}$ and the degrees of the elements of $\mathcal{A}(y)$ are bounded by $d'^{O(k)} d^{O(1)}$.
– Step 5: The number of elements in $\mathcal{A}$ is $s^{\ell+1} d'^{O(k\ell)} d^{O(\ell)}$, and the degrees of the elements of $\mathcal{A}$ are bounded by $d'^{O(k)} d^{O(1)}$. The complexity of this step is $s^{\ell(\ell+1)} d'^{O(k\ell)} d^{O(\ell)}$, using the complexity analysis of Algorithm 13.1 (Sampling).
– Step 6: For every ρ, the complexity is $s\, d'^{O(k\ell)} d^{O(\ell)}$. So the complexity of this step is $s^{\ell(\ell+1)+1} d'^{O(k\ell)} d^{O(\ell)}$ using the complexity analysis of Algorithm 12.19 (Triangular Sign Determination).

Finally the complexity is $s^{\ell(\ell+1)+1} d'^{O(k\ell)} d^{O(\ell)}$.

If $\mathrm{D} = \mathbb{Z}$, and the bitsizes of the coefficients of the polynomials are bounded by τ, then the bitsizes of the integers appearing in the intermediate computations and the output are bounded by $\tau d'^{O(k\ell)} d^{O(\ell)}$. □

We now define parametrized triangular Thom encodings.

A **parametrized triangular Thom encoding** of level k with parameters $Y = (Y_1, \ldots, Y_\ell)$ specified by $\mathcal{A}, \rho, \mathcal{T}, \sigma$ is

– a finite subset $\mathcal{A}$ of $\mathrm{R}[Y]$,
– a sign condition ρ on $\mathcal{A}$,
– a triangular system of polynomials $\mathcal{T}$, where $\mathcal{T}_i \in \mathrm{R}[Y, X_1, \ldots, X_i]$,
– a sign condition σ on $\mathrm{Der}(\mathcal{T})$ such that for every $y \in \mathrm{Reali}(\rho)$, there is a zero $z(y)$ of $\mathcal{T}(y)$ with triangular Thom encoding ρ.

Algorithm 14.7. **[Parametrized Triangular Thom Encoding]**

- **Structure**: an ordered integral domain D contained in a real closed field R.
- **Input**: a parametrized triangular system $\mathcal{T}$ with parameters $(Y_1, \ldots, Y_\ell)$ and variables $(X_1, \ldots, X_k)$.
- **Output**:
  – a finite set $\mathcal{A} \subset \mathrm{D}[Y]$, with $Y = (Y_1, \ldots, Y_k)$,
  – for every $\rho \in \mathrm{SIGN}(\mathcal{A})$, a list of sign conditions on $\mathrm{Der}(\mathcal{T})$ specifying for every $y \in \mathrm{Reali}(\rho)$, the list of triangular Thom encodings of the roots of $\mathcal{T}(y)$.
- **Complexity**: $d'^{O(k\ell)}$ where d' is a bound on the degrees of the polynomials in $\mathcal{T}$.
- **Procedure:** Apply Algorithm 14.6 (Parametrized Sign Determination) to $\mathcal{T}$ and $\mathrm{Der}(\mathcal{T})$.

Proof of correctness: Immediate. □

Complexity analysis: The complexity is $d'^{O(k\ell)}$, using the complexity of Algorithm 14.6 (Parametrized Sign Determination). The number of elements in $\mathcal{A}$ is $d'^{O(k\ell)}$, and the degrees of the elements of $\mathcal{A}$ are bounded by $d'^{O(k)}$. □

We follow the notations introduced in the last two sections.

Let $\mathcal{P} \subset \mathrm{R}[X_1, \dots, X_k, Y_1, \dots, Y_\ell]$ be finite and let Π denote a partition of the list of variables $X = (X_1, \dots, X_k)$ into blocks, $X_{[1]}, \dots, X_{[\omega]}$, where the block $X_{[i]}$ is of size $k_i, 1 \le i \le \omega$, $\sum_{1 \le i \le \omega} k_i = k$.

Recall that (Notations 14.8 and 14.10) for every $i = \omega - 1, \dots, 0$, the elements of $\mathrm{UR}_{\Pi,i}(\mathcal{P})$, are parametrized univariate representations in the variable T_{i+1}, contained in $\mathrm{D}[Y, X_{[1]}, \dots, X_{[i]}, \varepsilon_{i+1}, \delta_{i+1}][T_{i+1}]^{k_{i+1}+2}$. Let

$$u = (u_0, \dots, u_{\omega-1}) \in \mathcal{U} = \prod_{i=0}^{\omega-1} \mathrm{UR}_{\Pi,i}(\mathcal{P}),$$

with

$$u_{i-1} = (f^{[i]}, g_0^{[i]}, g_1^{[i]}, \dots, g_{k_i}^{[i]}).$$

Also recall that we denote,

$$\begin{aligned} \mathcal{T}_{u,i} &= (f^{[1]}(T_1), f_{u,1}^{[2]}(T_1, T_2), \cdots, f_{u,i-1}^{[i]}(T_1, T_2, \dots, T_i)), \\ \mathcal{T}_u &= (f^{[1]}(T_1), f_{u,1}^{[2]}(T_1, T_2), \cdots, f_{u,\omega-1}^{[\omega]}(T_1, T_2, \dots, T_\omega)). \end{aligned}$$

We now introduce the following notation which is used in the description of the algorithm below.

Notation 14.18. Let $u = (u_0, \dots, u_{j-1}) \in \mathcal{U}_i = \prod_{j=0}^{i-1} \mathrm{UR}_{\Pi,j}(\mathcal{P})$. We denote by $\mathcal{L}_{u,i}$ the set of all possible triangular Thom encodings of roots of $\mathcal{T}_{u,i}$ as y vary over $\mathrm{R}\langle \varepsilon_1, \delta_1, \dots, \varepsilon_\omega, \delta_\omega \rangle^\ell$. □

Algorithm 14.8. **[Local Quantifier Elimination]**

- **Structure:** an ordered domain D contained in a real closed field R.
- **Input:** a finite subset $\mathcal{P} \subset \mathrm{R}[X_1, \dots, X_k, Y_1, \dots, Y_\ell]$, a partition Π of the list of variables $X = (X_1, \dots, X_k)$ into blocks, $X_{[1]}, \dots, X_{[\omega]}$ and a $(\mathcal{P}, \Pi)$-formula $\Phi(Y)$.
- **Output:** a quantifier free formula, $\Psi(Y)$, equivalent to $\Phi(Y)$.
- **Complexity:** $s^{(k_\omega+1)\dots(k_1+1)} d^{\ell O(k_\omega)\dots O(k_1)}$ where s is a bound on the number of elements of $\mathcal{P}$, d is a bound on the degree of elements of $\mathcal{P}$, and k_i is the size of the block $X_{[i]}$.
- **Procedure:**
  - Initialize $\mathrm{B}_{Pi,\omega}(\mathcal{P}) := \mathcal{P}$.
  - Block Elimination Phase: Compute

$$\mathrm{B}_{\Pi,i}(\mathcal{P}) = \mathrm{BElim}_{X_{[i+1]}}(\mathrm{Bor}_{\Pi,i+1}(\mathcal{P})),$$

for $1 \le i \le \omega - 1$, applying repeatedly $\mathrm{BElim}_{X_{[i+1]}}$, using Algorithm 14.1 (Block Elimination).

Compute $\mathrm{UR}_{\Pi,i}(\mathcal{P})$, for every $i = \omega - 1, \ldots, 0$. The elements of $\mathrm{UR}_{\Pi,i}(\mathcal{P})$ are parametrized univariate representations in the variable T_{i+1}, contained in

$$\mathrm{D}[Y, X_{[1]}, \ldots, X_{[i]}, \varepsilon_{i+1}, \delta_{i+1}][T_{i+1}]^{k_{i+1}+2}.$$

- For every

$$u = (u_0, \ldots, u_{\omega-1}) \in \mathcal{U} = \prod_{i=0}^{\omega-1} \mathrm{UR}_{\Pi,i}(\mathcal{P}),$$

with

$$u_{i-1} = (f^{[i]}, g_0^{[i]}, g_1^{[i]}, \ldots, g_{k_i}^{[i]}),$$

compute the corresponding triangular system,

$$\mathcal{T}_u = (f^{[1]}(Y, T_1), f_{u,1}^{[2]}(Y, T_1, T_2), \ldots, f_{u,\omega-1}^{[\omega]}(Y, T_1, T_2, \ldots, T_\omega)).$$

(see Notation 14.10).

For $i = 0 \ldots \omega - 1$ compute the sets $\mathcal{L}_{u,i}$, using Algorithm 14.7 (Parametrized Triangular Thom Encoding) with input $\mathcal{T}_{u,i}$.

- Let

$$\Phi(Y) = (\mathrm{Qu}_1 X_{[1]}) \ldots (\mathrm{Qu}_\omega X_{[\omega]}) F(X, Y)$$

where $\mathrm{Qu}_i \in \{\forall, \exists\}$, $Y = (Y_1, \ldots, Y_\ell)$ and $F(X, Y)$ is a quantifier free $\mathcal{P}$-formula.

For every atom of the form $\mathrm{sign}(P) = \sigma, P \in \mathcal{P}$ occurring in the input formula F, and for every

$$u = (u_0, \ldots, u_{\omega-1}) \in \mathcal{U} = \prod_{i=0}^{\omega-1} \mathrm{UR}_{\Pi,i}(\mathcal{P}),$$

with

$$u_{i-1} = (f^{[i]}, g_0^{[i]}, g_1^{[i]}, \ldots, g_{k_i}^{[i]}),$$

and $\tau \in \mathcal{L}_{u,\omega}$ compute using Algorithm 14.5 (Quantifier Elimination) a quantifier-free formula $\phi_{u,\tau}$ equivalent to the formula

$$(\exists T_1, \ldots, T_\omega) \bigwedge_{i=1}^{\omega} \mathrm{SIGN}(\mathrm{Der}(f_{u,i-1}^{[i]})) = \tau_i \bigwedge \mathrm{SIGN}(P_{u,\omega}) = \sigma.$$

Let $F_{u,\tau}$ denote the quantifier-free formula obtained by replacing every atom ϕ in F by the corresponding formula $\phi_{u,\tau}$.

Also, for every

$$u = (u_0, \ldots, u_{\omega-1}) \in \mathcal{U} = \prod_{i=0}^{\omega-1} \mathrm{UR}_{\Pi,i}(\mathcal{P})),$$

with

$$u_{i-1} = (f^{[i]}, g_0^{[i]}, g_1^{[i]}, \ldots, g_{k_i}^{[i]}), \tau \in \mathcal{L}_{u,\omega}$$

and for every $j, 1 \le j \le \omega$, compute using Algorithm 14.5 (Quantifier Elimination) a quantifier-free formula $\psi_{u,\tau,j}$ equivalent to the formula,

$$(\exists T_1, \ldots, T_j) \bigwedge_{i=1}^{j} \mathrm{SIGN}(\mathrm{Der}(f^{[i]}_{u,i-1})) = \tau_i.$$

– For $u \in \mathcal{U}$ and $\tau \in \mathcal{L}_{u,\omega}$, let

$$\Phi_{\omega,u,\tau} = F_{u,\tau}.$$

Compute inductively for i from $\omega - 1$ to 0, and for every

$$u = (u_0, \ldots, u_{i-1}) \in \mathcal{U}_i = \prod_{j=0}^{i-1} \mathrm{UR}_{\Pi,j}(\mathcal{P}),$$

and $\tau \in \mathcal{L}_{u,i}$,

$$\begin{aligned} \Phi_{i,u,\tau} &= \bigwedge_{(v,\rho),\bar{v}=u,\bar{\rho}=\tau} (\psi_{v,\rho,i+1} \wedge \Phi_{i+1,v,\rho}) \text{ if } \mathrm{Qu}_{i+1} = \exists, \\ &= \bigwedge_{(v,\rho),\bar{v}=u,\bar{\rho}=\tau} (\psi_{v,\rho,i+1} \Longrightarrow \Phi_{i+1,v,\rho}) \text{ if } \mathrm{Qu}_{i+1} = \forall. \end{aligned}$$

Take $\overline{\Phi(Y)} = \Phi_0$.
– Output $\Psi(Y) = \mathrm{Remo}_{\varepsilon_1,\delta_1,\ldots,\varepsilon_\omega,\delta_\omega}(\overline{\Phi(Y)})$ (Notation 14.6).

Complexity analysis: It follows from the complexity analysis of Algorithm a14.1 (Block Elimination), Algorithm 14.7 (Parametrized Triangular Thom Encoding) and Algorithm 14.5 (Quantifier Elimination) that the complexity is bounded by $s^{(k_\omega+1)\ldots(k_1+1)} d^{\ell O(k_\omega)\ldots O(k_1)}$.

If $\mathrm{D} = \mathbb{Z}$, and the bitsizes of the coefficients of the polynomials are bounded by τ, then the bitsizes of the integers appearing in the intermediate computations and the output are bounded by $\tau d^{O(k_\omega)\cdots O(k_1)}$.

Note that the only improvement compared to Algorithm 14.5 (Quantifier Elimination) is that the exponent of s does not depend on the number of free variables ℓ. Note also that the total number of polynomials in Y appearing in the formula is $s^{(k_\omega+1)\ldots(k_1+1)} d^{\ell O(k_\omega)\cdots O(k_1)}$. Determining which are the realizable sign conditions on these polynomials would cost $s^{(\ell+1)(k_\omega+1)\ldots(k_1+1)} d^{\ell O(k_\omega)\ldots O(k_1)}$, but this computation is not part of the algorithm. □

We now give an application of Algorithm (Local Quantifier Elimination) 14.8 to the closure of a semi-algebraic set.

Let S be a semi-algebraic set described by a quantifier free $\mathcal{P}$-formula $\mathrm{F}(X)$, where $\mathcal{P}$ is a finite set of s polynomials of degree at most d in k variables. The closure of S is described by the following quantified formula $\Psi(X)$

$$\forall\, Z \quad \exists\, Y \quad \|X - Y\|^2 < Z^2 \wedge F(Y).$$

Note that $\Psi(X)$ is a first-order formula with two blocks of quantifiers, the first with one variable and the second one with k variables. Denote by $\mathcal{R}$ the set of polynomials in k variables obtained after applying twice Algorithm 14.1 (Block Elimination) to the polynomials appearing in the formula describing the closure of S in order to eliminate Z and Y. These polynomials have the property that the closure of S is the union of semi-algebraically connected components of sets defined by sign conditions over $\mathcal{R}$. According to Theorem 14.16 the set $\mathcal{R}$ has $s^{2k+1}d^{O(k)}$ polynomials and each of these polynomials has degree at most $d^{O(k)}$.The complexity for computing $\mathcal{R}$ is $s^{2(k+1)}d^{O(k)}$. Note that we cannot ensure that the closure of S is described by polynomials in $\mathcal{R}$. However, performing Algorithm 14.8 (Local Quantifier Elimination) gives a quantifier-free description of the closure of S in time $s^{2(k+1)}d^{O(k)}$ by $s^{2k+1}d^{O(k)}$ polynomials of degree at most $d^{O(k)}$.

14.4 Global Optimization

We describe an algorithm for finding the infimum of a polynomial on a semi-algebraic set as well as a minimizer if there exists one.

Algorithm 14.9. **[Global Optimization]**

- **Structure:** an ordered domain D contained in a real closed field R.
- **Input**: a finite subset $\mathcal{P} \subset \mathrm{D}[X_1, \ldots, X_k]$, a $\mathcal{P}$-semi-algebraic set S described by a quantifier free formula $\Phi(X)$ and $F \in \mathrm{D}[X_1, \ldots, X_k]$.
- **Output:** the infimum w of F on S, and a minimizer, i.e. a point $x \in S$ such that $F(x) = w$ if such a point exists.
- **Complexity:** $s^{2k+1} d^{O(k)}$, where s is a bound on the number of elements of $\mathcal{P}$ and d is a bound on degree of F and of the elements of $\mathcal{P}$.
- **Procedure:**
  - Let Y be a new variable and $G = Y - F \in \mathrm{D}[Y, X_1, \ldots, X_k]$. Denote by $S' \subset R^{k+1}$ the realization of $\Phi \wedge G = 0$.
  - Call Algorithm 14.1 (Block Elimination) with block of variables $X_1, \ldots,$ X_k and set of polynomials $\mathcal{P} \cup \{G\} \subset \mathrm{D}[Y, X_1, \ldots, X_k]$. Let $\mathcal{B} \subset \mathrm{D}[Y]$ denote $\mathrm{BElim}_X(\mathcal{P} \cup \{G\})$.
  - Call Algorithm 10.19 (Univariate Sample Points) with input $\mathcal{B}$ and denote by $\mathcal{C}$ the set of sample points so obtained. Each element of $\mathcal{C}$ is a Thom encoding (h, σ).
  - Fore each $y = (h, \sigma) \in \mathcal{C}$, the points associated to $\mathrm{UR}_X(\mathcal{P} \cup \{\mathcal{G}\})(y)$ intersect every semi-algebraically connected component of every realizable sign condition of the set $\mathcal{P} \cup \{\mathcal{G}\}(y)$ in $\mathrm{R}\langle\varepsilon, \delta\rangle^k$. Compute the subset $\mathcal{C}'$ of elements $y \in \mathcal{C}$ such that the set of ^points associated to $\mathrm{UR}_X(\mathcal{P} \cup \{\mathcal{G}\})(y)$ meets the extension of S' to $\mathrm{R}\langle\varepsilon, \delta\rangle$ using Algorithm 12.19 (Triangular Sign Determination).

- If there is no root y of a polynomial in $\mathcal{B}$ such that for all $y' \in \mathcal{C}'$, $y' \geqslant y$ holds, define w as $-\infty$. Otherwise, define w as the maximum $y \in \mathcal{C}$ which is a root of a polynomial in $\mathcal{B}$ and such that for all $y' \in \mathcal{C}'$, $y' \geqslant y$ holds.
- If $w = (h, \sigma) \in \mathcal{C}'$, pick $u = (f, g_0, \ldots, g_k) \in \mathrm{UR}_X(\mathcal{P} \cup \{\mathcal{G}\})(w)$ with associated point in the extension of S' to $\mathrm{R}\langle\varepsilon, \delta\rangle$. Replace δ and ε by appropriately small elements from the field of quotients of D using Algorithm 11.20 (Removal of Infinitesimals) with input f, its derivatives and the P_u, $P \in \mathcal{P}$ and using Remark 11.27. Then clear denominators to obtain univariate representation with entries in $\mathrm{D}[T]$.

Proof of correctness: Follows clearly from the correctness of Algorithm 14.1 (Block Elimination). □

Complexity analysis: The call to Algorithm 14.1 (Block Elimination) costs $s^{k+1}\, d^{O(k)}$. The call to Algorithm 10.19 (Univariate Sample Points) costs $s^{2k}\, d^{O(k)}$ since there are at most $s^k\, d^{O(k)}$ polynomials of degree at most $d^{O(k)}$. Each call to Algorithm 12.19 (Triangular Sign Determination) costs $s\, d^{O(k)}$ and there are $s^{2k}\, d^{O(k)}$ such calls. The call to Algorithm 11.20 (Removal of Infinitesimals) costs $(s + d^{O(k)})\, d^{O(k)}$ which is $s\, d^{O(k)}$. The total complexity is thus $s^{2k+1}\, d^{O(k)}$. □

14.5 Dimension of Semi-algebraic Sets

Let S be a semi-algebraic set described by a quantifier free $\mathcal{P}$-formula $\Phi(X)$

$$S = \{x \in \mathrm{R}^k \mid \Phi(x)\}$$

where $\mathcal{P}$ is a finite set of s polynomials in k variables with coefficients in a real closed field R. We denote by $\mathrm{SSIGN}(\mathcal{P})$ the set of **strict realizable sign conditions of** $\mathcal{P}$, i.e. the realizable sign conditions $\sigma \in \{0, 1, -1\}^{\mathcal{P}}$ such that for every $P \in \mathcal{P}$, $P \neq 0$, $\sigma(P) \neq 0$.

Proposition 14.19. *The dimension of S is k if and only is there exists $\sigma \in \mathrm{SSIGN}(\mathcal{P})$ such that* $\mathrm{Reali}(\sigma) \subset S$.

Proof: The dimension of S is k if and only if there exists a point $x \in S$ and $r > 0$ such that $B(x, r) \subset S$. The sign condition satisfied by $\mathcal{P}$ at such an x is necessarily strict. In the other direction, if the sign condition σ satisfied by $\mathcal{P}$ at such an x is strict, $\mathrm{Reali}(\sigma)$ is open, and contained in S since S is defined by a quantifier free $\mathcal{P}$-formula. □

It is reasonable to expect that the dimension of S is $\geq j$ if and only if the dimension of $\pi(S)$ is j, where π is a linear surjection of R^k to R^j.

Using results from Chapter 13, we are going to prove that using $j(k-j)+1$ well chosen linear surjections is enough. Recall that we have defined in Notation 13.26 a family

$$\mathcal{L}_{k,k-j}=\{V_i \mid 0\leq i\leq j(k-j)\}.$$

of $j\,(k-j)+1$ vector spaces such that any linear subspace T of R^k of dimension $k'\geq j$ is such that there exists $0\leq i\leq j(k-j)$ such that V_i and T span R^k (see Corollary 13.28). We denoted by $v_k(x)$ the Vandermonde vector

$$(1,x,\ldots,x^{k-1}).$$

and by V_ℓ the vector subspace of R^k generated by

$$v_k(\ell),v_k(\ell+1),\ldots,v_k(\ell+k-k'-1).$$

We also defined in Notation 13.26 a linear bijection $L_{j,i}$ such that $L_{j,i}(V_i)$ consists of vectors of R^k having their last j coordinates equal to 0. We denoted by $M_{k',\ell}=(d_{k-k',\ell})^{k'}L_{k',\ell}^{-1}$, with

$$d_{k-k',\ell}=\det(v_{k-k'}(\ell),\ldots,v_{k-k'}(\ell+k-k'-1)),$$

and remarked that $M_{k',\ell}$ plays the same role as the inverse of $L_{k',\ell}$ but is with integer coordinates.

We denote by π_j the canonical projection of R^k to R^j forgetting the first $k-j$ coordinates.

Proposition 14.20. *Let $0\leq j\leq k$. The dimension of S is $\geq j$ if and only if there exists $0\leq i\leq j(k-j)$ such that the dimension of $\pi_j(L_{j,i}(S))$ is j.*

Proof: It is clear that if the dimension of $\pi_j(L_{j,i}(S))$ is j, the dimension of S is $\geq j$. In the other direction, if the dimension of S is $k'\geq j$, by Proposition 5.53, there exists a smooth point x of S of dimension k' with tangent space denoted by T. By Corollary 13.28, there exists $0\leq i\leq j(k-j)$, such that V_i and T span R^k. Since $L_{j,i}(V_i)$ consists of vectors of R^k having their last j coordinates equal to 0, and $L_{j,i}(V_i)$ and $L_{j,i}(T)$ span R^k, $\pi_j(L_{j,i}(T))$ is R^j. Then the dimension of $\pi_j(L_{j,i}(S))$ is j. □

The idea for computing the dimension is simple: check whether the dimension of S is k or -1 (i.e. is empty) using Proposition 14.19. If it is not the case, try $k-1$ or 0 or, then $k-2$ or 1, etc.

Algorithm 14.10. **[Dimension]**

- **Structure:** an ordered domain D contained in a real closed field R.
- **Input**: a finite subset $\mathcal{P}\subset \mathrm{D}[X_1,\ldots,X_k]$, and a semi-algebraic set S described by a quantifier free $\mathcal{P}$-formula $\Phi(X)$.

- **Output:** the dimension k' of S.
- **Complexity:**

$$\begin{cases} s^{(k-k')k'} d^{O(k'(k-k'))} & \text{if } k' \geq k/2 \\ s^{(k-k'+1)(k'+1)} d^{O(k'(k-k'))} & \text{if } k' < k/2. \end{cases}$$

where s is a bound on the number of elements of $\mathcal{P}$ and d is a bound on their degree.
- **Procedure:**
  - Initialize $j := 0$.
  - ($\star$) Consider the block structure Π_{k-j} with two blocks of variables: $X_{j+1}, \ldots, X_k$ and $X_1, \ldots, X_j$.
  - For every $i = 0, \ldots, j(k-j)$ let $\mathcal{P}_{k-j,i} = \mathcal{P}(M_{k-j,i})$, using Notation 13.26 and

$$S_{k-j,i} = \{x \in \mathrm{R}^k \mid \Phi(M_{k-j,i}(x))\}.$$

    - Compute $\mathrm{SIGN}_{\Pi_{k-j}}(\mathcal{P}_{k-j,i})$ using Algorithm 14.2 (Block Structured Signs).
    - Defining $X_{\leq j} = X_1 \ldots, X_j$, compute

$$\mathrm{SSIGN}(\mathrm{BElim}_{X_{\leq j}}(\mathcal{P}_{k-j,i}))$$

    using Algorithm 13.2 (Sampling). Note, using Remark 14.12, that every sample point output by Algorithm 14.2 (Block Structured Signs) is above a sample point for $\mathrm{BElim}_{X_{\leq j}}(\mathcal{P}_{k-j,i})$ output by Algorithm 13.2 (Sampling).
    - Check whether one of the strict sign conditions in

$$\mathrm{SSIGN}(\mathrm{BElim}_{X_{\leq j}}(\mathcal{P}_{k-j,i}))$$

    is satisfied at some point of $\pi_{k-j}(S_{k-j,i})$.
    - If one of the strict sign conditions in

$$\mathrm{SSIGN}(\mathrm{BElim}_{X_{\leq j}}(\mathcal{P}_{k-j,i}))$$

    is satisfied at some point of $\pi_{k-j}(S_{k-j,i})$, output $k-j$.
  - Consider the block structure Π_j with two blocks of variables: $X_{k-j+1}, \ldots, X_k$ and $X_1, \ldots, X_{k-j}$.
  - For every $i = 0, \ldots, j(k-j)$ let $\mathcal{P}_{j,i} = \mathcal{P}(M_{j,i})$, using Notation 13.30 and

$$S_{j,i} = \{x \in \mathrm{R}^k \mid \Phi(M_{j,i}(x))\}.$$

    - Compute $\mathrm{SIGN}_{\Pi_j}(\mathcal{P}_{j,i})$ using Algorithm 14.2 (Block Structured Signs).
    - Defining $X_{\leq k-j} = X_1 \ldots, X_{k-j}$, compute

$$\mathrm{SSIGN}(\mathrm{BElim}_{X_{\leq k-j}}(\mathcal{P}_{j,i}))$$

using Algorithm 13.2 (Sampling). Note, using Remark 14.12, that every sample point output by Algorithm 14.2 (Block Structured Signs) is above a sample point for $\mathrm{BElim}_{X_{\leq k-j}}(\mathcal{P}_{j,i})$ output by Algorithm 13.2 (Sampling).

– Check whether one of the strict sign conditions in

$$\mathrm{SSIGN}(\mathrm{BElim}_{X_{\leq k-j}}(\mathcal{P}_{j,i}))$$

is satisfied at some point of $\pi_j(S_{j,i})$.

– If for every $i = 0 \ldots j(k-j)$ none of the strict sign conditions in

$$\mathrm{SSIGN}(\mathrm{BElim}_{X_{\leq k-j}}(\mathcal{P}_{j,i}))$$

is satisfied at some point of $\pi_j(S_{j,i})$, output $j-1$.

– Otherwise define $j := j+1$ and go to ($\star$).

Proof of correctness: Follows clearly from Proposition 14.19, Proposition 14.20, the correctness of of Algorithm 14.1 (Block Elimination), Algorithm 13.2 (Sampling). □

Complexity analysis: There are at most $(k+1)/2$ values of j considered in the algorithm.

For a given j, the complexity of the call to Algorithm 14.2 (Block Structured Signs) performed is $s^{(j+1)(k-j+1)} d^{O(j(k-j))}$, using the complexity analysis of Algorithm 14.2 (Block Structured Signs).

The call to Algorithm 13.2 (Sampling) for $\mathrm{BElim}_{X_{\leq j}}(\mathcal{P}_{k-j,i})$, has complexity $s^{(j+1)(k-j+1)} d^{O(j(k-j))}$, using the complexity analysis of Algorithm 14.1 (Block elimination) and 13.2 (Sampling), since the number of polynomials is $s^{j+1} d^{O(j)}$, their degrees are $d^{O(j)}$ and their number of variables is $k-j$.

Similarly, the call to Algorithm 13.2 (Sampling) for $\mathrm{BElim}_{X_{\leq k-j}}(\mathcal{P}_{j,i})$, has complexity $s^{(j+1)(k-j+1)} d^{O(j(k-j))}$, using the complexity analysis of Algorithm 14.1 (Block elimination) and 13.2 (Sampling), since the number of polynomials is $s^{k-j+1} d^{O(k-j)}$, their degrees are $d^{O(k-j)}$ and their number of variables is j.

Finally the total cost of the algorithm is

$$\begin{cases} s^{(k-k')k'} d^{O(k'(k-k'))} & \text{if } k' \geq k/2 \\ s^{(k-k'+1)(k'+1)} d^{O(k'(k-k'))} & \text{if } k' < k/2. \end{cases}$$

If $\mathrm{D} = \mathbb{Z}$, and the bitsizes of the coefficients of the polynomials are bounded by τ, then the bitsizes of the integers appearing in the intermediate computations and the output are bounded by $\tau d^{O(k'(k-k'))}$.

Note that this complexity result is output sensitive, which means that the complexity depends on the output of the algorithm. □

14.6 Bibliographical Notes

The idea of designing algorithms taking into account the block structure is due to Grigor'ev [76], who achieved doubly exponential complexity in the number of blocks for the general decision problem. It should be noted that for a fixed value of ω, this is only singly exponential in the number of variables. Heintz, Roy and Solerno [85] and Renegar [133] extended this result to quantifier elimination. Renegar's [133] algorithms solved the general decision problem in time $(s\, d)^{O(k_\omega)\cdots O(k_1)}$, and the quantifier elimination problem in time $(s\, d)^{O(k_\omega)\cdots O(k_1)O(\ell)}$.

Most of the results presented in this chapter are based on [13]. In terms of algebraic complexity (the degree of the equations), the complexity of quantifier elimination presented here is similar to [133]. However the bounds in this chapter are more precise in terms of combinatorial complexity (the dependence on the number of equations). Similarly, the complexity of Algorithm 14.10, coming from [19] improves slightly the result of [163] which computes the dimension of a semi-algebraic set with complexity $(s\, d)^{O(k'(k-k'))}$.

The local quantifier elimination algorithm is based on results in [12].

15

Computing Roadmaps and Connected Components of Algebraic Sets

In this chapter, we compute roadmaps and connected components of algebraic sets. Roadmaps provide a way to count connected components and to decide whether two points belong to the same connected component. Done in a parametric way the roadmap algorithm also gives a description of the semi-algebraically connected components of an algebraic set. The complexities of the algorithms given in this chapter are much better than the one provided by cylindrical decomposition in Chapter 11 (single exponential in the number of variables rather than doubly exponential).

We first define roadmaps. Let S be a semi-algebraic set. As usual, we denote by π the projection on the X_1-axis and set $S_x = \{y \in \mathrm{R}^{k-1} \mid (x, y) \in S\}$.

A **roadmap** for S is a semi-algebraic set M of dimension at most one contained in S which satisfies the following roadmap conditions:

- RM_1 For every semi-algebraically connected component D of S, $D \cap M$ is semi-algebraically connected.
- RM_2 For every $x \in \mathrm{R}$ and for every semi-algebraically connected component D' of S_x, $D' \cap M \neq \emptyset$.

The construction of roadmaps is based on the critical point method, using properties of pseudo-critical values provided in Section 15.1. In Section 15.2 we give an algorithm constructing a roadmap for $\mathrm{Zer}(Q, \mathrm{R}^k)$, for $Q \in \mathrm{R}[X_1, \ldots, X_k]$. As a consequence, we get an algorithm for computing the number of connected components (the zero-th Betti number) of an algebraic set, with single exponential complexity.

In Section 15.3 we obtain an algorithm giving a semi-algebraic description of the semi-algebraically connected components of an algebraic set. The idea behind the algorithm is simple: we perform parametrically the roadmap algorithm with a varying input point.

15.1 Pseudo-critical Values and Connectedness

We consider a semi-algebraic set S as the collection of its fibers S_x, $x \in \mathrm{R}$. In the smooth bounded case, critical values of π are the only places where the number of connected components in the fiber can change.

More precisely, we can generalize Proposition 7.6 to the case of a general real closed field.

Proposition 15.1. *Let* $\mathrm{Zer}(Q,\mathrm{R}^k)$ *be a non-singular bounded algebraic hypersurface,* $[a,b]$ *such that* π *has no critical value in* $[a,b]$, *and* $d \in [a,b]$.

a) The number of semi-algebraically connected components of $\mathrm{Zer}(Q,\mathrm{R}^k)_{[a,b]}$ *and* $\mathrm{Zer}(Q,\mathrm{R}^k)_d$ *are the same.*
b) Let S *be a semi-algebraically connected component of* $\mathrm{Zer}(Q,\ \mathrm{R}^k)_{[a,b]}$. *Then, for every* $d \in [a,b]$, S_d *is semi-algebraically connected.*

Proposition 15.1 immediately implies.

Proposition 15.2. *Let* $\mathrm{Zer}(Q,\mathrm{R}^k)$ *be a bounded non-singular algebraic hypersurface and* $[a,\ b]$ *such that* π *has no critical value in* $[a,\ b]$. *Let* S *be a semi-algebraically connected component of* $\mathrm{Zer}(Q,\ \mathrm{R}^k)_{[a,b]}$. *Then, for every* $d \in [a,b]$, S_d *is semi-algebraically connected.*

Proposition 15.3. *Let* $\mathrm{Zer}(Q,\mathrm{R}^k)$ *be a non-singular algebraic hypersurface and* S *a semi-algebraically connected component of* $\mathrm{Zer}(Q,\mathrm{R}^k)_{[a,b]}$. *If* $S_{[a,b)}$ *is not semi-algebraically connected then* b *is a critical value of* π *on* $\mathrm{Zer}(Q,\mathrm{R}^k)$.

Proof of Proposition 15.1: Over the reals (the case $\mathrm{R} = \mathbb{R}$), the two properties are true according to Proposition 7.6.

We now prove that Properties a and b hold for a general real closed field, using Theorem 5.46 (Semi-algebraic triviality) and the transfer principle (Theorem 2.80).

We first prove Property a.

Let $\{m_1, \ldots, m_N\}$ be a list of all monomials in the variables $x_1, \ldots x_k$ with degree at most the degree of Q. To an element $\mathrm{cof} = (c_1, \ldots, c_N)$ of R^N, we associate the polynomial

$$\mathrm{Pol}(\mathrm{cof}) = \sum_{i=1}^{N} c_i m_i.$$

Denoting by $\mathrm{cof}_i(Q)$ the coefficient of m_i in Q and by

$$\mathrm{cof}(Q) = (\mathrm{cof}_1(Q), \ldots, \mathrm{cof}_N(Q)),$$

we have $Q = \mathrm{Pol}(\mathrm{cof}(Q))$.

Consider the field $\mathbb{R}_{\mathrm{alg}}$ of real algebraic numbers and the subset $W \subset \mathbb{R}_{\mathrm{alg}}^{N+2+k}$ defined by

$$W = \{(\mathrm{cof}, a', b', x_1 \ldots, x_k) \mid a' \leq x_1 \leq b', \mathrm{Pol}(\mathrm{cof})(x_1, \ldots, x_k) = 0\}.$$

The set W can be viewed as the family of sets $\mathrm{Zer}(\mathrm{Pol}(\mathrm{cof}), \mathbb{R}_{\mathrm{alg}}^{N+2+k})_{[a',b']}$, parametrized by $(\mathrm{cof}, a', b') \in \mathbb{R}_{\mathrm{alg}}^{N+2}$. We also consider the subset $W' \subset \mathbb{R}_{\mathrm{alg}}^{N+1+k}$ defined by

$$W' = \{(\mathrm{cof}, d', x_1..., x_k) \mid \mathrm{Pol}(\mathrm{cof})(d', ..., x_k) = 0\}.$$

The set W' can be viewed as the family of sets $\mathrm{Zer}(\mathrm{Pol}(\mathrm{cof}), \mathbb{R}_{\mathrm{alg}}^{N+1+k})_{d'}$, parametrized by $(\mathrm{cof}, d') \in \mathbb{R}_{\mathrm{alg}}^{N+1}$. According to Theorem 5.46 (Hardt's triviality) applied to W (resp. W'), there is a finite partition $\mathcal{A}$ (resp. $\mathcal{B}$) of $\mathbb{R}_{\mathrm{alg}}^{N+2}$ (resp. $\mathbb{R}_{\mathrm{alg}}^{N+1}$) into semi-algebraic sets, and for every $A \in \mathcal{A}$ (resp. $B \in \mathcal{B}$) the sets $\mathrm{Zer}(\mathrm{Pol}(\mathrm{cof}), \mathbb{R}_{alg}^{N+2+k})_{[a',b']}$ (resp. $\mathrm{Zer}(\mathrm{Pol}(\mathrm{cof}), \mathbb{R}_{\mathrm{alg}}^{N+1+k})_{d'}$) are semi-algebraically homeomorphic as (cof, a', b') varies in A (resp. (cof, d') varies in B). Hence, they have the same number of bounded semi-algebraically connected components $\ell(A)$ (resp. $\ell(B)$).

Using the transfer principle (Theorem 2.80), for every real closed field R and every $(\mathrm{cof}, a', b') \in \mathrm{Ext}(A, \mathrm{R})$ (resp. $(\mathrm{cof}, d') \in \mathrm{Ext}(B, \mathrm{R})$), the set $\mathrm{Zer}(\mathrm{Pol}(\mathrm{cof}), \mathrm{R}^{N+2+k})_{[a',b']}$ has $\ell(A)$ (resp. $\mathrm{Zer}(\mathrm{Pol}(\mathrm{cof}), \mathrm{R}^{N+1+k})_{d'}$ has $\ell(B)$) bounded semi-algebraically connected components. Moreover, since the connected components of

$$W_A = \{(\mathrm{cof}, a', b', x_1, ..., x_k) \in W \mid (\mathrm{cof}, a', b') \in A\}$$

are semi-algebraic sets defined over $\mathbb{R}_{\mathrm{alg}}$, there exists, for every $A \in \mathcal{A}$, $\ell(A)$ quantifier free formulas

$$\Phi_1(A)(\mathrm{cof}, a', b', x_1, ..., x_k), ..., \Phi_{\ell(A)}(A)(\mathrm{cof}, a', b', x_1, ..., x_k),$$

such that for every real closed field R and for every $(\mathrm{cof}, a', b') \in \mathrm{Ext}(A, \mathrm{R})$ the semi-algebraic sets

$$C_j = \{(x_1..., x_k) \in \mathrm{R}^k \mid \Phi_j(A)(\mathrm{cof}, a', b', x_1, ..., x_k)\}$$

for $1 \le j \le \ell(A)$ are the bounded semi-algebraically connected components of $\mathrm{Zer}(\mathrm{Pol}(\mathrm{cof}), \mathrm{R}^{N+2+k})_{[a',b']}$.

Let A (resp. B) be the set of the partition $\mathcal{A}$ (resp. $\mathcal{B}$) such that $\mathrm{cof}(Q), a, b) \in \mathrm{Ext}(A, \mathrm{R})$ (resp. $(\mathrm{cof}(Q), d) \in \mathrm{Ext}(B, \mathrm{R})$), and let E be the semi-algebraic set of $(\mathrm{cof}, a', b', d') \in (\mathbb{R}_{\mathrm{alg}})^{N+3}$ such that $(\mathrm{cof}, a', b') \in A$, $(\mathrm{cof}, d') \in B$, $\mathrm{Zer}(\mathrm{Pol}(\mathrm{cof}), \mathbb{R}_{\mathrm{alg}}^{N+2+k})$ is a non-singular algebraic hypersurface, π has no critical value over $[a', b']$, and $a' < d' < b'$. Using the transfer principle (Theorem 2.80), the set E is non-empty since $\mathrm{Ext}(E, \mathrm{R})$ is non-empty, and hence $\mathrm{Ext}(E, \mathbb{R})$ is non-empty.

Given $(\mathrm{cof}, a', b', d') \in \mathrm{Ext}(E, \mathbb{R})$, the number of bounded connected components of $\mathrm{Zer}(\mathrm{Pol}(\mathrm{cof}), \mathbb{R}^{N+2+k})_{[a',b']}$ is equal to the number of bounded connected components of $\mathrm{Zer}(\mathrm{Pol}(\mathrm{cof}), \mathbb{R}^{N+2+k})_{d'}$, since Property 1 holds for the reals. It follows that $\ell(A) = \ell(B)$, so the number of bounded semi-algebraically connected components of $\mathrm{Zer}(Q, \mathrm{R}^k)_{[a,b]}$ is equal to the number of bounded semi-algebraically connected components of $\mathrm{Zer}(Q, \mathrm{R}^k)_d$.

To complete the proof of the proposition, it remains to prove Property b. According to the preceding paragraph, there exist j such that

$$S = \{(x_1 ..., x_k) \in \mathrm{R}^k \mid \Phi_j(A)(\mathrm{cof}(Q), a, b, x_1, ..., x_k)\}.$$

Since Property b is true over the reals, the formula expressing that for every $(\mathrm{cof}, a', b', d') \in \mathrm{Ext}(E, \mathbb{R})$ the set

$$\{(x_2, ..., x_k) \in \mathbb{R}^k \mid \Phi_j(A)(\mathrm{cof}(Q), a, b, d', ..., x_k)\}$$

is non-empty is true over the reals. Using the transfer principle (Theorem 2.80), this formula is thus true over any real closed field. Thus, S_d is non-empty. □

In the non-smooth case, we again consider X_1-pseudo-critical values introduced in Chapter 12. These pseudo critical-values will also be the only places where the number of connected components in the fiber can change. More precisely, generalizing Proposition 15.2 and Proposition 15.3, we prove the following two propositions, which play an important role for computing roadmaps.

Proposition 15.4. *Let* $\mathrm{Zer}(Q, \mathrm{R}^k)$ *be a bounded algebraic set and* S *a semi-algebraically connected component of* $\mathrm{Zer}(Q, \mathrm{R}^k)_{[a,b]}$. *If* $v \in (a, b)$ *and* $[a, b] \setminus \{v\}$ *contains no* X_1*-pseudo-critical value on* $\mathrm{Zer}(Q, \mathrm{R}^k)$, *then* S_v *is semi-algebraically connected.*

Proposition 15.5. *Let* $\mathrm{Zer}(Q, \mathrm{R}^k)$ *be a bounded algebraic set and let* S *be a semi-algebraically connected component of* $\mathrm{Zer}(Q, \mathrm{R}^k)_{[a,b]}$. *If* $S_{[a,b)}$ *is not semi-algebraically connected, then* b *is an* X_1*-pseudo-critical value of* $\mathrm{Zer}(Q, \mathrm{R}^k)$.

Before proving these two propositions, we need some preparation. Suppose that the polynomial $Q \in \mathrm{R}[X_1, ..., X_k]$, and $(d_1, ..., d_k)$ satisfy the following conditions:

- $Q(x) \geq 0$ for every $x \in \mathrm{R}^k$,
- $\mathrm{Zer}(Q, \mathrm{R}^k) \subset B(0, 1/c)$ for some $c \leq 1, c \in \mathrm{R}$,
- $d_1 \geq d_2 \cdots \geq d_k$,
- $\deg(Q) \leq d_1$, $\mathrm{tDeg}_{X_i}(Q) \leq d_i$, for $i = 2, ..., k$.

Let $\bar{d_i}$ be an even number $> d_i, i = 1, ..., k$, and $\bar{d} = (\bar{d}_1, ..., \bar{d}_k)$.

Let $G_k(\bar{d}, c) = c^{\bar{d}_1}(X_1^{\bar{d}_1} + \cdots + X_k^{\bar{d}_k} + X_2^2 + \cdots + X_k^2) - (2k - 1)$, and note that $\forall\, x \in B(0, 1/c) \quad G_k(\bar{d}, c)(x) < 0$.

Using Notation 12.35, we consider

$$\begin{aligned} \mathrm{Def}(Q, \zeta) &= \zeta G_k(\bar{d}, c) + (1 - \zeta)\, Q, \\ \mathrm{Def}_+(Q, \zeta) &= \mathrm{Def}(Q, \zeta) + X_{k+1}^2. \end{aligned}$$

The algebraic set $\mathrm{Zer}(\mathrm{Def}_+(Q,\zeta),\mathrm{R}\langle\zeta\rangle^{k+1})$ has the following property which is not enjoyed by $\mathrm{Zer}(\mathrm{Def}(Q,\zeta),\mathrm{R}\langle\zeta\rangle^k)$.

Lemma 15.6. *Let* $\mathrm{Zer}(Q, \mathrm{R}^k) \subset B(0, 1/c)$ *be a bounded algebraic set. For every semi-algebraically connected component* D *of* $\mathrm{Zer}(Q,\mathrm{R}^k)_{[a,b]}$ *there exists a semi-algebraically connected component* D' *of* $\mathrm{Zer}(\mathrm{Def}_+(Q,\zeta),\mathrm{R}\langle\zeta\rangle^{k+1})_{[a,b]}$ *such that* $\lim_\zeta (D') = D \times \{0\}$.

Proof: Let $y = (y_1, \ldots, y_k)$ be a point of $\mathrm{Ext}(D, \mathrm{R}\langle\zeta\rangle)$. Since $y \in B(0,1/c)$, we have $G_k(\bar{d}, c)(y) < 0$, hence $\mathrm{Def}(Q, \zeta)(y) < 0$. Thus, there exists a unique point $(y, f(y))$ in $\mathrm{Zer}(\mathrm{Def}_+(Q, \zeta), \mathrm{R}\langle\zeta\rangle^{k+1})$ for which $f(y) > 0$ and the mapping f is semi-algebraically continuous. Moreover for every z in D, $\mathrm{Def}(Q,\zeta)$ is infinitesimal, and hence $f(z) \in \mathrm{R}\langle\zeta\rangle$ is infinitesimal over R. So, $\lim_\zeta (z, f(z)) = (z, 0)$. Fix $x \in D$ and denote by D' the semi-algebraically connected component of $\mathrm{Zer}(\mathrm{Def}_+(Q,\zeta),\mathrm{R}\langle\zeta\rangle^{k+1})$ containing $(x, f(x))$. Since $\lim_\zeta (D')$ is connected (Proposition 12.43), contained in $\mathrm{Zer}(Q, \mathrm{R}^k)$, and contains x, it follows that $\lim_\zeta (D') \subset D$. Since f is semi-algebraic and continuous, and D is semi-algebraically path connected, for every z in D, the point $(z, f(z))$ belongs to the semi-algebraically connected component D' of $\mathrm{Zer}(\mathrm{Def}_+(Q, \zeta), \mathrm{R}\langle\zeta\rangle^{k+1})$ containing $(x, f(x))$. Since $\lim_\zeta (z, f(z)) = (z,0)$, we have $\lim_\zeta (D') = D \times \{0\}$. □

Exercise 15.1. Prove that for

$$Q = ((X+1)^2 + Y^2 - 1)((X-1)^2 + Y^2 - 1)((X-2)^2 + Y^2 - 4)$$

the statement of Lemma 15.6 is false if $\mathrm{Def}_+(Q,\zeta)$ is replaced by $\mathrm{Def}(Q,\zeta)$.

We are now able to prove Proposition 15.4 and Proposition 15.5.

Proof of Proposition 15.4: By Lemma 15.6, there exists D', a semi-algebraically connected component of $\mathrm{Zer}(\mathrm{Def}_+(Q, \zeta), \mathrm{R}\langle\zeta\rangle^{k+1})_{[a,b]}$ such that $D \times \{0\} = \lim_\zeta (D')$. Since $[a, b] \setminus \{v\}$ contains no X_1-pseudo-critical value, there exists an infinitesimal β such that the X_1-critical values on $\mathrm{Zer}(\mathrm{Def}_+(Q,\zeta),\mathrm{R}\langle\zeta\rangle^{k+1})$ in the interval $[a, b]$, if they exist, lie in the interval $[v-\beta, v+\beta]$.

We claim that $D'_{[v-\beta,v+\beta]}$ is semi-algebraically connected.

Let x, y be any two points in $D'_{[v-\beta,v+\beta]}$. We show that there exists a semi-algebraic path connecting x to y lying within $D'_{[v-\beta,v+\beta]}$. Since, D' itself is semi-algebraically connected, there exists a semi-algebraic path, $\gamma\colon [0,1] \to D'$, with $\gamma(0) = x$, $\gamma(1) = y$, and $\gamma(t) \in D'$, $0 \le t \le 1$. If $\gamma(t) \in D'_{[v-\beta,v+\beta]}$ for all $t \in [0,1]$, we are done. Otherwise, the semi-algebraic path γ is the union of a finite number of closed connected pieces γ_i lying either in $D'_{[a,v-\beta]}$, $D'_{[v+\beta,b]}$ or $D'_{[v-\beta,v+\beta]}$.

By Proposition 15.2 the connected components of $D'_{v-\beta}$ (resp. $D'_{v+\beta}$) are in 1-1 correspondence with the connected components of $D'_{[a,v-\beta]}$ (resp. $D'_{[v+\beta,b]}$) containing them. Thus, we can replace each of the γ_i lying in $D'_{[a,v-\beta]}$ (resp. $D'_{[v+\beta,b]}$) with endpoints in $D'_{v-\beta}$ (resp. $D'_{v+\beta}$) by another segment with the same endpoints but lying completely in $D'_{v-\beta}$ (resp. $D'_{v+\beta}$). We thus obtain a new semi-algebraic path γ' connecting x to y and lying inside $D'_{[v-\beta,v+\beta]}$.

It is clear that $\lim_\zeta (D'_{[v-\beta,v+\beta]})$ coincides with D_v. Since $D'_{[v-\beta,v+\beta]}$ is bounded, D_v is semi-algebraically connected by Proposition 12.43. □

Proof of Proposition 15.5: By Lemma 15.6, there exists D', a semi-algebraically connected component of $\mathrm{Zer}(\mathrm{Def}_+(Q,\zeta),\mathrm{R}\langle\zeta\rangle^{k+1})_{[a,b]}$ such that $D\times\{0\}=\lim_\zeta (D')$. According to Theorem 5.46 (Hardt's triviality), there exists $a'\in[a,\,b)$ such that for every $d\in[a',\,b)$, $D_{[a,d]}$ is not semi-algebraically connected. Hence, by Proposition 12.43, $D'_{[a,c]}$ is also not semi-algebraically connected for every $c\in\mathrm{R}\langle\zeta\rangle$ with $\lim_\zeta(c)=d$. Since D' is semi-algebraically connected, according to Proposition 15.3, there is an X_1-critical value c on $\mathrm{Zer}(\mathrm{Def}_+(Q,\zeta),\mathrm{R}\langle\zeta\rangle^{k+1})$, infinitesimally close to b. Hence b is an X_1-pseudo-critical value on $\mathrm{Zer}(Q,\mathrm{R}^k)$. □

15.2 Roadmap of an Algebraic Set

We describe the construction of a roadmap M for a bounded algebraic set $\mathrm{Zer}(Q,\,\mathrm{R}^k)$ which contains a finite set of points $\mathcal{N}$ of $\mathrm{Zer}(Q,\,\mathrm{R}^k)$. A precise description of how the construction can be performed algorithmically will follow.

We first construct X_2-pseudo-critical points on $\mathrm{Zer}(Q,\,\mathrm{R}^k)$ in a parametric way along the X_1-axis. This results in curve segments and their endpoints on $\mathrm{Zer}(Q,\mathrm{R}^k)$. The curve segments are continuous semi-algebraic curves parametrized by open intervals on the X_1-axis, and their endpoints are points of $\mathrm{Zer}(Q,\,\mathrm{R}^k)$ above the corresponding endpoints of the open intervals. Since these curves and their endpoints include, for every $x\in\mathrm{R}$, the X_2-pseudo-critical points of $\mathrm{Zer}(Q,\,\mathrm{R}^k)_x$, they meet every connected component of $\mathrm{Zer}(Q,\mathrm{R}^k)_x$. Thus the set of curve segments and their endpoints already satisfy RM_2. However, it is clear that this set might not be semi-algebraically connected in a semi-algebraically connected component, so RM_1 might not be satisfied (see Figure 15). We add additional curve segments to ensure that Mea is connected by recursing in certain distinguished hyperplanes defined by $X_1=z$ for distinguished values z.

The set of **distinguished values** is the union of the X_1-pseudo-critical values, the first coordinates of the input points $\mathcal{N}$ and the first coordinates of the endpoints of the curve segments. A **distinguished hyperplane** is an hyperplane defined by $X_1 = v$, where v is a distinguished value. The input points, the endpoints of the curve segments and the intersections of the curve segments with the distinguished hyperplanes define the set of **distinguished points** .

So we have constructed the distinguished values $v_1 < \cdots < v_\ell$ of X_1 among which are the X_1-pseudo-critical values. Above each interval (v_i, v_{i+1}), we have constructed a collection of curve segments $\mathcal{C}_i$ meeting every semi-algebraically connected component of $\mathrm{Zer}(Q, \mathrm{R}^k)_v$ for every $v \in (v_i, v_{i+1})$. Above each distinguished value v_i, we have constructed a set of distinguished points $\mathcal{N}_i$. Each curve segment in $\mathcal{C}_i$ has an endpoint in $\mathcal{N}_i$ and another in $\mathcal{N}_{i+1}$. Moreover, the union of the $\mathcal{N}_i$ contains $\mathcal{N}$.

We then repeat this construction in each distinguished hyperplane H_i defined by $X_1 = v_i$ with input $Q(v_i, X_2, \ldots, X_k)$ and the distinguished points in $\mathcal{N}_i$.

The process is iterated until for

$$I = (i_1, \ldots, i_{k-2}), 1 \leq i_1 \leq \ell, \ldots, 1 \leq i_{k-2} \leq \ell(i_1, \ldots, i_{k-3}),$$

we have distinguished values $v_{I,1} < \cdots < v_{I,\ell(I)}$ along the X_{k-1} axis with corresponding sets of curve segments and sets of distinguished points with the required incidences between them.

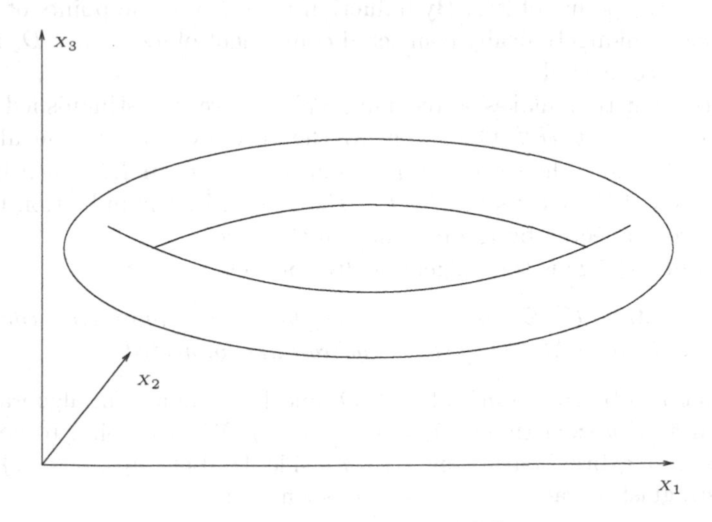

Fig. 15.1. A torus in R^3

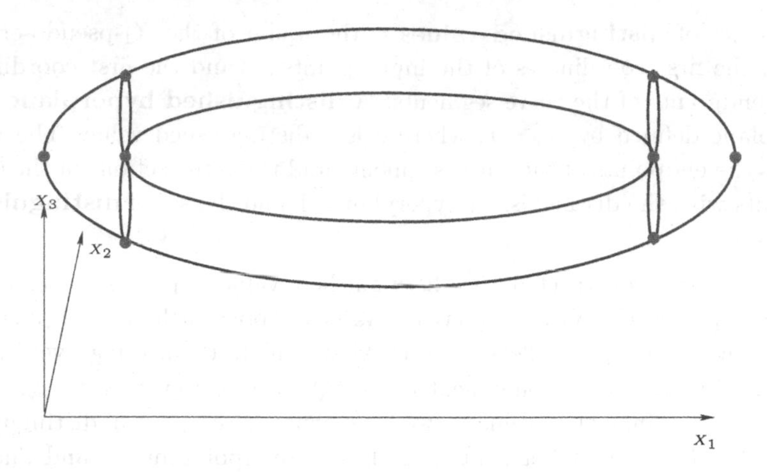

Fig. 15.2. The roadmap of the torus

Proposition 15.7. *The semi-algebraic set M obtained by this construction is a roadmap for* $\mathrm{Zer}(Q, \mathrm{R}^k)$.

The proof of Proposition 15.7 uses the following lemmas.

Lemma 15.8. *If $v \in (a, b)$ is a distinguished value such that $[a, b] \setminus \{v\}$ contains no distinguished value of π on* $\mathrm{Zer}(Q, \mathrm{R}^k)$ *and D is a semi-algebraically connected component of* $\mathrm{Zer}(Q, \mathrm{R}^k)_{[a,b]}$, *then $M \cap D$ is semi-algebraically connected.*

Proof: Since $[a, b] \setminus \{v\}$ contains no pseudo-critical value of the algebraic set $\mathrm{Zer}(Q, \mathrm{R}^k)$, we know, by Proposition 15.4, that D_v is semi-algebraically connected. Moreover, the points of $M \cap D$ are connected through curve segments to the points of $\mathcal{N}_v$. By induction hypothesis, the points of $\mathcal{N}_v$ are in the same semi-algebraically connected component of D_v, since D_v is semi-algebraically connected.

The construction makes a recursive call at every distinguished hyperplane and hence at H_v. The input to the recursive call is the algebraic set $\mathrm{Zer}(Q, \mathrm{R}^k)_v$ and the set of all distinguished points in H_v which includes the endpoints of the curves in $M \cap D \cap H_v$. Hence, by the induction hypothesis they are connected by the roadmap in the slice.

Therefore, $M \cap D$ is semi-algebraically connected. □

Lemma 15.9. *If D is a semi-algebraically connected component of* $\mathrm{Zer}(Q, \mathrm{R}^k)$, *then $M \cap D$ is semi-algebraically connected.*

Proof: Let x, y be two points of $M \cap D$, and let γ be a semi-algebraic path in D from x to y such that $\gamma(0) = x$, $\gamma(1) = y$. We are going to construct another semi-algebraic path from x to y inside M. Let $\{v_1 < \cdots < v_\ell\}$ be the set of distinguished values and choose u_i such that

$$u_1 < v_1 < u_2 < v_2 < \cdots < u_\ell < v_\ell < u_{\ell+1}.$$

There exist a finite number of points of γ, $x = x_0$, x_1, ..., $x_{N+1} = y$, with $\pi(x_i) = u_{n(i)}$, and semi-algebraic paths γ_i from x_i to x_{i+1} such that:

- $\gamma = \bigcup_{0 \leq i \leq N} \gamma_i$,
- $\gamma_i \subset D_{[u_{n(i)}, u_{n(i)+1}]}$ or $\gamma_i \subset D_{[u_{n(i)-1}, u_{n(i)}]}$.

Let D_i be the semi-algebraically connected component of $D_{[u_{n(i)}, u_{n(i)+1}]}$ (resp. $D_{[u_{n(i)-1}, u_{n(i)}]}$) containing γ_i. Since $D_{i-1} \cap D_i$ is a finite union of semi-algebraically connected components of $D_{\pi(x_i)}$, $M \cap D_{i-1} \cap D_i$ is not empty. Choose $y_0 = x, ..., y_i \in M \cap D_{i-1} \cap D_i, ..., y_{N+1} = y$. Then y_i and y_{i+1} are in the same semi-algebraically connected component of $M \cap D$ by Lemma 15.8. □

Proof of Proposition 15.7: We have already seen that M satisfies RM_2. We now prove that M satisfies RM_1.

The proof is by induction on the dimension of the ambient space. In the case of dimension one, the roadmap properties are obviously true for the set we have constructed. Now assume that the construction gives a roadmap for all dimensions less than k. That the construction gives a roadmap for dimension k follows from the following two lemmas. Lemma 15.8 and Lemma 15.9. □

We now describe precisely a way of performing algorithmically the preceding construction.

In our inductive construction of the roadmap, we are going to use the following specification describing points and curve segments:

A **real univariate triangular representation** $\mathcal{T}$, σ, u of level $i-1$ consists of:

- a triangular Thom encoding $\mathcal{T}, \sigma$ specifying $(z, t) \in \mathrm{R}^i$ with $z \in \mathrm{R}^{i-1}$
- a parametrized univariate representation

$$u(X_{<i}) = (\mathcal{T}_i(X_{<i}, T), g_0(X_{<i}, T), g_i(X_{<i}, T), ..., g_k(X_{<i}, T)),$$

  with parameters $X_{<i} = (X_1, ..., X_{i-1})$ (see Definition page 481).

The point associated to $\mathcal{T}, \sigma, u$ is

$$\left(z, \frac{g_i(z,t)}{g_0(z,t)}, ..., \frac{g_k(z,t)}{g_0(z,t)}\right).$$

A real univariate triangular representation $\mathcal{T}$, σ, u is **above** the triangular Thom encoding $\mathcal{T}', \sigma'$ if $\mathcal{T}' = \mathcal{T}_1, ..., \mathcal{T}_{i-1}, \sigma' = \sigma_1, ..., \sigma_{i-1}$.

It will be useful to compute the i-th projection of a point specified by a real univariate representation.

Algorithm 15.1. **[Projection]**

- **Structure:** a domain D contained in a field K.

- **Input:**
  a real univariate triangular representation $\mathcal{T}, \sigma, u$ of level $i-1$ with coefficients in D. We denote by z the root of $\mathcal{T}$ specified by σ and by x the point associated to $\mathcal{T}, \sigma, u$.
- **Output:** a Thom encoding $\mathrm{proj}_i(u), \mathrm{proj}_i(\tau)$ specifying the projection of the point associated to $\mathcal{T}, \sigma, u$ on the X_i axis.
- **Complexity:** $d^{O(i)}$, where d is a bound on the degree of the univariate representation and a bound on the degrees of the polynomials in $\mathcal{T}$.
- **Procedure:**
  - Compute the resultant $\mathrm{proj}_i(u)$ of $\mathcal{T}_i(X_{<i}, T)$, and

$$X_i\, g_0(X_{<i}, T) - g_i(X_{<i}, T)$$

  with respect to T, using Algorithm 8.21 (Signed subresultant).
  - Compute the Thom encoding of the root of $\mathrm{proj}_i(u)$ which is the i-th coordinate of x as follows: let d be the smallest even number not less than the degree of $\mathrm{proj}_i(u)$ with respect to X_i, and compute the sign of the derivatives of

$$g_0(X_{<i}, T)^d\, \mathrm{proj}_i(u)\left(\frac{g_i(X_{<i}, T)}{g_0(X_{<i}, T)}\right)$$

  with respect to T at the root z of $\mathcal{T}$ specified by σ. This gives the Thom encoding $\mathrm{proj}_i(\tau)$ of the i-th coordinate of x. This is done using Algorithm 12.19 (Triangular Sign Determination).

Proof of correctness: Immediate. □

Complexity analysis: The complexity is $d^{O(ki)}$ using the complexity of Algorithm 12.19 (Triangular Sign Determination).

If $\mathrm{D} = \mathbb{Z}$, and the bitsizes of the coefficients of the polynomials are bounded by τ, then the bitsizes of the integers appearing in the intermediate computations and the output are bounded by $\tau d^{O(i)}$. □

Let $\mathcal{V}_1, \tau_1, \mathcal{V}_2, \tau_2$ be two triangular Thom encodings above $\mathcal{T}, \sigma$. We denote by $z = (z_1, \dots, z_{i-1}) \in \mathrm{R}^{i-1}$ the point specified by $\mathcal{T}, \sigma$ and by (z, a), (z, b) the points specified by $\mathcal{V}_1, \tau_1$ and $\mathcal{V}_2, \tau_2$ (see Definition page 496).

A **curve segment representation** u, ρ **above** $\mathcal{V}_1, \tau_1, \mathcal{V}_2, \tau_2$ is:

- a parametrized univariate representation with parameters $(X_{\le i})$

$$u = (f(X_{\le i}, T), g_0(X_{\le i}, T), g_{i+1}(X_{\le i}, T), \dots, g_k(X_{\le i}, T)),$$

- a sign condition ρ on $\mathrm{Der}(f)$ such that for every $v \in (a, b)$ there exists a real root $t(v)$ of $f(z, v, T)$ with Thom encoding σ, ρ and $g_0(z, v, t(v)) \neq 0$.

The **curve segment associated to** u, ρ is the semi-algebraic function h which maps a point v of (a,b) to the point of R^k defined by

$$h(v)=\left(z,v,\frac{g_{i+1}(z,v,t(v))}{g_0(z,v,t(v))},\ldots,\frac{g_k(z,v,t(v))}{g_0(z,v,t(v))}\right).$$

It is a continuous injective semi-algebraic function.

The Curve Segments Algorithm will be the basic building block in our algorithm.

Algorithm 15.2. **[Curve Segments]**

- **Structure:** an ordered domain D contained in a real closed field R.
- **Input:**
  - a triangular Thom encoding $\mathcal{T}$, σ with coefficients in D specifying $z\in\mathrm{R}^{i-1}$,
  - a polynomial $Q\in\mathrm{D}[X_1\ldots,X_k]$, for which $\mathrm{Zer}(Q,\mathrm{R}^k)\subset B(0,1/c)$,
  - a finite set $\mathcal{N}$ of real univariate triangular representation above $\mathcal{T}$, σ with coefficients in D and associated points contained in $\mathrm{Zer}(Q,\mathrm{R}^k)$.
- **Output:**
  - an ordered list of triangular Thom encodings $\mathcal{V}_1,\tau_1,\ldots,\mathcal{V}_\ell,\tau_\ell$ above $\mathcal{T}$, σ specifying points $(z,v_1),\ldots,(z,v_\ell)$ with $v_1<\cdots<v_\ell$. The v_j are called distinguished values.
  - For every $j=1,\ldots,\ell$,
    - a finite set $\mathcal{D}_j$ of real univariate triangular representations representation above $\mathcal{V}_j,\tau_j$. The associated points are called distinguished points.
    - a finite set $\mathcal{C}_j$ of curve segment representations above $\mathcal{V}_j$, τ_j, $\mathcal{V}_{j+1}$, τ_{j+1}. The associated curve segments are called distinguished curves.
    - a list of pairs of elements of $\mathcal{C}_j$ and $\mathcal{D}_j$ (resp. $\mathcal{C}_{j+1}$ and $\mathcal{D}_j$) describing the adjacency relations between distinguished curves and distinguished points.

  The distinguished curves and points are contained in $\mathrm{Zer}(Q,\mathrm{R}^k)_z$. Among the distinguished values are the first coordinates of the points in $\mathcal{N}$ as well as the pseudo-critical values of $\mathrm{Zer}(Q,\mathrm{R}^k)_z$. The sets of distinguished values, distinguished curves, and distinguished points satisfy the following properties.
  - CS_1: For every $v\in\mathrm{R}$, the set of distinguished curve and distinguished points output intersect every semi-algebraically connected component of $\mathrm{Zer}(Q,\mathrm{R}^k)_{(z,v)}$.
  - CS_2: For each distinguished curve output over an interval with endpoint a given distinguished value, there exists a distinguished point over this distinguished value which belongs to the closure of the curve segment.

- **Complexity:** $d^{O(ik)}$, where d is a bound on the degree of Q and $O(d)^k$ is a bound on the degrees of the polynomials in $\mathcal{T}$, the degrees of the univariate representations in $\mathcal{N}$, and the number of these univariate representations.
- **Procedure:**
  - Step 1: Perform Algorithm 12.10 (Parametrized Multiplication Table) with input $\overline{\mathrm{Cr}}(Q^2,\zeta)$, (using Notation 12.46) and parameter $X_{\le i}$. Perform Algorithm 12.15 (Parametrized Limit of Bounded Points) and output $\mathcal{U}$.

    Consider for every $u=(f,\, g_0,\, g_{i+1},\, \ldots,\, g_k) \in \mathcal{U}$ the finite set $\mathcal{F}_u$ containing Q_u (Notation 13.8) and all the derivatives of f with respect to T, and compute

    $$\mathcal{D}_u = \mathrm{RElim}_T(f,\mathcal{F}_u) \subset \mathrm{D}[X_{\le i}],$$

    using Algorithm 11.19 (Restricted Elimination). Define $\mathcal{D} = \bigcup_{u\in\mathcal{U}} \mathcal{D}_u$.
  - Step 2: For every $\mathcal{T}', \tau, u \in \mathcal{N}$, compute $\mathrm{proj}_i(u), \mathrm{proj}_i(\tau)$ using Algorithm 15.1 (Projection), add to $\mathcal{D}$ the polynomial $\mathrm{proj}_i(u)$.
  - Step 3: Compute the Thom encodings of the zeroes of A, $A \in \mathcal{D}$ above $\mathcal{T}$, σ using Algorithms 12.20 (Triangular Thom Encoding), output their ordered list $A_1,\, \alpha_1,\, \ldots,\, A_\ell,\, \alpha_\ell$ and the corresponding ordered list $v_1 < \cdots < v_\ell$ of distinguished values using Algorithm 12.21 (Triangular Comparison of Roots). Define $\mathcal{V}_i, \tau_i = \mathcal{T}, A_i, \sigma, \alpha_i$.
  - Step 4: For every $j = 1,\, \ldots,\, \ell$ and every $(f,\, g_0,\, g_i,\, \ldots,\, g_k)$, $\tau \in \mathcal{N}$ such that $\mathrm{proj}_i(\tau) = \alpha_j$, append $(f,\, g_0,\, g_{i+1},\, \ldots,\, g_k)$, τ to $\mathcal{D}_j$, using Algorithm 12.19 (Triangular Sign Determination).
  - Step 5: For every $j=1,\ldots,\ell$ and every

    $$u=(f,g_0,g_{i+1},\ldots,g_k)\in\mathcal{U},$$

    compute the Thom encodings τ of the roots of f above $\mathcal{T}$, σ such that $\mathrm{proj}_i(\tau) = \alpha_j$, using Algorithm 12.20 (Triangular Thom Encoding). Append all pairs $(f,\, g_0,\, g_{i+1},\, \ldots,\, g_k)$, τ to $\mathcal{D}_j$ when the corresponding associated point belongs to $\mathrm{Zer}(Q,\mathrm{R}^k)_z$.
  - Step 6: For every $j=1,\ldots,\ell-1$ and every

    $$u=(f,g_0,g_{i+1},\ldots,g_k)\in\mathcal{U},$$

    compute the Thom encodings ρ of the roots of $f(z,v,T)$ over (v_j,v_{j+1}) using Algorithm 12.22 (Triangular Intermediate Points) and Algorithm 12.20 (Triangular Thom Encoding) and append pairs u, ρ to $\mathcal{C}_j$ when the corresponding associated curve is included in $\mathrm{Zer}(Q,\mathrm{R}^k)_z$.
  - Step 7: Determine adjacencies between curve segments and points. For every point of $\mathcal{D}_j$ specified by

    $$v' = (p, q_0, q_{i+1}, \ldots, q_k), \tau', \text{with } \{p, q_0, q_{i+1}, \ldots, q_k\} \subset \mathrm{D}[X_{\le i}][T]$$

and every curve segment representation of $\mathcal{C}_j$ specified by

$$v = (f, g_0, g_{i+1}, \ldots, g_k), \tau, \{f, g_0, g_{i+1}, \ldots, g_k\} \subset \mathrm{D}[X_{\le i}][T],$$

decide whether the associated point t is adjacent to the associated curve segment as follows: compute the first ν such that $(\partial^\nu g_0/\partial X_i^\nu)(v_j, t)$ is not zero and decide whether for every $\ell = i+1, \ldots, k$

$$\frac{\partial^\nu g_\ell}{\partial X_i^\nu}(v_j, t) q_0(t) - \frac{\partial^\nu g_0}{\partial X_i^\nu}(v_j, t) q_\ell(t)$$

is zero. This is done using Algorithm 12.19 (Triangular Sign Determination) above $\mathcal{T}, \sigma$.

Repeat the same process for every element of $\mathcal{D}_{j+1}$ and every curve segment representation of $\mathcal{C}_j$.

Proof of correctness: It follows from Proposition 12.42, the correctness of Algorithm 12.10 (Parametrized Multiplication Table), Algorithm 12.15 (Parametrized Limit of Bounded Points), Algorithm 11.19 (Restricted Elimination), Algorithm 15.1, Algorithm 12.22 (Triangular Intermediate Points), Algorithm 12.20 (Triangular Thom Encoding), Algorithm 12.21 (Triangular Comparison of Roots) and Algorithm 12.19 (Triangular Sign Determination). □

Complexity analysis:

- Step 1: This step requires $d^{O(i(k-i))}$ arithmetic operations in D, using the complexity analysis of Algorithm 12.10 (Parametrized Multiplication Table), Algorithm 12.15 (Parametrized Limit of Bounded Points), Algorithm 11.19 (Restricted Elimination). There are $d^{O(k-i)}$ parametrized univariate representations computed in this step and each polynomial in these representations has degree $O(d)^{k-i}$.
- Step 2: This step requires $d^{O(ik)}$ arithmetic operations in D, using the complexity analysis of Algorithm 15.1 (Projection).
- Step 3: This step requires $d^{O(ik)}$ arithmetic operations in D, using the complexity analysis of Algorithm 12.20 (Triangular Thom Encoding).
- Step 4: This step requires $d^{O(ik)}$ arithmetic operations in D, using the complexity analysis of Algorithm 12.19 (Triangular Sign Determination)
- Step 5: This step requires $d^{O(ik)}$ arithmetic operations in D, using the complexity analysis of Algorithm 12.20 (Triangular Thom Encoding).
- Step 6: This step requires $d^{O(ik)}$ arithmetic operations in D, using the complexity analysis of Algorithm 12.22 (Triangular Intermediate Points), Algorithm 12.20 (Triangular Thom Encoding).
- Step 7: This step requires $d^{O(ik)}$ arithmetic operations in D, using the complexity analysis of Algorithm 12.19 (Triangular Sign Determination).

Thus, the complexity is $d^{O(ik)}$. The number of distinguished values is bounded by $d^{O(k)}$.

If $D = \mathbb{Z}$, and the bitsizes of the coefficients of the polynomials are bounded by τ, then the bitsizes of the integers appearing in the intermediate computations and the output are bounded by $\tau d^{O(ik)}$. □

Given a polynomial Q and a set of real univariate representations $\mathcal{N}$, we denote by $\mathrm{RM}(\mathrm{Zer}(Q, \mathrm{R}^k), \mathcal{N})$ a roadmap of $\mathrm{Zer}(Q, \mathrm{R}^k)$ which contains the points associated to $\mathcal{N}$.

We now describe a recursive roadmap algorithm for bounded algebraic sets.

Algorithm 15.3. **[Bounded Algebraic Roadmap]**

- **Structure:** an ordered domain D contained in a real closed field R.
- **Input:**
  - a triangular Thom encoding $\mathcal{T}$, σ with coefficients in D specifying $z \in \mathrm{R}^i$,
  - a polynomial $Q \in \mathrm{D}[X_1, \ldots, X_k]$, for which $\mathrm{Zer}(Q, \mathrm{R}^k) \subset B(0, 1/c)$,
  - a finite set $\mathcal{N}$ of real univariate representation u, τ above $\mathcal{T}$, σ with coefficients in D with associated points contained in $\mathrm{Zer}(Q, \mathrm{R}^k)_z$.
- **Output:** a roadmap $\mathrm{RM}(\mathrm{Zer}(Q, \mathrm{R}^k)_z, \mathcal{N})$ which contains the points associated to$\mathcal{N}$.
- **Complexity:** $d^{O(k^2)}$, where d is a bound on the degree of Q and $O(d)^k$ is a bound on the degrees of the polynomials in $\mathcal{T}$, the degrees of the univariate representations in $\mathcal{N}$, and the number of these univariate representations.
- **Procedure:**
  - Call Algorithm 15.2 (Curve Segments), output ℓ and, for every $j = 1, \ldots, \ell$, $A_j, \alpha_j, \mathcal{D}_j$ and $\mathcal{C}_j$.
  - For every $j = 1, \ldots, \ell$, call Algorithm 15.3 (Bounded Algebraic Roadmap) recursively, with input $\mathcal{T}$, A_j, σ, α_j, specifying (z, v_j), Q and $\mathcal{D}_j$.

Proof of correctness: The correctness of the algorithm follows from Proposition 15.7 and the correctness of Algorithm 15.2 (Curve Segments). □

Complexity analysis: In the recursive calls to Algorithm 15.3 (Bounded Algebraic Roadmap), the number of triangular systems considered is at most $d^{O(k^2)}$ and the triangular systems involved have polynomials of degree $O(d)^k$. Thus the total number of arithmetic operations in D is bounded by $d^{O(k^2)}$ using the complexity analysis of Algorithm 15.2 (Curve Segments).

If $D = \mathbb{Z}$, and the bitsizes of the coefficients of the polynomials are bounded by τ, then the bitsizes of the integers appearing in the intermediate computations and the output are bounded by $\tau d^{O(k^2)}$. □

Since $\mathrm{RM}(\mathrm{Zer}(Q, \mathrm{R}^k)_z, \mathcal{N})$ contains $\mathrm{RM}(\mathrm{Zer}(Q, \mathrm{R}^k)_z)$, it is possible to extract from $\mathrm{RM}(\mathrm{Zer}(Q,\mathrm{R}^k)_z, \{u,\tau\})$ a path connecting the point p associated to u,τ to $\mathrm{RM}(\mathrm{Zer}(Q,\mathrm{R}^k)_z)$.

Algorithm 15.4. **[Bounded Algebraic Connecting]**

- **Structure:** an ordered domain D contained in a real closed field R.
- **Input:**
  - a triangular Thom encoding $\mathcal{T}$, σ with coefficients in D specifying $z \in \mathrm{R}^i$,
  - a polynomial $Q \in \mathrm{D}[X_1,\ldots,X_k]$ for which $\mathrm{Zer}(Q,\mathrm{R}^k) \subset B(0,1/c)$,
  - a real univariate triangular representation $\mathcal{V}$, τ, u above $\mathcal{T}$, σ with coefficients in D, with associated point p contained in $\mathrm{Zer}(Q,\mathrm{R}^k)_z$.
- **Output:** a path $\gamma(p) \subset \mathrm{Zer}(Q,\mathrm{R}^k)_z$ connecting p to a distinguished point of $\mathrm{RM}(\mathrm{Zer}(Q,\mathrm{R}^k)_z)$.
- **Complexity:** $d^{O(k^2)}$, where d is a bound on the degree of Q and $O(d)^k$ is a bound on the degrees of the polynomials in $\mathcal{T}$ and the degree of the real univariate triangular representation $\mathcal{V},\tau,u$.
- **Procedure:** Call Algorithm 15.3 (Bounded Algebraic Roadmap) with input Q, $\mathcal{T}$, σ and $\{\mathcal{V}, \tau, u\}$, and extract $\gamma(p)$ from $\mathrm{RM}(\mathrm{Zer}(Q, \mathrm{R}^k)$, $\{\mathcal{V},\tau,u\})$.

Proof of correctness: The correctness of the algorithm follows from the correctness of Algorithm 15.3 (Bounded Algebraic Roadmap). □

Complexity analysis: The total number of arithmetic operations in D is bounded by $d^{O(k^2)}$, using the complexity analysis of Algorithm 15.3 (Bounded Algebraic Roadmap).

If $\mathrm{D} = \mathbb{Z}$, and the bitsizes of the coefficients of the polynomials are bounded by τ, then the bitsizes of the integers appearing in the intermediate computations and the output are bounded by $\tau d^{O(k^2)}$. □

Remark 15.10. Note that the connecting path $\gamma(p)$ consists of two consecutive parts, $\gamma_0(p)$ and $\Gamma_1(p)$. The path $\gamma_0(p)$ is contained in $\mathrm{RM}(\mathrm{Zer}(Q,\mathrm{R}^k))$ and the path $\Gamma_1(p)$ is contained in $\mathrm{Zer}(Q, \mathrm{R}^k)_{p_1}$. The part $\gamma_0(p)$ consists of a sequence of sub-paths, $\gamma_{0,0}$, ..., $\gamma_{0,m}$. Each $\gamma_{0,i}$ is a semi-algebraic path parametrized by one of the co-ordinates $X_1, \ldots, X_k$, over some interval $[a_{0,i}, b_{0,i}]$ with $\gamma_{0,0}(a_{0,0}) = p$. The semi-algebraic maps, $\gamma_{0,0}, \ldots, \gamma_{0,m}$ and the end-points of their intervals of definition $a_{0,0}, b_{0,0}, \ldots, a_{0,m}, b_{0,m}$ are all independent of p (up to the discrete choice of the path $\gamma(p)$ in $\mathrm{RM}(\mathrm{Zer}(Q, \mathrm{R}^k), \{p\})$), except $b_{0,m}$ which depends on p_1.

Moreover, $\Gamma_1(p)$ can again be decomposed into two parts, $\gamma_1(p)$ and $\Gamma_2(p)$ with $\Gamma_2(p)$ contained in $\mathrm{Zer}(Q,\mathrm{R}^k)_{\bar{p}_2}$ and so on.

If $q = (q_1,\ldots,q_k) \in \mathrm{Zer}(Q,\mathrm{R}^k)$ is another point such that $p_1 \neq q_1$, then since $\mathrm{Zer}(Q,\mathrm{R}^k)_{p_1}$ and $\mathrm{Zer}(Q,\mathrm{R}^k)_{q_1}$ are disjoint, it is clear that

$$\mathrm{RM}(\mathrm{Zer}(Q,\mathrm{R}^k),\{p\}) \cap \mathrm{RM}(\mathrm{Zer}(Q,\mathrm{R}^k),\{q\}) = \mathrm{RM}(\mathrm{Zer}(Q,\mathrm{R}^k)).$$

Now consider a connecting path $\gamma(q)$ extracted from $\mathrm{RM}(\mathrm{Zer}(Q,\mathrm{R}^k),\{q\})$. The images of $\Gamma_1(p)$ and $\Gamma_1(q)$ are disjoint. If the image of $\gamma_0(q)$ (which is contained in $\mathrm{RM}(\mathrm{Zer}(Q,\mathrm{R}^k))$ follows the same sequence of curve segments as $\gamma_0(q)$ starting at p (that is, it consists of the same curves segments $\gamma_{0,0}, \ldots, \gamma_{0,m}$ as in $\gamma_0(p)$), then it is clear that the images of the paths $\gamma(p)$ and $\gamma(q)$ has the property that they are identical up to a point and they are disjoint after it. We call this the **divergence property**. □

Next we show how to handle the case when the input algebraic set $\mathrm{Zer}(Q,\mathrm{R}^k)$ is not bounded.

Algorithm 15.5. **[Algebraic Roadmap]**

- **Structure:** an ordered domain D contained in a real closed field R.
- **Input:** a polynomial $Q \in \mathrm{D}[X_1,\ldots,X_k]$ together with a finite set $\mathcal{N}$ of real univariate representations with coefficients in D.
- **Output:** a roadmap $\mathrm{RM}(\mathrm{Zer}(Q,\mathrm{R}^k),\mathcal{N})$ which contains $\mathcal{N}$.
- **Complexity:** $d^{O(k^2)}$, where d is a bound on the degree of Q and $O(d)^k$ is a bound on the degrees of the polynomials in $\mathcal{T}$, the degrees of the univariate representations in $\mathcal{N}$, and the number of these univariate representations.
- **Procedure:**
  - Introduce new variables X_{k+1} and ε and replace Q by the polynomial
  $$Q_\varepsilon = Q^2 + (\varepsilon^2 (X_1^2 + \cdots + X_{k+1}^2) - 1)^2.$$
  Replace $\mathcal{N} \subset \mathrm{R}^k$ by the set of real univariate representations specifying the elements of $\mathrm{Zer}(\varepsilon^2 (X_1^2 + \cdots + X_{k+1}^2) - 1, \mathrm{R}\langle\varepsilon\rangle^{k+1})$ above the points associated to $\mathcal{N}$ using Algorithm 12.11 (Univariate Representation).
  - Run Algorithm 15.3 (Bounded Algebraic Roadmap) without a triangular Thom encoding (i.e. with $i = 0$), Q_ε and $\mathcal{N}$ as input with structure $\mathrm{D}[\varepsilon]$. The algorithm outputs a roadmap of $\mathrm{RM}(\mathrm{Zer}(Q_\varepsilon,\mathrm{R}\langle\varepsilon\rangle^{k+1}),\mathcal{N})$ composed of points and curves whose description involves ε.
  - Denote by $\mathcal{L}$ the set of polynomials in $\mathrm{D}[\varepsilon]$ whose signs have been determined in the preceding computation and take $a = \min_{P \in \mathcal{L}} c'(P)$ (Definition 10.5). Replace ε by a in the polynomial Q_ε to get a polynomial Q_a. Replace ε by a in the output roadmap to obtain a roadmap $\mathrm{RM}(\mathrm{Zer}(Q_a,\mathrm{R}^{k+1}),\mathcal{N})$. When projected to R^k, this roadmap gives a roadmap for $\mathrm{RM}(\mathrm{Zer}(Q,\mathrm{R}^k),\mathcal{N}) \cap B(0,1/a)$.
  - In order to extend the roadmap outside the ball $B(0,\ 1/a)$ collect all the points $(y_1,\ \ldots,\ y_k,\ y_{k+1}) \in \mathrm{R}\langle\varepsilon\rangle^{k+1}$ in the roadmap $\mathrm{RM}(\mathrm{Zer}(Q_\varepsilon,\mathrm{R}\langle\varepsilon\rangle^{k+1}),\mathcal{N})$ which satisfies $\varepsilon(y_1^2 + \cdots + y_k^2) = 1$. Each such point is described by a real univariate representation involving ε. Add to the roadmap the curve segment obtained by first forgetting the last coordinate and then treating ε as a parameter which varies vary over $(0, a,]$ to get a roadmap $\mathrm{RM}(\mathrm{Zer}(Q,\mathrm{R}^k),\mathcal{N})$.

Proof of correctness: The choice of a guarantees that the roadmap for Q_ε just computed specializes to a roadmap for Q_a when ε is replaced by a. The correctness follows from the correctness of Algorithm 15.3 (Bounded Algebraic Roadmap). □

Complexity analysis: According to the complexity analysis of Algorithm 15.3 (Bounded Algebraic Roadmap), the number of arithmetic operations in the ring $\mathrm{D}[\varepsilon]$ is $d^{O(k^2)}$. Moreover, the degrees of the polynomials in ε generated by the algorithm do not exceed $d^{O(k^2)}$, using the complexity analysis of Algorithm 12.10 (Parametrized Special Multiplication Table). The complexity is thus $d^{O(k^2)}$ in the ring D, taking into account the complexity analyses of Algorithm 8.4 (Addition of multivariate polynomials), Algorithm 8.5 (Multiplication of Multivariate Polynomials), and Algorithm 8.6 (Exact Division of Multivariate Polynomials).

If $\mathrm{D} = \mathbb{Z}$, and the bitsizes of the coefficients of the polynomials are bounded by τ, then the bitsizes of the integers appearing in the intermediate computations and the output are bounded by $\tau d^{O(k^2)}$. □

Algorithm 15.6. **[Algebraic Connecting]**

- **Structure:** an ordered domain D contained in a real closed field R.
- **Input:**
  - a polynomial $Q \in \mathrm{D}[X_1, \ldots, X_k]$,
  - a real univariate representation u, τ with coefficients in D, with associated point p contained in $\mathrm{Zer}(Q, \mathrm{R}^k)$.
- **Output:** a path $\gamma(p, \mathrm{Zer}(Q, \mathrm{R}^k)) \subset \mathrm{Zer}(Q, \mathrm{R}^k)$ connecting p to a distinguished point of $\mathrm{RM}(\mathrm{Zer}(Q, \mathrm{R}^k))$.
- **Complexity:** $d^{O(k^2)}$, where d is a bound on the degree of Q and $O(d)^k$ is a bound on the degrees of u.
- **Procedure:** Call Algorithm 15.5 (Algebraic Roadmap) with input Q and (u, τ) and extract γ from $\mathrm{RM}(\mathrm{Zer}(Q, \mathrm{R}^k), \{u, \tau\})$.

Proof of correctness: The correctness of the algorithm follows from the correctness of Algorithm 15.5 (Algebraic Roadmap). □

Complexity analysis: The total number of arithmetic operations in D is bounded by $d^{O(k^2)}$, using the complexity analysis of Algorithm 15.5 (Algebraic Roadmap).

If $\mathrm{D} = \mathbb{Z}$, and the bitsizes of the coefficients of the polynomials are bounded by τ, then the bitsizes of the integers appearing in the intermediate computations and the output are bounded by $\tau d^{O(k^2)}$. □

We can now summarize our results on the complexity of the computation of the roadmap for an algebraic set.

Theorem 15.11. *Let* $Q \in \mathrm{R}[X_1, \ldots, X_k]$ *be a polynomial whose total degree is at most* d.

a) *There is an algorithm whose output is exactly one point in every semi-algebraically connected component of* $\mathrm{Zer}(Q, \mathrm{R}^k)$. *The complexity in the ring generated by the coefficients of* Q *is bounded by* $d^{O(k^2)}$. *In particular, this algorithm counts the number of semi-algebraically connected components of* $\mathrm{Zer}(Q, \mathrm{R}^k)$ *in time* $d^{O(k^2)}$. *If* $\mathrm{D} = \mathbb{Z}$, *and the bitsizes of the coefficients of the polynomials are bounded by* τ, *then the bitsizes of the integers appearing in the intermediate computations and the output are bounded by* $\tau d^{O(k^2)}$.
b) *Let* p *and* q *in* $\mathrm{Zer}(Q, \mathrm{R}^k)$ *be two points which are represented by real* k*-univariate real representation* u, σ v, τ *of degree* $O(d)^k$. *There is an algorithm deciding whether* p *and* q *belong to the same connected component of* $\mathrm{Zer}(Q, \mathrm{R}^k)$. *The complexity in the ring generated by the coefficients of* Q *and the coefficients of the polynomials in* u *and* v *is bounded by* $d^{O(k^2)}$. *If* $\mathrm{D} = \mathbb{Z}$, *and the bitsizes of the coefficients of the polynomials are bounded by* τ, *then the bitsizes of the integers appearing in the intermediate computations and the output are bounded by* $\tau d^{O(k^2)}$.

Proof: For a), proceed as follows: first compute $\mathrm{RM}(\mathrm{Zer}(Q, \mathrm{R}^k))$, then describe its connected components using the adjacencies between curve segments and points, and finally take one point in each of these connected components.

For b), use Algorithm 15.6 (Algebraic Connecting) for p and q. The points p and q are connected to points p' and q' of the roadmap. Use the first item to decide whether they belong to the same connected component or not. □

15.3 Computing Connected Components of Algebraic Sets

This section is devoted to the proof of the following result.

Theorem 15.12. *If* $\mathrm{Zer}(Q, \mathrm{R}^k)$ *is an algebraic set defined as the zero set of a polynomial* $Q \in \mathrm{D}[X_1, \ldots, X_k]$ *of degree* $\leq d$, *then there is an algorithm that outputs quantifier free formulas whose realizations are the semi-algebraically connected components of* $\mathrm{Zer}(Q, \mathrm{R}^k)$. *The complexity of the algorithm in the ring generated by the coefficients of* Q *is bounded by* $d^{O(k^3)}$ *and the degrees of the polynomials that appear in the output are bounded by* $O(d)^{k^2}$. *Moreover, if* $\mathrm{D} = \mathbb{Z}$, *and the bitsizes of the coefficients of the polynomials are bounded by* τ, *then the bitsizes of the integers appearing in the intermediate computations and the output are bounded by* $\tau d^{O(k^3)}$.

The proof is based on a parametrized version of the roadmap algorithm: we are going to find sign conditions on the parameters for which the description of the roadmap does not change.

For this purpose, we need parametrized versions of Algorithm 12.21 (Triangular Comparison of Roots) and Algorithm 12.22 (Triangular Intermediate Points). These algorithms will be based on Algorithm 14.6 (Parametrized Sign Determination).

Let $\mathcal{A} \subset \mathcal{B}$, ρ and $\bar{\rho}$ two sign conditions on $\mathcal{A}$ and $\mathcal{B}$. The sign condition $\bar{\rho}$ **refines** ρ if $\bar{\rho}(P) = \rho(P)$ for every $P \in \mathcal{A}$.

Notation 15.13. We denote by $\mathrm{SIGN}(\rho, \mathcal{B})$ the list of realizable sign conditions on $\mathcal{B}$ refining ρ. □

Algorithm 15.7. **[Parametrized Comparison of Roots]**

- **Structure**: an ordered integral domain D contained in a real closed field R.
- **Input**: a parametrized Thom encoding $\mathcal{A}$, ρ, $\mathcal{T}$, σ, of level $k-1$, with coefficients in D, two non-zero polynomials P and $Q \in \mathrm{D}[Y, X_1, \ldots, X_k]$.
- **Output**:
  - a finite set $\mathcal{B} \subset \mathrm{D}[Y]$ containing $\mathcal{A}$,
  - for every $\bar{\rho} \in \mathrm{SIGN}(\rho, \mathcal{B})$, a list of sign conditions on $\mathrm{Der}(\mathcal{T} \cup \{P\} \cup \{Q\})$ refining σ specifying for every $y \in \mathrm{Reali}(\rho)$ the ordered list of the triangular Thom encodings of the roots of P and Q above the point specified by σ.
- **Complexity:** $d^{O(k\ell)}$, where ℓ is the number of parameters and d is a bound on the degrees of the polynomials in $\mathcal{T}$, and the degree of P and Q.
- **Procedure:** Apply Algorithm 14.6 (Parametrized Sign Determination) to $\mathcal{T}, P$ and
$$\mathrm{Der}(\mathcal{T}) \cup \mathrm{Der}(P) \cup \mathrm{Der}(Q),$$
then to $\mathcal{T}, Q$ and
$$\mathrm{Der}(\mathcal{T}) \cup \mathrm{Der}(P) \cup \mathrm{Der}(Q)$$

Proof of correctness: Immediate. □

Complexity analysis: The complexity is $d^{O(k\ell)}$, using the complexity of Algorithm Algorithm 14.6 (Parametrized Sign Determination). The number of elements in $\mathcal{B}$ is $d^{O(k\ell)}$, and the degrees of the elements of $\mathcal{A}$ are bounded by $d^{O(k)}$.

If $\mathrm{D} = \mathbb{Z}$, and the bitsizes of the coefficients of the polynomials are bounded by τ, then the bitsizes of the integers appearing in the intermediate computations and the output are bounded by $\tau d^{O(k\ell)}$. □

Algorithm 15.8. **[Parametrized Intermediate Points]**

- **Structure**: an ordered integral domain D contained in a real closed field R.

- **Input**: a parametrized Thom encoding $\mathcal{A}$, ρ, $\mathcal{T}$, σ of level $k-1$, with coefficients in D, two non-zero polynomials P and Q in $\mathrm{D}[Y, X_1, \ldots, X_k]$ of degree bounded by p.
- **Output**:
  - a finite set $\mathcal{B} \subset \mathrm{D}[Y]$ containing $\mathcal{A}$
  - for every $\bar{\rho} \in \mathrm{SIGN}(\rho, \mathcal{B})$, a list of sign conditions on $\mathrm{Der}(\mathcal{T} \cup \{(PQ)'\})$ specifying for every $y \in \mathrm{Reali}(\bar{\rho})$ the triangular Thom encodings of a set of points intersecting all the intervals between two consecutive roots of P and Q.
- **Complexity:** $d^{O(k\ell)}$, where ℓ is the number of parameters and d is a bound on the degrees of the polynomials in $\mathcal{T}$, and the degree of P and Q.
- **Procedure:** Apply Algorithm 14.7 (Parametrized Thom Encoding) with input $\mathcal{T}$, P, $\mathcal{T}$, Q and $\mathcal{T}$, $P'Q$. Sort them using Algorithm 15.7 (Parametrized Comparison of Roots).

Proof of correctness: Immediate. □

Complexity analysis: The complexity is $d^{O(k\ell)}$, using the complexity of Algorithm Algorithm 14.6 (Parametrized Sign Determination). The number of elements in $\mathcal{A}$ is $d^{O(k\ell)}$, and the degrees of the elements of $\mathcal{B}$ are bounded by $d^{O(k)}$.

If $\mathrm{D} = \mathbb{Z}$, and the bitsizes of the coefficients of the polynomials are bounded by τ, then the bitsizes of the integers appearing in the intermediate computations and the output are bounded by $\tau d^{O(k\ell)}$. □

A **parametrized real univariate triangular representation** of level $i-1$ with parameters $Y = (Y_1, \ldots, Y_\ell)$ $\mathcal{T}, \sigma, u$ above $\mathcal{A}$, ρ is

- a parametrized triangular Thom encoding $\mathcal{T}, \sigma$ of level i,
- a parametrized representation $u = (\mathcal{T}_i, g_0, g_i, \ldots, g_k) \subset \mathrm{D}[Y, X_{\le i}, T]$ such that for every $y \in \mathrm{Reali}(\rho)$ there is a root $(z(y), t(y))$ of $\mathcal{T}$ with triangular Thom encoding σ.

A parametrized real univariate triangular representation $\mathcal{T}, \sigma, u$ is **above** the parametrized triangular Thom encoding $\mathcal{A}, \rho, \mathcal{T}', \sigma'$ if $\mathcal{T}, \sigma$ is above $\mathcal{A}, \rho$ and if $\mathcal{T}' = \mathcal{T}_1, \ldots, \mathcal{T}_{i-1}$, and $\sigma' = \sigma_1, \ldots, \sigma_{i-1}$.

Algorithm 15.9. **[Parametrized Projection]**

- **Structure:** a domain D contained in a field K.
- **Input:** a parametrized real univariate representation $\mathcal{T}$, u, σ above a parametrized triangular Thom encoding $\mathcal{A}$, ρ, with coefficients in D. For every $y \in \mathrm{Reali}(\rho)$, we denote by $z(y)$ the root of $\mathcal{T}(y)$ specified by τ and by $x(y)$ the point associated to $u(y, z(y))$.
- **Output:**
  - a finite set $\mathcal{B} \subset \mathrm{D}[Y]$ containing $\mathcal{A}$,

  - for every $\bar{\rho} \in \mathrm{SIGN}(\rho, \mathcal{B})$ a Thom encoding $(\mathrm{proj}_i(u), \mathrm{proj}_i(\tau))$ specifying, for every $y \in \mathrm{Reali}(\bar{\rho})$, the projection of the point associated to $x(y)$ on the X_i axis.
- **Complexity:** $d^{O(ki\ell)}$, where ℓ is the number of parameters, d is a bound on the degrees of on the degree of the univariate representation and of the polynomials in $\mathcal{T}$.
- **Procedure:**
  - Compute the resultant $\mathrm{proj}_i(u)$ of $f(Y, X_{<i}, T)$, and

$$X_i\, g_0(Y, X_{<i}, T) - g_i(Y, X_{<i}, T)$$

  with respect to T, using Algorithm 8.21 (Signed subresultant).
  - Use Algorithm 14.6 (Parametrized Sign Determination) with $\mathcal{T}$, f and the derivatives of

$$g_0(Y, X_{<i}, T)^d \mathrm{proj}_i(u)\left(\frac{g_i(Y, X_{<i}, T)}{g_0(Y, X_{<i}, T)}\right)$$

  with respect to T, where d is the smallest even number not less than the degree of $\mathrm{proj}_i(u)$ with respect to X_i. This gives a list of polynomials $\mathcal{B} \subset \mathrm{D}[Y]$ and for every $\bar{\rho} \in \mathrm{SIGN}(\rho, \mathcal{B})$ the Thom encoding $\mathrm{proj}_i(\tau)$ of the i-th coordinate of $x(y)$.

Prof of correctness: Immediate. □

Complexity analysis: The complexity is $d^{O(ki\ell)}$, using the complexity of Algorithm 14.6 (Parametrized Sign Determination).

If $\mathrm{D} = \mathbb{Z}$, and the bitsizes of the coefficients of the polynomials are bounded by τ, then the bitsizes of the integers appearing in the intermediate computations and the output are bounded by $\tau d^{O(ki\ell)}$. □

We now define parametrized curve segments.

Let $\mathcal{V}_1$, τ_1, $\mathcal{V}_2$, τ_2 be two parametrized triangular Thom encoding above $\mathcal{A}, \rho, \mathcal{T}, \sigma$. For every $y \in \mathrm{Reali}(\rho)$, we denote by $z(y) \in \mathrm{R}^{i-1}$ the point specified by $\mathcal{T}(y), \sigma$ and by $(z(y), a(y))$, $(z(y), b(y))$ the points specified by $\mathcal{V}_1(y)$, τ_1 and $\mathcal{V}_2(y)$, τ_2. A **parametrized curve segment representation** u, τ **above** $\mathcal{V}_1, \tau_1, \mathcal{V}_2, \tau_2$ is given by

- a parametrized univariate representation with parameters $(Y, X_{\leq i})$,

$$u = (f(Y, X_{\leq i}, T), g_0(Y, X_{\leq i}, T), g_{i+1}(Y, X_{\leq i}, T), \ldots, g_k(Y, X_{\leq i}, T)),$$

- a sign condition τ on $\mathrm{Der}(f)$ such that for every $y \in \mathrm{Reali}(\rho)$ and for every $v \in (a(y), b(y))$ there exists a real root $t(v)$ of $f(z(y), v, T)$ with Thom encoding σ, ρ, τ and $g_0(z(y), v, t(v)) \neq 0$.

Our aim is first to describe a parametrized version of Algorithm 15.2 (Curve Segments).

Algorithm 15.10. **[Parametrized Curve Segments]**

- **Structure:** an ordered domain D contained in a real closed field R.
- **Input:**
  - a parametrized Thom encoding $\mathcal{A}$, ρ, $\mathcal{T}$, σ with parameters $Y = (Y_1, \ldots, Y_\ell)$ of level $i-1$, with coefficients in D. For every $y \in \mathrm{Reali}(\rho)$, $(y, z(y))$ denotes the point specified by σ.
  - a polynomial $Q \in \mathrm{D}[Y, X_1 \ldots, X_k]$, for which $\mathrm{Zer}(Q, \mathrm{R}^k) \subset B(0, 1/c)$
  - a finite set $\mathcal{N}$ of parametrized real univariate triangular representation above $\mathcal{A}$, ρ, $\mathcal{T}$, σ with, for every $y \in \mathrm{Reali}(\rho)$, associated points contained in $\mathrm{Zer}(Q, \mathrm{R}^k)$.
- **Output:**
  - a finite set $\mathcal{B} \subset \mathrm{D}[Y]$ containing $\mathcal{A}$,
  - for every $\bar{\rho} \in \mathrm{SIGN}(\rho, \mathcal{B})$,
    - an ordered list of parametrized Thom encodings
$$\mathcal{V}_{\bar{\rho},1}, \tau_{\bar{\rho},1}, \ldots, \mathcal{V}_{\bar{\rho},\ell(\bar{\rho})}, \tau_{\bar{\rho},\ell(\bar{\rho})}$$
    above $\mathcal{B}, \bar{\rho}, \mathcal{T}, \sigma$
    - for every $i = 1, \ldots, \ell(\bar{\rho})$,
      - a finite set $\mathcal{N}_{\bar{\rho},i}$ of parametrized real univariate triangular representations above
$$\mathcal{B}, \bar{\rho}, \mathcal{V}_{\bar{\rho},j}, \tau_{\bar{\rho},j}$$
      - a finite set $\mathcal{C}_{\bar{\rho},j}$ of parametrized curve segments above
$$\mathcal{B}, \bar{\rho}, \mathcal{V}_{\bar{\rho},j}, \tau_{\bar{\rho},j}, \mathcal{V}_{\bar{\rho},j+1}, \tau_{\bar{\rho},j+1}$$
      - a list of pairs of elements of $\mathcal{C}_{\bar{\rho},j}$ and $\mathcal{N}_{\bar{\rho},j}$ (resp. $\mathcal{C}_{\bar{\rho},j+1}$ and $\mathcal{N}_{\bar{\rho},j}$) describing the adjacency relation.

  For every $y \in \mathrm{Reali}(\bar{\rho})$, this defines a set of curves and points contained in $\mathrm{Zer}(Q, \mathrm{R}^k)_{y,z(y)}$. The specifications of these points and curves is fixed for every point $y \in \mathrm{Reali}(\bar{\rho})$. These points and curves satisfy the properties of the output of Algorithm 15.2 (Curve Segments).
- **Complexity:** $d^{O(ki\ell)}$, where ℓ is the number of parameters, d is a bound on the degree of Q, $O(d)^k$ is a bound on the degrees of on the degree of the parametrized univariate representations in $\mathcal{N}$ and of the polynomials in $\mathcal{T}$.
- **Procedure:**
  - Step 1: Perform Algorithm 12.10 (Parametrized Multiplication Table) with input $\overline{\mathrm{Cr}}(Q^2, \zeta,)$, using Notation 12.46, and parameter $Y, X_{\leq i}$. Perform Algorithm 12.15 (Parametrized Limit of Bounded Points), and output a set $\mathcal{U}$ of parametrized univariate representations.

    Using Notation 13.8, consider for every $u = (f, g_0, g_{i+1}, \ldots, g_k) \in \mathcal{U}$ the finite set $\mathcal{F}_u$ containing $Q_u)$ and all the derivatives of f with respect to T, and compute $\mathcal{D}_u = \mathrm{RElim}_T(f, \mathcal{F}_u) \subset \mathrm{D}[Y, X_{\leq i}]$ using Algorithm 11.19 (Restricted Elimination).
  - Define $\mathcal{D} = \bigcup_{u \in \mathcal{U}} \mathcal{D}_u$.

- Step 2: Use Algorithm 15.9 (Parametrized Projection) with input $\mathcal{N}$ and output a finite set $\mathcal{B}_2 \subset \mathrm{D}[Y]$ containing $\mathcal{A}$, such that for every $\bar{\rho} \in \mathrm{SIGN}(\rho, \mathcal{B}_2)$ and every $u \in \mathcal{N}$ the Thom encoding $\mathrm{proj}_i(u)$, $\mathrm{proj}_i(\tau)$ specifying the projection of the associated point on the X_i axis is fixed for every $y \in \mathrm{Reali}(\bar{\rho})$. Add to $\mathcal{D}$ the polynomials $\mathrm{proj}_i(u)$.
- Step 3: Apply Algorithms 14.7 (Parametrized Thom Encoding), 15.7 (Parametrized Comparison of Roots) to the set $\mathcal{D}$. Denote by $\mathcal{B}_3 \subset \mathrm{D}[Y]$ the family of polynomials output, and for every $\bar{\rho} \in \mathrm{SIGN}(\rho, \mathcal{B}_3)$, denote by
$$A_{\bar{\rho},1}\alpha_{\bar{\rho},1}, \ldots, A_{\bar{\rho},\ell(\bar{\rho})}, \alpha_{\bar{\rho},\ell(\bar{\rho})}$$
the list of Thom encodings output. For every $y \in \mathrm{Reali}(\bar{\rho})$, these are the Thom encodings of the corresponding distinguished values
$$v_1(y, z(y)) < \cdots < v_\ell(y, z(y)).$$
Define $\mathcal{V}_i, \tau_i = \mathcal{T}, A_i$ and $\tau_i = \sigma, \alpha_i$.
- Step 4: For every $\bar{\rho} \in \mathrm{SIGN}(\rho, \mathcal{B}_3)$, every $j = 1, \ldots, \ell(\bar{\rho})$ and every $u = (f, g_0, g_i, \ldots, g_k), \tau \in \mathcal{N}$, use Algorithm 14.7 (Parametrized Triangular Thom Encoding) and output $\mathcal{B}_4(\bar{\rho}, j, u)$, containing $\mathcal{B}_3$. Append pairs $(f, g_0, g_{i+1}, \ldots, g_k)$, τ to $\mathcal{N}_{\rho_1,j}$ for every $\rho_1 \in \mathrm{SIGN}(\bar{\rho}, \mathcal{B}_4(\bar{\rho}, j, u, \tau))$ such that for every $y \in \mathrm{Reali}(\rho_1)$ $\mathrm{proj}_i(\tau)$ is the Thom encoding of a point of $\mathrm{Zer}(Q, \mathrm{R}^k)_z(y)$ with projection having Thom encoding α_j. Define $\mathcal{B}_4(\bar{\rho}) = \cup \mathcal{B}_4(\bar{\rho}, j, u, \tau)$.
- Step 5: For every $\bar{\rho} \in \mathrm{SIGN}(\rho, \mathcal{B}_3)$, every $j = 1, \ldots, \ell(\bar{\rho})$ and every $u = (f, g_0, g_i \ldots, g_k) \in \mathcal{U}$, use Algorithm 15.8 (Parametrized Intermediate Points) and Algorithm 14.7 (Parametrized Triangular Thom Encoding) and output $\mathcal{B}_5(\bar{\rho}, j, u)$, containing $\mathcal{B}_3$. Append pairs $(f, g_0, g_{i+1}, \ldots, g_k), \tau$ to $\mathcal{N}_{\rho_1,j}$ for every $\rho_1 \in \mathrm{SIGN}(\bar{\rho}, \mathcal{B}_5(\bar{\rho}, j, u))$ such that for every $y \in \mathrm{Reali}(\rho_1)$ $\mathrm{proj}_i(\tau)$ is the Thom encoding of a point of $\mathrm{Zer}(Q, \mathrm{R}^k)_z(y)$ with projection having Thom encoding α_j. Define $\mathcal{B}_5(\bar{\rho}) = \cup \mathcal{B}_5(\bar{\rho}, j, u)$.
- Step 6: For every $\bar{\rho} \in \mathrm{SIGN}(\rho, \mathcal{B}_3)$, every $j = 1, \ldots, \ell(\bar{\rho}) - 1$ and every $u = (f, g_0, g_i \ldots, g_k) \in \mathcal{U}$, use Algorithm 15.8 (Parametrized Intermediate Points) and Algorithm 14.7 (Parametrized Triangular Thom Encoding) and output a family $\mathcal{B}_6(\bar{\rho}, j, u)$ containing $\mathcal{B}_3$ such that for every sign condition ρ_1 on $\mathcal{B}_6$ and every $y \in \mathrm{Reali}(\rho_1)$ the Thom encodings τ of the roots of $f(y, z(y), v, T)$ over $(v_i(y), v_{i+1}(y))$ are fixed and the corresponding associated curves are contained in $\mathrm{Zer}(Q, \mathrm{R}^k)_z(y)$. Append all pairs $(f, g_0, g_{i+1} \ldots, g_k)$, τ to $\mathcal{C}_{\rho_3,i}$. Define $\mathcal{B}_6(\bar{\rho}) = \cup \mathcal{B}_6(\bar{\rho}, j, u)$.
- Step 7: Consider $\rho_1 \in \mathrm{SIGN}(\bar{\rho}, \mathcal{B}_4 \cup \mathcal{B}_5 \cup \mathcal{B}_6)$. For every $j = 1, \ldots, \ell(\bar{\rho}_1)$ and every parametrized real univariate triangular representation of $\mathcal{N}_{\rho_1,j}$ specified by
$$v' = (p, q_0, q_2, \ldots, q_k), \tau', \{p, q_0, q_2, \ldots, q_k\} \subset \mathrm{D}[Y, X_{\leq i}][T]$$

and every parametrized curve segment representation of $\mathcal{C}_{\rho_1,j}$ specified by

$$v=(f,g_0,g_2,\ldots,g_k),\tau,\{f,g_0,g_2,\ldots,g_k\}\subset \mathrm{D}[Y,X_{\le i}[T],$$

compute a family $\mathcal{B}_7(\rho_1, v', \tau', v, \tau)$ of polynomials containing $\mathcal{B}_4\cup\mathcal{B}_5\cup\mathcal{B}_6$ such that for every $\rho_2\in\mathrm{SIGN}(\rho_1,\mathcal{B}_7(\rho_1,v',\tau',v,\tau))$ and every $y\in\mathrm{Reali}(\rho_2)$ the algorithm deciding whether the corresponding point $t(y)$ is adjacent to the corresponding curve segment gives the same answer: compute the first ν such that $(\partial^\nu g_0/\partial X_i^\nu)(v_j,t)$ is not zero and decide whether for every $\ell=i+1,\ldots,k$

$$\frac{\partial^\nu g_\ell}{\partial X_i^\nu}(v_j,t)q_0(t)-\frac{\partial^\nu g_0}{\partial X_i^\nu}(v_j,t)q_\ell(t)$$

is zero, using Algorithm 14.6 (Parametrized Sign Determination).

Repeat the same process for every element of $\mathcal{N}_{\rho_1,i+1}$ and every curve segment of $\mathcal{C}_{\rho_1,i}$.

- Finally output $\mathcal{B}=\cup\mathcal{B}_7(\rho_1,v',\tau',v,\tau)$.

Proof of correctness: It follows from Proposition 12.42 and the correctness of Algorithm 12.10 (Parametrized Multiplication Table), Algorithm 12.15 (Parametrized Limit of Bounded Points), Algorithm 11.19 (Restricted Elimination), Algorithm 15.9 (Parametrized Projection), Algorithm 15.8 (Parametrized Intermediate Points), Algorithm 14.7 (Parametrized Thom Encoding), Algorithm 15.7 (Parametrized Comparison of Roots) and Algorithm 14.6 (Parametrized Sign Determination). □

Complexity analysis:

- Step 1: This step requires $d^{O((\ell+i)(k-i))}$ arithmetic operations in D, using the complexity analyses of Algorithm 12.10 (Parametrized Multiplication Table), Algorithm 12.15 (Parametrized Limit of Bounded Points), Algorithm 11.19 (Restricted Elimination). There are $d^{O(k-i)}$ parametrized univariate representations computed in this step and each polynomial in these representations has degree $O(d)^{k-i}$.
- Step 2: This step requires $d^{O((\ell+i)k)}$ arithmetic operations in D, using the complexity analysis of Algorithm 15.9 (Parametrized Projection).
- Step 3: This step requires $d^{O(\ell ik)}$ arithmetic operations in D, using the complexity analysis of Algorithm 14.7 (Parametrized Thom Encoding).
- Step 4: This step requires $d^{O(\ell ik)}$ arithmetic operations in D, using the complexity analysis of Algorithm 14.6 (Parametrized Sign Determination).
- Step 5: This step requires $d^{O(\ell ik)}$ arithmetic operations in D, using the complexity analysis of Algorithm 14.7 (Parametrized Thom Encoding).
- Step 6: This step requires $d^{O(\ell ik)}$ arithmetic operations in D, using the complexity analyses of Algorithm 15.8 (Parametrized Intermediate Points) and Algorithm 14.7 (Parametrized Thom Encoding).

– Step 7: This step requires $d^{O(\ell ik)}$ arithmetic operations, using the complexity analysis of Algorithm 14.6 (Parametrized Sign Determination).

Thus, the complexity is $d^{O(\ell ik)}$.

If $\mathrm{D} = \mathbb{Z}$, and the bitsizes of the coefficients of the polynomials are bounded by τ, then the bitsizes of the integers appearing in the intermediate computations and the output are bounded by $\tau d^{O(\ell ik)}$. □

Algorithm 15.11. **[Parametrized Bounded Algebraic Roadmap]**

- **Structure:** an ordered domain D contained in a real closed field R.
- **Input:**
  - a parametrized Thom encoding $\mathcal{A}$, ρ, $\mathcal{T}$, σ with parameters $Y = (Y_1, \ldots, Y_\ell)$ and variables $X_{\le i} = (X_1, \ldots, X_i)$, with coefficients in D. For every $y \in \mathrm{Reali}(\rho)$, $(y, z(y))$ denotes the point specified by σ,
  - a polynomial $Q \in \mathrm{D}[Y, X_1 \ldots, X_k]$, for which $\mathrm{Zer}(Q, \mathrm{R}^k) \subset B(0, 1/c)$,
  - a finite set $\mathcal{N}$ of parametrized real univariate triangular representations above $\mathcal{A}$, ρ, $\mathcal{T}$, σ with coefficients in D, with, for every $y \in \mathrm{Reali}(\rho)$, associated points contained in $\mathrm{Zer}(Q, \mathrm{R}^k)$.
- **Output:**
  - a subset $\mathcal{C}$ of $\mathrm{D}[Y]$ containing $\mathcal{A}$,
  - for every realizable sign condition τ on $\mathcal{C}$ refining ρ, a subset $\mathrm{RM}(\tau)$ such that, for every $y \in \mathrm{Reali}(\tau)$, $\mathrm{RM}(\tau)_y$ is a roadmap for $\mathrm{Zer}(Q, \mathrm{R}^k)_y$ that contains $\mathcal{N}_y$.
- **Complexity:** $d^{O(\ell k^2)}$, where ℓ is the number of parameters, $O(d)^k$ is a bound on the degrees of on the degree of the univariate representation and of the polynomials in $\mathcal{T}$.
- **Procedure:**
  - Call Algorithm 15.10 (Parametrized Curve Segments), output $\mathcal{B}$ and, for every realizable sign condition $\bar{\rho}$ on $\mathcal{B}$ refining ρ, $\ell(\rho)$. Output also, for every $j = 1, \ldots, \ell(\rho)$, $A_{\bar{\rho},i}$, $\alpha_{\bar{\rho},i}$, $\mathcal{N}_{\bar{\rho},i}$ and $\mathcal{C}_{\bar{\rho},i}$.
  - For every realizable sign condition $\bar{\rho}$ on $\mathcal{B}$ and for every i from 1 to $\ell(\bar{\rho})$, call Algorithm 15.11 (Parametrized Bounded Algebraic Roadmap) recursively, with input $\mathcal{B}, \bar{\rho}, \mathcal{T}, A_{\bar{\rho},j}, \sigma, \alpha_{\bar{\rho},j}$, Q and $\mathcal{N}_{\bar{\rho},j}$.

Proof of correctness: The correctness of the algorithm follows from Proposition 15.7 and the correctness of Algorithm 15.2 (Curve Segments). □

Complexity analysis: In the recursive calls to Algorithm 15.11 (Parametrized Bounded Algebraic Roadmap), the number of triangular systems considered is at most $d^{O(k^2)}$ and the triangular systems involved have polynomials of degree $O(d)^k$.

Thus, the total number of arithmetic operations in D is bounded by $d^{O(\ell k^2)}$ using the complexity analysis of Algorithm 15.10 (Parametrized Curve Segments).

If $\mathrm{D} = \mathbb{Z}$, and the bitsizes of the coefficients of the polynomials are bounded by τ, then the bitsizes of the integers appearing in the intermediate computations and the output are bounded by $\tau d^{O(\ell k^2)}$. □

We now want to obtain a parametrized connecting algorithm. We show how to obtain a covering of a given $\mathcal{P}$-closed semi-algebraic set contained in $\mathrm{Zer}(Q, \mathrm{R}^k)$ by a family of semi-algebraically contractible subsets. The construction is based on a parametrized version of the connecting algorithm: we compute a family of polynomials such that for each realizable sign condition σ on this family, the description of the connecting paths of different points in the realization, $\mathrm{Reali}(\sigma, \mathrm{Zer}(Q, \mathrm{R}^k))$, are uniform.

We first define parametrized paths. A parametrized path is a semi-algebraic set which is a union of semi-algebraic paths having the divergence property (see Remark 15.10).

More precisely,

Definition 15.14. A **parametrized path** γ is a continuous semi-algebraic mapping from $V \subset \mathrm{R}^{k+1} \to \mathrm{R}^k$, such that, denoting by $U = \pi_{1\ldots k}(V) \subset \mathrm{R}^k$, there exists a semi-algebraic continuous function $\ell\colon U \to [0, +\infty)$, and there exists a point a in R^k, such that

- $V = \{(x,t) \mid x \in U, 0 \leq t \leq \ell(x)\}$,
- $\forall\, x \in U,\ \gamma(x,0) = a$,
- $\forall\, x \in U,\ \gamma(x, \ell(x)) = x$,
- $\forall\, x \in U, \forall\, y \in U, \forall\, s, 0 \leqslant s \leqslant \ell(x), \forall\, t 0 \leqslant t \leqslant \ell(y)$
  $(\gamma(x,s) = \gamma(y,t) \Rightarrow s = t)$,
- $\forall\, x \in U, \forall\, y \in U, \forall\, s \in [0, \min(\ell(x), \ell(y))]$
  $(\gamma(x,s) = \gamma(y,s) \Rightarrow \forall\, t \leq s\ \gamma(x,t) = \gamma(y,t))$. □

Given a parametrized path, $\gamma\colon V \to \mathrm{R}^k$, we will refer to $U = \pi_{1\ldots k}(V)$ as its *base*. Also, any semi-algebraic subset $U' \subset U$ of the base of such a parametrized path, defines in a natural way the restriction of γ to the base U', which is another parametrized path, obtained by restricting γ to the set $V' \subset V$, defined by $V' = \{(x,t) \mid x \in U', 0 \leq t \leq \ell(x)\}$.

Proposition 15.15. *Let $\gamma\colon V \to R^k$ be a parametrized path such that $U = \pi_{1\ldots k}(V)$ is closed and bounded. Then, the image of γ is semi-algebraically contractible.*

Proof: Let $W = \mathrm{Im}(\gamma)$ and $M = \sup_{x \in U} \ell(x)$. We prove that the semi-algebraic mapping $\phi\colon W \times [0, M] \to W$ sending

$$(\gamma(x,t), s) \text{ to } \gamma(x,s) \text{ if } t \geq s,$$
$$(\gamma(x,t), s) \text{ to } \gamma(x,t) \text{ if } t < s$$

is continuous. Note that the map ϕ is well-defined, since

$$\gamma(x,t)=\gamma(x',t')\Rightarrow t=t',$$

by condition (4).Since ϕ satisfies

$$\begin{aligned}\phi(\gamma(x,t),0) &= a,\\ \phi(\gamma(x,t),M) &= \gamma(x,t)\end{aligned}$$

this gives a semi-algebraic continuous contraction from W to $\{a\}$.

Let $w\in W, s\in[0,M]$. Let $\varepsilon>0$ be an infinitesimal, and let

$$(w',s')\in \mathrm{Ext}(W\times[0,M],\mathrm{R}\langle\varepsilon\rangle)$$

be such that $\lim_\varepsilon(w',s')=(w,s)$. In order to prove the continuity of ϕ at w it suffices to prove that

$$\lim_\varepsilon \mathrm{Ext}(\phi,\mathrm{R}\langle\varepsilon\rangle)(w',s')=\phi(w,s).$$

Let $w=\gamma(x,t)$ for some $x\in U, t\in[0,\ell(x)]$, and similarly let $w'=(x',t')$ for some $x'\in\mathrm{Ext}(U,\mathrm{R}\langle\varepsilon\rangle)$ and $t'\in[0,\mathrm{Ext}(\ell,\mathrm{R}\langle\varepsilon\rangle)(x')]$. Note that $\lim_\varepsilon(x')\in U$ since U is closed and bounded and $\lim_\varepsilon t'\in[0,\ell(\lim_\varepsilon x')]$.

Now,

$$\begin{aligned}\gamma(x,t) &= w\\ &= \lim_\varepsilon(w')\\ &= \lim_\varepsilon \mathrm{Ext}(\gamma,\mathrm{R}\langle\varepsilon\rangle)(x',t')\\ &= \gamma(\lim_\varepsilon x',\lim_\varepsilon t').\end{aligned}$$

Condition (4) now implies that $\lim_\varepsilon t'=t$.

Without loss of generality let $t'\geq t$. The other case is symmetric. We have the following two sub-cases.

- Case $s'>t'$: Since $s,t\in\mathrm{R}$ and $\lim_\varepsilon s'=s$ and $\lim_\varepsilon t'=t$, we must have that $s\geq t$. In this case $\mathrm{Ext}(\phi,\mathrm{R}\langle\varepsilon\rangle)(w',s')=\mathrm{Ext}(\gamma,\mathrm{R}\langle\varepsilon\rangle)(x',t')$. Then,

$$\begin{aligned}\lim_\varepsilon \mathrm{Ext}(\phi,\mathrm{R}\langle\varepsilon\rangle)(w',s') &= \lim_\varepsilon \mathrm{Ext}(\gamma,\mathrm{R}\langle\varepsilon\rangle)(x',t')\\ &= \lim_\varepsilon w'\\ &= w\\ &= \phi(w,s).\end{aligned}$$

- Case $s'\leq t'$: Again, since $s,t\in\mathrm{R}$ and $\lim_\varepsilon s'=s$ and $\lim_\varepsilon t'=t$, we must have that $s\leq t$.

  In this case we have,

$$\begin{aligned}\lim_\varepsilon \phi(w',s') &= \lim_\varepsilon \mathrm{Ext}(\gamma,\mathrm{R}\langle\varepsilon\rangle)(x',s')\\ &= \gamma(\lim_\varepsilon x',\lim_\varepsilon s')\\ &= \gamma(\lim_\varepsilon x',s).\end{aligned}$$

Now,

$$\begin{aligned}\gamma(\lim_\varepsilon x', t) &= \gamma(\lim_\varepsilon x', \lim_\varepsilon t') \\ &= \lim_\varepsilon \mathrm{Ext}(\gamma, \mathrm{R}\langle\varepsilon\rangle)(x', t') \\ &= \lim_\varepsilon w' \\ &= w \\ &= \gamma(x, t).\end{aligned}$$

Thus, by condition (5) we have that $\gamma(\lim_\varepsilon x', s'') = \gamma(x, s'')$ for all $s'' \leq t$. Since, $s \leq t$, this implies,

$$\begin{aligned}\lim_\varepsilon \mathrm{Ext}(\phi, \mathrm{R}\langle\varepsilon\rangle)(w', s') &= \lim_\varepsilon \mathrm{Ext}(\gamma, \mathrm{R}\langle\varepsilon\rangle)(w', s') \\ &= \gamma(\lim_\varepsilon x', \lim_\varepsilon s') \\ &= \gamma(x, s) \\ &= \phi(w, s).\end{aligned}$$

This proves the continuity of ϕ, using Proposition 3.5. □

Algorithm 15.12. **[Parametrized Bounded Algebraic Connecting]**

- **Structure:** an ordered domain D contained in a real closed field R.
- **Input:**
  - a parametrized Thom encoding $\mathcal{A}$, ρ, $\mathcal{T}$, σ with parameters $Y = (Y_1, \ldots, Y_\ell)$ and variables $X_{\leq i} = (X_1, \ldots, X_i)$, with coefficients in D. For every $y \in \mathrm{Reali}(\rho)$, $(y, z(y))$ denotes the point specified by σ,
  - a polynomial $Q \in \mathrm{D}[Y, X_1 \ldots, X_k]$, for which $\mathrm{Zer}(Q, \mathrm{R}^k) \subset B(0, 1/c)$
  - a parametrized real univariate triangular representation above $\mathcal{A}$, ρ, $\mathcal{T}$, σ with coefficients in D, with, for every $y \in \mathrm{Reali}(\rho)$, associated point $p(y)$ contained in $\mathrm{Zer}(Q, \mathrm{R}^k)$.
- **Output:**
  - a subset $\mathcal{C}$ of $\mathrm{D}[Y]$ containing $\mathcal{A}$,
  - for every realizable sign condition τ on $\mathcal{C}$ refining ρ, a parametrized path $\gamma(\tau)$ such that, for every $y \in \mathrm{Reali}(\tau)$, $\gamma(\tau)(y)$ is a path connecting $p(y)$ to a distinguished point of $\mathrm{RM}(\mathrm{Zer}(Q, \mathrm{R}^k))$.
- **Complexity:** $d^{O(\ell k^2)}$, where ℓ is the number of parameters, $O(d)^k$ is a bound on the degrees of on the degree of the univariate representation and of the polynomials in $\mathcal{T}$.
- **Procedure:** Call Algorithm 15.11 (Parametrized Bounded Algebraic Roadmap) and extract γ from $\mathrm{RM}(\tau)$.

Proof of correctness: The correctness of the algorithm follows from the correctness of Algorithm 15.11 (Parametrized Bounded Algebraic Roadmap).It is easy to see that γ is a parametrized path (see Definition 15.14), using the divergence property of the paths $\gamma(y, \cdot\,)$ (see Remark 15.10). □

Complexity analysis: The total number of arithmetic operations in D is bounded by $d^{O(\ell k^2)}$, using the complexity analysis of Algorithm 15.11 (Parametrized Bounded Algebraic Roadmap).

If $\mathrm{D} = \mathbb{Z}$, and the bitsizes of the coefficients of the polynomials are bounded by τ, then the bitsizes of the integers appearing in the intermediate computations and the output are bounded by $\tau d^{O(\ell k^2)}$. □

Algorithm 15.13. **[Connected Components of an Algebraic Set]**

- **Structure:** an ordered domain D contained in a real closed field R.
- **Input:** a polynomial $Q \in \mathrm{D}[X_1, \ldots, X_k]$.
- **Output:** a subset $\mathcal{A}$ of $\mathrm{D}[X_1, \ldots, X_k]$ and for every semi-algebraically connected component S of $\mathrm{Zer}(Q, \mathrm{R}^k)$ a finite subset $\Sigma \subset \mathrm{SIGN}(\mathcal{A})$ such that $S = \bigcup_{\sigma \in \Sigma} \mathrm{Reali}(\sigma, \mathrm{Zer}(Q, \mathrm{R}^k))$.
- **Complexity:** $d^{O(k^3)}$, where d is a bound on the degree of the polynomial Q.
- **Procedure:**
  - Take $Q_\varepsilon = Q^2 + (\varepsilon^2 (X_1^2 + \cdots + X_k^2 + X_{k+1}^2) - 1)^2$.
  - Call Algorithm 15.11 (Parametrized Bounded Algebraic Roadmap) without parametrized triangular Thom encoding, Q_ε, and
  $$\mathcal{N} = \{(T - 1, 1, Y_1, \ldots, Y_k)\}.$$
  The output contains a family of polynomials $\mathcal{A}^\star \subset \mathrm{D}[\varepsilon][X]$ such that the realization of a non-empty sign condition ρ in $\mathcal{A}^\star$ is contained in a semi-algebraically connected component of $\mathrm{Zer}(Q_\varepsilon, \mathrm{R}\langle\varepsilon\rangle^{k+1})$.
  - Find a set $\mathcal{S}$ of sample points for every realizable sign condition on $\mathcal{A}^\star$ using Algorithm 13.1(Sampling). Compute $\mathrm{RM}(\mathrm{Zer}(Q_\varepsilon, \mathrm{R}\langle\varepsilon\rangle^{k+1})$ using Algorithm 15.3 (Bounded Algebraic Roadmap) and for every semi-algebraically connected component S' of $\mathrm{Zer}(Q_\varepsilon, \mathrm{R}\langle\varepsilon\rangle^{k+1})$, fix a point $y(S')$ of $S' \cap \mathrm{RM}(\mathrm{Zer}(Q_\varepsilon, \mathrm{R}\langle\varepsilon\rangle^{k+1})$. For every $x \in \mathcal{S}$ compute a roadmap $\mathrm{RM}(\mathrm{Zer}(Q_\varepsilon, \mathrm{R}\langle\varepsilon\rangle^{k+1}), x)$ of $\mathrm{Zer}(Q_\varepsilon, \mathrm{R}\langle\varepsilon\rangle^{k+1})$ containing x using Algorithm 15.3 (Bounded Algebraic Roadmap) and decide from $\mathrm{RM}(\mathrm{Zer}(Q_\varepsilon, \mathrm{R}\langle\varepsilon\rangle^{k+1}), x)$ whether x belongs to S'.
  - Output the description of S', i.e. the disjunction $\Phi(S')$ of realizable sign conditions on $\mathcal{A}^\star$ with a sample point belonging to S', for every semi-algebraically connected component S' of $\mathrm{Zer}(Q_\varepsilon, \mathrm{R}\langle\varepsilon\rangle^{k+1})$.
  - For every connected component S of $\mathrm{Zer}(Q, \mathrm{R}^k)$ there exists a connected component S' of $\mathrm{Zer}(Q_\varepsilon, \mathrm{R}\langle\varepsilon\rangle^{k+1})$, such that $\pi(S') \cap \mathrm{R}^k = S$, where π: $\mathrm{R}\langle\varepsilon\rangle^{k+1} \to \mathrm{R}\langle\varepsilon\rangle^k$ is the projection map forgetting the last coordinate.

  Consider the formula $\Phi(S')$ describing S' and, eliminating a quantifier, the formula Ψ describing $\pi(S')$. Then $\mathrm{Remo}_\varepsilon(\Psi(Y))$ (Notation 14.6) defines S.

Proof of correctness: All points satisfying the same sign condition on $\mathcal{A}$ can be connected by a semi-algebraic path in $\text{Zer}(Q, \text{R}^k)$ to some fixed curve segment of $\text{RM}(\text{Zer}(Q, \text{R}^k))$ and hence must belong to the same connected component of $\text{Zer}(Q, \text{R}^k)$. Which realizable sign conditions on $\mathcal{A}$ belong to the same semi-algebraically connected component of $\text{RM}(\text{Zer}(Q, \text{R}^k))$ follows from Step 2 and 3. We also use Proposition 14.7. □

Complexity analysis: The total number of arithmetic operations in D is bounded by $d^{O(k^3)}$, using the complexity analysis of Algorithm 15.11 (Parametrized Bounded Algebraic Roadmap). The degrees of the polynomials in $\mathcal{A}$ are bounded by $d^{O(k^2)}$.

If $\text{D} = \mathbb{Z}$, and the bitsizes of the coefficients of the polynomials are bounded by τ, then the bitsizes of the integers appearing in the intermediate computations and the output are bounded by $\tau d^{O(k^3)}$. □

So we have proved Theorem 15.12.

15.4 Bibliographical Notes

The problem of deciding connectivity properties of algebraic sets considered here is a base case for deciding connectivity properties of semi-algebraic sets, studied in Chapter 16.

The notion of a roadmap for a semi-algebraic set was introduced by Canny in [36].

We discuss in more details the various contributions to the roadmap problem and the computation of connected components at the end of Chapter 16.

It is interesting to remark that the complexity of computing the number of connected components of an algebraic set given in this chapter is significantly worse than that of the algorithm for computing the Euler-Poincaré characteristic of an algebraic set given in Chapter 12. Thus, currently we are able to compute the Euler-Poincaré characteristic of real algebraic sets (which is the alternative sum of the Betti numbers) more efficiently than any of the individual Betti numbers.

16

Computing Roadmaps and Connected Components of Semi-algebraic Sets

We compute roadmaps and connected components of semi-algebraic sets. The algorithms described in this chapter have complexity much better than the ones provided by cylindrical decomposition in Chapter 11 for the problem of deciding connectivity properties of semi-algebraic sets (single exponential in the number of variables rather than doubly exponential).

In Section 16.2, we study uniform roadmaps, which provide roadmaps in the realization of every weak sign condition obtained by the relaxation of a realizable sign condition of a finite set of polynomials $\mathcal{P}$. A key algorithm is the Connecting Algorithm which links a point to the uniform roadmap inside the same weak sign condition. The correctness of the uniform roadmap algorithm relies on properties of special values studied in Section 16.1.

In Section 16.3, using a parametrized version of the Connecting Algorithm we show how to compute descriptions of the semi-algebraically connected components of the realizations of sign conditions by quantifier free formulas.

In Section 16.4, we show how to compute descriptions of the semi-algebraically connected components of semi-algebraic sets by quantifier free formulas. In Section 16.5 we construct roadmaps for general semi-algebraic sets. Finally in Section 16.6 we give a single exponential complexity algorithm for computing the first Betti number of a semi-algebraic set.

16.1 Special Values

We want to prove a result similar to Proposition 15.4 for basic semi-algebraic sets. Unfortunately, the notion of pseudo-critical values is not strong enough to ensure this property, and this is why we define the technical notion of special value.

Let $\mathrm{Zer}(Q, \mathrm{R}^k)$ be bounded. Suppose that

- $Q \in \mathrm{D}[X_1, \ldots, X_k]$, of degree at most d is such that $\mathrm{Zer}(Q, \mathrm{R}^k) \subset B(0, 1/c)$,
- $d_1 \geq d_2 \cdots \geq d_k$,
- $\deg(Q) \leq d_1$, $\mathrm{tDeg}_{X_i}(Q) \leq d_i$, for $i = 2, \ldots, k$.

Let $\bar{d}_i = 2d_i + 2$, $i = 1, \ldots, k$, and $\bar{d} = (\bar{d}_1, \ldots, \bar{d}_k)$. Consider

$$\mathrm{Def}(Q^2, \zeta) = \zeta G_k(\bar{d}, c) + (1 - \zeta)\, Q^2,$$

using Notation 12.46.

An X_1**-special value** of $\mathrm{Zer}(Q, \mathrm{R}^k)$ is a $c \in \mathrm{R}$ for which there exists $y \in \mathrm{Zer}(\mathrm{Def}(Q^2, \bar{d}, c, \zeta), \mathrm{R}\langle\zeta\rangle^k)$ with $\lim_\zeta (\pi(y)) = c$, $g(y)$ infinitesimal and y a local minimum of g on $\mathrm{Zer}(\mathrm{Def}(Q^2, \zeta), \mathrm{R}\langle\zeta\rangle^k)$, where

$$g(X) = \frac{\sum_{i=2}^{k} \left(\frac{\partial \mathrm{Def}(Q^2, \zeta)}{\partial X_i}\right)^2}{\sum_{i=1}^{k} \left(\frac{\partial \mathrm{Def}(Q^2, \zeta)}{\partial X_i}\right)^2} \tag{16.1}$$

Note that any X_1-pseudo-critical value of $\mathrm{Zer}(Q, \mathrm{R}^k)$ is an X_1-special value of $\mathrm{Zer}(Q, \mathrm{R}^k)$.

Let S be a basic closed semi-algebraic set defined as

$$S = \{x \in \mathrm{R}^k \mid Q(x) = 0 \wedge \bigwedge_{P \in \mathcal{P}} P(x) \geq 0\}.$$

An X_1**-special value on** S is an X_1-special value on $\mathrm{Zer}(\mathcal{P}', \mathrm{R}^k)$ where $\mathcal{P}'$ is contained in $\{Q\} \cup \mathcal{P}$.

Using special values, a result similar to Proposition 15.4 holds for basic closed semi-algebraic sets.

Proposition 16.1. *Let* $\mathrm{Zer}(Q, \mathrm{R}^k)$ *be bounded, and let* S *be a basic closed semi-algebraic set defined as*

$$S = \{x \in \mathrm{R}^k \mid Q(x) = 0 \wedge \bigwedge_{P \in \mathcal{P}} P(x) \geq 0\}.$$

If C *is a semi-algebraic connected component of* $S_{[a,b]}$ *and* $[a, b] \setminus \{v\}$ *contains no* X_1*-special value of* S, $v \in (a, b)$, *then* C_v *is semi-algebraically connected.*

We first prove the following

Proposition 16.2. *If* $\mathrm{Zer}(Q, \mathrm{R}^k) \subset B(0, 1/c)$ *and* x *is a point of* $\mathrm{Zer}(Q, \mathrm{R}^k)_v$ *at which* $\mathrm{Zer}(Q, \mathrm{R}^k) \cap B(x, \varepsilon)_{<v}$ *is empty for a positive* ε, *then* v *is a* X_1*-special value of* π *on* $\mathrm{Zer}(Q, \mathrm{R}^k)$.

Proof: We first prove that the statement of the proposition can be translated into a formula of the language of ordered fields with coefficients in R. More precisely, we prove that the statement

$$\exists\, y \in \mathrm{Zer}(\mathrm{Def}(Q^2, \zeta), \mathrm{R}\langle\zeta\rangle^k) \quad \lim_{\zeta} (\pi(y)) = v \wedge \lim_{\zeta} (g(y)) = 0 \tag{16.2}$$

is equivalent to the formula

$$\forall\, \varepsilon > 0 \quad \exists\, \delta > 0 \quad \forall t \quad 0 < t < \delta \quad \exists\, y \quad \Phi(v),$$

with

$$\Phi(v) := \mathrm{Def}(Q^2, t)(y) = 0 \wedge g_t(y)^2 + (\pi(y) - v)^2 < \varepsilon,$$

where we write g_t for the rational fraction obtained after replacing ζ by t in the definition of g. Note that, for t small enough, g_t is well defined and the values of g_t are bounded by 1. Thus the limit g_0 of g_t, as t tends to 0, is well defined.

First observe that Equation (16.2) is equivalent to the fact that there exists a germ φ of semi-algebraic function represented by a semi-algebraic continuous function h defined on $[0,\ a]$ such that $\pi(h(0)) = v$ and $g_0(h(0)) = 0$ by Corollary 3.11, Proposition 3.18 and Lemma 3.21, since $\mathrm{Zer}(\mathrm{Def}(Q^2, \zeta), \mathrm{R}\langle\zeta\rangle^k)$ is bounded by Proposition 12.38. Equation (16.2) follows from the continuity of h, taking $y = h(t)$.

In the other direction, assume

$$\forall\, \varepsilon > 0 \quad \exists\, \delta > 0 \quad \forall t \quad 0 < t < \delta \quad \exists\, y \quad \Phi(v).$$

By Theorem 3.19 (Curve selection lemma), there exists $\varepsilon_0 > 0$ and a semi-algebraic continuous function d from $[0, \varepsilon_0]$ to R^k such that $d(0) = 0$ and for all $0 < \varepsilon < \varepsilon_0$, $d(\varepsilon) > 0$ and

$$\forall t \ \ 0 < t < d(\varepsilon) \quad \exists\, y \quad \mathrm{Def}(Q^2, t)(y) = 0 \wedge g_t(y)^2 + (\pi(y) - v)^2 < \varepsilon. \tag{16.3}$$

Since $d(\varepsilon_0) \in \mathrm{R}$ is positive, we can find, using Proposition 3.4, ε infinitesimal in $\mathrm{R}\langle\zeta\rangle$ such that $\mathrm{Ext}(d,\ \mathrm{R}\langle\zeta\rangle)(\varepsilon) = 2\zeta$. Choosing $t = \zeta$, there exists $y \in \mathrm{Zer}(\mathrm{Def}(Q^2, \zeta), \mathrm{R}\langle\zeta\rangle^k)$ such that $g(y)$ and $\pi(y) - v$ are infinitesimal. This proves Equation (16.2).

Considering all polynomials Q of fixed degree, the statement of the proposition can now be expressed by a sentence of the language of ordered fields with coefficients in $\mathbb{Z}$. By Theorem 2.80 (Tarski-Seidenberg principle), it thus suffices to prove the proposition over the reals, which is what we now proceed to do. The proposition for $\mathrm{R} = \mathbb{R}$ is an immediate consequence of the following two lemmas.

Let g be defined in Equation (16.1).

Lemma 16.3. *Suppose that* $\mathrm{Zer}(Q, \mathbb{R}^k) \subset B(0, 1/c)$ *and that* x *is a point of* $\mathrm{Zer}(Q, \mathbb{R}^k)_v$ *at which* $\mathrm{Zer}(Q, \mathbb{R}^k) \cap B(x, \varepsilon)_{<v}$ *is empty for some positive* ε, *then there is a point* $y \in \mathrm{Zer}(\mathrm{Def}(Q^2, \zeta), \mathbb{R}\langle\zeta\rangle^k) \cap B(x, \varepsilon)$ *for which* $\lim_\zeta(\pi(y)) = v$ *and* $\lim_\zeta(g(y)) = 0$.

Lemma 16.4. *If* y *is a point of* $\mathrm{Zer}(\mathrm{Def}(Q^2, \zeta), \mathbb{R}\langle\zeta\rangle^k) \cap B(x, \varepsilon)$ *at which* $\lim_\zeta\ (\pi(y)) = v$ *and* $\lim_\zeta\ (g(y)) = 0$ *then* v *is a* X_1*-special value of* π *on* $\mathrm{Zer}(Q, \mathbb{R}^k)$.

Proof of Lemma 16.3 : If there is a critical value of π on

$$\mathrm{Zer}(\mathrm{Def}(Q^2, \zeta), \mathbb{R}\langle\zeta\rangle^k)$$

infinitesimally close to v, we are done. Otherwise, suppose that there is no critical value of π on $\mathrm{Zer}(\mathrm{Def}(Q^2,\zeta),\mathrm{R}\langle\zeta\rangle^k)$ in an interval $(v-b,v+b)\subset\mathrm{R}\langle\zeta\rangle$ with $b\in\mathrm{R}$. We can suppose without loss of generality that $b>\varepsilon$.

We argue by contradiction and suppose that for every y at which

$$\mathrm{Def}(Q^2,\zeta)(y)=0\wedge\lim_{\zeta}(\pi(y))=c,$$

the value $g(y)$ is not infinitesimal.

Since $\mathrm{Zer}(Q,\mathrm{R}^k)\cap B(x,\varepsilon)_{<c}=\emptyset$, we know that for any

$$y\in\mathrm{Zer}(\mathrm{Def}(Q^2,\zeta),\mathrm{R}\langle\zeta\rangle^k)\cap B(x,\varepsilon)_{\leq v},$$

$\lim_\zeta(\pi(y))=v$ and thus $g(y)$ is not infinitesimal. Let $a\in\mathrm{R}$ be a positive number smaller than any value of g on

$$\mathrm{Zer}(\mathrm{Def}(Q^2,\zeta),\mathrm{R}\langle\zeta\rangle^k)\cap B(x,\varepsilon)_{\leq v}.$$

Let

$$U'=\{t\in\mathrm{R}\mid g_t<a\text{ on }\mathrm{Zer}(\mathrm{Def}(Q^2,t),\mathrm{R}^k)\cap B(x,\varepsilon)_{\leq v}\}.$$

Let U'' be the set of $t\in\mathrm{R}$ such that there is no critical value of π on $\mathrm{Zer}(\mathrm{Def}(Q^2,t),\mathrm{R}^k))$ in $(v-b,v+b)$ and $U=U'\cap U''$. The set U is semi-algebraic and its extension to $\mathrm{R}\langle\zeta\rangle$ contains ζ. Thus, it contains an interval $(0,t_0)$ by Proposition 3.17.

For every $t\in(0,t_0)$, let y_t be a point in

$$\mathrm{Zer}(\mathrm{Def}(Q^2,t),\mathrm{R}^k)\cap B(x,\varepsilon)_{\leq c}$$

whose last $k-1$ coordinates coincide with the last $k-1$ coordinates of x. Consider the curve γ_t on $\mathrm{Zer}(\mathrm{Def}(Q^2,t),\mathrm{R}^k)$ through y_t which at each of its points is tangent to the gradient of π on $\mathrm{Zer}(\mathrm{Def}(Q^2,t),\mathrm{R}^k)$. The gradient of π on $\mathrm{Zer}(\mathrm{Def}(Q^2,t),\mathrm{R}^k)$ at a point of $\mathrm{Zer}(\mathrm{Def}(Q^2,t),\mathrm{R}^k)$ is proportional to

$$G=\left(\sum_{i=2}^{k}\left(\frac{\partial\mathrm{Def}(Q^2,t)}{\partial X_i}\right)^2,\ldots,-\frac{\partial\mathrm{Def}(Q^2,t)}{\partial X_1}\frac{\partial\mathrm{Def}(Q^2,t)}{\partial X_k}\right)$$

(see page 240). For every point of γ_t, the vector G thus belongs to the half-cone $\mathcal{C}$ of center x, based on the $k-1$-sphere of radius $\sqrt{\frac{1-a}{a}}$ and center $(x_1-1,x_2,\ldots,x_k)$ in the hyperplane $X_1=x_1-1$. It follows that the curve γ_t is completely contained in $\mathcal{C}$. Since there is no critical value of π on $\mathrm{Zer}(Q_t,\mathrm{R}^k)$ in $(v-b,v+b)$, the curve γ_t is defined over $(v-b,v+b)$ and thus meets $S(x,\varepsilon)\cap\mathcal{C}$.

Since $\mathcal{C}\cap S(x,\varepsilon)\cap\mathrm{Zer}(\mathrm{Def}(Q^2,t),\mathrm{R}^k)\neq\emptyset$ is true for every $t\in(0,t_0)$ it follows from Proposition 3.17 that

$$\mathcal{C}\cap S(x,\varepsilon)\cap\mathrm{Zer}(\mathrm{Def}(Q^2,\zeta),\mathrm{R}\langle\zeta\rangle^k)\neq\emptyset.$$

Thus, taking $\lim_\zeta$ of the point so obtained, $B(x,\varepsilon)_{<v} \cap \mathrm{Zer}(Q,\mathbb{R}^k) \neq \emptyset$, which is a contradiction. □

Proof of Lemma 16.4 : If g is zero anywhere that the first coordinate is infinitesimally close to v, then v is a X_1-pseudo- critical value and we are done. Alternatively, we may assume that g is non-zero in any slab of infinitesimal width containing $X_1 = v$. Let y be given by our hypothesis, i.e. $\lim_\zeta(\pi(y)) = v$, $\lim_\zeta(g(y)) = 0$. We let C be the bounded semi-algebraically connected component of $\mathrm{Zer}(\mathrm{Def}(Q^2,\zeta), \mathbb{R}\langle\zeta\rangle^k)$ containing y. Define w by $\pi(y) = w$. Then g attains its minimum on C_w at some point $z \in C_w$. Let t be this minimum. It is clear that t is infinitesimal.

Consider the set $A = \{w \mid \min_{C_w}(g) \leq t\}$. This set A is closed, bounded, semi-algebraic, and thus a union of closed intervals $[a_1, b_1] \cup \ldots \cup [a_h, b_h]$ with $a_i \leq b_i < a_{i+1}$ Let $[a_i, b_i] = [a, b]$ be the interval containing w.

If a and b are both infinitesimally close to w take u and u' so that $b_{i-1} < u < a = a_i \leq b = b_i < u' < a_{i+1}$ with u and u' infinitesimally close to w. The minimum of g on $C_{[u,u']}$ occurs in the interior of the slab since it is smaller at C_w than its minimum both on C_u and $C_{u'}$. It follows that c is a X_1-special value on $\mathrm{Zer}(Q,\mathbb{R}^k)$.

Assume on the contrary that $[a,b]$ is such that a or b is not infinitesimally close to w. We are going to prove that this leads to a contradiction.

According to Theorem 5.46 (Semi-algebraic triviality), there exists a family ϕ_j of semi-algebraic curves parametrized by open segments (α_j, β_j) covering (a, b) (with the exception of a finite number of points) such that $g(\phi_j(x))$ is smaller than t. If $T_j(x) = (T_{j,1}(x), \ldots, T_{j,k}(x))$ is the tangent vector to ϕ_j at $(x, \phi_j(x))$, we have

$$\begin{aligned} -T_{j,1}\frac{\partial \mathrm{Def}(Q^2,t)}{\partial X_1} &= T_{j,2}\frac{\partial \mathrm{Def}(Q^2,t)}{\partial X_2} + \cdots + T_{j,k}\frac{\partial \mathrm{Def}(Q^2,t)}{\partial X_k} \\ T_{j,1}^2 &\leq \frac{t}{1-t}\|(T_{j,2},\ldots,T_{j,k})\|^2. \end{aligned}$$

Thus, at every point on each of these curves, $\left|\frac{T_{j,1}(x)}{T_{j,i}(x)}\right| < \sqrt{\frac{kt}{1-t}} = t'$ for some $2 \leq i \leq k$. Hence, we can suppose – subdividing further if needed and producing more curves – that on each of these curves, $\left|\frac{T_{j,1}(x)}{T_{j,i}(x)}\right| < t'$ for some $2 \leq i \leq k$.

Let N be the number of the curves so obtained. We prove now that the interval $(w, w + 2/cNt')$ contains w such that $\min_{C_w}(g) > t$. Suppose on the contrary that at every value $u \in (w, w + 2/cNt')$, $\min_{C_u}(g) \leq t$. Then there is an interval of length at least $2/ct'$ over which the curve $\phi_j(x)$ is differentiable and $\left|\frac{T_{j,1}(x)}{T_{j,i}(x)}\right|$ is less that t'. It follows from the mean value theorem that the projection of this curve to the X_i axis is bigger than $2/c$, which contradicts the fact that $C \subset B(0, 1/c)$. Similarly, the interval $(w - 2/cNt', w)$ contains u' such that $\min_{C_u}(g) > t$.

Note that both u and u' are infinitesimally close to v. This contradicts the fact that a or b is not infinitesimally close to w and ends the argument. □ □

The proof of Proposition 16.1 will use the following lemma.

Let C be a semi-algebraically connected component of $S_{[a,v]}$ and let $B_1, \ldots, B_h$ be the semi-algebraically connected components of $C_{[a,v)}$.

Lemma 16.5. *If $\bar{B}_1 \cap \bar{B}_2 \neq \emptyset$, then v is a X_1-pseudo-critical value on S.*

Proof: Suppose that $\overline{B_1} \cap \ldots \cap \overline{B_I} \neq \emptyset$ and that $1, \ldots, I$ is a maximal family with this property. Let x be a point of this intersection. Clearly, x belongs to the boundary of S and the set $\mathcal{P}' \subset \mathcal{P}$ of polynomials in $\mathcal{P}$ that vanish at x is not empty. According to Theorem 5.46 (Semi-algebraic triviality) there is $w \in [a, v)$ such that $\mathrm{Zer}(\mathcal{P}', \mathrm{R}^k)_{[w,v)}$ is semi-algebraically homeomorphic to $\mathrm{Zer}(\mathcal{P}', \mathrm{R}^k)_w \times [w, v)$ and $C_{[w,v)}$ is semi-algebraically homeomorphic to $C_w \times [w, v)$. Note that $C_{[w,v)}$ is not semi-algebraically connected. Let D be the connected component of $\mathrm{Zer}(\mathcal{P}', \mathrm{R}^k)_{[w,v]}$ containing x.

We consider two cases according to whether or not D_w is empty:

If D_w is empty, then v is an X_1-pseudo-critical value on $\mathrm{Zer}(\mathcal{P}', \mathrm{R}^k)$ by Proposition 16.2 and we have already noted that pseudo-critical values are special values.

If D_w is not empty, then some semi-algebraically connected component of $C_{[a,v)}$ intersects $\mathrm{Zer}(\mathcal{P}', \mathrm{R}^k)$ in any neighborhood of x. Suppose, without loss of generality that it is B_1. Consider a maximal subset of $\mathcal{P}$, say $\mathcal{P}''$, such that $\mathrm{Zer}(\mathcal{P}'', \mathrm{R}^k)$ intersects B_2 in any neighborhood of x. The set $\mathcal{P}''$ is non-empty and contained in $\mathcal{P}'$. According to Theorem 5.46 (Semi-algebraic triviality) there is a $w' \geq w$ such that $\mathrm{Zer}(\mathcal{P}'', \mathrm{R}^k)_{[w',v)}$ is semi- algebraically homeomorphic to $\mathrm{Zer}(\mathcal{P}'', \mathrm{R}^k)_{w'} \times [w', v)$. Let Z be the connected component of $\mathrm{Zer}(\mathcal{P}'', \mathrm{R}^k)_{[w',v]}$ containing x. By the maximality of $\mathrm{Zer}(\mathcal{P}'', \mathrm{R}^k)$, there is a connected component Z_1 of $Z_{[w',v)}$ contained in $B_{2_{[w',v)}}$. Since $\mathrm{Zer}(\mathcal{P}', \mathrm{R}^k) \subset \mathrm{Zer}(\mathcal{P}'', \mathrm{R}^k)$ and $\mathrm{Zer}(\mathcal{P}', \mathrm{R}^k)_{[w',v)}$ meets B_1, $\mathrm{Zer}(\mathcal{P}'', \mathrm{R}^k)_{[w',v)}$ is not semi-algebraically connected. We conclude by Proposition 15.5 that v is a X_1-pseudo-critical value on $\mathrm{Zer}(\mathcal{P}'', \mathrm{R}^k)$. □

Proof of Proposition 16.1: Suppose that C_v is empty. We take $d \in [a, b]$ such that C_d is non-empty and suppose that $v < d$ (the case $v > d$ can be treated similarly). We obtain a contradiction by proving that there is a X_1-special value on S in $(v, d]$. Since the set $\{w \in (v, d] | C_w \neq \emptyset\}$ is a closed semi-algebraic subset of $[v, d]$, it contains a smallest such value, say u. Choose an $x \in C_u$. Since x belongs to the boundary of S, the set $\mathcal{P}'$ of polynomials in $\mathcal{P}$ vanishing at x is non- empty. It is clear that $\mathrm{Zer}(Q, \mathrm{R}^k) \cap B(x, \varepsilon)_{<u} = \emptyset$ for ε small enough. Hence, by Proposition 16.2, u is an X_1-special value on $\mathrm{Zer}(\mathcal{P}', \mathrm{R}^k)$.

Suppose now that C_v is not semi-algebraically connected. Take $d \in [a, b]$ such that a semi-algebraically connected component of $C_{[v,d]}$ contains more than one connected component of C_v and suppose that $v < d$ (the case $v > d$ can be treated similarly). We obtain a contradiction by proving that there is a X_1-special value on S in $(v, d]$. Since the set of $w \in (v, d]$ for which $C_{[v,w]}$ contains more than one connected component of C_v is a closed semi-algebraic subset of $[v, b]$ by Theorem 5.46 (Semi-algebraic triviality), it contains a smallest such value, say u.

Consider a connected component B of $C_{[v,u]}$ containing more than one connected component of C_u. Let $B_1, \ldots, B_h$ be the connected components of $C_{[v,u)}$ contained in B, and let B_0 be the set of $x \in B_u$ such that $B(x, \varepsilon)_{<u} \cap C = \emptyset$ for ε small enough. Clearly, $B = B_0 \cup \overline{B_1} \cup \cdots \cup \overline{B_h}$.

We now prove that u is an X_1-special value on S whether or not $B_0 = \emptyset$.

If B_0 is non-empty, choose an $x \in B_0$ and let $\mathcal{P}'$ be the set of polynomials in $\mathcal{P}$ vanishing at x. Then $B(x, \varepsilon)_{<u} \cap \mathrm{Zer}(\mathcal{P}', \mathrm{R}^k)$ is empty and it follows from Proposition 16.2 that u is an X_1-special value on $\mathrm{Zer}(\mathcal{P}', \mathrm{R}^k)$.

Alternatively, if B_0 is empty we may assume, without loss of generality, that $\overline{B_1} \cap \overline{B_2} \neq \emptyset$. Thus by Lemma 16.5, u is an X_1-pseudo-critical value, hence a X_1-special value on S. □

We are going now to indicate how to compute special values. Consider the algebraic set Z defined by the $k+1$ polynomial equations in the $k+1$ variables $(X_1, \ldots, X_k, \lambda)$

$$\begin{aligned}
\mathrm{Def}(Q^2, \zeta) &= 0, \\
\frac{\partial \mathrm{Def}(Q^2, \zeta)}{\partial X_1} &= \lambda \frac{\partial g}{\partial X_1}, \\
&\vdots \\
\frac{\partial \mathrm{Def}(Q^2, \zeta)}{\partial X_k} &= \lambda \frac{\partial g}{\partial X_k}.
\end{aligned}$$

The local minima of g on $\mathrm{Zer}(\mathrm{Def}(Q^2, \zeta), \mathrm{R}\langle\zeta\rangle^k)$ are contained in the projection of Z to the first k coordinates.

Proposition 16.6. *If C' is a semi-algebraically connected component of Z on which g has an infinitesimal local minimum on* $\mathrm{Zer}(\mathrm{Def}(Q^2, \zeta), \mathrm{R}\langle\zeta\rangle^k)$ *then* $\lim_\zeta (C')$ *is a single point.*

Proof: Let x be a point of C' where g has an infinitesimal local minimum on $\mathrm{Zer}(\mathrm{Def}(Q^2, \zeta), \mathrm{R}\langle\zeta\rangle^k)$, and let $u = g(x)$. Note that g is constant on C'. The projection of C' to the X_1-axis, $\pi(C')$, is contained in $A = \{w \mid \min_{C_w}(g) \le u\}$ where C is the semi- algebraically connected component of $\mathrm{Zer}(\mathrm{Def}(Q^2, \zeta), \mathrm{R}\langle\zeta\rangle^k)$ containing x. Since $\pi(C')$ is semi-algebraically connected, following the proof of Lemma 16.4 we see that $\pi(C')$ is contained in an infinitesimal segment. □

Algorithm 16.1. **[Special Values]**

- **Structure:** an ordered domain D contained in a real closed field R.
- **Input:**
  - a triangular system $\mathcal{T}$ specifying $z \in \mathrm{R}^{i-1}$, with coefficients in D,
  - a polynomial $Q \in \mathrm{D}[X_1, \ldots, X_k]$, such that $\mathrm{Zer}(Q, \mathrm{R}^k) \subset B(0, 1/c)$.
- **Output:** a set of values containing the X_i-special values of $\mathrm{Zer}(Q, \mathrm{R}^k)_z$.
- **Complexity:** $d^{O(k)}$ where d is the degree of Q.
- **Procedure:**
  - Let $d_1 \geq d_2 \cdots \geq d_k$, $\deg(Q) \leq d_1$, $\mathrm{tDeg}_{X_i}(Q) \leq d_i$, for $i = 2, \ldots, k$, $\bar{d}_i = 2d_i + 2$, $i = 1, \ldots, k$, $\bar{d} = (\bar{d}_1, \ldots, \bar{d}_k)$, and

$$\mathrm{Def}(Q^2, \zeta) = \zeta G_{k-i}(\bar{d}, c) + (1 - \zeta) Q^2,$$

  using Notation 12.46. Denote by Z the algebraic set defined by the $k+1$ polynomial equations in the $k+1$ variables $(X_1, \ldots, X_k, \lambda)$,

$$\begin{aligned}
\mathcal{T}_j(X_1, \ldots, X_j) &= 0, \mathcal{T}_j \in \mathcal{T}, j = 1, \ldots, i-1, \\
\mathrm{Def}(Q^2, \zeta) &= 0, \\
\frac{\partial \mathrm{Def}(Q^2, \zeta)}{\partial X_i} &= \lambda \frac{\partial g}{\partial X_i}, \\
&\vdots \\
\frac{\partial \mathrm{Def}(Q^2, \zeta)}{\partial X_k} &= \lambda \frac{\partial g}{\partial X_k}.
\end{aligned}$$

  - Use Algorithm 12.16 (Bounded Algebraic Sampling) to find a set of points which meets every connected component of Z.
  - Compute $\lim_\zeta$ of these points using Algorithm 12.14 (Limit of Real Bounded Points).
  - Describe their k first coordinates using Algorithm 15.1 (Projection). Keep only the points whose $i-1$ first coordinates coincide with z.

Proof of correctness: By Proposition 16.6, we know that the X_i-special values of $\mathrm{Zer}(Q, \mathrm{R}^k)_z$ are the among the values computed by Algorithm 16.1 (Special Values). □

Complexity analysis: Using the complexity analysis of Algorithm 12.16 (Bounded Algebraic Sampling), Algorithm 12.14 (Limit of Real Bounded Points), and Algorithm 15.1 (Projection), we conclude that the complexity is $d^{O(k)}$. At most $O(d)^k$ univariate polynomials of degrees at most $O(d)^k$ whose real roots contain the X_1-special values of $\mathrm{Zer}(Q, \mathrm{R}^k)$ are computed.

If $\mathrm{D} = \mathbb{Z}$, and the bitsizes of the coefficients of the polynomials are bounded by τ, then the bitsizes of the integers appearing in the intermediate computations and the output are bounded by $\tau d^{O(k)}$. □

16.2 Uniform Roadmaps

We consider

- a polynomial $Q \in \mathrm{R}[X_1, \ldots, X_k]$ for which $\mathrm{Zer}(Q, \mathrm{R}^k)$ is bounded,
- a set of at most s polynomials $\mathcal{P}$ such that $\mathcal{P}$ is in strong ℓ-general position with respect to Q (Definition b).

We first indicate how to connect any point $x \in \mathrm{Zer}(Q, \mathrm{R}^k)$ to some roadmap of the zero set of the union of Q and a subset of $\mathcal{P}$.

Denote by $\sigma(x)$ the sign condition on $\mathcal{P}$ at x. Let

$$\mathrm{Reali}(\bar{\sigma}(x), \mathrm{Zer}(Q, \mathrm{R}^k)) = \{x \in \mathrm{Zer}(Q, \mathrm{R}^k) \mid \bigwedge_{P \in \mathcal{P}} \mathrm{sign}(P(x)) \in \bar{\sigma}(x)(P)\},$$

where $\bar{\sigma}$ is the relaxation of σ (Definition 5.32). We say that $\bar{\sigma}(x)$ is the weak sign condition defined by x on $\mathcal{P}$. We denote by $\mathcal{P}(x)$ the union of $\{Q\}$ and the set of polynomials in $\mathcal{P}$ vanishing at x.

Algorithm 16.2. **[Bounded Connecting]**

- **Structure:** an ordered domain D contained in a real closed field R.
- **Input:**
  - a polynomial $Q \in \mathrm{D}[X_1, \ldots, X_k]$ such that $\mathrm{Zer}(Q, \mathrm{R}^k) \subset B(0, 1/c)$,
  - a finite set of polynomials $\mathcal{P} \subset \mathrm{D}[X_1, \ldots, X_k]$ in strong ℓ-general position with respect to Q,
  - a point $p \in \mathrm{Zer}(Q, \mathrm{R}^k)$ described by a real univariate representation u, τ with coefficients in D.
- **Output:** a subset $\mathcal{P}' \subset \mathcal{P}$ and a semi-algebraic path Γ which connects p to $\mathrm{RM}(\mathrm{Zer}(\mathcal{P}' \cup \{Q\}, \mathrm{R}^k))$ inside $\mathrm{Reali}(\bar{\sigma}(x), \mathrm{Zer}(Q, \mathrm{R}^k))$.
- **Complexity:** $\ell\, s\, d^{O(k^2)}$, where s is a bound on the number of polynomials in $\mathcal{P}$, d is a bound on the degree of Q and the polynomials in $\mathcal{P}$ and $O(d)^k$ is a bound on the degree of of the univariate representation u.
- **Procedure:**
  - Initialize $\Gamma = \emptyset$, $q := p$, $u', \tau' := u, \tau$.
  - ($\star$) Construct a path γ connecting q to $\mathrm{RM}(\mathrm{Zer}(\mathcal{P}(q), \mathrm{R}^k), \{u', \tau'\})$, using Algorithm 15.4 (Bounded Algebraic Connecting).
  - If a polynomial of $\mathcal{P} \setminus \mathcal{P}(q)$ vanishes somewhere on γ, let $t \in (0, 1)$ such that no polynomial in $\mathcal{P} \setminus \mathcal{P}(q)$ vanishes on $\gamma((0, t))$ and there are polynomials in $\mathcal{P} \setminus \mathcal{P}(q)$ vanishing on $\gamma(t)$. Add $\gamma_{|(0,t]}$ to the end of Γ. Call the algorithm recursively returning to ($\star$) with input $q := \gamma(t)$, taking as u', τ' the real univariate representation describing $\gamma(t)$.
  - If γ is such that no polynomial of $\mathcal{P} \setminus \mathcal{P}(q)$ vanishes on γ, add γ to the end of Γ.

Proof of correctness : Follows clear from the correctness of Algorithm 15.4 (Bounded Algebraic Connecting). □

Complexity analysis: Since $\mathcal{P}$ is in strong ℓ-general position with respect to Q, the algorithm terminates after $\ell' \leq \ell$ iterations. The degrees of the univariate representations representing the ℓ' successive values of p are bounded by $d^{O(k^2)}$. Thus the complexity of the Bounded Connecting Algorithm is clearly $\ell\, s\, d^{O(k^2)}$. The number of different curve segments in the connecting semi-algebraic path is at most $\ell\, d^{O(k^2)}$.

If $\mathrm{D} = \mathbb{Z}$, and the bitsizes of the coefficients of the polynomials are bounded by τ, then the bitsizes of the integers appearing in the intermediate computations and the output are bounded by $\tau d^{O(k^2)}$. □

A **uniform roadmap of** $(Q, \mathcal{P})$ is a union of open curve segments and points satisfying the following two conditions:

- URM_1: The signs of the polynomials $P \in \mathcal{P}$ are constant on each curve segment,
- URM_2: The intersection of this set with any basic closed semi-algebraic set

$$\mathrm{Reali}(\sigma, \mathrm{Zer}(Q, \mathrm{R}^k)) = \{x \in \mathrm{R}^k \mid Q(x) = 0 \wedge \mathrm{sign}(P(x)) \in \sigma(P)\},$$

  where $\sigma \in \{\{0\}, \{0, 1\}, \{0, -1\}\}^{\mathcal{P}}$ is a weak sign condition on $\mathcal{P}$, is a roadmap for $\mathrm{Reali}(\sigma, \mathrm{Zer}(Q, \mathrm{R}^k))$.

As a first step we describe an algorithm which, given a polynomial Q, a set of polynomials $\mathcal{P}$, and a point p in $\mathrm{Zer}(Q, \mathrm{R}^k)$ constructs a finite number of continuous semi-algebraic curves starting at p so that every semi-algebraically connected component of every realizable sign condition of $\mathcal{P}$ in $\mathrm{Zer}(Q, \mathrm{R}^k)$ sufficiently near and to the left of p contains one of these curves without the point p.

If $p \in \mathrm{Zer}(Q, \mathrm{R}^k)$, denote by $\mathrm{SIGN}(\mathcal{P}, p)$ the set of sign conditions σ such that

$$\mathrm{Reali}(\sigma, \mathrm{Zer}(Q, \mathrm{R}^k)) \cap B(p, r)_{<v}$$

is non-empty for all sufficiently small $r > 0$.

Algorithm 16.3. **[Linking Paths]**

- **Structure:** an ordered domain D contained in a real closed field R.
- **Input:**
  - a polynomial $Q \in \mathrm{D}[X_1, \ldots, X_k]$ such that $\mathrm{Zer}(Q, \mathrm{R}^k) \subset B(0, 1/c)$,
  - a finite set of polynomials $\mathcal{P} \subset \mathrm{D}[X_1, \ldots, X_k]$, in strong ℓ-general position with respect to Q,
  - a point $p \in \mathrm{Zer}(Q, \mathrm{R}^k)_z$, described by a real univariate representation of degree at most $d^{O(k)}$ with coefficients in D.

- **Output:** a finite set of semi-algebraic paths starting at p such that for some sufficiently small r and for every σ in $\mathrm{SIGN}(\mathcal{P}, p)$ every connected component of $\mathrm{Reali}(\sigma, \mathrm{Zer}(Q, \mathrm{R}^k)) \cap B(p,r)_{<v}$ contains one of these semi-algebraic paths (without the endpoint p).
- **Complexity:** $(s+2^\ell)\, d^{O(k)}$, where s is a bound on the number of polynomials in $\mathcal{P}$ and d is a bound on the degree of Q and the polynomials in $\mathcal{P}$.
- **Procedure:**
  - Let $\mathcal{P}(p)$ be the set of polynomials in $\mathcal{P}$ (possibly empty) that are zero at p and let $B(p, \varepsilon)$ be a ball of radius ε and center p, where ε is a new variable. Using Algorithm 13.1 (Sampling) with the polynomials defining $B(p,\varepsilon)_{<v}$ along with the polynomials $\mathcal{P}(p)$ as input and structure $\mathrm{D}[\varepsilon] \subset \mathrm{R}\langle\varepsilon\rangle$, find a finite set of points $\mathcal{S}(\varepsilon)$ intersecting every semi-algebraically connected component of every realizable sign condition of the polynomials in $\mathcal{P}$ in $B(p,\varepsilon)_{<v}$.
  - For every $u = (f, g_0, g_1, \ldots, g_k) \in \mathcal{S}(\varepsilon)$, apply Algorithm 11.20 (Removal of Infinitesimals) with input f and $\mathcal{P}(p)_u$, output $t(u)$ and define $t_0 = \min_{u \in \mathcal{S}(\varepsilon)} t(u)$. Replacing ε by $t \in (0, t_0]$ defines for each $u(\varepsilon) \in \mathcal{S}(\varepsilon)$, with associated point $q(\varepsilon)$, a semi-algebraic path $\gamma(u)$ such that $\gamma(0) = p$.

Proof of correctness: The semi-algebraic paths $\gamma(u)$, $u \in \mathcal{S}(\varepsilon)$, join p to points in every semi-algebraically connected component of the realizable sign conditions of $\mathcal{P}$ intersected with $\mathrm{Zer}(Q, \mathrm{R}^k) \cap B(p,\varepsilon)_{<v}$. □

Complexity analysis: Since at most ℓ polynomials can be zero at p, the complexity is $(s+2^\ell)\, d^{O(k)}$, using Remark 13.3.

If $\mathrm{D} = \mathbb{Z}$, and the bitsizes of the coefficients of the polynomials are bounded by τ, then the bitsizes of the integers appearing in the intermediate computations and the output are bounded by $\tau d^{O(k)}$. □

We are going to describe now the Uniform Roadmap Algorithm. The algorithm will call itself recursively. In each recursive call the number of variables will strictly decrease. The base case when $k=1$ or $\ell=0$ are easy.

Note that if $\ell = 0$, then a roadmap $\mathrm{RM}(\mathrm{Zer}(Q, \mathrm{R}^k))$ is a uniform roadmap for $(Q, \mathcal{P})$ since on every semi-algebraically connected component of $\mathrm{Zer}(Q, \mathrm{R}^k)$ the signs of the polynomials in $\mathcal{P}$ are fixed.

If $k=1$, the zeroes of Q are isolated, the roadmap consists of the zeroes of Q.

Algorithm 16.4. **[Uniform Roadmap]**

- **Structure:** an ordered domain D contained in a real closed field R.
- **Input:**
  - a polynomial $Q \in \mathrm{D}[X_1, \ldots, X_k]$ such that $\mathrm{Zer}(Q, \mathrm{R}^k) \subset B(0, 1/c)$,

– a finite set $\mathcal{P} \subset \mathrm{D}[X_1, \ldots, X_k]$,
– a natural number ℓ such that $\mathcal{P}$ is in strong ℓ-general position with respect to Q.

- **Output:** a semi-algebraic set $\mathrm{URM}(Q, \mathcal{P})$ satisfying conditions URM_1 and URM_2. Moreover, $\mathrm{URM}(Q, \mathcal{P})$ is described by real univariate representations and curve segments representations, and each of these representations is labeled by a subset $\mathcal{R} \subset \mathcal{P}$ such that its associated point or curve segment is contained in $\mathrm{Zer}(\mathcal{R} \cup \{Q\}, \mathrm{R}^k)$.
- **Complexity:** $s^{\ell+1} d^{O(k)}$, where s is a bound on the number of elements of $\mathcal{P}$ and d is a bound on the degrees of Q and the elements of $\mathcal{P}$.
- **Procedure:**

– Initialize $i := 1, \mathcal{T}, \sigma := \emptyset, S := Q^2, \mathcal{R} := \mathcal{P}, m := \ell$.
– Step 1: ($\star$) For each $\mathcal{R}' \subset \mathcal{R}, \#(\mathcal{R}') \leq m$.
  – Step 1 a): If $\#(\mathcal{R}') = m$, describe the isolated zeroes of $\mathcal{T}, S + \sum_{P \in \mathcal{R}'} P^2$. These points are placed in a set of distinguished points for $\mathcal{R}'$. A distinguished value for $\mathcal{R}'$ is the i-th coordinate of a distinguished point for $\mathcal{R}'$.
  – Step 1 b): If $\#(\mathcal{R}') < m$, run Algorithm 15.2 (Curve Segments) with $\mathcal{T}, \sigma, S + \sum_{P \in \mathcal{Q}'} P^2$, and an empty set of univariate representations as input.

    Label each curve segment by $\mathcal{R}'$. The endpoints of these curve segments are labeled by $\mathcal{R}'$ and are placed in a set of distinguished points for $\mathcal{R}'$. A distinguished value for $\mathcal{R}'$ is the i-th coordinate of a distinguished point for $\mathcal{R}'$.
  – Step 1 c): Run Algorithm 16.1 (Special Values) for $\mathcal{T}, \sigma, S + \sum_{P \in \mathcal{R}'} P^2$ and intersect the curve segments obtained in Step 1 a) with the corresponding special hyperplanes. Add these points to the set of distinguished points for $\mathcal{R}'$. Append their i-th coordinates to the set of distinguished values for $\mathcal{R}'$.
  – Step 1 d): Compute the intersection of each curve segment output in Step 1 a) with $\mathrm{Zer}(P, \mathrm{R}^k)$ for each $P \in \mathcal{R}$. Note that the intersection of a curve segment with the zero set of a polynomial, is either the segment itself, or a finite set of points (possibly empty), This is checked by substituting the parametrized univariate representation of the curve into each polynomial in $\mathcal{R}$ and checking whether the resulting univariate polynomial vanishes identically or not.

    If the intersection is the curve segment itself, ignore this intersection. Otherwise, the points of intersection yield a partition of the curve segment. Add these points to the set of distinguished points for $\mathcal{R}' \cup \{P\}$. Append their i-th coordinates to the set of distinguished values for $\mathcal{R}' \cup \{P\}$. Store the sign vector of the set of polynomials $\mathcal{R}$ on each curve segment and point computed above.

- Step 2: For every distinguished point p with label $\mathcal{R}'$ and i-th coordinate v, add to the distinguished points for $\mathcal{R}'$ the intersections of the hyperplane H_v with the curves constructed for $\mathcal{R}''$ in Step 1, where $\mathcal{R}' \subset \mathcal{R}'' \subset \mathcal{R}$, $\#(\mathcal{R}'') \leq m - \#(\mathcal{R}') - 1$.
- Step 3: For all distinguished value v specified by A, α, call the algorithm recursively, returning to $(\star)$ with input

$$\begin{aligned} i &:= i+1, \\ \mathcal{T}, \sigma &:= \mathcal{T}, A, \sigma, \alpha, \\ S &:= S + \sum_{P \in \mathcal{P}'} P^2, \\ \mathcal{R} &:= \mathcal{R} \setminus \mathcal{R}', \\ m &:= m - \#(\mathcal{R}'. \end{aligned}$$

  Denote by $\mathrm{URM}_0(Q, \mathcal{P})$ the output so obtained.
- Step 4: For each distinguished point p and the corresponding distinguished hyperplane, use Algorithm 16.3 (Linking Paths) to construct semi-algebraic paths joining p to points in every semi-algebraically connected component of every realizable sign condition of the set of polynomials $\mathcal{P}$ intersected with $\mathrm{Zer}(Q, \mathrm{R}^k) \cap B(p, r)_{<\pi(p)}$, for some small enough r. Let the other endpoints of these curves be a finite set $\mathcal{S}$. Connect the points of $\mathcal{S}$ to some $\mathrm{Zer}(\mathcal{P}', \mathrm{R}^k)$ using Algorithm 16.2 (Bounded Connecting).
- Output all the curve segments and distinguished points, each labeled by the sign condition it satisfies. This is the set $\mathrm{URM}(Q, \mathcal{P})$.

The proof of correctness of Algorithm 16.4 (Uniform Roadmap) is based on the following results.

Let S be the semi-algebraic set defined by $Q = 0$, $P \geq 0$, $P \in \mathcal{P}$, and let $\mathrm{RM}(S) = S \cap \mathrm{URM}(Q, \mathcal{P})$.

Proposition 16.7. *Thc set* $\mathrm{RM}(S)$ *is a roadmap for the set* S.

Proof: We first show that $\mathrm{RM}(S)$ satisfies RM_2.

For any $x \in \mathrm{R}$ such that S_x is non-empty, and for any semi- algebraically connected component C of S_x, there exists a semi-algebraically connected component C' of a non-empty algebraic set, $\mathrm{Zer}(\mathcal{P}', \mathrm{R}^k)_x$ such that $C' \subset C$ (see Proposition 13.1). Since, in Step 1 b of the algorithm we construct curves using the Algorithm 15.2 (Curve Segments) on all non-empty algebraic sets of the form $\mathrm{Zer}(\mathcal{P}', \mathrm{R}^k)$, it is clear that $\mathrm{RM}(S)$ intersects C. Thus $\mathrm{RM}(S)$ satisfies RM_2.

We next show that $\mathrm{RM}(S)$ satisfies condition RM_1 as well. This is the content of the following two lemmas. Let $v(1), \ldots, v(\ell)$ be the set of distinguished values computed by the algorithm.

Lemma 16.8. *For* $1 \leq i \leq \ell$, *if* $\mathrm{RM}(S)_{\leq v(i)}$ *satisfies condition* RM_1 *for the set* $S_{\leq v(i)}$ *then,* $\mathrm{RM}(S)_{< v(i+1)}$ *satisfies condition* RM_1 *for the set* $S_{< v(i+1)}$.

Proof: Let C be a semi-algebraically connected component of $S_{<v(i+1)}$ and let Γ be a semi-algebraically connected component of $\mathrm{RM}(S) \cap C_{[v(i),v(i+1))}$. The set $\Gamma_{v(i)}$ is non-empty since there is no distinguished value in $(v(i), v(i+1))$. It is then clear that $\mathrm{RM}(S) \cap C_{\leq v(i)} \cup \Gamma$ is semi-algebraically connected. Since $\mathrm{RM}(S) \cap C_{\leq v(i)}$ is semi-algebraically connected, the conclusion follows. □

Lemma 16.9. *For* $1 \leq i \leq \ell$, *if* $\mathrm{RM}(S)_{<v(i)}$ *satisfies condition* RM_1 *for the set* $S_{<v(i)}$, *then* $\mathrm{RM}(S)_{\leq v(i)}$ *satisfies condition* RM_1 *for the set* $S_{\leq v(i)}$.

Proof: Let C be a semi-algebraically connected component of $S_{\leq v(i)}$. We prove that $\mathrm{RM}(S) \cap C$ is semi-algebraically connected.

Let B_1, ..., B_h be the semi-algebraically connected components of $S \cap C_{<v(i)}$. Then, by Lemma 16.3, $C = C_1 \cup C_2 \cdots \cup C_N$, where each C_i is either $\bar{B}_j$ or a semi-algebraically connected component of $\mathrm{Zer}(\mathcal{P}', \mathrm{R}^k) \cap S_{v(i)}$, for some $I \subset \{1, ..., s\}$, where $v(i)$ is an X_1-special value of $\mathrm{Zer}(\mathcal{P}', \mathrm{R}^k)$.

Let $\Gamma = \mathrm{RM}(S) \cap C$ and $\Gamma(i) = \mathrm{RM}(S) \cap C_i$ for $1 \leq i \leq N$. Then $\Gamma = \bigcup_i \Gamma(i)$.

First, we claim that each $\Gamma(j)$ is semi-algebraically connected. If C_j is a semi-algebraically connected component of $\mathrm{Zer}(\mathcal{P}') \cap S_{v(i)}$ for some $I \subset \{1, ..., s\}$ containing Q, then, since $v(i)$ is an X_1-special value for this algebraic set, $\Gamma(j)$ is semi-algebraically connected by Step 4 of the algorithm.

Else, by the hypothesis of the lemma, we know that $\Gamma(j)_{<v(i)}$ is semi-algebraically connected. Thus, $\Gamma(j)$ can have at most one semi-algebraically connected component whose intersection with $(\mathrm{R}^k)_{<v(i)}$ is non-empty, and all the other semi-algebraically connected components of $\Gamma(j)$ must lie in $\pi^{-1}(v(i))$. Hence each of these must contain a distinguished point. But, by Step 4 of the algorithm, the distinguished points get connected to $\Gamma(j)_{<v(i)}$. Thus, $\Gamma(j)$ can have only one semi-algebraically connected component.

Moreover, if $C_j \cap C_j' \neq \emptyset$, then $\Gamma(j)$ and $\Gamma(j')$ are connected in $\mathrm{RM}(S)$. This is so since, according to Lemma 16.4, $C_j \cap C_j'$ intersects an algebraic set which has $v(i)$ as an X_1-pseudo-critical value and thus contains a distinguished point which gets linked to both $\Gamma(j)$ and $\Gamma(j')$.

It follows that Γ is semi-algebraically connected. This proves the lemma. □

The proposition now follows by induction on i. □

Proof of correctness: Note that Step 2 b and Step 3 of the algorithm make it evident that $\mathrm{RM}(Q, \mathcal{P})$ satisfies condition URM_1.That it satisfies condition URM_2 follows from Proposition 16.7. □

Complexity analysis: When $\ell = 1$, it follows from the analysis of the algebraic case that the number of arithmetic operations is $s\, d^{O(k^2)}$. When $k = 1$, the number of arithmetic operations is $s\, d^{O(1)}$.

In Step 1 a) the total number of arithmetic operations in D is $s\binom{s}{\ell}\, d^{O(k)} = s^{\ell+1}\, d^{O(k)}$.

In Step 1 b, the total number of calls to Algorithm 15.2 (Curve Segments) is $\sum_{j=1}^{\ell-1} \binom{s}{j}$, and each call costs $d^{O(k)}$. Thus, the total cost of the calls to Algorithm 15.2 (Curve Segments) is bounded by $\sum_{j=1}^{\ell-1} \binom{s}{j}\, d^{O(k)}$ arithmetic operations in D.

In Step 1 c, the cost of each call to Algorithm 16.1 (Special Values) as well as the cost of computing the intersection of each curve segment with the special hyperplanes are bounded by $d^{O(k)}$, and hence the total cost of this step is bounded by $\sum_{j=1}^{\ell-1} \binom{s}{j}\, d^{O(k)}$

In Step 1d, the cost of computing the intersection of each curve segment computed with the zero sets of each polynomial in $\mathcal{P}$ is bounded by $s\, d^{O(k)}$, and the total cost of Step 1 d, for all I considered, is $s \sum_{j=1}^{\ell-1} \binom{s}{j}\, d^{O(k)}$

In Step 2, the cost is $\sum_{j=1}^{\ell-1} \binom{s}{j}\, 2^j\, d^{O(k)}$, since $\sum_{j=1}^{\ell-1} \binom{s}{j}\, 2^j$ is the number of pairs $(\mathcal{R}', \mathcal{R}'')$ considered.

Note that the combinatorial level in the recursive call is at most $\ell - \#(\mathcal{P}') - 1$ and the number of variables is $k - 1$.

We now count the recursive calls. For each j, $0 \leq j \leq \ell - 1$, we make $\sum_{j=1}^{\ell-1} \binom{s}{j}\, d^{O(k)}$ recursive calls to the algorithm with combinatorial level $\ell - j$ and ambient space dimension $k - 1$.

Let $T(s, d, \ell, k)$ denote the complexity of the algorithm with these parameters. Since at any depth of the recursion the cost of a single arithmetic operations is bounded by $d^{O(k^2)}$ arithmetic operations in D, we ignore the fact that the ring changes as we go down in the recursion. Thus, we have the following recurrence,

$$
\begin{aligned}
T(s,d,\ell,k) \;&\leq\; \sum_{j=1}^{\ell-1} \binom{s}{j}\, d^{O(k)}\, T(s-j,\, d,\, \ell-j,\, k-1) + s \sum_{j=0}^{\ell} \binom{s}{j} d^{O(k)}, \\
&\qquad \ell > 0, k > 1, \\
T(s,d,0,k) \;&=\; s\, d^{O(k^2)}, \quad k > 1, \\
T(s,d,\ell,1) \;&=\; s\, d^{O(1)}.
\end{aligned}
$$

This recurrence solves to $T(s, d, \ell, k) = s^{\ell+1}\, d^{O(k^2)}$.

In Step 4 the total cost of the calls to the Algorithm 16.3 (Linking Paths) and Algorithm 16.2 (Bounded Connecting) is bounded by $s^{\ell+1}\, d^{O(k)}$.

It follows immediately that the total cost is still bounded by $s^{\ell+1}\, d^{O(k)}$.

If $\mathrm{D} = \mathbb{Z}$, and the bitsizes of the coefficients of the polynomials are bounded by τ, then the bitsizes of the integers appearing in the intermediate computations and the output are bounded by $\tau d^{O(k^2)}$. □

16.3 Computing Connected Components of Sign Conditions

For complexity reasons, the formulas describing the semi-algebraically connected components of a given semi-algebraic set produced by our algorithm will not necessarily be written as disjunctions of sign conditions. This differs from some our previous algorithms (such as eliminating quantifiers, or describing the semi-algebraically connected components of algebraic sets).

Notation 16.10. Let $\phi(Y)$ be a quantifier free formula,

$$\mathcal{T}, \sigma, u = (\mathcal{T}_\ell, g_0, g_\ell, \ldots, g_k)$$

a parametrized real univariate triangular representation with parameters $Y = (Y_1, \ldots, Y_k)$. With $T = (T_1, \ldots, T_\ell)$, we denote by $\phi_u(Y)$ the formula

$$(\forall T)\left(\bigwedge_{1 \le i \le \ell}\left(\mathcal{T}_i(Y, T_1, \ldots, T_i) = 0 \wedge \bigwedge_{h \in \mathrm{Der}(\mathcal{T})} \mathrm{sign}(h(Y, T)) = \sigma(h)\right)\right) \Rightarrow \phi(u)$$

where $\phi(u)$ is obtained by replacing Y_j by $g_j(Y, T_1, \ldots, T_\ell)/g_0(Y, T_1, \ldots, T_\ell)$, $j \ge \ell$ and clearing denominators. □

Algorithm 16.5. **[Parametrized Bounded Connecting]**

- **Structure:** an ordered domain D contained in a real closed field R.
- **Input:**
  - a polynomial $Q \in \mathrm{D}[X_1, \ldots, X_k]$, such that $\mathrm{Zer}(Q, \mathrm{R}^k) \subset B(0, 1/c)$,
  - a finite set of polynomials $\mathcal{P} \subset \mathrm{D}[X_1, \ldots, X_k]$ in strong k-general position with respect to Q.
- **Output:**
  - a finite set of polynomials $\mathcal{A}$ containing $\mathcal{P}$,
  - a finite set Θ of $\mathcal{A}$-quantifier free formulas such that for every semi-algebraically connected component S of the realization of every weak sign condition on $\mathcal{P}$ on $\mathrm{Zer}(Q, \mathrm{R}^k)$, there exists a subset $\Theta(S) \subset \Theta$ such that

  $$S = \bigcup_{\theta \in \Theta(S)} \mathrm{Reali}(\theta, \mathrm{Zer}(Q, \mathrm{R}^k)),$$

  - for every $\theta \in \Theta$, a parametrized path $\Gamma(\theta) \subset \mathrm{R}^{2k}$ such that $\Gamma(\theta)_y$ is a semi-algebraic set of dimension at most one, that connects for every $y \in \mathrm{Reali}(\theta)$ the point y to some roadmap $\mathrm{RM}(\mathrm{Zer}(\mathcal{P}' \cup \{Q\}, \mathrm{R}^k))$ where $\mathcal{P}' \subset \mathcal{P}$, staying inside

  $$\mathrm{Reali}(\overline{\sigma}(y), \mathrm{Zer}(Q, \mathrm{R}^k)).$$

  Moreover, for every $y \in \mathrm{Reali}(\theta, \mathrm{Zer}(Q, \mathrm{R}^k))$, the description of $\Gamma(\theta)_y$ is fixed and the endpoint of $\Gamma(\theta)_y$, described by the real univariate representation $w(\theta)$ is independent of y.

- **Complexity:** $s^{\ell+1} d^{O(k^4)}$, where s is a bound on the number of elements of $\mathcal{P}$ and d is a bound on the degrees of Q and the elements of $\mathcal{P}$.
- **Procedure:**
  - Step 1: Fix an ordered tuple of indices I with elements in $\{1, \ldots, s\}$, such that $\#(I) \leq \ell$, and denote by $\mathcal{P}_I$ the set of polynomials $\{Q\} \cup \{P_i \in \mathcal{P} | i \in I\}$. If $\mathrm{Zer}(\mathcal{P}_I, \mathrm{R}^k) \neq \emptyset$, compute using Algorithm 15.12 (Parametrized Bounded Algebraic Connecting) a family of polynomials $\mathcal{A}(I) \subset \mathrm{D}[Y_1, \ldots, Y_k]$, and for every $\rho \in \mathrm{SIGN}(\mathcal{A}(I), \mathcal{P}_I)$ a semi-algebraic set $\Gamma(\rho) \subset \mathrm{R}^{2k}$ such that, for every $y \in \mathrm{Reali}(\rho)$, $\Gamma(\rho)_y$ connects the point y to a distinguished point of $\mathrm{RM}(\mathrm{Zer}(\mathcal{P}_I, \mathrm{R}^k))$, described by the real univariate representation $w(\rho)$. Moreover, for every $y \in \mathrm{Reali}(\rho)$, the description of $\Gamma(\rho)_y$ is fixed.
  - Step 2: Fix an ordered tuple of indices I with elements in $\{1, \ldots, s\}$, such that $\#(I) \leq \ell$, and $j \in \{1, \ldots, s\} \setminus I$. Compute a family of polynomials whose signs control the manner in which $\mathrm{Zer}(P_j, \mathrm{R}^k)$ intersects $\Gamma(\rho)$. More precisely, compute a family of polynomials $\mathcal{B}(\rho, j)$ containing $\mathcal{A}(I)$ and the subset $\Sigma(\rho, j)$ of elements ρ' of $\mathrm{SIGN}(\rho, \mathcal{B}(\rho, j))$ such that for every $y \in \mathrm{Reali}(\rho')$,
    - the intersection of $\Gamma(\rho)(y)$ with $\mathrm{Zer}(P_j, \mathrm{R}^k)$ is non-empty,
    - the Thom encodings describing the various points of intersection of $\Gamma(\rho)_y$ with $\mathrm{Zer}(P_j, \mathrm{R}^k)$ remain constant.

    In order to achieve this, we first use Algorithm 14.7 (Parametrized Triangular Thom Encoding) as follows. Each curve segment of $\Gamma(\rho)$ is described by:
    - a parametrized triangular Thom encoding $\mathcal{T}(Y, X_{<i}), \sigma$,
    - a parametrized univariate representation with parameters $(X_{\leq i})$,

    $$u = (f(Y, X_{\leq i}, T), g_0(Y, X_{\leq i}, T), g_{i+1}(Y, X_{\leq i}, T), \ldots, g_k(Y, X_{\leq i}, T)),$$

    - a sign condition on $\mathrm{Der}(f)$.

    For each such curve segment in $\Gamma(\rho)$, first compute

    $$\mathrm{RElim}_{\mathcal{T}}(P_{j,u}, f) \subset \mathrm{D}[Y, X_{\leq i}]$$

    using Algorithm 11.19. Then call Algorithm 14.7 (Parametrized Triangular Thom Encoding) with input $\mathcal{T} \cup \{P\}$ for each $P \in \mathrm{RElim}_{\mathcal{T}}(P_{j,u}, f)$.

    The output is:
    - a finite set $\mathcal{B}' \subset \mathrm{D}[Y]$,
    - for every $\rho' \in \mathrm{SIGN}(\mathcal{B}')$, a list of sign conditions on $\mathrm{Der}(\mathcal{T})$ specifying, for every $y \in \mathrm{Reali}(\rho')$, the list of triangular Thom encodings of the roots of $\mathcal{T}(y) \cup \{P(y)\}$.

    Let $\mathcal{B}(\rho, j)$ be the union of all the $\mathcal{B}'$ obtained above along with $\mathcal{A}(I)$. Now use Algorithm 15.7 (Parametrized Comparison of Roots) to the order the various points of intersections and compute $\Sigma(\rho, j)$.

– Step 3: Fixing an ordered tuple of indices I with elements in $\{1,...,s\}$, such that $\#(I) \leq \ell$, denote by $\Phi(I)$ the set of formulas

$$\mathcal{F}(\rho) := \left(Q = 0 \wedge \bigwedge_{i \in I} P_i(x) = 0 \wedge \bigwedge_{A \in \mathcal{A}(I)} \operatorname{sign}(A)(x) = \rho(A) \right)$$

for all $\rho \in \mathrm{SIGN}(\mathcal{A}(I))$. Similarly, fixing an ordered tuple of indices I with element in $\{1,...,s\}$, with $\#(I) \leq \ell - 1$, and $j \in \{1,...,s\} \setminus I$, denote by $\Phi(I, j)$ the set of formulas

$$\mathcal{F}(\rho) \wedge \bigwedge_{B \in \mathcal{B}(\rho, j)} \operatorname{sign}(B)(x) = \rho'(B),$$

for all $\rho \in \mathrm{SIGN}(\mathcal{A}(I))$ and $\rho' \in \Sigma(\rho, j)$.

For every $\phi \in \Phi(I, j)$ and every $y \in \mathrm{Reali}(\phi)$, the first point of intersection of $\Gamma(\phi)_y$ with $\mathrm{Zer}(P_j, \mathrm{R}^k)$, $F(\phi)(y)$, is described by a real parametrized univariate representation with parameters Y, denoted by $u(\phi)(Y), \tau(\phi)$.

Denote by $\gamma(\phi)_y$ the part of the semi-algebraic path $\Gamma(\phi)_y$, starting at y and ending at $F(\phi)(y)$, and by $\gamma(\phi) \subset \mathrm{R}^{2k}$ the union of $\{y\} \times \gamma(\phi)_y$ for $y \in \mathrm{Reali}(\phi)$.

Compose the functions $F(\phi)$ inductively, as follows. Fix an ordered tuple of indices I with elements in $\{1, ..., s\}$, such that $\#(I) \leq \ell$ and $\mathrm{Zer}(\mathcal{P}_I, \mathrm{R}^k) \neq \emptyset$ and initialize $\Psi(I) := \Phi(I)$ and for every $\psi \in \Phi(I)$ associated to ρ, $v(\psi)(Y) := Y$, $w(\psi) := w(\rho)$, $\Gamma(\psi) = \emptyset$.

Fix an ordered tuple of indices I with elements in $\{1, ..., s\}$, such that $\#(I) \leq \ell - 1$. We will compute for every J, $1 \leq \#(J) \leq \ell - \#(I)$ a finite set of quantifier free formulas $\Psi(I, J)$ and for every $\psi \in \Psi(I, J)$, a parametrized real univariate triangular representation $\mathcal{T}(\psi)$, $\sigma(\psi)$, as well as $v(\psi)$, $w(\psi)$, $\Gamma(\psi)$.

Let J be an ordered tuple of indices with elements in $\{1,...,s\}$ such that $1 \leq \#(J) \leq \ell - \#(I)$ and suppose that a finite set of quantifier free formulas $\Psi(I, J)$, as well as for every $\psi \in \Psi(I, J)$, parametrized real univariate triangular representation $\mathcal{T}(\psi), \sigma(\psi), v(\psi)$, $w(\psi)$,$\Gamma(\psi)$, have already been computed.

Let $\bar{J} = J \cdot j$, $j \in \{1, ..., s\} \setminus I \cdot J$, and define the set $\Psi(I, \bar{J})$ as follows.

For each $\psi \in \Psi(I, J)$, each $\phi_1 \in \Phi(I \cdot J, j)$, and each $\phi_2 \in \Phi(I \cdot J \cdot j)$, let $v = u(\phi_1)_{v(\psi)}$. Compute, using Algorithm 14.5 (Quantifier Elimination), a quantifier-free formula $\overline{\phi_{1,v(\psi)} \wedge \phi_{2,v}}$ equivalent to $\phi_{1,v(\psi)} \wedge \phi_{2,v}$. Include in $\Psi(I, \bar{J})$ all $\psi \wedge \overline{\phi_{1,v(\psi)} \wedge \phi_{2,v}}$ which are realizable using Algorithm 13.1 (Computing Realizable Sign Conditions), with input the family of polynomials appearing in $\psi \wedge \overline{\phi_{1,v(\psi)} \wedge \phi_{2,v}}$.

For every $\psi' = \psi \wedge \overline{\phi_{1,v(\psi)} \wedge \phi_{2,v}} \in \Psi(I, \bar{J})$, define $v(\psi')$ as $v = (\mathcal{T}_{i+1}, \bar{g}_0, \ldots, \bar{g}_k)$, and define a new triangular system $\mathcal{T}(\psi')$ obtained by appending $\mathcal{T}_{i+1}$ to $\mathcal{T}(\psi)$ and a new list of sign vectors $\sigma(\psi')$ by appending $\tau(\phi_1)$ to the list $\sigma(\psi)$. Finally, let $w(\psi') := w(\phi_2)$.

Define $\Gamma(\psi') = \Gamma(\psi) \cup \gamma(\phi_{1,v(\psi)})$.

- Step 4: For an ordered tuple of indices I with elements in $\{1, \ldots, s\}$ such that $\#(I) \leq \ell - 1$, and an ordered tuple of indices J such that $1 \leq \#(J) \leq \ell - \#(I)$ with elements in $\{1, \ldots, s\}$ and a formula $\psi \in \Psi(I, J)$ the semi-algebraic path $\Gamma(\psi) \cup \Gamma(\rho)(v(\psi))$, where ρ is the sign condition on $\mathcal{A}_{I \cdot J}$ satisfied at $v(\psi)$, may or may not be a valid connecting semi-algebraic path depending on whether any polynomials in $\mathcal{P} \setminus \mathcal{P}_{I \cdot J}$ vanish on any one of its segments.

  Compute the formula $\theta(\psi, j)$ expressing the conditions on Y ensuring that P_j does not vanish on $\Gamma(\psi) \cup \Gamma(\rho)(v(\psi))$, using Algorithms 14.7 (Parametrized Triangular Thom Encoding) and 15.7 (Parametrized Comparison of Roots) for all the real parametrized univariate representations describing $\Gamma(\psi) \cup \Gamma(\rho)(v(\psi))$.

  Define the set $\Theta(I, J)$ of formulas $\psi \wedge \bigwedge_{j \notin I \cdot J} \theta(\psi, j)$ with $\psi \in \Psi(I, J)$, and, for $\theta \in \Theta(I, J)$,
$$w(\theta) = w(\psi),\ \Gamma(\theta) = \Gamma(\psi) \cup \Gamma(\rho)(v(\psi))$$
  $w(\theta) = w(\psi)$, $\Gamma(\theta) = \Gamma(\psi) \cup \Gamma(\rho)(v(\psi))$, where ρ is the sign condition on $\mathcal{A}_{I \cdot J}$ satisfied at $v(\psi)$.

  Since formulas of $\Theta(I, J)$ are refinements of formulas in $\Psi(I, J)$, every $\theta \in \Theta(I, J)$ defines a subset $\Gamma(\theta)$ such that $\Gamma(\theta)(y)$ is a path connecting y to the point of the roadmap for $\mathcal{P}_{I \cdot J}$ inside $\mathrm{Reali}(\sigma(y), \mathrm{Zer}(Q, \mathrm{R}^k))$, described by the real univariate representation $w(\theta)$.

  Define Θ as the union for every ordered tuple of indices I with elements in $\{1, \ldots, s\}$ such that $\#(I) \leq \ell - 1$, and every ordered tuple of indices J with elements in $\{1, \ldots, s\} \setminus I$ such that $1 \leq \#(J) \leq \ell - \#(I)$ of $\Theta(I, J)$.

  Define $\mathcal{A} \subset \mathrm{D}[X_1, \ldots, X_k]$ to be the set of polynomials appearing in the formulas of Θ.

Proof of correctness : It is clear from the algorithm that each formula θ obtained in Step 4 has the property that every $y \in \mathrm{Reali}(\theta)$ gets connected to a unique distinguished point of some algebraic set $\mathrm{Zer}(\mathcal{P}_I)$ by $\Gamma(\theta)_y$ inside $\mathrm{Reali}(\overline{\sigma}(y), \mathrm{Zer}(Q, \mathrm{R}^k))$. Thus, each $\mathrm{Reali}(\theta)$ must be fully contained in some connected component of a realizable weak sign condition of $\mathcal{P}$. Moreover, clearly $\bigcup_{\theta \in \Theta} \mathrm{Reali}(\theta) = \mathrm{Zer}(Q, \mathrm{R}^k)$. It is easy to see that $\Gamma(\theta)$ is a parametrized path, by the correctness of Algorithm 15.12 (Parametrized Bounded Algebraic Connecting). □

Complexity analysis: Using the complexity analysis of Algorithm 15.12 the complexity of Step 1 is bounded by $\sum_{i=1}^{\ell}\binom{s}{i} d^{O(k^3)}$. The number and degrees of the polynomials in the various $\mathcal{A}(I)$ are bounded by $d^{O(k^2)}$. Note that the number of elements of $\Phi(I)$ coincides with the number of non-empty sign conditions on $\mathcal{A}(I)$ and is bounded by $d^{O(k^3)}$.

Similarly, using the complexity analysis of Algorithms 14.7 and 15.7, the complexity of Step 2 is bounded by $\sum_{i=1}^{\ell-1}(s-i)\binom{s}{i} d^{O(k^3)}$. The number and degrees of the polynomials in the various $\mathcal{B}(\rho, j)$ are bounded by $d^{O(k^2)}$. Note that the number of elements of $\Phi(I,\, j)$ coincides with the number of non-empty sign conditions on $\mathcal{B}(I, j)$ and is bounded by $d^{O(k^3)}$.

In Step 3 the complexity for an ordered tuple I of indices of length p is $\sum_{i=1}^{\ell-p}\binom{s}{i} d^{O(k^4)}$ since there are $\sum_{i=1}^{\ell-p}\binom{s}{i}$ choices for J. The degrees of the polynomials appearing in the formulas of $\Psi(I, J)$ are bounded by $d^{O(k^3)}$. Thus, the total complexity of Steps 3 is bounded by $s^{\ell}\, d^{O(k^4)}$. The number of elements of Θ is $s^{\ell}\, d^{O(k^3)}$. Moreover, for every $\theta \in \Theta$, since $w(\theta)$'s is a distinguished point of the roadmap of some $\mathrm{Zer}(P_I, \mathrm{R}^k)$, the triangular system defining $w(\theta)$ has polynomials of degree at most $d^{O(k)}$.

Finally, the complexity of Step 4 is bounded by $s^{\ell+1}\, d^{O(k^4)}$.

The complexity of the algorithm is $s^{\ell+1} d^{O(k^4)}$. The number of polynomials in the family $\mathcal{A}$ is also $s^{\ell+1}\, d^{O(k^4)}$.

If $\mathrm{D} = \mathbb{Z}$, and the bitsizes of the coefficients of the polynomials are bounded by τ, then the bitsizes of the integers appearing in the intermediate computations and the output are bounded by $\tau d^{O(k^4)}$. □

Algorithm 16.6. **[Basic Connected Components]**

- **Structure:** an ordered domain D contained in a real closed field R.
- **Input:** a finite set $\mathcal{P} \subset \mathrm{D}[X_1, \ldots, X_k]$ of polynomials of degree at most d.
- **Output:** quantifier free formulas whose realizations are the semi-algebraically connected components of $\mathrm{Reali}(\sigma,\, \mathrm{R}^k)$, for the realizable sign conditions $\sigma \in \{0, 1, -1\}^{\mathcal{P}}$.
- **Complexity:** $s^{k+1} d^{O(k^4)}$, where s is a bound on the number of elements of $\mathcal{P}$ and d is a bound on the degrees of Q and the elements of $\mathcal{P}$.
- **Procedure:**
  - Take $Q = \varepsilon^2(X_1^2 + \cdots + X_k^2 + X_{k+1}^2) - 1$.
  - Replace the set $\mathcal{P}$ by the family $\mathcal{P}^\star$ defined by

$$\begin{aligned} P_i^\star &= \{(1-\delta)\, P_i + \delta\, H_k(d', i), (1-\delta)\, P_i - \delta\, H_k(d', i),\\ &\quad (1-\delta)\, P_i + \delta\, \gamma\, H_k(d', i), (1-\delta)\, P_i - \delta\, \gamma\, H_k(d', i)\}\\ \mathcal{P}^\star &= \{P_1^\star, \ldots, P_s^\star\} \end{aligned}$$

for $0 \le i \le s$, where $H_k(d', i) = (1 + \sum_{j=1}^{k} i^j X_j^{d'})$ and $d' > d$.

- For every non-empty sign condition σ on $\mathcal{P}$

$$\begin{array}{ll} P_i=0, & i\in I\subset\{1,\ldots,s\} \\ P_i>0, & i\in J\subset\{1,\ldots,s\}\setminus I \\ P_i<0, & i\in\{1,\ldots,s\}\setminus(I\cup J), \end{array}$$

  let $\sigma^\star$ be the weak sign condition on Q and $\mathcal{P}^\star$ defined by

$$Q=0.$$

$$-\gamma\,\delta\,H_k(d',i)\leq(1-\delta)\,P_i\leq\gamma\,\delta\,H_k(d',i),\ i\in I,$$

$$(1-\delta)\,P_i\geq\delta\,H_k(d',i),\ i\in J$$

$$(1-\delta)\,P_i\leq-\,\delta\,H_k(d',i),\ i\in\{1,\ldots,s\}\setminus(I\cup J).$$

  Apply Algorithm 16.5 (Parametrized Bounded Connecting) with input Q and $\mathcal{P}^\star$ and output Θ. Compute the set Θ_σ of $\theta\in\Theta$ such that $w(\theta)$ belongs to the realization of $\sigma^\star$. Using Algorithm 16.4 (Uniform Roadmap) with input Q, $\mathcal{P}^\star$ and the $w(\theta)$, $\theta\in\Theta_\sigma$, partition $\{w(\theta)\mid\theta\in\Theta_\sigma\}$ into subsets $W_1,\ldots,W_r$ such that all points of W_i belong to the same semi-algebraically connected component of the realization of $\sigma^\star$. Compute $\Phi_1,\ldots,\Phi_r$ with $\Phi_i=\bigvee_{\{\theta\in\Theta_\sigma\mid w(\theta)\in W_i\}}\theta$.
- For every semi-algebraically connected component C of the realization of σ there exists a semi-algebraically connected component C' of the realization of $\sigma^\star$ such that $\pi(C')\cap\mathrm{R}^k=C$. Consider the formula $\Phi_i(X_1,\ldots,X_k,X_{k+1})$ describing C' and denote by $\Psi_i(X_1,\ldots,X_k)$ the formula obtained by replacing each atom $F(X_1,\ldots,X_{k+1})$ of Φ_i by the quantifier free formula equivalent to $(\exists X_{k+1})\,X_{k+1}<0\wedge F(X_1,\ldots,X_{k+1})$ using Algorithm 14.5 (Quantifier Elimination). The formula Ψ_i describes $\pi(C')$. Then $\mathrm{Remo}_{\varepsilon,\delta,\gamma}(\Psi_i(Y))$ (Notation 14.6) defines C.

Proof of correctness: According to Proposition 13.7, every semi-algebraically connected component C of every strict sign condition of the original family $\mathcal{P}$ corresponds to a semi-algebraically connected component C' of a weak sign condition on $\mathcal{P}^\star$. Moreover, $C=\pi(C')\cap\mathrm{R}^k$. Now use Proposition 14.7. □

Complexity analysis: The family $\mathcal{P}^\star$ has combinatorial level k by Proposition 13.6. Using the complexity analysis of Algorithm 16.5 (Parametrized Bounded Connecting), the complexity of computing Θ is $s^{k+1}d^{O(k^4)}$. Applying Algorithm 16.4 costs $s^{k+1}d^{O(k^2)}$, and the points $w(\theta),\theta\in\Theta$ are distinguished points of this uniform roadmap. For every atom, the quantifier elimination performed costs $d^{O(k^4)}$, since there is one variable to eliminate, k free variables and one polynomial of degree $d^{O(k^3)}$, according to the complexity of Algorithm 14.5 (Quantifier Elimination). The total number of the atoms to consider is $s^kd^{O(k^4)}$.

So the total complexity is $s^{k+1}d^{O(k^4)}$. The degrees of the polynomials that appear in the output are bounded by $d^{O(k^3)}$.

If $\mathrm{D} = \mathbb{Z}$, and the bitsizes of the coefficients of the polynomials are bounded by τ, then the bitsizes of the integers appearing in the intermediate computations and the output are bounded by $\tau d^{O(k^4)}$. □

Theorem 16.11. *Let $\mathcal{P} = \{P_1, \ldots, P_s\} \subset \mathrm{D}[X_1, \ldots, X_k]$ with $\deg(P_i) \leq d$, for $1 \leq i \leq s$. There exists an algorithm that outputs quantifier-free semi-algebraic descriptions of all the semi-algebraically connected components of every realizable sign condition of the family $\mathcal{P}$. The complexity of the algorithm is bounded by $s^{k+1}d^{O(k^4)}$. The degrees of the polynomials that appear in the output are bounded by $d^{O(k^3)}$. Moreover, if the input polynomials have integer coefficients whose bitsize is bounded by τ the bitsize of coefficients output is $\tau d^{O(k^3)}$.*

16.4 Computing Connected Components of a Semi-algebraic Set

We first construct data for adjacencies for $\mathcal{P}$ on $\mathrm{Zer}(Q, \mathrm{R}^k)$, ensuring that if the union of two semi-algebraically connected components of two different sign conditions for $\mathcal{P}$ on $\mathrm{Zer}(Q, \mathrm{R}^k)$ is semi-algebraically connected, a path starting in a sign condition and ending in the other is constructed.

A set N of **data for adjacencies** for $\mathcal{P}$ on $\mathrm{Zer}(Q, \mathrm{R}^k)$ is a set of triples (p, q, γ), where $p, q \in \mathrm{Zer}(Q, \mathrm{R}^k)$, and γ is semi-algebraic path joining p to q inside $\mathrm{Zer}(Q, \mathrm{R}^k)$, such that for any two semi-algebraically connected components, C and D of $\mathrm{Reali}(\sigma, \mathrm{Zer}(Q, \mathrm{R}^k))$ and $\mathrm{Reali}(\tau, \mathrm{Zer}(Q, \mathrm{R}^k))$ where $\sigma, \tau \in \{-0, 1, -1\}^{\mathcal{P}}$, with $\bar{C} \cap D \neq \emptyset$, there exists $(p, q, \gamma) \in N$, such that $q \in D$ and $\gamma \setminus \{q\} \in C$.

Thus, if C and D are two semi-algebraically connected components of two distinct sign conditions whose union is semi-algebraically connected then there exists $(p, q, \gamma) \in N$ such that γ connects the point $p \in C$ with the point $q \in D$ through a semi-algebraic path lying in $C \cup D$.

We first describe the algorithm constructing data for adjacencies and then prove its correctness.

Algorithm 16.7. **[Data for Adjacencies]**

- **Structure:** an ordered domain D contained in a real closed field R.
- **Input:** a polynomial $Q \in \mathrm{D}[X_1, \ldots, X_k]$ such that $\mathrm{Zer}(Q, \mathrm{R}^k) \subset B(0, 1/c)$, and $\mathrm{Zer}(Q, \mathrm{R}^k)$ is of real dimension k', a finite set $\mathcal{P} \subset \mathrm{D}[X_1, \ldots, X_k]$.
- **Output:** a set N of data for adjacencies for $\mathcal{P}$ on $\mathrm{Zer}(Q, \mathrm{R}^k)$, described by real univariate representations and parametrized real univariate representations.

- **Complexity:** $s^{k'+1} d^{O(k)}$ where s is a bound on the number of elements of $\mathcal{P}$ and d us a bound on the degrees of Q and the elements of $\mathcal{P}$.
- **Procedure:**
  - Introduce a new variable β and define $\mathcal{P}' = \{\{P, P+\beta, P-\beta\}, P \in \mathcal{P}\}$.
  - Call Algorithm 13.3 (Sampling on an Algebraic Set) with input Q and $\mathcal{P}'$ and structure $\mathrm{D}[\beta] \subset \mathrm{R}\langle\beta\rangle$ to obtain a set of real univariate representations.
  - For each associated point $p(\beta)$, compute $q = \lim_\beta(p(\beta))$, using Algorithm 12.14 (Limit of Bounded Points). The point $p(\beta)$ is represented as a real k-univariate representation (u, σ) with

    $$u = (f(\beta, T), g_0(\beta, T), \ldots, g_k(\beta, T)).$$

    Replacing β in u by a small enough $t_0 \in \mathrm{R}$ using Algorithm 11.20 (Removal of Infinitesimals). Call Algorithm 11.20 (Removal of Infinitesimals) with input the polynomial f as well as the family of polynomials $\{P_u | P \in \mathcal{P}\}$ (see Notation 13.8) to obtain $t_0 \in \mathrm{R}$ replacing β. Letting t vary over the interval $[0, t_0]$ gives a semi-algebraic path γ joining $p(t_0)$ to q. Include the triple $(q, p(t_0), \gamma)$ in the set N.

The proof of correctness uses the following lemma.

Lemma 16.12. *Let* $Q \in \mathrm{R}[X_1, \ldots, X_k]$ *with* $\mathrm{Zer}(Q, \mathrm{R}^k) \subset B(0, 1/c)$. *Let* $\mathcal{P} \subset \mathrm{R}[X_1, \ldots, X_k]$ *be a finite set of polynomials and*

$$\mathcal{P}' = \{\{P, P+\beta, P-\beta\}, P \in \mathcal{P}\}.$$

Suppose that σ *and* τ *are distinct realizable sign conditions on* $\mathcal{P}$ *and that* C *and* D *are two semi-algebraically connected components of* $\mathrm{Reali}(\sigma, \mathrm{Zer}(Q, \mathrm{R}^k))$, *and* $\mathrm{Reali}(\tau, \mathrm{Zer}(Q, \mathrm{R}^k))$ *respectively such that* $\bar{C} \cap D \neq \emptyset$. *Then there is a semi-algebraically connected component* C', *of a realizable sign condition* σ' *of* $\mathcal{P}'$ *on* $\mathrm{Zer}(Q, \mathrm{R}^k)$ *such that* $C' \subset \mathrm{Ext}(C, \mathrm{R}\langle\beta\rangle)$ *and* $\lim_\beta(C') \subset D$.

Proof: Let $\mathcal{P} = \{P_1, \ldots, P_s\}$. Suppose without loss of generality that σ is

$$P_1 = \cdots = P_\ell = 0, P_{\ell+1} > 0, \ldots, P_s > 0.$$

After a possible re-ordering of the indices, τ is

$$P_1 = \cdots = P_m = 0, P_{m+1} > 0, \ldots, P_s > 0$$

with $m > \ell$. This is clear since a point $p \in \bar{C} \cap D$ must satisfy

$$P_1 = \cdots = P_\ell = 0, P_{\ell+1} \geq 0, \ldots, P_s \geq 0.$$

Consider the set defined by the formula σ'

$$\begin{gathered} P_1 = \cdots = P_\ell = 0, \\ 0 \leq P_{\ell+1} \leq \beta, \ldots, 0 \leq P_m \leq \beta, \\ P_{m+1} > 0, \ldots, P_s > 0. \end{gathered}$$

Let us prove first that the realization of σ' is non-empty. Let $x \in \bar{C} \cap D$. According to Theorem 3.19 (Curve Selection Lemma), there is a semi-algebraic path γ such that $\gamma(0) = x$, $\gamma((0, 1]) \subset C$. Since at $\gamma(1)$ we have $P_{\ell+1} > 0, \ldots, P_m > 0$ and at $\gamma(0)$ we have $P_{\ell+1} = \cdots = P_m = 0$, there exists $t \in \mathrm{R}\langle\beta\rangle$ such that $0 < P_{\ell+1} \leq \beta, \ldots, 0 < P_m \leq \beta$ on $\gamma((0, t])$ (use Exercise 3.1 part 3).

It is clear that the realization of σ' is contained in the extension of σ to $\mathrm{R}\langle\beta\rangle$. Consider the semi-algebraically connected component C' of the realization of σ' that contains $y = \gamma(1)$. It is clear that $C' \subset \mathrm{Ext}(C, \mathrm{R}\langle\beta\rangle)$. Moreover, $\lim_\beta(C')$ satisfies sign condition τ and contains $x \in D$.

Since, $\lim_\beta$ maps semi-algebraically connected sets to semi-algebraically connected sets, we see that $\lim_\beta(C) \subset D$. □

Proof of correctness: We need to show that the set of triples computed above is set of data for adjacencies for $\mathcal{P}$. This is an immediate consequence of Lemma 16.12.

It is clear that if $p(\beta)$ is a point in C' $q = \lim_\beta(p)$, and γ is the semi-algebraic path obtained by replacing β by a small enough $t > t_0$ in $p(\varepsilon)$ then $p(t_0) \in C$, $q \in D$ and γ is a semi-algebraic path joining $p(t_0)$ and q contained in C except at the endpoint q. □

Complexity analysis: The complexity of the whole computation is $s\sum_{j \leq k'} 4^j \binom{s}{j} d^{O(k)}$ in $\mathrm{D}[\beta]$, using the complexity analyses of Algorithm 13.3 (Sampling on a Bounded Algebraic Set) (with the extra remark that P, $P - \beta$ and $P + \beta$ have no common zeroes) and Algorithm 11.20 (Removal of Infinitesimals). Since the degree in β of the intermediate computations is also bounded by $d^{O(k)}$, the complexity in D is finally $s\sum_{j \leq k'} 4^j \binom{s}{j} d^{O(k)} = s^{k'+1} d^{O(k)}$.

If $\mathrm{D} = \mathbb{Z}$, and the bitsizes of the coefficients of the polynomials are bounded by τ, then the bitsizes of the integers appearing in the intermediate computations and the output are bounded by $\tau d^{O(k)}$. □

We can now describf the semi-algebraically connected components of a semi-algebraic set.

Algorithm 16.8. **[Connected Components of a Semi-algebraic Set]**

- **Structure:** an ordered domain D contained in a real closed field R.
- **Input:** a finite set $\mathcal{P} \subset \mathrm{D}[X_1, \ldots, X_k]$, a $\mathcal{P}$-semi-algebraic set Sa.
- **Output:** a description of the semi-algebraically connected components of S.
- **Complexity:** $s^{k+1} d^{O(k^4)}$ where s is a bound on the number of polynomials in $\mathcal{P}$ and d is a bound on their degree.

- **Procedure:** Using Algorithms 16.7 (Data for Adjacencies) and 16.6 (Basic Connected Components), compute the equivalence classes of the transitive closure of the adjacency relation between semi-algebraically connected components of the realizations of realizable sign condition, and take the union of the corresponding equivalence classes.

Proof of correctness: Follows from the correctness of Algorithms 16.7 (Data for Adjacencies) and 16.6 (Basic Connected Components). □

Complexity analysis: The complexity of the algorithm is bounded by $s^{k+1}d^{O(k^4)}$ using the preceding results on the complexity of Algorithm 16.7 (Data for Adjacencies) and Algorithm 16.6 (Basic Connected Components). The degrees of the polynomials that appear in the output are bounded by $d^{O(k^3)}$.

If $\mathrm{D} = \mathbb{Z}$, and the bitsizes of the coefficients of the polynomials are bounded by τ, then the bitsizes of the integers appearing in the intermediate computations and the output are bounded by $\tau d^{O(k^4)}$. □

We have proved the following theorem:

Theorem 16.13. *Let* $\mathcal{P} = \{P_1, \ldots, P_s\} \subset \mathrm{D}[X_1, \ldots, X_k]$ *with* $\deg(P_i) \leq d$, *for* $1 \leq i \leq s$ *and a semi-algebraic set* S *defined by a* $\mathcal{P}$ *quantifier-free formula. There exists an algorithm that outputs quantifier-free semi-algebraic descriptions of all the semi-algebraically connected components of* S. *The complexity of the algorithm is bounded by* $s^{k+1} d^{O(k^4)}$. *The degrees of the polynomials that appear in the output are bounded by* $d^{O(k^3)}$. *Moreover, if the input polynomials have integer coefficients whose bitsize is bounded by* τ *the bitsize of coefficients output is* $\tau d^{O(k^3)}$.

16.5 Roadmap Algorithm

Our aim in this section is to construct a roadmap of a semi-algebraic defined by a $\mathcal{P}$-quantifier free formula on an algebraic set $\mathrm{Zer}(Q, \mathrm{R}^k)$ of dimension k'. We use the construction of approximating varieties described in Section 13.3 in order to achieve better complexity for our algorithm.

Let S be an arbitrary semi-algebraic set defined by a finite set of polynomials $\mathcal{P}$ which is contained in a bounded algebraic set $\mathrm{Zer}(Q, \mathrm{R}^k)$ of real dimension k'.

We first assume that $\mathrm{Zer}(Q, \mathrm{R}^k)$ is bounded. The idea is to construct uniform roadmaps for a perturbed finite set of polynomials which are in general position over approximating varieties (see Chapter 13 page 523) which are close to $\mathrm{Zer}(Q, \mathrm{R}^k)$ and of dimension k'.

We then take the limits of the curves obtained when the parameter of deformation tends to 0, i.e.the images of the curves so constructed under a lim map.

We first describe this limit process. The idea is to modify Algorithm 15.2 (Curve Segments) so that the limit of the curve segments when the parameter of deformation tends to 0 is also output.

Algorithm 16.9. **[Modified Curve Segments]**

- **Structure:** an ordered domain D contained in a real closed field R.
- **Input:**
  - a polynomial $Q \in \mathrm{D}[X_1, X_2, \ldots, X_k]$, such that $\mathrm{Zer}(Q, \mathrm{R}^k) \subset B(0, 1/c)$,
  - $\varepsilon = (\varepsilon_1, \ldots, \varepsilon_m)$
  - a polynomial $\bar{Q} \in \mathrm{D}[\varepsilon, X_1, X_2, \ldots, X_k]$, for which
  $$\lim_{\varepsilon} (\mathrm{Zer}(\bar{Q}, \mathrm{R}\langle\varepsilon\rangle^k)) \subset \mathrm{Zer}(Q, \mathrm{R}^k)$$
  - a triangular Thom encoding $\mathcal{T}, \sigma$ specifying $z \in \mathrm{R}\langle\varepsilon\rangle^{i-1}$ with coefficients in $\mathrm{D}[\varepsilon]$,
  - a triangular Thom encoding $\mathcal{T}', \sigma'$ specifying $\lim_{\varepsilon} (z) \in \mathrm{R}^{i-1}$, with coefficients in D
  - a set of at most m points, $\mathcal{N} \subset \mathrm{Zer}(\bar{Q}, \mathrm{R}\langle\varepsilon\rangle^k)$, where each point of $\mathcal{N}$ is defined by a real k-univariate representation u, σ with coefficients in $\mathrm{D}[\varepsilon]$, above $\mathcal{T}, \sigma$.
- **Output:**
  - An ordered list of Thom encodings A_1, α_1, ..., A_ℓ, α_ℓ above $\mathcal{T}$, σ specifying points $(z, v_1), \ldots, (z, v_\ell)$ with $v_1 < \cdots < v_\ell$.
  - An ordered list of Thom encodings B_1, β_1, ..., B_ℓ, β_ℓ above $\mathcal{T}'$, σ' specifying the image under $\lim_\varepsilon$ of these distinguished values:
  - For every $j = 1, \ldots, \ell$,
    - a finite set $\mathcal{D}_j$ of real univariate representation above $\mathcal{T}, A_j, \sigma, \alpha_j$. The associated points are called distinguished points.
    - a finite set $\mathcal{D}'_j$ of real univariate representation above $\mathcal{T}', B_j, \sigma', \beta_j$. The associated points are the image under $\lim_\varepsilon$ of the distinguished points of $\mathcal{D}_j$.
    - a finite set $\mathcal{C}_j$ of curve segment representations above
    $$\mathcal{T},\ \sigma,\ A_j,\ \alpha_j,\ A_{j+1},\ \alpha_{j+1}.$$
    The associated curve segments are called distinguished curves.
    - a finite set $\mathcal{C}'_j$ of curve segment representations
    $$\mathcal{T}',\ \sigma',\ B_j,\ \beta_j,\ B_{j+1},\ \beta_{j+1}.$$
    with associated curve segments the image under $\lim_\varepsilon$ of the curve segments in $\mathcal{C}_j$.
    - a list of pairs of elements of $\mathcal{C}_j$ and $\mathcal{D}_j$ (resp. $\mathcal{C}_{j+1}$ and $\mathcal{D}_j$) describing the adjacency relations between distinguished curves and distinguished points.

The distinguished curves and points are contained in $\mathrm{Zer}(\bar{Q}, \mathrm{R}\langle\varepsilon\rangle^k)_z$. Among the distinguished values are the first coordinates of the points in $\mathcal{N}$ as well as the pseudo-critical values of $\mathrm{Zer}(\bar{Q}, \mathrm{R}\langle\varepsilon\rangle^k)_z$. The sets of distinguished values, distinguished curves and distinguished points satisfy the following properties.

  - CS_1: For every $v \in \mathrm{R}\langle\varepsilon\rangle$ the set of distinguished curve and distinguished points output intersect every semi-algebraically connected component of $\mathrm{Zer}(\bar{Q}, \mathrm{R}\langle\varepsilon\rangle^k)_{z,v}$.
  - CS_2: For each distinguished curve output over an interval with endpoint a given distinguished value, there exists a distinguished point over this distinguished value which belongs to the closure of the curve segment.

- **Complexity:** $d^{O(ik)}$, where d is a bound on the degree of Q and $\bar{Q}$, and $O(d)^k$ is a bound on the degree of the polynomials in $\mathcal{T}$, on the degree of the univariate representations in $\mathcal{N}$ and on the number of these univariate representations.
- **Procedure:**
  - Step 1: Perform Algorithm 12.10 (Parametrized Multiplication Table) with input $\overline{\mathrm{Cr}}(\bar{Q}^2, \zeta)$, (using Notation 12.46) and parameter $X_{\leq i}$. Perform Algorithm 12.15 (Parametrized Limit of Bounded Points) and output $\mathcal{U}$.
  - Step 2: For every $u, \tau \in \mathcal{N}$, compute $\mathrm{proj}_i(u)$, $\mathrm{proj}_i(\tau)$ using Algorithm 15.1 (Projection), add to $\mathcal{D}$ the polynomial $\mathrm{proj}_i(u)$.
  - Step 3: Compute the Thom encodings of the zeroes of $\mathcal{T}, A$, $A \in \mathcal{D}$ above $\mathcal{T}, \sigma$ using Algorithms 12.20 (Triangular Thom Encoding) and output their ordered list A_1, α_1, ..., A_ℓ, α_ℓ and the corresponding ordered list $v_1 < \cdots < v_\ell$ of distinguished values using 12.21.

    Compute the Thom encoding of $\lim_\varepsilon (v_1) \leq \ldots \leq \lim_\varepsilon (v_\ell)$.
  - Step 4: For every $j = 1, \ldots, \ell$ and every $(f, g_0, g_i, \ldots, g_k), \tau \in \mathcal{N}$ such that $\mathrm{proj}_i(\tau) = \alpha_j$, append $(f, g_0, g_{i+1}, \ldots, g_k), \tau$ to $\mathcal{D}_j$.
  - Step 5: For every $j = 1, \ldots, \ell$ output a finite set of univariate representations $\mathcal{D}_j$ such that the set of associated points contains the set of X_i-pseudo-critical points of $\mathrm{Zer}(\bar{Q}, \mathrm{R}\langle\varepsilon\rangle^k)_{v_i}$ as well as a set of univariate representations $\mathcal{D}'_j$ with associated points the $\lim_\varepsilon$ image of the points associated to $\mathcal{D}_j$.

    For every $j = 1, \ldots, \ell$ and every $u = (f, g_0, g_i, \ldots, g_k) \in \mathcal{U}$, compute the Thom encodings τ of the roots of $\mathcal{T}$, f such that $\mathrm{proj}_i(\tau) = \alpha_j$, using Algorithm 12.20 (Triangular Thom Encoding) and append all pairs $(f, g_0, g_{i+1}, \ldots, g_k)$, τ to $\mathcal{D}_j$ when the corresponding associated point belongs to $\mathrm{Zer}(\bar{Q}, \mathrm{R}\langle\varepsilon\rangle^k)_z$.

    For every $u \in \mathcal{D}_j$, such that $o(f) = o(u)$, put

$$\hat{u}(X_{\leq i}, T) = \lim_{\varepsilon} (\varepsilon^{-o(f)} u(\varepsilon, X_{\leq i}, T)).$$

    with coefficients in $\mathrm{D}[X_{\leq i}, T]$ in $\mathcal{D}'_j$.

- Step 6: Output on each open interval (v_j, v_{j+1}) a finite set of curve segments $\mathcal{C}_j$ such that for every $v \in (v_j, v_{j+1})$ the set of associated points contains the set of X_i-pseudo-critical points of $\mathrm{Zer}(Q, \mathrm{R}^k)_v$.

  For every $j = 1, \ldots, \ell - 1$ and every $u = (f, g_0, g_{i+1}, \ldots, g_k) \in \mathcal{U}$, compute the Thom encodings ρ of the roots of $f(z, v, T)$ over (v_j, v_{j+1}) using Algorithm 12.22 (Triangular Intermediate Points) and Algorithm 12.20 (Triangular Thom Encoding). Append pairs u, ρ to $\mathcal{C}_j$ when the corresponding associated curve is included in $\mathrm{Zer}(\bar{Q}, \mathrm{R}\langle\varepsilon\rangle^k)_z$.

  For every $u \in \mathcal{C}_j$, such that $o(f) = o(u)$, put

$$\bar{u}\,(X_{\le i}, T) = \lim_{\varepsilon} (\varepsilon^{-o(f)} u(\varepsilon, X_{\le i}, T))$$

  with coefficients in $\mathrm{D}[X_{\le i}, T]$ in $\mathcal{C}'_j$.
- Step 7: Determine adjacencies between curve segments and points. For every point of $\mathcal{D}_j$ specified by

$$v' = (p, q_0, q_{i+1}, \ldots, q_k), \tau', \{p, q_0, q_{i+1}, \ldots, q_k\} \subset \mathrm{D}[X_{\le i}][T]$$

  and every curve segment representation of $\mathcal{C}_j$ specified by

$$v = (f, g_0, g_{i+1}, \ldots, g_k), \tau, \{f, g_0, g_{i+1}, \ldots, g_k\} \subset \mathrm{D}[X_{\le i}][T],$$

  decide whether the corresponding point t is adjacent to the corresponding curve segment as follows: compute the first ν such that $\partial^\nu g_0 / \partial X_i^\nu (v_j, t)$ is not zero and decide whether for every $\ell = i + 1, \ldots, k$

$$\frac{\partial^\nu g_\ell}{\partial X_i^\nu}(v_j, t) q_0(t) - \frac{\partial^\nu g_0}{\partial X_i^\nu}(v_j, t) q_\ell(t)$$

  is zero. This is done using Algorithm 12.19 (Triangular Sign Determination).

  Repeat the same process for every element of $\mathcal{D}_{j+1}$ and every curve segment representation of $\mathcal{C}_j$.

Proof of correctness: It follows from Proposition 12.42, the correctness of Algorithm 12.10 (Parametrized Multiplication Table), Algorithm 12.15 (Parametrized Limit of Bounded Points), Algorithm 11.19 (Restricted Elimination), Algorithm 15.1 (Projection), Algorithm 12.22 (Triangular Intermediate Points), Algorithm 12.20 (Triangular Thom Encoding), Algorithm 12.21 (Triangular Comparison of Roots) and Algorithm 12.19 (Triangular Sign Determination). □

Complexity analysis: Step 1: This step requires $d^{O(i(k-i))}$ arithmetic operations in D, using the complexity analysis of Algorithm 12.10 (Parametrized Multiplication Table), Algorithm 12.15 (Parametrized Limit of Bounded Points), Algorithm 11.19 (Restricted Elimination). There are $d^{O(k-i)}$ parametrized univariate representations computed in this step and each polynomial in these representations has degree $O(d)^{k-i}$.

Step 2: This step requires $d^{O(ik)}$ arithmetic operations in D, using the complexity analysis of Algorithm 15.1 (Projection).

Step 3: This step requires $d^{O(ik)}$ arithmetic operations in D, using the complexity analysis of Algorithm 12.20 (Triangular Thom Encoding).

Step 4: This step requires $d^{O(ik)}$ arithmetic operations in D, using the complexity analysis of Algorithm 12.19 (Triangular Sign Determination).

Step 5: This step requires $d^{O(ik)}$ arithmetic operations in D, using the complexity analysis of Algorithm 12.20 (Triangular Thom Encoding).

Step 6: This step requires $d^{O(ik)}$ arithmetic operations in D, using the complexity analysis of Algorithm 12.22 (Triangular Intermediate Points), Algorithm 12.20 (Triangular Thom Encoding).

Step 7: This step requires $d^{O(ik)}$ arithmetic operations in D, using the complexity analysis of Algorithm 12.19 (Triangular Sign Determination).

Thus, the complexity is $d^{O(ik)}$. The number of distinguished values is bounded by $d^{O(k)}$.

If $\mathrm{D} = \mathbb{Z}$, and the bitsizes of the coefficients of the polynomials are bounded by τ, then the bitsizes of the integers appearing in the intermediate computations and the output are bounded by $\tau d^{O(ik)}$. □

We describe the construction of a set, L, such L meets every semi-algebraically connected component of every realizable weak sign condition of $\mathcal{P}$ on $\lim_\eta(Z_j) \cap \lim_\eta(Z_\ell)$, where Z_j and Z_ℓ are the approximating varieties defined in Notation 13.30, for every $0 \le j \le k'(k-k')$ and $0 \le \ell \le k'(k-k')$,

Algorithm 16.10. **[Linking Points]**

- **Structure:** an ordered domain D contained in a real closed field R.
- **Input:**
  - $Q \in \mathrm{D}[X_1, \ldots, X_k]$ such that $\mathrm{Zer}(Q, \mathrm{R}^k) \subset B(0, 1/c)$ is of real dimension k',
  - a finite set $\mathcal{P} \subset \mathrm{D}[X_1, \ldots, X_k]$.
- **Output:** a set of points L such that for every $0 \le j \le k'(k-k')$ and $0 \le \ell \le k'(k-k')$, L meets every semi-algebraically connected component of every realizable weak sign condition of $\mathcal{P}$ on $\lim_\eta(Z_j) \cap \lim_\eta(Z_\ell)$.
- **Complexity:** $s^{k'+1} d^{O(k^2)}$, where s is a bound on the number of elements of $\mathcal{P}$ and d is a bound on the degrees of Q and the elements of $\mathcal{P}$.
- **Procedure:**
  - For every $0 \le j \le k'(k-k')$, denote by $\mathcal{R}_j$ the set of polynomials in $k+1$ variables obtained after two steps of Algorithm 14.1 (Block Elimination) applied to the polynomials appearing in the formula

$$(\exists\,(X,T))\;\|(X,T)-Y\|^2 < Z^2 \wedge T > 0 \wedge \bigwedge_{P \in \mathrm{App}(Q_j,T)} P(X) = 0$$

describing the closure of the set

$$\{(x,t)\in \mathrm{R}^{k+1} \mid t>0 \wedge \bigwedge_{P\in \mathrm{App}(Q_j,t)} P(x)=0\},$$

in order to eliminate Z and X,T. Denote by $\mathcal{P}_j$ the set of polynomials in k variables obtained by substituting 0 for T in $M_{k',j}(\mathcal{R}_j)$ (see Notation 13.26).

– For every $0 \le j \le k'(k-k')$ and $0 \le \ell \le k'(k-k')$, apply Algorithm 13.3 (Sampling on a Bounded Algebraic Set), with input $\mathrm{Zer}(Q,\mathrm{R}^k)$, $\mathcal{P}\cup\mathcal{P}_\ell\cup\mathcal{P}_j$ to obtain the set $L_{\ell,j}$. The set L is the union of the $L_{\ell,j}$.

Proof of correctness: Note that,

$$\lim_\eta\,(\mathrm{Zer}(\mathrm{App}(Q_\ell,\eta),\mathrm{R}\langle\eta\rangle^k))$$

is the closure of

$$\{(x,t)\in \mathrm{R}^{k+1} \mid t>0 \wedge \bigwedge_{P\in \mathrm{App}(Q_\ell,t)} P(x)=0\})\cap\{t=0\}.$$

The polynomials $\mathcal{R}_j$ have the property that the closure of

$$\{(x,t)\in \mathrm{R}^{k+1} \mid t>0 \wedge \bigwedge_{P\in P\in \mathrm{App}(Q_j,t)} P(x)=0\}$$

is the union of semi-algebraically connected components of sets defined by sign conditions over $\mathcal{R}_j$ (see page 556).

For every $0\le j\le k'(k-k')$ and $1\le \ell\le k'(k-k')$, L meets every semi-algebraically connected component of every realizable weak sign condition of $\mathcal{P}$ on $\lim_\eta(Z_j)\cap\lim_\eta(Z_\ell)$. □

Complexity analysis: According to the complexity of Algorithm 14.1 (Block Elimination), the set $\mathcal{R}_j$ has $d^{O(k)}$ polynomials and each of these polynomials has degree at most $d^{O(k)}$.

According to the complexity of Algorithm 13.3 (Sampling on a Bounded Algebraic Set), the set $L_{i,j}$ consists of $\sum_{j=0}^{k'}\binom{s}{j}4^j d^{O(k^2)}$ points defined by polynomials of degree at most $d^{O(k^2)}$. The complexity is

$$s\sum_{j=0}^{k'}\binom{s}{j}4^j d^{O(k^2)} = s^{k'+1}d^{O(k^2)}.$$

If $\mathrm{D}=\mathbb{Z}$, and the bitsizes of the coefficients of the polynomials are bounded by τ, then the bitsizes of the integers appearing in the intermediate computations and the output are bounded by $\tau d^{O(k^2)}$. □

In order to ensure that the roadmaps constructed on the various approximating varieties take into account connectivity in the original algebraic set, we need to add points in the various roadmaps for approximating varieties.

Algorithm 16.11. **[Touching Points]**

- **Structure:** an ordered domain D contained in a real closed field R.
- **Input:**
  - $Q \in \mathrm{D}[X_1, \ldots, X_k]$ such that $\mathrm{Zer}(Q, \mathrm{R}^k) \subset B(0, 1/c)$ is of real dimension k',
  - a real univariate representation u describing a point $p \in \mathrm{Zer}(Q, \mathrm{R}^k)$.
- **Output:** for every $0 \leq j \leq k'(k-k')$ such that Z_j is infinitesimally close to p, a set of real univariate representations describing points meeting every semi-algebraically connected component of Z_j infinitesimally close to p.
- **Complexity:** $s^{k'+1} d^{O(k^2)}$ where s is a bound on the number of elements of $\mathcal{P}$, d is a bound on the degrees of Q and the elements of $\mathcal{P}$ and $d^{O(k^2)}$ is a bound on the degrees of u.
- **Procedure:**
  - Let $u = (f, g_0, \ldots, g_k), \sigma$. For every $0 \leq j \leq k'(k-k')$ proceed as follows. Let β be a new variable and let $P_p(T, X_1, \ldots, X_k)$ be the system
$$\{f(T), \sum_{i=1}^{k} (g_0(T) X_i - g_i(T))^2 - g_0(T)^2 \beta^2\}$$
    Call Algorithm 13.3 (Sampling on a Bounded Algebraic Set) with input $\mathrm{App}(Q_j, \eta)$, P_p and $\mathrm{Der}(f)$ in the ring $\mathrm{D}[\beta, \eta]$. For each real univariate representation obtained, keep all those corresponding to points q at which the sign of P_p is negative and such that the sign condition satisfied by $\mathrm{Der}(f)$ at $\lim_{\beta,\eta}(q)$ is σ and discard the rest. Denote by $\mathcal{U}_j$ the real univariate representations representing points of Z_j obtained by applying $M_{k',j}$ (see Notation 13.26)to the real univariate representation associated to q.
  - Output the set $\mathcal{U} = \bigcup_{j=0}^{k'(k-k')} \mathcal{U}_j$ of real univariate representations so obtained. The touching points are the points associated to the elements of $\mathcal{U}$.

Proof of correctness: Immediate. □

Complexity analysis: The number of arithmetic operations in D for computing the set of touching points is $s \sum_{j=0}^{k'} \binom{s}{j} 4^j d^{O(k^2)} = s^{k'+1} d^{O(k^2)}$. This follows from the complexity of Algorithm 13.3 (Sampling on a Bounded Algebraic Set).

If $\mathrm{D} = \mathbb{Z}$, and the bitsizes of the coefficients of the polynomials are bounded by τ, then the bitsizes of the integers appearing in the intermediate computations and the output are bounded by $\tau d^{O(k^2)}$. □

We now describe the roadmap algorithm in the bounded case.

Algorithm 16.12. **[Bounded Roadmap]**

- **Structure:** an ordered domain D contained in a real closed field R.
- **Input:**
  - a polynomial $Q \in \mathrm{D}[X_1, \ldots, X_k]$ such that $\mathrm{Zer}(Q, \mathrm{R}^k) \subset B(0, 1/c)$ is of real dimension k',
  - a semi-algebraic subset S of $\mathrm{Zer}(Q, \mathrm{R}^k)$ defined by a $\mathcal{P}$-quantifier-free formula where $\mathcal{P} \subset \mathrm{D}[X_1, \ldots, X_k]$.
- **Output:** a roadmap for S.
- **Complexity:** $s^{k'+1} d^{O(k^2)}$ where s is a bound on the number of elements of $\mathcal{P}$, and d is a bound on the degree of Q and of the polynomials in $\mathcal{P}$.
- **Procedure:**
  - Let $d' = 2(d+1)$.
  - For every $0 \le \ell \le k'(k-k')$, define

$$\bar{Q}_\ell = Q_\ell^2 + (\varepsilon^2 (X_1^2 + \cdots + X_k^2 + X_{k+1}^2) - 1)^2,$$

    and define $\mathrm{App}(\bar{Q}_\ell, \eta)$ and $\mathcal{P}_\ell^\star$, using Notation 13.30 and Notation 13.32. Use a modified version of Algorithm 16.4 (Uniform Roadmap) with input $(\mathrm{App}(\bar{Q}_\ell, \eta), \mathcal{P}_\ell^\star)$ using Algorithm 16.9 (Modified Curve Segments) rather than Algorithm 15.2 (Curve Segments).
  - Call Algorithm 16.7 (Data for Adjacencies) and Algorithm 16.10 (Linking Points). For each element of $\bar{N} \cup L$, obtained above apply Algorithm 16.11 (Touching points). This defines a set $\mathcal{A}_\ell$ of real univariate representations. Connect the points associated to the elements of $\mathcal{A}_\ell$ to the uniform roadmap for $(\mathrm{App}(\bar{Q}_\ell, \eta), \mathcal{P}_\ell^\star)$ using a modified version of Algorithm 16.2 (Bounded Connecting), using Algorithm 16.9 (Modified Curve Segments) rather than Algorithm 15.2 (Curve Segments).
  - Output the image of the segments and points constructed above under the $\lim_{\gamma,\eta}$ map, using the computation done in the calls to Algorithm 16.9 (Modified Curve Segments) and retain only those portions which are in the given set S.

Proof of correctness: The correctness follows from the correctness of Algorithm 16.4 (Uniform Roadmap), Algorithm 16.9 (Modified Curve Segments), Algorithm 16.7 (Data for Adjacencies), Algorithm 16.10 (Linking Points) and Algorithm 16.11 (Touching points),as well as Proposition 13.33 and Proposition 13.35. □

Complexity analysis: The number of arithmetic operations for computing the set of added points $\mathcal{A}_\ell$ is $s \sum_{j=0}^{k'} \binom{s}{j} 4^j d^{O(k^2)}$ in D, using the complexity analysis of Algorithm 16.7 (Data for Adjacencies), Algorithm 16.10 (Linking Points) and Algorithm 16.11 (Touching points).

Since the set $\mathcal{P}_\ell^\star$ is in k'-general position with respect to $\mathrm{App}(\bar{Q}_\ell, d', \varepsilon, \eta)$ according to Proposition 13.33, using the complexity bound of Algorithm 16.4 (Uniform Roadmap), we see that the complexity is bounded by $s^{k'+1} d^{O(k^2)}$ in D.

Similarly, using the complexity bounds for Algorithm 16.2 (Bounded Connecting), the complexity of connecting a point x described by polynomials of degree at most $d^{O(k)}$ to the roadmap is $k' s\, d^{O(k^2)}$ in D.

If $\mathrm{D} = \mathbb{Z}$, and the bitsizes of the coefficients of the polynomials are bounded by τ, then the bitsizes of the integers appearing in the intermediate computations and the output are bounded by $\tau d^{O(k^2)}$. □

Now we show how to modify Algorithm 16.12 (Bounded Roadmap) to handle the case when the input algebraic set $\mathrm{Zer}(Q, \mathrm{R}^k)$ is not bounded.

Algorithm 16.13. **[General Roadmap]**

- **Structure:** an ordered domain D contained in a real closed field R.
- **Input:**
  - a polynomial $Q \in \mathrm{D}[X_1, \ldots, X_k]$ such that $\mathrm{Zer}(Q, \mathrm{R}^k)$ is of real dimension k',
  - a semi-algebraic subset S of $\mathrm{Zer}(Q, \mathrm{R}^k)$ described by a finite set $\mathcal{P} \subset \mathrm{R}[X_1, \ldots, X_k]$.
- **Output:** a roadmap for S.
- **Complexity:** $s^{k'+1} d^{O(k^2)}$ where s is a bound on the number of elements of $\mathcal{P}$, and d is a bound on the degree of Q and of the polynomials in $\mathcal{P}$.
- **Procedure:**
  - Step 1: Introduce new variables X_{k+1} and ε and replace Q by the polynomial $Q^\star = Q^2 + (\varepsilon^2 (X_1^2 + \cdots + X_{k+1}^2) - 1)^2$. Let $S^\star \in R\langle\varepsilon\rangle^{k+1}$ be the set defined by the same formula as S but with Q replaced by $Q^\star$. Run Algorithm 16.12 (Bounded Roadmap) with input $Q^\star$ and $S^\star$ and output a roadmap for $\mathrm{RM}(S^\star)$, composed of points and curves whose description involves ε.
  - Step 2: Denote by $\mathcal{L}$ be the set of all polynomials in $\mathrm{D}[\varepsilon]$ whose signs were determined in the various calls to the Multivariate Sign Determination Algorithm in Step 1. Replace ε by

$$a = \min_{P \in \mathcal{L}} c(P)$$

  (Definition 10.5) in the output roadmap to obtain a roadmap $\mathrm{RM}(S_a)$. When projected on R^k, this gives a roadmap $\mathrm{RM}(S) \cap B(0, 1/a)$.
  - Step 3: Collect all the points $(y_1, \ldots, y_k)$ in the roadmap which satisfies $\varepsilon^2 (y_1^2 + \cdots + y_k^2) = 1$. Each such point is described by a univariate representation involving ε. Add to the roadmap the curve segment obtained by treating ε as a parameter and letting ε vary over $(0, a,]$, to get a roadmap $\mathrm{RM}(S)$.

Proof of correctness: Follows from the correctness of Algorithm 16.12 (Bounded Roadmap). □

Complexity analysis: The complexity is bounded by $s^{k'+1}d^{O(k^2)}$ in D and coincides with the complexity of Algorithm 16.12 (Bounded Roadmap).

If $\mathrm{D} = \mathbb{Z}$, and the bitsizes of the coefficients of the polynomials are bounded by τ, then the bitsizes of the integers appearing in the intermediate computations and the output are bounded by $\tau d^{O(k^2)}$. □

Using the preceding algorithms we can now prove.

Theorem 16.14. *Let $Q \in \mathrm{R}[X_1, \ldots, X_k]$ with $\mathrm{Zer}(Q, \mathrm{R}^k)$ of dimension k' and let $\mathcal{P} \subset \mathrm{R}[X_1, \ldots, X_k]$ be a set of at most s polynomials for which the degrees of the polynomials in $\mathcal{P}$ and Q are bounded by d. Let S be a semi-algebraic subset of $\mathrm{Zer}(Q, \mathrm{R}^k)$ defined by a $\mathcal{P}$-quantifier-free formula.*

a) *Let $p \in \mathrm{Zer}(Q, \mathrm{R}^k)$ a point which is represented by a k-univariate representation with specified Thom encoding (u, σ) of degree $d^{O(k)}$. There is an algorithm whose output is a semi-algebraic path connecting p to $\mathrm{RM}(S)$. The complexity of the algorithm in the ring D generated by the coefficients of Q, u and the elements of $\mathcal{P}$ is bounded by $k' s\, d^{O(k^2)}$. If $\mathrm{D} = \mathbb{Z}$, and the bitsizes of the coefficients of the polynomials are bounded by τ, then the bitsizes of the integers appearing in the intermediate computations and the output are bounded by $\tau d^{O(k^2)}$.*
b) *There is an algorithm whose output is exactly one point in every semi-algebraically connected component of S. The complexity in the ring generated by the coefficients of Q and $\mathcal{P}$ is bounded by $s^{k'+1}d^{O(k^2)}$. In particular, this algorithm counts the number semi-algebraically connected component of S in time $s^{k'+1}d^{O(k^2)}$ in the ring D generated by the coefficients of Q and the coefficients of the elements of $\mathcal{P}$. If $\mathrm{D} = \mathbb{Z}$, and the bitsizes of the coefficients of the polynomials are bounded by τ, then the bitsizes of the integers appearing in the intermediate computations and the output are bounded by $\tau d^{O(k^2)}$.*
c) *Let p and q be two points that are represented by real k-univariate real representation u and v, of degree $d^{O(k)}$ belonging to S. There is an algorithm deciding whether p and q belong to the same connected component of S. The complexity in the ring D generated by the coefficients of Q, u, v and the coefficients of the polynomials in $\mathcal{P}$. is bounded by $s^{k'+1}d^{O(k^2)}$. If $\mathrm{D} = \mathbb{Z}$, and the bitsizes of the coefficients of the polynomials are bounded by τ, then the bitsizes of the integers appearing in the intermediate computations and the output are bounded by $\tau d^{O(k^2)}$.*

Proof: a) In order to connect a point x to the roadmap in the bounded case, chose a $0 \leq j \leq k'(k - k')$ such that $x \in \lim_{\gamma,\eta} (Z_j)$ and construct a point x_j infinitesimally close to x in Z_j using Algorithm 13.3 (Sampling on an Algebraic Set) and Algorithm 16.11 (Touching Points). This point x_j is connected to the uniform roadmap $\mathrm{RM}(\mathrm{App}(Q_\ell, \eta), \mathcal{P}_j^*)$ using a modified version of Algorithm 16.2 (Bounded Connecting) using Algorithm 16.9 (Modified Curve Segments) instead of Algorithm 15.2 (Curve Segments). Then output the image of the connecting curves under the map $\lim_{\eta,\gamma}$ using the computations done in the calls to Algorithm 16.9 (Modified Curve Segments). In the unbounded case, we modify the preceding method using the same method as in Step 3 of Algorithm 16.13 (General Roadmap).

b) and c) are clear after a). □

16.6 Computing the First Betti Number of Semi-algebraic Sets

Our aim in this section is to compute the first Betti number of a $\mathcal{P}$-closed semi-algebraic set.

We first describe an algorithm for computing closed a contractible coverings of a $\mathcal{P}$-closed semi-algebraic set in single exponential time when the family $\mathcal{P}$ is in general position. This algorithm computes parametrized connecting paths using Algorithm 16.5 (Parametrized Bounded and uses them to construct a contractible covering.

We are given a polynomial $Q \in \mathrm{D}[X_1, \ldots, X_k]$ such that $\mathrm{Zer}(Q, \mathrm{R}^k)$ is bounded and a finite set of polynomials $\mathcal{P} \subset \mathrm{D}[X_1, \ldots, X_k]$ in strong k-general position with respect to Q.

We fix a closed semi-algebraic set S contained in $\mathrm{Zer}(Q, \mathrm{R}^k)$. We follow the notations of Algorithm 16.5. and let $\#\mathcal{A} = t$. We denote by $\mathrm{SIGN}(S)$ the set of realizable sign conditions of $\mathcal{A}$ on $\mathrm{Zer}(Q, \mathrm{R}^k)$ whose realizations are contained in S, remembering that $\mathcal{P} \subset \mathcal{A}$. For each $\sigma \in \mathrm{SIGN}(S)$ $\mathrm{Reali}(\sigma, \mathrm{Zer}(Q, \mathrm{R}^k))$ is contained in $\mathrm{Reali}(\theta, \mathrm{Zer}(Q, \mathrm{R}^k))$ for some $\theta \in \Theta$. We denote by $\gamma(\sigma)$ the restriction of $\gamma(\theta)$ to the base $\mathrm{Reali}(\sigma, \mathrm{Zer}(Q, \mathrm{R}^k))$. Since $\gamma(\theta)$ is a parametrized path, $\gamma(\sigma)$ is also a parametrized path. However, since $\mathrm{Reali}(\sigma, \mathrm{Zer}(Q, \mathrm{R}^k))$ is not necessarily closed and bounded, we cannot use Proposition 15.15, and $\mathrm{Im}\gamma(\sigma)$ might not be contractible. In order to ensure contractibility, we restrict the base of $\gamma(\sigma)$ to a slightly smaller set which is closed, using infinitesimals.

We introduce infinitesimals

$$\varepsilon_{2t} \gg \varepsilon_{2t-1} \gg \cdots \gg \varepsilon_2 \gg \varepsilon_1 > 0.$$

For $i = 1, \ldots, 2t$ we denote by D_i the ring $\mathrm{D}[\varepsilon_{2t}, \cdots, \varepsilon_i]$, and by R_i the field $\mathrm{R}\langle\varepsilon_{2t}\rangle \cdots \langle\varepsilon_i\rangle$.

For $\sigma \in \mathrm{SIGN}(S)$ we define the level of σ by,

$$\mathrm{level}(\sigma) = \#\{P \in \mathcal{A} \mid \sigma(P) = 0\}.$$

Given $\sigma \in \mathrm{SIGN}(S)$, with $\mathrm{level}(\sigma) = j$, we denote by $\mathrm{Reali}(\sigma_-)$ the set defined on $\mathrm{Zer}(Q, \mathrm{R}_{2j}^k)$ by the formula σ_- obtained by taking the conjunction of

$$\begin{array}{rl} P = 0, & \text{for each } P \in \mathcal{A} \text{such that } \sigma(P) = 0, \\ P \geq \varepsilon_{2j}, & \text{for each } P \in \mathcal{A} \text{such that } \sigma(P) = 1, \\ P \leq -\varepsilon_{2j}, & \text{for each } P \in \mathcal{A} \text{such that } \sigma(P) = -1. \end{array}$$

Notice that $\mathrm{Reali}(\sigma_-) \subset \mathrm{Reali}(\sigma, \mathrm{Zer}(Q, \mathrm{R}_{2j}^k))$ is closed and bounded. Proposition 15.15 implies,

Proposition 16.15. The set $\gamma(\sigma)(\mathrm{Reali}(\sigma_-))$ *is semi-algebraically contractible.*

Note that the sets $\gamma(\sigma)(\mathrm{Reali}(\sigma_-))$ do not necessarily cover S. So we are going to enlarge them, preserving contractibility, to obtain a covering of S.

Given $\sigma \in \mathrm{SIGN}(S)$, with $\mathrm{level}(\sigma) = j$, we denote by $\mathrm{Reali}(\sigma_-^+)$ the set defined on $\mathrm{Zer}(Q, \mathrm{R}_{2j-1}^k)$, by the formula σ_-^+ obtained by taking the conjunction of

$$\begin{array}{rl} -\varepsilon_{2j-1} \leq P \leq \varepsilon_{2j-1} & \text{for each } P \in \mathcal{A} \text{ such that } \sigma(P) = 0, \\ P \geq \varepsilon_{2j}, & \text{for each } P \in \mathcal{A} \text{ such that } \sigma(P) = 1, \\ P \leq -\varepsilon_{2j}, & \text{for each } P \in \mathcal{A} \text{ such that } \sigma(P) = -1. \end{array}$$

with the formula ϕ defining S. Let $C(\sigma)$ be the set defined by,

$$C(\sigma) = \gamma(\sigma)(\mathrm{Reali}(\sigma_-)) \cup \mathrm{Reali}(\sigma_-^+)).$$

We now prove that

Proposition 16.16. $C(\sigma)$ *is semi-algebraically contractible.*

Let C be a closed and bounded semi-algebraic set contained in $\mathrm{R}\langle\varepsilon\rangle^k$. We can suppose without loss of generality that C is defined over $\mathrm{R}[\varepsilon]$ by Proposition 2.82. We denote by $C(t)$ the semi-algebraic subset of R^k defined by replacing ε by t in the definition of C. Note that $C(\varepsilon)$ is nothing but C.

We are going to use the following lemma.

Lemma 16.17. *Let B be a closed and bounded semi-algebraic set contained in* R^k *and let C be a closed and bounded semi-algebraic set contained in* $\mathrm{R}\langle\varepsilon\rangle^k$. *If there exists t_0 such that for every $t < t' < t_0$, $C(t) \subset C(t')$, and* $\lim_\varepsilon (C) = B$, *then* $\mathrm{Ext}(B, R\langle\varepsilon\rangle)$ *has the same homotopy type as C.*

Proof: Hardt's Triviality Theorem (Theorem 5.46) implies that there exists $t_0 > 0$, and a homeomorphism

$$\phi_{t_0}: C(t_0) \times (0, t_0] \to \cup_{0 < t \leq t_0} C_t$$

which preserves $C(t_0)$. Replacing t_0 by ε gives a homeomorphism

$$\phi(\varepsilon): C \times (0,\varepsilon] \to \cup_{0<t\leq\varepsilon} C(t).$$

Defining

$$\psi: C \times [0,\varepsilon] \to C$$

by

$$\begin{array}{lcll} \psi(x,s) & = & \pi_{1\ldots k} \circ \phi(x,s), & \text{if } s>0 \\ \psi(x,0) & = & \lim_{s\to 0_+} \pi_{1\ldots k} \circ \phi(x,s), & \end{array}$$

it is clear that ψ is a semi-algebraic retraction of C to $\mathrm{Ext}(B, \mathrm{R}\langle\varepsilon\rangle)$. □

We now prove Proposition 16.16.

Proof of Proposition 16.16: Apply Lemma 16.17 to C_σ and

$$\mathrm{Ext}(\gamma(\sigma)(\mathrm{Reali}(\sigma_-)), \mathrm{R}_{2j-1}):$$

thus $C(\sigma)$ can be semi-algebraically retracted to $\mathrm{Ext}(\gamma(\sigma)(\mathrm{Reali}(\sigma_-)), \mathrm{R}_{2j-1})$.

Since $\mathrm{Ext}(\gamma(\sigma)(\mathrm{Reali}(\sigma_-)), \mathrm{R}_{2j-1})$ is semi-algebraically contractible, so is $C(\sigma)$. □

We now prove that the sets $\mathrm{Ext}(C(\sigma), \mathrm{R}_1)$ form a covering of $\mathrm{Ext}(S, \mathrm{R}_1)$.

Proposition 16.18. [Covering property]

$$\mathrm{Ext}(S, \mathrm{R}_1) = \bigcup_{\sigma \in SIGN(S)} \mathrm{Ext}(C(\sigma), \mathrm{R}_1).$$

The proposition is an immediate consequence of the following stronger result.

Proposition 16.19.

$$\mathrm{Ext}(S, \mathrm{R}_1) = \bigcup_{\sigma \in SIGN(S)} \mathrm{Reali}(\sigma_-^+, \mathrm{R}_1^k).$$

Proof: By definition,

$$\mathrm{Ext}(S, \mathrm{R}_1) \supset \bigcup_{\sigma \in \mathrm{SIGN}(S)} \mathrm{Reali}(\sigma_-^+, \mathrm{R}_1^k).$$

We now prove the reverse inclusion. Clearly, we have that

$$S = \bigcup_{\sigma \in \mathrm{SIGN}(S)} \mathrm{Reali}(\sigma, \mathrm{R}).$$

Let $x \in \mathrm{Ext}(S, \mathrm{R}_1)$ and σ be the sign condition of the family $\mathcal{A}$ at x and let level$(\sigma) = j$. If $x \in \mathrm{Reali}(\sigma_-^+, \mathrm{R}_1^k)$, we are done. Otherwise, there exists $P \in \mathcal{A}$, such that x satisfies either $0 < P(x) < \varepsilon_{2j}$ or $-\varepsilon_{2j} < P(x) < 0$. Let $\mathcal{B} = \{P \in \mathcal{A} \mid \lim_{\varepsilon_{2j}} P(x) = 0\}$. Clearly $\#\mathcal{B} = j' > j$. Let $y = \lim_{\varepsilon_{2j}} x$. Since, $\mathrm{Ext}(S, \mathrm{R}_1)$ is closed and bounded and $x \in \mathrm{Ext}(S, \mathrm{R}_1)$, y is also in $\mathrm{Ext}(S, \mathrm{R}_1)$. Let τ be the sign condition of $\mathcal{A}$ at y with level$(\tau) = j' > j$. If $x \in \mathrm{Reali}(\tau_-^+, \mathrm{R}_1^k)$ we are done. Otherwise, for every $P \in \mathcal{A}$ such that $P(y) = 0$, we have that $-\varepsilon_{2j'-1} \leq P(x) \leq \varepsilon_{2j'-1}$, since $\lim_{\varepsilon_{2j}} (P(x)) = P(y) = 0$ and $\varepsilon_{2j'-1} \gg \varepsilon_{2j}$. So there exists $P \in \mathcal{A}$ such that x satisfies either $0 < P(x) < \varepsilon_{2j'}$ or $-\varepsilon_{2j'} < P(x) < 0$, and we replace $\mathcal{B}$ by $\{P \in \mathcal{A} \mid \lim_{\varepsilon_{2j'}} P(x) = 0\}$, and y by $y = \lim_{\varepsilon_{2j'}} x$. This process must terminate after at most t steps. □

Algorithm 16.14. **[Covering by Contractible Sets]**

- **Structure:** an ordered domain D contained in a real closed field R.
- **Input:**
  - a finite set of s polynomials $\mathcal{P} \subset \mathrm{D}[X_1, \ldots, X_k]$ in strong k-general position on R^k, with $\deg(P_i) \leq d$ for $1 \leq i \leq s$,
  - a $\mathcal{P}$-closed semi-algebraic set S, contained in the sphere of center 0 and radius r, defined by a $\mathcal{P}$-closed formula ϕ.
- **Output:** a set of formulas $\{\phi_1, \ldots, \phi_M\}$ defined by ^polynomials in $\mathrm{D}_1[X_1, \ldots, X_k]$ such that
  - each $\mathrm{Reali}(\phi_i, \mathrm{R}_1^k)$ is semi-algebraically contractible, and
  - $\bigcup_{1 \leq i \leq M} \mathrm{Reali}(\phi_i, \mathrm{R}_1^k) = \mathrm{Ext}(S, \mathrm{R}_1)$.
- **Complexity:** The complexity of the algorithm is bounded by $s^{(k+1)^2} d^{O(k^5)}$.
- **Procedure:**
  - Step 1: Let $Q = X_1^2 + \ldots + X_{k+1}^2 - r^2$. Call Algorithm 16.5 (Parametrized Bounded Connecting) with input $Q, \mathcal{P}$. Let $\mathcal{A}$ be the family of polynomials output.
  - Step 2: Compute the set of realizable sign conditions SIGN(S) using Algorithm 13.1 (Computing Realizable Sign Conditions) .
  - Step 3: Using Algorithm 14.21 (Quantifier Elimination), eliminate one variable to compute the image of the semi-algebraic map γ_{σ_-}. Finally, output the set of formulas $\{\phi_\sigma \mid \sigma \in \mathrm{SIGN}(\mathcal{A}, S)\}$ describing the semi-algebraic set $C(\sigma)$.

Proof of correctness: The correctness of the algorithm is a consequence of Proposition 16.16, Proposition 16.18 and the correctness of Algorithm 16.5 (Parametrized Bounded Connecting), as well as the correctness of Algorithm 13.1 (Computing Realizable Sign Conditions) and Algorithm 14.5 (Quantifier Elimination). □

Complexity analysis: The complexity of Step 1 of the algorithm is bounded by $s^{k+1}d^{O(k^4)}$, where s is a bound on the number of elements of $\mathcal{P}$ and d is a bound on the degrees of the elements of $\mathcal{P}$, using the complexity analysis of Algorithm 16.5 (Parametrized Bounded Connecting). The number of polynomials in $\mathcal{A}$ is $s^{k+1}d^{O(k^4)}$ and their degrees are bounded by $d^{O(k^3)}$. Thus the complexity of computing SIGN(S) is bounded by $s^{(k+1)^2}d^{O(k^5)}$ using Algorithm 13.1 (Computing Realizable Sign Conditions). In Step 3 of the algorithm there is a call to Algorithm 14.5 (Quantifier Elimination). There are two blocks of variables of size k and 2 respectively. The number and degrees of the input polynomials are bounded by $s^{k+1}d^{O(k^4)}$ and $d^{O(k^3)}$ respectively. Moreover, observe that even though we introduced $2s$ infinitesimals, each arithmetic operation is performed in the ring D adjoined with at most $O(k)$ infinitesimals since the polynomials $\{P, P\pm\varepsilon_{2j}, P\pm\varepsilon_{2j-1}, P\in\mathcal{P}, 1\le j\le s\}$ are in strong general position. Thus, the complexity of this step is bounded by $s^{(k+1)^2}d^{O(k^5)}$ using the complexity analysis of Algorithm 14.5 (Quantifier Elimination) and the fact that each arithmetic operation costs at most $d^{O(k^5)}$ in terms of arithmetic operations in the ring D. □

We now want to compute the first Betti number of a $\mathcal{P}$-closed semi-algebraic set S when $\mathcal{P}$ is not necessary in general position. We first replace S by a $\mathcal{P}^\star$-closed and bounded semi-algebraic set, where the elements of $\mathcal{P}^\star$ are slight modifications of the elements of $\mathcal{P}$, and the family $\mathcal{P}^\star$ is in general position and $b_i(S^\star)=b_i(S), 0\le i\le k$.

Define

$$H_i = 1 + \sum_{1\le j\le k} i^j X_j^{d'}.$$

where d' is the smallest number strictly bigger than the degree of all the polynomials in $\mathcal{P}$. Using arguments similar to the proof of Proposition 13.6 it is easy to see that the family $\mathcal{P}^\star$ of polynomials $P_i - \delta H_i$, $P_i + \delta H_i$, with $P_i\in\mathcal{P}$. is in general position in $\mathrm{R}\langle\delta\rangle^k$.

Lemma 16.20. *Denote by $S^\star$ the set obtained by replacing any $P_i\ge 0$ in the definition of S by $P_i\ge -\delta H_i$ and every $P_i\le 0$ in the definition of S by $P_i\le\delta H_i$. If S is bounded, the set* $\mathrm{Ext}(S,\mathrm{R}\langle\delta\rangle^k$ *is semi-algebraically homotopy equivalent to $S^\star$.*

Proof: The claim follows by Lemma 16.17. Note that S is closed and bounded, $\lim_\delta S^\star = S$, and $S^\star(t)\subset S^\star(t')$ for $t<t'$. □

Algorithm 16.15. **[First Betti Number in the $\mathcal{P}$-closed case]**

- **Structure:** an ordered domain D contained in a real closed field R
- **Input:**
  - a finite set of polynomials $\mathcal{P}\subset \mathrm{D}[X_1,\ldots,X_k]$,

  - a formula defining a $\mathcal{P}$-closed semi-algebraic set, S.
- **Output:** the first Betti number $b_1(S)$.
- **Complexity:** $(s\,d)^{k^{O(1)}}$, where $s = \#\mathcal{P}$ and $d = \max_{P \in \mathcal{P}} \deg(P)$.
- **Procedure:**
  - Step 1: Let ε be an infinitesimal. Replace S by the semi-algebraic set T defined as the intersection of the cylinder $S \times \mathrm{R}\langle\varepsilon\rangle$ with the upper hemisphere defined by $\varepsilon^2(X_1^2 + \ldots + X_k^2 + X_{k+1}^2) = 1, X_{k+1} \geq 0$.
  - Step 2: Replace T by $T^\star$ using the notation of Lemma 16.20.
  - Step 3: Use Algorithm 16.14 (Covering by Contractible Sets) with input $\varepsilon^2(X_1^2 + \ldots + X_k^2 + X_{k+1}^2) - 4$ and $\mathcal{P}^\star$, to compute a covering of $T^\star$ by closed, bounded and contractible sets, T_i, described by formulas ϕ_i.
  - Step 4: Use Algorithm 16.13 (General Roadmap) to compute exactly one sample point of each connected component of the pairwise and triplewise intersections of the T_i's. For every pair i, j and every k compute the incidence relation between the connected components of $T^\star_{ijk}$ and $T^\star_{ij}$ as follows: compute a roadmap of $T^\star_{ij}$, containing the sample points of the connected components of $T^\star_{ijk}$ using Algorithm 16.13 (General Roadmap).
  - Step 5: Using linear algebra compute

$$b_1(T^\star) = \dim(\mathrm{Ker}(\delta_2)) - \dim(\mathrm{Im}(\delta_1)),$$

with

$$\prod_i H^0(T_i^\star) \xrightarrow{\delta_1} \prod_{i<j} H^0(T_{ij}^\star) \xrightarrow{\delta_2} \prod_{i<j<\ell} H^0(T_{ij\ell}^\star)$$

Proof of correctness: First note that T is closed and bounded and has the same Betti numbers as S, using the local conical structure at infinity. It follows from Lemma 16.20 that T and $T^\star$ have the same Betti numbers. The correctness of the algorithm is a consequence of the correctness of Algorithm 16.14 (Covering by Contractible Sets), Algorithm 16.13 (General Roadmap), and Theorem 6.9. □

Complexity analysis: The complexity of Step 3 of the algorithm is bounded by $s^{(k+1)^2} d^{O(k^6)}$ using the complexity analysis of Algorithm 16.14 (Covering by Contractible Sets) and noticing that each arithmetic operation takes place a ring consisting of D adjoined with at most k infinitesimals. Finally, the complexity of Step 4 is also bounded by $(s\,d)^{k^{O(1)}}$, using the complexity analysis of Algorithm 16.13 (General Roadmap). □

Now we describe the algorithm for computing the first Betti number of a general semi-algebraic set. We first replace the given set by a closed and bounded one, using Theorem 7.45. We then apply Algorithm 16.15.

Algorithm 16.16. **[First Betti Number of a $\mathcal{P}$-Semi-algebraic Set]**

- **Structure:** an ordered domain D contained in a real closed field R.

- **Input:**
  - a finite set of polynomials $\mathcal{P} \subset \mathrm{D}[X_1, \ldots, X_k]$,
  - a formula defining a $\mathcal{P}$-semi-algebraic set, S.
- **Output:** the first Betti number $b_1(T)$.
- **Complexity:** $(s\,d)^{k^{O(1)}}$, where $s = \#\mathcal{P}$ and $d = \max_{P \in \mathcal{P}} \deg(P)$.
- **Procedure:**
  - Step 1: Let ε be an infinitesimal. Define $\tilde{S}$ as the intersection of $\mathrm{Ext}(S, \mathrm{R}\langle\varepsilon\rangle)$ with the ball of center 0 and radius $1/\varepsilon$. Define $\mathcal{Q}$ as
  $$\mathcal{P} \cup \{\varepsilon^2(X_1^2 + \ldots + X_k^2 + X_{k+1}^2) - 4, X_{k+1}\}.$$
  - Replace $\tilde{S}$ by the $\mathcal{Q}$- semi-algebraic set S defined as the intersection of the cylinder $\tilde{S} \times \mathrm{R}\langle\varepsilon\rangle$ with the upper hemisphere defined by $\varepsilon^2(X_1^2 + \ldots + X_k^2 + X_{k+1}^2) = 4, X_{k+1} \geq 0$.
  - Step 2: Using Definition 7.44, replace T by a $\mathcal{Q}'$-closed set, T', where
  $$\mathcal{Q}' = \{P \pm \varepsilon_i \mid P \in \mathcal{Q}, i = 1, \ldots, 2\,s\}.$$
  - Step 3: Use Algorithm 16.15 to compute the first Betti number of T'.

Proof of correctness: The correctness of the algorithm is a consequence of Theorem 7.45 and the correctness of Algorithm 16.15. □

Complexity analysis: In Step 2 of the algorithm the cardinality of $\mathcal{Q}'$ is $2(s+1)^2$ and the degrees of the polynomials in $\mathcal{Q}'$ are still bounded by d. The complexity of Step 3 of the algorithm is then bounded by $(s\,d)^{k^{O(1)}}$ using the complexity analysis of Algorithm 16.15. □

16.7 Bibliographical Notes

A motivation for deciding connectivity of semi-algebraic sets comes from robot motion planning [146]. This is equivalent to deciding whether the two corresponding points in the free space are in the same connected component of the free space. The solution by Schwartz and Sharir [146] using Collin's method of cylindrical algebraic decomposition. The complexity of their solution is thus polynomial in d and s and doubly exponential in k.

Canny introduced the notion of a roadmap for a semi-algebraic set and gave an algorithm [36] which after subsequent modifications [38] constructed a roadmap for a semi-algebraic set defined by polynomials whose sign invariant sets give a stratification of R^k and whose complexity is $s^k\,(\log s)\,d^{O(k^4)}$. For an arbitrary semi-algebraic set he perturbs the defining polynomials and is then able to decide if two points are in the same semi-algebraically connected component with the same complexity. However, this algorithm does not give a path joining the points. A Monte Carlo version of this algorithm has complexity $s^k(\log s)d^{O(k^2)}$.

Grigor'ev and Vorobjov [78] gave an algorithm with complexity $(s\,d)^{k^{O(1)}}$, for counting the number of connected components of a semi-algebraic set. Heintz, Roy, and Solerno [86] and Gournay and Risler [75] gave algorithms which compute a roadmap for any semi-algebraic set whose complexity was also $(s\,d)^{k^{O(1)}}$.

Unlike the complexity of Canny's algorithm, the complexities of these algorithms are not separated into a combinatorial part (the part depending on s) and an algebraic part (the part depending on d). Since the given semi-algebraic set might have $(s\,d)^k$ different connected components, the combinatorial complexity of Canny's algorithm is nearly optimal. Canny's algorithm makes use of Thom's isotopy lemma for stratified sets and consequently requires the use of generic projections, as well as perturbations to put the input polynomials into general position in a very strong sense. In order to do this in a deterministic fashion, $O(s + k^2)$ different transcendental are introduced, requiring the algebraic operations to be performed over an extended ring. This raises the algebraic complexity of the deterministic algorithm to $d^{O(k^4)}$.

In [16] a deterministic algorithm constructing a roadmap for any semi-algebraic set contained in an algebraic set $\mathrm{Zer}(Q, \mathrm{R}^k)$ of dimension k' with complexity $s^{k'+1} d^{O(k^2)}$ is given. In robot motion planning, the configuration space of a robot is often embedded as a lower dimensional algebraic set in a higher dimensional real Euclidean space (see [103]), so it is of interest to design algorithms which take advantage of this fact and whose complexity reflects the dimension of this algebraic set rather than the dimension of the ambient space. The combinatorial complexity of this algorithm is nearly optimal. The algorithm uses only a fixed number of infinitesimal quantities which reduces the algebraic complexity to $d^{O(k^2)}$. The algorithm also computes a semi-algebraic path between the input points if they happen to lie in the same connected component and hence solves the full version of the problem.

A single exponential bound $(s\,d)^{k^{O(1)}}$ for computing the connected components of a semi-algebraic set is due to Canny, Grigor'ev, Vorobjov and Heintz, Roy and Solernò [39, 87]. The results presented here are significantly more precise.

References

1. J. Abdeljaoued, H. Lombardi, *Méthodes matricielles: Introduction à la Complexité Algébrique*, Mathématiques & Applications, Springer (2004).
2. J.W. Alexander, *A proof of the invariance of certain constants of analysis situs*, Trans. Amer. Math. Soc., 16 148–154 (1915).
3. N. Alon, *Tools from higher algebra*, Handbook of combinatorics,1749–1783, Elsevier, Amsterdam (1995).
4. M. E. Alonso, E. Becker, M.-F. Roy, T. Wörmann, *Zeroes, Multiplicities and Idempotents for Zerodimensional Systems*, Algorithms in algebraic geometry and applications, Progress in Mathematics, vol 143. Birkhauser 1–16 (1996).
5. C. Andradas, L. Bröcker, J. Ruiz, *Constructible sets in Real Geometry*, Ergeb. Math. Grenzgeb. (3) 33 Berlin etc.: Springer-Verlag (1996).
6. E. Artin, *Über die Zerlegung definiter Funktionen in Quadrate*, Hamb. Abh. **5**, 100-115 (1927). The collected papers of Emil Artin, 273–288. Reading: Addison-Wesley (1965).
7. E. Artin, O. Schreier, *Algebraische Konstruktion reeller Körper*, Hamb. Abh. 5 8(-99 (1925). The collected papers of Emil Artin, 258-271. Addison-Wesley (1965).
8. P. Aubry, F. Rouillier, M. Safey El Din, *Real solving for positive dimensional systems*, Journal of Symbolic Computation, 34(6), 543–560 (2002).
9. E. H. Bareiss, *Sylvester's Identity and Multistep Integer-Preserving Gaussian Elimination*, Math. Comp. 22 565–578 (1968).
10. S. Basu, *On different bounds on different Betti numbers*, Discrete and Computational Geometry, 30:1, 65–85, (2003).
11. S. Basu, *On Bounding the Betti Numbers and Computing the Euler Characteristics of Semi-algebraic Sets*, Discrete and Computational Geometry, 22 1–18 (1999).
12. S. Basu, *New Results on Quantifier Elimination over Real Closed Fields and Applications to Constraint Databases,* Journal of the ACM, 46 (4), 537–555 (1999).
13. S. Basu, R. Pollack, M.-F. Roy, *On the Combinatorial and Algebraic Complexity of Quantifier Elimination*, Journal of the ACM, 43 1002–1045, (1996).
14. S. Basu, R. Pollack, M.-F. Roy, *On the number of cells defined by a family of polynomials on a variety*, Mathematika, 43 120–126 (1996).

15. S. BASU, R. POLLACK, M.-F. ROY, *On Computing a Set of Points meeting every Semi-algebraically Connected Component of a Family of Polynomials on a Variety*, Journal of Complexity, March 1997, 13 (1), 28–37.
16. S. BASU, R. POLLACK, M.-F. ROY, *Computing Roadmaps of Semi-algebraic Sets on a Variety*, Journal of the AMS, 3 (1) 55–82 (1999).
17. S. BASU, R. POLLACK, M.-F. ROY *On the Betti Numbers of Sign Conditions*, Proc. Amer. Math. Soc. 133, 965–974 (2005).
18. S. BASU, R. POLLACK, M.-F. ROY, *Computing Euler-Poincaré characteristic of sign conditions*, Journal of Computational Complexity, 14, 53–71 (2005).
19. S. BASU, R. POLLACK, M.-F. ROY, *Computing the Dimension of a Semi-Algebraic Set,* Zap. Nauchn. Semin. POMI 316 42–54 (2004).
20. S. BASU, R. POLLACK, M.-F. ROY, *Algorithms in real algebraic geometry*, Springer-Verlag (2003).
21. S. BASU, R. POLLACK, M.-F. ROY, *Computing the first Betti number and describing the connected components of semi-algebraic sets*, preprint (2005), [arXiv:math.AG/0603248].
22. R. BENEDETTI, J.-J. RISLER, *Real algebraic and semi-algebraic sets*, Actualités Mathématiques. Hermann, Paris (1990).
23. M. BEN-OR, D. KOZEN, J. REIF, *The complexity of elementary algebra and geometry*, J. of Computer and Systems Sciences, 18:251–264, (1986).
24. E. BÉZOUT, *Recherche sur les degrés de l'équation résultant de l'évanouissement des inconnues*, Histoire de l'académie royale des sciences, 288–338, (1764).
25. G.D. BIRKHOFF, *Collected Mathematical Papers*, Vol II, Amer. Math. Soc., New York (1950).
26. J. BOCHNAK, M. COSTE, M.-F. ROY, *Géométrie algébrique réelle,* Springer-Verlag (1987). *Real algebraic geometry*, Springer-Verlag (1998).
27. A. BOREL, J. C. MOORE, *Homology theory for locally compact spaces*, Mich. Math. J., 7: 137–159, (1960).
28. H. BRAKHAGE, *Topologische Eigenschaften algebraischer Gebilde über einen beliebigen reell-abgeschlossenen Konstantenkörper*, Dissertation, Univ. Heidelberg (1954).
29. E. BRIAND, F. CARRERAS, L. GONZALEZ–VEGA, N. GONZALEZ–CAMPOS, I. NECULA, H. PERDRY, N. DEL RIO, C. TANASESCU, F. ROUILLIER, M.-F. ROY, A. SEIDL, *Creating an Electronic Book for Algorithms in Real Algebraic Geometry: a first experiment,* ISSAC Poster (2004). http://www0.risc.uni-linz.ac.at/issac2004/poster-abstracts/abstract24.pdf
30. C. BROWN, *QEPCAD B: a program for computing with semi-algebraic sets using CADs* ACM SIGSAM 37 (4): 97–108 (2003). http://www.cs.usna.edu/~qepcad/B/QEPCAD.html
31. W. D. BROWNAWELL *Local diophantine Nullstellen identities*, J. AMS 1, 311–322 (1988).
32. B. BUCHBERGER, *An Algorithm for Finding the Basis Elements in the Residue Class Ring Modulo a Zero Dimensional Polynomial Ideal*, PhD Thesis, Mathematical Institute, University of Innsbruck, Austria, (1965).
33. B. BUCHBERGER, *Gröbner bases: an algorithmic method in polynomial ideal theory*, Recent trends in multidimensional systems theory, Reider ed. Bose, (1985).

34. F. BUDAN DE BOISLAURENT, *Nouvelle méthode pour la résolution des équations numériques d'un degré quelconque*, (1807), 2nd edition, Paris (1822).
35. L. CANIGLIA, A. GALLIGO, J. HEINTZ, *Borne simplement exponentielle pour les degrés dans le théorème des zéros sur un corps de caractéristique quelconque*, C. R. Acad. Sci. Paris 307, 255–258 (1988).
36. J. CANNY, *The Complexity of Robot Motion Planning*, MIT Press (1987).
37. J. CANNY, *Some Algebraic and Geometric Computations in PSPACE*, Proc. Twentieth ACM Symp. on Theory of Computing, 460–467, (1988).
38. J. CANNY, *Computing road maps in general semi- algebraic sets*, The Computer Journal, 36: 504–514, (1993).
39. J. CANNY, D. GRIGOR'EV, N. VOROBJOV, *Finding connected components of a semi-algebraic set in subexponential time*, Appl. Algebra Eng. Commun. Comput., 2 (4), 217–238 (1992).
40. F. CARUSO, *SARAG: Some Algorithms in Real Algebraic Geometry*, (2005).
41. A. CAUCHY, *Calcul des indices des fonctions*, Journal de l'Ecole Polytechnique, vol. 15, Cahier 25, 176–229 (1832).
42. A. CHISTOV, H. FOURNIER, L. GURVITS, P. KOIRAN, *Vandermonde Matrices, NP-Completeness and Transversal Subspaces,* Foundations of Computational Mathematics, 3 (4) 421–427 (2003).
43. P. J. COHEN, *Decision procedures for real and p-adic fields*, Comm. Pure. Appl. Math. 22, 131–151 (1969).
44. G. E. COLLINS, *Subresultants and Reduced Polynomial Remainder Sequences*, Journal of the ACM 14 128–142 (1967).
45. G. COLLINS, *Quantifier elimination for real closed fields by cylindric algebraic decomposition*, In Second GI Conference on Automata Theory and Formal Languages. Lecture Notes in Computer Science, vol. 33, 134–183, Springer- Verlag, Berlin (1975).
46. G. E. COLLINS, H. HONG, *Partial Cylindrical Algebraic Decomposition for Quantifier Elimination*, Journal of Symbolic Computation, 12, 299–328 (1991).
47. M. COSTE, *An introduction to semi-algebraic geometry*, Dip. Mat. Univ. Pisa, Dottorato di Ricerca in Matematica, Istituti Editoriali e Poligrafici Internazionali, Pisa (2000).
48. M. COSTE, *An introduction to o-minimal geometry*, Dip. Mat. Univ. Pisa, Dottorato di Ricerca in Matematica, Istituti Editoriali e Poligrafici Internazionali, Pisa (2000).
49. M. COSTE, T. LAJOUS-LOEZA, H. LOMBARDI, M.-F. ROY, *Generalized Budan-Fourier theorem and virtual roots*, Journal of Complexity, 21, 478–486, (2005).
50. M. COSTE, M.-F. ROY, *Thom's lemma, the coding of real algebraic numbers and the topology of semi-algebraic sets.* Journal of Symbolic Computation 5, No.1/2, 121–129 (1988).
51. D. COX, J. LITTLE, D. O'SHEA, *Ideals, varieties and algorithms: an introduction to computational algebraic geometry and commutative algebra*, Undergraduate Texts in Mathematics, Springer-Verlag, New York (1997).
52. J. H. DAVENPORT, J. HEINTZ, *Real quantifier elimination is doubly exponential*, Journal of Symbolic Computation 5, No.1/2, 29–35 (1988).
53. R. DESCARTES, *Géométrie* (1636). A source book in Mathematics, 90–31. Harvard University press (1969).

54. C. L. Dogdson, *Condensation of determinants, being a new and brief method for computing their numerical values*, Proc. Royal. Soc. Lond. 15 150–155 (1866).
55. A. Eigenwillig, V. Sharma, C. Yap, *Almost tight complexity bounds for Descartes method*, preprint (2006).
56. L. Euler, *Démonstration sur le nombre de points où deux lignes d'ordre quelconque peuvent se couper*, Mémoires de l'Académie des Sciences de Berlin, 4, 234–248, (1750).
57. G. Farin, *Curves and surfaces for Computer Aided Design*, Academic Press (1990).
58. J.-C. Faugère, *A new efficient algorithm for computing Gröbner basis (F 4)* Journal of Pure and Applied Algebra, 139 61–88 (1999).
59. J.-C. Faugère, *FGb (Gröbner basis computations)* http://fgbrs.lip6.fr/Software/
60. J. Fourier, *Analyse des équations déterminées*, F. Didot, Paris (1831).
61. F. G. Frobenius, *Uber das Traegheitsgesetz des quadratishen Formen,* S-B Pruss. Akad. Wiss. 241-256 (1984), 403–431 (1884).
62. A. Gabrielov, N. Vorobjov, *Betti Numbers for Quantifier-free Formulae.* Discrete and Computational Geometry, 33 395–401 (2005).
63. F. R. Gantmacher, *Theory of matrices, Vol I.* AMS-Chelsea (2000).
64. J. von zur Gathen, J.Gerhard *Modern computer algebra*, Cambridge University Press (1999).
65. C. F. Gauss *Demonstratio Nova Altera Theorematis Omnem Funct. Alg.*, Commentationes societatis regieae scientiarum Gottingensis recentiores, 3, 107–134 (1816). Werke III 31-56 (1876).
66. L. Gonzalez Vega, *La sucesión de Sturm–Habicht y sus aplicaciones al Algebra Computacional*, Doctoral Thesis, Universidad de Cantabria (1989).
67. L. Gonzalez Vega, M. El Kahoui, *An improved upper complexity bound for the topology computation of a real algebraic curve*, J. Complexity 12 527–544 (1996).
68. L. Gonzalez Vega, H. Lombardi, L. Mahé, *Virtual roots of real polynomials* Journal of Pure and Applied Algebra, 124, 147–166 (1998).
69. L. Gonzalez Vega, H. Lombardi, T. Recio, M.-F. Roy, *Spécialisation de la suite de Sturm et sous-résultants I*, Informatique théorique et applications 24 561–588 (1990).
70. L. Gonzalez Vega, H. Lombardi, T. Recio, M.-F. Roy, *Spécialisation de la suite de Sturm et sous-résultants II*, Informatique théorique et applications 28 1-24 (1994).
71. L. Gonzalez-Vega, I. Necula, *Efficient topology determination of implicitly defined algebraic plane curves*, Comput. Aided Geom. Design 19, no. 9, 719–743 (2004).
72. L. Gonzalez Vega, F. Rouillier, M.-F. Roy, *Symbolic Recipes for Polynomial System Solving*, In: Some tapas of computer algebra, A. Cohen et al. ed. Algorithms and Computation in Mathematics, vol. 4, 34–64, Springer.
73. L. Gonzalez Vega, F. Rouillier, M.-F. Roy, G. Trujillo *Symbolic Recipes for Real Solutions*, In: Some tapas of computer algebra, A. Cohen et al. ed. Algorithms and Computation in Mathematics, vol. 4, 121–167, Springer.
74. P. Gordan, *Verlesungen über Invarientheorie- Ester Band: Determinanten*, B. G. Teubner, Leipzig (1885).

75. L. GOURNAY, J. J. RISLER, *Construction of roadmaps of semi-algebraic sets*, Appl. Algebra Eng. Commun. Comput. 4, No.4, 239–252 (1993).
76. D. GRIGOR'EV, *The Complexity of deciding Tarski algebra*, Journal of Symbolic Computation 5 65–108 (1988).
77. D. GRIGOR'EV, N. VOROBJOV, *Solving Systems of Polynomial Inequalities in Subexponential Time*, Journal of Symbolic Computation, 5 37–64 (1988).
78. D. GRIGOR'EV, N. VOROBJOV, *Counting connected components of a semi-algebraic set in subexponential time*, Comput. Complexity 2, No.2, 133–186 (1992).
79. W. HABICHT, *Eine Verallgemeinerung des Sturmschen Wurzelzählverfahrens*, Comm. Math. Helvetici 21, 99–116 (1948).
80. J. HADAMARD, *Résolution d'une question relative au déterminant*, Bull. Sci. Math. 17, 240–246 (1893).
81. HAM, LE, *Un theoreme de Zariski du type de Lefschetz*, Ann. Sc. Ec. Norm. Sup., (3) vol 6. 317–355 (1973).
82. HAM, LE, *Lefschetz theorems on quasi-projective varieties*, Bull. Soc. Math. France, 113, 123–142 (1985).
83. R. M. HARDT, *Semi-algebraic Local Triviality in Semi-algebraic Mappings*, Am. J. Math. 102, 291–302 (1980).
84. J. HEINTZ, M.-F. ROY, P. SOLERNÓ , *On the Complexity of Semi-Algebraic Sets*, Proc. IFIP 89, San Francisco. North- Holland 293–298 (1989).
85. J. HEINTZ, M.-F. ROY, P. SOLERNÒ, *Sur la complexité du principe de Tarski-Seidenberg*, Bull. Soc. Math. France 118 101–126 (1990).
86. J. HEINTZ, M.-F. ROY, P. SOLERNÒ, *Single exponential path finding in semi-algebraic sets II: The general case*, Bajaj, Chandrajit L. (ed.), Algebraic geometry and its applications. Collections of papers from Shreeram S. Abhyankar's 60th birthday conference held at Purdue University, West Lafayette, IN, USA, June 1-4, 1990. New York: Springer-Verlag, 449–465 (1994).
87. J. HEINTZ, M.-F. ROY, P. SOLERNÒ, *Description of the Connected Components of a Semialgebraic Set in Single Exponential Time*, Discrete and Computational Geometry, 11, 121–140 (1994).
88. G. HERMANN, *Die Frage der endlich vielen Schritte in der Theorie der Polynomialideale*, Math. Annalen 95, 736–788 (1926).
89. C. HERMITE, *Remarques sur le théorème de Sturm*, C. R. Acad. Sci. Paris 36, 52–54 (1853).
90. C. HERMITE, *Sur l'indice des fractions rationnelles*, Bull. Soc. Math. France, tome 7, 128–131 (1879).
91. D. HILBERT, *Über die Theorie der algebraischen Formen*, Math. Annalen 36, 473–534 (1890).
92. L. HÖRMANDER, *The analysis of linear partial differential operators*, vol. 2. Berlin etc.: Springer-Verlag (1983).
93. A. HURWITZ, *Über die Bedingungen, unter welchen eine Gleichnung nur Wurzeln mit negativen reellen Theilen besitzt,* Math. Annalen, vol. 46, 273–284 (1895).
94. N. V. ILYUSHECKIN, *On some identities for the elements of a symmetric matrix*, Zapiski Nauchnyh Seminarov POMI Vol. 303, Investigations on Linear Operators and Function Theory. Part 31, editor S.V.Kislyakov (1997), http://www.pdmi.ras.ru/znsl/2003/v303.html.

95. A. KHOVANSKY, *Fewnomials*, Transl. Math. Monogr. 88, Providence, RI: American Mathematical Society (1991).
96. M. KNEBUSCH, C. SCHEIDERER, *Einführung in die reelle Algebra*, Vieweg-Studium 63, Aufbaukurs Mathematik, Vieweg (1989).
97. J. KOLLAR, *Effective Nullstellensatz for arbitrary ideals*, J. Eur. Math. Soc. 1 no. 3, 313–337 (1999).
98. W. KRANDICK, K. MEHLHORN, *New bounds for the Descartes method*, Journal of Symbolic Computation, vol. 41, 49–66 (2006).
99. M. G. KREIN AND M. A. NAIMARK, *The method of symmetric and hermitian forms in the theory of separation of roots of algebraic equations*, Kharkov 1936 (in Russian), English translation in: Linear and Multi-linear Algebra 10 265–308 (1981).
100. L. KRONECKER, *Werke*, Vol 1. Leipzig, Teubner (1895).
101. J.-M. LANE, R. F. RIESENFELD, *Bounds on a polynomial*, 21 112–117 (1981).
102. S. LANG, *Algebra*, Reading: Addison-Wesley (1971).
103. J.-C. LATOMBE, *Robot Motion Planning*, The Kluwer International Series in Engineering and Computer Science. 124. Dordrecht: Kluwer Academic Publishers Group (1991).
104. D. LAZARD, *Gröbner-Bases, Gaussian elimination and resolution of systems of algebraic equations*. EUROCAL 146–156 (1983).
105. S. LEFSCHETZ, *Algebraic geometry*. Princeton University Press, Princeton, N. J., (1953).
106. U. LE VERRIER *Sur les variations séculaires des éléments elliptiques des sept planètes principales: Mercure, Vénus, La Terre, Mars, Jupiter, Saturne et Uranus*, J. Math. Pures Appli., (4), 220–254 (1840).
107. T. LICKTEIG, M.-F. ROY, *Sylvester-Habicht sequences and fast Cauchy index computation*, Journal of Symbolic Computation, 31 315–341 (2001).
108. A. LIÉNARD, M. H. CHIPART *Sur le signe de la partie réelle des racines d'une équation algébrique*, J. Math. Pures Appl. (6) 10 291–346 (1914).
109. Z. LIGATSIKAS, M.-F. ROY, *Séries de Puiseux sur un corps réel clos*. C. R. Acad. Sci. Paris 311 625–628 (1990).
110. S. LOJASIEWICZ *Ensembles semi-analytiques*. Inst. Hautes Etudes Sci., (preprint) (1964).
111. S. LOJASIEWICZ *Triangulation of semi-analytic sets*. Ann. Scuola Norm. Sup. Pisa, Sci. Fis. Mat. (3) **18**, 449–474 (1964).
112. H. LOMBARDI, M.-F. ROY, M. SAFEY, *New structure theorems for subresultants*, Special Issue Symbolic Computation in Algebra, Analysis, and Geometry Journal of Symbolic Computation, 29 663–690 (2000).
113. Y. MATIYASEVICH *Hilbert's tenth problem*. Translated from the 1993 Russian original by the author. With a foreword by Martin Davis. Foundations of Computing Series. MIT Press, Cambridge, MA (1993).
114. E. W. MAYR *Some complexity results for polynomial ideals*, Journal of Complexity, 13(3):303–325 (1997).
115. E. MENDELSON *Introduction to mathematical logic*, Princeton, N.J., Van Nostrand (1964).
116. M. MIGNOTTE, D. STEFANESCU *Polynomials, an algorithmic approach*, Springer Verlag, Singapore (1999).
117. J. MILNOR *Morse Theory*, Annals of Mathematical Studies, Princeton University Press (1963).

118. J. MILNOR, *On the Betti numbers of real varieties*, Proc. AMS 15, 275–280 (1964).
119. R. T. MOENCK *Fast computation of GCDs*, Proc. STOC '73, 142–151 (1973).
120. L. G. MONK, *Elementary-recursive decision procedures*, Thesis (1976).
121. M. MORSE, *Relations between the critical points of a real function of n independent variables*, Trans. Amer. Math. Soc., 27 345–396 (1925).
122. B. MOURRAIN, M. N. VRAHATIS, J.-C. YAKHOUBSON *On the Complexity of Isolating Real Roots and Computing with Certainty the Topological Degree*, Journal of Complexity, 182, 612–640 (2002).
123. I. NEWTON, *The mathematical papers of Isaac Newton*, Cambridge University Press (1968,1971,1976).
124. O. A. OLEINIK, *Estimates of the Betti numbers of real algebraic hypersurfaces*, Mat. Sb. (N.S.), 28 (70): 635–640 (Russian) (1951).
125. O. A. OLEINIK, I. B. PETROVSKII, *On the topology of real algebraic surfaces*, Izv. Akad. Nauk SSSR 13, 389–402 (1949).
126. A. OSTROWSKI, *Notes sur les produits de séries normales*, Bulletin de la Société Royale des Sciences de Liège 8 458–467 (1939).
127. C. PAPADIMITRIOU, *Computational Complexity*, Addison-Wesley (1994).
128. P. PEDERSEN, *Counting real zeroes of polynomials*, PhD Thesis, Courant Institute, New York University (1991).
129. P. PEDERSEN, M.-F. ROY, A. SZPIRGLAS, *Counting real zeroes in the multivariate case*, Computational algebraic geometry, Eyssette et Galligo ed. Progress in Mathematics 109, 203–224, Birkhauser (1993).
130. H. POINCARE, *Analysis Situs*, Oeuvres, vol VI, Gauthier-Villars, Paris, 193-288 (1953).
131. R. POLLACK, M.-F. ROY, *On the number of cells defined by a set of polynomials*, C. R. Acad. Sci. Paris 316 573–577 (1993).
132. F. PREPARATA, D. SARWATE *An improved parallel processor bound in fast matrix inversion*, Inf. Proc. Letters, (7)/3, 148–150 (1978).
133. J. RENEGAR. *On the computational complexity and geometry of the first order theory of the reals*, Journal of Symbolic Computation, 13: 255–352 (1992).
134. F. ROUILLIER, *Solving Polynomial Systems through the Rational Univariate Representation,* Applicable Algebra in Engineering Comunication and Computing, 9 (5) 433–461 (1999).
135. F. ROUILLIER, *RS (Real roots of systems with a finite number of complex solutions)*, http://fgbrs.lip6.fr/Software/
136. F. ROUILLIER, *On solving zero-dimensional polynomial systems with rational coefficients*, preprint (2005).
137. F. ROUILLIER, M.-F. ROY, M. SAFEY EL DIN, *Finding at least one point in each connected component of a real algebraic set defined by a single equation*, Journal of Complexity 16, 716–750 (2000).
138. F. ROUILLIER, P. ZIMMERMANN, *Efficient Isolation of a Polynomial Real Roots*, Rapport de Recherche INRIA 4113 (2001).
139. M.-F. ROUTH, *Stability of a given state of motion,* London (1877), The advanced part of a treatise of the system of rigid bodies Dover, New York (1955).
140. M.-F. ROY, *Basic algorithms in real algebraic geometry: from Sturm theorem to the existential theory of reals,* Lectures on Real Geometry in memoriam of Mario Raimondo, de Gruyter Expositions in Mathematics, 1–67 (1996).

141. M.-F. ROY, A. SZPIRGLAS, *Complexity of the computations with real algebraic numbers*, Journal of Symbolic computation 10 39–51 (1990).
142. M.-F. ROY, N. VOROBJOV, *Computing the Complexification of a Semi-algebraic Set*, Math. Zeitschrift 239, 131–142 (2002).
143. M. SAFEY EL DIN, *RAG'Lib*. http://fgbrs.lip6.fr/Software/
144. M. SAFEY EL DIN, E. SCHOST, *Properness defects of projections and computation of one point in each connected component of a real algebraic set*, Discrete and Computational Geometry, 32 (3), 417–430 (2004).
145. A. SCHONHAGE, *Schnelle Berechnung von Kettenbruchentwicklungen*, Acta Informatica 1,139–144, (1971).
146. J. SCHWARTZ, M. SHARIR, *On the 'piano movers' problem II. General techniques for computing topological properties of real algebraic manifolds*, Adv. Appl. Math. 4, 298–351 (1983).
147. I.R. SHAFAREVITCH, *Basic algebraic geometry*, Springer (1974).
148. A. SEIDENBERG, *A new decision method for elementary algebra*, Annals of Mathematics, 60:365–374, (1954).
149. V. SHARMA, C. YAP, *Sharp Amortized Bounds for Descartes' and de Casteljau's Methods for Real Root Ioslation*, preprint (2005).
150. E. H. SPANIER, *Algebraic Topology*, McGraw-Hill Book Company (1966).
151. A. STREZEBONSKI, *Solving Systems of Strict Polynomial Inequalities*, Journal of Symbolic Computation 29 (3) 471–480 (2000).
152. C. STURM, *Mémoire sur la résolution des équations numériques.* Inst. France Sc.Math. Phys.6 (1835).
153. J. J. SYLVESTER, *On a theory of syzygetic relations of two rational integral functions, comprising an application to the theory of Sturm's function.* Trans. Roy. Soc. London (1853).
154. A. TARSKI, *Sur les ensembles définissables de nombres réels*, Fund. Math. 17, 210-239 (1931).
155. A. TARSKI, *The completeness of elementary algebra and geometry*, 1939, Preprint Institut Blaise Pascal, CNRS (1967).
156. A. TARSKI, *A Decision method for elementary algebra and geometry*, University of California Press (1951).
157. R. THOM, *Sur l'homologie des variétés algébriques réelles*, Differential and Combinatorial Topology, 255–265. Princeton University Press, Princeton (1965).
158. J.V. USPENSKY, *Theory of equations*, MacGraw Hill (1948).
159. B. L. VAN DER WAERDEN, *Modern Algebra, Volume II*, F. Ungar Publishing Co. (1950).
160. B. L. VAN DER WAERDEN, *Topologische Begründung des Kalküls der abzählenden Geometrie*, Math. Ann. 102, 337–362 (1929).
161. L. VIETORIS, *Uber die Homologiegruppen der Vereinigung zweier Komplexe*, Monatsh. fur Math. u. Phys., 37 159–162 (1930).
162. A.J.H. VINCENT, *Sur la résolution des équations numériques*, Journal de Mathématiques Pures et Appliquées, 341–372 (1836).
163. N. N. VOROBJOV. *Complexity of computing the dimension of a semi-algebraic set*. J. of Symbolic Comput., 27: 565–579 (1999).
164. R. J. WALKER, *Algebraic Curves*, Princeton University Press (1950).

165. H.E. Warren, *Lower bounds for approximations by non-linear manifolds*, Trans. Amer. Math. Soc. 1333, 167–178, (1968).
166. A. Zapletal, *Gröbner bases bibliography*. http://www.ricam.oeaw.ac.at/Groebner-Bases-Bibliography/index.php

Index of Notation

Chapter 1

Chapter 2

Chapter 3

Chapter 4

Chapter 5

Chapter 6

Chapter 9

Chapter 10

Chapter 11

Chapter 12

Chapter 13

Chapter 14

Chapter 15

Chapter 16

Index